Lecture Notes in Physics

The Lecture Notes in Physics

The series Lecture Notes in Physics (LNP), founded in 1969, reports new developments in physics research and teaching – quickly and informally, but with a high quality and the explicit aim to summarize and communicate current knowledge in an accessible way. Books published in this series are conceived as bridging material between advanced graduate textbooks and the forefront of research and to serve three purposes:

- to be a compact and modern up-to-date source of reference on a well-defined topic
- to serve as an accessible introduction to the field to postgraduate students and nonspecialist researchers from related areas
- to be a source of advanced teaching material for specialized seminars, courses and schools

Both monographs and multi-author volumes will be considered for publication. Edited volumes should, however, consist of a very limited number of contributions only. Proceedings will not be considered for LNP.

Volumes published in LNP are disseminated both in print and in electronic formats, the electronic archive being available at springerlink.com. The series content is indexed, abstracted and referenced by many abstracting and information services, bibliographic networks, subscription agencies, library networks, and consortia.

Proposals should be sent to a member of the Editorial Board, or directly to the managing editor at Springer:

Christian Caron
Springer Heidelberg
Physics Editorial Department I
Tiergartenstrasse 17
69121 Heidelberg / Germany
christian.caron@springer.com

A. Mulero (Ed.)

Theory and Simulation of Hard-Sphere Fluids and Related Systems

Springer

Ángel Mulero
Universidad de Extremadura
Dept. Física Aplicada
Avenida de Elvas, s/n
06071 Badajoz
Spain
mulero@unex.es

Mulero, A (Ed.), *Theory and Simulation of Hard-Sphere Fluids and Related Systems*, Lect. Notes Phys. 753 (Springer, Berlin Heidelberg 2008), DOI 10.1007/978-3-540-78767-9

ISBN: 978-3-642-09758-4 e-ISBN: 978-3-540-78767-9

DOI 10.1007/978-3-540-78767-9

Lecture Notes in Physics ISSN: 0075-8450

Softcover reprint of the hardcover 1st edition 2008

Cover design: eStudio Calamar S.L., F. Steinen-Broo, Pau/Girona, Spain

Printed on acid-free paper

9 8 7 6 5 4 3 2 1

springer.com

To Ana, Alejandro, and Alberto, my wife and children.

Preface

In our days, it seems natural using a system of hard bodies perturbed by intermolecular potentials for the modeling of the thermodynamic properties of fluids. In fact, it seems so natural that one hardly notices the misuse of the term ‘intermolecular potential’ in the previous phrase to refer to the ‘inter-hard-body-potential’ purported to exist for the modeling of fluids. The reason for this tacit acceptance is that, in our mental models for fluids, we identify molecules with hard bodies, almost as a primitive concept. The question then naturally arises: Was it always this way? Or, perhaps, was there a time when scientists did not recognize the existence of molecules? Could it be possible that there was a time when scientists would not accept the use of a system formed by hard bodies as the supporting structure for the modeling of fluids? It is not easy to imagine that less than one hundred years ago, the existence of atoms and molecules was still under discussion. As commented by Einstein’s biographer A. Pais [1], it was only in 1908 that Friederich Wilhelm Ostwald recognized the existence of atoms and molecules, and Ernst Mach died in 1926 still unconvinced of their existence. When one gains this perspective, one really understands the value of the contribution of Johannes Diderik van der Waals who, in 1873, took a gigantic first step with the introduction of hard bodies representing atoms and molecules for the modeling of the pressure-volume-temperature behaviour of fluids. Van der Waals innovative thinking is so remarkable that it deserves special attention at the beginning of this volume on the physics of hard-sphere (HS) systems.

We all know that a model is just an approximation to reality and that models should never be confused with reality itself. However, the closer a model captures the essential elements of reality the better is its predictive ability and, thus, its usefulness. Hence, the main problem in the modeling of a physical system is precisely the step of recognizing what are its essential aspects. This is certainly not an easy task.

When van der Waals presented his famous equation of state in 1873, he was not the old Professor that we see in many postal stamps celebrating the Nobel Prize that he obtained in 1910. In 1873, he was just a 36-year-old man, who had studied part time at the university from age 25–28 years and had afterwards prepared his Doctoral Dissertation 'On the Continuity of the Gas and Liquid State' without any known mentor. His doctoral work was primarily motivated by the results of T. Andrews experiments, who in 1869 reported the existence of critical temperatures above which gases could not be liquefied by an increase in pressure only [2]. As a way to understand better the major step made by van der Waals, in connection with the subject of this monograph, let us digress for a moment and consider the nature of the problem he wanted to solve. As a high school teacher of mathematics and physics [2], van der Waals obviously knew that the equation for the isothermal change of pressure as a function of volume had to have an inflection point at the critical point. This was necessary in order to give two roots for volume at temperatures below the critical isotherm, and a single root for volume at temperatures above the critical isotherm. Thus, in mathematical terms, he knew that the equation of state for the pressure of a fluid should be a cubic equation in terms of the volume. This being the case, one wonders why he did not solve his problem using the simplest possible cubic equation of state in the form of a truncated virial expansion:

$$P = \frac{RT}{v} + \frac{\alpha}{v^2} + \frac{\beta}{v^3}$$

This equation obviously satisfies all the conditions required by the problem of continuity of the phases that van der Waals was considering. In fact, following van der Waals own procedure of equating the first and second derivatives of the pressure with respect to the volume to zero at the critical point, this equation gives a critical compressibility factor equal to 1/3 [3]. Instead, van der Waals proposed the following, seemingly unnecessarily complex, form:

$$P = \frac{RT}{v - b} - \frac{a}{v^2}$$

Following the 'van der Waals' procedure to evaluate both constants, this form gives a critical compressibility factor equal to 3/8. Why choosing a more complex form that gives a higher compressibility factor? The reason for this can be found in the Nobel Lecture presented by van der Waals, when he said: 'It will be perfectly clear that in all my studies I was quite convinced of the real existence of molecules, that I never regarded them as a figment of my imagination, nor even as mere centres of force effects. I considered them to be actual the bodies, thus what we term "body" in daily speech ought better to be called "pseudo-body". It is an aggregate of bodies and empty space.' Thus, van der Waals equation was not just an analytical expression able to fit data, it was a model of reality. It introduced two physical approximations in one single step: the idea that the molecules exist and occupy part of the volume of the system, represented by the parameter 'b', and that a cohesive potential, represented by the parameter 'a', acts upon the molecules. In retrospect, we have to admire not only the genius of van der Waals but also his bravery. In another paragraph of his speech accepting the Nobel Prize, referring to the 'real' existence of molecules, van der Waals said: 'When I began my studies I had the feeling that I was almost alone in holding that view. And when, as it occurred already in my 1873 treatise, I determined their number in one gram-mol, their size and the nature of their action, I was strengthened in my opinion, yet there still often arose within me the question whether in the final analysis a molecule is a figment of the imagination and the entire molecular theory too.' Think for a moment on this. Being van der Waals what we would now call 'a graduate student', he had the courage to go against the views of well established scientists and propose a model incorporating hard spheres to represent the body of molecules. We should get more graduate students like him these days!

Once the ideas of van der Waals are accepted, the 'van der Waals model' can be directly generalized as a tool to help understanding the molecular assumptions implicit in other equations of state [4]. Empirical applications have been mostly concerned with temperature and volumetric corrections to the cohesive energy term. More fundamental studies have concentrated on the modeling of the free volume available for the movement of the molecules in the system. This latter aspect is underlying the topic of this book, which is basic for the understanding of liquids, liquid mixtures and other condensed states of matter.

After van der Waals innovative work, the field experienced a long period of stagnation. In fact, the necessary theoretical and experimental tools were not available at the time for the study of the contribution of repulsive effects to the thermodynamic properties of fluids. In an attempt to provide the necessary perspective, I will briefly call the attention here to what it seems to be the most important steps in the historical development of HS fluids and related systems.

From the theoretical point of view, the next qualitative change after the work of van der Waals was introduced by John G. Kirkwood in three seminal papers: 'Quantum Statistics of Almost Classical Assemblies' [5], 'Statistical Mechanics of Fluid Mixtures' [6] and 'Molecular Distribution in Liquids' [7]. The origins of the well-known equations of state for HS systems of Percus–Yevick–Frisch [8] and of Percus–Yevick–Thiele [9], can be directly traced back to the work of Kirkwood. The successful equation of state for hard spheres of Carnahan–Starling [10] is a direct interpolation between the two other equations mentioned above. Most notably, Thiele in his derivation clearly observed that the equation of state for hard spheres he was presenting failed to give the correct limit for ordered close-packed hard spheres. This failure of the equations of state to meet the high-pressure limiting behaviour, common to the three equations of state mentioned above, was addressed only 10 years ago [11].

On the experimental side, the exact contribution of the repulsive potentials is not directly accessible from actual fluid data. Here it was necessary to wait for the advent of computers to generate 'experimental data' for HS systems. The first reliable computational results were those obtained by Monte Carlo calculations by Rosenbluth and Rosenbluth [12] and by Wood and Jacobson [13]. One year after Wood and Jacobson, Alder and Wainwright reported results obtained by molecular dynamics [14]. These two latter sets of results were found to be in agreement 'with respect to their calculated equation of state' [8].

From there on, the development of the field advanced at a vertiginous speed both in theory and in the generation of computer-simulated experimental results. Although much progress has been made in the last few decades, much still needs to be done, mainly in the study of mixtures consisting of hard bodies of different shapes and sizes. This book presents the state of the art of our knowledge in the field at the dawn of the second millennium. Reading these manuscripts that are at the frontier of knowledge, it is sobering to think that this fascinating field originated from an idea of a single man, working almost in isolation, who dared to challenge the strong prejudices existing in his time.

Canada, January 2007 *Juan H. Vera*

References

1. A. Pais: *Subtle is the Lord. The Science and Life of Albert Einstein* (Oxford University Press, Oxford, 1982)
2. *Nobel Lectures, Physics 1901-1921*, (Elsevier Publishing Company, Amsterdam, 1967) *This material can also be found at http://nobelprize.org/nobel_prizes/physics/*
3. H. Orbey and J.H. Vera: Chem. Eng. Sci. **45**, 3319 (1990)
4. J.H. Vera and J.M. Prausnitz: Chem. Eng. J. **3**, 1 (1972)
5. J.G. Kirkwood: Phys. Rev. **44**, 31 (1933)
6. J.G. Kirkwood: J. Chem. Phys. **3**, 300 (1935)
7. J.G. Kirkwood: J. Chem. Phys. **7**, 919 (1939)
8. H. Reiss, H.L. Frisch, and J.L. Lebowitz: J. Chem. Phys. **31**, 369 (1959)
9. E. Thiele: J. Chem. Phys. **39**, 474 (1963)
10. N.F. Carnahan and K.E. Starling: J. Chem. Phys. **51**, 635 (1969); **53**, 600 (1970)
11. W. Wang, M.K. Khoshkbarchi, and J.H. Vera: Fluid Phase Equil. **115**, 25 (1996)
12. M.N. Rosenbluth and A.W. Rosenbluth: J. Chem. Phys. **22**, 881 (1954)
13. W.W. Wood and J.D. Jacobson: J. Chem. Phys. **27**, 1207 (1957)
14. B.J. Alder and T.E. Wainwright: *Contract Reports, Radiation Laboratory at Livermore* (University of California, 1958); J. Chem. Phys. **33**, 1439 (1960)

Introduction

HARD SPHERES (HS) are a paradigmatic system. Despite the simplicity of that system, and that it only exists in computers, it has served as the basis for the advance of science in the fields of general liquids, amorphous solids, liquid crystals, colloids, granular matter, etc. HS systems exhibit a surprisingly rich structural and thermodynamic behaviour (phase transitions, metastable states, demixing, etc.) that call for careful study. Also surprisingly, this behaviour is not completely understood and is an object of current research. The same situation holds for related systems: hard disks, hard diatomics, spherocylinders, etc.

This book reviews some important progress in the study of HS and related systems (hard disks and mixtures of hard systems, mainly). Their structure, thermodynamic properties, and kinetic and transport properties are considered from different and complementary points of view. In some cases, the study is extended to other simple systems such as square-well or Lennard–Jones fluids. Therefore, this book can be an important tool for graduate students and teachers, as well as for researchers in statistical mechanics, physics of liquids, physical chemistry, chemical physics, and chemical engineering. In this Introduction, I shall briefly describe the main characteristics of the HS systems, and try to guide the readers so that they can properly use the book.

As is well known, a HS system is simply defined as that in which the molecules interact with an infinitely strong mutual repulsion whenever their centres come within the distance of the particle diameter. The interest for the study of these systems is due to the fact that the structure of real fluids is mainly determined by repulsive forces. Therefore, the HS model represents the simplest statistical mechanical system with liquid-like properties at high densities, and it is commonly used as a reference system for perturbation approaches. In fact, the idea of representing a liquid by a system of HS can already be found in the work of van der Waals: his famous equation of state was derived using essentially this principle.

The simple systems of hard bodies (rods, disks, spheres, hyper-spheres, some other anisotropic non-spherical molecules, and mixtures of them) have been studied in a great number of computer experiments, and the results have been compared with the predictions of many theoretical approximations. Indeed, they have been a continual source of help to improve those approximations. In fact the HS model for fluids is one of the oldest, and most studied systems in statistical mechanics. Some of the most important of these approaches and the main results obtained from them are considered in this book, as explained below.

In Chaps. 1–3 and 5–7 the main structural and thermodynamic properties of HS and related systems are considered. The main theoretical approaches used are the integral equation theories, the Rational Function approach, and the Density Functional theory, which are combined with or compared to results from computer simulations. In Chap. 4 the application of the HS and hard-disk equations of state to the study of the purely repulsive part of the Lennard–Jones system is developed. Chapter 8 is devoted to the study of ensemble effects in small HS systems. Finally, some kinetic and dynamic properties of several hard particle systems, and also other models (for instance, square-well, Lennard–Jones fluids, and granular mixtures) are discussed in Chaps. 9 and 10.

It is important to take into account that, as in the case of the ideal gas, hard spheres are an 'athermal' system: the canonical partition function and any structural properties are independent of the temperature. Its free energy derives entirely from the entropy. The finite particle size of hard spheres introduces no additional energy terms because the energy is infinite if the particles overlap and zero otherwise. There is no energy scale. The particles have only excluded volume interactions, and equilibrium is determined by maximizing the number of accessible states. Therefore, the study of its statistical properties is essentially a geometry problem, what could be called 'statistical geometry': to 'count' the number of ways in which non-overlapping spheres may be arranged under different constraints, such as fixed volume and number of particles to obtain the equation of state, or around a sphere to obtain the radial distribution function. As the reader will verify throughout this book, the geometry of the problem leads sometimes to almost 'magical relationships' between the most important parameters, and in fact some of the most successful approaches have been those with 'harmonious' analytical expressions.

Unlike the ideal gas, the HS system has not one but two length scales: the average interparticle space, commonly recast as the number density, ρ, and the particle diameter, σ. These parameters can be combined to form the so-called packing fraction, $\eta = \rho\pi\sigma^3/6$, which is the only relevant dimensionless parameter of the bulk system. The extension of η to a local packing fraction in inhomogeneous systems is the key element of the exact density functional free energy for dimensional hard-rods and for the most advanced density functional approaches for hard spheres.

The study of the main thermodynamic properties of HS and some related systems is an important basis for the knowledge of the thermodynamic behaviour of real fluids. Some of these properties are included in several chapters of this book from different and sometimes complementary points of view, the main objective being the calculation of the compressibility factor, i.e. the equation of state of the system.

Although there are no exact analytical formulae for the equation of state of hard-body systems beyond the one-dimensional hard-rod case, there are several useful expressions that agree well with data from computer simulations and that contain the basic description of the thermodynamic behaviour of these systems. A great number of equations of state (EOS) for HS and related systems are presented and used in this book, providing what is perhaps the broadest collection of analytical expressions published to date.

A basic procedure to obtain the equation of state of a fluid is the use of a virial expansion of the compressibility factor in powers of the density. Indeed, if all the exact virial coefficients were known, the infinite virial series would formally represent the exact EOS for the fluid phase. The virial coefficients are then very important from a theoretical point of view because they represent the non-ideal behaviour of real gases. Indeed, the importance of the virial coefficients lies also in the fact that they are related directly to the interactions between clusters of molecules. Thus, the second virial coefficient represents the departure from ideal behaviour due to the interactions between pairs of molecules; the third virial coefficient gives the effects of interactions of molecular triplets, and so on.

Efforts have been made to calculate the virial coefficients even for the simplest systems, but unfortunately, the difficulty increases sharply with increasing order of the virial coefficient. In the case of hard-body mono-component fluids, the virial coefficients do not depend on the temperature, and they are simply pure numbers. But even in this case, calculations of coefficients higher than the fourth must be done numerically, and a large amount of computation is needed. Some numerical methods to calculate them, as well as the values of the first ten virial coefficients and extrapolated values for some higher ones, are given in Chap. 2 for both hard disks and hard spheres. Values for virial coefficients of mixtures are given in Chap. 5 and considered again in Chap. 6.

The virial expansion and other procedures followed in order to obtain analytical expressions for the EOS of pure hard systems are summarized in the introduction of Chap. 3. As indicated there, since the virial series converges slowly, several types of approximants, such as those of Padé and of Levin, are used in order to accelerate the convergence. This procedure leads to EOSs, which are more accurate at high densities than the truncated virial series, and in some cases they include the correct limit behaviour at very high densities.

The connection between the thermodynamic properties (EOS) and the molecular structure of a fluid comes through the definition and study of the radial distribution function (RDF). As is well known, the RDF is a statistical measure of the average structure in the fluid, and its knowledge allows one to determine the thermodynamic properties of systems interacting by pair-wise potentials. The RDF thus plays a central role in the study of properties of fluids, and great efforts have been devoted to calculating it from different theoretical approaches once the intermolecular potential is known. As the details of the intermolecular forces in real fluids are not known, the consideration of models including the main characteristics of real fluids and defined for a simple intermolecular potential, as is the case of the HS model, is of great theoretical relevance in understanding the properties of fluids. An important point here is that the relationship between EOS and RDF is not unique. In fact, the EOS can be obtained following the virial or the compressibility route, and the same (exact) results should be obtained if one could use the exact RDF. That gives a measure of the quality of any approximate theory as reflected in the difference in the EOS obtained by the two routes. It also provides a goal in the development of approximations to reduce that inconsistency, or even avoid it for uniform bulk fluids.

Researches working on the EOS of HS usually accept the virial equation as the most relevant EOS, which is then given by a simple expression containing the contact value of the radial distribution function within a given approximate scheme. On the contrary, workers in Density Functional theory use the compressibility route as the most natural connection between the correlation structure and the thermodynamic properties. Yet an alternative 'energy route' may be used for systems with interactions beyond the simple hard-core models, so that the RDF may be used to obtain the internal energy from the pair interaction potential.

In any case, the calculation of the RDF from theoretical procedures is the most important step in the modern theory of dense fluids, and in Chap. 1 of this book the basic concepts about the structure of HS and related systems are introduced, and then the calculation of the RDF for pure hard spheres is considered. The low- and the high-density regions are considered separately. At low densities, the correlation functions can be obtained from a density expansion, in which the coefficients include integrals that are represented by 'diagrams'. Both analytical and computer simulation results for the diagrams are presented and discussed in that chapter.

A crucial element of the diagrammatic series is the fact that the whole series could be formally expressed in terms of a reduced set of 'connected diagrams', which represent the expansion of what is called the direct correlation function (DCF). The total correlation function and hence the RDF would come out of the DCF through the Ornstein–Zernike (OZ) equation (see Sect. 1.3.2 in Chap. 1), which reconstructs all the diagrams in the expansion of the RDF as convolutions of those in the DCF. The OZ equation may be considered as just a formal definition of the DCF, but it opened a whole, and most successful, class of theoretical approaches, the so-called integral equation theories to get the properties of fluids at high density, beyond the practical applicability of the virial expansions. That is done through what is called 'a closure relation', some physically based assumption on the form of the DCF and/or on its relation to the RDF, which may be used together with the OZ to close a system of integral equations. The most important of these relations are the hypernetted chain (HNC) and Percus–Yevick (PY) closures, which are briefly described in Sect. 1.3.3 of Chap. 1. As indicated there, one advantage of considering the HS fluid is that the PY approach gives an analytical solution for the OZ equation, which can be used as a sound starting point for the study of more complex systems. Thus, in the case of HS, these analytical solutions lead to simple EOSs commonly used as referents for the development of more complex expressions (see Chap. 3).

For both pure HS fluids and binary mixtures, the PY approach gives generally more satisfactory results than the HNC. In particular, exact solutions are obtained using the PY approach for the one dimensional hard rod system. Because of that, it is commonly used in the development of other theoretical approximations and in the study of more complex systems. Clear examples are presented in Chap. 7, which is devoted to reviewing the Density Functional theory, and where the OZ equation is considered again under a different light.

For some complex hard-body systems no analytical solutions are known, so numerical procedures to solve the OZ equation are needed. In Chap. 1 some numerical methods are analyzed and several references are given to guide the reader on how to apply the methods for the metastable phases or for the mixtures. As there are several closures that one can choose, the readers can find at the end of Chap. 1 several recommendations about what closure should be used for different hard-body fluids.

Another question to take into account is the specific design of closures which give consistent results for the EOS by the different routes described above. In particular, in Sect. 1.3.3 of Chap. 1, several improvements to the classical closures are briefly described, including self-consistent, phenomenological, and other closures. Moreover, in Sect. 6.3 of Chap. 6, the so-called Rational Function approximation is presented as an analytical approximation constructed to be thermodynamically consistent and thus being a clear alternative to the integral equation approaches.

The results obtained by using the aforementioned theoretical approximations can be applied to obtain the thermodynamic properties of fluids. In fact, it is very important to pay attention to those EOS expressions which have a sound theoretical basis, as are those based on the solution of the PY approach, in the use of the concept of excluded volume, or in the scaled-particle theory (SPT). In particular, the SPT is a clear example in which the 'statistical geometry' considerations are used. Thus, the SPT allows the approximate determination of the radial distribution function at contact by considering the reversible work necessary to create a spherical cavity in the fluid. The resulting EOSs for odd-dimensional hard systems are the same as those obtained by the PY approximation using the compressibility route, and this coincidence can be considered a consequence of the geometry. The mentioned purely theoretical EOSs have served as referents to construct a large number of other analytical expressions. Nevertheless, the latter are analytically more complex and in most cases lack a theoretical basis. Hence, the EOSs based on purely theoretical arguments are commonly used as a fundamental tool in the development of other theories. A clear example, shown in this book, is the use of the PY approximation in the development of some Density Functional approaches (see Sect. 7.4 of Chap. 7).

As noted above, the PY EOS obtained using the virial route is different from that obtained using the compressibility route. As an attempt to solve this inconsistency, Carnahan and Starling proposed which is perhaps the most used EOS for HS fluids, and also as a reference part in EOSs for real fluids. This EOS can be obtained as a linear combination of the PY-virial and PY-compressibility EOSs, but also by considering the first virial coefficients to be integer numbers and then resumming the series. As is shown in Chap. 3, it constitutes perhaps the simplest EOS for HS fluids that significantly agrees with computer simulation data, at least at low and intermediate densities. Thus, the Carnahan–Starling EOS has been widely used, and the reader will find it in several parts of this book. For instance, it is used to study the density profile of a HS fluid against a planar wall and to describe the HS crystal using Density Functional theories (see Sect. 7.4 of Chap. 7). This EOS is also used in some versions of the Fundamental Measure theory (see Sect. 7.5.7 in Chap. 7).

Some improvements to the Carnahan–Starling EOS have been made in order to give the correct close-packed value or to improve its accuracy at high densities, and most of them can be found in Chap. 3. Nevertheless, this improvement also means using a more complex analytical expression, and thus encountering clear difficulties in the theoretical developments.

Another important matter to bear in mind is that the HS system is not always a fluid, but can also be a solid. Almost from the beginning of the study of this system, mainly by using computer simulations, the appearance of a fluid–solid phase transition was considered. The study of the nature and location of this phase transition, as well as the packing of spheres and the existence of a metastable region and a glass transition, has generated much scientific work over the years, and there are still some discrepancies between the results obtained from different approaches. Those approaches go from the application of the idea of excluded volume to the consideration of Density Functional theories.

The basic phase diagram of pure HS is presented in Fig. 3.5 of Chap. 3. As can be seen, there are four important branches shown in the phase diagram. There is a fluid branch that ends at a freezing point, where the phase diagram splits into two parts. One part is a metastable extension of the fluid branch which follows continuously from the previous one and is conjectured to end at a point known as random close packing. The other branch represents the thermodynamically stable part of the phase diagram. Along the horizontal portion of this branch both fluid and solid can coexist until the melting point is reached. The portion of the curve which continues above the melting point is referred to as the solid or ordered branch, ending at the close-packed crystal.

The study of the different parts of the phase diagram can require different techniques or approaches. In particular, different EOSs must be defined and used on each branch. In any case, there have been several attempts to build an EOS that is valid for both the fluid and metastable regions, whereas a different expression must be used for the solid phase (see Chap. 3). With respect to the HS crystallization, several Density Functional approximations have been developed aimed mainly at trying to characterize it (see Sect. 7.4.2 in Chap. 7).

The different procedures and the existence of different phases have led to proposals of a great number of analytical expressions for the EOS of pure hard-body fluids. For that reason, in Chap. 3 most of the EOSs for hard disks and HS, based mainly on the use of approximants, on different theories, and on the use of computer simulation data, have been collected. Moreover, the main computer simulations devoted to the study of thermodynamic properties of these systems are summarized. Then some selected data are used to test the accuracy of several equations of state in order to reproduce the calculated compressibility factor.

In addition to the study of the compressibility factor, in this book attention is given to the chemical potential and, to a lesser degree, to the isothermal compressibility of HS. As is well-known, knowledge of the chemical potential is needed in the study of phase equilibria. Hence, in Chap. 2 some methods of obtaining it from computer simulations are considered, and in Chap. 3 the calculation is performed analytically through the EOS. The isothermal compressibility is also considered in this third chapter. The accuracy of the analytical expressions obtained for the two properties is then checked.

As remarked above, the HS system serves as a referent for the study of more complex fluids. In particular, it is the fundamental basis for the development of perturbation theories of fluids. Perturbation theories, in a similar way to the van der Waals equation of state, are based on the separation of the contributions of repulsive and attractive intermolecular forces. The main idea is that the repulsive forces strongly determine the behaviour of dense fluids, and the attractive forces are regarded as perturbations. If the properties of a hard-body system are adequately modified to introduce the temperature dependence, which is done through a scaling procedure of the molecular diameter, then it can be used to model the properties of the repulsive reference system. This scaling procedure, applied to the study of Lennard–Jones fluids, is extensively analyzed in Chaps. 4 and 9. In particular, Chap. 4 includes the application of the HS and hard-disk EOSs to the study of the Weeks–Chandler–Andersen reference system. In Chap. 9, the perturbative approach is used to study dynamic properties of fluids. Also, the application of the Rational Function approximation to perturbation theories is considered in Sect. 6.5 of Chap. 6.

In the preceding paragraphs, we have been considering mainly the properties of pure fluids, but it is also important to consider those of mixtures. An adequate theoretical basis for the study of real mixtures requires knowledge of properties of hard-body mixtures. Of course, sources of knowledge about these mixtures are similar to those for pure hard-body fluids. Virial coefficients, computer simulation data, and theories of mixtures of HS and of some related systems are reviewed in several chapters of this book. First of all, it is important to distinguish between mixtures of additive HS and mixtures of non-additive HS. Additive mixtures are particularly interesting because of their resemblance to real colloidal suspensions, whereas non-additive mixtures are mainly considered when dealing with metallic systems and with mixtures of colloidal particles and polymers. The procedures to study them are similar, but the difficulties and hence the advances in theoretical developments are different. Additive mixtures are considered in Chaps. 5–7, and non-additive mixtures in Chaps. 6 and 7.

The study of additive HS mixtures as the simplest models of colloidal suspensions is clearly interesting, with the possible existence of a demixing transition as a phenomenon to consider. Chapter 5 is devoted specifically to the study of the main properties of these binary mixtures using both theory and computer simulations. The reader can find there a summary of the main computer simulation methods and their results. An important topic considered in Chap. 5 is the calculation of virial coefficients of mixtures. In particular, for additive HS mixtures, the second and third virial coefficients are known analytically, while several of the higher order virial coefficients have been determined numerically or through approximate analytical expressions. The methods used and their results are then summarized in that chapter. As in the case of pure fluids, the virial expansion can be rescaled in order to accelerate the convergence. The main methods and results of this approach are summarized briefly in Sect. 5.7.

Chapter 5 includes the study of the consistency conditions, i.e., the exact conditions that the equation of state, the contact values of the partial radial distribution functions, and other thermodynamic properties of HS additive mixtures must satisfy. These conditions are summarized and are then used to derive improved analytical expressions and density expansions for the EOS and the contact values of the partial RDFs.

A great amount of computational effort has been devoted to determining the phase transitions in these mixtures. In Chap. 5, special attention is given to the crystallization and demixing phenomena. Demixing is a phase transition due only to the asymmetric interactions between the particles of the different components in the mixture. This phenomenon has been widely studied but is still a controversial subject. In addition to the aspects considered in Chap. 5, in Sect. 6.2.4 of Chap. 6 some routes that have been taken to try to clarify the main aspects of this subject are presented.

In Chap. 6, another way to obtain the structural and thermodynamic properties of simple fluids, the so-called Rational Functional approximation, is presented. As noted above, it constitutes a clear alternative to the integral equation theories, and it leads by construction to thermodynamic consistency. The main steps and alternatives in the application of the method are carefully explained, and then results for pure HS and multicomponent mixtures are given. Moreover, some proposals are presented for the contact values of the RDF and the EOS of additive mixtures in d-dimensions, with especial relevance given to the three-dimensional case. Non-additive mixtures are also considered, and a simple proposal for the EOS of these systems is presented in Sect. 6.2.3.

In this Chap. 6, comparisons with computer simulation results and other theoretical approaches are carefully made in order to show the accuracy and applicability of the procedure. Finally, its application to other systems such as sticky hard spheres, square-well fluids, and Lennard–Jones fluids, in this case in connection with perturbation theories, is also considered.

Chapter 7 is devoted to Density Functional theories. The development of these theories, which permit one to obtain the equilibrium density distribution of non-homogeneous systems, has been in great part through the consideration of hard-core models. The reader can find in this seventh chapter what is perhaps the most complete and up-to-date review of these theories, including the study of HS and disks, HS mixtures, and non-spherical (anisotropic) hard particles, including a useful tutorial introduction to the Density Functional (DF) formalism. It is important to note that the results for uniform hard-body systems, presented in the other chapters of this book, are fundamental to the construction of the DF approximations. The application of the DF formalism when the correlation structure is known, and vice versa, i.e., the use of the DF approximation to obtain the correlation structure, is clearly explained in Sect. 7.2.2, with a more physically intuitive interpretation of the direct correlation function, and hence of the OZ equation. Similarly to the virial expansion used to get the thermodynamic and structural properties of fluids, Sect. 7.2.3 shows how to apply the virial expansion to the excess free energy in the DF scheme. Finally, in this second section the Local Density Approximation, the Gradient Expansion, and the Mean Field Approximation are briefly considered as generic DF approximations, and a first analysis of their advantages and limitations is made.

The exact free energy DF for hard rods, presented first by Percus in 1976, is considered in Sect. 7.3 as the departure point for the application of the DF approach to hard disks and spheres. Efforts devoted to improving the quality of the DF approximations have led to excellent results, mainly thanks to the development of the Weighted Density Approximation (WDA) and of the Fundamental Measure Theory (FMT), described in Sects. 7.4 and 7.5, respectively. In particular, the WDA is used here to describe the density profile of a HS fluid against a planar wall, and also the main properties of HS crystals. The FMT, first proposed by Rosenfeld in 1989 but subsequently improved by other authors, is regarded as the most successful DF approximation, although it is perhaps more difficult to apply than other approaches. In Sect. 7.5 of Chap. 7, the main steps in applying both the original Rosenfeld version and the so-called Dimensional Interpolation version, proposed by Tarazona in 2000, are clearly explained. The application to some important problems, such as the study of zero-dimensional cavities, the density profiles near a hard wall, and the HS crystal, are then considered, including comparisons with the WDA.

The application of DF approaches to mixtures, which can not be based solely on the consideration of results for pure substances, is considered in Sect. 7.6. Different versions, mainly the WDA and the FMT, are presented and compared. Their use for polydisperse systems and the extensions to HS non-additive mixtures are also considered.

In the last Sects. of Chap. 7, the reader can find the development of Density Functionals for hard anisotropic, i.e., non-spherical hard-core, particles. As can be seen there, there are no good approximations for the correlation functions and equation of state of hard anisotropic fluids, and therefore most of the functionals described need to use the thermodynamic and structural properties of the HS fluid, which have already been described in this and previous chapters.

As is well known, anisotropic particles have both positional and orientational degrees of freedom, and new phases, different from those of isotropic particles, can be defined (the so-called nematic, smectic, and columnar phases, and also some other crystalline phases). The application of DF or other approaches must take into account this rich behaviour and then must try to explain it. In Sect. 7.10.1 the Mean Field approach, which lays stress on the attractive contributions to the anisotropic interaction potential, is considered. In Sect. 7.10.2 Density Functionals based on purely hard-core interactions are presented. Finally, in Sect. 7.11 the FMT for hard anisotropic particles is developed. Applications to the study of parallel hard cubes, the Zwanzig model, and freely rotating hard anisotropic particles are considered. Added to the theoretical developments, phase diagrams of some of these anisotropic hard-body mixtures are presented.

The use of new materials in industry that involve very small systems has led to great interest in the study of systems consisting of a reduced number of particles. Computer simulation of simple models such as HS is a fundamental tool in the study of these small systems. In these simulations, it is very important to study the finite size effects, which are reviewed in Chap. 8.

These small systems are clear examples of inhomogeneous systems, so that the DF theory can be applied. Nevertheless, as explained in Chap. 8, the traditional formulation of the DF is inadequate, and hence new approaches have been developed to study such small systems. In this sense, Chap. 8 clearly serves to extend some of the approaches previously considered in Chap. 7, but with especial attention given to the finite size effects on both kinetic and structural properties.

The last part of the book, Chaps. 9 and 10, is devoted to the study of transport and kinetic properties. In particular, transport properties are extremely important in both research and industry, and their accurate modeling is very desirable in various applications. An important aspect to take into account in the modeling procedure is the molecular interpretation of transport processes and hence their properties. Such coefficients describe the relaxation to equilibrium which a system undergoes from an initially perturbed state. For instance, viscosity measures the resistance experienced by adjacent layers of a fluid flowing at different velocities, which results from the momentum exchange occurring when molecules collide among them. Diffusion refers to the process by which particles intermingle as a result of their microscopic motion. Finally, thermal conductivity is interpreted as the result of energy transportation that molecules pass to each other under mutual collisions.

Accuracy of the molecular description of transport processes depends primarily on the intermolecular potential function adopted. Chapter 9 is devoted to simple systems, starting with HS but including also square-well, one-component plasma, soft-spheres, Lennard–Jones and purely repulsive WCA reference systems (as already mentioned, the thermodynamic properties of the latter are considered in Chap. 4). In particular, the calculation of self-diffusion coefficient, shear viscosity, and thermal conductivity by computer simulations is briefly described in Sect. 9.2.

Section 9.3 is devoted to the most important theoretical approaches. In the region of low densities, where collisions among three and more molecules can be neglected, explicit relations for transport quantities can be derived for the hard sphere system from theoretical approaches, such as the Enskog kinetic theory. Nonetheless, for dense fluids, at present no strict theory can provide reliable predictions. Hence, HS properties must be obtained by correcting Enskog theory with computer simulation data. More realistic model systems may be approached by adequately modifying results obtained for HS. As explained in Chaps. 4 and 9, this may be done through a scaling procedure in which a temperature- and possibly density-dependent effective HS diameter is used. In both chapters, several proposals for that diameter are considered and carefully analyzed.

Another very interesting approach is to relate transport coefficients with fluid structural properties. Thus, in Sect. 9.3.7 of Chap. 9, the relationship with the excess entropy is focused. The next section is devoted to the study of the self-diffusion coefficient, including the tracer diffusivity, for HS and other simple systems. The extension to real fluids leads to the proposal of specific values of molecular parameters that characterize each fluid. Several tables have been included where the reader can find those values for a large number of substances.

Sections 9.5 and 9.6 of Chap. 9 concentrate on viscosity and thermal conductivity, respectively. In the case of viscosity, several models for the HS system are considered, and special attention is dedicated to free-volume models. Lennard–Jones and real fluids are also reviewed. Section 9.5 finalizes with a brief discussion and some relevant conclusions.

With respect to the thermal conductivity, Sect. 9.6 starts with a compilation of HS expressions. Subsequently, real substances are studied on the basis of the free-volume theory, modified Enskog theory, and van der Waals theory. Excess entropy-scaling law approach is also briefly discussed. The chapter ends with some conclusions and final remarks about the modeling of the transport properties.

As is well known, several properties of granular media can be modeled through a fluid of HS with inelastic collisions. The last chapter of this book, Chap. 10, is devoted to the study of the kinetic theory of inelastic hard spheres. Transport properties, instabilities, non-Newtonian behaviour, etc., are presented from both theory and computer simulation perspectives. In particular, the theoretical description is based mainly on the Boltzmann kinetic equation, valid at low densities. Recent results obtained for binary granular mixtures are then carefully reviewed. A hydrodynamic description is subsequently considered from the solution to the Boltzmann equation using the Chapman–Enskog method. The Navier–Stokes hydrodynamic equations are obtained and then compared with numerical solutions from computer simulations. In particular, the results for the tracer diffusion coefficient and shear viscosity are considered.

The second part of Chap. 10 is devoted to some applications of the Navier–Stokes equations. In particular, an analysis of the validity of the Einstein relation, which is a consequence of the fluctuation-dissipation theorem, and of Onsager's relations, related to the time reversal symmetry of the microscopic equations of motion, for granular mixtures is given in Sects. 10.6 and 10.7, respectively.

A linear stability analysis of the non-linear Navier–Stokes hydrodynamic equations is carried out in Sect. 10.8 to determine the conditions for stability of the homogeneous cooling state in terms of the wave vector, the dissipation, and the parameters of the mixture. The Navier–Stokes description is used again in Sect. 10.9 to study the segregation due to thermal diffusion. It is important to mention that a new criterion for segregation is found, which is consistent with recent experimental results. Finally, Sect. 10.10 deals with one of the most interesting problems in granular fluids: uniform shear flow. This is a non-equilibrium situation in which the rheological properties of the system cannot be obtained from the Navier–Stokes description, so that more complex constitutive equations are needed. Theoretical developments and computer simulation results are presented in this section in order to clarify the main aspects of this phenomenon.

As explained above, much research has been devoted to the study of hard-body fluids and their mixtures, and some of the most important theoretical developments and computer simulation results are summarized in this book. Nevertheless, it must be pointed out that there are still many unsolved problems, and other problems have only partial solutions that must be improved. Several of these have also been described in different chapters of the book. The reader can thus find in this book a useful reference tool with which to start to work on some of those problems.

I acknowledge the hard work and dedication of all the authors who have contributed to this book. My deepest gratitude for their help and cooperation, and for the great amount of work they put in. Special thanks are due to Drs I. Cachadiña and M.I. Parra for their help in the editing process of this book, and to Dr Tarazona for his helpful suggestions to improve this Introduction.

The authors of this book will be pleased if the readers enjoy "playing with marbles"....

Badajoz (Spain), October 2007 *A. Mulero*

Contents

1

Structure of Hard Spheres and Related Systems

A. Malijevský[1,2] and J. Kolafa[1]

[1] Department of Physical Chemistry, Institute of Chemical Technology, Prague, Technická 5, 166 28 Praha 6, Czech Republic
anatol.malijevsky@vscht.cz, jiri.kolafa@vscht.cz

[2] Center for Biomolecules and Complex Molecular Systems, Prague

The fluid of hard spheres, impenetrable molecules of spherical shape with diameter σ and no attraction, defined by the intermolecular potential

$$u(r) = \begin{cases} \infty & \text{for } r < \sigma \,, \\ 0 & \text{for } r > \sigma \,, \end{cases} \tag{1.1}$$

where r is the intermolecular distance, has served as the simplest nontrivial model of fluid structure for more than half a century. Similarly, the simplest nontrivial model of fluid mixtures is a mixture of additive hard spheres characterized by the mole fractions of components and sphere diameter ratios. There are a number of models of hard nonspherical molecules, "children of hard spheres", such as hard spherocylinders or diatomics whose pair intermolecular potential is a generalization of Eq. (1.1):

$$u(r, \boldsymbol{\omega}) = \begin{cases} \infty & \text{overlap} \,, \\ 0 & \text{nonoverlap} \,, \end{cases} \tag{1.2}$$

where $\boldsymbol{\omega}$ is the vector of angles defining the mutual orientation of molecules. Another generalization of the model are hard-body fluids in an external field, for example, hard spheres at a hard wall or in pores of different geometry. In the next section some important quantities used for characterization of fluid structure of hard spheres and related models are introduced. The main body of the chapter is divided into two sections. In Sect. 1.2 density expansions of correlation functions are discussed, and in Sect. 1.3 approaches based on the Ornstein–Zernike (OZ) integral equation are considered.

1.1 Basic Concepts

The fundamental quantity describing the structure of pure fluids is the pair correlation function, $g(\boldsymbol{r}_1, \boldsymbol{\omega}_1, \boldsymbol{r}_2, \boldsymbol{\omega}_2)$. It represents the normalized probability

Malijevský, A., Kolafa, J.: *Structure of Hard Spheres and Related Systems*. Lect. Notes Phys. **753**, 1–26 (2008)
DOI 10.1007/978-3-540-78767-9_1

of finding molecule 1 in a position given by $\boldsymbol{r}_1$ and orientation $\boldsymbol{\omega}_1$ and molecule 2 at $\boldsymbol{r}_2$ and $\boldsymbol{\omega}_2$. In the canonical NVT ensemble it is defined as

$$g(\boldsymbol{r}_1,\boldsymbol{\omega}_1,\boldsymbol{r}_2,\boldsymbol{\omega}_2) = V^2\frac{N-1}{N}\frac{\int \exp[-\beta U_N(\boldsymbol{r}^N,\boldsymbol{\omega}^N)]\,\mathrm{d}\boldsymbol{r}_3\cdots\mathrm{d}\boldsymbol{r}_N\mathrm{d}\boldsymbol{\omega}_3\cdots\mathrm{d}\boldsymbol{\omega}_N}{z}\;. \tag{1.3}$$

Here z is the configurational integral

$$z = \int \exp[-\beta U_N(\boldsymbol{r}^N,\boldsymbol{\omega}^N)]\,\mathrm{d}\boldsymbol{r}_1\cdots\mathrm{d}\boldsymbol{r}_N\mathrm{d}\boldsymbol{\omega}_1\cdots\mathrm{d}\boldsymbol{\omega}_N\;, \tag{1.4}$$

where N is the number of particles, $\beta = 1/(k_\mathrm{B}T)$, V is the volume of the system, and $\boldsymbol{r}^N$ stands for N vectors $\boldsymbol{r}_1,\ldots,\boldsymbol{r}_N$. Throughout this and the next chapter we will consider the pair additivity approximation of the total potential energy, i.e.,

$$U_N(\boldsymbol{r}^N,\boldsymbol{\omega}^N) = \sum_{i<j} u_{i,j}(\boldsymbol{r}_i,\boldsymbol{\omega}_i,\boldsymbol{r}_j,\boldsymbol{\omega}_j)\;. \tag{1.5}$$

For the homogeneous fluid of spherically symmetric molecules the pair correlation function becomes the radial distribution function (RDF), $g(r)$, where r is the particle separation distance between molecules 1 and 2; similarly, the pair potential becomes $u_{i,j}(\boldsymbol{r}_i,\boldsymbol{\omega}_i,\boldsymbol{r}_j,\boldsymbol{\omega}_j) = u(r)$. The RDF can be generalized to the *cavity distribution function* (also called background correlation function), $y(r)$, which for hard spheres describes the (normalized) probability of finding a pair of spherical cavities r apart. It holds

$$g(r) = y(r)\exp[-\beta u(r)]\;. \tag{1.6}$$

For hard-body systems (the systems for which the pair intermolecular potential is either zero or infinity) the RDF is zero in the overlap region while $y(r)$ is not. Mechanical (nonentropic) thermodynamic quantities may be calculated from the RDF. The pressure expressed conveniently via the compressibility factor $Z = \beta P/\rho$ can be obtained using the pressure (virial) equation

$$Z = 1 - \frac{2}{3}\pi\beta\rho\int_0^\infty \frac{\mathrm{d}u(r)}{\mathrm{d}r} g(r) r^3 \mathrm{d}r \tag{1.7}$$

or after integration from the compressibility equation

$$\frac{1}{\beta}\left(\frac{\partial\rho}{\partial P}\right)_T = 1 + 2\pi\rho\int_0^\infty h(r) r^2 \mathrm{d}r\;, \tag{1.8}$$

where $h(r)$ is the total correlation function,

$$h(r) = g(r) - 1\;. \tag{1.9}$$

Note that the use of Eq. (1.8) requires either a grandcanonical ensemble or a thermodynamic limit in Eq. (1.3) taken before the upper bound limit in integral (1.8) because formal integration of $h(\mathbf{r})$ over a box gives $\partial\rho/\partial P = 0$ in accordance with the $V = \text{const}$ condition in the NVT ensemble.

The excess internal energy with respect to the ideal gas (also called the residual internal energy[1]) is given by

$$U^{\text{ex}} = 2\pi\rho \int_0^\infty u(r)g(r)r^2 \mathrm{d}r \ . \qquad (1.10)$$

For hard spheres, their mixtures, and generally hard body systems the correlation functions and thermodynamic quantities do not depend on temperature, and the energy equation does not play role. Instead for hard spheres there are two equations relating the background correlation function and thermodynamic quantities called the first and the second zero separation theorems. The first is the Hoover–Poirier relation [1]

$$\beta\mu^{\text{ex}} = \ln y(0) \ , \qquad (1.11)$$

where μ^{ex} is the excess chemical potential. The second is the Meeron–Siegert relation [2]

$$\left(\frac{\mathrm{d}\ln y}{\mathrm{d}r}\right)_{r=0} = -6\eta y(\sigma) \ , \qquad (1.12)$$

where $y(\sigma) = g(\sigma+)$ is the contact value of the RDF and η is the packing fraction (the volume occupied by hard spheres to the total volume)

$$\eta = \frac{\pi}{6}\frac{N\sigma^3}{V} \ . \qquad (1.13)$$

Finally, for hard spheres the pressure equation, Eq. (1.7), simplifies to

$$Z = 1 + 4\eta y(\sigma) \ . \qquad (1.14)$$

In further text we set $\sigma = 1$ for hard spheres.

Besides the RDF, the three-body distribution function is important. For hard spheres (more generally for spherically symmetric molecules) it is a function of $\boldsymbol{r}_1, \boldsymbol{r}_2, \boldsymbol{r}_3$, or equivalently three distances between molecules 1, 2, and 3:

$$g_3(r_{12}, r_{13}, r_{23}) = V^3 \frac{N-1}{N}\frac{N-2}{N} \times \frac{\int \exp[-\beta U_N(\boldsymbol{r}^N, \boldsymbol{\omega}^N)]\, \mathrm{d}\boldsymbol{r}_4 \cdots \mathrm{d}\boldsymbol{r}_N \mathrm{d}\boldsymbol{\omega}_4 \cdots \mathrm{d}\boldsymbol{\omega}_N}{z} \ . \qquad (1.15)$$

[1] The residual quantities (with respect to the ideal gas at the system pressure) and excess quantities (with respect to the ideal gas at standard pressure) generally differ.

For systems with linear molecules the pair correlation function depends on the center–center distance and three angles

$$g(\boldsymbol{r}_1, \boldsymbol{\omega}_1, \boldsymbol{r}_2, \boldsymbol{\omega}_2) = g(r, \theta_1, \theta_2, \phi_{12}) \; ,$$

where $\phi_{12} = \phi_1 - \phi_2$. It may be expanded in normalized spherical harmonics [3],

$$g(r, \theta_1, \theta_2, \phi_{12}) = 4\pi \sum_{\ell_1=0}^{\infty} \sum_{\ell_2=0}^{\infty} \sum_{m=-n}^{n} g_{\ell_1 \ell_2 m}(r) Y_{\ell_1 m}(\theta_1, \phi_{12}) Y_{\ell_2 \overline{m}}(\theta_2, 0) \; , \tag{1.16}$$

where $n = \min(\ell_1, \ell_2)$ and $\overline{m} = -m$. Spherical harmonics $Y_{\ell m}$ are

$$Y_{\ell m}(\theta, \phi) = (-1)^m \left[\frac{2\ell+1}{4\pi} \frac{(\ell-m)!}{(\ell+m)!} \right]^{1/2} \exp(im\phi) P_{\ell m}(\cos\theta)$$

for $|m| \leq \ell$ and zero otherwise. In the above formula, $P_{\ell,m}$ is the associated Legendre polynomial, see Appendix A.1 in [3].

The harmonic coefficients are given by

$$g_{\ell_1 \ell_2 m}(r) = \frac{1}{2} \int_{-1}^{1} \int_{-1}^{1} \int_{-\pi}^{\pi} g(r, \theta_1, \theta_2, \phi_{12}) \times Y_{\ell_1 m}(\theta_1, \phi_{12}) Y_{\ell_2 \overline{m}}(\theta_1, 0) \mathrm{d}\cos\theta_1 \mathrm{d}\cos\theta_2 \mathrm{d}\phi_{12} \; . \tag{1.17}$$

In the next section we will discuss density expansions of the pair correlation and related functions that are useful at low and medium densities. In Sect. 1.3 the correlation functions will be discussed for high-density gas and liquid regions.

1.2 Low-Density Region

This section deals with density (diagram) expansions of the pair distribution function, related correlation functions, and the bridge function for one-component hard spheres. Analytical and computer simulation results for the diagrams are shown and discussed.

1.2.1 Density Expansions of the Pair Correlation Functions

The RDF may be expanded in powers of density

$$g(r) = e(r)\, y(r) = e(r) \left[1 + y_1(r)\rho + y_2(r)\rho^2 + \cdots \right] , \tag{1.18}$$

where $e(r) = \exp[-\beta u(r)]$, coefficients $y_i(r)$ in the expansion of the background correlation function are sums of multifold integrals where the integrands are products of the Mayer function

$$f(r) = e(r) - 1 = \exp[-\beta u(r)] - 1 \,. \tag{1.19}$$

For hard spheres $f(r) = -1$ for $r < 1$ and $f(r) = 0$ for $r > 1$. More generally, it holds for hard-body systems and their mixtures $f(r) = -1$ when particles overlap and $f(r) = 0$ otherwise. The integrals in Eq. (1.18) are conventionally depicted by diagrams [4].

The diagrams contributing to $y(r)$ are the connected diagrams with two root (white) points (see Fig. 1.1) separated by distance r and n field (black) points without the articulation points and without a direct root–root bond. The integration is performed over the field points while the bonds connecting points i and j stand for $f(r_{ij})$ bonds. The first two terms are

$$y_1 = y_1(r_{13}) = \int f(\boldsymbol{r}_{12}) f(\boldsymbol{r}_{23}) \mathrm{d}\boldsymbol{r}_2 =$$ 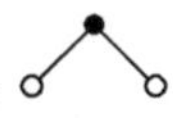 , (1.20)

$y_2 =$ 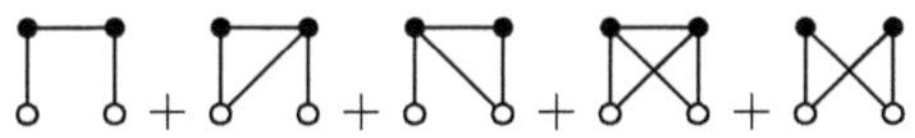 , (1.21)

where because of symmetry $\delta_1(r) = \delta_2(r)$. The above diagrams may be divided into the chain (series) diagrams (at least one field point is the nodal point)

$$\gamma(r) = \gamma_1(r)\rho + \gamma_2(r)\rho^2 + \cdots \,, \tag{1.22}$$

where $\gamma_1 = c_1$, $\gamma_2 = c_2 + \delta_1 + \delta_2$; the parallel diagrams (with the root articulation pair)

$$\pi(r) = \pi_2\rho^2 + \cdots \,, \tag{1.23}$$

where $\pi_2 = \frac{1}{2}\gamma_1^2$ etc., and the elementary (bridge) diagrams (without the root articulation pair and the nodal point)

$$B(r) = E_2(r)\rho^2 + E_3(r)\rho^3 + \cdots \,. \tag{1.24}$$

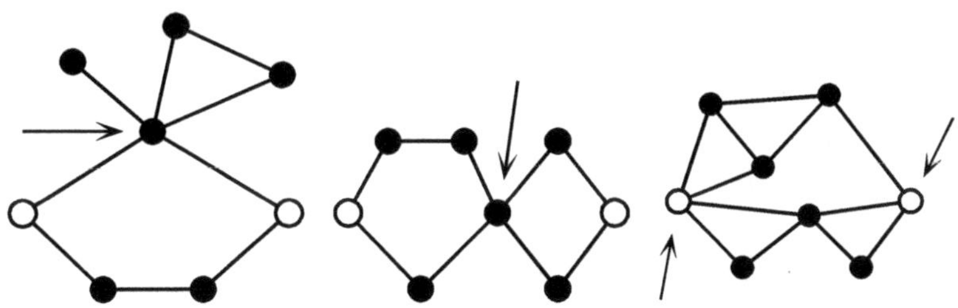

Fig. 1.1. *Left*: if an *articulation point* is removed, the graph splits into components so that at least one of them contains a field (*black*) point. *Middle*: if a *nodal point* is removed, the graph splits into components so that each root (*white*) point appears in a different component. *Right*: if a *root articulation pair* is removed, the graph splits into components

1.2.2 Diagrams Contributing to the RDF

All diagrams in the first two terms in the density expansion of the RDF, Eqs. (1.20) and (1.21), are known analytically [5],

$c_1 =$ 

$$= \begin{cases} \frac{\pi}{12}(r+4)(r-2)^2 & \text{for } 0 \le r < 2\,, \\ 0 & \text{for } r \ge 2\,, \end{cases} \tag{1.25}$$

$c_2 =$ 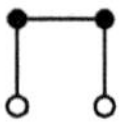

$$= \begin{cases} \frac{\pi^2}{630}(r^6 - 63r^4 + 315r^2 - 525) & \text{for } 0 \le r < 1\,, \\ -\frac{\pi^2}{1260r}(r^3 + 12r^2 + 27r - 6)(r-3)^4 & \text{for } 1 \le r < 3\,, \\ 0 & \text{for } r \ge 3\,, \end{cases} \tag{1.26}$$

$$\delta_2 = \begin{cases} \frac{-\pi^2}{1260}(r^6 - 63r^4 - 70r^3 + 315r^2 + 525r - 1050) & \text{for } 0 \le r < 1\,, \\ \frac{\pi^2}{1260r}(r^5 + 4r^4 - 51r^3 - 10r^2 + 479r - 81) \\ \qquad \times (r-2)^2 & \text{for } 1 \le r < 2\,, \\ 0 & \text{for } r \ge 2\,, \end{cases}$$

$E_2 =$ 

$$= \begin{cases} X_0(r) & \text{for } 0 \le r\,, \\ -\frac{1}{2}\gamma_1^2(r) + \frac{\pi}{2}X_1(r) & \text{for } 1 \le r < \sqrt{3}\,, \\ -\frac{1}{2}\gamma_1^2(r) & \text{for } \sqrt{3} \le r \le 2\,, \\ 0 & \text{for } r > 2\,, \end{cases} \tag{1.27}$$

$$\begin{aligned} X_0(r) = \frac{\pi}{2}\Bigg[& \pi\left(-\frac{3r^6}{560} + \frac{r^4}{15} - \frac{r^3}{9} - \frac{r^2}{2} + \frac{22r}{15} - \frac{5}{6} - \frac{9}{35r}\right) \\ & + \left(-\frac{3r^4}{280} + \frac{41r^2}{420}\right)(3-r^2)^{1/2} \\ & + \left(-\frac{23r}{15} + \frac{36}{35r}\right)\arccos\frac{r}{(12-3r^2)^{1/2}} \\ & + \left(\frac{3r^6}{560} - \frac{r^4}{15} + \frac{r^2}{2}\arccos\frac{r^2-2}{4-r^2}\right) \\ & + \left(\frac{2}{15}r - \frac{9}{35r}\right)\arccos\frac{-12+11r^2-2r^4}{12-3r^2}\Bigg]\,, \end{aligned}$$

$$\begin{aligned} X_1(r) = & \left(-\frac{3}{280}r^4 + \frac{41}{420}r^2\right)(3-r^2)^{1/2} \\ & + \left(-\frac{23}{15}r + \frac{36}{35r}\right)\arccos\frac{r}{(12-3r^2)^{1/2}} \\ & + \left(\frac{3}{560}r^6 - \frac{r^4}{15} + \frac{r^2}{2} + \frac{2r}{15} - \frac{9}{35r}\right)\arccos\frac{r^2+r-3}{(12-3r^2)^{1/2}} \\ & + \left(\frac{3}{560}r^6 - \frac{r^4}{15} + \frac{r^2}{2} - \frac{2r}{15} + \frac{9}{35r}\right)\arccos\frac{-r^2+r+3}{(12-3r^2)^{1/2}}\,. \end{aligned}$$

The third term in Eq. (1.18), $y_3(r)$, was calculated numerically using the Monte Carlo (MC) integration [6]. Recently, the higher order terms up to $y_6(r)$ have been obtained and fitted to splines [7]. The calculations were based on the Ree–Hoover representation of the diagrams [8, 9], symbolic algebra programming and automatic code generation (see, e.g., [10, 11, 12]), and the MC integration [13] utilizing the spanning diagrams. The results are shown in Fig. 1.2. Figure 1.3 shows convergence of the truncated series $g_n(r)$ to the simulated RDF at the medium fluid density $\rho = N\sigma^3/V = 0.5$.

Due to the discontinuous nature of the hard-sphere potential, functions $y_n(r)$ are discontinuous at higher derivatives. The most important are those at $r = 1, 2, \ldots, \mathrm{n}+1$ (n + 1 denotes the distance beyond which $y_n(r)$ becomes zero); at $r = 2$ the second derivative is discontinuous, and other discontinuities are of higher order. In addition, there are discontinuities at noninteger points. The most important is at $r = \sqrt{3}$ which is of $\frac{9}{2}$th order (term $\propto (r - 3^{1/2})^{9/2}$), as can be derived from Eq. (1.27).

We are not aware of any y_n, $n > 3$ data on hard-sphere mixtures and on nonspherical hard-body systems. This gap is caused by a larger number of degrees of freedom. For example, for one-component hard spheres one has the only diagram contributing to $y_1(r)$, see Eq. (1.20). For a binary mixture of hard spheres there are five diagrams, and the number of diagrams rapidly increases with the increasing order of the diagram. For one-component fluids of axially symmetric molecules the number of diagrams is the same as for pure

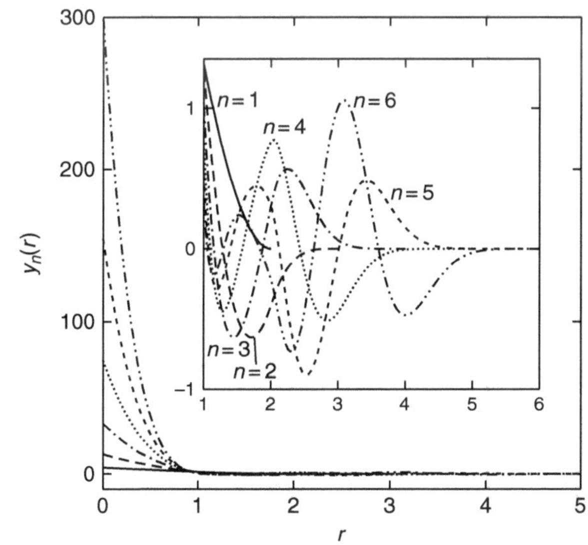

Fig. 1.2. The expansion coefficients $y_n(r)$ of the background correlation function, Eq. (1.18), for hard spheres

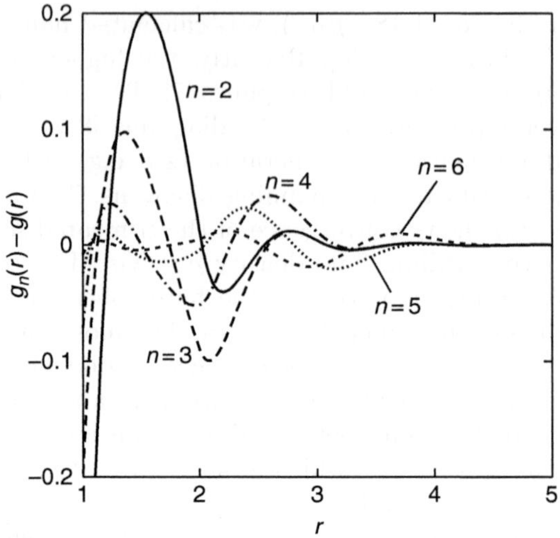

Fig. 1.3. Convergence of the expansion to the RDF at $\rho = 0.5$ [7]. The differences of the partial sums from the simulation results are shown

hard spheres. However, y_n are functions not only of the particle separation but also of three angles defining mutual orientation of the molecules, $y_n = y_n(r, \theta_1, \theta_2, \phi)$. In the case of nonlinear hard-body rigid molecules the situation is even worse; y_n are functions of the distance and five angles.

1.2.3 Elementary Diagrams

The elementary diagrams, the terms in the density expansion of the bridge function (1.24), play a key role in theories based on the OZ equation, see Sect. 1.3.2. The first two terms in the density expansion are zero. The second-order term is known analytically (1.27). The higher coefficients must be calculated numerically. E_3 was calculated by Attard and Patey [14]. The terms E_3 to E_5 were calculated by Labík et al. [15]. Recently, E_6 was calculated and the lower order terms precisely recalculated by Kolafa and Labík [7].

Figure 1.4 shows the expansion coefficients $E_n(r)$ for $n = 2$ to 6. It is interesting that while $E_2(r)$ is negative in the whole range of particle separations, the higher order terms change sign, see also a discussion in Sect. 1.3.5. Figure 1.5 shows convergence of the truncated series $B_n(r)$ to the simulated $B(r)$ at the medium fluid density $\rho = N\sigma^3/V = 0.5$.

The only results for binary mixtures of hard spheres are E_2 and E_3 diagrams calculated by Attard and Patey [14]. There are no literature results for E_n of nonspherical hard-body systems.

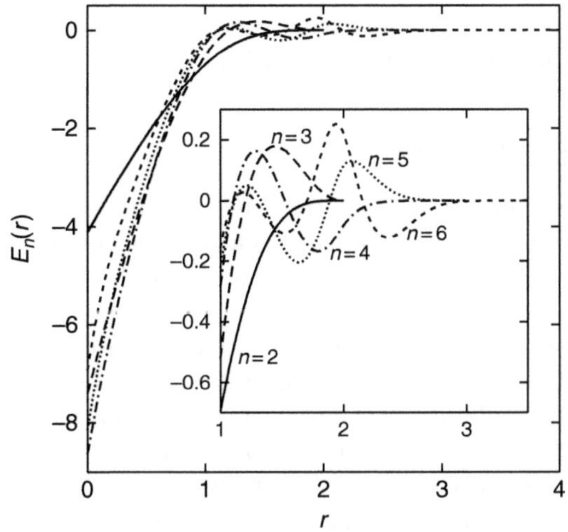

Fig. 1.4. The expansion coefficients $E_n(r)$ of the bridge function for hard spheres

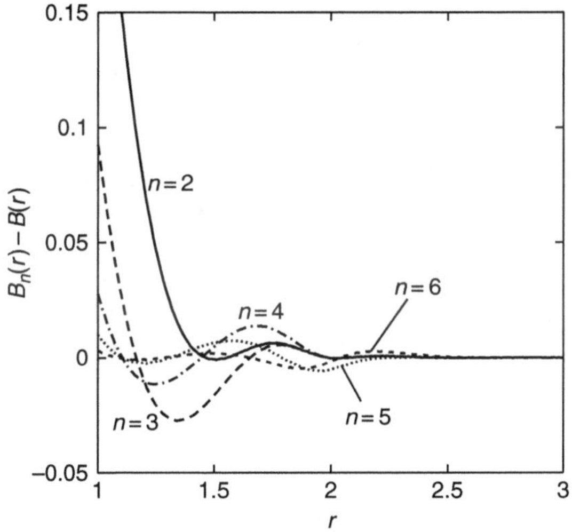

Fig. 1.5. Convergence of the expansion to the bridge function at $\rho = 0.5$. The differences of the partial sums from the bridge function are shown. The values of the bridge function are from [16]

1.3 High-Density Region

Besides computer simulations, there are a number of theoretical approaches of studying fluid structure. In this section we will concentrate on the methods based on the OZ equation. Theories starting from the Born–Green–Yvon hierarchy [4] and related approaches [17, 18, 19] are not considered here.

Chapter 6 deals with an alternative rational function approximation and Chap. 7 with the density functional theories.

1.3.1 Computer Simulations

The RDF can be simulated using the MC and molecular dynamics (MD) methods [13]. The first simulations go back to fifties of the last century [20]. Since then a number of simulations of the RDF have been performed for hard-sphere fluids (for a review on early results see, e.g., [21, 22]). For a long time and even now the most frequently cited and used are those of Barker and Henderson [23, 24]. However, their precision is low because of the low power of computers available in seventies of the last century: only 108 particles in a simulation box, only 10^6 configurations generated, only separation step size 0.02, and only reduced distances up to 2.27 were considered. At present, much more powerful computers and new algorithms are available allowing calculations with by two-order better precision, at least. The present state-of-the-art demonstrate recent data of Kolafa et al. [16], see Sect. 1.3.5 for details.

There are a number of simulations on the pair correlation function of fluids with hard linear molecules such as hard spherocylinders, hard ellipsoids of revolution, hard diatomics, or hard linear triatomics. The data are often presented in the form of the spherical harmonic coefficients $g_{\ell_1\ell_2m}(r)$, Eq. (1.17). Less often, radial slices through $g(r,\theta_1,\theta_2,\phi)$ and various angle averages are used [22]. Fluids with nonlinear hard molecules are studied rarely, see, e.g., [25, 26].

The background correlation function for hard spheres in the overlap region was simulated by Patey and Torrie [27], Ballance and Speedy [28], and Labík et al. [29, 30, 31]. The three-body distribution function was studied using computer simulations rarely [32, 33].

1.3.2 The OZ Equation

The OZ integral equation defines the direct correlation function c in terms of the total correlation function h given by Eq. (1.9). For homogeneous one-component systems the OZ equation has the form

$$h(1,2) = c(1,2) + \rho \int c(1,3)h(3,2)\mathrm{d}3 \,. \tag{1.28}$$

Symbol d3 is an abbreviation for $\mathrm{d}\boldsymbol{r}_3\mathrm{d}\boldsymbol{\omega}_3$ and 1 stands for $\boldsymbol{r}_1, \boldsymbol{\omega}_1$ in arguments of the correlation functions. If the system is isotropic, the equation becomes

$$h(r) = c(r) + \rho \int c(|\boldsymbol{r}-\boldsymbol{s}|)h(s)\mathrm{d}\boldsymbol{s} \tag{1.29}$$

or using the bipolar coordinates (see, e.g., [34])

$$h(r) = c(r) + \frac{2\pi\rho}{r}\int_0^\infty sh(s)\int_{|r-s|}^{r+s} tc(t)\mathrm{d}t\,\mathrm{d}s\,. \tag{1.30}$$

Baxter proposed two different (but equivalent) forms of Eq. (1.30), see [35, 36], which have some advantages but are not often used.

On taking the Fourier transform of Eq. (1.28), we obtain

$$\tilde{h}(k) = \tilde{c}(k) + \rho\tilde{h}(k)\tilde{c}(k)\,, \tag{1.31}$$

where

$$\tilde{h}(k) = \frac{4\pi}{k}\int_0^\infty h(r)\sin(kr)r\mathrm{d}r\,. \tag{1.32}$$

For axially symmetric systems the OZ equation in Fourier space is

$$\tilde{h}(k) = \tilde{c}(k) + \rho\int \tilde{c}(1,3)\tilde{h}(3,2)\mathrm{d}\boldsymbol{\omega}_3 \tag{1.33}$$

or in terms of the harmonic coefficients

$$\tilde{h}_{\ell_1\ell_2m}(k) = \tilde{c}_{\ell_1\ell_2m}(k) + (-1)^m\rho\sum_{\ell_3=m}^{\infty}\tilde{c}_{\ell_1\ell_3m}(k)\tilde{h}_{\ell_3\ell_2m}(k)\,. \tag{1.34}$$

The multicomponent generalization of Eq. (1.28) can be written as

$$h_{ij}(1,2) = c_{ij}(1,2) + \rho\sum_{k=1}^{n}x_k\int c_{ik}(1,3)h_{kj}(3,2)\mathrm{d}3\,. \tag{1.35}$$

where n is the number of components and x_k is the mole fraction of component k. For example, for a binary mixture of spherically symmetric molecules the above equation is (note that $c_{ij} = c_{ji}$ and $h_{ij} = h_{ji}$)

$$\begin{aligned}
h_{11}(r) &= c_{11}(r) + \rho x_1\int c_{11}(|\boldsymbol{r}-\boldsymbol{s}|)h_{11}(s)\mathrm{d}\boldsymbol{s} + \rho x_2\int c_{12}(|\boldsymbol{r}-\boldsymbol{s}|)h_{21}(s)\mathrm{d}\boldsymbol{s}\,,\\
h_{12}(r) &= c_{12}(r) + \rho x_1\int c_{11}(|\boldsymbol{r}-\boldsymbol{s}|)h_{12}(s)\mathrm{d}\boldsymbol{s} + \rho x_2\int c_{12}(|\boldsymbol{r}-\boldsymbol{s}|)h_{22}(s)\mathrm{d}\boldsymbol{s}\,,\\
h_{22}(r) &= c_{22}(r) + \rho x_1\int c_{21}(|\boldsymbol{r}-\boldsymbol{s}|)h_{12}(s)\mathrm{d}\boldsymbol{s} + \rho x_2\int c_{22}(|\boldsymbol{r}-\boldsymbol{s}|)h_{22}(s)\mathrm{d}\boldsymbol{s}\,.
\end{aligned} \tag{1.36}$$

For inhomogeneous (nonuniform) fluids, the simplest example is a fluid of hard spheres at a hard wall, two approaches can be used. The first is a generalization of Eq. (1.28) to a fluid in the presence of an external field

$$h(1,2) = c(1,2) + \int \rho(3)c(1,3)h(3,2)\mathrm{d}3\,, \tag{1.37}$$

where $\rho(3)$ is the local density. The pair correlation function depends on the interparticle distance and on distances of particles from a surface, e.g., $g = f(r_{12}, z_1, z_2)$.

The second and a simpler form (the singlet nonuniform OZ equation) has been proposed by Henderson et al. [37], see also [38, 39, 40] and [14].

Consider, for example, a binary mixture of large and small hard spheres in the limit $\sigma_{\text{large}} \to \infty$, $x_{\text{large}} \to 0$. In this limit the set of three Eqs. (1.36) decouples to an equation for the bulk–bulk interaction (solved using standard methods) and the bulk–wall interaction (giving the particle-surface density profile $g(z)$)

$$h(z) = c(z) + \rho_{\text{b}} \int c_{\text{b}}(|\boldsymbol{r} - \boldsymbol{s}|)h(s)\text{d}\boldsymbol{s} \ , \tag{1.38}$$

where ρ_{b} and c_{b} are the bulk density and direct correlation function, respectively, and z is the distance from the surface. The third equation describes the surface–surface interaction.

1.3.3 The Closures

Application of the OZ equation requires a closure, i.e., another relation between h and c. Generally, it holds for spherically symmetric systems

$$h(r) = \exp[-\beta u(r) + \gamma(r) + B(r)] - 1 \ , \tag{1.39}$$

where

$$\gamma(r) = h(r) - c(r) \tag{1.40}$$

is the indirect (chain) correlation function and $B(r)$ is the bridge function. Equation (1.39) does not yet provide a closure. Although the bridge function is exactly defined as a sum of elementary diagrams, see Eq. (1.24), it is in fact an unknown function of interparticle distance. From this point approximations are introduced for $B(r)$, as described below.

Classical Closures

The simplest closure is the hypernetted chain (HNC) approximation

$$B_{\text{HNC}}(r) = 0 \tag{1.41}$$

postulating the sum of elementary diagrams to vanish. Another classical closure is the Percus–Yevick (PY) closure relating the bridge and the chain functions

$$B_{\text{PY}}(r) = \gamma(r) - \ln[\gamma(r) + 1] \ . \tag{1.42}$$

It is based on an idea that the sums of elementary and parallel diagrams cancel. The HNC and PY theories can also be derived in another, perhaps a more elegant, way—from a first-order functional Taylor expansions [4].

The HNC and PY theories were first tested on computer simulation data on the pair distribution function and pressure of pure hard spheres. Results are excellent at low densities, agree well with simulations at medium densities, but are poor at high densities, see, e.g., [41].

Second-Order PY and HNC

Many attempts have been made to improve the performance of the PY and the HNC closures. Probably, the most systematic way to do it is to extend the functional expansions including second-order terms [42, 43, 44]. The solutions are much more computer time demanding [45] than those of the first-order OZ theories.

Truncated Series of the Elementary Diagrams

The direct way to a theoretically based closure is to use the definition of the bridge function as a series of the elementary diagrams, Eq. (1.24), and to truncate it at some order in density. Unfortunately, the series converges slowly at high densities, see, e.g., [7, 15]. To accelerate convergence, Stell [46] proposed to replace f bonds in Eq. (1.24) by h bonds. This proposal has been recently tested by Kwak and Kofke [47]. The authors calculated modified (density-dependent) elementary diagrams up to order ρ^4. Their results at high densities are not better than those using f-bond expansion and terms up to order ρ^6.

Self-Consistent Theories

Another approach to improving performance of the classical theories is called self-consistent. It is based on two facts. First, approximative OZ theories give different pressures when the pressure Eq. (1.7) or the compressibility Eq. (1.8) is applied. Second, the pressure PY equation for pure hard spheres underestimates and the compressibility equation overestimates computer simulation results. The converse holds for the HNC solutions. These two facts led Rowlinson [48] to the proposal of a closure

$$B_{\mathrm{R}}(r) = \alpha B_{\mathrm{PY}}(r) + (1-\alpha) B_{\mathrm{HNC}}(r)\,, \tag{1.43}$$

where B_{PY} and $B_{\mathrm{HNC}}(r)$ are given by Eqs. (1.42) and (1.41), respectively, and α is a density-dependent parameter obtained in such a way that the consistency between the pressure and the compressibility equation is guaranteed. A number of other self-consistent (or nearly self-consistent) closures have been proposed [49, 50, 51, 52, 53, 54, 55, 56, 57, 58, 59, 60].

Phenomenological Closures

This group of theories is based on different arguments. Probably, the most known is the Martynov–Sarkisov (MS) [61] closure

$$B_{\mathrm{MS}}(r) = 1 + \gamma(r) - [1 + 2\gamma(r)]^{1/2} \,. \tag{1.44}$$

This closure, while as simple as the PY or the HNC ones, gives much better results. Its disadvantage is that it loses solution when the function under the square root becomes negative which typically happens at high densities.[2] To overcome it, Ballone et al. [64] proposed a modification of the equation. Another modification of Eq. (1.44) was proposed by Vompe and Martynov [54].

Verlet proposed for hard spheres a closure [65]

$$B_{\mathrm{V}}(r) = \frac{1}{2}\frac{\gamma^2}{1 + A\gamma} \,, \tag{1.45}$$

where $A = 4/5$ was taken ad hoc. This closure was empirically modified by Labík et al. [66] by making A state dependent on η, $A = 1.1 - \frac{2\eta}{\pi}$ (called Verlet modified or shortly VM), and by Henderson et al. [67] with $A = \frac{17}{120\eta} + 0.515 - 0.221\eta$. The former was used also for nonspherical hard-body fluids, with η being the packing fraction, while the latter was proposed for hard spheres only. These modified Verlet closures give slightly better results than the original Verlet closure, and the MS closure and its extensions.

The RHNC Closure

The reference hypernetted chain approximation (RHNC) postulates that the bridge function of a system under study is the same as the bridge function of a reference system [68, 69, 70, 71, 72, 73]

$$B = B_{\mathrm{ref}} \,. \tag{1.46}$$

The simplest nontrivial system of pure hard spheres is chosen, as a rule, as the reference system

$$B_{\mathrm{ref}} = B_{\mathrm{HS}}(d) \,, \tag{1.47}$$

where d is an effective sphere diameter, which is usually obtained using a consistency condition.

---

[2] One of the authors, G. A. Martynov [62, 63] believes that this is an advantage of the closure; losing solution is an indication of phase transition from fluid to solid. However, this view is not generally shared.

1.3.4 Numerical Solutions of the OZ Equation

The OZ equation can be rarely solved analytically. One of exceptions is the PY solution for hard spheres [74, 75, 76, 77]. In most cases one has to resign to numerical solutions.

A number of numerical methods have been developed. Let us start from the OZ equation expressed in the form of Eq. (1.30) coupled with a chosen closure. We have an integral equation for an unknown function $\gamma(r)$. The first thing in its numerical implementation is replacing infinity in the upper bound of the integral by a suitably chosen $r_{\max}$; typically, $r_{\max} \approx 10$ for hard spheres. The next step is a discretization of integrals using, say, the trapezoidal rule with a number of grid points N and step size Δr, $N\Delta r = r_{\max}$ (typically $\Delta r \approx 0.01$ or less). In this way we obtain a set of N nonlinear equations for N unknown values of $\gamma(r_i)$.

The simplest way of solving the set is the Picard (or direct iteration) method [78]. Starting from a suitably chosen initial approximation (either a low density limit, $\gamma^0(r_i) = 0$, or a nearest lower density result) in the right-hand-side (rhs) of the equation a new left-hand-side result is obtained and the procedure is repeated until convergence is reached,

$$\gamma^{j+1} = \mathrm{rhs}(\gamma^j) \ .$$

An advantage of this numerical method is its simplicity. Its disadvantage is a bad convergence or even divergence at medium and high densities, which can be improved by mixing (damping) the iterations with the previous value,

$$\gamma^{j+1} = \omega\gamma^j + (1-\omega)\mathrm{rhs}(\gamma^j) \ .$$

At high densities the dumping parameter ω is usually close to unity and the convergence is slow.

A simple variant of the damped iterations method based on smart automatic choice of the damping parameter in each step has been proposed by Kolafa [79]. It is surprisingly efficient even at high densities.

To overcome the slow convergence problem, the Newton–Raphson method can be used instead [80]. A disadvantage of it is a solution of large matrixes $N \times N, N \approx 1000$, at each iteration step. Gillan [81] proposed a compromise between the Picard and the Newton–Raphson methods that partly suppresses their shortcomings.

The slow calculation of the convolution (N^2 operations) can be improved by evaluating the necessary convolutions in the k space ($N \log N$ operations) or most advantageously by rewriting the whole OZ equation to the Fourier form (1.31). The Picard iteration scheme then becomes [82]

$$\gamma(r) \longrightarrow c(r) \longrightarrow \tilde{c}(k) \longrightarrow \tilde{\gamma}(k) \longrightarrow \gamma(r) \ .$$

The convergence is much faster and robust than when using the Picard iterations in the configuration space, but still there are problems with convergence at high densities.

A number of other sophisticated numerical algorithms for solving the OZ equation have been proposed [83, 84, 85, 86, 87, 88]. We have practical experience with the LMV method of Labík et al. [84] and will describe it in more detail:

1. Initial approximation $\gamma^0(r)$. Either low density limit $\gamma^0(r) = 0$ or a nearest lower density result.
2. Linearization of the closure (1.39), where $c = h - \gamma$ with respect to γ
$$c(r_i) = c^0(r_i) + \left(\frac{\mathrm{d}c(r_i)}{\mathrm{d}\gamma(r_i)}\right)^0 [\gamma(r_i) - \gamma^0(r_i)] \ , \ i = 0, \ldots, N \ .$$
3. Fourier transform of the linearized closure
$$\tilde{c}(k_j) = \tilde{c}^0(k_j) + \sum_{\ell=0}^{N} D^0(k_j, k_\ell)[\tilde{\gamma}(k_\ell) - \tilde{\gamma}^0(k_\ell)] \ , \ j = 0, \ldots, N \ .$$
4. Solution of the set of quadratic equations in Fourier space
$$\tilde{\gamma}(k_j) = \rho\tilde{c}(k_j)[\tilde{c}(k_j) + \tilde{\gamma}(k_j)] \ , \ j = 0, \ldots, N,$$
for N unknown $\tilde{\gamma}(k_j)$. For the first M (the dominant) components $M \approx 100$, the set is solved using the Newton–Raphson method; the remaining components are calculated by the Picard iterations.
5. New values of $\gamma(r_i)$ are obtained by the inverse Fourier transform and the procedure is repeated until convergence is considered to be obtained using the criterion
$$\left(\frac{1}{N}\sum_{i=0}^{N}[\gamma(r_i) - \gamma^0(r_i)]^2\right)^{1/2} < \epsilon \ .$$

The algorithm is both efficient (i.e., rapidly convergent and thus not much demanding on the computer time) and robust (i.e., insensitive to an initial approximation). To test the robustness, we solved the OZ equation hard spheres at densities beyond the density of crystallization, i.e., in the metastable and unstable regions [89]. The algorithm converged even at densities far beyond the random close packing density.

The algorithm can be straightforwardly extended to mixtures of hard spheres [90] (more generally to fluid mixtures of spherical molecules) and to nonuniform fluids within the Henderson–Abraham–Barker [37] form of the OZ equation for density profiles [91, 92, 93]. The extension of it to fluids of linear molecules is more difficult and it is described in detail in [94].

A disadvantage of the LMV algorithm is that it must be reprogrammed for systems in different geometries, which is demanding on a programmer's time. In some cases it may be more convenient to use general packages for solution of sets of nonlinear equations [95, 96, 97].

The LMV method has not been extended to the second-order OZ equation, nor to the nonuniform OZ equation. For the latter case a method of numerical solution was suggested by Sokołowski [98, 99].

1.3.5 The Bridge Function for Hard Spheres

Although the bridge function plays a fundamental role in theories based on the OZ equation, see Sect. 1.3.3, it has not been accurately known until recently even for the simplest model (after ideal gas) of one-component hard spheres. The closures to the OZ equation are of little use for this purpose because of their approximative nature. The direct and accurate way to determine $B(r)$ is the sum of elementary diagrams, Eq. (1.24). The diagrams only up to the order ρ^6 are known, see Sect. 1.2.3. This is sufficient for determining $B(r)$ precisely at low and medium densities (for, say, $\rho \leq 0.5$, [7]), see Fig. 1.5. At high densities the series (1.24) converges slowly and calculation of higher order terms is necessary. However, this is too painstaking. Attempts to accelerate convergence by using Padé approximants [14] or by replacing f bonds by h bonds in the elementary diagrams [47] do not improve convergence much. The bridge function can be obtained from simulated RDF via the OZ equation. Early attempts in this direction [100, 101] were of low precision due to a low accuracy of $g(r)$ simulation data. Later Malijevský and Labík [102, 103] proposed an empirical formula (ML)

$$B_{\mathrm{ML}}(r) = -b^2(r) , \tag{1.48}$$

where

$$b(r) = \begin{cases} (a_1 + a_2 x)(x - a_3)(x - a_4)/(a_3\, a_4) & \text{for } x \leq a_4 , \\ A_1 \exp[-a_5(x - a_4)] \sin[A_2(x - a_4)]/r & \text{for } x > a_4 , \end{cases} \tag{1.49}$$

and $x = r - 1$. Free parameters a_i as functions of density were fitted to computer simulation data on $g(r)$. (A_1 and A_2 are combinations of a_i that guarantee continuity of $b(r)$ and its first derivative at $x = a_4$.) The ML formula precisely reproduces RDF and thermodynamic simulation data for densities up to $\rho = 0.86$. It extrapolates them accurately to higher fluid densities, see [104] where it was extended to the overlap region. It even predicts (with an unexpectedly good accuracy) simulation RDF data in the vicinity of the random close packing density [89]. While it is reasonably accurate, it has one principal defect: it postulates that $B(r)$ does not change sign. First, the main reason for introducing this approximation was to increase the numerical stability of solving a set of highly nonlinear equations to obtain parameters a_i. Second, the approximation seemed to be reasonable because (i) the known closures to the OZ equation for hard spheres with the exception of the second-order PY theory [45] predict nonpositive values of $B(r)$, and (ii) the first elementary diagram $E_2(r)$ is also nonpositive. The postulate of nonpositivity of the bridge function was questioned by Rast et al. [105] and independently by Yuste et al. [106, 107]. Of course, these findings are inconclusive due to the approximations involved. However, the doubts are further supported by the fact that the higher elementary diagrams, E_3 to E_6, also change sign. Thus, the problem of true dependence of the bridge function has been recently reopened.

There is a direct and free of any approximation way of determination $B(r)$ from simulated values of the RDF, $g(r)$. It consists of four steps:

1. Calculation of the Fourier transform, $\tilde{h}(k)$, of the computer simulation values of $h(r) = g(r) - 1$.
2. Calculation of $\tilde{c}(k)$ from the Fourier transform of the OZ Eq. (1.31).
3. Calculation of $c(r)$ using the inverse Fourier transform.
4. Calculation of $B(r)$ from Eq. (1.39).

Why this simple noniterative method has not been used earlier, but rather painstaking methods have been developed? Because this method is numerically unstable as we will see. The success of the approaches combining the OZ equation with approximate closures is due to the relative insensitivity of the pair distribution function to the bridge function. For example, the simplest HNC theory postulating incorrectly $B(r) = 0$ gives reasonably good values of $g(r)$ and moreover predicts exactly the second and third virial coefficients. Therefore, it is not a surprise that if the way $B(r) \longrightarrow g(r)$ is stable the reverse way $g(r) \longrightarrow B(r)$ may be unstable. To prove it, the following numerical experiment was performed [16]. The propagation of small errors in $g(r)$ to $B(r)$ can be expressed as

$$\delta B(r) = \int_1^{r_{\max}} \frac{\partial B(r)}{\partial g(r')} \delta g(r') \mathrm{d}r' ,$$

where $r_{\max}$ is a maximal distance for which $g(r)$ is known from simulations, and $\delta g(r)$ is an uncertainty of simulated $g(r)$. Assuming that $\delta g(r)$ is a small constant, $\delta g = 0.001$ on interval $r \in [1, r_{\max}]$, we obtain for density $\rho = 0.94$

$$\delta B(r) = 2900 \delta g .$$

This alarming result indicates that a systematic uncertainty in $g(r)$ of 0.001 (very accurate simulation data!) causes an uncertainty in $B(r)$ of 2.9, which is almost twice its contact value at the given density. Let us conclude that the above-described method becomes useless unless highly accurate computer simulation data are used and corrected for any systematic errors.

Detailed analysis [16] shows that there are four sources of errors of the computer simulation data, which are described below.

1. **Finite size errors.** These errors are caused by a finite number of particles in the system and can be divided into the *ensemble error*, which is caused by fluctuations of measured quantities, and the *periodic error*. The ensemble error is of the order of $1/N$ (so that a very large number of particles would be needed to obtain accurate results) and vanishes for hard bodies in the grandcanonical μNT ensemble (simulations in this ensemble are possible, but inefficient). The most efficient approach is to simulate

a moderately large system in the NVT ensemble and to recalculate the results to the μVT via

$$g_{\mu\mathrm{VT}}(r) = g_{\mathrm{NVT}}(r) + \frac{1}{2N\beta}\frac{\partial^2 \rho^2 g(r)}{\partial \rho^2}\left(\frac{\partial \rho}{\partial p}\right)_\beta + \mathcal{O}(N^{-2}) \, .$$

The periodic errors are caused by the influence of the periodic boundary conditions. Their correction is difficult. Fortunately, the periodic errors decay exponentially with the number of particles in a basic simulation box and may be neglected for moderately large values of N.

2. **Grid size errors.** The $g(r)$ data are collected as average values over the intervals $[r_i - \Delta r/2, r_i + \Delta r/2]$. Too large Δr causes error in the numerical calculation of

$$\tilde{h}(k) = \frac{4\pi}{k}\int_0^\infty rh(r)\sin(kr)\,\mathrm{d}r \, .$$

Too small Δr increases statistical errors in $g(r_i)$. The best strategy, we believe, is to use sophisticated methods of the numerical quadrature, and to set an "optimal" Δr by trial and error.

3. **Tail errors.** The RDF is simulated up to a cut-off distance r_{max} that depends on the number of particles in a basic box and on density. Setting $g(r) = 1$ beyond this distance results in two errors in $B(r)$. First, it systematically influences its values in the vicinity of r_{max}. Second, which is much more important, it causes a severe error in the evaluation of $\tilde{h}(k)$. At high densities the upper bound should be truncated at distances 20–30, which are much larger distances than the typical values of r_{max}. The tail error may be suppressed by using a suitable extrapolation formula for $h(r)$ [63] and fitting their constants to the simulation data:

$$h_{\mathrm{tail}}(r) = \exp(T_0 - T_1 r)\cos(T_2 + T_3 r)/r \, .$$

4. **Statistical errors.** They may be suppressed by using a large number of generated configurations and by choosing a suitable Δr. Fortunately, the statistical errors are random (at least in principle). So, they do not influence $B(r)$ as drastically as the systematic errors.
The up-to-date data [16] contain uncertainties in the resulting values of $g(r)$ of about 0.002 in the vicinity of contact and 0.0001 at large particle separations. Uncertainties in $B(r)$ are of the order of 0.001.

Figure 1.6 compares the bridge function for hard spheres obtained in this way with the ML formula, Eqs. (1.48) and (1.49), at $\rho = 0.94$.

The bridge function changes sign in the regions of maxima. For instance, at $\rho = 0.94$ ($\eta = 0.4922$) the first maximum is positive (0.004 ± 0.001), which is not visible on the scale of Fig. 1.6. The positive maxima at larger separations are well seen.

The above-described calculations have been extended [108] to the whole range of fluid densities from very low ones up to a deep metastable region

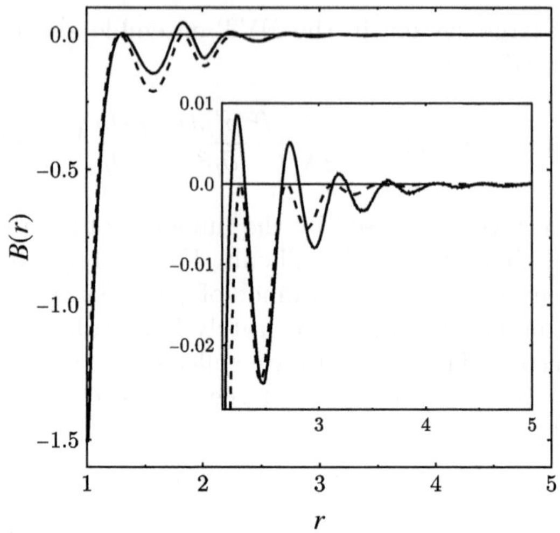

Fig. 1.6. The bridge function obtained from approximate formulas and computer simulations [16] at $\rho = 0.94$. *Solid line*: simulations, *dashed line*: approximation (1.48). Note that (1.50) is indistinguishable from the simulations within the scale of the graph

($\rho = 1.02$). The data on $B(r)$ were fitted as a function of interparticle separation distance

$$B(r) = \begin{cases} \sum_{i=0}^{13} a_i T_i(r) & \text{for } 1 \le r \le 2\,, \\ \sum_{i=0}^{9} b_i T_i(r) & \text{for } 2 < r < 3\,, \\ \frac{1}{r^2}\{\exp(-1.25r)\,[c_0\cos(13.75r) + c_1 \sin(13.75r)] & \\ \quad + c_2 \exp(-1.5r)\} & \text{for } r > 3\,. \end{cases} \tag{1.50}$$

Finally, constants a_i, b_i, and c_i were accurately fitted as functions of density.

The numerically accurate bridge function allows testing of approximative closures of Sect. 1.3.3. All the closures known in literature are based either on the assumption

$$B(r) = f[\gamma(r)], \tag{1.51}$$

or on a more general assumption

$$B(r) = f[\gamma(r), \rho], \tag{1.52}$$

or they do not postulate any analytical relation between the bridge function and the series function. The PY, HNC, and MS closures belong to the first category, and all the self-consistent approximations and the VM closure to the second. Typical examples of the third category are closures based on truncated series of elementary diagrams.

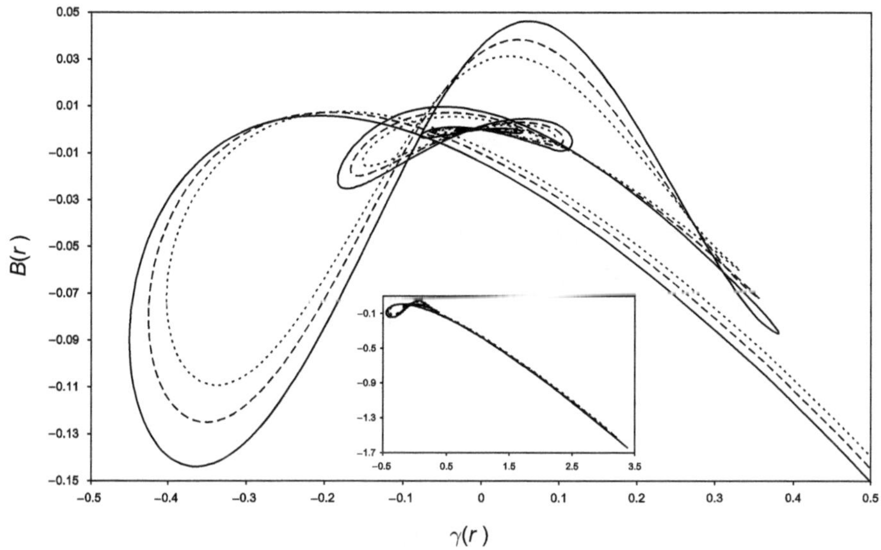

Fig. 1.7. Dependence of the bridge function on the series function at three selected densities: $\rho = 0.90$ (*dotted line*), $\rho = 0.92$ (*dashed line*), $\rho = 0.94$ (*solid line*)

Figure 1.7 plots $B(r)$ against $\gamma(r)$ at three selected (high) densities. It is seen that assumption (1.51) about the functional dependence is wrong in principle, as B is not a function of γ. Assumption (1.52) is incorrect, too.

1.3.6 What Closure to Use?

The previous subsection ended somewhat pessimistically—none of currently available closures agrees with the simulated bridge function for pure hard spheres. The exceptions may be truncated series of elementary diagrams either with f bonds or with h bonds. Unfortunately, they converge slowly and are of little use in the most interesting region of high densities. However, the situation is not as bad as it seems from Fig. 1.7.

The closures to the OZ equation have been used for decades not only for hard spheres but also for more complex fluids. They structural and thermodynamic predictions agree well and sometimes even excellently with computer simulation data. So, what is wrong? Consider Fig. 1.8. It compares the bridge function extracted from the simulations with the older ML formula and the bridge function predicted by the PY, MS, and VM theories at $\rho = 0.94$. There are two distinct regions in the figure (and also in Fig. 1.7): the region of large γ roughly from $\gamma = 0.5$ to 3.5 and the region of small γ from -0.5 to 0.5. In the first, "regular", region $B(r)$ is a monotonous function of $\gamma(r)$. The ML results almost perfectly agree with the simulated bridge function. Also the MS and VM closures are in an excellent agreement. The PY values of the bridge function are low. In the "irregular" region of small values of γ the

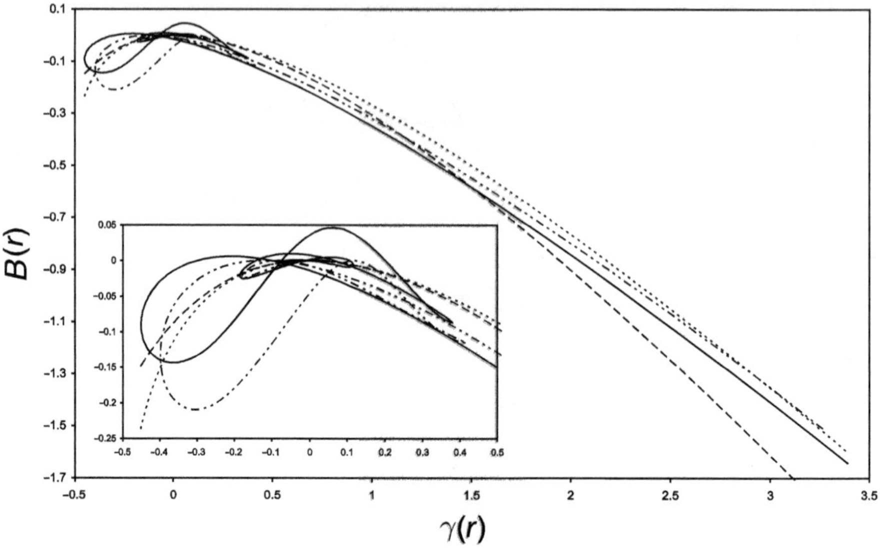

Fig. 1.8. Comparison of the simulated bridge function with several closures at $\rho = 0.94$. *Solid line*: Eq. (1.50), *dash-dot-dot line*: Eq. (1.48) (the ML formula), *dashed line*: PY, *dotted line*: MS

behavior is very complicated and the ML results differ considerably from the simulated ones. Of course, such complicated behavior cannot be described by closures such as the PY, MS, and VM at all. The regular region is much more important—high values of γ and B appear in the vicinity of contact and at high densities. This explains the success of the OZ equation–based theories.

There are two aspects that decide what closure to apply. Certainly, the first is the expected accuracy of results and the second is the amount of numerical work necessary to reaching it. It is not easy to answer the first question unambiguously. Different closures were tested on computer simulation data of different precision in different ranges of state variables by different authors. The answer to the second question is easier.

Pure Hard Spheres

Computationally, the most convenient are the PY, HNC, and the phenomenological closures. More costly are the self-consistent closures and the most arduous are the second-order theories.

The PY and HNC theories do not give good results at high densities, i.e., in the vicinity and beyond the density of freezing. For example, the PY theory (while better than the HNC) even predicts negative (!) values of the RDF in the metastable region. The phenomenological theories give much better results. From them, the VM closure is likely the most successful. The MS closure is almost of the same accuracy, but it loses solution in the metastable region. The self-consistent approaches are better than the PY and HNC but worse than the phenomenological ones at medium and high densities, see, e.g., [66] where the Rogers–Young closure is compared. The second-order theories are not much investigated [45, 109]. There is still space for improving them by proposing more sophisticated closures.

Hard-Sphere Mixtures

The closures listed in Sect. 1.3.3 for one-component hard spheres can be extended to their mixtures, see, e.g., [90, 110]. Moreover, they can be combined. For example, in the case of a binary mixture, one closure can be used for B_{11}, another for B_{12}, and still another for B_{22}. However, such combinations of closures usually do not bring better results.

The RHNC theory that becomes identity for pure hard spheres can be used for their mixtures. Recently, Amokrane et al. [111] have shown that their new variant of the RHNC theory is superior to all the other closures investigated in literature.

Nonspherical Hard-Body Fluids

The closures proposed for hard spheres can be used directly for fluids of hard nonspherical molecules such as fused hard spheres, hard spherocylinders, or hard ellipsoids. Using some of them is computationally almost prohibitively arduous, though. Let us realize that for pure hard spheres one OZ equation is solved, for hard-sphere binary mixtures a set of three coupled equations, and for axially symmetric molecules a set of at least 30 coupled integral equations (1.34) for Fourier transforms of the harmonic coefficients must be considered. Numerical solution for fluids with hard nonlinear molecules is at the edge of current computer facilities and we are not aware of any attempts in this direction.

For all models considered the VM closure is the superior to the PY, HNC, and MS closures [66, 112, 113, 114, 115, 116]. It is also better than different variants of the RHNC theory [66, 112, 113, 117, 118].

Nonuniform Fluids of Hard Spheres

For the singlet OZ equation the same closures as for bulk hard spheres may be applied. In this case the bulk OZ equation is solved first, followed by the singlet OZ Eq. (1.38). Thus, one closure may be applied for bulk fluid and another for a fluid in external field. However, the combined closures such as PY/HNC or HNC/PY are much worse than the noncombined ones [91, 92, 93]. For hard spheres near a hard wall [91] and for hard spheres in a slit pore [92] variants of the RHNC theory proposed in the papers are superior to other closures. The MS and MV closures give very good results, too.

Acknowledgement

This work was supported by The Ministry of Education, Youth and Sports of the Czech Republic under the projects LC512 (Center for Biomolecules and Complex Molecular Systems) and 604 613 7307.

References

1. W. G. Hoover, J. C. Poirier: J. Chem. Phys. **37**, 1041 (1962).
2. E. Meeron, A. J. F. Siegert: J. Chem. Phys. **48**, 3139 (1968).
3. C. G. Gray, K. E. Gubbins: *Theory of Molecular Fluids* (Clarendon Press, Oxford 1984).
4. J. P. Hansen, I. R. McDonald: *Theory of Simple Liquids* (Academic Press, Oxford 1986).
5. B. R. A. Nijboer, L. van Hove: Phys. Rev. **85**, 777 (1952).
6. Y. T. Lee, F. H. Ree, T. Ree: J. Chem. Phys. **26**, 3506 (1968).
7. J. Kolafa, S. Labík: Mol. Phys. **104**, 1915 (2006).
8. F. H. Ree, W. G. Hoover: J. Chem. Phys. **40**, 939 (1964).
9. F. H. Ree, W. G. Hoover: J. Chem. Phys. **46**, 4181 (1967).
10. E. J. Janse van Rensburg, G. M. Torrie: J. Phys. A: Math. Gen. **26**, 943 (1992).
11. E. J. Janse van Rensburg: J. Phys. A: Math. Gen. **26**, 4805 (1993).
12. A. Y. Vlasov, X.-M. You, A. J. Masters: Mol. Phys. **100**, 3313 (2002).
13. M. P. Allen, D. J. Tildesley: *Computer Simulation of Liquids* (Clarendon Press, Oxford 1987).
14. P. Attard, G. N. Patey: J. Chem. Phys. **49**, 4970 (1990).
15. S. Labík, H. Gabrielová, J. Kolafa, A. Malijevský: Mol. Phys. **101**, 1139 (2003).
16. J. Kolafa, S. Labík, A. Malijevský: Mol. Phys. **100**, 2629 (2002).
17. S. Labík, A. Malijevský, W. R. Smith: Mol. Phys. **83**, 983 (1994).
18. W. R. Smith, S. Labík, A. Malijevský, J. Šedlbauer: Mol. Phys. **83**, 1223 (1994).
19. J. Šedlbauer, S. Labík, A. Malijevský, W. R. Smith: Mol. Phys. **85**, 61 (1995).
20. M. N. Rosenbluth, A. W. Rosenbluth: J. Chem. Phys. **22**, 881 (1954).
21. D. Levesque, J. J. Weis, J. P. Hansen: *Topics in Current Physics*, vol. 36 (Springer, Berlin 1984), p. 37.
22. I. Nezbeda, S. Labík, A. Malijevský: Collect. Czech. Chem. Commun. **54**, 1137 (1989).
23. J. A. Barker, D. Henderson: Mol. Phys. **21**, 187 (1971).
24. J. A. Barker, D. Henderson: Ann. Rev. Phys. Chem. **23**, 439 (1972).
25. S. Labík, I. Nezbeda: Mol. Phys. **48**, 97 (1983).
26. I. Nezbeda, H. L. Voertler: Mol. Phys. **57**, 909 (1986).
27. G. N. Patey, G. M. Torrie: Mol. Phys. **34**, 1623 (1977).
28. J. A. Ballance, R. J. Speedy: Mol. Phys. **54**, 1035 (1985).
29. S. Labík, A. Malijevský: Mol. Phys. **53**, 381 (1984).
30. S. Labík, W. R. Smith: J. Chem. Phys. **88**, 1223 (1988).
31. S. Labík, W. R. Smith, R. Pospíšil, A. Malijevský: Mol. Phys. **69**, 649 (1990).
32. Y. Uehara, Y. T. Lee, T. Ree, F. H. Ree: J. Chem. Phys. **68**, 1275 (1979).
33. W. J. McNeil, W. G. Madden, A. D. J. Haymet, S. A. Roce: J. Chem. Phys. **78**, 388 (1983).

34. T. Boublík, I. Nezbeda, K. Hlavatý: *Statistical Thermodynamics of Liquids and Liquid Mixtures* (Elsevier, Amsterdam 1980).
35. R. J. Baxter: J. Chem. Phys. **41**, 553 (1964).
36. R. J. Baxter: Phys. Rev. **154**, 170 (1967).
37. D. Henderson, F. F. Abraham, J. A. Barker: Mol. Phys. **31**, 1291 (1976).
38. Y. Zhou, G. Stell: Mol. Phys. **66**, 767 (1989).
39. Y. Zhou, G. Stell: J. Chem. Phys. **92**, 5535 (1990).
40. Y. Zhou, G. Stell: Mol. Phys. **92**, 5544 (1990).
41. A. Malijevský, S. Labík: Czech. J. Phys. **B38**, 250 (1988).
42. L. Verlet: Physica **30**, 95 (1964).
43. L. Verlet: Physica **31**, 959 (1965).
44. M. S. Wertheim: J. Math. Phys. **8**, 927 (1967).
45. D. Henderson, S. Sokołowski: J. Chem. Phys. **92**, 7541 (1995).
46. G. Stell: *The Equilibrium Theory of Classical Fluids*. Editors H. L. Frisch, J. L. Lebowitz (Benjamin, New York 1964).
47. S. K. Kwak, D. A. Kofke: J. Chem. Phys. **122**, 104508 (2005).
48. J. S. Rowlinson: Mol. Phys. **9**, 217 (1965).
49. P. Huchinson, Q. R. Conkie: Mol. Phys. **21**, 881 (171).
50. P. Huchinson, Q. R. Conkie: Mol. Phys. **24**, 567 (1972).
51. D. S. Hall, W. R. Conkie: Mol. Phys. **40**, 907 (1980).
52. F. J. Rogers, D. A. Young: Phys. Rev. A **30**, 999 (1984).
53. G. Zerah, J. -P. Hansen: J. Chem. Phys. **84**, 2336 (1986).
54. A. G. Vompe, G. A. Martynov: J. Chem. Phys. **100**, 5249 (1994).
55. Y. Rosenfeld: Phys. Rev. E **54**, 2827 (1996).
56. L. L. Lee: J. Chem. Phys. **103**, 9388 (1995).
57. L. L. Lee: J. Chem. Phys. **110**, 7589 (1999).
58. J. M. Bomont, J. L. Bretonnet: J. Chem. Phys. **114**, 4141 (2001).
59. J. M. Bomont, J. L. Bretonnet: J. Chem. Phys. **119**, 2188 (2003).
60. R. Fatoni, G. Pastore: J. Chem. Phys. **119**, 3810 (2003).
61. G. A. Martynov, G. N. Sarkisov: Mol. Phys. **49**, 1495 (1983).
62. G. A. Martynov: private communication.
63. G. A. Martynov: *Fundamental Theory of Liquids* (Adam Hilger, Bristol 1992).
64. P. Ballone, G. Pastore, G. Galli, D. Gazillo: Mol. Phys. **59**, 275 (1986).
65. L. Verlet: Mol. Phys. **41**, 183 (1980).
66. S. Labík, A. Malijevský, W. R. Smith: Mol. Phys. **73**, 87 (1991).
67. D. Henderson, K. Y. Chan, L. Degreve: J. Chem. Phys. **101**, 6975 (1994).
68. F. Lado: Phys. Rev. A **8**, 2548 (1973).
69. F. Lado: Mol. Phys. **31**, 1117 (1976).
70. F. Lado: Phys. Lett. **89A**, 196 (1982).
71. Y. Rosenfeld, N. W. Ashcroft: Phys. Rev. A **20**, 1208 (1979).
72. F. Lado, S. M. Foiles, N. W. Ashcroft: Phys. Rev. A **28**, 2374 (1983).
73. S. M. Foiles, N. W. Ashcroft, L. Reatto: J. Chem. Phys. **80**, 4441 (1984).
74. E. Thiele: J. Chem. Phys. **39**, 474 (1963).
75. M. S. Wertheim: Phys. Rev. Lett. **19**, 321 (1963).
76. M. S. Wertheim: J. Math. Phys. **5**, 643 (1964).
77. R. J. Baxter: Aust. J. Phys. **21**, 563 (1968).
78. A. A. Broyles: J. Chem. Phys. **33**, 456 (1960).
79. J. Kolafa: Mol. Phys. **75**, 577 (1992).
80. R. O. Watts: *Specialist Periodical Report of Chemical Society. Statistical Mechanics.* Vol. 1 (British Chemical Society, London 1973).

81. M. J. Gillan: Mol. Phys. **38**, 1781 (1979).
82. F. Lado: J. Chem. Phys. **47**, 4828 (1967).
83. G. Zerah: J. Comput. Phys. **61**, 280 (1985).
84. S. Labík, A. Malijevský, P. Voňka: Mol. Phys. **56**, 709 (1985).
85. M. Kinoshita, M. Harada: Mol. Phys. **65**, 599 (1988).
86. E. Lomba: Mol. Phys. **68**, 87 (1989).
87. L. Mier y Terán, E. Díaz-Herrera, M. Lozada-Cassou: J. Comput. Phys. **84**, 326 (1989).
88. M. J. Booth, A. G. Schlijper, L. E. Scales, A. D. J. Haymet: Comp. Phys. Comm. **119**, 122 (1999).
89. A. Malijevský, S. Labík, W. R. Smith: Mol. Phys. **72**, 193 (1991).
90. M. Barošová, A. Malijevský, S. Labík, W. R. Smith: Mol. Phys. **87**, 423 (1996).
91. A. Malijevský, R. Pospíšil, W. R. Smith, S. Labík: Mol. Phys. **72**, 199 (1991).
92. R. Pospíšil, A. Malijevský, P. Jech, W. R. Smith: Mol. Phys. **78**, 1461 (1993).
93. R. Pospíšil, A. Malijevský: Mol. Phys. **76**, 1423 (1992).
94. S. Labík, R. Pospíšil, A. Malijevský, W. R. Smith: J. Comput. Phys., **115** 12 (1994).
95. M. J. Both, A. G. Schlijper, L. E. Scales, A. D. J. Haymet: Comp. Phys. Commun. **119**, 122 (1999).
96. C. T. Kelley, B. M. Pettitt: J. Comp. Phys. **197**, 49 (2004).
97. P. Morávek, J. Kolafa, T. Hujo, S. Labík: J. Mol. Phys. **125**, 22 (2006).
98. S. Sokołowski: J. Chem. Phys. **73**, 3507 (1980).
99. S. Sokolowski: Mol. Phys. **49**, 1481 (1983).
100. E. W. Grundke, D. Henderson: Mol. Phys. **24**, 269 (1972).
101. R. D. Groot, N. M. Faber, J. P. van der Eerden: Mol. Phys. **62**, 861 (1987).
102. A. Malijevský, S. Labík: Mol. Phys. **60**, 663 (1987).
103. A. Malijevský, S. Labík: **67**, 431 (1989).
104. S. Labík, A. Malijevský: Mol. Phys. **67**, 431 (1989).
105. S. Rast, P. H. Fries, H. Krienke: Mol. Phys. **96**, 1543 (1999).
106. S. B. Yuste, A. Santos: Phys. Rev. A **43**, 5418 (1991).
107. S. B. Yuste, M. López de Haro, A. Santos: Phys. Rev. E **53**, 4820 (1996).
108. M. Francová, A. Malijevský, J. Kolafa, S. Labík: submitted to Phys. Chem. Chem. Phys.
109. S. C. Attard, G. Stell: Chem. Phys. Lett. **189**, 128 (1992).
110. D. Henderson, A. Malijevský, S. Labík, K. Y. Chan: Mol. Phys. **87**, 273 (1996).
111. S. Amokrane, A. Ayadim, J. G. Malherbe: J. Chem. Phys. **123**, 174508 (2005).
112. S. Labík, A. Malijevský, W. R. Smith: Mol. Phys. **73**, 495 (1991).
113. R. Pospíšil, A. Malijevský, S. Labík, W. R. Smith: Mol. Phys. **75**, 253 (1991).
114. S. Labík, A. Malijevský, R. Pospíšil, W. R. Smith: Mol. Phys. **74**, 261 (1991).
115. J. Šedlbauer, S. Labík, A. Malijevský: Phys. Rev. E **49**, 3179 (1994).
116. R. Pospíšil, A. Malijevský, W. R. Smith: Mol. Phys. **79**, 1011 (1993).
117. F. Lado: Mol. Phys. **47**, 299 (1982).
118. S. Labík, W. R. Smith, R. Pospíšil, A. Malijevský: Mol. Phys. **69**, 649 (1990).

2

Introduction to the Thermodynamics of Hard Spheres and Related Systems

A. Malijevský[1,2] and J. Kolafa[1]

[1] Department of Physical Chemistry, Institute of Chemical Technology, Prague, Technická 5, 166 28 Praha 6, Czech Republic
anatol.malijevsky@vscht.cz, jiri.kolafa@vscht.cz
[2] Center for Biomolecules and Complex Molecular Systems, Prague

This chapter, similarly as the previous one, is divided into two sections corresponding approximately to low- and high-density regions. In Sect. 2.1 we consider density (virial) expansion of pure hard-spheres and hard disks; hard-sphere mixtures are postponed to Chap. 5. In Sect. 2.2 we concentrate on computer simulation and mixed methods of determining the chemical potentials of dense hard-sphere fluids, nonspherical hard-body fluids, and their mixtures. Equations of state are described in detail in Chap. 3.

2.1 Virial Coefficients of Pure Hard Spheres and Disks

2.1.1 Known Virial Coefficients

Virial coefficients are cornerstones of theory of fluids at low and medium densities. They are the coefficients in the density expansion of the equation of state (EOS) expressed via the compressibility factor

$$Z = \frac{\beta P}{\rho} = 1 + \sum_{i=2}^{\infty} B_i \eta^{i-1} , \tag{2.1}$$

where $\beta = 1/(k_B T)$ is the inverse temperature, P the pressure, ρ the number density, and we use the packing fraction

$$\eta = \rho \mathcal{V} , \tag{2.2}$$

where $\mathcal{V}$ stands for sphere volume or disk area. The dimensionless packing fraction thus represents the portion of space occupied by the molecules.

The virial coefficients B_i are defined by exact formulas in terms of integrals whose integrands are products of Mayer functions [1]. For hard spheres, more generally hard-body one-component fluids, the integrals are numbers (they do not depend on temperature).

Malijevský, A., Kolafa, J.: *Introduction to the Thermodynamics of Hard Spheres and Related Systems*. Lect. Notes Phys. **753**, 27–36 (2008)
DOI 10.1007/978-3-540-78767-9_2

For hard spheres and hard disks the terms up to B_4 are known analytically. For hard spheres [2, 3]

$$B_2 = 4\ , \tag{2.3}$$

$$B_3 = 10\ , \tag{2.4}$$

$$B_4 = \frac{2707\pi + [438\sqrt{2} - 4131\arccos(1/3)]}{70\pi} = 18.364768\ldots\ , \tag{2.5}$$

and for hard disks [4, 5]

$$B_2 = 2\ , \tag{2.6}$$

$$B_3 = \frac{16}{3} - \frac{4\sqrt{3}}{\pi} = 3.128018\ldots\ , \tag{2.7}$$

$$B_4 = 16 - 36\frac{\sqrt{3}}{\pi} + \frac{80}{\pi^2} = 4.257854\ldots\ . \tag{2.8}$$

The higher virial coefficients must be calculated numerically. The fifth virial coefficients for hard spheres and disks were calculated by Rosenbluth and Rosenbluth [6] and by Kratky [7, 8, 9, 10]; the sixth by Ree and Hoover [11]; the seventh virial coefficients by Ree and Hoover [12], by Kim and Henderson [13], and by Janse van Rensburg and Torrie [14]; the eighth virial coefficients by Janse van Rensburg [15] and by Vlasov et al. [16]; the ninth virial coefficients by Labík et al. [17]; and the tenth by Clisby and McCoy [18, 19]. As a rule, when the higher virial coefficients were calculated, the lower ones were more precisely recalculated.

Virial coefficients are sums of irreducible (cluster) integrals represented by diagrams. It is a task for a high school student to determine the second virial coefficient for hard spheres or disks. However, dimensionality of the irreducible integrals and their numbers increase rapidly with the increasing order of the virial coefficient. For hard disks the integrals in B_n have dimensionality up to $2n-3$ and for hard spheres it is $3n-6$. For example, the tenth virial coefficient for hard spheres contains 24-dimensional integrals.

Numbers of integrals (no general formula is known for the number of irreducible diagrams based on n) increase even much more rapidly than the dimensionality of the integrals. This is seen in Table 2.1. Some of them are topologically equivalent, which means that their integrands differ only in numbering of variables. In the table, heading "Unlabeled" denotes the total number of irreducible diagrams and "Labeled" the number of topologically different diagrams with numbered (labeled) black points. To reduce these numbers, Ree and Hoover in their pioneering work [11] replaced Mayer diagrams with f bonds by generalized Ree–Hoover diagrams with f bonds and e bonds using identity $f(r) + e(r) = 1$. Besides the reduction in the number of topologically different diagrams, there is another, and now even more important, advantage of the Ree–Hoover approach: computer codes using them are more efficient than those using the Mayer approach.

Table 2.1. Number of unlabeled and labeled Mayer and Ree–Hoover diagrams, and dimensionality d of corresponding cluster integrals for hard spheres

n	Mayer		Ree–Hoover		d
	Unlabeled	Labeled	Unlabeled	Labeled	
2	1	1	1	1	1
3	1	1	1	1	3
4	3	10	2	4	6
5	10	238	5	68	9
6	56	11 368	23	3 053	12
7	468	1 014 888	171	297 171	15
8	7 123	166 537 616	2 606	56 671 216	18
9	194 066	50 680 432 112	81 564	21 286 987 064	21

The number of Ree–Hoover diagrams is lower than the number of Mayer diagrams, as is seen in the table. But for higher virial coefficients, say $n > 6$, it is impossible within a lifetime of an explorer to determine the diagrams and their weights using "pencil and paper" avoiding "human factor" errors. Topological analysis must be done automatically on a computer. Different algorithms utilizing symbolic algebra programming and automatic code generation have been proposed recently [16, 17, 18, 19].

Another problem is to evaluate the multifold integrals. This can be done by random-shooting Monte Carlo integration, but it may spend too much computer time. A more efficient way is to start from the so-called *spanning diagrams*—the integrals that can be calculated analytically. Configurations of the spanning diagrams are sampled either using standard Metropolis Monte Carlo method or (for the linear chains) by the so-called *reptation* [20]. The values of the diagrams of interest are then evaluated by Monte Carlo integration.

The state-of-the-art values of the virial coefficients for hard spheres and disks are listed in Table 2.2. See Chap. 5 for a review of values of hard mixtures.

Table 2.2. Summary of recommended virial coefficients B_n (n = 5–10) for hard spheres and hard disks

n	Hard spheres	Hard disks
5	28.22445 (10)	5.33689664 (64)
6	39.81545 (34)	6.363026 (11)
7	53.3418 (15)	7.352080 (28)
8	68.534 (88)	8.318668 (62)
9	85.805 (58)	9.27236 (29)
10	105.8 (4)	10.2161 (41)

2.1.2 Extrapolation of Virial Coefficients

It is prohibitively difficult to calculate virial coefficients beyond B_{10} because numbers of diagrams increase enormously. There are $2^{\binom{n}{2}}$ of diagrams with n field points that must be analyzed. It is 2^{45} for $n = 10$, 2^{55} for $n = 11$, 2^{66} for $n = 12$, and so on. The number of diagrams increases by more than by three orders with increasing n by 1. It is thus practical to estimate higher virial coefficients for $n > 10$ rather than trying to calculate them.

There are several approaches to extrapolate virial coefficients based on (unjustified) assumption that the higher ones depend on the lower ones. In the theory of fluids the most popular are Padé approximants, see [21] and references therein. They are based on the assumption that the compressibility factor may be expressed in the form

$$Z = \frac{P_n(\eta)}{Q_m(\eta)} , \tag{2.9}$$

where P_n and Q_m are polynomials of order n and m, respectively, whose coefficients are combinations of known virial coefficients. By expanding Eq. (2.9) into the Taylor series, the higher virial coefficients are obtained.

An extension of Padé approximants are differential approximants [22]. The first-order differential approximants have the form

$$P_n(\eta)\frac{\mathrm{d}Z}{\mathrm{d}\eta} + Q_m(\eta)Z(\eta) = R_m(\eta) , \tag{2.10}$$

where P_n, Q_m, and R_m are polynomials; for $P_n(\eta) = 0$ Padé approximants are revealed. The differential approximants have not been used in theories of fluids until recently [23].

These two approaches utilize only a part of available information, namely, the lower virial coefficients. Another source of information are computer simulation data on the compressibility factors. They can be utilized by proposing an EOS some of whose parameters are determined from known virial coefficients and the remaining constants fitted to simulation data. For example, Kolafa et al. [24] used an EOS in the form proposed by Barboy and Gelbart [25]

$$Z = 1 + \sum_{i=1}^{n} a_i x^i , \quad x = \frac{\eta}{1-\eta} , \tag{2.11}$$

where a_i are combinations of known B_i. By expanding the equation in powers of η, the higher virial coefficients may be estimated.

All the above extrapolation methods suffer from the assumption that the values of the lower virial coefficients are known precisely. This is, certainly, not true, see table 2.2. To be accurate, uncertainties in the lower virial coefficients (as well as in EOS simulation data) should be taken into account [26, 27].

It was suggested [23, 27] that the safest way of utilizing both the lower virial coefficients data and the EOS data is to minimize the objective function

Table 2.3. Extrapolated virial coefficients for hard spheres [23] and hard disks [27]

n	Hard spheres	Hard disks
11	130 (2)	11.172 (10)
12	150 (5)	12.132 (30)
13	175 (7)	13.097 (60)
14	205 (10)	14.053 (80)
15	225 (25)	14.94 (21)
16	–	15.7 (4)

$$S = \sum_{i=2}^{n} \left(\frac{B_i - B_i^{\text{exp}}}{\sigma_i} \right)^2 + \sum_{j=1}^{k} \left(\frac{Z_j - Z_j^{\text{exp}}}{\sigma_j} \right)^2 , \tag{2.12}$$

where n is the number of known virial coefficients, σ_i are their standard deviations, k is the number of simulated state points, and σ_j are their standard deviations. Z_j is the value of the compressibility factor at density η_j given by a chosen compressibility factor in the form of Eq. (2.9) (optimized Padé approximants), or in the form of Eq. (2.10) (optimized differential approximants), in the form of Eq. (2.11), or in another form. In other words, the higher virial coefficients appear in Eq. (2.9) etc. as free parameters.

While the statistical errors in the predicted higher virial coefficients can be easily evaluated, this cannot be said about the systematic errors introduced by a particular functional form. The only way to assess the systematic errors is to compare as many functional forms (Padé indices, powers in polynomials, etc.) as possible. Then subjectively (after canceling of evidently defective values) estimate average values of them and their uncertainties. This is more art than science. Our estimates of higher virial coefficients for hard spheres obtained by several methods [23] and hard disks by Eq. (2.11) [27] including the estimated uncertainties are given in Table 2.3. They all have been obtained using known virial coefficients B_2 to B_{10}, their estimated uncertainties, and the EOS data for hard spheres [24] and hard disks [27]. Different above-mentioned extrapolation methods based on Eqs. (2.9), (2.10), and (2.11) were used to guess their values and uncertainties.

2.2 Chemical Potentials of Hard Spheres and Related Systems

The chemical potential plays the main role in studying the phase and chemical equilibria. It can be obtained from an EOS by standard thermodynamic methods, see Chap. 3. However, it is worthwhile to obtain it directly by computer simulations. This is the aim of this section.

Widom in his pioneering work [28] derived a formula

$$\beta\mu^{\rm ex} = -\log \mathcal{P}(\sigma) \ , \tag{2.13}$$

where $\beta = 1/(k_{\rm B}T)$, $\mu^{\rm ex} = \mu - \mu^{\rm id}$ is the difference from the ideal gas (at the same pressure; also called the residual chemical potential), and $\mathcal{P}(\sigma)$ the probability of inserting an extra hard sphere of diameter σ to the system of N hard spheres. A simulation algorithm is simple: try to insert an extra particle to the system. If the extra particle overlaps with one or more particles, call it an "unsuccessful" attempt, otherwise call it a "successful" attempt. After a number of attempts probability $\mathcal{P}$ and its uncertainty can be estimated using standard statistical methods. What can be simpler?

The problem arises at high densities. Most attempts become unsuccessful. For example, for pure hard spheres at densities near the crystallization density the probability of a single successful inserting an extra sphere is of the order of 1 to 10 million. Billions of attempts are needed to obtain a good statistic.

Many methods have been suggested to overcome the above shortcoming [29, 30, 31, 32, 33, 34, 35, 36, 37, 38, 39, 40, 41]. We have a practical experience with a method originally proposed by Labík and Smith for hard spheres [42] and called the scaled particle Monte Carlo (SP-MC) method by the authors. The SP-MC method was later extended to hard-sphere mixtures [43] and to nonspherical hard-body fluids [44], and modified [45]. The principle of the SP-MC method consists of measuring of the probabilities of successful insertion of smaller copies of molecules of the fluid during conventional Monte Carlo simulations, combined with certain exact theoretical constraints. It is described in detail below.

2.2.1 SP-MC Method for Pure Hard Spheres

Consider a fluid of hard spheres of diameter σ at the packing fraction η and a test particle of diameter d. Equation (2.13) applied to inserting this particle into the system becomes

$$\beta\mu^{\rm ex}(d) = -\log \mathcal{P}(d) \ , \tag{2.14}$$

where $\mu^{\rm ex}(d)$ is the excess chemical potential of the d particle at infinite dilution in the hard-sphere fluid and $\mathcal{P}(d)$ is the probability of inserting a smaller particle of diameter $d < \sigma$ into the system. This probability can be easily and accurately determined in computer simulations (even at high densities) over a range of such d values. Assuming that $\mathcal{P}(d)$ is a smooth function of d, it can be expressed as a polynomial

$$\log \mathcal{P}(d) = \sum_{i=0}^{n} a_i d^i \ . \tag{2.15}$$

The probability of insertion a point particle ($d = 0$) is exactly given by the spare volume fraction

$$\mathcal{P}(0) = 1 - \eta \, . \tag{2.16}$$

Its first two derivatives are also known exactly [46]

$$\left(\frac{\partial \log \mathcal{P}(d)}{\partial d}\right)_{d=0} = -\frac{1}{\sigma}\frac{3\eta}{1-\eta} \, , \tag{2.17}$$

$$\left(\frac{\partial^2 \log \mathcal{P}(d)}{\partial d^2}\right)_{d=0} = -\frac{3}{\sigma^2}\frac{\eta(2+\eta)}{(1-\eta)^2} \, . \tag{2.18}$$

Moreover, the probability derivative at $d = \sigma$ is exactly related to the compressibility factor of pure hard spheres [47]

$$\left(\frac{\partial \log \mathcal{P}(d)}{\partial d}\right)_{d=\sigma} = -3(Z-1) \, . \tag{2.19}$$

In addition to information on $\mathcal{P}(d)$ gathered by the test insertions of small particles and the above theoretical constraints, one may gain additional information by measuring the conditional probability $\mathcal{P}^*(\sigma \to d)$ that a randomly selected sphere can be enlarged to a diameter $d > \sigma$ [35, 39]. This conditional probability is related to $\mathcal{P}(d)$ and $\mathcal{P}(\sigma)$ by

$$\mathcal{P}(d) = \mathcal{P}(\sigma)\mathcal{P}^*(\sigma \to d) \, , \quad d > \sigma \, . \tag{2.20}$$

Combining this equation with Eq. (2.15) gives

$$\log \mathcal{P}^*(d) = \sum_{i=1}^{n} a_i (d^i - \sigma^i) \, . \tag{2.21}$$

Finally, constants a_i in Eq. (2.15) are obtained by fitting this equation and Eq. (2.21) together with the constraints (2.16) to (2.19) to the simulation data on inserting test particles for $d < \sigma$ and the data on enlarging spheres to $d > \sigma$. Following the regression, Eqs. (2.13) and (2.15) are used to determine the chemical potential from the calculated value of $\mathcal{P}(\sigma)$. The method remains efficient even in a deep metastable region [47].

2.2.2 SP-MC Method for Hard-Sphere Mixtures

Accurate determination of the chemical potentials for hard-sphere fluid mixtures is of particular interest. It is because of an attractive and yet unsolved hypothesis of a possible existence of the fluid–fluid phase equilibrium at which the chemical potentials of each component must be equal in the both phases. The phase separation—if it exists—is expected for highly asymmetric mixtures at high densities and low concentrations of large spheres, i.e., in the region where computer simulations are extremely demanding. Therefore, sophisticated methods of determining component chemical potentials are highly desirable. The Widom's test particle insertion method, Eq. (2.13), can be generalized to

$$\beta\mu_i^{\mathrm{ex}} = -\log \mathcal{P}(\sigma_i)\ , \quad i = 1, 2, \ldots, k\ , \tag{2.22}$$

where μ_i is the chemical potential of component i, k the number of mixture components, and $\mathcal{P}(\sigma_i)$ the probability of successful insertion of particle of diameter σ_i into the mixture.

For simplicity of notation, we will limit the next discussion to binary (additive) mixtures of large hard spheres with diameter σ_{L} and small hard spheres with σ_{S}. The system is characterized by the diameter ratio $\gamma = \sigma_{\mathrm{S}}/\sigma_{\mathrm{L}}$, the mole fractions of the components, x_{L} and x_{S}, $x_{\mathrm{L}} + x_{\mathrm{S}} = 1$, and by the packing fraction

$$\eta = \frac{\pi}{6}\frac{N}{V}(x_{\mathrm{L}}\sigma_{\mathrm{L}}^3 + x_{\mathrm{S}}\sigma_{\mathrm{S}}^3)\ . \tag{2.23}$$

To illustrate the impracticality of the Widom's method at certain state conditions, consider, for example, the state point characterized by diameter ratio $\gamma = 0.3$, mole fraction of the large spheres $x_{\mathrm{L}} = 0.0625$, and packing fraction $\eta = 0.49$. The probability of inserting a large sphere is $\mathcal{P}(\sigma_{\mathrm{L}}) \approx 1.1 \times 10^{-29}$, see [43]. If each attempted particle insertion required only 1 ns of computer time, and we ran the simulations on a cluster of 100 dual-core computers, then a single successful insertion would occur once in 14 billion years (the age of the Universe).

The SP-MC method can be readily extended to the case of binary hard-sphere mixtures [43]. An assumption of the functional form (2.15) and constraint (2.16) with η given by Eq. (2.23) remain unchanged. Measuring of the probability of inserting smaller test particles is the same as in the case of pure fluid, too. Equations (2.17) and (2.18) are generalized to

$$\left(\frac{\partial \log \mathcal{P}(d)}{\partial d}\right)_{d=0} = -\frac{3\eta}{1-\eta}\frac{x_{\mathrm{L}}\sigma_{\mathrm{L}}^2 + x_{\mathrm{S}}\sigma_{\mathrm{S}}^2}{x_{\mathrm{L}}\sigma_{\mathrm{L}}^3 + x_{\mathrm{S}}\sigma_{\mathrm{S}}^3}\ , \tag{2.24}$$

$$\left(\frac{\partial^2 \log \mathcal{P}(d)}{\partial d^2}\right)_{d=0} = -\frac{3\eta}{1-\eta}\left[2\frac{x_{\mathrm{L}}\sigma_{\mathrm{L}} + x_{\mathrm{S}}\sigma_{\mathrm{S}}}{x_{\mathrm{L}}\sigma_{\mathrm{L}}^3 + x_{\mathrm{S}}\sigma_{\mathrm{S}}^3}\right. \tag{2.25}$$

$$\left. + \frac{3\eta}{1-\eta}\left(\frac{x_{\mathrm{L}}\sigma_{\mathrm{L}}^2 + x_{\mathrm{S}}\sigma_{\mathrm{S}}^2}{x_{\mathrm{L}}\sigma_{\mathrm{L}}^3 + x_{\mathrm{S}}\sigma_{\mathrm{S}}^3}\right)^2\right]\ . \tag{2.26}$$

Instead of Eq. (2.19) we have two equations

$$\left(\frac{\partial \log \mathcal{P}(d)}{\partial d}\right)_{d=\sigma_{\mathrm{L}}} = -12\eta_{\mathrm{L}}g_{\mathrm{L}}/\sigma_{\mathrm{L}} - 12\eta_{\mathrm{S}}g_{\mathrm{LS}}\sigma_{\mathrm{LS}}^2/\sigma_{\mathrm{S}}^3\ , \tag{2.27}$$

$$\left(\frac{\partial \log \mathcal{P}(d)}{\partial d}\right)_{d=\sigma_{\mathrm{S}}} = -12\eta_{\mathrm{S}}g_{\mathrm{S}}/\sigma_{\mathrm{S}} - 12\eta_{\mathrm{L}}g_{\mathrm{LS}}\sigma_{\mathrm{LS}}^2/\sigma_{\mathrm{L}}^3\ , \tag{2.28}$$

where $\eta_i = \frac{\pi}{6}\rho x_i\sigma_i^3$, and g_{L}, g_{LS}, and g_{S} are the contact values of the radial distribution functions for $L-L$, $L-S$, and $S-S$ interactions, respectively. Instead of measuring a single conditional probability $\mathcal{P}^*(\sigma \to d)$, two conditional probabilities $\mathcal{P}^*(\sigma_L \to d), d > \sigma_L$, and $\mathcal{P}^*(\sigma_S \to d), d > \sigma_S$, are to be determined from computer simulations.

We remark that in the case of pure hard spheres there are four constraints available, Eqs. (2.16) to (2.19). In the case of binary hard-sphere mixtures five constraints are to one's disposal. Moreover, two conditional probabilities can be measured. For an n-component mixture, $n+3$ constraints and n measured conditional probabilities are available. One can thus anticipate that the SP-MC results combining computer simulations and the exact constraints may be even more accurate for mixtures than in the one-component case.

It is difficult (or almost impossible) to judge the efficiency of the SP-MC method described in detail above with alternative methods proposed in literature [31, 32, 33, 34, 35, 36, 37, 38, 39, 40, 41]. It depends on too many factors:

- System chosen (pure hard spheres, their mixtures, nonspherical hard-body systems, their mixtures, etc.)
- State variables such as density, composition and diameter ratio for hard-sphere mixtures, and elongation for diatomics. The simplest Widom method works efficiently for low densities and/or small solutes only
- Precision chosen vs. the computer time spent
- Computer codes developed including specific programmer's tricks
- Computers used
- Whether one is interested in one state point only or a whole range; in the latter case, thermodynamic integration based on a series of standard simulations can be efficiently used.

Thus, in sum, one has to rely on his/her intuition. The chemical potential simulations/calculation are just an example of a general problem of accuracy:efficiency:robustness.

Acknowledgement

This work was supported by The Ministry of Education, Youth and Sports of the Czech Republic under the projects LC512 (Center for Biomolecules and Complex Molecular Systems) and 604 613 7307.

References

1. T. L. Hill: *Statistical Mechanics* (McGraw-Hill, New York 1956).
2. B. R. A. Nijboer, L. van Hove: Phys. Rev. **85**, 777 (1952).
3. I. Lyberg: e-print cond-mat/0410080.
4. M. Metropolis, A. W. Rosenbluth, M. N. Rosenbluth, A. H. Teller, E. Teller: J. Chem. Phys. **21**, 1087 (1953).
5. N. Clisby, B. McCoy: J. Stat. Phys. **114**, 1343 (2004).

6. M. N. Rosenbluth, A. W. Rosenbluth: J. Chem. Phys. **33**, 1439 (1953).
7. K. W. Kratky: Physica A **85**, 607 (1976).
8. K. W. Kratky: Physica A **87**, 548 (1977).
9. K. W. Kratky: J. Stat. Phys. **27**, 533 (1982).
10. K. W. Kratky: J. Stat. Phys. **29**, 129 (1982).
11. F. H. Ree, W. G. Hoover: J. Chem. Phys. **40**, 939 (1964).
12. F. H. Ree, W. G. Hoover: J. Chem. Phys. **46**, 4181 (1967).
13. S. Kim, D. Henderson: Phys. Lett. **27 A**, 379 (1968).
14. E. J. Janse van Rensburg, G. M. Torrie: J. Phys. A: Math. Gen. **26**, 943 (1992).
15. E. J. Janse van Rensburg: J. Phys. A: Math. Gen. **26**, 4805 (1993).
16. A. Y. Vlasov, X. -M. You, A. J. Masters: Mol. Phys. **100**, 3313 (2002).
17. S. Labík, J. Kolafa, A. Malijevský: Phys. Rev. E **71**, 021105 (2005).
18. N. Clisby, B. M. McCoy: J. Stat. Phys. **122**, 15 (2006).
19. N. Clisby, B. M. McCoy: Pramana-J. Phys. **4**, 3609 (2005).
20. M. P. Allen, D. J. Tildesley: *Computer Simulation of Liquids* (Clarendon Press, Oxford 1987).
21. I. C. Sanchez: J. Chem. Phys. **101**, 7003 (1994).
22. A. J. Guttmann: *Phase Transitions and Critical Phenomena*, Vol. 3, Chap. 1. Eds. C. Domb, J. Lebowitz (Academic Press, New York 1989).
23. M. Ončák, A. Malijevský, J. Kolafa, S. Labík: to be published.
24. J. Kolafa, S. Labík, A. Malijevský: Phys. Chem. Chem. Phys. **6**, 2335 (2004).
25. B. Barboy, W. M. Gelbart: J. Chem. Phys. **72**, 3035 (1979).
26. J. J. Erpenbeck, W. W. Wood: J. Stat. Phys. **3**, 321 (1984).
27. J. Kolafa, M. Rottner: Mol. Phys. **104**, 3435 (2006).
28. B. Widom: J. Chem. Phys. **39**, 2808 (1964).
29. D. J. Adams: Mol. Phys. **28**, 1241 (1974).
30. J. P. Valleau, G. M. Torrie: *Statistical Mechanics, Part A*, Chap. 5, Ed. B. J. Berne (Plenum Press, New York 1977).
31. C. Seiter, B. J. Alder: J. Solut. Chem. **7**, 73 (1978).
32. K. S. Shing, K. E. Gubbins: Mol. Phys. **43**, 717 (1981).
33. M. Fixman: J. Chem. Phys. **78**, 4223 (1983).
34. K. E. Gubbins: Mol. Simul. **2**, 223 (1989).
35. G. L. Dietrick, L. E. Scriven, H. T. Davis: J. Chem. Phys. **90**, 2370 (1989).
36. I. Nezbeda, J. Kolafa: Mol. Simul. **5**, 391 (1991).
37. J. Talbot, P. Bereolos, K. C. Chao: J. Chem. Phys. **98**, 1531 (1993).
38. P. Attard: J. Chem. Phys. **98**, 2225 (1993).
39. L. E. S. Souza, A. Stamatopoulou, D. Ben-Amotz: J. Chem. Phys. **100**, 1456 (1994).
40. I. Nezbeda, M. Strnad: Mol. Phys. **85**, 91 (1995).
41. D. A. Kofke, P. T. Cummings: Mol. Phys. **92**, 973 (1997).
42. S. Labík, W. R. Smith: Mol. Simul. **12**, 23 (1994).
43. M. Barošová, A. Malijevský, S. Labík, W. R. Smith: Mol. Phys. **87**, 423 (1996).
44. S. Labík, V. Jirásek, A. Malijevský, W. R. Smith: Chem. Phys. Lett. **247**, 227 (1995).
45. S. Labík, V. Jirásek, A. Malijevský, W. R. Smith: Mol. Phys. **94**, 385 (1998).
46. H. Reiss, P. Schaft: J. Chem. Phys. **91**, 2514 (1989).
47. W. R. Smith, S. Labík: Mol. Phys. **80**, 1561 (1994).

3

Equations of State for Hard Spheres and Hard Disks

A. Mulero[1], C.A. Galán[1], M.I. Parra[2], and F. Cuadros[1]

[1] Departamento de Física Aplicada, Universidad de Extremadura, 06071 Badajoz, Spain
mulero@unex.es, cgalango@unex.es, cuadros1@unex.es

[2] Departamento de Matemáticas, Universidad de Extremadura, 06071 Badajoz, Spain
mipa@unex.es

The equation of state (EOS) of a system is perhaps its most important thermodynamic relationship as it allows one to calculate most of its thermodynamic properties. Thus, it is essential to have an adequate analytical expression for the EOS of a hard-sphere system, which can be used in perturbation theories or as the foundation for the construction of EOSs of real fluids. A great variety of proposals can be found in the literature. This chapter collects and reviews most of the equations of state for single-component hard spheres and hard disks in order to test their accuracy in reproducing the available computer simulation data for the compressibility factor of the system. Some of these equations are also used to derive the chemical potential and the isothermal compressibility of hard spheres. Their application to the Weeks–Chandler–Andersen perturbation theory is presented in Chap. 4.

3.1 Introduction

As is well known, the hard sphere (HS) (or hard disk, HD) system is defined by an interaction potential that considers only the repulsive forces among molecules [1]. The simplicity of this model allows one to calculate its thermodynamic properties by obtaining analytical solutions of some theories [2, 3, 4] or by performing computer simulations. Nevertheless, the results given by the various methods can be analytically and numerically different. In particular, the equation of state (EOS) of a system is perhaps its most important thermodynamic relationship. Unfortunately, there is no exact theoretical solution for the EOS of these systems, except for the one-dimensional case. As a consequence, a great variety of expressions for the HS EOS can be found in the literature. Most were obtained from knowledge of the virial coefficients and/or by directly fitting computer simulation data. The use of different computer

Mulero, A., et al.: *Equations of State for Hard Spheres and Hard Disks.* Lect. Notes Phys. **753**, 37–109 (2008)
DOI 10.1007/978-3-540-78767-9_3

simulation data sets and of different analytical forms for the equations have led to the wide range of proposals which will be reviewed in this chapter.

The HS system can be regarded as the most widely studied, although this does not mean that all its properties are definitively and exactly known. Theoretically, attention has been paid to the calculation of the radial distribution function, virial coefficients (only the first 10 are known with certain accuracy), the EOS, some thermodynamic properties, and the random packing of spheres, as well as to the nature and location of the solid–fluid transitions. Here we focus our attention exclusively on the different expressions proposed for the EOS, in order to give some useful criteria to make a suitable choice (in any case, references on the other interesting subjects mentioned are also included).

The choice of an analytical expression for the EOS of the HS is an important step in the development of perturbation theories and of EOSs for real fluids used in physical chemistry or in chemical engineering calculations [5, 6, 7, 8, 9, 10, 11, 12, 13, 14, 15, 16, 17, 18, 19, 20, 21, 22, 23, 24, 25, 26, 27, 28, 29, 30, 31, 32, 33]. In a more general sense and according to Song et al. [34] "understanding the packing of hard spheres at moderate densities turned out to be the key to understanding simple liquids". In fact, the packing of spheres is a classical matter of discussion, with several applications in science, engineering, and medicine [35, 36, 37].

In the case of two-dimensional fluids, the HD properties serve as a basis in the development of simple theories or equations explaining some simple adsorption processes [38, 39, 40, 41, 42, 43, 44]. Moreover, computer simulations can be performed with a greater number of particles than in the three-dimensional case, and hence, the effects of the small finite size of the sample are clearly reduced.

The study of HS systems with more than three dimensions (hard hyperspheres) has the interest of being a general test for theoretical and computational techniques. See, for instance, [34, 45, 46, 47, 48, 49, 50, 51, 52, 53, 54, 55, 56, 57] and other references therein.

For the study of anisotropic molecules, an appropriate reference system is a fluid of anisotropic hard particles, e.g., rods, ellipsoids, dumbbells, and spherocylinders. The theoretical study of these systems [58] has been based mainly on the scaled particle approach originally developed by Reiss et al. [2, 3] for HS fluids. Attention has also been devoted to the calculation of virial coefficients, computer simulations, phase diagrams, and the proposal of EOSs [58, 59, 60, 61, 62, 63, 64, 65, 66, 67, 68, 69, 70, 71, 72, 73, 74, 75, 76, 77] and references therein.

As is well known, the EOS of a pure substance is defined as the mathematical relation between pressure (P), volume, and temperature (T). It is usually an explicit function of pressure, which is given through the density dependence of the compressibility factor Z:

$$Z = \frac{P}{\rho T} \tag{3.1}$$

ρ being the density of the system. All the previous magnitudes are given in reduced Lennard–Jones units, which are related to real magnitudes as:

$$P_{\text{real}} = P\frac{(\varepsilon/k)R}{N_{\text{a}}\sigma^D} \qquad (\text{Pa}) \tag{3.2}$$

$$\rho_{\text{real}} = \frac{\rho}{N_{\text{a}}\sigma^D} \qquad (\text{mol m}^{-3}) \tag{3.3}$$

$$T_{\text{real}} = T(\varepsilon/k) \qquad (\text{K}) \tag{3.4}$$

where N_{a} is Avogadro's number, R is the perfect gas constant, D is the dimension of the system, k is Boltzmann´s constant, and ε/k and σ are the parameters that characterize the Lennard–Jones interaction potential (the minimum of the potential and the distance for which the potential becomes zero). Note that ε/k and σ are expressed in kelvin and meters, respectively. Values for these two parameters can be taken from several works in the literature (see, for instance, [78, 79]).

The problem of obtaining an EOS for any fluid could be solved exactly if one could determine the coefficients in the infinite virial series expansion of the compressibility factor in powers of the density, i.e.,

$$Z = 1 + \sum_{n=2}^{\infty} B_n \rho^{n-1} \tag{3.5}$$

B_n being the so-called virial coefficients (see Chap. 2, as well as recent references for latest results [46, 47, 48, 80, 81]).

Unfortunately, only the first few virial coefficients of this expansion are known at best, and, since the virial series converges slowly, frequent use is made of several types of approximants in order to accelerate the convergence of the series. They are functions that depend on a certain number of parameters, which are determined through the condition that the series expansion of the approximant must reproduce a given number of the known virial coefficients (usually all of them). The most widely used are the Padé and the Levin approximants, which are quotients of polynomials in the density or in the so-called packing fraction. This procedure leads to EOSs based on the virial series expansion, which are more accurate at high densities than the truncated virial series itself. However, there are many other additional interesting proposals that must be taken into account.

In sum, the main ways followed in order to obtain analytical expressions for the EOS of HSs are (i) expressions based on theoretical developments; (ii) empirical or semiempirical expressions, based on the knowledge of virial coefficients and/or the fit of simulation data; (iii) expressions derived from Padé, Levin, Tová, and other approximants with poles at different density values; (iv) volume-explicit EOSs, where the compressibility factor is expressed in terms of pressure and temperature; and (v) nonanalytical EOSs based mainly on the solution to a nonlinear differential equation.

Each of those methods has specific advantages and disadvantages. Theoretical EOSs usually have a limited accuracy, but they do not contain adjustable coefficients and do permit some data to be predicted. Empirical or semiempirical EOSs are generally very accurate, but they cannot be improved in a systematic way. Approximants allow the exact reproduction of the known virial coefficients, but sometimes they can only be used at low densities. Moreover, it is found that sometimes they do not agree with the computer simulation values for the EOS over the whole density range. Also the number of calculated coefficients is sometimes very large, and hence, their handling can be difficult for practical purposes. These practical difficulties are even greater in the case of volume-explicit or nonanalytical expressions. Nevertheless, they can be quite useful for particular purposes.

In this chapter, approximately 80 analytical expressions for the HS EOS and more than 30 for the HD system are collected. The main features or some interesting aspects of each of these expressions are reported and, for a significant number of them, some studies to test their accuracy to reproduce the compressibility factor of the system by comparing with available computer simulation data are described in detail. For that purpose, several simulation data sets were selected as a reference, including thermodynamic states near the phase transitions. The main conclusions are listed and, as a result, some of the analytical expressions are recommended for accuracy and/or simplicity in their analytical form.

Finally, the use of some of those analytical expressions to reproduce values of the chemical potential and the isothermal compressibility of HSs is also described [13, 14].

3.2 Equations of State for Hard Disks

Some features of the HD system are particularly interesting for researchers, such as their use in the study of monolayer repulsive interactions, as well as the nature and exact location of the solid–fluid transition. For these purposes, many computer simulations have been carried out and reported, giving rise to a great number of works with important results and conclusions. We focus here only on those simulations giving the HD EOS.

Although the HD system can be regarded as being very simple, completely satisfactory results in the development of very accurate and simple EOSs have not yet been obtained [80, 82, 83, 84, 85]. There are many different analytical expressions to calculate the HD EOS. They are usually expressed as a function of the packing fraction η, which is defined as the ratio between the volume occupied by the particles and the total volume. In particular, for two-dimensional systems it is given by:

$$\eta = \pi\rho/4 \tag{3.6}$$

with ρ being the number of particles per unit area (in reduced units). The maximum value of η is the so-called closest packing fraction (η_c), which is defined as the highest possible packing fraction which can be calculated from the geometric properties of the molecules. In particular, for the two-dimensional fluid one has:

$$\eta_c = \frac{\pi\sqrt{3}}{6} \tag{3.7}$$

which corresponds to a density $\rho_c = 2/\sqrt{3} \approx 1.1547$. At this value, the HD system is a crystal where the disks constitute a regular triangular lattice, each disk in contact with six neighbors [86, 87]. As will be explained in Sect. 3.2.2, the closest-packing fraction has usually been taken as a reference to build EOSs by imposing suitable analytical conditions to be met at high densities.

It is also important to know that freezing takes place at $\eta \approx 0.7$ [88] and that random close packing, i.e., a configuration with no statistically significant short- or long-range order, is also possible at higher densities. Although the experimental value of the density at which this packing occurs is not well known, the oldest estimations regard this point as being located at around 0.89 to 0.92 times the regular close packing value ($\eta_{\mathrm{rcp}} = 0.82 \pm 0.02$ [89]). Moreover, Song et al. [34] showed that, within the accuracy of the computer data available at that time, the divergence of pressure at this point is characterized by a fractional critical exponent. More recently, Torquato et al. [36] observed that this point cannot be well defined mathematically, and therefore, they introduced a new concept, the maximally random jammed state, in order to be more precise. Other studies of the statistical geometry of HDs can be found in [86, 87, 90] and references therein.

To build an analytical expression for the EOS of a monocomponent HD system, the values of the virial coefficients and the computer simulation data are the main sources of information. For the HD system, the first 10 virial coefficients are known, although important statistical uncertainties must be taken into account beyond the fourth [46, 47, 48, 81, 85]. Estimations of the virial coefficients from the 11th to the 16th have been performed recently [80] (see Chap. 2). With respect to computer simulations, many have been carried out in order to obtain values of the compressibility factor for different density ranges. Data can be found from the pioneering simulations in 1953 [91] to the most recent ones [80]. A summary of most of those computer simulations is presented in the following section. Section 2.2 lists most of the analytical proposals for the EOS of HD fluids and solids. In Sect. 3.2.3 a group of them are selected to establish a comparison with some of the most important computer simulation data available in the literature. Finally, the conclusions are presented in Sect. 3.2.4.

3.2.1 Summary of Computer Simulations

Despite the simplicity of the HD model, the statistical mechanics involved has not been solved exactly, though many numerical and approximate analytical calculations have been carried out. Indeed, the Monte Carlo simulation technique was pioneered by Metropolis et al. [91]. They studied a system of 224 HDs with periodic boundary conditions and calculated its EOS. Their results were in agreement with free volume theory at high densities and with a four-term virial expansion at lower densities. The other simulation technique, known as molecular dynamics, was used by Alder and Wainwright [92] and by Hoover and Alder [93] in order to study the liquid–solid transition and the effect of the system size. The first computer simulation data available for the solid phase were reported by Hoover and Ree [94] and by Alder et al. [95].

Other simulations including the calculation of the EOS will be briefly summarized here in chronological order. Thus, Chae et al. [96] performed another Monte Carlo simulation in order to obtain the radial distribution function at densities $\rho/\rho_c = 0.4$, 0.5, and 0.6. The first virial coefficients of pressure were also evaluated. A Padé approximant was proposed for the EOS in order to reproduce the computer simulation data (see the next subsection).

Monte Carlo techniques with the isothermal–isobaric ensemble, NPT, were developed by Wood [97, 98] and initially applied to a system of only 12 HDs. Results for the EOS and radial distribution function in the high-density range were obtained, and were in agreement with those of the previous molecular dynamics [92, 93] and Monte Carlo [96] studies.

Two main computer simulations were performed in the seventies. The first [99] was a molecular dynamics simulation considering a system of 1600 particles, but the results were obtained only at a reduced density of 0.85. The second [100] was a Monte Carlo simulation for a small system of 32 particles. Calculations were made in the reduced density range from 0.05 to 0.9. The values generated, together with a new proposal for the EOS, were used to develop a perturbation theory for the two-dimensional Lennard–Jones fluid.

Erpenbeck and Luban [101] performed a combined Monte Carlo–molecular dynamics simulation in the fluid phase and found excellent agreement with results obtained from a Levin approximant to the first six terms of the virial series (see the next subsection). Simulations using 5822 and 1512 HDs were carried out in the ranges $0.1 \leq \rho/\rho_c \leq 0.7143$ and $\rho/\rho_c \leq 0.05$, respectively. This computer simulation has for a long time been regarded as the most accurate and has therefore been widely used as a validity test of HD EOSs [42, 44, 83, 84, 85].

The Monte Carlo computer simulation of Fraser et al. [65] for the HD fluid was performed using the NPT ensemble via a new method, based on the Voronoi tessellation, to keep track of nearest neighbors. Data for a 408 HD system were obtained in both the fluid and the transition regions (high densities).

Kolafa and Rottner [80] have recently given new molecular dynamic results in the density range from 0.4 to 0.9. Finite effects were discussed in detail and taken into account before generating their values for the compressibility factor. Those authors used from 4000 to 50 000 disks. They noted that their results in the region close to the phase transition agree quite accurately with

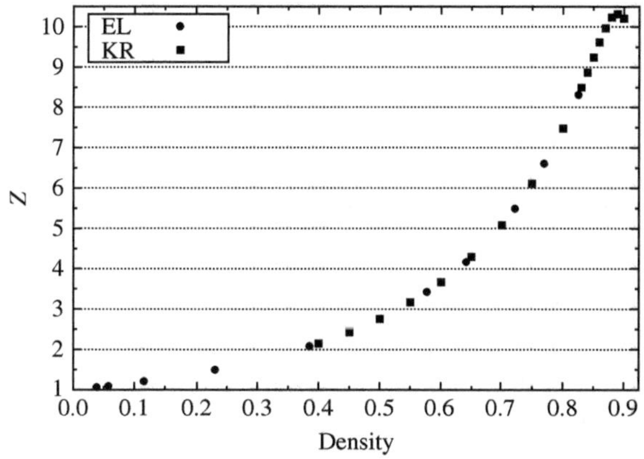

Fig. 3.1. Computer simulation data for the compressibility factor obtained by Erpenbeck and Luban (EL) [101] and by Kolafa and Rottner (KR) [80]

other extensive Monte Carlo simulation data [102, 103], except at density 0.9. As can be seen in Fig. 3.1, a higher value for the compressibility factor should have been obtained at the highest density. In relation to this, the authors pointed out that this point could be affected by finite size effects. At intermediate densities very good agreement with Erpenbeck and Luban's data is found.

Much computational effort has also been devoted to the study of the solid–fluid transition, and this subject is still an interesting field of research. Some of the most interesting related papers are included in [65, 94, 102, 103, 104, 105, 106, 107, 108, 109, 110, 111, 112, 113, 114, 115, 116, 117, 118, 119]. Some of these references deal with the nature of the melting transition in two-dimensional systems, which is still a subject of debate. Let us note, in particular, that in the recent work of Mak [103] large-scale computer simulations with more than four million particles were performed to study the melting transition in an HD fluid. The van der Waals loop previously observed in the pressure–density relation of smaller simulations was shown to disappear systematically as the sample size was increased, but even with such a large number of particles, the freezing transition still exhibits what appears to be a weakly first-order behavior.

We must finally mention that computer simulations of anisotropic hard particles in two dimensions have also used hard ellipses, discorectangles, needles, and dumbbells. More information can be found in [59, 60, 62, 64, 120, 121, 122, 123] and references therein.

3.2.2 Analytical Expressions

We shall now give a brief description of the main analytical expressions proposed for the EOS of the HD fluid, following a chronological order.

The first expression was derived by Helfand et al. [124]:

$$Z = \frac{1}{(1-\eta)^2} \tag{3.8}$$

It was based on the scaled particle theory (SPT), and presents the simplest analytical form. According to those authors, this equation is valid for $\eta \leq \eta_c (\sim 0.907)$ and reproduces accurately the oldest computer simulation results. As pointed out by Santos et al. [83], Eq. (3.8) works extremely well at all densities, despite its analytical simplicity.

Alder et al. [95] proposed an expression which is perhaps the first EOS for the HD solid phase. It is based on the expansion of the pressure in powers of the relative free volume $\alpha = \eta_c/\eta - 1$, and is given by:

$$Z = \frac{2}{\alpha} + 1.90 + 0.67\alpha \tag{3.9}$$

The coefficients given by the authors were in good agreement with those obtained from cell models including correlations between neighboring particles.

The third expression for the EOS is a Padé approximant proposed by Chae et al. [96] in order to fit their own computer simulation results:

$$Z = 1 + \frac{2\eta - 0.8084\eta^2 + 0.0448\eta^3}{1 - 1.9682\eta + 0.9716\eta^2} \tag{3.10}$$

The authors indicated that Eq. (3.10) is valid for $\eta \leq 0.691$ with a mean deviation of around 2% with respect to their own reference data.

In 1975 Henderson [125] proposed a slight modification of the SPT EOS, Eq. (3.8), to give:

$$Z_{\mathrm{H75}} = \frac{1 + \eta^2/8}{(1-\eta)^2} \tag{3.11}$$

This equation, which will be referred here as H75, improved the results at low and intermediate densities, but presented poorer behavior in the high density range. For that reason, in 1977 he proposed a further modification (H77) by including an additional term in order to fit the computer simulation results more accurately over the whole density range [100]:

$$Z_{\mathrm{H77}} = Z_{\mathrm{H75}} - \frac{0.043\eta^4}{(1-\eta)^3} \tag{3.12}$$

Andrews [126] proposed a new equation based on a simple physical interpretation of the statistical mechanical expression for the reciprocal of the activity of a classical fluid:

$$Z = \frac{3\eta}{\left(1 - \frac{\eta}{\eta_c}\right)(1 - \alpha_A \eta)} - \frac{1}{\eta}\ln(1-\eta) - \frac{3\eta_c}{\alpha_A \eta}\ln(1 - \alpha_A \eta) + \frac{3\eta_c^2}{(1 - \alpha_A \eta_c)\eta}\ln\frac{1 - \frac{\eta}{\eta_c}}{1 - \alpha_A \eta} \tag{3.13}$$

with α_A defined as:

$$\alpha_A = \frac{1}{2}\left(b_3 - \frac{1}{3}\right) - \eta_c^{-1} = \frac{5}{2} - \frac{4\sqrt{3}}{\pi} \cong 0.2947 \tag{3.14}$$

where $b_3 = B_3/(B_2/2)^2 \approx 3.128$. Andrews' equation contains a singularity at close packing and reproduces the first three virial coefficients by construction. Nevertheless, in general it does not perform better than Eq. (3.8) [83].

Woodcock [99] used a generalization of free volume theory and proposed an expression that explicitly contains up to the sixth virial coefficient:

$$Z = \frac{1 + \dfrac{3\eta}{\eta_c}}{1 - \dfrac{\eta}{\eta_c}} + \sum_{n=2}^{6} (b_n - 4)\left(\frac{\eta}{\eta_c}\right)^{n-1} \tag{3.15}$$

with

$$b_n = \frac{B_n}{(B_2/2)^{n-1}} \tag{3.16}$$

being reduced values of the virial coefficients. Comparison with the computer simulation data from Chae et al. [96] at $\rho = 0.7$ and also with his own reference data at $\rho = 0.85$ showed a high degree of agreement [99].

Kratky [127] pointed out that EOS H75, Eq. (3.11), leads to slightly high values for the compressibility factor at certain densities, and therefore proposed the following modified expression to obtain a better overall fit to computer simulation results:

$$Z = \frac{1 + 0.112\eta^2}{(1-\eta)^2} \tag{3.17}$$

Taking into account the sixth and seventh virial coefficients, Kratky extended Eq. (3.17) to obtain the new equation:

$$Z = \frac{1 + 0.12802\eta^2}{(1-\eta)^2} - 0.03003\left(\frac{\eta}{1-\eta}\right)^3 \tag{3.18}$$

The values of the compressibility factor given by Eqs. (3.17) and (3.18) are listed in [127], together with predictions for the 6th to the 10th virial coefficients.

Baram and Luban [128] used Tová approximants to accelerate the convergence of the virial series, and proposed the following HD EOS:

$$\begin{aligned} Z = {} & \frac{\eta_c^4 (6b_6\eta_c - 5b_5)}{1 - \eta/\eta_c} + \frac{30\eta_c^5 (b_6\eta_c - b_5)}{\eta} \ln(1 - \eta/\eta_c) \\ & + \sum_{n=1}^{4} \left\{ (\eta/\eta_c)^{n-1} \left[b_n \eta_c^{n-1} - \eta_c^4 (6b_6\eta_c - 5b_5) + \frac{30\eta_c^5 (b_6\eta_c - b_5)}{n} \right] \right\} \end{aligned} \tag{3.19}$$

As can be seen, this EOS explicitly includes the first six virial coefficients in reduced units defined in Eq. (3.16). It also takes into account that, according to those authors, the virial expansion diverges at the closest packing density with critical exponent equal to 1.

Aguilera-Navarro et al. [129] proposed Padé approximants that reproduce the first six virial coefficients, and that diverge at the closest packing density, but also near the random close-packing density (between 0.89 and 0.92 times the regular close packing value, approximately). Unfortunately, coefficients for those approximants were not given in [129], and therefore their analytical expression is not presented here.

Devore and Schneider [130] proposed two new Padé approximants using the first six virial coefficient values and containing an explicit divergence at η_c:

$$Z = \frac{1 - 0.3802607\xi + 0.1960393\xi^2}{(1-\xi)(1 - 1.1940601\xi + 0.4088725\xi^2 - 0.0297987\xi^3)} \tag{3.20}$$

$$Z = \frac{1 - 0.3064720\xi + 0.1654813\xi^2 + 0.0120603\xi^3}{(1-\xi)(1 - 1.1202714\xi + 0.3182654\xi^2)} \tag{3.21}$$

with $\xi = \eta/\eta_c$. Comparison with the EOSs of Woodcock, Eq. (3.15), and of Baran and Lubam, Eq. (3.19), as well as estimation of the 7th to the 10th virial coefficients were given [130].

Verlet and Levesque [131] developed a new HD EOS based on a pressure-consistent integral equation for the radial distribution function:

$$Z = \frac{1 + 0.125\eta^2}{(1-\eta)^2} - \left(\frac{2^{-5}\eta}{1-\eta}\right)^4 \tag{3.22}$$

Note the similarity to the proposals of Henderson [100] and of Kratky [127], Eqs. (3.12) and (3.17). The authors indicated that Eq. (3.22) agreed with computer simulation data within the statistical errors of the latter (around 1%). A comparison with other expressions was carried out by Baus and Colot [45].

Hoste and van Dael [132] proposed a new EOS based on the use of the first seven virial coefficients [127], and also on the existence of a divergence located at a fixed density and with a certain slope as supported by theoretical arguments. Two adjustable parameters were included in their expression in order to fit computer simulation data [96, 98, 133]. The final expression is:

$$Z = 1 + \frac{2\eta/\eta_c}{1 - \eta/\eta_c} + \frac{7.893817\eta^2}{(1 - 0.6790401\eta)^2} + \sum_{q=2}^{7} \beta_q \eta^{q-1} \tag{3.23}$$

with

$$\beta_2 = -0.2053156 \quad \beta_3 = -7.197508 \quad \beta_4 = -9.143925$$
$$\beta_5 = -8.539116 \quad \beta_6 = -6.780652 \quad \beta_7 = -4.639083$$

Another correction to equation H77 [100] was proposed by Singh and Sinha [134]:

$$Z = \frac{1 + 0.128\eta^2}{(1-\eta)^2} - \frac{0.043\eta^4}{(1-\eta)^3} - \frac{1.47021\eta^2(1 - 7\eta/16)^2}{(1-\eta)^4} \tag{3.24}$$

In this equation, however, the correction term plays only a very minor role, as noted by Cuadros and Mulero [135].

Erpenbeck and Luban [101] proposed a Levin approximant from the use of the first seven virial coefficients and that reproduced quite accurately their own computer simulation results. It will be referred to as EL85, and is expressed as:

$$Z = \frac{\sum_{n=0}^{4} p_n \eta^n}{\sum_{n=0}^{5} q_n \eta^n} \tag{3.25}$$

where the coefficients are given by:

$$q_n = (-1)^n \binom{6}{n} \left(1 - \frac{n}{6}\right)^5 \frac{b_6}{b_{6-n}} \tag{3.26}$$

$$p_n = \sum_{m=0}^{n} b_{n+1-m} q_m \tag{3.27}$$

b_n being defined in Eq. (3.16). As reported by the authors, their EOS contains five simple poles at real positive densities. As will be noted below, Eq. (3.25) has been widely used as a reference to study the performance of new proposals.

A nonanalytical EOS for HDs was proposed by Alexanian [136], which is the solution to the nonlinear differential equation:

$$\frac{\mathrm{d}(Z\rho)}{\mathrm{d}B_2} = \eta^2 \frac{1 + AZ\rho + C(Z\rho)^2 + E(Z\rho)^3}{1 + BZ\rho + D(Z\rho)^2 + F(Z\rho)^3} \tag{3.28}$$

The author used the values of the first eight virial coefficients and obtained good agreement with Hoover and Alder's computer simulation data [93] (see Fig. 3.1 in [136]) for constants $A = 2.38787, B = 2.82387, C = 1.38276$, $D = 2.27327$, $E = 0.112033$, and $F = 0.467241$.

Baus and Colot [45] proposed a rescaling of the virial series by writing the density expansion of the compressibility factor as:

$$Z = \frac{1 + \sum_{n=1}^{6} c_n \eta^n}{(1 - \eta)^2} \tag{3.29}$$

where $c_1 = 0$, $c_2 = 0.128$, $c_3 = 0.018$, $c_4 = -0.0507$, $c_5 = -0.0533$, and $c_6 = -0.0410$. In this EOS the singularity is maintained at $\eta = 1$ following the idea of the SPT and of most earlier proposed EOSs. Values of c_n were obtained from the virial coefficients, and the method is extendable to hard rods, spheres, and hyperspheres. In particular, in two dimensions, they found that their proposal agreed with Erpenbeck and Luban's data [101] more accurately than the EOS of Verlet and Levesque [131], Eq. (3.22), and that its accuracy was as similar to the EOSs proposed by Henderson [100], by Kratky [127], and by Erpenbeck and Luban [101], Eqs. (3.12), (3.18), and (3.25) respectively (see Fig. 3.1 in [45]). Let us finally note that Brunner et al. [137] found good agreement between their experimental results for a two-dimensional liquid of charged colloidal particles suspended in water and Eq. (3.29), and therefore concluded that this liquid behaves as a two-dimensional fluid of HDs over a wide density range.

Song et al. [34] showed that the divergence of pressure at the random closest packing density (ρ_{rcp}) has a fractional critical exponent (s), and gave a general semiempirical expression for the EOS of the "hard fluid" in any number of dimensions:

$$Z = 1 + b\rho \frac{\sum_{n=1}^{4} c_n (b\rho)^{n-1}}{\left(1 - \frac{\rho}{\rho_{\mathrm{rcp}}}\right)^{s}} \tag{3.30}$$

where b is the so-called van der Waals co-volume, and where the coefficients c_n are not adjustable but are fixed by the known values of s, ρ_{rcp}, and the virial coefficients. In particular, for the two-dimensional case, they proposed:

$$Z = 1 + 2\eta \frac{\sum_{n=1}^{4} c_n (2\eta)^{n-1}}{\left(1 - \frac{\eta}{\eta_{\mathrm{rcp}}}\right)^{0.84}} \tag{3.31}$$

with $\eta_{\mathrm{rcp}} = 0.82$ [89] and $c_n = 1, 0.2698, 0.1067$, and 0.03552 (in numerical order). Results obtained with this EOS were compared with those of the SPT and Woodcock [99] EOSs, finding a high degree of accuracy over a wide range of densities.

Luban and Michels [49] presented an expression providing an accurate representation, for all densities, of computer simulation data for the fluid phase of hard disks, hard spheres, and hard four- and five-dimensional hyperspheres. They incorporated the exact values of the first four virial coefficients and included two adjustable constants determined from computer simulation data. The analytical expression has the form:

$$Z = \frac{B_2^* \xi \left\{1 + \left[(B_3^*/B_2^*) - \Psi(\xi)(B_4^*/B_3^*)\right]\xi\right\}}{1 - \Psi(\xi)(B_4^*/B_3^*\xi + \left[\Psi(\xi) - 1\right](B_4^*/B_2^*)\xi^2} \tag{3.32}$$

where $\xi = \eta/\eta_{\mathrm{c}}$, B_n^* are parameters given in terms of the virial coefficients as $B_n^* = B_n \rho_{\mathrm{c}}^{n-1}$, and $\Psi(\xi)$ is an arbitrary function containing the two adjustable coefficients. In particular, for the two-dimensional case the function selected was $\Psi(\xi) = 1.5317 - 0.0627\xi$. The authors performed a detailed comparison with results obtained from the Levin approximant in EOS EL85, Eq. (3.25), from two Padé approximants and also from the computer simulation values of Erpenbeck and Luban [101]. They noted that values obtained from their EOS agree extremely accurately with those of the Levin approximant.

Maeso et al. [50, 138] proposed a more general type of expression that consisted of a Padé approximant divided by the term $(1 - \eta)^n$, where n is an integer. In particular, for the HD system the authors showed that the expression

$$Z_{\mathrm{MSAV}} = \frac{10^3 - 947.989\eta + 128.018\eta^2 - 113.987\eta^3 - 52.9722\eta^4 - 1.580596\eta^5}{(1 - 0.947989\eta)(1-\eta)^2 10^3} \tag{3.33}$$

which was obtained from the first seven virial coefficients known at that time [127], gives quite accurate results when compared with the computer simulation data of Erpenbeck and Luban [101]. They reported an average absolute deviation (AAD) of 0.05%, which is a slightly higher value [82] than that obtained using the H77 [100] expression, Eq. (3.12).

Sanchez [139] used the value of the eighth virial coefficient and more accurate values of the lower order ones [140] in order to propose a new Padé approximant:

$$Z_{\mathrm{S}} = \frac{1 + 1.03961\eta - 0.4768481\eta^2 + 0.30164\eta^3}{1 - 0.960399\eta - 1.68407\eta^2 + 2.41607\eta^3 - 0.811998\eta^4} \tag{3.34}$$

He investigated the singularities of that expression, showing that it has a pole at $\eta = 0.9002$, which is near the closest packing value ($\eta_{\mathrm{c}} = 0.9069$). No comparison was made with data from computer simulations.

Santos et al. [83] carried out a classification of the earlier types of EOSs. On the one hand, there is a set of EOSs aimed at achieving good accuracy either when reproducing virial coefficients—like for instance those in Eqs. (3.15), (3.18) to (3.21), (3.29), (3.31), (3.33), and (3.34)– or when fitting computer simulation results—like Eqs. (10), (17), and (25). On the other hand, there are proposals for which accuracy is sacrificed in favor of analytical simplicity or the inclusion of only a reduced number of fitting parameters, like Eqs. (3.8), (3.11), and (3.12). Following with the latter aim, Santos et al. [83, 141] proposed the simple EOS:

$$Z_{\mathrm{SHY}} = \left(1 - 2\eta + \frac{2\eta_{\mathrm{c}} - 1}{\eta_{\mathrm{c}}^2}\eta^2\right)^{-1} \tag{3.35}$$

which will be referred to as the SHY EOS, and which does not contain any adjustable coefficients. It was obtained by merely imposing an exact fit to the second virial coefficient and the existence of a pole singularity at the close-packing fraction. As can be seen, it has another pole at $\eta_{\mathrm{c}}/(2\eta_{\mathrm{c}} - 1)$. According to the authors, the compressibility factor is accurately reproduced by Eq. (3.35), with relative errors always below 1.5% when compared with Erpenbeck and Luban's computer simulation results. The authors also compared the first ten virial coefficients obtained with their proposal with those obtained from the EOSs given in Eqs. (3.8), (3.11), (3.13), (3.15), (3.19), (3.25), and (3.34) and with the values known in 1995 [140]. Table 3.1 lists the same comparison, but now taking as reference the values for the virial coefficients given in [47, 48]. It can be seen that Andrews' EOS, Eq. (3.13), yields the least accurate results, whereas the expression of Erpenbeck and Luban, Eq. (3.25), shows the best agreement. The results in Table 3.1 confirm the conclusion of Santos et al. that their EOS provides reasonable estimates of the virial coefficients, performing better than some of the more complex expressions. Santos et al. also noted that the value that Eq. (3.35) predicts for

Table 3.1. Reference values for the reduced virial coefficients b_n (as defined in Eq. (3.16)) obtained by Clisby and McCoy [47, 48], and percentage deviations of the values predicted by several EOSs. Cells labeled with "–" indicate that the corresponding virial coefficient is directly included in the analytical expression of the EOS

n	b_n	SPT Eq. (3.8)	H75 Eq. (3.11)	Andrews Eq. (3.13)	Woodcock Eq. (3.15)	BL Eq. (3.19)	EL Eq. (3.25)	Sanchez Eq. (3.34)	SHY Eq. (3.35)
3	3.128017748	−4.1	–	–	–	–	–	–	−3.7
4	4.257854456	−6.1	0.04	−8.1	–	–	–	–	−5.1
5	5.33689664	−6.3	0.9	−14.4	–	–	–	–	−4.3
6	6.36296	−5.7	2.3	−18.2	–	–	–	–	−2.4
7	7.3518592	−4.8	3.9	−20.3	−2.2	0.5	−0.1	–	0.3
8	8.319104	−3.8	5.4	−21.1	−4.7	1.7	−0.2	–	3.3
9	9.2721408	−2.9	6.7	−21.0	−5.7	3.5	−0.4	0.8	6.8
10	10.2162944	−2.1	7.9	−20.2	−5.7	5.9	−0.4	2.1	10.5

the seventh virial coefficient is more accurate than that from Eq. (3.15) [99] or Eq. (3.19) [128], which explicitly includes up to the sixth coefficient in their construction.

Hamad [142] proposed volume-explicit EOSs for HDs, HSs, and mixtures of HSs. As he indicated, the volume-explicit expressions are very convenient to find the correct volume roots in chemical engineering calculations. In the two-dimensional case, the proposal takes the following analytical form:

$$Z = 1 + 2Z\eta + \frac{4}{3}\left(\frac{\sqrt{3}}{\pi} - \frac{1}{3}\right) Z\eta \ln \frac{5 + Z\eta}{5 + 16Z\eta} \tag{3.36}$$

This expression gives the exact second and third virial coefficients and predicts higher ones with a maximum error of 1.2% up to the eighth coefficient [140]. The author compared his results with the computer simulation data of Erpenbeck and Luban [101], finding a high degree of accuracy. Also, the value obtained for the limiting packing fraction at infinite pressure was 0.837, slightly above the random close packing, which is estimated as 0.82 approximately [89].

A nonanalytical HD EOS was proposed by Edgal and Huber [143], which is the solution of the differential equation:

$$\frac{\mathrm{d}u}{\mathrm{d}x} = 2\left(\sqrt{u} + 1\right)\left\{-\sqrt{u} + 2e^{x}(1 - qe^{x})^{-m}\left[1 + \frac{mqe^{x}}{1 - qe^{x}} - \frac{\mathrm{d}m}{\mathrm{d}x}\ln\left(1 - qe^{x}\right)\right]\right\} \tag{3.37}$$

where $u = (Z-1)^2$, $x = \ln \eta$, $q = 1/\eta_c$, and $m(\eta)$ is a function of η. Edgal and Huber used a virial development for $m(\eta)$ and achieved a solution of Eq. (3.37) involving 15 coefficients (see Table 3.1 in [143]). The authors showed that their proposal agrees quite successfully with Erpenbeck and Luban's data [101].

Wang [144] proposed the following van der Waals–Tonks type EOS:

$$Z = \frac{4\eta}{\pi}\left(\frac{4.08768}{1-\eta/\eta_{\mathrm{c}}} - 3.08768 - 2.50732\eta - 1.84204\eta^2 - 1.222376\eta^3 - 0.70592\eta^4 - 0.29728\eta^5\right) \tag{3.38}$$

which agrees with the correct close-packing limit. He showed that adequate results were obtained for the first nine virial coefficients and also for the compressibility factor at low densities, but not at high densities. We note that comparison was made only using old computer simulation results instead of the more recent Erpenbeck and Luban data [101].

Rusanov [18, 19, 44] derived a new theory for EOSs based on the concept of the exclusion factor (volume or area), which had already been used by Boltzmann and van der Waals. As is well known, the exclusion volume is defined as the volume that must be subtracted from the actual volume of a system due to the fact that particles in the system have their own volumes. In fact, Rusanov [19, 44] proposed a family of equations for the HD fluid, with the form:

$$Z^{(n)} = \frac{1 - \dfrac{2nk-2k-4}{(n-1)(n-2)k^2\eta}\left[(1-k\eta)^{n-1} - 1 + (n-1)k\eta\right]}{(1-k\eta)^{(n-1)}}, n \geq 3 \tag{3.39}$$

where n is an integer and k an adjustable constant that depends on n. Thus, for instance, for $n = 3$ and $k = 1$ the SPT EOS, Eq. (3.8), is recovered. Rusanov showed that taking $n = 3$ and $k = 1.0145$ gives better agreement with Erpenbeck and Luban's [101] data, with a maximum deviation of around 1%. If the equation is developed until the fourth approximation, i.e., with $n = 4$, then $k \approx 0.876677$ and the maximum deviation is reduced to 0.46%. Rusanov noted that the fifth approximation fits a little worse so that higher degrees should not be considered [19, 44]. This author compared the values of the compressibility factor obtained from his fourth-degree approach with those given by the SHY and EL85 EOSs, Eqs. (3.25) and (3.35). According to his results, his proposal appeared to be clearly more accurate than the SHY EOS, but slightly less accurate than the EL85 EOS, except for the highest density data, for which Eq. (3.39) with $n = 4$ is even superior to the expression EL85. Unfortunately, no comparison with more recent computer simulation results including higher densities was made.

Clisby and McCoy [47, 48] evaluated the 9th and 10th virial coefficients for HD (and also for HS and hard hyperspheres). They proposed the use of two Padé approximants at low densities:

$$Z = \frac{1 + 0.69939247\,(2\eta) - 0.33033017\,(2\eta)^2 + 0.11294457\,(2\eta)^3 - 0.012320562\,(2\eta)^4}{1 - 0.30060752\,(2\eta) - 0.81172709\,(2\eta)^2 + 0.62751627\,(2\eta)^3 - 0.17862580\,(2\eta)^4 + 0.021359218\,(2\eta)^5} \tag{3.40}$$

$$Z = \frac{\begin{array}{c}1 - 0.062894522\,(2\eta) + 0.13851476\,(2\eta)^2 + 0.0067699403\,(2\eta)^3 \\ +0.0039942056\,(2\eta)^4 + 0.00047760798\,(2\eta)^5\end{array}}{\begin{array}{c}1 - 1.0628945\,(2\eta) + 0.41940485\,(2\eta)^2 + \\ -0.11367848\,(2\eta)^3 + 0.021846467\,(2\eta)^4\end{array}} \tag{3.41}$$

Nevertheless, they pointed out that the variation in the polynomial coefficients in those approximants when the uncertainty in the virial coefficients is taken into account is of the same order as the coefficients themselves. Hence, one should not ascribe too much importance to the exact value of the Padé coefficients as they will change as future improvements are made in the accuracy of the virial coefficients. Estimates of the values of the 11th to 18th virial coefficients were also made.

Solana [84, 85] compared the virial equation obtained by considering the first 10 coefficients given by Clisby and McCoy [47, 48] with several computer simulation data sets. He found very good agreement in the approximate range $0 \leq \eta \leq 0.55$. Moreover, he showed that the results obtained with the Padé approximant given by Clisby and McCoy, Eq. (3.41), agree very accurately with the computer simulation data, although more simple analytical expressions could also work as accurately. In particular, Solana proposed the following two new expressions:

$$Z = \frac{1 + 5\eta^2/64}{(1-\eta)^2} \tag{3.42}$$

$$Z = \frac{1 + \eta^2/8 - \eta^4/10}{(1-\eta)^2} \tag{3.43}$$

The first arises as the sum of Eqs. (3.8) and (3.11), with weighting coefficients 3/8 and 5/8, respectively. It reproduces computer simulation data [101] with an absolute error below 0.04 at all densities, being therefore of similar accuracy to the clearly more complex Clisby and McCoy EOSs. The second, Eq. (3.43), is a proposal designed from a generalized Padé approximant, and has shown excellent accuracy for its apparent simplicity. The greatest absolute difference with respect to Erpenbeck and Luban's data is only about 0.03. Equation (3.43) also improves the prediction of the 5th to 10th virial coefficients when compared with Eq. (3.11).

Kolafa and Rottner [80] used their own computer simulation data together with the values of the first 10 virial coefficients to build new expressions for the HD EOS with the analytical form:

$$Z = \sum_{i=0}^{k} C_i \left(\frac{\eta}{1-\eta}\right)^i \tag{3.44}$$

where C_0–C_4 were determined so that the first five virial coefficients were reproduced exactly. Parameters for higher values of i are adjustable, but some of them may be zero. The authors then presented three developments of Eq. (3.44) differing in the maximum density and number of fitted parameters.

Table 3.2. Coefficients for Eq. (3.44) obtained by Kolafa and Rottner [80]

i	$\rho \leq 0.88$	$\rho \leq 0.89$	$\rho \leq 0.90$
0	1	1	1
1	2	2	2
2	1.12801775	1.12801775	1.12801775
3	0.00181895291	0.00181895291	0.00181895291
4	−0.0526134737	−0.0526134737	−0.0526134737
5	0.0504951668	0.0504963915	0.0504960168
6	−0.0325433846	−0.0325578581	−0.0325537792
7	0.0133946531	0.0134816028	0.0134578632
8	0.00174265604	0.0012918784	0.00140888182
9	−0.00944632202	−0.00808881628	−0.00834273601
10	0.00851111768	0.00669011963	0.00694127367
11	−0.0035963525	−0.00250795961	−0.00262254723
12	0.000577345106	0.000336036442	0.000355746352
19	$-1.06399127 \times 10^{-7}$		
22		$-5.15282664 \times 10^{-9}$	$-5.24672938 \times 10^{-9}$
57		$5.57730095 \times 10^{-23}$	$5.88054639 \times 10^{-23}$

The coefficients obtained are listed in Table 3.2. They noted that the proposals with maximum density greater than or equal to 0.89 predict to some extent a loop at the critical fluid–hexatic point.

Finally, let us note that some EOSs for anisotropic two-dimensional hard particles have also been developed. In this sense, several extensions of Eq. (3.8) have been proposed for planar convex particles and dumbbells, and the Percus–Yevick equation has been solved numerically for ellipses and dumbbells [62, 63, 120, 121, 122, 123, 145].

3.2.3 Test of Accuracy for a Set of Equations of State

Given the great number of proposals for the HD EOS, some investigators have focused their efforts on the study of their applicability and/or accuracy in reproducing certain computer simulation data sets, as well as on performing comparisons among them in order to achieve as much accuracy as possible for particular purposes. In particular, we focus here on the works of Santos et al. [83], Mulero et al. [82], and Solana [85], which may be regarded as being the most relevant in this sense. We also add a final comparison of some selected EOSs with the most recent simulation of Kolafa and Rottner [80].

In particular, in the work of Santos et al. [83] the results for the compressibility factor obtained with the EOSs given in Eqs. (3.8), (3.11), (3.13), (3.15), (3.19), (3.25), (3.34), and (3.35) were compared with the computer simulation data of Erpenbeck and Luban [101]. The corresponding percentage deviations

Table 3.3. Percentage deviations (%), in absolute value, of the values of the compressibility factor calculated with different expressions for the EOS of the HD fluid, taking as reference the computer simulation data of Erpenbeck and Luban [101]. Deviations 0.00 should be regarded as being lower than 0.005%

η_c/η	SPT Eq. (3.8)	H75 Eq. (3.11)	Andrews Eq. (3.13)	Woodcock Eq. (3.15)	BL Eq. (3.19)	EL Eq. (3.25)	Sanchez Eq. (3.34)	SHY Eq. (3.35)
1.4	2.95	2.26	8.20	0.19	1.91	0.45	1.24	0.64
1.5	3.20	1.34	7.29	0.51	0.84	0.02	0.42	0.75
1.6	3.04	0.94	6.21	0.42	0.47	0.02	0.19	1.26
1.8	2.62	0.54	4.49	0.25	0.18	0.01	0.06	1.55
2.0	2.23	0.34	3.31	0.13	0.08	0.01	0.02	1.53
3.0	1.11	0.04	0.97	0.00	0.00	0.00	0.00	0.92
5.0	0.42	0.05	0.22	0.02	0.02	0.02	0.02	0.35
10.0	0.11	0.00	0.02	0.00	0.00	0.00	0.00	0.10
20.0	0.01	0.00	0.00	0.00	0.00	0.01	0.01	0.01
30.0	0.00	0.00	0.00	0.00	0.00	0.01	0.01	0.00

are listed in Table 3.3, where values have been presented to two decimal digits. It can be seen that their proposal, Eq. (3.35), does a good job over the whole density range and is clearly superior to the other simple, and even to some of the complex, equations for the higher densities (i.e., lower values of η_c/η). It is worthwhile to stress that the relative error incurred through the use of this equation is always less than 2.5% in the interval of densities considered. Note that beyond $\eta_c/\eta = 1.6$ the relative error decreases dramatically. In sum, the authors regarded their proposal for the SHY EOS, Eq. (3.35), as presenting impressive behavior in view of its simple analytical form [83].

Mulero et al. [82] performed calculations of the compressibility factor of the HD system with six of the EOSs referred to in Sect. 3.2.2 (in particular, Eqs. (3.8), (3.11), (3.12), and (3.33) to (3.35)), and carried out a detailed comparison with the computer simulation data of Erpenbeck and Luban [101] which were taken as the most accurate at that date. The authors obtained the AADs between the calculated results from the EOSs and those from the computer simulation (see Table 3.4). Since theoretical EOSs (like SPT, Eq. (3.8)) fail when reproducing the simulation data at high densities [50, 83], they considered three different ranges for the packing fraction. For the whole range ($0.030 \leq \eta \leq 0.648$) different accuracies for the simple (Eqs. (3.8), (3.11), and (3.35)) and the more complex equations were found. Obviously, in the first case accuracy is sacrificed in favor of analytical simplicity, which is even more evident at high densities. Despite this, equation H77 [100], Eq. (3.12) (which contains only two adjustable parameters), gives the lowest average deviation with respect to simulation results, whereas the EOS proposed by Sanchez [139] (Eq. (3.34), with seven coefficients) gives greater deviations than Henderson's. Although the SHY EOS, Eq. (3.35), represents a good improvement with respect to SPT, especially at high densities, it was found to be not as good as H75 [125], Eq. (3.11), in particular at lower densities. Finally, the EOS given

Table 3.4. Average of percentage mean deviations (%) between the results obtained from the EOSs and computer simulation data of [101] for HD systems. Different ranges of η are considered

	SPT Eq. (3.8)	H75 Eq. (3.11)	H77 Eq. (3.12)	Maeso et al. Eq. (3.33)	Sanchez Eq. (3.34)	SHY Eq. (3.35)
$0.45 < \eta < 0.65$	2.81	1.09	6.2×10^{-2}	8.7×10^{-2}	0.39	1.14
$0.03 < \eta < 0.30$	0.33	2.1×10^{-2}	6.8×10^{-3}	6.2×10^{-3}	7.9×10^{-3}	0.28
$0.03 < \eta < 0.65$	1.57	0.55	3.5×10^{-2}	4.7×10^{-2}	0.20	0.71

in Eq. (3.33) [50] was reported to reach an accuracy similar to that of H77, Eq. (3.12).

Solana [85] compared the EOSs SPT, H75, the Padé approximant proposed by Clisby and McCoy, Eqs. (3.8), (3.11), and (3.41), and his own two proposals, Eqs. (3.42) and (3.43), with available computer simulation data. He first showed that, as expected, Eq. (3.41) gives excellent results when compared with Erpenbeck and Luban's simulation data. Therefore, the values yielded by Eq. (3.41) were taken as reference to test the accuracy of the other selected expressions. Figure 3.2 shows the difference between the compressibility factor calculated with several EOSs and that given by Eq. (3.41). Differences with respect to computer simulation data [93, 96, 98, 101] are also shown. As can be seen, the second proposal of Solana [84], Eq. (3.43), still very simple in its formal structure, gives the most accurate results. Differences with Erpenbeck and Luban's data are in the third decimal digit, except at the highest density value (for which this difference is approximately 0.03).

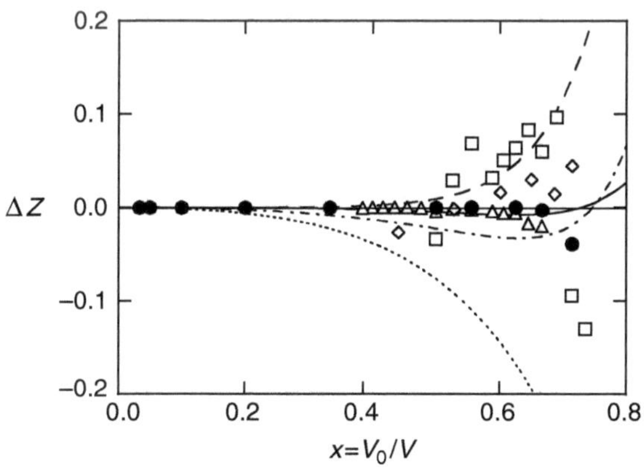

Fig. 3.2. Difference between calculated compressibility factors and the values given by Eq. (3.41) (taken from [84]). *Points*: computer simulation values of [93, 96, 98, 101] (*square*, *triangle*, *rhombus*, and *circle*, respectively). *Curves*: values obtained from EOSs in Eqs. (3.8), (3.11), (3.42), and (3.43) (*dotted*, *dashed*, *dotted-dashed*, and *continuous lines*, respectively)

Table 3.5. Average of percentage mean deviations (%) between the results obtained from the EOSs and computer simulation data of [80] for HD systems. Different density ranges are considered. The Rusanov proposal considered is Eq. (3.39) with $n = 4$. CM1 and CM2 were proposed by Clisby and McCoy [47]

	Density range		
EOS	0.4–0.75	0.8–0.89	0.4–0.89
EL, Eq. (3.25)	0.005	3.27	1.64
BC, Eq. (3.29)	0.181	3.82	2.00
Maeso et al., Eq. (3.33)	0.056	3.00	1.53
Sanchez, Eq. (3.34)	0.076	4.45	2.26
SHY, Eq. (3.35)	1.299	4.36	2.83
Rusanov, Eq. (3.39)	0.350	2.01	1.18
CM1, Eq. (3.40)	0.002	3.11	1.56
CM2, Eq. (3.41)	0.003	3.22	1.61
Solana, Eq. (3.43)	0.117	3.25	1.68

Given that accurate computer simulation data have been published recently [80], we include here a new validity test for the selection of the most accurate EOSs. The results are given in Table 3.5 for three different density ranges.

As can be seen in Table 3.5, the Solana EOS, Eq. (3.43), despite its simplicity, gives very accurate results for reduced densities lower than 0.75, being in clearly better agreement than the SHY EOS, Eq. (3.35), which is another simple expression. With respect to the Rusanov proposal, Eq. (3.39) with $n = 4$, we note that despite having an adjustable parameter, it deviates from the computer simulation data more than the simple Solana EOS. Figure 3.3a shows the percentage deviations of some EOSs in the density range from 0.4 to 0.75. As can be seen, the Rusanov EOS always overpredicts the data in this range, whereas the Solana EOS underpredicts them, although with lower deviations. Both the Baus and Colot (BC) and the Sanchez expressions increase in deviation as the density increases.

In this range of intermediate densities the Clisby and McCoy and the EL EOSs, which are analytically very similar, give the best results. There are only slight differences between their predictions, which can be seen in Fig. 3.3b, where the percentage deviations are shown. As can be seen, the CM2 EOS, Eq. (3.41), gives excellent results for densities lower than 0.7, but gives the greatest deviations (of the three EOSs considered here) for 0.75. The CM1 expression, Eq. (3.40), is more regular, and although its deviations increase at 0.7 (with an extremely low deviation of only −0.006%), it gives very good results at 0.75. Finally, we note that the EL underpredicts the data except for $\rho = 0.75$.

Table 3.5 also gives the AADs in the high density range from 0.8 to 0.89. As expected, the deviations obtained are clearly greater than in the previous

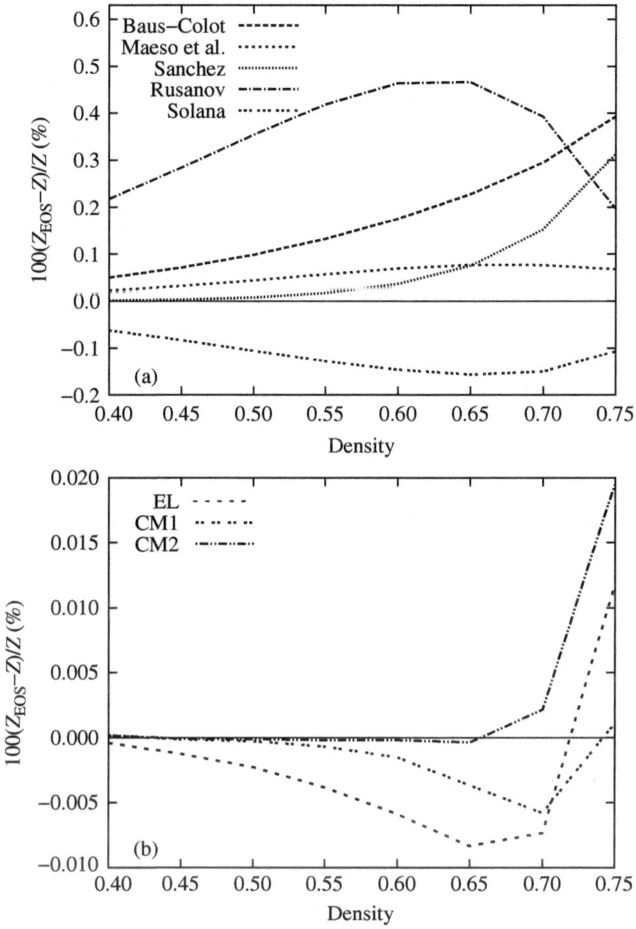

Fig. 3.3. Percentage deviations between compressibility factor values given by several EOSs and those given in Kolafa and Rottner's [80] computer simulation for $\rho \leq 0.75$

range. As can be seen, the Rusanov EOS, Eq. (3.39) with $n = 4$, gives the lowest AAD. This success is probably due to the inclusion of an adjustable coefficient ($k \approx 0.876677$), although it must be taken into account that densities which were not very high were considered in the calculation of that coefficient. The good results obtained with the simple Solana EOS, Eq. (3.43), must also be noticed. The AAD obtained is similar to that given by the EL and CM2 EOSs and lower than that given by the BC, Sanchez, and SHY EOSs. The Maeso et al. EOS also gives very good results despite being slightly simpler than other proposals. Percentage deviations obtained with several EOSs in the high density range are shown in Fig. 3.4. As can be seen, the Rusanov EOS gives deviations lower than 0.3% in the density range from 0.75 to 0.85.

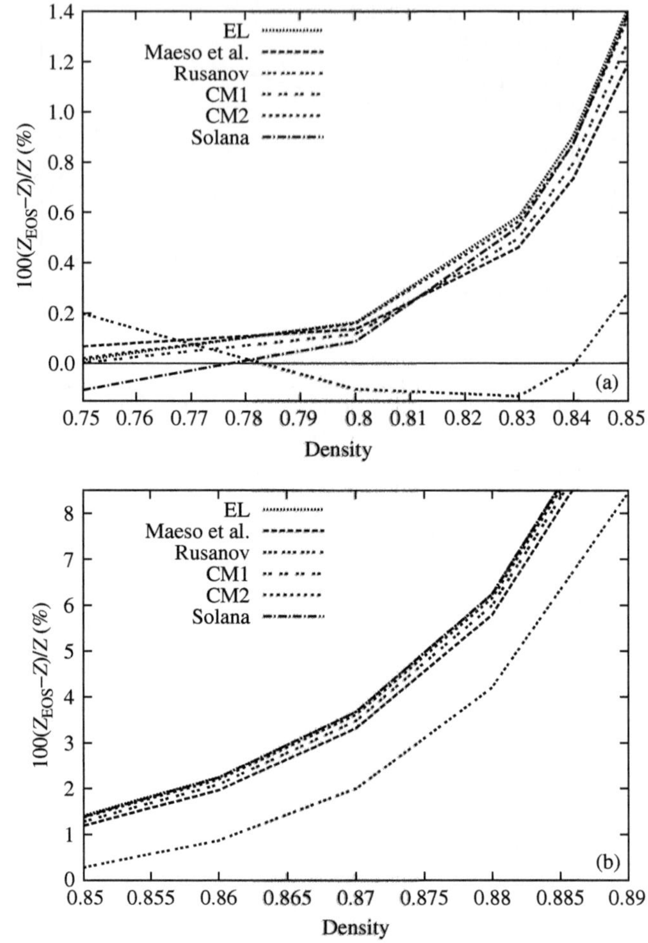

Fig. 3.4. Percentage deviations between compressibility factor values given by several EOSs and those given in Kolafa and Rottner's [80] computer simulation at high densities

Moreover, it is the only one giving deviations below 8.5% over the whole density range. One can also see that for $\rho \geq 0.8$ all the EOSs overpredict the Z values, with the only exception of the Rusanov EOS in the density range 0.79 to 0.84. Mainly due to this agreement at high densities, the Rusanov EOS gives the lowest AAD value when the whole density range is considered.

Finally, it must be noted that the validity of these EOSs has been investigated only in reproducing computer simulation results for the compressibility factor. Obviously, one can consider other criteria to make a choice depending on different applications.

3.2.4 Summary and Conclusions

Let us now make some concluding remarks about the main aspects referred to in this section.

A brief review of the most important computer simulations for the HD system available in the literature was given in Sect. 3.2.1. They ranged from the pioneering work of Metropolis et al. [91] in 1953 to the recent simulation of Kolafa and Rottner [80], and data were generated following both Monte Carlo and molecular dynamics computational techniques. The main features or any specific relevant aspect was noted for each of them. Some references relating to the study of the solid–fluid transition in two-dimensional systems were also included in that section.

Section 2.2 listed more than 30 proposals for the HD EOS. In particular, the dependence of the compressibility factor on the packing fraction–therefore on density– was given explicitly in most cases (see, for instance, Eqs. (3.8), (3.11), (3.13), (3.22), (3.29), and (3.34)). Some other expressions appear as the solution to a differential equation, like those proposed by Alexanian [136] and by Edgal and Huber [143], Eqs. (3.28) and (3.37), respectively. Some of these expressions were selected to carry out a detailed analysis of their accuracy in reproducing the compressibility factor of HD fluids. In this sense, the important work of Santos et al. [83], Mulero et al. [82], and Solana [85] were described in Sect. 3.2.3. In the three cases, Erpenbeck and Luban's [101] computer simulation data were taken as reference. A new comparison with respect to the more recent computer simulation data of Kolafa and Rottner [80] was also presented.

Santos et al. [83] checked the accuracy of the EOSs given by Eqs. (3.8), (3.11), (3.13), (3.15), (3.19), (3.25), (3.34), and (3.35). They proved that their proposal, Eq. (3.35), works extremely accurately at all densities in spite of its analytical simplicity. Moreover, its analytical structure agrees with the divergence at close packing. The authors emphasized that one should be cautious, however, in thinking that taking the singularity at close packing into account is enough to devise an accurate EOS.

Mulero et al. [82] selected six analytical expressions for the EOS. They considered four of them as being analytically "simple" (Eqs. (3.8), (3.11), (3.12), and (3.35)), and the other two (Eqs. (3.33) and (3.34)) as having a "complex" structure. The main conclusion the authors arrived at was that the simple H75, Eq. (3.11), and the complex Eq. (3.33) both give the best approximations to the computer simulation results of Erpenbeck and Luban [101] among the EOSs considered in their work. The simplest analytical equation, SPT (Eq. (3.8)), yielded the least accurate values, especially at high densities. According to those authors, the SHY EOS, Eq. (3.35), does not represent such a good improvement with respect to other simple expressions like SPT and H75, Eqs. (3.8) and (3.11), respectively.

After his detailed studies, Solana [84, 85] stated that if as much accuracy as possible is needed for the HD EOS, the two Padé approximants proposed by

Clisby and McCoy [48], Eqs. (3.40) and (3.41), should be used. Nevertheless, the expression H75, Eq. (3.11), is much simpler in its analytical form and offers enough accuracy for many purposes. When even more accurate values are needed, but analytical simplicity should also be considered, then Eqs. (3.42) and (3.43) [84] must be regarded as being excellent alternatives.

Comparison with the Kolafa and Rottner [80] data confirms that Eq. (3.43) as proposed by Solana is a simple expression giving accurate results over the whole temperature range. The Erpenbeck and Luban and the Clisby and McCoy proposals reproduce the data at densities below 0.8 with excellent accuracy. Nevertheless, at high densities the Rusanov proposal, Eq. (3.29), which contains an adjustable parameter, gives the smallest deviations.

Let us finally emphasize the great variety of proposals for the EOS of the HD system available in the literature –and even more so for the three-dimensional system. This allows one to make a suitable choice depending on the physical property to be calculated, on the analytical complexity one can accept, and on the specific working density range.

3.3 Equations of State for Hard Spheres

The HS fluid is perhaps the most widely studied physical system. There are thousands of papers devoted to the analysis of its geometry, structure, dynamic and thermodynamic properties, etc. In the particular case of EOSs, efforts have mainly focused on the search for an appropriate expression that allows one to reproduce the virial coefficients and/or the computer simulation data over some specific -or even over the whole- density range. Those expressions are currently used in the development of perturbation theories and of EOSs for real fluids used in physical chemistry or in chemical engineering calculations [5, 6, 7, 8, 9, 10, 11, 12, 13, 14, 15, 16, 17, 18, 19, 20, 21, 22, 23, 24, 25, 26, 27, 28, 29, 30, 31, 32, 33].

In both computer simulations and analytical expressions for HS systems, the compressibility factor Z is calculated as a function of the packing fraction, η, or the packing fraction ratio, ξ. The packing fraction in three dimensions is defined as:

$$\eta = \frac{\pi\rho}{6} \tag{3.45}$$

whereas the ratio between the packing fraction occupied by the spheres and its maximum value is:

$$\xi = \frac{\eta}{\eta_c} = \frac{\rho}{\sqrt{2}} \tag{3.46}$$

where η_c is the closest-packing fraction $\eta_c = \pi\sqrt{2}/6$.

In the following subsections, the main computer simulations and the most important proposals for the EOS of the HS system will be presented. But first we shall comment briefly on the shape of the phase diagram for hard spheres

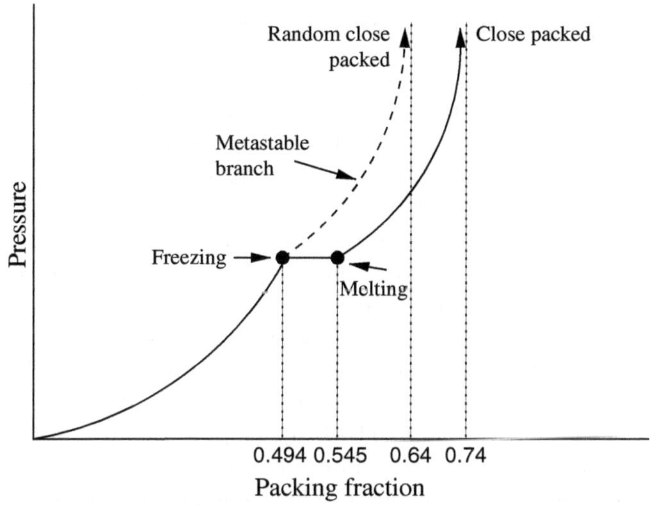

Fig. 3.5. Main branches of the phase diagram of an HS fluid. The diagram was constructed using the data of [148]

(Fig. 3.5). As is well known, despite the simplicity of the HS interaction potential, there is strong numerical evidence for the existence of even a first-order disorder/order phase transition. The phase diagram shows four important branches when pressure is plotted versus η (see Fig. 3.5).

There is a fluid branch that ends in a freezing point at approximately $\eta = 0.494$ ($\rho = 0.943$). At this point, the phase diagram splits into two parts. One is a metastable extension of the fluid branch [146], which follows continuously from the previous branch and is conjectured to end at a point known as random close packing. This state can be defined as the maximum packing fraction over all ergodic ensembles at which the mean nearest neighbor distance is equal to the sphere diameter. Extensive simulation work indicated that this occurs at $\eta = 0.644$ [89, 147, 148]. Torquato et al. [36] have pointed out that the traditional picture of the random close packing phenomena cannot be made mathematically precise, and they suggested using the new concept of a maximally random jammed state, which can be made precise.

The other branch that splits off from the freezing point represents the thermodynamically stable part of the phase diagram. Along the horizontal portion of this branch both fluid and solid can coexist until the melting point $\eta \approx 0.545$ is reached. The portion of the curve which continues above the melting point is referred to as the solid or ordered branch, ending at the close packed face-cubic-centered crystal at $\eta_c = 0.7405$, which is the maximum value of the packing fraction (closest-packing fraction). We note also that Williams et al. [149] found that HS fluids undergo a first-order transition to a crystalline solid at a density slightly above the melting value.

3.3.1 Summary of Computer Simulations

Computer simulation techniques represent the most usual method to test the accuracy of the theories and EOSs. A great number of computer simulations of the HS system have been performed in the last few decades, most of which will be referred to in the next paragraphs, following a chronological order – except for some specific exceptions.

The molecular dynamics computer simulation of Alder and Wainwright [150] is the oldest reporting a large number of data for the compressibility factor of the HS system. Their aim was to study how many particles are required to describe the equilibrium behavior of a real system. The authors presented results for systems ranging in size from 4 to 500 particles. The high density range $0.885 \leq \rho \leq 1.18$ was considered. It includes a range (from the normal freezing density, $\rho = 0.943$, up to $\rho = 1.085$) where the system remains fluid at metastable equilibrium, and also a range ($\rho > 1.085$) where the system crystallizes spontaneously. These computer simulation data have quite often been used as a reference to test the validity of numerous proposals for the EOS (see, for instance, [16, 21, 132, 151, 152, 153]). Further data for the solid phase were reported by other authors [94, 95, 133].

A more extensive simulation was performed by Barker and Henderson [154] using the Monte Carlo method. Values of Z were obtained from the radial distribution function of the fluid. The authors used a system of 108 particles and obtained results for $\eta \leq 0.484$ ($\rho \leq 0.925$). These computer simulation data have been used [12, 22, 23, 155] as a reference to test the accuracy of several analytical EOSs.

Another Monte Carlo computer simulation was performed by Adams [156]. Results were obtained for sets of 32 and 256 particles in the reduced density range from 0.1 to 0.8. Another run for $\rho = 0.7$ was carried out using 864 particles. He found clear discrepancies with the data reported by Barker and Henderson [154] due to the use of different numerical methods and approaches. Since new and more accurate data for Z in the same range were obtained later by other authors, neither of these simulations will be considered as reference in this chapter. Nevertheless, the sets of data obtained by Adams [156] for the chemical potential were considered by Mulero et al. [13] to test the accuracy of the analytical expressions derived from several EOSs (see Sect. 3.4.2).

The molecular dynamics computer simulations of Woodcock [99, 157, 158, 159] provide results for systems of 500 and 512 particles and for a greater number of states. The most widely used data set [159] corresponds to the range $0.942 \leq \rho \leq 1.085$, i.e., the metastable region. Several authors have considered Woodcock's computer simulation data as reference to test the accuracy of EOSs (see, for instance, [152, 160]).

Labík and Malijevský [161] made new Monte Carlo simulations for the radial distribution function. They used 864 particles, and worked in the range $0.3 \leq \rho \leq 0.86$, obtaining the compressibility factor Z from the contact value of the radial distribution function. By performing a comparison with

Carnahan–Starling's EOS and with the computer simulation data of [154, 156], those authors concluded that their results were considerably less scattered than those of the aforementioned references.

The simulation of Erpenbeck and Wood [162] includes results for a system of 4000 particles in the low density range $0.057 \leq \rho \leq 0.884$ (indeed, for $\rho = 0.3536$ only 1372 particles were considered). They stated that their results differ from others in two aspects: (i) they are statistically more precise, having statistical uncertainties at least an order of magnitude smaller than other previous results; and (ii) a correct dependence on the system's size is taken into account. In fact, Erpenbeck and Wood's data, together with those of Alder and Wainwright [150], have been considered as reference in a vast number of studies on the accuracy of different proposals for the HS EOS [18, 45, 152, 153, 160, 163]. We note that the computer simulation data of [161, 162] as well as their uncertainties were listed by Iglesias-Silva and Hall [164] and by Ayala and Lonngi [165]. In the last reference, molecular dynamics results for the density range from 0.1 to 0.943 for a system of 108 particles can also be found.

Speedy [166, 167] performed a new molecular dynamics simulation for a system of 500 HSs in the range $0.73 \leq \xi \leq 0.87$ (density range $1.03 \leq \rho \leq 1.23$), also including the formation of a glass at around $\xi \approx 0.75$. New simulations, using fluids with 1372 and 4000 HSs, were reported in the metastable density range $0.625 \leq \xi \leq 0.77$ [168]. Speedy pointed out that his values agree with those of Woodcock for $\xi \leq 0.74$. At higher densities Woodcock's pressures are lower, suggesting that his fluid and glass were partially frozen. For higher densities the results obtained were less accurate and were averaged over several short runs. Finally, let us note that these computer simulation data have recently been used by Liu [160] in order to test the accuracy of several expressions for the EOS of the HS fluid.

Rintoul and Torquato [148] presented results of large-scale molecular dynamics and Monte Carlo simulations along the disordered, metastable branch of the phase diagram, from the freezing point to the random close packing volume. They pointed out that many previous simulations contained deficiencies caused by crystallization and finite size effects. Their results indicated that in the vicinity of the random close packing fraction, the pressure scales as $(\eta - 0.644)^{-1}$. Unlike other studies [166, 167, 169], the authors found no evidence of a thermodynamic glass transition. The computer simulation data of Rintoul and Torquato have recently been compared with some other simulations [21] and used by Liu [160] to develop a new expression for the EOS.

The simulation data of Barošová [170] were computed using Monte Carlo's method and cover the density range $0.57 \leq \rho \leq 0.94$, although only five values were generated [171]. Nevertheless, the simulation results of Barošová are associated with large systematic errors, and do not include corrections for the size of the system.

Labík [172] carried out a Monte Carlo simulation with 13500 particles for a density range $0.40 \leq \rho \leq 0.94$. Kolafa [172] performed a similar simulation

(13500 particles, $0.60 \leq \rho \leq 0.95$, almost the same range as Barošová's simulation) using the method of molecular dynamics. Corrections due to finite size effects were included in both cases. Standard deviations were of the order of 10^{-4} for the simulation of Kolafa [172], even at high densities, and 10^{-3} at densities in the simulation of Labík [172]. More recently, new accurate data were obtained by highly optimized molecular dynamics calculations performed by Kolafa et al. [173], simulating 13500 HSs in the microcanonical ensemble in the density range 0.20–1.03. The relative inaccuracy at the 95% confidence level was better than 4×10^{-5} except for the last deeply metastable point. Finite size effects were taken into account in this computer simulation.

Wu and Sadus [21] reported new Monte Carlo molecular simulation data covering the isotropic liquid ($\rho \leq 0.95$), metastable fluid ($0.98 \leq \rho \leq 1.21$), and solid ($0.96 \leq \rho \leq 1.4$) ranges of density. The size of the simulated systems varied from 500 to 32000 particles. A detailed comparison with the data of Woodcock [99], Alder and Wainwright [150], Barker and Henderson [154], Adams [156], Labík and Malijevský [161], and Erpenbeck and Wood [162] was performed. Wu and Sadus reported a clear disagreement among those values only in the metastable region $\eta > 0.62$. We have compared Wu and Sadus' data with those of Kolafa et al. [173] for $\rho < 1.05$, finding very good agreement (see Fig. 3.6).

Let us also note that the data set of Wu and Sadus [21] has been used by Nasrifar and Bolland [17], Waziri and Hamad [174], and Liu [160] to test the accuracy of several expressions for the EOS of hard spheres (see the following subsection).

Rambaldi et al. [175] performed new molecular dynamics simulations in which values of Z were obtained from the mean free path. They used careful management of the list of collisions and were able to consider a system of 104–106 particles.

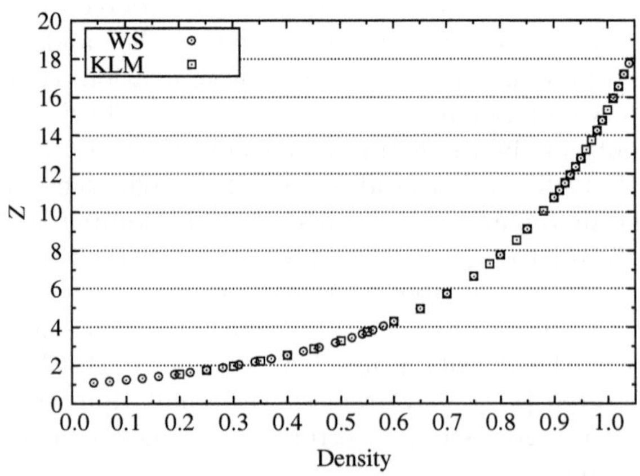

Fig. 3.6. Plot of the values of the compressibility factor obtained in the computer simulations performed by Wu and Sadus (WS) [21] and by Kolafa et al. (KLM) [173]

Finally, Kolafa [176] carried out a molecular dynamics computer simulation in the high density range $1.02 \leq \rho \leq 1.09$, where runs of 1000, 2000, 4000, and 13500 particles were combined. Good agreement was found with respect to Wu and Sadus' [21] simulation. Kolafa used his own data to propose a new nonanalytical EOS. Liu [160] used some of Kolafa's data to test a new proposal for the EOS. As has been shown, a long list of computer simulation data sets are available in the literature. Therefore, one should take specific criteria (range, number of simulated states, particular statistical indicators) into account when choosing suitable computer simulation values as references for comparison. In particular, this should be taken into account if several sets of data are used simultaneously in order to cover a wider range in the phase diagram. This is the case for the accuracy test for several EOSs performed by Mulero et al. [152] that will be described in Sect. 3.3.3, in which six different sets of computer simulation data were selected [99, 150, 162, 170, 172]. Despite the fact that they had not been performed following a single method, and that they covered different ranges of densities, the degree of agreement among them could be regarded as being quite satisfactory (see Fig. 3.7).

3.3.2 Analytical Expressions

A well-behaved HS EOS might be interpreted as satisfying various requirements [160]: (i) It should be very accurate, as is necessary for many applications. For instance, in perturbation theories for real fluids, the HS system is used as a reference and additional uncertainty will be created by the theory. Deviations in HS EOSs will be carried over into the real system and cause confusion. (ii) An HS EOS should reproduce as many as possible exact or accurate virial coefficients. This is because the virial EOS is the only equation with a sound theoretical background, and can produce the most accurate compressibility data and other thermodynamic properties in the stable fluid region given sufficient virial terms. (iii) The EOS should be fairly simple for mathematical manipulations since other thermodynamic properties will be algebraically derived from it. Also, the location of the poles should be taken into account when designing an expression for the EOS, especially at unity, random packing, or closest packing.

Following some of the considerations presented in the previous paragraphs, a vast number of analytical expressions for the EOS of hard spheres have been proposed since van der Waals proposed the simple form:

$$Z = \frac{1}{1 - 4\eta} \tag{3.47}$$

Approximately 80 expressions will be collected here, Eqs. (3.47) to (3.61), (3.63) to (3.123), and (3.125). Among them, those listed in Table 3.11 (22 expressions) were considered by Mulero et al. [152, 178] in order to test their accuracy in reproducing available computer simulation data of the

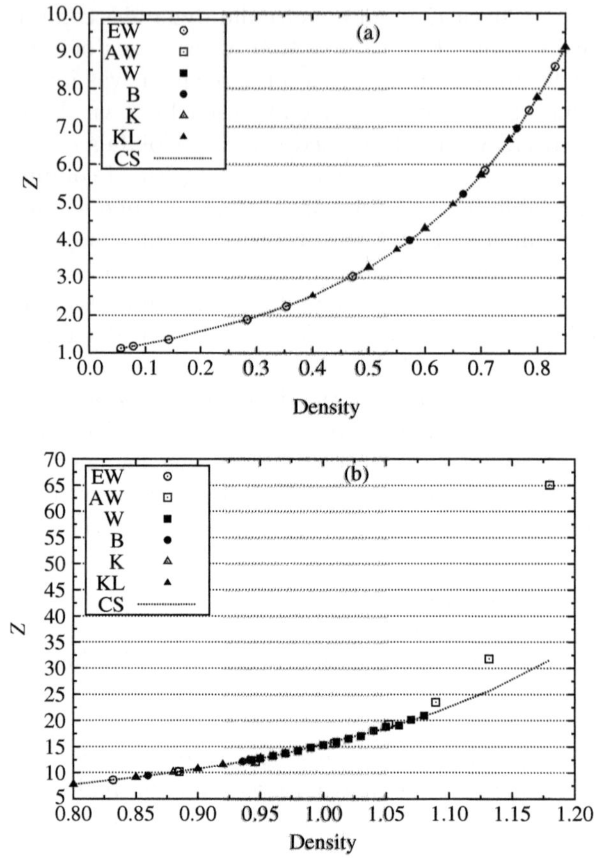

Fig. 3.7. Computer simulation data for the compressibility factor (Z) of the HS system plotted vs density (ρ), together with the results obtained with the EOS of Carnahan and Starling [177] (CS), Eq. (3.54). Notation: EW, Erpenbeck and Wood [162]; AW, Alder and Wainwright [150]; W, Woodcock [99]; B, Barošová [170]; K, Kolafa [172]; L, Labík [172]

compressibility factor. The rest were specifically collected for this chapter. We shall now briefly describe each of the EOSs in chronological order.

A pioneering study of the EOS of hard spheres was made by Tonks [179] who proposed a set of EOSs for one-, two-, and three-dimensional HS systems based on cell theory. The three-dimensional expression of this equation has the form of a Padé approximant:

$$Z = \frac{1 + 2.9619\xi + 5.483\xi^2}{1 - 0.8517\xi^3 - 0.1483\xi^4} \tag{3.48}$$

Although this equation satisfies the correct limit for the volume dependence (i.e., Z diverges to infinity as the volume approaches the close-packed value), it suffers from two major shortcomings. It can only reproduce up to the third virial coefficient well and represents poorly the machine-calculated values of the compressibility factor. In addition, the use of the Padé approximant in the denominator does not correspond to the thermodynamic framework of cell theory, in terms of the probability of finding a certain cell filled with an HS.

Together with van der Waals equation, one of the simplest and oldest expressions for the HS EOS is that derived from SPT [2, 180], which is the same as the expression derived from the Percus–Yevick compressibility equation:

$$Z = \frac{1+\eta+\eta^2}{(1-\eta)^3} \tag{3.49}$$

It has been proved that this equation can only be applied for low densities [45, 162], and not for the metastable or high density ranges [138, 153]. Recent extensions of SPT [181, 182] have led to a marked improvement in the theoretical prediction of the properties of HS fluids.

Thiele [4] derived the following EOS from the Percus–Yevick virial equation:

$$Z = \frac{1+2\eta+3\eta^2}{(1-\eta)^2} \tag{3.50}$$

This expression does not show any improvement with respect to the previous one. In fact, it has been shown that it gives worse results for the compressibility factor [45] and the virial coefficients [153] at any density. Thiele [4] observed that Eq. (3.50) must fail at densities near that of closest packing, since its singularity corresponds to the physically impossible density at which all the space is filled.

With the aim of reproducing the theoretical values of the virial coefficients up to the fifth, Ree and Hoover [183] proposed the expression:

$$Z = \frac{1+1.75399\eta+2.31704\eta^2+1.108928\eta^3}{1-2.246004\eta+1.301056\eta^2} \tag{3.51}$$

The analytical form of its denominator suggests a square of a binomial term. The authors showed that this expression reproduces Erpenbeck and Wood's computer simulation results better than Eqs. (3.49) and (3.50), although it gives a poor representation of the limiting behavior at the close-packed volume.

A quite simple and moderately accurate HS equation was derived by Guggenheim [184] in order to improve the algebraic simplicity of Eqs. (3.49) and (3.50):

$$Z = \frac{1}{(1-\eta)^4} \tag{3.52}$$

It has been used, for instance, to predict the critical properties of a diverse range of binary mixtures [20].

The proposal of Alder et al. [95] is perhaps the first EOS for the solid HS phase:

$$Z = 3/\alpha + 2.56 + 0.56\alpha \tag{3.53}$$

It is based on the expansion of the pressure in powers of the relative free volume $\alpha = \eta_c/\eta - 1$. An extension of this proposal, considering up to the third term, was later suggested by Young and Alder [133] from their own computer simulation results.

Carnahan and Starling [177] approximated the first virial coefficients to their nearest integers and then added the virial series to obtain their well-known expression for the EOS:

$$Z = \frac{1 + \eta + \eta^2 - \eta^3}{(1 - \eta)^3} \tag{3.54}$$

This can also be obtained from Eqs. (3.49) and (3.50), by multiplying the first by 2/3 and the second by 1/3 and then adding the two resulting expressions. It has been shown that it behaves very accurately at low densities [45] as well as for the metastable fluid region [138]. Nevertheless, it fails to reproduce the virial coefficients beyond the third, as well as failing in the closest-packing limit [153]. Equation (3.54) is actually the best well-known HS EOS and has been widely used in the construction of empirical and theoretical equations (see, for instance, [5, 8, 9, 10, 11]) to describe the repulsive contribution to the pressure of fluids.

The following expressions were proposed by Scott, as quoted by Marsh et al. [185]:

$$Z = \frac{1 + 3\eta + 4\eta^2}{(1 - 2\eta)(1 + \eta)} \tag{3.55}$$

$$Z = \frac{1 + 2\eta + 3\eta^2}{(1 - \eta)^2} \tag{3.56}$$

They give natural number values for the virial coefficients, although only the first two are reproduced correctly. As was shown by Miandehy and Modarress [155], both EOSs can only be applied at very low densities.

The equation given by Hall [186] is an empirical modification of expression CS, and has a pole at the regular closest-packing volume:

$$Z = \frac{1 + \eta + \eta^2 - 0.67825\eta^3 - \eta^4 - 0.5\eta^5 - 1.7\eta^6}{1 - 3\eta + 3\eta^2 - 1.04305\eta^3} \tag{3.57}$$

This equation fails in the prediction of the seventh virial coefficient, although it is in good agreement with several computer simulation results [20, 178]. Frenkel and Smit [187] used this equation to represent the liquid curve in solid–liquid equilibrium calculations for the HS system.

The expression of Le Fevre [188] was proposed to satisfy the idea that a pole must exist at $\eta = 0.6366$:

$$Z = \frac{1 - 0.8824\eta}{1 - 4.8823\eta + 9.5291\eta^2 - 7.6583\eta^3 - 3.2567\eta^4 + 12.8117\eta^5 - 8.7529\eta^6} \tag{3.58}$$

It was noted later [138] that this pole seems to be incorrect. Equation (3.58) works satisfactorily when reproducing virial coefficients but is expected not to reproduce those beyond the seventh. It was also shown that it behaves less accurately than the CS equation, Eq. (3.54), when compared with computer simulation data at low and intermediate densities, although the opposite is the case at high densities [138, 178].

Andrews [189] proposed a simple physical interpretation of the statistical–mechanical expression for the reciprocal of the activity, and used that expression to derive the following equation:

$$\begin{aligned} Z = {} & \frac{5.1834\xi}{1 - 1.5340\xi + 0.5340\xi^2} - \frac{1.3504}{\xi} \ln(1 - 0.7405\xi) \\ & - \frac{4.8534}{\xi} \ln\left(1 - 1.5340\xi + 0.5340\xi^2\right) - \frac{15.977}{\xi} \ln\left(\frac{1 - 0.5340\xi}{1 - \xi}\right) \end{aligned} \tag{3.59}$$

After a comparison with the results from the CS equation, Eq. (3.54), Andrews concluded that his proposal represents a significant improvement over Eq. (3.49) for $\xi^{-1} > 1.4$ (i.e., in the stable fluid region).

Woodcock [99] proposed an expression based on the use of the values of the first seven virial coefficients (B_i) in the expansion of Z in powers of the relative density ξ:

$$Z = 1 + \frac{9\xi}{1 - \xi} + \sum_{i=2}^{7} (B_i - 9)\, \xi^{i-1} \tag{3.60}$$

He carried out a comparison with Monte Carlo computer simulation results of Adams [156] and of Barker and Henderson [154], and also with his own molecular dynamics values at high densities. This comparison also included the previous EOSs Eqs. (3.49) to (3.52), (3.54), and (3.58). He concluded that his proposal was more accurate than all the aforementioned expressions.

Barboy and Gelbart [58] proposed an EOS with a clearly different structure from those referred to so far:

$$Z = \sum_{i=1}^{n} A_i \left(\frac{\eta}{1 - \eta}\right)^i \tag{3.61}$$

It has a theoretical background based on the SPT developed by Reiss et al. [2]. The coefficients A_i are obtained from the virial coefficients (B_i) as:

$$A_i = \sum_{i=0}^{k} \binom{k-1}{i} (-1)^{k-i-1} B_{i+1} \tag{3.62}$$

In 1979, 10 virial coefficients were known [190], although only the six first with sufficient accuracy. Barboy and Gelbart [58] compared the values of the compressibility factor obtained with Eq. (3.61) with the molecular dynamics data of Alder and Wainwright [150], finding good accuracy (similar to that of the CS equation) for $n = 4$. As noted by Wang et al. [153], Eq. (3.61) has the shortcoming of being extremely sensitive to the uncertainty of the values of the virial coefficients used. These authors compared the Barboy–Gelbart EOS obtained by using the first seven virial coefficients (i.e., taking $n = 7$) with the computer simulation data of Alder and Wainwright [150], and also with those of Erpenbeck and Wood [162]. The corresponding deviations were always greater than 9% at high densities (Alder and Wainwright's data), being clearly lower at low densities.

In 1981, Woodcock [159] proposed a second expression involving the ratio of two quartic polynomials:

$$Z = 1 + \frac{3.55173\eta}{1 - 2.6213\eta + 2.0485\eta^2 - 0.23782\eta^3 - 0.15085\eta^4} \tag{3.63}$$

This expression has recently been used by Liu [160] as a reference for the development of a new EOS valid for the whole density range, as will be described below.

Aguilera-Navarro et al. [129] derived a set of Padé approximants that reproduce the first six virial coefficients, and that diverge at the closest-packing density and also near the random close-packing density proposed by Scott and Kilgour [191] (0.8597 times the regular close-packing value). These EOSs contain information about the metastable liquid, and have the general form:

$$Z_{i,j} = \frac{1 + \sum p_i \xi^i}{1 + \sum q_j \xi^j} \tag{3.64}$$

The authors reported the best three approximants to be $Z_{3,3}$, $Z_{2,4}$, and $Z_{2,3}$, whose coefficients are listed in Table 3.6.

Table 3.6. Coefficients for the Padé approximants proposed by Aguilera-Navarro et al. [129], Eq. (3.64)

i,j	3,3	2,4	2,3
p_1	0.6178954155	0.6787391837	0.6714698708
p_2	0.6279348203	0.6687895775	0.6808674091
p_3	−0.0414265388		
q_1	−2.3440265434	−2.283182775	−2.290452088
q_2	2.0877644956	1.948285220	1.981894189
q_3	−0.8286676181	−0.7080817076	−0.7677703833
q_4		−0.0467140432	

Devore and Schneider [130] reviewed the proposal of Aguilera-Navarro et al. [129] and indicated that the molecular dynamics study of Woodcock and Angell [192] gave a slightly higher value for the random close-packing density ($0.8826\rho_c$). After this study, those authors proposed two new Padé approximants containing an explicit divergence at η_c:

$$Z = \frac{1 + 0.6300440\xi + 0.7498612\xi^2}{(1-\xi)\left(1 - 1.3318780\xi + 0.8417103\xi^2 - 0.2666873\xi^3\right)} \tag{3.65}$$

$$Z = \frac{1 + 1.1403861\xi + 1.1440680\xi^2 + 0.3624862\xi^3}{(1-\xi)\left(1 - 0.8216133\xi + 0.2348176\xi^2\right)} \tag{3.66}$$

The coefficients for these two equations were obtained using the first six virial coefficients. Estimates for the 7th to 10th virial coefficients were given by Devore and Schneider [130]. Finally, those authors proposed a fully empirical equation with a simple analytical form:

$$Z = \frac{1 + \frac{3}{2}\xi + \frac{3}{2}\xi^2 + \frac{1}{3}\xi^3}{(1-\xi)\left(1 - \frac{1}{2}\xi\right)} \tag{3.67}$$

which reproduces quite accurately Barker and Henderson's Monte Carlo computer simulation values, but which is not consistent with the virial coefficients.

Hoste and van Dael [132] proposed new EOSs for the crystal, glass, and fluid phases. For the crystal phase, they used the same analytical expression given by Young and Alder [133] but with new values for the adjustable coefficients, which were obtained by fitting computer simulation data selected from the simulations of Alder et al. [95], Hoover and Ree [94], and Young and Alder [133]. Their final expression was:

$$Z = \frac{3}{\alpha} + 2.81 + 0.47\alpha - 1.36\alpha^2 + 6.41\alpha^3 \tag{3.68}$$

with $\alpha = \eta_c/\eta - 1$. Equation (3.68) gave a root mean squared deviation of only 0.075% with respect to the aforementioned simulation data.

For the glass phase, Hoste and van Dael took Woodcock's simulation data [159] as reference to compare with, together with the analytical expression proposed by Alder et al. [95], Eq. (3.53). The proposal of Hoste and van Dael [132] takes the form:

$$Z = \frac{3}{\alpha_a} - 2.944 + 29.43\alpha_a \tag{3.69}$$

where $\alpha_a = \eta_a/\eta - 1$, with $\eta_a = 0.653839$ being the value of the packing fraction for which the glass branch diverges, which is taken to be another adjustable parameter. Finally, for the fluid phase, the proposed expression is:

$$Z = 1 + \frac{3\eta/\eta_c}{1-\eta/\eta_c} + \frac{5.372804\eta^2}{(1-1.041709\eta)^2} - 0.05142342\eta - 0.8441475\eta^2 - 0.2179363\eta^3 + 0.7549648\eta^4 + 2.063786\eta^5 + 6.282366\eta^6 \tag{3.70}$$

which reproduces the first seven virial coefficients and contains two adjustable parameters in order to ensure good agreement with computer simulation data. The value of the MD is less than those obtained for the EOSs given by Eqs. (3.54), (3.57), (3.58), and (3.65) to (3.67).

Erpenbeck and Wood [162] fitted their own computer simulation data to the following Padé approximant:

$$Z = \frac{1 + 1.7227128\eta + 2.2532688\eta^2 + 0.89244864\eta^3 - 0.34302926\eta^4}{1 - 2.2772872\eta + 1.32624176\eta^2} \tag{3.71}$$

which does not produce a singularity for any real value of η. They also proved that this EOS gives a better approximation in this range than Eqs. (3.51), (3.54), (3.59), and some others. Nevertheless, it has been shown [153, 178] that Eq. (3.71) does not show accurate behavior when compared with Alder and Wainwright's [150] computer simulation results.

A nonanalytical EOS was proposed by Alexanian [136] which, as in the two-dimensional case, is the solution to the nonlinear differential equation in Eq. (3.28). The best results, when compared with Erpenbeck and Wood's [162] data, were obtained for the constants $A = 2.46754$, $B = 3.21754$, $C = 1.39279$, $D = 2.94510$, $E = 0.0861991$, and $F = 0.695249$.

A new equation was proposed by Kolafa (quoted in [193]):

$$Z = \frac{1 + \eta + \eta^2 - \frac{2}{3}\left(\eta^3 + \eta^4\right)}{(1-\eta)^3} \tag{3.72}$$

It is similar in simplicity to the CS expression, Eq. (3.54), although it reproduces the HS virial coefficients and exact values of the compressibility factor somewhat more satisfactorily. In particular, it gives excellent results with respect to Erpenbeck and Wood's [162] simulation data, showing somewhat better behavior than Eq. (3.54) [45, 152, 153, 178]. Also, in the case of the metastable fluid region, Eq. (3.72) behaves more accurately than Eq. (3.54) [138]. With respect to Alder and Wainwright's [150] computer simulation data, both equations behaved poorly [152, 153, 178]. Some detailed comments will be given in Sect. 3.3.3. Finally, we note that a simple theoretical description of the derivation of the EOSs in Eqs. (3.47), (3.49), (3.50), (3.54), and (3.72) can be found in [194].

The following expression, proposed by Kim et al. [195], was also developed as a simplification of the CS EOS in order to represent the repulsive term in a cubic chain-of-rotators EOS. Its analytical expression is very simple:

$$Z = \frac{1 + 3.08\eta}{1 - 1.68\eta} \tag{3.73}$$

Miandehy and Modarress [155] showed that this equation is not as accurate as others (CS, for instance) when compared with Alder and Wainwright's [150] computer simulation data. If the simulation results of Barker and Henderson

[154] are considered as reference, then the AAD deviation is 3.3% [12], being 0.9% if the CS EOS is used.

Ma and Ahmadi [151] proposed a new equation with a pole at the random close-packed value $\eta = 0.64356$, and with the form:

$$Z = 1 + 4\eta \left\{ \frac{1 + 2.5\eta + 4.5904\eta^2 + 4.515439\eta^3}{\left[1 - (\eta/0.64356)^3\right]^{0.67802}} \right\} \tag{3.74}$$

The coefficients were obtained using the virial coefficient values given by Ree and Hoover [183]. A comparison with Alder and Wainwright's [150] computer simulation data, as well as with the results obtained with the EOSs in Eqs. (3.51) and (3.54), was performed. They observed that this new equation is remarkably accurate even at extremely high densities.

Baus and Colot [45] obtained the following two simple expressions for the EOS by rescaling the exact low density results and taking into account the values of the virial coefficients up to the seventh:

$$Z = \frac{1 + \eta + \eta^2 - 0.6352\eta^3 - 0.8697\eta^4}{(1-\eta)^3} \tag{3.75}$$

$$Z = \frac{1 + \eta + \eta^2 - 0.6352\eta^3 - 0.8697\eta^4 + 0.2543\eta^5 + 2.9231\eta^6}{(1-\eta)^3} \tag{3.76}$$

They compared the results with the computer simulation data of Erpenbeck and Wood [162], concluding that Eq. (3.75) behaves more accurately than Eq. (3.76), and that its behavior approaches that of Eqs. (3.54), (3.71), and (3.72) in that range. A comparison at very high densities will be presented in the next subsection.

Goldman and White [196] suggested two expressions to describe the HS system, which are obtained by interpolating between gas and close packed configurations. Their analytical expressions are:

$$Z = \frac{1 - 1.63523\eta^3}{(1-\eta)^4} \tag{3.77}$$

$$Z = \frac{1 + 2.649526\eta + 4.598102\eta^2 + 4.860055\eta^3 + 3.498\eta^4}{1 - \xi} \tag{3.78}$$

These expressions indicated some instability and possible phase transition in the same region as that in Alder and Wainwright's [150] computer simulation. The first has a pole at $\xi = 1$ and gives exact values for the first four virial coefficients. The pole of the second is located at the close-packed density. Comparison with more recent computer simulation results has not been carried out so far.

Zhou and Stell [197] proposed an EOS obtained from a linear combination of those derived from the Percus–Yevick compressibility and virial approximations, Eqs. (3.49) and (3.50), respectively. Their proposal can be written as follows:

$$Z = \frac{1+\eta+\eta^2-3\eta^3\theta}{(1-\eta)^3} \tag{3.79}$$

where θ is, in general, a function of the packing fraction. The particular cases of $\theta = 1/3$ and $\theta = 2(1+\eta)/9$ reproduce the CS and Kolafa EOSs, Eqs. (3.54) and (3.72), respectively. After comparison with Erpenbeck and Wood's data, Zhou and Stell proposed taking $\theta = 5/23 + (3/13)\eta$, finding significantly more accurate values for the compressibility factor than those calculated with the CS and Kolafa EOSs.

Song et al. [34] observed that the data at high densities diverge at the density of random closest packing, and that this divergence is characterized by a fractional critical exponent. The aforementioned EOS derived by Ma and Ahmadi [151], Eq. (3.74), is a clear example of the use of a fractional power to characterize such divergence. In particular, Song et al. [34] proposed the value 0.76 for the fractional power instead of the value 0.678 used in Eq. (3.74). By fixing this fractional power value and the value of the random closed packing density ($\eta = 0.6435$), the authors reached an expression similar to that of Eq. (3.74), although it includes a larger number of coefficients obtained from the virial coefficients. This new equation takes the form:

$$Z = 1 + \frac{\begin{array}{c} 4\eta + 0.3298\,(4\eta)^2 + 0.08867\,(4\eta)^3 + 0.01472\,(4\eta)^4 + \\ 10^{-3} \times (0.5396\,(4\eta)^5 - 0.3574\,(4\eta)^6 - 0.5705\,(4\eta)^7 \\ -0.1212\,(4\eta)^8 - 0.1151\,(4\eta)^9) \end{array}}{(1-\rho/\rho_{\mathrm{rcp}})^{0.76}} \tag{3.80}$$

Very good agreement was found with respect to the computer simulation values of Alder and Wainwright [150] and of Erpenbeck and Wood [162]. Song et al. [34] pointed out that their proposal may be useful for arbitrary dimensionality, and some results were presented for four and five dimensions.

Luban and Michels [49] proposed a new analytical representation for the HS EOS by incorporating the exact values of the first four virial coefficients [190] and two constants determined from a fit to the results obtained with Erpenbeck and Wood's EOS [162], Eq. (3.71). The resulting expression is:

$$Z = \frac{\sqrt{2}B_2\xi\left\{1 + \left[\frac{B_3}{B_2} - (1.43544 - 0.10283\xi)\,\frac{B_4}{B_3}\right]\sqrt{2}\xi\right\}}{1 - (1.43544\xi - 0.10283\xi^2)\,\frac{B_4}{B_3}\sqrt{2} + (0.87088\xi^2 - 0.20566\xi^3)\,\frac{B_4}{B_2}} \tag{3.81}$$

where the B_i represent the virial coefficients. They also proposed expressions for hard four- and five-dimensional hyperspheres. No comparison with more recent data from computer simulations has been carried out so far.

Maeso et al. [138] tested the accuracy of a large number of the EOSs already referred to in the metastable fluid region. They noted that, although this region is particularly sensitive to the quality of the expression for the EOS, it is frequently ignored. A large number of a new general type of approximants were proposed and discussed. According to Maeso et al., the two most accurate are the following:

$$Z = \frac{1 + 2\eta + 3\eta^2 + 2.364799\eta^3 + 1.494402\eta^4 + 1.750397\eta^5 + 4.672003\eta^6}{(1-\eta)^3} \tag{3.82}$$

$$Z = \frac{1 + 20.15439\eta + 18.09291\eta^2 + 31.55595\eta^3}{(1 + 19.15439\eta - 2.061472\eta^2 + 15.09823\eta^3)(1-\eta)^3} \tag{3.83}$$

which were based on the use of the seven first virial coefficients [190]. They showed that both expressions present accuracy similar to that of the CS EOS, Eq. (3.54), for both the metastable and the stable fluid regions (taking as reference the computer simulation data of Erpenbeck and Wood [162]), their main goal being to reproduce the first seven virial coefficients within their uncertainties. A more detailed comparison with several computer simulation data sets will be presented in Sect. 3.3.3.

Ayala and Lonngi [165] proposed a new method for obtaining the analytical expression of the HS EOS. It was first based on Thiele's formula that permits one to recover exactly an arbitrary rational function from tabulated values. In particular, they used as reference their own computer simulation data as well as those of Erpenbeck and Wood [162], and of Labík and Malijevský [198]. Then, they obtained six different rational functions for the compressibility factor, which are Newton–Padé approximants with the same degree in the numerator and denominator. After that they divided out from the numerator and denominator of the Newton–Padé approximants the factor associated with the closely lying roots. Six irreducible rational functions were then obtained, which differ in the data used as reference. Moreover, the authors constructed another approximant from the values of the first five virial coefficients, and yet one more by considering simultaneously all the simulation data. Hence, a total of eight expressions were given by Ayala and Lonngi [165], expressed as:

$$Z = \frac{\sum_i a_i \rho^i}{\sum_i b_i \rho^i} \tag{3.84}$$

with coefficients listed in Table 3.7. Comparison with the CS and Kolafa EOSs, Eqs. (3.54) and (3.72), as well as predictions for the first 10 virial coefficients were performed. The results showed that some simple irreducible rational approximants constructed by using two cubic or even two quadratic polynomials in the density have similar or even better accuracy than the EOSs given by Eqs. (3.54) and (3.72).

Table 3.7. Coefficients for the set of EOSs represented by Eq. (3.84)

i	a_i	b_i	i	a_i	b_i
0	1	1	0	1	1
1	1.3777578	−0.7166212	1	1.5434490	−0.5510603
2	1.0181363	−0.2227314	2	1.1584598	−0.4296627
3	0.3947018	0.1879037	3	0.2551780	0.4832663
0	1	1	0	1	1
1	0.742510	−1.351485	1	0.7694829	−1.3249120
2	0.414829	0.499168	2	0.4391413	0.4724746
0	1	1	0	1	1
1	0.871472	−1.222923	1	0.848995	−1.245400
2	0.593020	0.384102	2	0.546424	0.402013
0	1	1	0	1	1
1	0.781844	−1.312551	1	0.69864	−1.39576
2	0.455055	0.462600	2	0.38663	0.56834

Solana [199] described an improvement in and generalization of the CS method of obtaining EOSs for HS fluids in d-dimensions, and proposed two new expressions. The first is:

$$Z = \frac{1 + \eta + \eta^2 - 0.6352\eta^3}{(1-\eta)^3} \tag{3.85}$$

which is very similar to Eq. (3.75), and which was obtained in order to calculate the virial coefficients up to the fourth. Nevertheless, that equation reproduces computer simulation results at low and metastable densities worse than the CS equation [152]. The second includes a larger number of coefficients:

$$Z = \frac{1 - \eta - 1.6352\eta^3 + 1.4005\eta^4 + 1.1764\eta^5}{(1-\eta)^5} \tag{3.86}$$

It reproduces seven virial coefficients and presents a behavior similar to that of Eq. (3.85) at low densities, yielding less accurate results at metastable densities. The author concluded that in the three-dimensional case no improvement in the EOSs can be achieved by following the CS method, in contrast to the two-dimensional case.

Two Padé approximants were proposed by Sanchez [139],

$$Z = \frac{1 + 1.024385\eta + 1.104537\eta^2 - 0.4611472\eta^3 - 0.7430382\eta^4}{1 - 2.975615\eta + 3.007000\eta^2 - 1.097758\eta^3} \tag{3.87}$$

$$Z = \frac{1 + 1.7758\eta + 2.32413\eta^2 + 1.117943\eta^3}{1 - 2.22420\eta + 1.22094\eta^2 - 0.172934\eta^3 - 0.278645\eta^4} \tag{3.88}$$

which were obtained by considering eight virial coefficients. Equation (3.87) was considered by Bravo-Yuste et al. [200] to study a rational-function approximation method, which permits one to derive analytically the radial distribution function and the structure factor of a HS metastable fluid (see Chap. 6).

Iglesias-Silva and Hall [164] noted that the previous EOSs did not adequately represent hard-convex body behavior. Then they used the values for the first four virial coefficients given by Boublik for hard-convex body fluids [61] to derive a couple of expressions with the form of Padé approximants:

$$Z = \frac{1 - 0.71239\eta - 1.44735\eta^2 - 3.02606\eta^3 + 1.3\pi\eta^5\left(1 + 1.5\eta\right)}{\left(1 - \frac{\pi\eta}{2}\right)^3} \quad (3.89)$$

$$Z = \frac{1 + 2.4292\eta + 3.71681\eta^2 + 2.65693\eta^3 + 6\eta^4\left(-0.42699 - 22.5\eta^4\right)}{1 - \frac{\pi\eta}{2}} \quad (3.90)$$

which have a pole at approximately the random close-packed density $(2/\pi)$. Each expression contains one adjustable coefficient, which was obtained by fitting Erpenbeck and Wood's [162] data. One of the advantages of their equations over many of the others is that they work both for HS systems and for prolate spherocylinders. Nevertheless, as Khoshkbarchi and Vera [163] have pointed out, Eq. (3.89) exhibits anomalous behavior at medium to high densities, and Eq. (3.90) can only reproduce correctly the first four virial coefficients. Moreover, they both give negative values for Z at values of η greater than $2/\pi$.

Speedy [166, 167] gave a very simple empirical relationship that is valid only in the metastable region:

$$Z = \frac{2.67}{1 - 1.54269\eta} \quad (3.91)$$

As will be seen below in this subsection, this expression was subsequently improved [169].

Mohsen-Nia et al. [201] proposed a very simple expression in order to serve as reference in EOSs for real fluids while maintaining their cubic character [202, 203]. Its analytical expression is:

$$Z = \frac{1 + 2.48\eta}{1 - 1.88\eta} \quad (3.92)$$

Nevertheless, this simple proposal fails both at the close-packed density and at reproducing the virial coefficients [155]. In fact, it reproduces accurately the computer simulation values of the compressibility factor only at very low densities [155].

A new volume dependence for Z was proposed by Wang et al. [153] to agree with both the low density and the closest-packing limits. The new volume dependence was based on the definition of a new cell geometry which fills up all the space and, as a result, the compressibility factor tends to infinity when the hard spheres are in the close-packed state. The new cell, in contrast

Table 3.8. Coefficients for the EOSs of Wang et al. [153], Eqs. (3.93) to (3.95)

i	c_i	d_i	D_i
1	1	1	1
2	0.96192	−0.03808	1.96192
3	0.55927	−0.40265	0.55927
4	−0.54794	−1.10721	−1.10721
5	−0.94405	−0.39611	0.55626
6	−0.64773	0.29632	−0.11923
7	−0.00006	0.64715	0.00954
8	−5.13137	−2.37792	
9	0.30736	0.75175	
10	7.45126	2.83837	

to the spherical cell, has a tetrakaidecahedral geometry with centers placed at the corners of a rhombohedron. The three proposals (with coefficients in Table 3.8) are expressed in terms of ξ as follows:

$$Z = \frac{\sum_{i=1} c_i \xi^{i-1}}{(1-\xi)^2} \tag{3.93}$$

$$Z = \frac{\sum_{i=1} d_i \xi^{i-1}}{(1-\xi)^3} \tag{3.94}$$

$$Z = \frac{1}{\xi} \sum_{i=1} D_i \left(\frac{\xi}{1-\xi} \right)^i \tag{3.95}$$

which have a singularity at $\xi = 1$, corresponding to the physically realistic case in which the HSs are at their closest-packing state but not all space is filled.

Equations (3.93) and (3.94) reproduce exactly seven virial coefficients, whereas Eq. (3.95) does so with considerably poorer accuracy. Comparison with computer simulations shows that Eqs. (3.93) and (3.94) reproduce better the values of Erpenbeck and Wood [162], whereas Eq. (3.95) behaves significantly more accurately in the range of very high densities (where simulation data of Alder and Wainwright [150] were taken as reference). Comparison with other previous proposals was also made by Wang et al. [153]. The behavior of these equations in the metastable region will be analyzed in more detail in Sect. 3.3.3.

Following Kim et al. [195], Lin et al. [12] proposed a new simplification of the CS EOS, Eq. (3.54), in order to represent a cubic perturbed hard-chain EOS for real fluids:

$$Z = \frac{1 + 3.453667\eta}{1 - 1.610016\eta} \tag{3.96}$$

The AAD with respect to the computer simulation data of Barker and Henderson [154] was found to be 2.2%, a value lower than that corresponding to the Kim et al. EOS, Eq. (3.73), but clearly higher than that obtained with the CS EOS (0.9%). In any case, Kim et al. noted that both Eqs. (3.73) and (3.96) were developed mainly as the repulsive part of an EOS for real fluids. In this case, the use of Eqs. (3.73) and (3.96) avoids some mathematical problems found when the CS EOS is taken as [195].

Khoshbarchi and Vera [163] developed a simplified expression based on the volume dependence proposed by Wang et al. [153], Eq. (3.94), and therefore having the correct limit for the close-packed density:

$$Z = \frac{1 - \frac{1}{25}\xi - \frac{2}{5}\xi^2 - \frac{5}{4}\xi^3 + \frac{9}{50}\xi^5 + \frac{71}{50}\xi^{12}}{(1-\xi)^3} \tag{3.97}$$

As they noted, Wang et al.'s proposals, Eqs. (3.93) to (3.95), are quite accurate over a wide range of densities, but their analytical form is rather too complex to obtain analytical solutions for some integral equations, such as that of Percus–Yevick. They therefore proposed a simpler expression which they showed improved the prediction of the values of the first eight virial coefficients [190] compared with the expressions given in Eqs. (3.54), (3.89), and (3.90). Moreover, it reproduces accurately the computer simulation data of Alder and Wainwright [150] and of Erpenbeck and Wood [162]. This EOS has been used to extend to mixtures the radial distribution function of Percus–Yevick type HS EOSs [204], and also to study the performance of three mixing rules for HSs [205].

Hamad [142] developed a new EOS, but using a different approach in which the analytical expression is volume-explicit:

$$Z = 1 + 4Z\eta + \frac{3}{4}Z\eta \ln\left(\frac{3 + Z\eta}{1 + 25Z\eta}\right) \tag{3.98}$$

This expression gives the exact values for the second and third virial coefficients. Higher coefficients up to the eighth [140] are represented with a maximum deviation of 1.4%. Comparison with the computer simulation data of Labík and Malijevský [161] and of Erpenbeck and Wood [162] showed that the accuracy was very similar to the CS EOS, Eq. (3.54). The limit value for η is 0.631, which is near the random close-packing. In a more recent work [174] the comparison was extended to the computer simulation values of Wu and Sadus [21]. In particular, for the isotropic phase, Eq. (3.98) gave an AAD of 0.623%, a value higher than those obtained with the CS or Kolafa EOSs, Eqs. (3.54) and (3.72) (around 0.3%). Extension to mixtures of hard spheres was subsequently performed [142, 174]. We note that although volume-explicit expressions are not very common (see [206]), some authors have stressed some of their advantages [142, 174].

Speedy [168] used the known values for the first eight coefficients and his own computer simulation results in the metastable region to propose a new Padé approximant:

$$Z = 1 + \frac{\xi + 0.076014\xi^2 + 0.019480\xi^3}{1 - 0.548986\xi + 0.075647\xi^2} \quad (3.99)$$

which is valid only for low densities. In particular, he showed that it reproduces Erpenbeck and Wood's data to within twice the very small errors reported by them, and that it fits the new data to within their errors up to $\xi = 0.68$, which is just above the equilibrium freezing density ($\xi = 0.663$). He also noted that other more complex Padé approximants give similar quantitative accuracy. For higher densities, $0.663 < \xi \leq 0.75$, he added a correction term to Eq. (3.99), to give the following expression:

$$Z = 1 + \frac{\xi + 0.076014\xi^2 + 0.019480\xi^3}{1 - 0.548986\xi + 0.075647\xi^2} + 400\,(\xi - 0.663)^3 \quad (3.100)$$

Speedy [168] also noted that a possible interpretation of Eq. (3.100) was that the Padé approximants based on the known virial coefficients do not sum the tail of the series accurately, so that when more virial coefficients are known, the Padé approximants will change and they may then account for the high-density data. A second interpretation is that the Padé approximants sum the tail of the virial series accurately, but that the virial expansion is not valid above the freezing density.

Speedy [169] also proposed an expression valid only for the solid phase, with the same analytical form as one of his earlier proposals, Eq. (3.91):

$$Z = \frac{C_0}{1 - C_1\eta} \quad (3.101)$$

where in this case $C_0 = 2.8$ and $C_1 = 1.5474$, although other values may also be valid.

Edgal and Huber [143] proposed a nonanalytical EOS which is the solution to the differential equation in Eq. (3.37). In this case, the authors used a virial expansion for $m(\eta)$:

$$\begin{aligned} m(\eta) = {} & 2.4065615915 - 0.9468575772\eta - 0.8747161956\eta^2 \\ & +1.0053182969\eta^3 - 0.6677809217\eta^4 \end{aligned} \quad (3.102)$$

Good agreement was found with respect to the simulation data of Erpenbeck and Wood [162]. Edgal and Huber [143] noted that their approach, which uses a five-term series, gives clearly better results than the virial equation obtained by considering seven virial coefficients (which were used as reference to obtain Eq. (3.102)). A discussion about the extrapolation of this proposal to the high density region was also given by those authors.

Malijevský and Veverka [171] proposed an EOS with the form of the Padé approximant of the rescaled virial series, using the first seven virial coefficients. This equation has the form:

$$Z = \frac{1 + 1.056\eta + 1.6539\eta^2 + 0.3262\eta^3}{(1-\eta)^3\,(1 + 0.056\eta + 0.5979\eta^2 + 0.3076\eta^3)} \quad (3.103)$$

We must note that this equation has the same analytical form as Eq. (3.83), but with different coefficients due to a more recent source for the values of the first seven virial coefficients [140]. The results obtained with Eq. (3.103) were compared to the computer simulation data of Barošová [170], as well as to the results derived from the EOSs given in Eqs. (3.54), (3.71), and (3.72). It was shown that Eq. (3.103) is the most accurate in reproducing those simulation data. Nevertheless, Wu and Sadus [21] have pointed out that for $\rho \leq 0.95$ the improvement of Eqs. (3.72) and (3.103) with respect to the CS EOS is not significant. A detailed comparison with other computer simulation data will be given in Sect. 3.3.3 [152].

Yelash et al. [22] proposed the following two expressions:

$$Z = \frac{3 + 5\eta + 6\eta^2}{(1-\eta)(3-4\eta)} \tag{3.104}$$

$$Z = \frac{3 + 8\eta + 14\eta^2}{3 - 4\eta} \tag{3.105}$$

The first was obtained by multiplying the equation of Carnahan and Starling, Eq. (3.54), by the factor $(3-4\eta)/(3-4\eta)$ and then shifting the value of the coefficient of the η^2 term in the resulting equation from 4 to 6. Equation (3.104) has therefore a pole at $\eta = 0.75$, which is near the actual close-packed density. The second proposal was derived starting from the van der Waals equation repulsion term and extending it with respect to the limit value $\eta = 0.75$. A comparison with the computer simulation values of Barker and Henderson [154] showed that Eq. (3.104) overestimates those values (as well as those predicted by Eq. (3.54)), while Eq. (3.105) clearly underestimates them. Both equations were then used to design new EOSs for real fluids by adding the van der Waals attractive term. Equation (3.104) was also used, together with the EOSs given in Eqs. (3.47), (3.49), (3.50), (3.54), and (3.72) to describe the properties of HS chains [194].

Nasrifar et al. [16] pointed out that the EOSs that give accurate results and that yield the correct close-packed limit, like those proposed by Wang et al. [153] or by Khorskhbarchi and Vera [163], include excessively high powers of the density, and therefore are not very useful to serve as references for EOSs of real fluids. Nasrifar et al. [16] therefore first proposed three different expressions including only quadratic or cubic terms. Although the cubic expression gave the best predictions for the virial coefficients, it was valid only for densities less than around 0.67, and gave negative values at high densities. Therefore, the authors finally selected a ratio of second-order polynomials, with the functional form:

$$Z = 1 + \frac{2.962\xi - 0.4407\xi^2}{(1-\xi)^2} \tag{3.106}$$

This equation satisfies the ideal gas and close-packed limits, exactly reproduces the first three virial coefficients, and is in reasonable agreement with the fourth and fifth. Nevertheless, in general, it overestimates the values for higher coefficients [17]. When compared with the computer simulation values of Alder and Wainwright [150] and of Wu and Sadus [21], there is also overestimation at intermediate densities [17]. Nevertheless, at high densities Eq. (3.106) gives more accurate results than other expressions like Eqs. (3.54), (3.72), (3.104), and (3.105) [17].

Yelash and Kraska [23] developed a generic expression for the EOS, which provided the basis of a hierarchy of terms for different levels of simplification. This basis was a postulated relationship between the limiting slope of the logarithm of the virial coefficients and the high density limit. The final selected expression can be taken to be an extension of that originally proposed by Yelash et al. [22], Eq. (3.105), and has the form:

$$Z = \frac{3 + 8\eta + 14\eta^2 + 14\eta^3 + \frac{40}{3}\eta^4}{3 - 4\eta} \tag{3.107}$$

This equation reproduces accurately the values of the first 10 virial coefficients and is in good agreement with the computer simulation data generated by Barker and Henderson [154].

Ghotbi and Vera [207] proposed two new equations satisfying the correct close packed limit:

$$\begin{aligned} Z = {} & 1 + 2.9619\xi + 5.4831\xi^2 + 7.4564\xi^3 + 8.4856\xi^4 \\ & + \frac{8.85\xi^5}{1-\xi} - \frac{0.62\xi^7}{(1-\xi)^2} + \frac{0.04\xi^{10}}{(1-\xi)^3} \end{aligned} \tag{3.108}$$

$$\begin{aligned} Z = {} & 1 + 2.9619\xi + 5.4831\xi^2 + 7.4564\xi^3 + 8.4856\xi^4 \\ & + \frac{8.9\xi^5 - 2.8\xi^8}{1-\xi} \end{aligned} \tag{3.109}$$

Both were based on the fact that when the virial equation is expressed in terms of ξ, the corresponding fifth to eighth virial coefficients have a small variation. Therefore, Eqs. (3.108) and (3.109) include the first five virial coefficients [190] plus some other terms that include the correct pole and that are obtained by taking into account the computer simulation results of Alder and Wainwright [150]. Ghotbi and Vera [207] showed that their Eq. (3.109) gives values for the first eight virial coefficients in better agreement with the values reported by Kratky [190] than those obtained from Eqs. (3.54), (3.72), and (3.97). Their equations were extended to mixtures of hard spheres [150] and also to hard spheres of different diameters [37].

Wang [144] developed the following van der Waals–Tonks type expression from the first 10 virial coefficients:

$$\begin{aligned} Z = {} & \frac{8.8854}{(1 - \eta/\eta_c)} - 7.8854 - 8\eta - 6.2057\eta^2 \\ & -3.52\eta^3 - 1.3312\eta^4 + 2.048\eta^6 \end{aligned} \tag{3.110}$$

This equation was in agreement with the values predicted by Eqs. (3.51) and (3.54) in the low-density fluid regime. Nevertheless, it does not agree with the simulation results at high densities. Wang noted that its most important feature is that it can be a useful guide when treating more complex interaction potentials.

Eu and Ohr [208] applied the Wiener–Hopf technique to solve the Ornstein–Zernike equation for HS fluids. A thermodynamically consistent analytical expression for the EOS was obtained by considering the correlation range as an adjustable parameter. The final expression proposed is:

$$Z = \frac{\eta\left(1+\eta+\eta^2\right)}{\left(1-\eta\right)^3} - 2.7\ln\left(1-\eta\right) - \frac{2.7\eta}{\left(1-\eta\right)^5}\left(1-\frac{9}{2}\eta+\frac{47}{6}\eta^2-\frac{77}{12}\eta^3+\frac{157}{60}\eta^4-\frac{1}{3}\eta^5\right) \quad (3.111)$$

This EOS was found to be numerically comparable to that of Carnahan and Starling, Eq. (3.54), and is consistent when either the virial or the isothermal compressibility route is used. Nevertheless, recently, Wu and Sadus [21] compared it with their own simulation data and found a significantly poorer behavior than that of Eqs. (3.54), (3.72), (3.103), and (3.107) for any density range. Waziri and Hamad [174] repeated this comparison, but also including Eqs. (3.98) and (3.91) for the stable and metastable phases, respectively. They found that for the isotropic phase the EOS of Malijevský and Veverka, Eq. (3.103), gives the smallest deviation (only 0.228%). Nevertheless, in the metastable region the best results were obtained with the EOS of Hamad, Eq. (3.98) which gives a deviation of only 2.8%, whereas Speedy's EOS, Eq. (3.91) (which is valid only in that range), gives an average deviation greater than 10% (being around 30%–40% for the rest of the EOSs considered).

Miandehy and Modarress [155] presented an EOS in which the numerator was simply fixed as the virial expansion using the values of the first four virial coefficients. The denominator was a polynomial obtained by using the first seven virial coefficients and adding an extra term (in particular, the term $29.55\eta^8$) to fit the computer simulation results generated by Alder and Wainwright [150]. The expression has the following form:

$$Z = \frac{1+4\eta+10\eta^2+18.36\eta^3}{1-28.22\eta^4+73.05\eta^5-66.11\eta^6+29.55\eta^8} \quad (3.112)$$

The behavior of this new proposal was compared with the simulation data of Barker and Henderson [154] and of Alder and Wainwright [150], as well as with the results yielded by Eqs. (3.47), (3.49), (3.50) to (3.52), (3.54) to (3.56), (3.72), (3.73), (3.92), (3.104), and (3.105). The results showed that Eq. (3.112) behaves very similarly to the CS EOS, Eq. (3.54), in the reproduction of the compressibility factor. Virial coefficients higher than the seventh are predicted only roughly, although more accurately than some other expressions. Extension to mixtures was also performed by Miandehy and Modarress [155].

Kolafa et al. [173] obtained highly optimized molecular dynamics computer simulation calculations in the range of reduced densities 0.20–1.03. Following the idea of Barboy and Gelbart [58], these data were fitted to power series in $\eta/(1-\eta)$ in order to obtain an analytical expression for the compressibility factor in both the stable and metastable regions. The first six coefficients were determined to reproduce the first six virial coefficients (the fifth and sixth virial coefficients were specifically recalculated by Kolafa et al. [173]), whereas the others were considered to be adjustable parameters. As was noted by those authors, some of these coefficients can be zero. Two different expressions were proposed with the coefficients listed in Table 3.9, sharing the same analytical form:

$$Z = \sum_i C_i \left(\frac{\eta}{1-\eta} \right)^i \tag{3.113}$$

The coefficients in the first column of Table 3.9 are valid only for densities $\rho \leq 0.98$, i.e., the corresponding EOS describes the stable fluid as well as a moderately metastable region. For the whole range of densities, $\rho \leq 1.03$, more terms are needed and the new coefficients in column II of Table 3.9 should be used.

Rusanov [18, 19] proposed a new theory based on the concept of the exclusion factor (the basis is the same as for the aforementioned expression in Sect. 3.2.2 for the HD fluid). As in the two-dimensional case, his proposal was actually a set of expressions in which n stands for integer values and k is an adjustable parameter that takes different values for each value of n:

$$Z = \frac{1}{(1-k\eta)^{n-1}} \left\{ 1 + \frac{8-2k\,(n-1)}{(n-1)\,(n-2)\,k^2\eta} \left[(1-k\eta)^{n-1} - 1 + (n-1)\,k\eta \right] \right\} \tag{3.114}$$

Table 3.9. Coefficients for the two EOSs proposed by Kolafa et al. [173], based on Eq. (3.113). (I) $\rho \leq 0.98$; (II) $\rho \leq 1.03$

i	Eq. (3.113) (I)	Eq. (3.113) (II)
0	1	1
1	4	4
2	6	6
3	2.3647684	2.3647684
4	−0.8698551	−0.8698551
5	1.1062803	1.1062803
6	−1.2105096	−1.095049
7	1.1356305	0.637614
8	−0.5947886	0
10	0	−0.2279397
12	0.0849805	0
14	0	0.1098948
22	0	−0.00906797

Thus, for example, for $n = 5$ and $k = 1$ the equation of Guggenheim, Eq. (3.52), is recovered. Values of the parameter k for $n = 2, 3 \cdots, 8$ (taking the computer simulation of Alder and Wainwright [150] and of Erpenbeck and Wood [162] as reference) were reported by Rusanov [19]. The greatest accuracy was achieved for $n = 7$, and with $k = 0.717260$ and $k = 0.715955$, when the Alder and Wainwright or the Erpenbeck and Wood data were used, respectively. For $n = 6$ and after a formal simplification, Rusanov proposed the surprisingly simple expression:

$$Z = \frac{1}{(1 - k\eta)^5} \tag{3.115}$$

with k approximately equal to 0.803, no matter which of the two computer simulation data sets was used. In contrast with Eq. (3.114), this approach does not reproduce the second virial coefficient exactly. When compared with the computer simulation results of Erpenbeck and Wood [162], Eq. (3.114) with $n = 7$ gives a maximum deviation of only 0.2%, while Eq. (3.115) gives 0.45% (note that the CS EOS gives 0.13%, and 0.59% that of Woodcock, Eq. (3.60)).

Polishuk and Vera [209] introduced a new EOS that satisfies the close-packed limit and that includes an exponential term in its numerator:

$$Z = \frac{e^{3.078\eta} - 0.4285\eta - 0.1389\eta^2}{1 - \eta/\eta_c} \tag{3.116}$$

It reproduces the first nine virial coefficients better than the equations of Carnahan and Starling and of Yelash and Kraska, Eqs. (3.54), (3.104), and (3.105). Nevertheless, Nasrifar and Bolland [17] have shown that Eq. (3.116) reproduces the compressibility factor values obtained by computer simulation [21] at intermediate densities clearly worse than other expressions, like Eqs. (3.54), (3.72), and (3.106). Equation (3.116) was subsequently used, together with an attractive term, to reproduce the properties of real fluids.

Clisby and McCoy [48] have recently evaluated the first 10 virial coefficients for hard spheres in dimensions from 2 to 8. This allowed them to propose the two following Padé approximants:

$$Z = \frac{1 + 2.03995984\eta + 3.3399824\eta^2 + 2.332827008\eta^3 + 0.900518016\eta^4}{1 - 1.96004012\eta + 1.180142976\eta^2 - 1.152111936\eta^3 + 1.4787069952\eta^4 - 0.5607328768\eta^5} \tag{3.117}$$

$$Z = \frac{1 + 3.21634468\eta + 7.4659272\eta^2 + 9.0866976\eta^3 + 7.613060608\eta^4 + 2.8869096448\eta^5}{1 - 0.78365528\eta + 0.600548464\eta^2 - 3.843711104\eta^3 + 3.149556992\eta^4} \tag{3.118}$$

The authors noted that these expressions are valid only at low densities. Finally, they focused their research on the location of the singularities.

Rambaldi et al. [175] performed new computer simulations and proposed two new expressions for the EOS of the HS system. Their derivation is an extension of the work of Hoover and Alder [93] who showed that the compressibility factor can be measured through the collision rate. The first expression approximates uniformly their numerical solution for the whole fluid branch, and has the form:

$$Z = 1 + \frac{4\eta}{1 - 2.5\eta + 1.658808\eta^2} \tag{3.119}$$

The second approximates the solid branch and was derived following the method developed by Hall [186] and introducing the variable $\delta = (\eta_c - \eta)$. It is expressed as follows:

$$Z = 2.22144415\delta^{-1} - 0.44234 + 0.67689946\delta + 5.14275116\delta^2 \tag{3.120}$$

The results yielded by these two equations have as yet not been compared with those obtained by other expressions or with computer simulation data.

After a detailed analysis of the advantages of the equations proposed by Yelash et al. [22] and by Nasrifar et al. [16], Eqs. (3.104), (3.105), and (3.106), Nasrifar and Bolland [17] proposed an expression which, as the previous ones, consisted of a ratio of second-order polynomials that satisfies the ideal gas and close-packed density limits:

$$Z = 1 + \frac{2\eta}{1-\eta}\left(2 + \frac{3\eta}{1 - \eta/\eta_c}\right) \tag{3.121}$$

The coefficients of this equation were determined by matching the second and third virial coefficients, and it also reproduces closely the fourth and the fifth, being in reasonable agreement with the sixth, the seventh, and the eighth. In any case, the authors pointed out that both the CS and Kolafa EOSs, Eqs. (3.54) and (3.72), reproduce those coefficients more accurately. In relation to the calculation of the compressibility factor, the accuracy of Eq. (3.121) was tested via comparison with the computer simulation data of Wu and Sadus [21]. Very good agreement was found for $\xi \le 0.5$, whereas in the range $0.5 < \xi \le 0.67$ the accuracy was found to be poorer than for Eqs. (3.54) and (3.72). In the metastable region, $0.69 \le \xi \le 0.82$, the EOSs given in Eqs. (3.104), (3.105), and (3.121) yield the most accurate results, while Eq. (3.106) was found to present the most accurate behavior for $0.82 < \xi \le 0.86$.

Miandehy et al. [210] obtained a slightly modified version of Eq. (3.93) of Wang et al. [153], with the form:

$$Z = \frac{1 + 0.9619\xi + 0.5593\xi^2 - 0.5499\xi^3 - 0.9415\xi^4 - 0.647\xi^5 - 0.7324\xi^7}{(1-\xi)^2} \tag{3.122}$$

This expression reproduces the first eight virial coefficients, and was used to obtain excess properties of mixtures by using the perturbation theory of Leonard–Henderson–Barker [210]. As was noted by Vera and Mulero [211] the main difference between the treatments of Miandehy et al. and of Wang et al. is the method used to evaluate the parameters in Eq. (3.95). Miandehy et al. showed a comparison with the simulation data of Alder and Wainwright [150] and of Erpenbeck and Wood [162], and also with the EOSs proposed by van der Waals, Eq. (3.47), Wertheim, Eq. (3.49), Thiele, Eq. (3.50), Guggenheim, Eq. (3.52), CS, Eq. (3.54), Scott, Eqs. (3.55) and (3.56), and Kolafa, Eq. (3.72). Although these authors concluded that Eq. (3.122) appears to be the most accurate in reproducing those data, Vera and Mulero noted that some other EOSs previously referred to in this section as well as more recent computer simulation data should be considered in order to obtain a definite conclusion about the accuracy of Eq. (3.122). They also noted that the recently calculated 11th and 12th virial coefficients [47, 48, 81] should also be taken into account.

Kolafa [176] used some expressions derived from the simplest version of the classical nucleation theory in combination with the virial expansion to develop an EOS which is not analytical at the freezing point. He indicated that the aim of his work was to test a theoretically based term, rather than to compete with accurate many-constant equations. His proposal tried to sew together both the low-density region and the anomalous part valid near the freezing point. The final expression is a simple sum of a polynomial in the density and an anomalous term:

$$Z_m = \sum_{i=1}^{m} C_i \rho^{i-1} + \beta \Psi_3 \left[\alpha(\rho - \rho_\mathrm{f}) \right] \tag{3.123}$$

where C_i $(i = 1, \ldots, m)$ are coefficients determined to reproduce the virial coefficients up to B_m. Two adjustable parameters, α and β, were fitted to molecular dynamics computer simulation data [173]. The value of the reduced freezing density $\rho_\mathrm{f} = 0.9435$ was taken from [94]. Finally, Ψ_3 is a nonanalytical function defined as:

$$\Psi_n(u) = \int_0^\infty x^n \exp\left(\frac{x^3}{3} u - \frac{x^2}{2} \right) \mathrm{d}x \tag{3.124}$$

Values for α and β for different values of m have been obtained by Kolafa, and the resulting EOS was compared with his own computer simulation results at very high densities, as well as with those obtained in the simulation of Kolafa et al. [173]. The author found that for $\rho \leq 0.95$, the nonanalytical EOSs improve the results of the expression of Kolafa et al. [173], Eq. (3.113), for $m > 5$ [173]. The smallest deviations were achieved for $m = 7$, $\alpha = 0.3877$, and $\beta = 40.56$. When considering a wider density range, $\rho \leq 1.09$, the author found that the best results with his nonanalytical EOS were obtained for $m = 10$, whereas Eq. (3.113) presents quite irregular behavior when new terms are added.

Finally, Liu [160] developed an analytical EOS for the entire stable and metastable regions. He used a potential energy landscape analysis combined with Woodcock's EOS, Eq. (3.60), to derive the following expression:

$$Z = Z_V + \frac{0.31416\eta}{1 - 1.573357\eta} + 4.1637 \times 10^{10}\eta^{40} \\ -2.3452 \times 10^{11}\eta^{42} + 3.6684 \times 10^{11}\eta^{44} \quad (3.125)$$

where Z_V is defined as:

$$Z_V = 1 + \frac{3.68584\eta}{1 - 2.5848\eta + 1.9499\eta^2 - 0.172284\eta^3 - 0.16012\eta^4} \quad (3.126)$$

The denominator included in the second term gives a random close-packed density of 1.21389 ($\eta = 0.635591$), which is consistent with the values obtained by other means [36, 191]. Liu built two databases. For the stable region (64 data), he used the data from [21, 162, 173]. For the metastable region (81 data), values from [21, 148, 159, 168, 173, 176] were selected. A comparison with the values calculated with several EOSs was performed, and the main results are listed in Table 3.10.

As can be seen, the EOS of Liu behaves as accurately as Kolafa's in the stable region. In the metastable region, the overall AAD obtained with the equation of Liu is below 1%, thus improving significantly the results achieved by the equations of Speedy or Le Fevre. In sum, the EOS of Eq. (3.125) reproduces the compressibility factor for the complete region with high accuracy, and can be extended to real systems (including supercooled liquids and glasses) by using conventional approaches, such as van der Waals theory or perturbation theory [160] (see Chap. 4).

As a final comment, we would note that an important feature of the EOSs is the location of poles (values of density or of the packing fraction for which the compressibility factor becomes infinite). As was already observed by Thiele [4], the simplest equations fail at densities near that corresponding to the closest-packing ($\eta_c \approx 0.7405$), since their singularity ($\eta = 1$) corresponds to the density at which the whole space would be filled ($\rho \approx 1.91$). However, Williams et al. [149] later showed that HS fluids undergo a first-order transition to a crystalline solid at a density slightly above the melting density. Therefore, it seems that the close-packed density is a matter of convenience. Nevertheless, as Nasrifar and Bolland [17] pointed out, HS EOSs with a singularity at the

Table 3.10. AADs of the values of the compressibility factor calculated with several EOSs, with respect to the reference values mentioned in the text [160]

EOS	Stable region (64 points)	Metastable region (81 points)
Liu, Eq. (3.125)	0.0841	0.98
Kolafa, Eq. (3.123)	0.0834	–
Carnahan and Starling, Eq. (3.54)	0.196	–
Le Fevre, Eq. (3.58)	0.592	8.36
Speedy, Eq. (3.91)	–	4.70

close-packed limit will normally reproduce the compressibility factor of HS fluids in the metastable region more accurately than those with a singularity at $\eta = 1$. In this sense, many of the proposed EOSs satisfy the close-packed limit condition. Some of them fulfill this condition by including the term $(1-\xi)$ in the denominator, with Eqs. (3.65) to (3.67), (3.78), (3.93) to (3.95), (3.97), (3.106), (3.116), and (3.122) being clear examples. Others, like Eqs. (3.89), (3.90), (3.104), (3.105), and (3.107), do not diverge exactly when $\eta = \eta_c$, although they can be regarded as approximately reproducing the close-packed limit. Finally, there are other proposals, like Eqs. (3.58) and (3.115), in which the pole is located at a random close-packed density.

The preceding paragraphs have provided a detailed review of the main expressions for the EOS of the HS fluid. The most important features of each of those equations were presented and some comments were added concerning the location of the poles or the analytical structure of the equation.

This review has illustrated the wide dispersion of results and comparisons that can be found in the literature. There are many expressions that have not been tested for some density ranges or have ever been compared among themselves. As an attempt to provide some explicit calculations in that sense, in the following section we describe what should be taken as perhaps the most detailed comparison among EOSs and available computer simulation data sets, referring to the calculation of the compressibility factor. Only a portion of the computer simulations and analytical expressions proposed until 2001 were considered in detail, whereas for more recent comparisons we will only summarize briefly some of the most important results. In any case, we think that the following subsection can be regarded as being an adequate representation of the state-of-the-art, and that the conclusions it leads to can serve as a starting point for future work in the development and testing of EOSs. Moreover, we include a study of other derived thermodynamic properties: the isothermal compressibility and the chemical potential.

3.3.3 Test of Accuracy for a Set of Equations of State

In order to establish some conclusions about the validity of the analytical expressions proposed for the compressibility factor of the HS fluid, Mulero et al. [152] considered results taken from six different computer simulations, which were performed by Alder and Wainwright [150], Woodcock [99], Erpenbeck and Wood [162], Barošová [170], Labík [172], and Kolafa [172], in chronological order. These results were compared with those obtained with the 22 analytical EOSs listed in Table 3.11, all described in Sect. 3.3.2. Obviously, the accuracy of these expressions and their validity for reproducing the virial coefficients of the HS system were already studied by their own authors.

As a first attempt to classify these EOSs in terms of complexity, we will consider as "simple" those in which the denominator has the form $(1-\eta)^n$ ($n = 2, 3, 4, 5$), and "complex" those in which denominator includes a more complicated polynomial. Hence, expressions SPT, PYT, G, CS, K, BC4, BC6,

Table 3.11. Set of EOSs taken from those presented in Sect. 3.3.2. Note that an abbreviation is associated with each for clarity of nomenclature

Author(s)	Abbreviation	Equation	Reference
Reiss–Frisch–Lebowitz (1959)	SPT	(3.49)	[2]
Thiele (1963)	PYT	(3.50)	[4]
Ree–Hoover (1964)	RH	(3.51)	[183]
Guggenheim (1965)	G	(3.52)	[184]
Carnahan–Starling (1969)	CS	(3.54)	[177]
Hall (1972)	H	(3.57)	[186]
Le Fevre (1972)	LF	(3.58)	[188]
Andrews (1975)	A	(3.59)	[189]
Erpenbeck–Wood (1984)	EW	(3.71)	[162]
Kolafa (1986)	K	(3.72)	[193]
Baus–Colot (1987)	BC4	(3.75)	[45]
Baus–Colot (1987)	BC6	(3.76)	[45]
Maeso–Solana–Amorós–Villar (1991)	MSAV1	(3.82)	[138]
Maeso–Solana–Amorós–Villar (1991)	MSAV2	(3.83)	[138]
Solana (1992)	SOL1	(3.85)	[199]
Solana (1992)	SOL2	(3.86)	[199]
Sanchez (1994)	S1	(3.87)	[139]
Sanchez (1994)	S2	(3.88)	[139]
Wang–Khoshkbarchi–Vera (1996)	WKV1	(3.93)	[153]
Wang–Khoshkbarchi–Vera (1996)	WKV2	(3.94)	[153]
Wang–Khoshkbarchi–Vera (1996)	WKV3	(3.95)	[153]
Malijevský–Veverka (1999)	MV	(3.103)	[171]

MSAV1, SOL1, and SOL2 are taken as having a simple analytical form. On the other hand, Andrews' EOS is considered to be the most complex, because it contains logarithmic terms. Expressions WKV1, WKV2, and WKV3 are intermediate between "simple" and "complex", since they use a different variable and in some cases a novel analytical form.

Mulero et al. [152] made a detailed analysis of the accuracy of the EOSs listed in Table 3.11 by a separate comparison with each of the six aforementioned computer simulation data sets. Table 3.12 lists the AADs of the results using the equations of state (Z_{EOS}) with respect to the computer simulation data (Z_{sim}):

$$\mathrm{AAD} = \frac{1}{n} \sum_{i=1}^{n} \frac{100 \times |Z_{\mathrm{EOS}_i} - Z_{\mathrm{sim}_i}|}{Z_{\mathrm{sim}_i}} \tag{3.127}$$

where n is the number of thermodynamic states.

The computer simulation results of Alder and Wainwright [150], which cover the range of very high densities (only seven data) including the metastable region, seem to be the most difficult to reproduce using analytical EOSs (see Table 3.12). Only the expressions WKV1, WKV2, WKV3, and LF give AADs $< 6\%$. As was concluded by Wang et al. [153], expression WKV3 is the

Table 3.12. AADs obtained from the EOSs listed in Table 3.11, applied to the HS system, with respect to the computer simulation values of Alder and Wainwright (AW) [150], Woodcock (W) [99], Erpenbeck and Wood (EW) [162], Barošová (B) [170], Kolafa (K) [172], and Labík (L) [172]

EOS	AW (1960) ($0.885 \le \rho \le 1.180$)	W (1976) ($0.942 \le \rho \le 1.085$)	EW (1984) ($0.057 \le \rho \le 0.814$)	B (1996) ($0.573 \le \rho \le 0.936$)	K (2001) ($0.5 \le \rho \le 0.97$)	L (2001) ($0.4 \le \rho \le 0.94$)
SPT	12.79	8.52	2.04	4.15	7.72	3.75
PYT	27.97	18.20	4.47	9.24	16.28	8.18
RH	14.13	2.46	0.22	0.60	1.18	0.47
G	23.95	27.57	5.74	12.03	22.67	10.72
CS	12.39	0.60	0.13	0.31	0.28	0.23
H	11.02	0.72	0.05	0.13	0.19	0.09
LF	1.39	2.05	0.34	0.86	1.83	0.70
A	9.78	6.01	0.79	1.11	2.70	1.13
EW	12.81	0.67	0.14	0.15	0.30	0.20
K	12.66	0.70	0.01	0.10	0.08	0.03
BC4	12.93	1.25	0.11	0.34	0.56	0.24
BC6	11.15	3.52	0.45	0.95	2.24	0.89
MSAV1	12.68	0.64	0.13	0.17	0.44	0.21
MSAV2	12.28	0.63	0.11	0.16	0.49	0.20
SOL1	11.73	2.86	0.66	1.32	2.64	1.22
SOL2	9.89	5.00	0.41	0.94	2.48	0.88
S1	10.00	3.28	0.26	0.58	1.61	0.57
S2	109.12	51.55	4.63	10.94	24.62	9.67
WKV1	5.54	1.95	0.64	1.30	2.05	1.08
WKV2	3.14	2.38	0.60	1.36	2.61	1.13
WKV3	0.94	2.97	1.40	2.80	4.50	2.40
MV	12.41	0.56	0.04	0.05	0.18	0.07

most accurate of those included in Table 3.11 to calculate the compressibility factor in this range. Mulero et al. [152] added that equation LF, which was not included in the study of Wang et al. [153], is also a good alternative, giving smaller deviations than WKV1 or WKV2. As can be seen in Table 3.12, some other expressions (BC4, BC6, MSAV1, MSAV2, SOL1, and MV) give AADs > 11% for this range, i.e., they are not really appropriate for these very high densities. The least accurate EOSs in Table 3.11 are S2 and PYT. In relation to other EOSs not included in Table 3.11, Wang et al. [153] reported an AAD of 29% when the EOS of Barboy and Gelbart [58], Eq. (3.61), is used with $n = 7$. Also, Ma and Ahmadi [151] showed that their proposal, Eq. (3.74), is clearly more accurate in this range than the EOSs of Ree and Hoover and CS, Eqs. (3.51) and (3.54), respectively.

Mulero et al. [152] noted that the above results depend strongly on the last two values of the computer simulation data set of Alder and Wainwright [150], i.e., for $\xi^{-1} = 1.2496$ and $\xi^{-1} = 1.1987$, which are located near the closest-packing limit. If these two points were ignored, then all the equations, except SPT, PYT, and SOL2, would give AADs < 4.5%. As shown in Fig. 3.8, only the equations WKV3 and LF among those listed in Table 3.11 are valid at those two particular densities.

When considering the computer simulation results of Woodcock [99], Mulero et al. [152] emphasized the accuracy of one of the simplest EOSs, namely, CS, as was also stressed by Maeso et al. [138]. As observed in Table 3.12, other simple expressions like K and MSAV1 also give excellent results. Woodcock [99] showed that his proposal, Eq. (3.60), reproduces these data more accurately than the CS EOS and some others (Eqs. (3.49) to (3.52), (3.54), and (3.58), in particular). A small deviation was also found when using

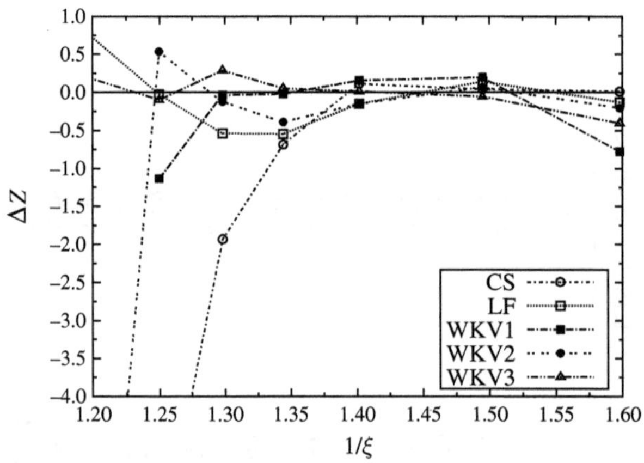

Fig. 3.8. Deviations ($\triangle Z = Z_{\mathrm{EOS}} - Z_{\mathrm{AW}}$) of the results obtained with some of the expressions of Table 3.12 with respect to the computer simulation values of Alder and Wainwright [150]

equation MV (see Table 3.12). Nevertheless, it must be pointed out that these improvements were not very significant, and therefore, it became clear that the use of complex or more recent expressions is not needed in that particular density range.

Expressions WKV1 and WKV2 improved the results with respect to the range covered by the computer simulation of Alder and Wainwright [150], but WKV3 (as well as LF) yielded clearly less accurate values. Also the very complex equation A does not give adequate results in that range. Finally, Mulero et al. [152] pointed out that the use of expression H in the study of the liquid–solid equilibrium as performed by Frenkel and Smit [187] can produce poorer results than a simpler one, such as equation CS, which gives more accurate results near the liquid transition density ($\rho \approx 0.94$).

The computer simulation results of Erpenbeck and Wood [162] have commonly been used as a reference to test the accuracy of EOSs. As can be seen in Table 3.12, the simple equation K and the complex H and MV were the best in this range [152]. Very good results were obtained with most of the EOSs listed in Table 3.11: an AAD $< 0.5\%$ was found for 13 expressions, including some as simple as CS, BC4, BC6, MSAV1, and SOL2, whereas deviations between 0.5% and 1% were found for four equations (one of which, SOL1, can be considered as simple in its analytical form). As in previous cases SPT, PYT, G, and S2 yielded the least accurate values for the compressibility factor. With respect to other EOSs not included in Table 3.11, we note that Wang et al. [153] reported an AAD of 2.3% when the EOS of Barboy and Gelbart, Eq. (3.61), is used with $n = 7$. We also note that the expression of Zhou and Stell, Eq. (3.79), reproduces Erpenbeck and Wood's data more accurately than the CS or Kolafa EOSs. Finally, Rusanov [19] showed that in this range his proposal, Eq. (3.114) with $n = 7$, gives a maximum deviation of only 0.2%. For the simplest of his expressions, Eq. (3.115), he found a maximum deviation of 0.45%, being 0.31% and 0.59% when the CS and Woodcock expressions were used.

According to Mulero et al. [152], the computer simulation data of Barošová [170] had been compared only with results given by expressions CS, EW, K, and MV at that date. Table 3.12 lists that deviations corresponding to those four EOSs, as well to the simple MSAV1 and BC4, and to the complex H and MSAV2, are below 0.35%. As in previous cases, the use of complex EOSs did not imply a significant improvement in the results, except perhaps in the case of equation MV. In any case, the simple equation K seemed to be quite accurate. As reported by Mulero et al. [152], in general deviations were greater than those obtained when taking the computer simulation data of Erpenbeck and Wood [162] as reference, increasing significantly for the three EOSs proposed by Wang et al. [153] (WKV1, WKV2, and WKV3).

Table 3.12 also includes the deviations with respect to the molecular dynamics simulation data of Kolafa [172]. These data cover almost the same density range as the simulation of Barošová [170], but includes a greater number of values, some of which belong to the metastable region. The simple equations

CS and K were the only ones reproducing the computer simulation values of Kolafa [172] more accurately than those of the simulation of Barošová [170]. In particular, the expression K achieved the best agreement. The complex H, EW, and MV, as well as the simple CS, also presented accurate behavior.

Mulero et al. [152] also considered the unpublished Monte Carlo computer simulation data of Labík (L) [172], for almost the same range as the two previous sets. Again, equations K and MV led to very good agreement. In general, all deviations were less than they were for the computer simulation set of Kolafa [172]. The simple CS, BC4, and MSAV1 also yielded excellent results, more accurate than those obtained with many other complex proposals.

The great influence of the metastable region in the deviations of the last two columns of Table 3.12 was stressed by Mulero et al. [152]. It was also noticeable that some equations gave more accurate results for the compressibility factor in that region than in the stable range. It seemed to be clear that the coefficients given by Wang et al. [153] should be revised in order to reproduce more accurately recent computer simulation data instead of the old values generated by Alder and Wainwright [150]. Unfortunately, the only available revision of Wang et al.'s EOSs, carried out by Miandehy et al. [210], does not include new data [211].

By comparing the behavior of the EOSs with the six computer simulation data sets, Mulero et al. [152] concluded that the LF and WKV3 expressions could be regarded as the most "homogeneous", giving adequate results over any density range. However, those authors did not confirm the conclusion of Wang et al. [153] that good results at high densities are obtained with their expressions "without sacrificing" the precision in the low-density region. As Mulero et al. [152] showed, WKV3 did not produce adequate values in the metastable region when Woodcock's [99] data set was taken as reference instead of Alder and Wainwright's [150]. Moreover, values of AAD for the three expressions of Wang et al. [153] were always greater than 1% with respect to the most recent computer simulation data, a significantly greater percentage than those obtained with simple EOSs, such as CS (AAD $\leq$ 0.31%) or K (AAD $\leq$ 0.1%).

To sum up, Mulero et al. [152] noted that the use of the simple equations CS, K, or MSAV1 is highly recommended to calculate the compressibility factor of HS systems, except for the case when very high densities are considered when WKV3 is the most accurate expression (of course, among those listed in Table 3.11). In any case, one should take into account that the data in this range are quite old, and their associated statistical uncertainties are not small. If calculations for both low and very high densities are required simultaneously, equation LF seemed to be an adequate choice. Expressions SPT, PYT, G, and S2 should not be used in any case. Neither was the use of equation A recommended by Mulero et al. [152], given its analytical complexity and that it does not improve the results obtained with simpler proposals. Many equations proposed between 1987 and 1996 (BC4, BC6, SOL1, SOL2, S1, S2, WKV1, and WKV2) had not led to a significant improvement with respect to previous simple and well-known proposals. The complex H, EW, MSAV2, and especially MV gave good results except when compared with the computer simulation data of Alder and Wainwright [150]. Nevertheless, their analytical complexity does not necessarily imply more accurate results than those obtained with simpler equations. In most cases, therefore, the use of more complex expressions does not give any significant improvement as there is no direct correlation between the simplicity or complexity and the accuracy of the EOSs. Indeed, the results achieved by Mulero et al. [152] justify the use of the very simple equation CS over a wide range of densities (see Fig. 3.7).

Finally, Mulero et al. [152] stressed that they investigated the validity of the EOSs listed in Table 3.11 exclusively to reproduce computer simulation results for the compressibility factor. Obviously, one can consider other testing criteria such as the calculation of virial coefficients with different applications in mind. But their aim in that work was to apply those expressions to the calculation of the pressure of a more realistic model, and to do so only the compressibility factor was needed. The validity of those HS EOSs for calculating the chemical potential and the isothermal compressibility have been considered elsewhere [13, 14].

Other more recent studies of the accuracy of EOSs when compared with computer simulation data have been reported by Miandehy and Modarress [155], Wu and Sadus [21], Waziri and Hamad [174], Nasrifar and Bolland [17], and Liu [160]. Some of their main results will be briefly described in the following paragraphs.

Miandehy and Modarress [155] compared the accuracy of Eqs. (3.47), (3.49), (3.50) to (3.52), (3.54) to (3.56), (3.72), (3.73), (3.92), (3.104), (3.105), and (3.112) in reproducing the computer simulation data of Barker and Henderson [154] and of Alder and Wainwright [150]. They found out that their proposal, Eq. (3.112), behaves very similarly to the CS EOS, Eq. (3.54), in the specific density range considered.

Wu and Sadus [21] compared the expression of Eu and Ohr, Eq. (3.111), with their own simulation data and found a significantly poorer behavior than that of Eqs. (3.54), (3.72), (3.103), and (3.107) for any density range. Waziri and Hamad [174] performed the same comparison, but now including Eq. (3.98) for the stable phase and Eq. (3.91) for the metastable phase. They found that for the isotropic phase the EOS of Malijevský and Veverka, Eq. (3.103), gives the smallest deviation (only 0.228%). The CS and Kolafa expressions give AADs around 0.3%, while the expression of Hamad, Eq. (3.98), gives an AAD equal to 0.623%. Nevertheless, in the metastable region the most accurate results were obtained by using the equation of Hamad, Eq. (3.98), which gives a deviation of only 2.8%, whereas Speedy's proposal, Eq. (3.91) (valid only in that range), gave an AAD greater than 10% (between approximately 30% and 40% for the rest of the EOSs considered).

Nasrifar and Bolland [17] tested the behavior of several EOSs when compared with the computer simulation data of Alder and Wainwright [150], and of Wu and Sadus [21]. First, they indicated that Polishuk and Vera's EOS,

Eq. (3.116), gives clearly poorer results than other currently used and simpler expressions. Second, they compared their own EOS, Eq. (3.121), with Wu and Sadus [21] data, finding very good agreement for $\xi \leq 0.5$. In the range $0.5 < \xi \leq 0.67$ the accuracy of Eq. (3.121) was found not to be as good as that of Eqs. (3.54) and (3.72). In the metastable region, $0.69 \leq \xi \leq 0.82$, the EOSs given by Eqs. (3.104), (3.105), and (3.121) yielded the most accurate results, while the expression of Nasrifar et al., Eq. (3.106), was found to be the most accurate for the high density range $0.82 < \xi \leq 0.86$.

Liu [160] tested several EOSs versus two sets of computer simulation data taken from different sources. The results are listed in Table 3.10. He found that his own proposal, Eq. (3.125), behaves as accurately as Kolafa's, Eq. (3.123), in the stable region. In the metastable region, the AAD obtained is below 1%, therefore improving significantly the results achieved with the EOSs of Speedy or Le Fevre. In sum, the EOS shown in Eq. (3.125) reproduces highly accurately the compressibility factor for the complete region.

3.4 Other Properties of the Hard-Sphere Fluid from Equations of State

The important study of Mulero et al. [152] on the test of accuracy of a set of HS EOSs in the reproduction of the compressibility factor was referred to in detail in the previous section. However, there are some other interesting thermodynamic properties with direct industrial or research applications, such as the isothermal compressibility and the chemical potential.

In most cases, the parameters needed in the EOSs are calculated from fitting computer simulation data for the pressure and/or the potential energy. This implies that an equation can reproduce accurately a certain physical property but not others related to it [8, 9, 10, 13]. Moreover, the available data for the thermodynamic properties are usually given for different thermodynamic ranges and were generated in computer simulations with different particular computational features. For this reason, no direct relationship between the accuracy achieved in reproducing the pressure and that in reproducing other thermodynamic properties can be found. Nevertheless, suitably designed analytical expressions must give accurate values for other thermodynamic properties. In particular, in this chapter we will focus on the calculation of the isothermal compressibility and of the chemical potential using different proposals for the EOS.

It is well known that isothermal compressibility is a relevant property directly related to the pressure, the heat capacity at constant pressure, the coefficient of thermal expansion [8, 212], and the structure of the system. The knowledge of the chemical potential of a system is needed in certain important calculations, such as the determination of the phase equilibrium properties for which analytical expressions must be considered [13]. We shall comment on some aspects of these two properties in the following two subsections.

3.4.1 Isothermal Compressibility

The isothermal compressibility, χ_T, is defined as:

$$\chi_T = \frac{1}{\rho}\frac{\partial \rho}{\partial P} \tag{3.128}$$

where P is the pressure, and where all properties are expressed in reduced Lennard–Jones units. In order to obtain an analytical expression for χ_T from Eq. (3.128), the dependence of P on density, i.e., an analytical expression for the EOS, is required. Some examples of this can be found in reviews [6] and books [212] using only the simplest EOSs for the HS system. An important study by Mulero et al. [14] of the accuracy of a set of 19 EOSs in the reproduction of the isothermal compressibility will be described in this subsection. Their results seemed to be very helpful for the choice of the most appropriate analytical expression, especially for calculations where knowledge of this property is needed [8].

Mulero et al. [14] calculated the isothermal compressibility of the HS fluid using Eq. (3.128) and all the EOSs listed in Table 3.11 except G, S2, and MV. Table 3.13 lists the values of $(\chi_T T)$ obtained with the equation CS,

Table 3.13. Values of the isothermal compressibility for HS calculated with the equation CS [14]

ρ	χ_T
0.1	6.6110
0.2	2.2027
0.3	0.9828
0.4	0.4935
0.5	0.2634
0.6	0.1453
0.7	0.0814
0.8	0.0457
0.82	0.0407
0.84	0.0362
0.844	0.0354
0.86	0.0323
0.88	0.0287
0.90	0.0255
0.92	0.0226
0.94	0.0201
0.943	0.0197
0.95	0.0189
0.97	0.0167
1.0	0.0139
1.07	0.0089
1.4142	6.6×10^{-3}

which is the most used and best-known expression, at densities in the stable region $0.1 \leq \rho \leq 0.943$ and in the metastable region $0.95 \leq \rho \leq 1.07$. For the close-packed density ($\rho_c = \sqrt{2}$) the isothermal compressibility must be zero, since the pressure diverges to infinity, so that this point is also included in Tables 3.13 and 3.14 for the CS and the other equations, respectively, as a test of their accuracy in this density limit. Table 3.14 also gives the mean deviation between the isothermal compressibility calculated with the other 18 expressions with respect to that obtained from CS, for different ranges of densities. As was shown by Mulero et al. [178], equation CS yielded adequate values for the pressure of the HS fluid except in the range $0.885 < \rho < 0.943$, for which WKV3 gave the most accurate results.

As can be seen in Table 3.14, at low densities $\rho < 0.6$, no expression led to a large deviation. In this range the equation K was the most accurate in the case of the pressure, the mean deviation being very small (0.19%), and the CS values always being greater than the K values. For these low densities, other simple (like MSAV1) or complex (like EW or WKV1) expressions also gave small deviations. As in the case of the calculation of the pressure, expressions PYT and SPT gave the highest deviations.

For $0.6 \leq \rho \leq 0.8$ equations SPT, PYT, and WKV3 gave the greatest deviations with respect to CS. These equations also showed major discrepancies when their values for the compressibility factor were compared to available computer simulation data [152, 178]. Seven EOSs gave mean deviations between 1% and 3%, whereas eight led to deviations below 1%, the smallest corresponding to expressions H and BC4.

For the highest stable densities, $0.8 < \rho \leq 0.943$, greater differences were found between the results obtained with the various EOSs, with only seven expressions giving deviations below 2% with respect to CS. For this region, the accuracy of equation CS was questionable because of the poor results found in the calculation of the pressure. Better results might be obtained using other expressions, such as WKV3.

For the metastable region, $0.943 < \rho \leq 1.07$, expressions H, EW, K, BC4, MSAV1, and MSAV2 led to results which were similar to those calculated with the equation CS (which presented accurate behavior in the calculation of pressure in this range). Mean deviations greater than 10% were found for five expressions, including WKV3. The only conclusion arising from the analysis performed in this region was that there was major dispersion in the results, so that it was not clear which expression should be regarded as the most accurate.

Therefore, the aforementioned results showed that, for both the stable and metastable ranges of HS fluids, the use of expressions EW, K, and MSAV2 led practically to the same results (mean deviations below 1%) for the isothermal compressibility as when the simpler expression CS was considered. Thus, the use of more complex EOSs was only justified if calculations were to be performed at higher densities.

Bravo-Yuste et al. [200] pointed out that the isothermal compressibility calculated from equation S1 was greater than that obtained from equation

Table 3.14. Mean deviations (%) between the isothermal compressibility for HSs calculated with several EOSs and that obtained with the equation CS. The value of $\chi_T T$ at the close-packed density is also given

EOS	$(0.1 \leq \rho < 0.6)$	$(0.6 \leq \rho \leq 0.8)$	$(0.8 < \rho \leq 0.943)$	$(0.934 < \rho \leq 1.07)$	$\chi_T T(\rho_c = \sqrt{2})$
SPT	1.17	6.02	9.45	11.86	5.21×10^{-4}
PYT	−2.49	−14.91	−15.39	−37.06	1.46×10^{-3}
RH	0.15	−0.49	−2.39	−5.20	1.26×10^{-3}
H	0.16	0.11	0.54	3.17	4.51×10^{-12}
LF	0.20	−0.94	−3.89	5.36	1.13×10^{-2}
A	−0.17	1.44	6.31	13.87	-6.67×10^{-13}
EW	0.32	0.56	0.04	−0.94	9.78×10^{-4}
K	0.19	0.25	−0.24	−0.87	7.0×10^{-4}
BC4	0.16	−0.12	−1.06	−2.12	7.29×10^{-4}
BC6	0.25	1.73	4.00	6.69	4.89×10^{-4}
MSAV1	0.22	0.68	0.49	−0.26	7.81×10^{-4}
MSAV2	0.20	0.62	0.79	0.95	6.48×10^{-4}
SOL1	0.43	2.29	3.67	4.68	6.03×10^{-4}
SOL2	0.21	1.58	5.13	11.37	1.84×10^{-4}
S1	0.20	1.10	3.42	8.27	1.50×10^{-7}
WKV1	0.06	−2.08	−1.82	9.88	9.39×10^{-17}
WKV2	0.13	−2.86	−4.05	7.48	1.44×10^{-21}
WKV3	−0.26	−4.76	−4.13	12.03	5.83×10^{-40}

SPT in the stable and metastable regions. However, the two values were almost equal at $\rho \approx 1.07$, i.e., at the density associated with the glass transition [166]. Also, the value given by equation S1 was lower than that given by equation SPT at higher densities. According to these authors, this means that the appearance of a different structure in the system can be predicted theoretically and that the method they proposed will not be applicable to Sanchez's EOS beyond the stable or metastable fluid regions. Mulero et al. [14] also noted that none of the expressions considered in their work gave the same value for the isothermal compressibility as equation SPT at $\rho = 1.07$. For instance, the value yielded by equation WKV3 was some 18% lower than that given by expression SPT at that density.

Finally, as can be seen in Tables 3.13 and 3.14, most of the EOSs led to small but finite values of the isothermal compressibility at the close-packed density. Only the expressions proposed by Wang et al. [153] (namely, WKV1, WKV2, and WKV3) gave values below 10^{-10} (reduced Lennard–Jones units), the WKV3 value being the nearest to zero.

3.4.2 Chemical Potential

The residual or configurational chemical potential, μ, i.e., the excess chemical potential of the fluid relative to a perfect gas, can be obtained analytically as:

$$\mu = T\left[\int \left(\frac{Z-1}{\rho}\right) d\rho + Z - 1\right] \tag{3.129}$$

The evaluation of the chemical potential of a fluid by computer simulation methods is not straightforward and different techniques have been proposed for the purpose. The test particle or potential distribution method derived by Widom [213] or some modification of it was widely used in Monte Carlo [156, 214, 215, 216, 217] and molecular dynamics [218, 219, 220] computer simulations.

In order to obtain an analytical expression for the chemical potential from Eq. (3.129), the dependence of the compressibility factor on density and temperature (i.e., an analytical expression for the EOS) is required. One of the first works on the evaluation of the accuracy of HS EOSs in reproducing computer simulation data for the chemical potential was carried out by Adams [156], considering equations CS and LF in the range $0.1 \leq \rho \leq 0.8$. Both EOSs yielded accurate results, expression LF giving a slightly closer agreement with respect to the Monte Carlo computer simulation values.

Mulero et al. [13] calculated the residual chemical potential for the HS fluid using Eq. (3.129) and 18 EOSs (those listed in Table 3.11 except G, A, S2, and MV), and compared the results with those generated by Adams [156] in a Monte Carlo computer simulation. As can be seen in Table 3.15, when the whole density range is considered, the average absolute deviations are less than 1% except for the old expressions SPT and PYT, and surprisingly for

Table 3.15. AADs of the values for the residual chemical potential of HSs given by several EOSs with respect to computer simulation values from [156]

EOS	AAD (%) $(0.1 \leq \rho \leq 0.8)$
SPT	1.5
PYT	3.9
RH	0.5
CS	0.6
II	0.5
LF	0.5
EW	0.4
K	0.5
BC4	0.5
BC6	0.5
MSAV1	0.4
MSAV2	0.4
SOL1	0.6
SOL2	0.5
S1	0.4
WKV1	0.8
WKV2	0.7
WKV3	1.3

WKV3. Excellent agreement was found for the rest of the equations, without meaningful differences in accuracy. Therefore, the authors concluded that there seemed to be no direct correlation between the simplicity or complexity of the EOSs and their accuracy in reproducing computer simulation data for the chemical potential.

Despite the good agreement between the EOSs and the computer simulation data, none of the expressions considered by those authors gave deviations below 1% for all densities [13]. Thus, deviations greater than 1% were always found at $\rho = 0.3$, except for the equation SPT. Moreover, for SPT and WKV3 these deviations were always greater than 1% at high densities ($\rho > 0.5$). For $\rho > 0.6$, expressions MSAV1 and MSAV2 yielded excellent results. On the contrary, at the highest density considered ($\rho = 0.8$), expressions SPT, PYT, SOL1, WKV1, WKV2, and WKV3 gave deviations greater than 1% [13].

Mulero et al. [14] stressed, as in the case of the calculation of the compressibility factor and of the isothermal compressibility referred to in previous paragraphs, that the use of simple modifications of the equation CS or of more complex EOSs does not imply a significant increase in accuracy in the calculation of the chemical potential. In particular, the use of the most recent equations considered in their study (namely, WKV1, WKV2, and WKV3) led to less accurate results than those calculated with most of the other expressions, at least for stable densities. Although some other simple or complex

Table 3.16. AADs of the values for the chemical potential of HSs given by several equation of state EOSs with respect to computer simulation values in [156, 181, 221]

EOS	AAD (%)
Liu, Eq. (3.125)	0.64
CS, Eq. (3.54)	0.71
Kolafa, Eq. (3.72)	0.64
Le Fevre, Eq. (3.58)	1.07

expressions could have been used, CS was sufficiently accurate and simple to be considered an adequate choice to calculate the chemical potential of HSs. Therefore, the well-known expression CS can be regarded as a fairly good representation of the contribution of repulsive forces to the chemical potential in EOSs [8, 9, 10, 13].

Another relevant work on the calculation of the chemical potential of HS fluids was that recently carried out by Liu [160]. As a further test for any proposal for the EOS, as suggested by Mulero et al. [13], Liu calculated the chemical potential at 35 thermodynamic states using his own expression, Eq. (3.125), in Eq. (3.129), and compared the results with those from the computer simulations in [156, 181, 221]. The AADs are listed in Table 3.16, which also includes results for some other expressions for the EOS. As for the case of the compressibility factor, he found the values yielded with his proposal to be highly accurate, and noted that such a high accuracy is necessary for an HS EOS because it is demonstrated elsewhere that it is crucial for glass transition studies [222]. It can be seen that very accurate values are also obtained with some simple proposals for the EOS, such as those of CS [177] and of Kolafa [193].

3.5 Final Remarks

The EOS of an d-dimensional hard fluid appears to perhaps be the most important analytical relationship required to study its thermodynamic behavior.

A wide variety of analytical and nonanalytical expressions for the EOS of monocomponent HD and HS fluids are available in the literature. Each one shows particular advantages and/or disadvantages in its handling. The analytical structure, the validity range in the corresponding phase diagram, and the location of poles must be regarded as the three main features of an EOS, and should therefore be borne in mind in order to achieve accurate results for a specific physical magnitude.

There are several ways of deriving an adequate expression for the EOS of the hard fluid. In this sense, the literature includes expressions based on theoretical developments, empirical or semiempirical expressions based on the virial coefficients and/or on the fit of computer simulation data, expressions derived from approximants with poles located at specific density values, volume-explicit EOSs, and nonanalytical equations obtained as the solution to a nonlinear differential equation.

In particular, more than 30 and 80 different expressions for the EOSs of the HD and HS fluids, respectively, have been reviewed in this chapter.

The accuracy of most of the proposed expressions in the calculation of the compressibility factor of the HD and HS systems was analyzed by comparison with available computer simulation data and/or virial coefficients.

The choice of a set of computer simulation values of a specific physical magnitude is crucial in order to have an adequate reference to compare with. In this sense, there have been numerous studies based on either molecular dynamics or Monte Carlo techniques, although many of them do not satisfy appropriate conditions in terms of number of simulated states, number of particles of the system, number of time steps, width of the validity range in the phase diagram, etc. Thus, a detailed analysis of the performance of the reference values is crucial for further studies of the accuracy of any physical property.

As a general conclusion, there seemed to be no direct relationship between the analytical complexity and the accuracy of the EOSs. This conclusion was found to apply for both two- and three-dimensional cases. Therefore, for HSs the simple and well-known expression proposed by Carnahan and Starling in 1969 was regarded as adequate in a great number of cases. However, it must be noted that most comments about the accuracy of the EOSs presented refer to the calculation of the compressibility factor. Any further study about some other thermophysical property should take specific criteria into account in order to make coherent statements about the accuracy of the results yielded by the equations.

As a further test of the analytical expressions collected here for the EOS of the HS system, calculations to obtain the isothermal compressibility and the residual chemical potential were performed, and the results were compared with available computer simulation data.

Acknowledgement

This work was supported by project FIS2006-02794 of the "Ministerio de Educación y Ciencia" of Spain and the "Fondo Europeo de Desarrollo Regional" (FEDER) of the European Union.

References

1. J.P. Hansen, I.R. McDonald: *Theory of Simple Liquids* (Academic Press, London 1976)
2. H. Reiss, H.L. Frisch, J.L. Lebowitz: J. Chem. Phys. **31**, 369 (1959)
3. H. Reiss: *Statistical Mechanics and Statistical Methods in Theory and Application* (U. Landman, Plenum, New York 1977)

4. E. Thiele: J. Chem. Phys. **39**, 474 (1963)
5. G. Aly, I. Ashour: Fluid Phase Equil. **101**, 137 (1994)
6. J.A. Barker, D. Henderson: Rev. Mol. Phys. **48**, 587 (1976)
7. T. Boublik, I. Nezbeda: Coll. Czech. Chem. Commun. **51**, 2301 (1986)
8. F. Cuadros, W. Ahumada: Thermochim. Acta **297**, 109 (1997)
9. F. Cuadros, A. Mulero, W. Ahumada: Thermochim. Acta **277**, 85 (1996)
10. F. Cuadros, A. Mulero, W. Okrasiński, W. Ahumada: Int. Rev. Phys. Chem. **16**, 141 (1997)
11. M.D. Donohue, J.M. Prausnitz: AIChE J. **24**, 849 (1978)
12. H.M. Lin, J.T. Chen, M.J. Lee: Fluid Phase Equil. **126**, 29 (1996)
13. A. Mulero, C.A. Faúndez, F. Cuadros: Mol. Phys. **97**, 453 (1999)
14. A. Mulero, C.A. Faúndez, M.I. Parra, F. Cuadros: Termochim. Acta **334**, 1 (1999)
15. A. Mulero, F. Cuadros, C.A. Faúndez: Aust. J. Phys. **52**, 101 (1999)
16. Kh. Nasrifar, Sh. Ayatollahi, M. Moshfeghian: Canadian J. Chem. Eng. **78**, 1111 (2000)
17. Kh. Nasrifar, O. Bolland: Chem. Eng. Comm. **193**, 1277 (2006)
18. A.I. Rusanov: J. Chem. Phys. **121**, 1873 (2004)
19. A.I. Rusanov: Chem. Rev. **74**, 117 (2005)
20. Y.S. Wei, R.J. Sadus: AIChE J. **46**, 169 (2000)
21. G. Wu, R.J. Sadus: AIChE J. **51**, 309 (2005)
22. L.V. Yelash, T. Kraska, U.K. Deiters: J. Chem. Phys. **110**, 3079 (1999)
23. L.V. Yelash, T. Kraska: Phys. Chem. Chem. Phys. **3**, 3114 (2001)
24. T. Boublik: Fluid Phase Equil. **1**, 37 (1977)
25. F. Cuadros, A. Mulero: Chem. Phys. **156**, 33 (1991)
26. F. Cuadros, A. Mulero: Chem. Phys. **160**, 375 (1992)
27. F. Cuadros, A. Mulero, P. Rubio: J. Chem. Educ. **71**, 959 (1994)
28. F. Cuadros, A. Mulero, W. Okrasiński, C.A. Faúndez, M.I. Parra: Rev. Chem. Eng. **19**, 387 (2003)
29. A. Mulero, I. Cachadiña: Int. J. Thermophys. **27**, 1435 (2006)
30. E.A. Müller, K.E. Gubbins: Ind. Eng. Chem. Res. **40**, 2193 (2001)
31. F.L. Román, A. Mulero, F. Cuadros: Phys. Chem. Chem. Phys. **6**, 5402 (2004)
32. J.R. Solana, J. Amorós, E. Villar: Physica B **160**, 352 (1990)
33. J.D. Weeks, D. Chandler, H.C. Andersen: J. Chem. Phys. **54**, 5237 (1971)
34. Y. Song, R.M. Stratt, E.A. Mason: J. Chem. Phys. **88**, 1126 (1988)
35. J.H. Conway, N.J.A. Sloane: *Sphere Packings, Lattices and Groups* (Springer-Verlag, New York 1998)
36. S. Torquato, T.M. Truskett, P.G. Debenedetti: Phys. Rev. Lett. **84**, 2064 (2000)
37. C. Ghotbi, J.H. Vera: Ind. Eng. Chem. Res. **41**, 1122 (2002)
38. F. Cuadros, A. Mulero: Analysis of Steele's theory monolayer adsorption, in *Fluid Physics. Lecture Notes of Summer Schools*, ed by M.G. Velarde, C.I. Christov (World Scientific, Singapore 1995) pp 135–141
39. F. Cuadros, A. Mulero, L. Morala, V. Gómez-Serrano: Langmuir **17**, 1576 (2001)
40. A. Mulero, F. Cuadros: Chem. Phys. **205**, 379 (1996)
41. A. Mulero, F. Cuadros: Langmuir **12**, 3265 (1996)
42. A. Mulero, F. Cuadros: J. Coll. Interf. Sci. **186**, 110 (1997)

43. A. Mulero, L. Morala, V. Gómez-Serrano, F. Cuadros: Two dimensional models for physisorption at surfaces. In *Encyclopedia of Surface and Colloid Science*, ed. A. Hubbard, M. Dekker (United States, New York 2002)
44. A.I. Rusanov: J. Chem. Phys. **120**, 10736 (2004)
45. M. Baus, J.L. Colot: Phys. Rev. A **36**, 3912 (1987)
46. N. Clisby, B.M. McCoy: Pramana **64**, 775 (2005)
47. N. Clisby, B.M. McCoy. arXiv:cond-mat/0503525 available at URL http://arxiv.org/archive/cond-mat (2005)
48. N. Clisby, B.M. McCoy: J. Statistical Phys. **122**, 15 (2006)
49. M. Luban, J.P.J. Michels: Phys. Rev. A **41**, 6796 (1990)
50. M.J. Maeso, J.R. Solana, J. Amorós, E. Villar: Materials Chem. Phys. **30**, 39 (1991)
51. M. Bishop, A. Masters, J.H.R. Clarke: J. Chem. Phys. **110**, 11449 (1999)
52. M. Bishop, A. Masters, A.Y. Vlasov: J. Chem. Phys. **121**, 6884 (2004)
53. M. Bishop, P.A. Whitlock: J. Chem. Phys. **123**, 14507 (2005)
54. L. Lue: J. Chem. Phys. **122**, 44513 (2005)
55. M. Robles, M. López de Haro, A. Santos: J. Chem. Phys. **120**, 9113 (2004); Erratum 125, 219903(E) (2006)
56. M. Robles, M. López de Haro, A. Santos: J. Chem. Phys. **126**, 16101 (2007)
57. A. Santos: J. Chem. Phys. **112**, 10680 (2000)
58. B. Barboy, W.M. Gelbart: J. Chem. Phys. **71**, 3053 (1979)
59. M.P. Allen, G.T. Evans, D. Frenkel, B.M. Mulder: Adv. Chem. Phys. **86**, 1 (1993)
60. M.P. Allen, D.J. Tildesley: *Computer Simulations of Liquids.* Oxford Science Publications, (Oxford, United Kingdom 1987)
61. T. Boublik: J. Chem. Phys. **63**, 4084 (1975)
62. T. Boublik: Mol. Phys. **29**, 421 (1975)
63. T.K. Dey, H.D. Singh, S.S. Karki, S.K. Sinha: Indian J. Phys. **79**, 959 (2005)
64. A. Donev, S. Torquato, F.H. Stillinger: J. Computational Phys. **202**(2), 765 (2005)
65. D.P. Fraser, M.J. Zuckermann, O.G. Mouritsen: Phys. Rev. A **42**, 3186 (1990)
66. A. Avazpour, M. Moradi: Physica B: Condensed Matter **392**(1–2), 242 (2007)
67. T. Boublik: Mol. Phys. **42**, 209 (1981)
68. L.E.S. de Souza, S. Canuto: Phys. Chem. Chem. Phys. **4**(6), 922 (2002)
69. A. Esztermann, H. Reich, M. Schmidt: Phys. Rev. E **73**, 11409 (2006)
70. S.C. Gay, J.C. Rainwater, P.D. Beale: J. Chem. Phys. **112**(22), 9849 (2000)
71. T. Koda, S. Ikeda: Mol. Cryst. Liq. Cryst. Sci. Technol., Sect. A **346**, 127 (2000)
72. M.J. Maeso, J.R. Solana, J. Amorós: Materials Chem. Phys. **33**, 134 (1993)
73. S.C. McGrother, D.C. Williamson, G. Jackson: J. Chem. Phys. **104**(17), 6755 (1996)
74. S.V. Savenko, M. Dijkstra: Phys. Rev. E **72**, 1 (2005)
75. D.J. Tildesley, W.B. Streett: Mol. Phys. **41**, 85 (1980)
76. A.Y. Vlasov, X.M. You, A.J. Masters: Mol. Phys. **100**(20), 3313 (2002)
77. X.M. You, A.Y. Vlasov, A.J. Masters: J. Chem. Phys. **123**(3), 1 (2005)
78. F. Cuadros, A. Mulero, I. Cachadiña, W. Ahumada: Int. Rev. Phys. Chem. **14**, 205 (1995)
79. F. Cuadros, I. Cachadiña, W. Ahumada: Molecular Eng. **6**, 319 (1996)
80. J. Kolafa, M. Rottner: Mol. Phys. **104**, 3435 (2006)

81. S. Labík, J. Kolafa, A. Malijevský: Phys. Rev. E **71**, 21105 (2005)
82. A. Mulero, F. Cuadros, C.A. Galán: J. Chem. Phys. **107**, 6887 (1997)
83. A. Santos, M.L. Haro, S.B. Yuste: J. Chem. Phys. **103**, 4622 (1995)
84. J.R. Solana: Ecuación de estado de un sistema de discos duros: avances recientes. In: *La Investigación del Grupo Especializado de Termodinámica de las R.S.E.F. y R.S.E.Q., Año 2006*, ed. J.M. Ortiz de Zárate, M. Khayet (R.S.E.F., Madrid 2006) pp 291–305
85. J.R. Solana: Ecuaciones de estado precisas y sencillas para el fluido de discos duros a partir de aproximantes. In: *La Investigación del Grupo Especializado de Termodinámica de las R.S.E.F. y R.S.E.Q., Año 2006*, ed. J.M. Ortiz de Zárate, M. Khayet (R.S.E.F., Madrid 2006) pp 306–313
86. D.K. Stillinger, F.H. Stillinger, S. Torquato, T.M. Truskett, P.G. Debenedetti: J. Stat. Phys. **100**, 49 (2000)
87. D.K. Stillinger, F.H. Stillinger, S. Torquato, T.M. Truskett, P.G. Debenedetti: J. Stat. Phys. **113**, 10186 (2000)
88. C.F. Tejero, J.A. Cuesta: Phys. Rev. E **47**, 490 (1993)
89. J.G. Berryman: Phys. Rev. A **27**, 1053 (1983)
90. E. Eisenberg, A. Baram: Phys. Rev. E **73**, 25104 (2006)
91. M. Metropolis, A.W. Rosenbluth, M.N. Rosenbluth, A.N. Teller, E. Teller: J. Chem. Phys. **21**, 1087 (1953)
92. B.J. Alder, T.E. Wainwright: Phys. Rev. **127**, 359 (1962)
93. W.G. Hoover, B.J. Alder: J. Chem. Phys. **46**, 686 (1967)
94. W.G. Hoover, F.H. Ree: J. Chem. Phys. **49**, 3609 (1968)
95. B.J. Alder, W.G. Hoover, D.A. Young: J. Chem. Phys. **49**, 3688 (1968)
96. D.G. Chae, F.H. Ree, T. Ree: J.Chem.Phys. **50**, 1581 (1969)
97. W.W. Wood: J. Chem. Phys. **48**, 415 (1968)
98. W.W. Wood: J. Chem. Phys. **52**, 729 (1970)
99. L.V. Woodcock: J. Chem. Soc. Faraday Trans. II **72**, 731 (1976)
100. D. Henderson: Mol. Phys. **34**, 301 (1977)
101. J.J. Erpenbeck, M. Luban: Phys. Rev. A **32**, 2920 (1985)
102. A. Jaster: Phys. Rev. E **59**, 2594 (1999)
103. C.H. Mak: Phys. Rev. E **73**, 65104 (2006)
104. J.J. Alonso, J.F. Fernández: Phys. Rev. E **59**, 2659 (1999)
105. K. Binder, S. Sengupta, P. Nielaba: J. Phys.: Condens. Matt. **14**, 2323 (2002)
106. M.A.A. da Silva, A. Caliri, B.J. Mokross: Phys. Rev. Lett. **58**, 2312 (1987)
107. J.F. Fernández, J.J. Alonso, J. Stankiewicz: Phys. Rev. Lett. **75**, 3477 (1995)
108. J.F. Fernández, J.J. Alonso, J. Stankiewicz: Phys. Rev. E **55**, 750 (1997)
109. A. Jaster: Europhys. Lett. **42**, 277 (1998)
110. A. Jaster: Phys. Lett. A **330**, 120 (2004)
111. J. Lee, K.J. Strandburg: Phys. Rev. B **46**, 11190 (1992)
112. A.C. Mitus, H. Weber, D. Marx: Phys. Rev. E **55**, 6855 (1997)
113. A.C. Mitus, A.Z. Patashinski, A. Patrykiejew, S. Sokolowski: Phys. Rev. B **66**, 184202 (2006)
114. T.M. Truskett, S. Torquato, S. Sastry, P.G. Debenedetti, F.H. Stillinger: Phys. Rev. E **58**, 3083 (1998)
115. E. Velasco, L. Mederos: Phys. Rev. B **56**, 2432 (1997)
116. H. Watanabe, S. Yukawa, Y. Ozeki, N. Ito: Phys. Rev. E **69**, 045103 (2004)
117. H. Weber, D. Marx, K. Binder: Phys. Rev. B **51**, 14636 (1995)
118. J.A. Zollweg, G.V. Chester, P.W. Leung: Phys. Rev. B **39**, 9518 (1989)

119. J.A. Zollweg, G.V. Chester: Phys. Rev. B **46**, 11186 (1992)
120. T. Boublik: Mol. Phys. **63**, 685 (1988)
121. G. Lasher: J. Chem. Phys. **53**, 4141 (1970)
122. D.A. Ward, F. Lado: Mol. Phys. **63**, 623 (1988)
123. D.A. Ward, F. Lado: Mol. Phys. **64**, 1185 (1988)
124. E. Helfand, H.L. Frisch, J.L. Lebowitz: J. Chem. Phys. **34**, 1037 (1961)
125. D. Henderson: Mol. Phys. **30**, 971 (1975)
126. F.C. Andrews: J. Chem. Phys. **64**, 1941 (1976)
127. K.W. Kratky: J. Chem. Phys. **69**, 2251 (1978)
128. A. Baram, M. Luban: J. Phys. C: Solid State Phys. **12**, 659 (1979)
129. V.C. Aguilera-Navarro, M. Fortes, M. de Llano, A. Plastino: J. Chem. Phys. **76**, 749 (1982)
130. J.A. Devore, E. Schneider: J. Chem. Phys. **77**, 1067 (1982)
131. L. Verlet, D. Levesque: Mol. Phys. **46**, 969 (1982)
132. B.R. Hoste, W. van Dael: J. Chem. Soc. Faraday Trans. II **80**, 477 (1984)
133. D.A. Young, B.J. Alder: J. Chem. Phys. **70**, 473 (1979)
134. T.P. Singh, S.K. Sinha: Pramana J. Phys. **25**, 733 (1985)
135. F. Cuadros, A. Mulero: Chem. Phys. **156**, 33 (1991)
136. M. Alexanian: Phys. Rev. B **31**, 7160 (1985)
137. M. Brunner, C. Bechinger, U. Herz, H.H. von Grünberg: Europhys. Lett. **63**, 791 (2003)
138. M.J. Maeso, J.R. Solana, J. Amorós, E. Villar: J. Chem. Phys. **94**, 551 (1991)
139. I.C. Sanchez: J. Chem. Phys. **101**, 7003 (1994)
140. E.J. Janse van Rensburg: J. Phys. A **26**, 4805 (1993)
141. M. López de Haro, A. Santos, S. Bravo-Yuste: Eur. J. Phys. **19**, 281 (1998)
142. E.Z. Hamad: Ind. Eng. Chem. Res. **36**, 4385 (1997)
143. U.F. Edgal, D.L. Huber: J. Chem. Phys. **108**, 1578 (1998)
144. X.Z. Wang: Phys. Rev. E **66**, 31203 (2002)
145. A. Chamoux, A. Perera: Phys. Rev. E **58**, 1933 (1998)
146. M.D. Rintoul, S. Torquato: Phys. Rev. E **58**, 532 (1998)
147. S. Torquato: Phys. Rev. E **51**, 3170 (1995)
148. M.D. Rintoul, S. Torquato: J. Chem. Phys. **105**, 9258 (1996), Erratum **107**, 2698 (1997)
149. S.R. Williams, I.K. Snook, W. van Megen: Phys. Rev. E **64**, 21506 (2001)
150. B.J. Alder, T.E. Wainwright: J. Chem. Phys. **33**, 1439 (1960)
151. D. Ma, G. Ahmadi: J. Chem. Phys. **84**, 3449 (1986)
152. A. Mulero, C.A. Galán, F. Cuadros: Phys. Chem. Chem. Phys. **3**, 4991 (2001)
153. W. Wang, M.K. Khoshkbarchi, J.H. Vera: Fluid Phase Equil. **115**, 25 (1996)
154. J.A. Barker, D. Henderson: Mol. Phys. **21**, 187 (1971)
155. M. Miandehy, H. Modarress: J. Chem. Phys. **119**, 2716 (2003)
156. D. Adams: Mol. Phys.**28**, 1241 (1974)
157. L.V. Woodcock: J. Chem. Soc. Faraday Trans. II **72**, 1667 (1976b)
158. L.V. Woodcock: J. Chem. Soc. Faraday Trans. II **73**, 11 (1977)
159. L.V. Woodcock: Ann. (N.Y.) Acad. Sci. **371**, 274 (1981)
160. H. Liu arXiv:cond-matt/0605392 available at URL http://arxiv.org/archive/cond.mat
161. S. Labík, A. Malijevský: Mol. Phys. **42**, 739 (1981)
162. J.J. Erpenbeck, W.W. Wood: J. Stat. Phys. **35**, 321 (1984)
163. M.K. Khoshkbarchi, J.H. Vera: Fluid Phase Equil. **130**, 189 (1997)

164. G.A. Iglesias-Silva, K.R. Hall: Fluid Phase Equil. **98**, 49 (1994)
165. D. Ayala, P.A. Lonngi: Mol. Phys. **73**, 763 (1991)
166. R.J. Speedy: J. Chem. Phys. **100**, 6684 (1994)
167. R.J. Speedy: Mol. Phys. **83**, 591 (1994)
168. R.J. Speedy: J. Phys.: Cond. Matt. **9**, 8591 (1997)
169. R.J. Speedy: Mol. Phys. **95**, 169 (1998)
170. M. Barošová: Statistical thermodynamic study of binary mixtures of hard spheres. PhD Thesis, Institute of Chemical Technology, Czech Republic (1996)
171. A. Malijevský, J. Veverka: Phys. Chem. Chem. Phys. **1**, 4267 (1999)
172. J. Kolafa: Personal communication (2001)
173. J. Kolafa, S. Labík, A. Malijevský A: Phys. Chem. Chem. Phys. **6**, 2335 (2004)
174. S.M. Waziri, E.Z. Hamad: Ind. Eng. Chem. Res. **45**, 7251 (2006)
175. S. Rambaldi, G. Salustri, C. Benedetti: Physica A **361**, 180 (2006)
176. J. Kolafa: Phys. Chem. Chem. Phys. **8**, 464 (2006)
177. N.F. Carnahan, K.E. Starling: J. Chem. Phys. **51**, 635 (1969)
178. A. Mulero, C.A. Galán, F. Cuadros: J. Chem. Phys. **111**, 1 (1999)
179. L. Tonks: Phys. Rev. **50**, 955 (1936)
180. M.S. Wertheim: Phys. Rev. Lett. **10**, 321 (1963)
181. M. Heying, D.S. Corti: J. Phys. Chem. B **108**, 19756 (2004)
182. F.H. Stillinger, P.G. Debenedetti, S. Chatterjee: J. Chem. Phys. **125**, 204504 (2006)
183. F.H. Ree, W.G. Hoover: J. Chem. Phys. **40**, 939 (1964)
184. E.A. Guggenheim: Mol. Phys. **9**, 199 (1965)
185. K. N. Marsh, M.L. McGlashan, C. Warr: Trans. Faraday Soc. **66**, 2453 (1970)
186. K.R. Hall: J. Chem. Phys. **57**, 2252 (1972)
187. D. Frenkel, B. Smit: *Understanding Molecular Simulation*, (Academic Press, New York 1996)
188. E.J. Le Fevre: Nature (London) Phys. Sci. **235**, 20 (1972)
189. F.C. Andrews: J. Chem. Phys. **62**, 272 (1975)
190. K.W. Kratky: Physica A, **87**, 584 (1977)
191. G.D. Scott, D.M. Kilgour: J. Phys. D **2**, 863 (1969)
192. L.V. Woodcock, C. A. Angell: Phys. Rev. Lett. 47, 1129 (1981)
193. T. Boublik: Mol. Phys. **59**, 371 (1986)
194. I.H. Kim, Y.C. Bae: Fluid Phase Equil. **167**, 187 (2000)
195. I.H. Kim, H.M. Lin, K.C. Chao: Ind. Eng. Chem. Fundam. **25**, 75 (1986)
196. J.I. Goldman, J.A. White: J. Chem. Phys. **89**, 6403 (1988)
197. Y. Zhou, G. Stell: Int. J. of Thermophys. **9**, 953 (1988)
198. S. Labík, A. Malijevský: Czech. J. Phys. B **33**, 128 (1983)
199. J.R. Solana: Personal communication (1992)
200. S. Bravo-Yuste, M. López de Haro, A. Santos: Phys. Rev. E **53**, 4820 (1986)
201. M. Mohsen-Nia, H. Modarress, G.A. Mansoori: Chem. Eng. Comm. **131**, 15 (1995)
202. M.M. Khalkhali, H. Modarress, M. Mohsen-Nia: Chem. Eng. Comm. **192**, 725 (2005)
203. M. Mohsen-Nia, H. Modarress, G.A. Mansoori: Fluid Phase Equil. **206**, 27 (2003)
204. M.K. Khoshkbarchi, J.H. Vera: Fluid Phase Equil. **142**, 131 (1998)
205. C. Ghotbi, J.H. Vera: Fluid Phase Equil. **321**, 187 (2001)
206. I. Nezbeda, K. Aim, J. Kolafa: Z. Phys. Chem. (Leipzig) **270**, 533 (1989)

207. C. Ghotbi, J.H. Vera: Canadian J. Chem. Eng. **79**, 678 (2001)
208. B.C. Eu, Y.G. Ohr: J. Chem. Phys. **118**, 2264 (2003)
209. I. Polishuk, J.H. Vera: J. Phys. Chem. B **109**, 5977 (2005)
210. M. Miandehy, H. Modarress, M.R. Dehghani: Fluid Phase Equil. **239**, 91 (2006)
211. J.H. Vera, A. Mulero: Fluid Phase Equil. **253**, 176 (2007)
212. T.M. Reed, K.E. Gubbins: *Applied Statistical Mechanics* (McGraw-Hill, New York 1978)
213. B. Widom: J. Chem. Phys. **39**, 2808 (1963)
214. D. Adams: Mol. Phys.**32**, 647 (1976)
215. D. Adams: Mol. Phys.**37**, 211 (1979)
216. K.S. Shing, K.E. Gubbins: Mol. Phys. **43**, 717 (1981)
217. B. Smit, D. Frenkel: Mol. Phys. **68**, 951 (1989)
218. D.M. Heyes: Chem. Phys. **82**, 285 (1983)
219. A. Lotfi, J. Vrabec, J. Fischer: Mol. Phys. **76**, 1319 (1992)
220. J.G. Powles, W.A.B. Evans, N. Quirke: Mol. Phys. **46**, 1347 (1982)
221. P. Attard: J. Chem. Phys. **98**, 2225 (1993)
222. H. Liu arXiv:cond-matt/0605358
available at URL http://arxiv.org/archive/cond.mat

4

Application of Hard-Sphere and Hard-Disk Equations of State to the Weeks–Chandler–Andersen Reference System

A. Mulero, C.A. Galán, I. Cachadiña, and F. Cuadros

Departamento de Física Aplicada,
Universidad de Extremadura,
06071 Badajoz, Spain
mulero@unex.es, cgalango@unex.es, icacha@unex.es, cuadros1@unex.es

Perturbation theories in statistical mechanics are based on the separation of the contributions of repulsive and attractive intermolecular forces. The former can be modeled by an appropriate modification of the hard-sphere or hard-disk properties through a scaling procedure, for which a temperature-dependent molecular diameter is needed. In this chapter, we analyze such a scaling procedure when applied to the well-known Weeks–Chandler–Andersen perturbation theory. Calculations are performed not only for the whole range of temperatures and densities, but attention is also focused on certain zones of the phase diagram of special interest. The best choices for the combination of equations of state and molecular diameters are given, following accuracy and/or simplicity criteria.

4.1 Introduction

Perturbation theories were born some decades ago, and can be seen as a modern version of the old van der Waals theory [1, 2, 3, 4, 5]. They are based on the separation of the repulsive and attractive effects of the intermolecular interaction potential, considering that the structure of dense fluids is strongly determined by its repulsive part. In this way, the effect of the attractive forces is regarded as a perturbation. These theories allowed one to study the properties of fluids even at thermodynamic states close to the triple point, and hence, they have become an extremely significant contribution to the development of the molecular physics of fluids [1, 2, 3, 4, 5]. Moreover, many equations of state (EOSs), arising from the well-known van der Waals equation, are based on the separation of repulsive and attractive molecular interactions [6, 7, 8, 9, 10, 11, 12].

Mulero, A., et al.: *Application of Hard-Sphere and Hard-Disk Equations of State to the Weeks–Chandler–Andersen Reference System.* Lect. Notes Phys. **753**, 111–132 (2008)
DOI 10.1007/978-3-540-78767-9_4

One of the most widely used perturbation theories is that of Weeks, Chandler, and Andersen (WCA) [1, 2, 3, 4, 5], which states as a basic hypothesis that the structure of dense fluids practically does not depend on the attractive forces. As a direct consequence, it is possible to approach the thermodynamic study of a dense fluid from the knowledge of the structure and thermodynamic properties of the reference system (RS) [1, 2, 3, 4, 5, 6, 13, 14, 15, 16, 17, 18, 19, 20, 21, 22, 23, 24, 25, 26, 27, 28, 29, 30, 31, 32] whose molecules interact with a repulsive potential. From a practical point of view, the properties of this RS can be obtained from computer simulations or by suitably modifying the properties of a hard system including a temperature-dependent molecular diameter [5, 13, 14, 15, 16, 17, 18, 19, 20, 21, 22, 23, 24, 25, 26]. The choice of an analytical expression for the repulsive contribution is a major step in the development of perturbation theories and of EOSs for real fluids to be used in physical chemistry or chemical engineering calculations [5, 6, 7, 8, 9, 10, 11, 13, 14, 20]. In the case of two-dimensional fluids, the properties of the RS [13, 15, 16, 17, 18, 22, 29, 32] serve as the basis in the development of simple theories or equations explaining some simple adsorption processes [33, 34, 35, 36, 37, 38].

The study of the WCA RS and its possible applications is hence interesting for both students and researchers in statistical mechanics, chemical physics, physical chemistry, chemical engineering, and related areas. This chapter gives an extensive review of the scaling procedure from the properties of hard-sphere (HS) and hard-disk (HD) systems. For that, the validity test performed by Mulero et al. [22, 25, 26] of a set of analytical expressions for the EOS of the hard fluids to reproduce the pressure and potential energy of the WCA RS in two [15, 22] and three [25, 26] dimensions is presented. To do so, the expressions of Verlet–Weis [39] and of Kolafa–Nezbeda [40] for the temperature dependence of the molecular diameter are considered. The results are compared with those obtained through molecular dynamics computer simulation [5, 14].

The best choices for the combination of EOSs and molecular diameters in the calculation of the pressure and potential energy are given. It is shown how that choice may differ for different ranges of temperatures and densities, and how there is no direct relationship between the simplicity or complexity of the analytical expressions and their accuracy. As will be noted, this general conclusion applies for both two and three dimensions.

Although the study of other properties (such as chemical potential, isothermal compressibility, and critical point) and their applications (simple adsorption properties, for example) is not included in this chapter, there are appropriate references in the literature [5, 14, 17, 18, 19, 20, 23, 24, 29, 32, 33, 35, 36, 37, 38].

The chapter is organized as follows. Section 4.2 presents the main analytical expressions that will be needed in the calculations performed to obtain the pressure and the potential energy of the WCA RS. The three-dimensional case will be considered in Sects. 4.3 (calculation of pressure using a selection of EOSs, and considering either the whole fluid range in the phase diagram or some specific ranges) and 4 (calculation of the potential energy). Section 4.5 presents results for the pressure and the potential energy in the two-dimensional case. Finally, the most relevant conclusions will be given in Sect. 4.6.

4.2 Analytical Expressions

As is well known, in the WCA theory the RS is defined by the short-range repulsive forces, whereas the perturbation part contains all the long-range attractive forces. When the Lennard–Jones potential is used, the reference repulsive potential is given by

$$U_0(r) = \begin{cases} 4(r^{-12} - r^{-6}) + 1 & r \leq r_\mathrm{m} = 2^{1/6} \\ 0 & r > r_\mathrm{m} \end{cases} \tag{4.1}$$

where r is the interparticle distance and r_m the minimum of the potential (the distance at which the force changes from repulsive to attractive). Note that in this and in the following equations all the properties are expressed in reduced Lennard–Jones units, as defined in Chap. 3. The properties of the RS can be obtained using the expressions for an HS (or HD) system, where the HS (or HD) diameter, σ, is replaced by an appropriate diameter, d, in the calculation of the packing fraction η:

$$\eta_\mathrm{RS} = \frac{\pi \rho d^3}{6} \tag{4.2}$$

$$\eta_\mathrm{RS} = \frac{\pi \rho d^2}{4} \tag{4.3}$$

in three and two dimensions, respectively. In these expressions ρ is the reduced density.

Several criteria have been proposed in order to adequately calculate the effective diameter, and they are briefly summarized in Sect. 9.3.3 of Chap. 9. In particular, in the Barker–Henderson criteria the molecular diameter d is defined as [41]:

$$d(T) = \int_0^{r_\mathrm{m}} \left[1 - e^{-U_0(r)/T}\right] \mathrm{d}r \tag{4.4}$$

where T is the reduced temperature. Nevertheless, an analytical expression for d is much more useful for practical applications. In this sense, it is well known that the dependence of the molecular diameter on density is much weaker than that on temperature. For that reason, only the latter is commonly considered. Analytical expressions of d as a function of the temperature are summarized in Table 9.1 of Chap. 9. We use here which is perhaps the most widely used and easiest expression for d, proposed by Verlet and Weis (VW) [39]:

$$d_\mathrm{VW} = \frac{0.3837T + 1.068}{0.4293T + 1} \tag{4.5}$$

which gives an error less than 2×10^{-4} with respect to Eq. (4.4). This expression has been used in many previous works [5, 14, 15, 21, 22, 23, 24, 25, 42, 43, 44, 45, 46, 47, 48], since Cuadros and Mulero [15] proved that it gives good results in the calculation of the compressibility factor of the WCA repulsive Lennard–Jones system.

Another analytical expression for $d(T)$ that we consider here is the one proposed by Kolafa and Nezbeda (KN) [40]:

$$\begin{aligned} d_{\mathrm{KN}} =& 0.011117524\,T^{-1} - 0.076383859\,T^{-1/2} + 1.080142248 \\ &+ 0.000693129\,T^{1/2} - 0.063920968\ln(T) \end{aligned} \tag{4.6}$$

which fits the molecular diameter with a maximum error of 9.05×10^{-5} and its derivative with a maximum error of 1.67×10^{-4}.

We note that the dependence of the effective molecular diameter on density should be taken into account at high densities. Other approximations for this dependence were proposed elsewhere [49, 50, 51], and are summarized in Table 9.2 of Chap. 9. Nevertheless, it has been proved that its influence can be neglected for practical purposes [15, 49].

The properties of the WCA RS to be discussed in this chapter are the pressure (P) and the potential energy (E), given by the following expressions:

$$P = Z\rho T \tag{4.7}$$

$$E = -2T^2 \frac{Z-1}{d(T)} \frac{\partial d(T)}{\partial T} \tag{4.8}$$

where Z is the compressibility factor.

4.3 Pressure of the Three-Dimensional WCA RS

Mulero et al. [25, 26] reported two detailed studies of the accuracy of a broad set of EOSs in reproducing the pressure of the WCA RS of Lennard–Jones fluids. They were carried out for three specific zones of the phase diagram with a set of 22 EOSs [26] and also considering the whole fluid range for a set of 6 EOSs [25]. Their main results will be presented in the following two subsections. The computer simulation data sets used in each case will also be outlined.

4.3.1 Results for Certain Zones of the Phase Diagram

Mulero et al. [26] collected 22 expressions for the HS EOS in order to reproduce computer simulation values of the pressure of the WCA RS in three specific zones of importance in the phase diagram: (Zone I) thermodynamic states near the fluid–solid coexistence curve; (Zone II) thermodynamic states near the vapor–liquid coexistence curve; and (Zone III) thermodynamic states near the Lennard–Jones critical point. The authors considered as reference the molecular dynamics computer simulation results obtained by Cuadros et al. [5, 14] for the WCA RS. The computational system contained $N = 504$ particles with initial velocities randomly selected from a Maxwell distribution. The repulsive part of the Lennard–Jones potential was used, the interparticle forces having been computed by means of the particle neighbor table method. The classical Newtonian equations of motion were then numerically integrated using Verlet's algorithm, under the usual periodic boundary conditions. The computer simulation values for pressure were calculated as

$$P_{\text{sim}} = \rho T - \frac{\rho}{3N} \left\langle \sum_{i<j} r_{ij} \frac{\mathrm{d}U(r_{ij})}{\mathrm{d}r_{ij}} \right\rangle \tag{4.9}$$

where r_{ij} is the intermolecular distance (less than r_m), "<>" means averaging over time, and $U(r_{ij})$ is defined in Eq. (4.1).

The authors considered only the results obtained at thermodynamic states near the Lennard–Jones liquid curve in the vapor–liquid equilibrium (the critical point located at $T_\mathrm{c} = 1.316$, $\rho_\mathrm{c} = 0.304$, in reduced units [52]) or near the fluid side in the fluid–solid coexistence equilibrium, including the Lennard–Jones triple point ($T_\mathrm{t} = 0.7$, $\rho_\mathrm{t} = 0.844$ [1, 53]). The results obtained for the 32 selected states are listed in Table 4.1 (note that the state $\rho = 0.844$, $T = 0.7$ belongs to Zones I and II simultaneously), for the aforementioned zones of the phase diagram.

Mulero et al. [26] calculated the values for the pressure of the RS, Eq. (4.7), using the expressions for the HS EOSs from Table 3.11 in Chap. 3, where the packing fraction, Eq. (4.2), was obtained using expression VW for the molecular diameter [39], Eq. (4.5). The authors compared their results with the computer simulation values listed in Table 4.1, and obtained the average absolute deviations ($\mathrm{AAD_P}$) given in Table 4.2.

As can be seen, for thermodynamic states at the triple point density (Zone I), none of the EOSs gives $\mathrm{AAD_P} < 3\%$. In fact, only the old PYT expression (which is one of the simplest) and WKV1, WKV2, and WKV3 are suitable to reproduce the computer simulation data with deviations below 5.5%, the most accurate being WKV3. Nevertheless, it was found that a very simple equation like PYT gives a deviation smaller than that obtained with WKV1, WKV2, and all the other simple or complex expressions (except WKV3). Also, expressions CS and BC4 gave a deviation similar to that obtained with more complex equations (RH, H, or LF), and produced better results than the other simple equations (except PYT). Equation MV, however, did not lead to better agreement.

Figure 4.1 shows the individual absolute percentage deviations ($\mathrm{D_P}$) for the thermodynamic states near the liquid curve in the vapor–liquid equilibrium

Table 4.1. Pressure for the three-dimensional WCA RS at thermodynamic states selected for the three zones of the phase diagram: (Zone I) near the fluid–solid coexistence curve; (Zone II) near the vapor–liquid coexistence curve; and (Zone III) near the Lennard–Jones critical point

Zone	ρ	T	P_{sim}	Zone	ρ	T	P_{sim}
I	0.844	0.7	6.191	II	0.844	0.7	6.191
	0.844	0.8	6.802		0.800	0.8	5.646
	0.844	0.9	7.411		0.800	0.9	6.143
	0.844	1.0	7.989		0.750	1.0	5.305
	0.844	1.1	8.512		0.700	1.1	4.567
	0.844	1.2	9.028		0.650	1.2	3.947
	0.844	1.3	9.573		0.550	1.3	2.710
	0.844	1.4	10.055	III	0.500	1.4	2.280
	0.844	1.5	10.592		0.450	1.4	1.805
	0.844	1.6	10.982		0.400	1.4	1.416
	0.844	1.7	11.438		0.350	1.4	1.086
	0.844	1.8	11.875		0.300	1.4	0.825
	0.844	1.9	12.305		0.250	1.4	0.610
	0.844	2.0	12.788				
	0.844	2.1	13.220				
	0.844	2.2	13.613				
	0.844	2.3	14.053				
	0.844	2.4	14.490				
	0.844	2.5	14.886				
	0.844	2.6	15.299				

(Zone II). Note that WKV3 gave the best results, with an important difference with respect to the rest of the expressions for the EOS. The only others with $\text{AAD}_\text{P} \leq 4.1\%$ are LF, WKV1, and WKV2 (see Table 4.2). The authors observed that the well-known equation CS gave better results than other simple and complex expressions, including the most recent MV. For all the EOSs the poorest results were obtained near the triple point, with deviations always greater than or equal to 4% (see Fig. 4.1). On the contrary, it was observed that near the critical point all the EOSs gave practically the same good results (deviations near 1%), except SPT, PYT, and G (note that results for these equations are not shown in Fig. 4.1).

For the six selected states near the Lennard–Jones critical point (Zone III), deviations with respect to the computer simulated pressures were found to be equal to or less than 1%, except for expressions SPT, G, and S2 (see Table 4.2). Expression A showed the best agreement with computer simulation results, although other simpler expressions may also be considered as good choices. For example, the old equations PYT and CS produced practically the same or even smaller deviations than WKV1, WKV2, WKV3, and MV.

Table 4.2. Average absolute percentage deviations (AAD_P) of the results for the pressure of the three-dimensional WCA RS calculated with the set of expressions for the HS EOSs given in Table 3.11 of Chap. 3. Note that expression VW for the effective molecular diameter was used

EOS	Zone I	Zone II	Zone III	EOS	Zone I	Zone II	Zone III
SPT	12.96	9.34	1.49	BC6	8.73	5.90	0.93
PYT	4.76	4.89	0.73	MSAV1	7.66	5.13	0.92
RH	6.58	4.32	0.89	MSAV2	7.63	5.10	0.91
G	23.56	17.64	2.41	SOL1	9.21	6.33	1.03
CS	7.05	4.60	0.77	SOL2	8.66	5.83	0.91
H	7.19	4.76	0.89	S1	8.17	5.48	0.91
LF	6.11	4.05	0.91	S2	21.99	15.85	1.44
A	8.33	5.34	0.24	WKV1	5.34	3.43	0.86
EW	7.57	5.10	1.00	WKV2	5.33	3.46	0.89
K	7.28	4.84	0.90	WKV3	3.17	1.75	0.71
BC4	6.93	4.58	0.89	MV	7.42	4.94	0.90

Mulero et al. [26] concluded that the approximation of using HS EOSs together with the VW formula for the molecular diameter is very accurate at temperatures and densities near the Lennard–Jones critical point. At densities close to the triple point or to the liquid curve in the liquid–vapor equilibrium, the agreement is adequate only if an appropriate expression for the HS EOS is used.

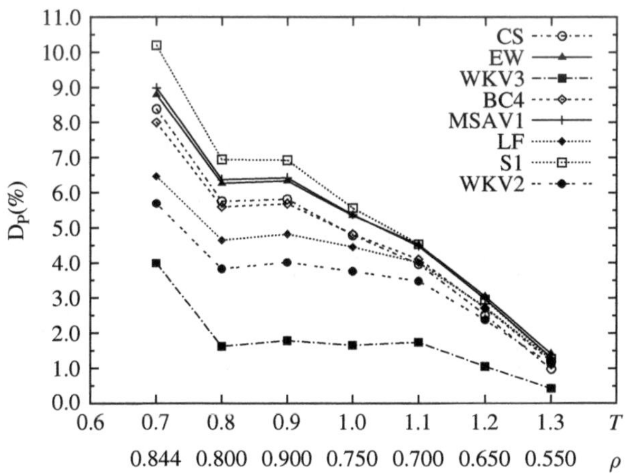

Fig. 4.1. Individual absolute percentage deviations in the calculation of pressure for the three-dimensional WCA RS performed with a selection of EOSs, at each thermodynamic state of Zone II (see Table 4.1)

4.3.2 Results over the Whole Fluid Range in the Phase Diagram

Let us now consider the results of Cuadros et al. [5, 14] for the compressibility factor of the WCA RS of Lennard–Jones fluids and for a total of 269 thermodynamic states over the wide range of temperatures and densities shown in Fig. 4.2.

Taking the aforementioned computer simulation values as reference, Mulero et al. [25] carried out a complete analysis of the validity of six analytical expressions for the EOS of an HS system in order to reproduce the pressure (and potential energy, as will be explained in the next section) of the WCA RS of Lennard–Jones fluids. Their main results are presented in the following paragraphs.

For each thermodynamic state the values of P (and E) were calculated [25] using the expressions CS, K, S1, WKV1, WKV2, and WKV3 from Table 3.11 in Chap. 3, where the corresponding packing fractions were obtained using the VW and the KN expressions for the molecular diameter, Eqs. (4.5) and (4.6), respectively. The values the authors found for $\mathrm{AAD_P}$ are listed in Table 4.3, not only for the whole range, but also for certain specific temperature ranges (all densities being considered) or density ranges (all temperatures being considered).

As can be seen, all the expressions for the EOS yielded good results, $\mathrm{AAD_P} < 3\%$ for the total range (all temperatures and densities shown in Fig. 4.2), with WKV1, WKV2, and WKV3 being the most accurate. Also the very simple CS EOS gave a deviation smaller than those obtained with more complex equations (such as K or S1) in all cases. The authors pointed out that there is no direct correlation between complexity and accuracy for the EOSs.

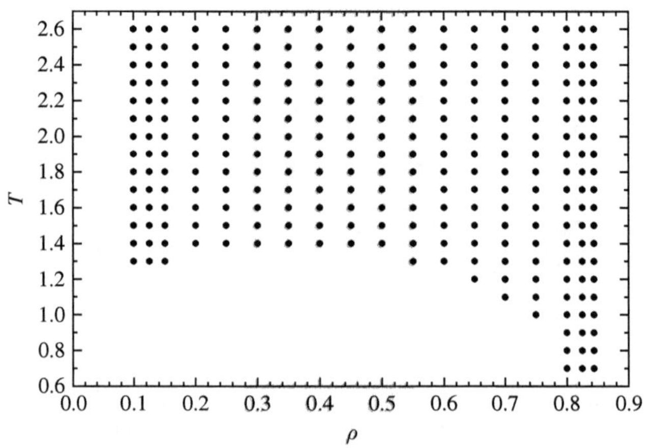

Fig. 4.2. Thermodynamic states selected in the computer simulation of the three-dimensional WCA RS [5, 14]

Table 4.3. Average absolute percentage deviations (AAD_P) of the values of pressure for the three-dimensional WCA RS obtained from EOSs with respect to those obtained by computer simulation. Different ranges of temperatures and densities were considered

Range	CS	K	S1	WKV1	WKV2	WKV3
Total (VW)	2.45	2.62	2.82	2.02	2.11	1.26
Total (KN)	2.70	2.87	3.07	2.27	2.35	1.50
$\rho \leq 0.35$ (VW)	0.37	0.39	0.39	0.39	0.39	0.38
$0.35 < \rho \leq 0.65$ (VW)	1.47	1.70	1.73	1.49	1.58	1.03
$\rho > 0.65$ (VW)	5.41	5.68	6.23	4.16	4.31	2.36
$T \leq 1.2$ (VW)	6.18	6.40	7.32	4.52	4.45	2.41
$T > 1.2$ (VW)	2.08	2.25	2.38	1.78	1.88	1.15

Mulero et al. [25] noted that, as one could expect, although the VW diameter gave smaller deviations, no significant differences were found when the KN molecular diameter was used. This conclusion will also apply for two-dimensional systems, as will be described in Sect. 4.5.

Table 4.3 also shows that there were significant differences in accuracy for different density ranges, the deviations being greater at high densities. As expected, for very low densities ($\rho \leq 0.35$), all the EOSs gave excellent results ($AAD_P \leq 0.4\%$, when the VW diameter was used) without meaningful differences. For intermediate densities ($0.35 < \rho \leq 0.65$), the WKV3 and CS equations yielded the most accurate results. The authors found the most significant deviations at high densities, for which WKV3 gave the most accurate results, with a significant difference with respect to the others.

Some authors [39, 54] suggested that results at high densities could be improved by including also the density dependence in the expression for the molecular diameter. Nevertheless, this dependence is too small to change the results significantly [15]. Moreover, including the density dependence in the EOSs would lead to analytical expressions that are excessively complex to be used for practical purposes.

With respect to the influence of temperature on the accuracy of the EOSs, Mulero et al. [25] observed that at low temperatures ($T \leq 1.2$) the deviations obtained are greater than 5% except for WKV1, WKV2, and WKV3, finding the best results for WKV3. At higher temperatures, this expression was again the most accurate; other good choices were the other two equations proposed by the same authors (WKV1 and WKV2), and even the simple and well-known CS equation. The fact that this high temperature range corresponds approximately to the supercritical range of a Lennard–Jones fluid makes it of special interest, since having a good reference in this range becomes important for further applications.

More recently, Liu [21] pointed out that accurate results for the three-dimensional WCA RS can also be achieved by using the Lado theory diameter, which includes a density dependence in its analytical form [50], instead of the

VW diameter. Nevertheless, he noted that in the analysis of these results it must be taken into account that the error due to the use of an approximate EOS may be canceled by that due to the use of an approximate expression for the effective diameter, hence leading to accurate final values.

4.4 Potential Energy of the Three-Dimensional WCA RS

We report in this section the most significant results achieved by Mulero et al. [25] in their study of the validity of six EOSs (the same as those considered in Sect. 4.3.2) to reproduce computer simulation data of potential energy of the WCA RS of Lennard–Jones fluids, taking as reference the computer simulation values also referred to in that section [5, 14].

As in the case of the pressure, values for the average absolute percentage deviations for the potential energy (AAD_E) taken from [25] are listed in Table 4.4.

As can be seen, great differences were found when expressions VW or KN for the effective molecular diameter were used [25]. For all ranges of densities (all temperatures being considered) or low and intermediate temperature ranges, the use of the KN molecular diameter was reported as leading to more accurate results. This is obviously due to the fact that it also fits the derivative of the molecular diameter, whereas the derivative of the VW diameter succeeds only at high temperatures. Nevertheless, for the highest temperatures ($T > 2.2$), both expressions for the molecular diameter show similar accuracy, with VW giving the smallest deviation (see Table 4.4).

In particular, when the complete range is considered, none of the selected EOSs gave $AAD_E < 3\%$ if the VW diameter was used, whereas KN gave deviations below that value for all the EOSs. Expressions WKV3, WKV1, and WKV2 showed, in this order, the smallest deviations. Mulero et al. [25]

Table 4.4. Average absolute percentage deviations of the values of potential energy (AAD_E) for the three-dimensional WCA RS obtained from EOSs with respect to those obtained by computer simulation. Different ranges of temperatures and densities were considered

Range	CS	K	S1	WKV1	WKV2	WKV3
Total (VW)	4.22	4.44	4.60	3.91	4.04	3.39
Total (KN)	1.98	2.21	2.43	1.66	1.76	1.65
$\rho < 0.6$ (KN)	1.98	2.13	2.13	2.08	2.11	1.91
$0.6 \le \rho \le 0.7$ (KN)	1.72	2.09	2.21	1.38	1.64	0.76
$\rho > 0.7$ (KN)	2.13	2.42	3.12	1.00	1.17	1.65
$T \le 1.3$ (KN)	1.32	1.54	2.32	0.70	0.75	2.41
$1.3 < T \le 2.2$ (KN)	1.98	2.31	2.37	1.69	1.80	1.56
$T > 2.2$ (VW)	1.84	2.07	2.18	1.61	1.73	1.40
$T > 2.2$ (KN)	2.29	2.52	2.63	2.05	2.18	1.46

reported CS as being a good choice, whereas values calculated with equation S1 were the least accurate.

Table 4.4 also lists $\mathrm{AAD_E}$ for different ranges of densities. In this case, only the KN diameter was considered [25]. At low densities ($\rho < 0.6$), all the EOSs gave $\mathrm{AAD_E}$ near 2%, without significant differences. In the range $0.6 \leq \rho \leq 0.7$, expression WKV3 was the only one with $\mathrm{AAD_E} < 1\%$. The WKV1 equation and the simple CS were reported as being adequate choices, whereas expressions K and S1 gave again the greatest deviations. For high densities, WKV1, WKV2, and WKV3 gave the best results (in particular WKV1, for which $\mathrm{AAD_E} = 1\%$). All the other EOSs gave deviations greater than 2%, CS being one of the most accurate among them.

According to Mulero et al. [25], when ranges of constant temperature were considered, it was observed that the best results were achieved for low temperatures ($T \leq 1.3$), using the KN molecular diameter. In this range, only equations WKV1 and WKV2 gave $\mathrm{AAD_E} < 1\%$. The simple CS and K gave good results, whereas WKV3 surprisingly gave the greatest deviation in this range. At intermediate temperatures ($1.3 < T \leq 2.2$), the KN diameter was also recommended. In this range, expressions WKV1, WKV2, WKV3, and CS gave $\mathrm{AAD_E} < 2\%$, the best results corresponding to WKV3. Finally, for high temperatures, expression VW for the molecular diameter produced lower average deviations than KN. The best results were also obtained using equation WKV3 with either of the two expressions for the molecular diameter.

4.5 Pressure and Potential Energy of the Two-Dimensional WCA RS

The scaling procedure from the HD system to the two-dimensional WCA RS was first reviewed by Bishop [43]. He used his own molecular dynamics simulation results for the compressibility factor as reference to test the accuracy of the SPT and the seven-term virial EOSs. The VW effective diameter was used in both cases. Nevertheless, only data for reduced temperatures near 0.95 were considered, and therefore, the possible temperature dependence of the procedure was not taken into account. Results for dimensions higher than two and three were also obtained by Bishop et al. [42, 44, 45].

Mulero et al. [22] performed a more detailed study of the validity of a set of six EOSs for the HS system to reproduce computer simulation data for both the pressure and potential energy of the WCA RS of two-dimensional Lennard–Jones fluids over a wide range of temperatures and densities. They computed their own molecular dynamic simulation data as a reference, and then calculated those properties by using a scaling procedure through the temperature dependence of the effective molecular diameter, considering the expressions VW and KN, Eqs. (4.5) and (4.6), respectively. Their main results will be described in this section.

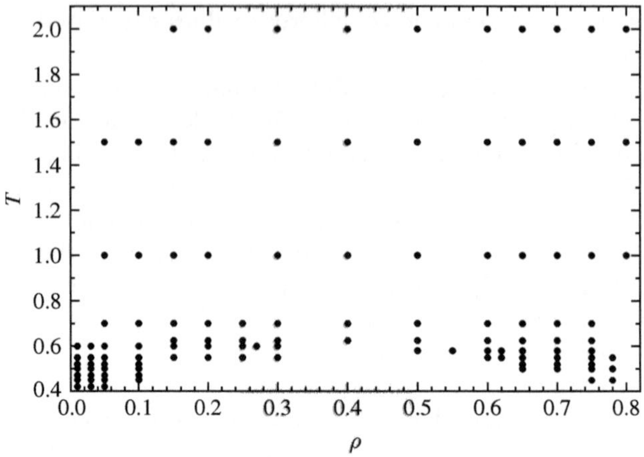

Fig. 4.3. Thermodynamic states selected for the computer simulation of the two-dimensional WCA RS [15, 22]

Computer simulation data were obtained for the wide range of temperatures and densities shown in Fig. 4.3 [15, 22]. The computational system was isochoric and adiabatic, and hence isoenergetic (isolated). It contained $N = 256$ particles (16×16) with initial velocities randomly selected from a Maxwell distribution. The repulsive part of the Lennard–Jones potential (Eq. (4.1)) was used, having been computed for the interparticle forces by the particle neighbor table method [55]. The classical Newtonian equations of motion were then numerically integrated using Verlet's algorithm. The usual periodic boundary conditions were used and, finally, the average values of compressibility factor and potential energy were calculated as

$$Z_{\text{sim}} = 1 + 6T^{-1}N^{-1}\left\langle \sum_{i \neq j}(2r_{ij}^{-12} - 6r_{ij}^{-6}) \right\rangle \tag{4.10}$$

$$E_{\text{sim}} = \frac{1}{2N}\left\langle \sum_{i \neq j} U(r_{ij}) \right\rangle \tag{4.11}$$

Results for the 89 thermodynamic states shown in Fig. 4.3 are listed in Table 4.5.

For each thermodynamic state, Mulero et al. [22] calculated values for Z and E of the two-dimensional Lennard–Jones WCA RS, using expressions SPT, H75, H77, Maeso et al. (MSAV), Sanchez (S), and SHY (Eqs. (4.8), (4.11), (4.12), and (4.33) to (4.35), respectively, in Chap. 3, Sect. 3.2.2). The results were compared with the computer simulation values by calculating the average of the percentage deviations (D) and the average of the absolute percentage deviations (AAD) for Z and E, respectively. Values of AAD were calculated for the complete range and also for fixed temperature ranges (all

Table 4.5. Computer simulation results of the pressure and the potential energy of the two-dimensional WCA RS [15, 22]

T	ρ	P_{sim}	E_{sim}	T	ρ	P_{sim}	E_{sim}
2.0	0.15	0.383	0.047	0.625	0.25	0.248	0.023
2.0	0.20	0.557	0.066	0.625	0.30	0.338	0.032
2.0	0.30	1.027	0.121	0.625	0.40	0.572	0.052
2.0	0.40	1.676	0.187	0.625	0.50	0.947	0.082
2.0	0.50	2.638	0.282	0.625	0.60	1.524	0.123
2.0	0.60	4.124	0.424	0.625	0.65	1.955	0.156
2.0	0.65	5.151	0.517	0.625	0.70	2.549	0.200
2.0	0.70	6.532	0.647	0.625	0.75	3.370	0.261
2.0	0.75	8.229	0.800	0.60	0.10	0.072	0.0073
2.0	0.80	10.36	0.985	0.60	0.15	0.118	0.012
1.5	0.10	0.177	0.021	0.60	0.20	0.174	0.017
1.5	0.15	0.290	0.034	0.60	0.25	0.242	0.023
1.5	0.20	0.424	0.049	0.60	0.30	0.322	0.030
1.5	0.30	0.776	0.087	0.58	0.50	0.868	0.073
1.5	0.40	1.295	0.139	0.58	0.55	1.122	0.093
1.5	0.50	2.067	0.213	0.58	0.60	1.423	0.114
1.5	0.60	3.255	0.320	0.58	0.62	1.571	0.125
1.5	0.65	4.072	0.391	0.58	0.65	1.860	0.147
1.5	0.70	5.164	0.487	0.58	0.70	2.397	0.185
1.5	0.75	6.581	0.609	0.58	0.75	3.145	0.238
1.5	0.80	8.457	0.769	0.55	0.10	0.065	0.0064
1.0	0.10	0.118	0.013	0.55	0.15	0.108	0.010
1.0	0.15	0.196	0.022	0.55	0.20	0.160	0.015
1.0	0.20	0.284	0.030	0.55	0.25	0.221	0.020
1.0	0.30	0.528	0.055	0.55	0.30	0.299	0.027
1.0	0.40	0.884	0.088	0.55	0.60	1.371	0.109
1.0	0.50	1.434	0.136	0.55	0.62	1.510	0.119
1.0	0.60	2.294	0.209	0.55	0.65	1.774	0.139
1.0	0.65	2.908	0.259	0.55	0.70	2.301	0.175
1.0	0.70	3.713	0.323	0.55	0.75	3.044	0.228
1.0	0.75	4.796	0.410	0.55	0.78	3.602	0.267
1.0	0.80	6.236	0.523	0.52	0.10	0.059	0.0057
0.70	0.10	0.083	0.0083	0.52	0.65	1.682	0.129
0.70	0.15	0.137	0.014	0.52	0.70	2.182	0.164
0.70	0.20	0.201	0.020	0.52	0.75	2.896	0.213
0.70	0.25	0.282	0.028	0.50	0.10	0.059	0.0055
0.70	0.30	0.374	0.036	0.50	0.65	1.638	0.124
0.70	0.40	0.638	0.059	0.50	0.70	2.113	0.157
0.70	0.50	1.036	0.091	0.50	0.75	2.798	0.204
0.70	0.60	1.680	0.141	0.50	0.78	3.307	0.238
0.70	0.65	2.179	0.180	0.47	0.10	0.056	0.0054
0.70	0.70	2.792	0.225	0.45	0.10	0.054	0.0051
0.70	0.75	3.656	0.290	0.45	0.75	2.571	0.183
0.625	0.15	0.123	0.012	0.45	0.78	3.066	0.215
0.625	0.20	0.181	0.018				

Table 4.6. Average absolute percentage deviations (AAD) and mean percentage deviations (D) for the values obtained from some of the EOSs considered in Sect. 3.2.2 of Chap. 3, and from expressions VW and KN for the effective molecular diameter (Eqs. (4.5) and (4.6)), considering the complete temperature and density range ($0.45 \leq T \leq 2.0$ and $0.1 \leq \rho \leq 0.8$). The molecular dynamics computer simulation data given in Table 4.5 were taken as reference

Diameter		SPT	H75	H77	MSAV	S	SHY
Verlet–Weis (VW)	D_Z	−0.41	1.99	1.39	1.34	1.59	0.68
	AAD_Z	0.88	2.15	1.56	1.51	1.77	1.32
Kolafa–Nezbeda (KN)	D_Z	−0.11	2.31	1.71	1.65	1.92	1.00
	AAD_Z	0.86	2.44	1.84	1.79	2.05	1.52
Verlet–Weis (VW)	D_E	−10.1	−7.24	−7.87	−7.93	−7.67	−8.96
	AAD_E	10.9	9.69	9.90	9.92	9.82	10.3
Kolafa–Nezbeda (KN)	D_E	−1.25	1.98	1.26	1.20	1.50	0.10
	AAD_E	2.65	2.31	1.74	1.71	1.92	1.93

densities being considered) or density ranges (all temperatures being considered). The values obtained for those deviations are listed in Tables 4.6–4.8.

If the whole range is considered, Table 4.6, one observes that, although all the EOSs gave average absolute percentage deviations for Z (AAD_Z) less than the computer simulation errors, the simple equation SPT gave the best approximation to the computer simulation values, with the AAD_Z being less than 1% [22]. Moreover, the mean percentage deviations for Z (D_Z) are very small. No significant differences are found for the expressions H77, MSAV, S, and SHY, although the last one also gave good accuracy, whereas expression H75 yielded the greatest AAD_Z. From this, the authors pointed out that there seemed to be no direct relationship between simplicity and accuracy among the EOSs studied in their work. As a consequence of the accuracy of the analytical expressions, no significant differences were found with respect to the use of the VW and KN effective molecular diameters.

The results of Mulero et al. [22] also showed that the accuracy of the EOSs in the calculation of Z may be different for particular temperature and density ranges. For example, the greatest deviations were found at $T = 1$ and $\rho = 0.8$, being near 3.5% for expression SPT (the computer simulation error being equal to 2.1%), whereas deviations from 6% to 9% were found for the rest of the EOSs. Also, taking for instance $T = 1$ and $\rho = 0.6$, the deviation of equation SHY with respect to the computer simulation value was less than 0.05%, being between 1% and 2% for the rest of the EOSs. Table 4.7 lists the AAD_Z for different temperature or density ranges. As can be seen, the values were of the same order for low and high temperatures, whereas major differences in accuracy were found for different density ranges (see Fig. 4.4). At the lowest densities ($\rho < 0.3$), where the computer simulation errors for Z are always below 1.5%, all the EOSs behaved quite accurately ($AAD_Z < 0.45\%$).

Table 4.7. Average absolute percentage deviations in the calculation of Z (AAD_Z) for different temperature or density ranges

Range	$d(T)$	SPT	H75	H77	MSAV	S	SHY
$2.0 \geq T \geq 1.0$	VW	0.92	2.39	1.90	1.85	2.05	1.52
	KN	0.93	2.49	1.99	1.95	2.14	1.60
$0.70 \geq T \geq 0.45$	VW	0.87	2.01	1.37	1.32	1.61	1.21
	KN	0.82	2.41	1.76	1.70	2.00	1.48
$0.1 \leq \rho < 0.3$	VW	0.43	0.40	0.40	0.40	0.40	0.42
	KN	0.42	0.40	0.40	0.40	0.40	0.42
$0.3 \leq \rho < 0.6$	VW	1.24	0.51	0.46	0.46	0.46	0.95
	KN	1.13	0.55	0.49	0.48	0.49	0.85
$0.6 \leq \rho < 0.7$	VW	1.13	2.24	1.61	1.55	1.71	0.49
	KN	0.80	2.63	1.99	1.93	2.09	0.55
$0.7 \leq \rho < 0.8$	VW	0.92	5.20	3.62	3.48	4.27	3.18
	KN	1.15	5.94	4.32	4.18	5.00	3.92

At intermediate densities (from 0.3 to 0.6), SPT and SHY gave the poorest accuracy, whereas all the other equations presented similar deviations (see Fig. 4.4a). At high densities ($0.6 \leq \rho < 0.7$), equation SHY showed the most accurate behavior in the calculation of Z, with a significant difference with respect to the others (Fig. 4.4b). At very high densities ($0.7 \leq \rho < 0.8$), expression SPT yielded the most accurate results (Fig. 4.4c). Finally, Mulero et al. [22] noted that in the case of equation SPT the deviation is practically the same for high or very high densities, whereas for expression SHY the deviation is seven times greater at high densities than at very high ones.

To sum up, in order to calculate Z in the two-dimensional Lennard–Jones WCA RS, Mulero et al. [22] proposed using the SPT equation, except if the calculation is to be done only at intermediate densities, for which expressions H75, H77, M, and S behave more accurately. The use of equation SHY is regarded as being also a good choice for all the ranges considered, especially at high (but not for very high) densities. The aforementioned authors found no significant differences in the choice of expressions VW or KN for the effective molecular diameter, and hence proposed the first for the obvious criterion of simplicity.

With respect to the calculation of potential energy through Eq. (4.8), great differences were found when VW or KN diameters were used (see Table 4.8), except for the range of high temperatures [22]. Obviously, this is due to the fact that the KN diameter fits the derivative of the molecular diameter and, as shown in Fig. 4.5, the derivative of the VW diameter behaves adequately only at high temperatures ($T \geq 1$).

In particular, when the VW diameter was used, and all the ranges were considered, none of the selected EOSs gave mean deviations less than the computer simulation error. The values obtained with Eq. (4.8) were always lower than those generated in the computer simulation. Good agreement was

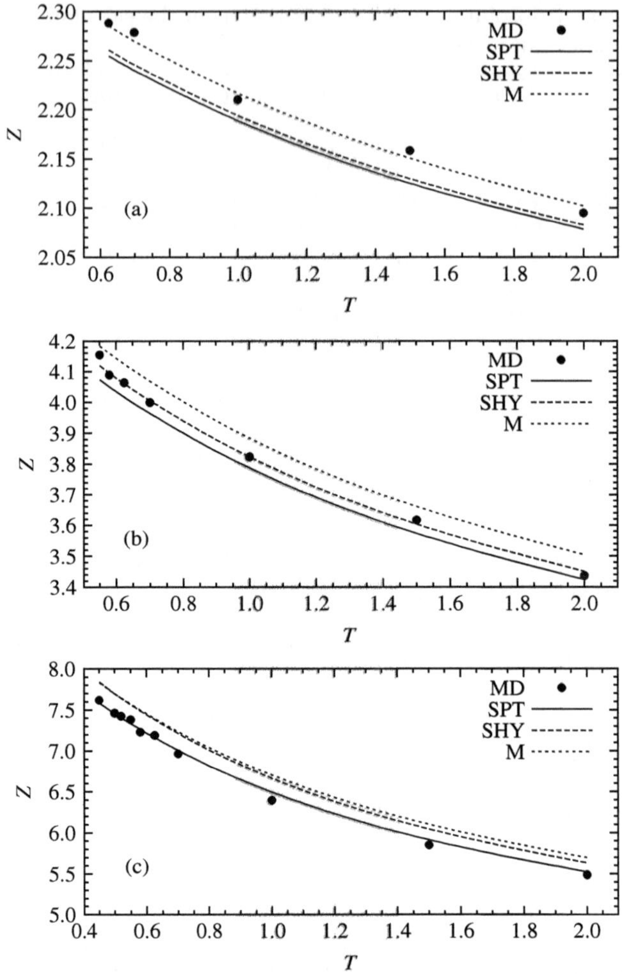

Fig. 4.4. Compressibility factor vs temperature obtained from molecular dynamics (MD) computer simulation (*dots*) and from some of the studied analytical EOSs (*lines*). Note that the VW effective molecular diameter was used in all cases. (**a**) $\rho = 0.4$; (**b**) $\rho = 0.6$; (**c**) $\rho = 0.75$

found only at high temperatures $T \geq 1$, for which all the AAD_E were less than 4%, the smallest deviations being for the SPT and SHY equations. Despite this, deviations greater than 4% were found for some particular states in this range, even for the equation SPT (for $T = 1.5$ and $\rho = 0.1$, for example). At low temperatures ($T \leq 0.7$), none of the selected EOSs gave acceptable results. When density ranges were considered (for all temperatures), the lowest deviations were found for intermediate densities ($0.3 \leq \rho < 0.6$), the AAD_E being quite similar for all the selected EOSs (and always below 8.5%). At low

Table 4.8. Average absolute percentage deviations in the calculation of E (AAD_E) for different temperature or density ranges

Range	$d(T)$	SPT	H75	H77	MSAV	S	SHY
$2.0 \geq T \geq 1.0$	VW	2.04	3.50	2.96	2.92	3.12	2.33
	KN	2.69	1.41	0.98	0.97	1.12	1.75
$0.70 \geq T \geq 0.45$	VW	15.9	13.2	13.8	13.9	13.6	14.7
	KN	2.63	2.81	2.16	2.12	2.37	2.03
$0.1 \leq \rho < 0.3$	VW	10.1	9.68	9.69	9.69	9.69	10.0
	KN	2.66	3.16	3.15	3.15	3.15	2.69
$0.3 \leq \rho < 0.6$	VW	8.17	7.29	7.36	7.37	7.36	7.96
	KN	2.46	1.05	1.05	1.05	1.05	2.01
$0.6 \leq \rho \leq 0.8$	VW	12.7	10.8	11.2	11.2	11.0	11.5
	KN	2.73	2.33	1.15	1.09	1.53	1.42

densities ($\rho < 0.3$), the mean deviations were close to 10%. Finally, at high densities, the values of AAD_E were always greater than 10%, the greatest values corresponding to the highest densities. These results were due to the fact that Verlet and Weis [39] did not consider the derivative of the molecular diameter in their fit. Because of this, Mulero et al. [22] concluded that expression VW for the molecular diameter should not be used to calculate the potential energy in the WCA RS of two-dimensional Lennard–Jones fluids.

When the KN molecular diameter is used, Eq. (4.6), the average deviations with respect to computer simulation results were always less than 3.5%, the greatest deviation being near 9.5% for $T = 0.5$ and for all the EOSs. The lowest D_E was found for expression SHY, whereas the lowest AAD_E was

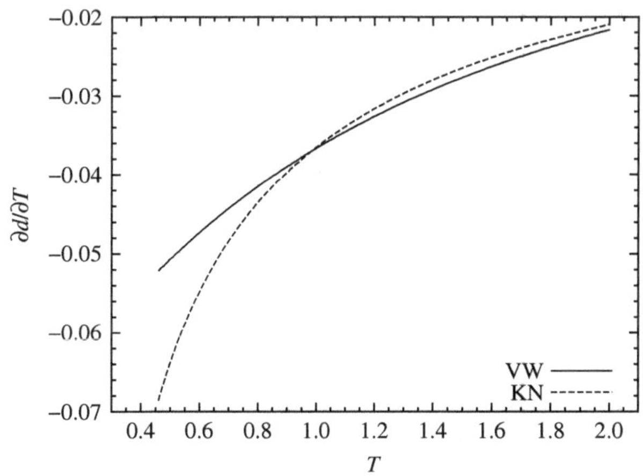

Fig. 4.5. Derivative of the effective molecular diameter vs temperature for expressions VW and KN

that given by equations MSAV or H77. Unfortunately, the SPT EOS, despite its successful behavior in the calculation of Z, gave the worst result in the determination of E.

At high temperatures, the results were significantly better than in the low temperature case, except for equations SPT and SHY. Hence, the use of expressions MSAV or H77 was suggested at high temperatures, whereas for low temperatures (when deviations are similar for the different EOSs) Mulero et al. [22] suggested the use of equation SHY because of both its analytical simplicity and the successful results achieved.

The use of equations SPT or SHY gave the smallest deviations for low densities, but the largest at intermediate densities. For these intermediate values of the density, all the other EOSs could be used with little error. At high or very high densities, equations MSAV and H77 were reported as being the best behaved.

In conclusion, expression KN for the effective molecular diameter should be used to calculate the potential energy of the WCA RS [22]. Equation MSAV appears to be the best choice for calculations where different ranges of temperature or density need to be considered. If calculations are confined to low temperatures or low densities, then equation SHY is regarded as yielding the most accurate results.

Sometimes Z and E need to be determined simultaneously. When complete the range of temperatures and densities was considered, equations SHY, SPT, H77, and MSAV showed similar accuracies, although the use of the first two was suggested because of their analytical simplicity. At high temperatures, Mulero et al. [22] suggested equations H77 and MSAV, together with the KN diameter, or equation SPT with the VW diameter. For other ranges, the KN molecular diameter should always be used. Equation SPT was recommended at low temperatures, and also at low or very high densities. Equation SHY was a good choice for low temperatures, as also for low or high (not very high) densities. Expressions MSAV and H77 gave the best results at high temperatures and also at intermediate or high densities. Finally, the use of equations H75 or S was not recommended [22], since they gave accurate results only for particular values of temperatures or densities.

4.6 Conclusions

We shall now describe the most significant conclusions obtained in the study of the validity of several expressions for the EOS of the HS system to reproduce the pressure of the WCA RS of Lennard–Jones fluids [26].

In the calculation of this property using HS EOSs together with VWs' diameter, significant differences are found for each zone of the phase diagram. Thus, the greatest deviations are found at very high densities. For supercritical states near the critical point, practically all the EOSs gave good results without meaningful differences. Equation WKV3 (using expression VW for the effective molecular diameter) seemed to be the most suitable to obtain an accurate representation of the thermodynamic behavior of the WCA repulsive Lennard–Jones system for states near the phase transitions. The use of expressions WKV1 and WKV2, or any more complex equations, is not needed in these ranges. In particular, the complex equation A only gave good results over a small density range and, due to its analytical form, Mulero et al. [26] do not recommend its use unless if very accurate results are needed in that specific density range.

As a general conclusion, Mulero et al. [26] noted that the use of most complex analytical expressions in phase-transition calculations is not necessary. They observed that there does not seem to be a direct relationship between the complexity and the accuracy of the equations. Thus, the simple equations of Carnahan–Starling (CS) turned out to be adequate in many cases. The equation WKV3 is highly recommended for the calculation of the pressure of the WCA RS. However, certain simpler equations also gave good results. More complex modifications of that equation did not improve the results significantly (except, maybe, for particular temperature and density ranges). Finally, the approximation of using an HS EOS together with VWs, formula for the effective molecular diameter led to satisfactory results (deviations below 3%) only if the appropriate expression for the compressibility factor was selected.

Mulero et al. [25] proposed the use of expression WKV3, together with VW's molecular diameter, Eq. (4.5), to calculate the pressure of the WCA RS, at least if calculations are needed for ranges of $T < 2.7$ and $\rho < 0.85$. According to these authors, none of the other EOSs should be used at low temperatures ($T \leq 1.2$) and high densities ($0.844 \geq \rho > 0.65$) if accuracy is required. When only low or intermediate densities are considered ($\rho < 0.65$), the simple and well-known CS EOS can be used successfully giving good results also for high temperatures. The use of more complex expressions is not needed in these ranges. In particular, the complex equation S1 gives good results only for low densities, but in general gives the poorest results [25] of the EOSs considered in Sect. 4.3.2.

According to the calculation of the potential energy of the WCA RS of Lennard–Jones fluids, Mulero et al. [25] proposed the use of equation WKV3 together with expression KN for the molecular diameter, except at low temperatures ($T \leq 1.3$) or at high densities ($0.7 < \rho < 0.844$), when WKV1 is more accurate. Equation CS must be regarded as being a good choice as well, especially at low densities where it gives practically the same deviation than WKV3, or at low temperatures ($T \leq 1.3$), where it is more accurate than WKV3. The use of some modified expressions of CS, such as expression K, or more complex equations, like S1, does not imply an improvement in the calculation of the potential energy. Mulero et al. [25] recommended expression KN for the molecular diameter, except at very high temperatures ($2.2 \leq T \leq 2.6$). Nevertheless, expression WKV3 becomes a good choice together with either the VW or KN expressions for the diameter in this range without significant differences.

It must be pointed out as a general conclusion that the use of some complex analytical EOSs can only be justified to compute certain thermodynamic properties, or when calculations have to be done in fixed ranges of temperatures and densities. There seemed to be no direct relationship between complexity and accuracy in the EOSs. It was also shown that the volume dependence of the HS EOSs given by expressions WKV1, WKV2, and WKV3 improves the results significantly, especially those for pressure at high densities ($0.844 \geq \rho > 0.65$) or low temperatures ($T \leq 1.2$). The advantage of using one of the other expressions for the molecular diameter in the calculation of pressure and potential energy, respectively, of the WCA RS of Lennard–Jones fluids is clearly justified.

Let us now focus on the calculation of the compressibility factor (pressure) and the potential energy of the WCA RS of two-dimensional Lennard–Jones fluids. Again, according to the study of Mulero et al. [22], there is no direct correlation between the simplicity of the analytical expressions and their accuracy. In general, good results for the compressibility factor (pressure) were found throughout the studied range, the smallest deviations corresponding to low temperatures ($T < 0.7$) or low densities ($\rho < 0.3$), and independently of the expression used for the molecular diameter. Nevertheless, when the VW diameter is used, good results for E were found only at high temperatures ($T \geq 1$), whereas excessively high deviations were found for most of the thermodynamic states considered. The use of the KN diameter gave significantly better results, with deviations less than the computer simulation errors. Mulero et al. [22] noted that in the calculation of E through Eq. (4.8) not only has the expression for the effective molecular diameter great relevance, but also its derivative with respect to temperature.

When the complete range of temperatures and densities was considered, calculation with equation SPT gave the best approximation to the compressibility or the pressure of the system, whereas expressions MSAV or H77 gave the best results for the potential energy. When all the properties were calculated simultaneously, equation SHY also became a good choice. In this case, Mulero et al. [22] indicated that the use of equations MSAV or H77 had the advantage of giving the same deviations in Z and in E.

When particular ranges of temperatures and/or densities are considered, the accuracy of the EOSs may differ. In this sense, equation SHY, which is a very good approximation for the calculation of Z and E in the complete range, gave the best approximation for low or high densities, although noticeable results at the highest densities. Equations H77, MSAV, and S presented similar accuracy, giving the best results at intermediate densities ($0.3 \leq \rho < 0.6$). Finally, equation H75 must be used only at low or intermediate densities ($\rho < 0.6$), where it gave deviations of the same order as the other EOSs.

Finally, Mulero et al. [22] reported that expressions SPT or SHY yielded, in general, more accurate results than S to calculate the properties of the WCA RS. Again, as was the case for the WCA RS of three-dimensional Lennard–Jones fluids, the large deviations found for SPT or other EOSs at the highest densities can be reduced if an appropriate dependence of the molecular diameter on density is used.

Acknowledgement

This work was supported by project FIS2006-02794 of the "Ministerio de Educación y Ciencia" of Spain and the "Fondo Europeo de Desarrollo Regional" (FEDER) of the European Union.

References

1. J. P. Hansen, I. R. McDonald: *Theory of Simple Liquids* (Academic Press, London 1976)
2. J. D. Weeks, D. Chandler, H. C. Andersen: J. Chem. Phys. **54**, 5237 (1971)
3. J. D. Weeks, D. Chandler, H. C. Andersen: J. Chem. Phys. **55**, 5422 (1971)
4. F. Cuadros, A. Mulero, P. Rubio: J. Chem. Educ. **71**, 959 (1994)
5. F. Cuadros, A. Mulero, W. Okrasinski, W. Ahumada: Int. Rev. Phys. Chem. **16**, 141 (1997)
6. F. Cuadros, A. Mulero, J. O. Valderrama: Chem. Eng. Sci. **48**, 513 (1993)
7. Y. S. Wei, R. J. Sadus: AIChE J. **46**, 169 (2000)
8. E. A. Müller, K. E. Gubbins: Ind. Eng. Chem. Res. **40**, 2193 (2001)
9. B. E. Poling, J. M. Prausnitz, J. P. O'Connell: *The Properties of Gases and Liquids.* (McGraw-Hill, New York 2001)
10. J. O. Valderrama: Ind. Eng. Chem. Res. **42**, 1603 (2003)
11. F. L. Román, A. Mulero, F. Cuadros: Phys. Chem. Chem. Phys. **6**, 5402 (2004)
12. A. Mulero, I. Cachadiña: Int. J. Thermophys. **28**, 279 (2007)
13. F. Cuadros, A. Mulero: Chem. Phys. **177**, 53 (1993)
14. F. Cuadros, A. Mulero, W. Ahumada: Termochim. Acta **277**, 85 (1996)
15. F. Cuadros, A. Mulero: Chem. Phys. **156**, 33 (1991)
16. F. Cuadros, A. Mulero: Chem. Phys. **160**, 375 (1992)
17. F. Cuadros, A. Mulero: J. Chem. Phys. **99**, 419 (1995)
18. F. Cuadros, A. Mulero, W. Ahumada: Mol. Phys. **85**, 207 (1995)
19. F. Cuadros, A. Mulero, C.A. Faúndez: Mol. Phys. **98**, 899 (2000)
20. F. Cuadros, A. Mulero, W. Okrasinski, C.A. Faúndez, M.I. Parra: Rev. Chem. Eng. **19**, 387 (2003)
21. H. Liu: arXiv:cond-matt/0605392 available at URL http://arxiv.org/archive/cond.mat (2006)
22. A. Mulero, F. Cuadros, C. A. Galán: J. Chem. Phys. **107**, 6887 (1997)
23. A. Mulero, C. A. Faúndez, F. Cuadros: Mol. Phys. **97**, 453 (1999)
24. A. Mulero, C. A. Faúndez, M. I. Parra, F. Cuadros: Thermochim. Acta **334**, 1 (1999)

25. A. Mulero, C. A. Galán, F. Cuadros: J. Chem. Phys. **111**, 1 (1999)
26. A. Mulero, C. A. Galán, F. Cuadros: Phys. Chem. Chem. Phys. **3**, 4991 (2001)
27. J. D. Weeks, J. Q. Broughton: J. Chem. Phys. **78**, 4197 (1983)
28. F. Cuadros, W. Ahumada, A. Mulero: Chem. Phys. **204**, 41 (1996)
29. F. Cuadros, A. Mulero, W. Okrasinski: Physica A **214**, 162 (1995)
30. L. F. Rull, F. Cuadros, J. J. Morales: Phys. Lett. A **123**, 217 (1987)
31. J. O. Valderrama, A. Mulero, F. Cuadros: Chem. Phys. Lett. **166**, 437 (1990)
32. F. Cuadros, A. Mulero, W. Okrasinski: Physica A **223**, 321 (1996)
33. A. Mulero, F. Cuadros: J. Colloid and Interface Sci. **186**, 110 (1997)
34. F. Cuadros, A. Mulero: *Fluid Physics. Lecture Notes of Summer Schools.* Ed. M. G. Velarde, C. I. Christov (World Scientific, Singapore 1995)
35. A. Mulero, F. Cuadros: Chem. Phys. **205**, 379 (1996)
36. A. Mulero, F. Cuadros: Langmuir **12**, 3265 (1996)
37. F. Cuadros, A. Mulero, L. Morala, V. Gómez-Serrano: Langmuir **17**, 1576 (2001)
38. A. Mulero, L. Morala, V. Gómez-Serrano, F. Cuadros: *Encyclopedia of Surface and Colloid Science.* Ed. Dr. A. Hubbard (Marcel Dekker, New York 2002)
39. L. Verlet, J. J. Weis: Phys. Rev. A **5**, 939 (1972)
40. J. Kolafa, I. Nezbeda: Fluid Phase Equil. **100**, 1 (1994)
41. J. A. Barker, D. J. Henderson: Chem. Phys. **47**, 4714 (1967)
42. M. Bishop: Am. J. Phys. **52**, 158 (1984)
43. M. Bishop: J. Chem. Phys. **84**, 535 (1986)
44. M. Bishop: J. Chem. Phys. **88**, 5779 (1988)
45. M. Bishop, A. Masters, J. H. R. Clarke: J. Chem. Phys. **110**, 11449 (1999)
46. A. Mulero, F. Cuadros, C.A. Faúndez: Aust. J. Phys. **52**, 101 (1999)
47. F. Cuadros, W. Ahumada: Thermochim. Acta **297**, 109 (1997)
48. F. Cuadros, W. Okrasinski: Chem. Eng. Sci. **51**, 4189 (1996)
49. W. A. Steele: J. Chem. Phys. **65**, 5256 (1976)
50. F. Lado: Mol. Phys. **53**, 363 (1984)
51. T. P. Singh, S. K. Sinha: Pramana J. Phys. **25**, 733 (1985)
52. A. Lotfi, J. Vrabec, J. Fischer: Mol. Phys. **76**, 1319 (1992)
53. J. A. Barker, D. Henderson: Rev. Mod. Phys. **48**, 587 (1976)
54. H. S. Kang, C. S. Lee, T. Ree, F. H. Ree: J. Chem. Phys. **82**, 414 (1985)
55. M. P. Allen, D. J. Tildesley: *Computer Simulation of Liquids.* (Ed. Oxford University Press, Oxford 1987)

5

Binary Mixtures of Additive Hard Spheres. Simulations and Theories

C. Barrio[1] and J.R. Solana[2]

[1] Departamento de Matemáticas, Universidad Carlos III, 28911 Leganés (Madrid), Spain.
cbarrio@math.uc3m.es
[2] Departamento de Física Aplicada, Universidad de Cantabria, 39005 Santander, Spain.
ramon.solana@unican.es

In this chapter we review our present knowledge on the thermodynamic and structural properties of additive hard-sphere fluid mixtures, as revealed by computer experiments and some exact relationships, with special emphasis on the equation of state and other related properties. In the first place, we summarize the available simulation data for these properties. The phase diagram and the demixing transition, as revealed by computer simulations, are analyzed with some detail. Next we review a number of consistency conditions, that is exact conditions that must satisfy the equation of state, the contact values of the partial radial distribution functions, and other thermodynamic properties. One section is devoted to the virial coefficients, including exact results, numerical data, and approximate expressions derived on the basis of several consistency conditions. Two more sections deal with some applications of the consistency conditions. One of them illustrates the usefulness of the consistency conditions to derive improved analytical expressions for the equation of state and the contact values of the partial radial distribution functions by discussing several proposals reported in the literature. Another section uses the consistency conditions to derive density expansions of the contact values of the partial radial distribution functions on the basis of the approximate analytical virial expansions previously discussed. Then, as these density expansions converge slowly, two ways of re-scaling them are analyzed. The last section summarizes the main conclusions.

5.1 Introduction

For many years the properties of the monodisperse hard-sphere (HS) fluid have been profusely studied from theory as well as from computer simulation. One of the reasons is that it is perhaps the simplest non-ideal fluid whereas

Barrio, C., Solana, J.R.: *Binary Mixtures of Additive Hard Spheres. Simulations and Theories*. Lect. Notes Phys. **753**, 133–182 (2008)
DOI 10.1007/978-3-540-78767-9_5

shares in common with more realistic fluids some properties, especially if the molecules of the real fluid interact with each other by means of a spherically symmetric potential. Moreover, in this case, the properties of the latter kind of fluid can be accurately obtained by means of perturbation theories which use the HS fluid as the reference system. Therefore, at present, most of the properties of the monodisperse HS fluid are well known.

Similarly, the polydisperse HS fluid presents a considerable interest because it shares in common certain properties with more realistic mixtures, for which constitutes the natural reference system. Moreover, the binary mixtures of additive hard spheres with extreme diameter ratios between the spheres, belonging to different species and the bigger spheres present in low concentrations, are particularly interesting because of their resemblance with real colloidal dispersions. In addition, the binary mixtures of additive hard spheres, namely those for which the minimum distance between the centers of particles i and j is $\sigma_{ij} = (\sigma_{ii} + \sigma_{jj})/s$, where σ_{ii} is the diameter of particle i, might undergo a demixing transition, purely entropic in nature. There is no need to emphasize here the interest that present colloidal suspensions and the possible existence of a demixing transition for many industrial applications. Therefore, it is easy to understand the interest in the study of the additive HS fluid mixtures as the simplest model of such dispersions.

Because of the preceding considerations, it is not surprising the fact that this kind of fluid mixtures have been the subject of a considerable amount of research for many years. Among other subjects, part of this research has been devoted to perform computer simulations of the thermodynamic and structural properties [1, 2, 3, 4, 5, 6, 7, 8, 9, 10, 11, 12, 13, 14, 15, 16, 17, 18]; to obtain consistency conditions [19, 20, 21, 22, 23, 24, 25, 26, 27, 28]; to the exact or numerical calculation of the virial coefficients [29, 30, 31, 32, 33, 34, 35, 36][1] or to obtain approximate but accurate analytical expressions for the virial coefficients beyond the third [37, 38]; to derive theories leading to analytical expressions of the contact values of the radial distribution functions (RDFs) and the equation of state, such as the scaled particle theory (SPT) [19] or the Percus-Yevick (PY) theory [39], or semiempirical corrections to them [40, 41]; or to develop further refinements to the SPT-based expressions on the basis of the use of consistency conditions leading to improved analytical expressions for the RDFs at contact distance and the equation of state [28, 42, 43, 44, 45, 46, 47]; and to derive accurate analytical equations of state on a semiempirical basis [48, 49, 50, 51, 52].

A question which is deserving much attention from some time now is the above-mentioned possible existence of a demixing transition in binary mixtures of additive hard spheres for extreme mole fractions x_1 and diameter

[1] There is a misprint in Table 5.4 of [35]. The right partial contributions to the seventh virial coefficient of binary mixtures of additive hard spheres with size ratio $R = 10$ are as follows: $B_7(1,6) = 7.2(3)10^{-16}$, $B_7(2,5) = 4.6(2)10^{-13}$, $B_7(3,4) = 2.6(2)10^{-10}$, $B_7(4,3) = 1.28(6)10^{-7}$, $B_7(5,2) = 5.0(1)10^{-5}$, and $B_7(6,1) = 1.27(2)10^{-2}$.

ratios $R = \sigma_{11}/\sigma_{22}$, where subscript 1 refers to the bigger spheres. This is reflected in a considerable number of papers on this subject published in recent years [14, 53, 54, 55, 56, 57, 58, 59, 60, 61]. However, results are controversial since certain theories predict a demixing transition under certain conditions, whereas other does not. On the other hand, computer simulations pose a formidable challenge because usual simulation methods are not applicable for the conditions for which the demixing transition is expected to occur, so that, at present, this way neither can provide a conclusive answer. Several experiments carried out with colloidal suspensions of particles resembling additive HS mixtures [62, 63, 64] seem to indicate that there is not a demixing transition in such mixtures. Instead, a phase coexistence takes place between a fluid phase in which the smaller spheres are dominant and a solid phase in which the bigger spheres are dominant. To obtain more insight on the subject, very accurate analytical expressions for the equation of state and the contact values $g_{ij}(\sigma_{ij})$ of the RDFs are needed.

In this review, which is intended appropriate for postgraduate students, we will attempt to give a survey on the state of the art of the subject. We will focus mainly on "exact" and "experimental" results and on the information we can extract from them. By "exact" we mean certain consistency conditions and the analytical or numerical virial coefficients known. By "experimental" we mean the computer simulation results.

5.2 Computer Simulations

In this Section we will analyze the information on the HS binary mixtures available from computer simulation. We will start summarizing the simulation data available in the literature about the thermodynamic and structural properties of these mixtures. Then, we will discuss the nature of the depletion forces arising in this kind of mixtures and the reliability of treating the larger particles in the actual mixture as an effective monodisperse fluid with an effective potential due to the depletion forces. Finally, we will discuss the information on the phase diagram of binary HS mixtures arising from direct computer simulations as well as from the effective one-fluid description.

5.2.1 Thermodynamic and Structural Properties

Apart from some pioneering works [1, 2, 3], computer simulations for the compressibility factor $Z = pV/Nk_{\mathrm{B}}T$ and the contact values $g_{ij}(\sigma_{ij})$ of the partial RDFs for binary mixtures of additive hard spheres are available in the literature from several authors. Thus, some data for these quantities at high densities for $R = 2$ and 3 and several mole fractions were reported in [5]. The same functions were determined in [8] for $R = 5/3$, 3, 5, and 20 for several compositions and a limited number of densities. More abundant are the data in [9], for the diameter ratios $R = 1/0.9$, $1/0.6$, and $1/0.3$, and the mole fractions $x_1 = 0.125$, 0.25, 0.50, and 0.75, at moderate to high densities. Values of $g_{ij}(\sigma_{ij})$ and Z are available too in [10] for $R = 3/2$, 2, and 3, a wide range of densities, and compositions $x_1 = 0.25$, 0.50, and 0.75 and in [11] for the same diameter ratios and compositions $x_1 = 0.0625$, 0.125, and 0.25. All these authors use the Monte Carlo (MC) method in the *NVT* ensemble. The same technique is used in [12] to obtain the contact values of the solute-solvent RDF $g_{12}(\sigma_{12})$ for low values of x_1, diameter ratios R in the range 4–20 and high densities. Monte Carlo and molecular dynamics (MD) *NVT* simulation data for the partial pressures p_{ij} and the compressibility factor Z of mixtures with $R = 10$, mole fractions $x_1 = 0.002, 0.005, 0.01, 0.02, 0.05$, and 0.1, and moderate to high densities were reported in [13]. The same authors reported later additional MD data for p_{ij} for the same diameter ratio and $x_1 = 0.05$ [14]. A few additional MD data on $g_{ij}(\sigma_{ij})$ for $R = 10$ and $x_1 = 0.002$ have been reported in [15]. Partial excess pressures p_{12}^{e}, and p_{22}^{e} for diameter ratios R ranging from 1 to 2.2, mole fraction $x_1 = 0.00926$ and several densities are available in [16]. Finally, MC *NVT* data for the RDFs and the compressibility factor for $R = 5$, mole fractions x_1 of the order of 10^{-2} and high densities are available in [17].

On the other hand, a few references contain simulation data for other thermodynamic properties. Thus, [9] contains simulation data for the excess chemical potentials μ_i^{e}, [13] includes simulation data for the excess chemical potential μ_2^{e} of the smaller spheres, and [16] reports computer simulations for the excess chemical potentials μ_i^{e} and the excess entropies S_i^{e}.

The structure of additive HS fluid mixtures has been analyzed by means of extensive MC simulation in [4, 10, 18]. In the first of these references, results were reported in tabular form for several diameter ratios, compositions, and densities. In the latter reference it was found that mixtures with diameter ratio $R = 1/0.3$ and low concentrations of the larger spheres presented unusual maxima in the RDFs $g_{ij}(r)$ at distances $r \approx \sigma_{ij} + \sigma_{22}$, and the position of the maxima was found to be nearly independent on density. This was attributed to the preference of arrangements in which two larger spheres are separated by a smaller one.

A selection of references for the available simulation data is summarized in Table 5.1 in which $\rho^* = \rho\sigma_{\text{mix}}^3$ is the reduced density, with $\sigma_{\text{mix}} = (\sum_{i,j} x_i \sigma_{ii}^3)^{1/3}$, $\eta = (\pi/6)\rho^*$ is the packing fraction of the mixture, and d is the ratio of the diameter of the solute spheres to that of the solvent spheres, so that $d = R$ for $d > 1$ and $d = 1/R$ for $d < 1$.

It is worth mentioning here that the simulation of binary mixtures of additive hard spheres with high values of the diameter ratio, say $R > 3$, and/or low mole fractions x_1 of the larger spheres, say $x_1 < 0.1$, poses considerable difficulties at high densities. This is due to the decay in the mobility of the larger spheres which may prevent the system to explore the phase space ergodically. To overcome this difficulty, several techniques have been developed [7, 13]. In the former, the Procedure consists in looking for small particles that can be interchanged one to one with a large particle without producing overlaps and then performing the interchange selecting randomly

Table 5.1. Selected references for the simulation data on the thermodynamic properties for binary mixtures of additive hard spheres. See the text for explanations

Ref.	Functions	R or d	x_1	ρ^*, η
[5]	$g_{ij}(\sigma_{ij}), Z$	$R = 2$	0.10156	$\eta = 0.45, 0.50, 0.55, 0.59$
			0.05,0.19922	$\eta = 0.45, 0.50, 0.55$
		$R = 3$	0.062	
			0.01978	$\eta = 0.45$
[8]	$g_{ij}(\sigma_{ij}), Z$	$R = 5/3$	0.5	$\rho^* = 0.2, 0.4, 0.6, 0.7, 0.8,$ $0.8495, 0.9515, 1.0067$
		$R = 3$	0.5	$\rho^* = 0.5884$
			0.0648	$\rho^* = 0.3820, 0.8623$
		$R = 5$	0.5	$\rho^* = 0.5760$
		$R = 20$	0.037	$\rho^* = 1.0636$
			0.1019	$\rho^* = 0.9889, 1.0636$
			0.5	$\rho^* = 0.5737$
[9]	$g_{ij}(\sigma_{ij}), Z,$ μ_i^e	$R = 1/0.9$	0.25,0.50,0.75	$\eta = 0.30, 0.35, 0.40, 0.45,$ 0.49
		$R = 1/0.6$	0.125,0.25,0.50,0.75	
		$R = 1/0.3$	0.0625,0.125,0.25,0.50,0.75	
[10]	$g_{ij}(\sigma_{ij}), Z$	$R = 3/2, 2, 3$	0.25,0.50,0.75	$\eta = 0.20, 0.25, 0.30, 0.35,$ 0.40,0.45
[11]	$g_{ij}(\sigma_{ij}), Z$	$R = 3/2, 2, 3$	0.0625,0.125,0.25	$\eta = 0.20, 0.25, 0.30, 0.35,$ 0.40,0.45.0.50
[12]	$g_{12}(\sigma_{12})$	$R = 4, 5, 7, 9,$ 11,13,15	0.0020,0.0039,0.0092	$\rho^* = 0.8, 0.9$
		R=20	0.020	
[13]	p_{ij}, Z, μ_2^e	$R = 10$	0.1,0.05,0.02,0.01	$\eta = 0.35, 0.40, 0.45, 0.50,$ 0.55
			0.005	$\eta = 0.35, 0.40, 0.45, 0.50,$ 0.55,0.60
	p_{ij}, Z		0.002	$\eta = 0.75$
[14]	p_{ij}	$R = 10$	0.05	$\eta = 0.35, 0.40, 0.45, 0.50,$ 0.55,0.60,0.62,0.63, 0.64,0.65,0.655
[15]	$g_{ij}(\sigma_{ij})$	$R = 10$	0.001,0.002,0.005,0.01,0.02	$\eta = 0.3$
			0.002	$\eta = 0.4$
[16]	p_{12}^e, p_{22}^e	$d = 0 - 3.4$	1/108	$\rho\sigma_{22}^3 = 0.1$
	S_1^e, S_2^e	$d = 0 - 3.2$		$\rho\sigma_{22}^3 = 0.4$
	μ_1^e, μ_2^e	$d = 0 - 2.4$		$\rho\sigma_{22}^3 = 0.8$
[17]	$g_{ij}(\sigma_{ij}), Z$	$R = 5$	0.00311	$\rho^* = 0.60, 0.75, 0.85$
			0.00415	$\rho^* = 0.6, 0.75, 0.9, 1.0$
			0.00466	$\rho^* = 0.929$
			0.005	$\rho^* = 0.6, 0.75, 0.85$
			0.00588	$\rho^* = 0.658$

between the small particles fulfilling that condition. Obviously this procedure becomes increasingly inefficient as density and/or diameter ratio increase. A more sophisticated technique has been devised [13] in which the smaller spheres are moved in the usual way, whereas in each attempted move of a larger sphere two situations may arise: 1) the displaced sphere overlaps another large sphere and 2) the displaced sphere intersects other smaller spheres. In the first case, the attempted move is rejected. In the second case, the intersected small spheres are attempted to move to the "cavity" left by the displaced large sphere by inverting their positions with respect to the midpoint of the line connecting the old and new positions of the large sphere. If this attempted move leads to any overlap of one of the displaced small spheres with other spheres, the move is rejected and otherwise is accepted.

5.2.2 Depletion Forces and Effective One-Fluid Description

It was first shown by Asakura and Oosawa [65, 66] that between two bodies immersed in a solution of macromolecules appears an interaction, purely entropic in nature, denoted "depletion" interaction. This kind of interaction appears in any situation in which one has a solute consisting in large particles dispersed in a solvent consisting in smaller particles. This is the case of a mixture of additive hard spheres with different diameters. The effective attraction between the solute particles arises as a consequence of the excluded volume when the separation between the surfaces of two solute particles is lower than the diameter of the solvent particles. The exclusion of the latter kind of particles from a certain region situated between the larger particles gives rise to an effective attractive interaction between the latter due to an anisotropy in the local pressure (see Fig. 5.1). If we consider that the two kind of particles are spherical in shape and that the diameter of the smaller ones is σ_{22}, the effective interaction between the larger particles is attractive for a separation between their surfaces $\delta < \sigma_{22}$ and monotonically increasing with δ and zero for $\delta \geq \sigma_{22}$; moreover, it is proportional to the packing fraction $\eta_2 = (\pi/6)x_2\rho\sigma_{22}^3$ of the small particles, that is, to the fraction of the total volume occupied by them.

On the basis of simple geometrical arguments, Asakura and Oosawa [65, 66] determined the depletion force F_d acting between two large hard spheres of diameter σ_{11}, with their centers separated a distance r, immersed in a fluid of small hard spheres with diameters σ_{22} in the form

$$
\begin{aligned}
F_\mathrm{d} = -\frac{1}{4}\pi\rho k_\mathrm{B}T\left[(\sigma_{11}+\sigma_{22})^2-(\sigma_{11}+\delta)^2\right] \ , \quad & \delta \leq \sigma_{22} \\
0 \qquad\qquad\qquad\qquad\qquad\qquad\qquad\qquad \ , \quad & \delta > \sigma_{22} ,
\end{aligned}
\qquad (5.1)
$$

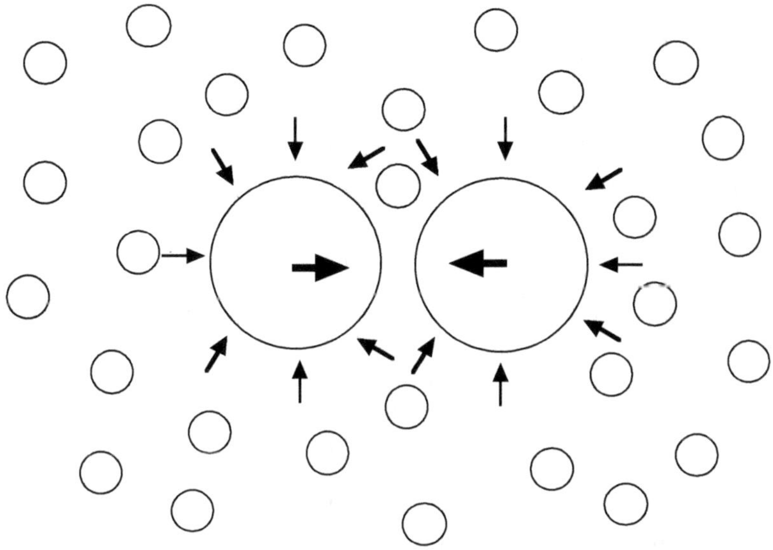

Fig. 5.1. Illustration of the depletion forces arising when a solute consisting in large particles is suspended in a solvent consisting in much smaller particles. The *thin arrows* indicate the directions in which the pressure forces act exerting the smaller particles on the larger ones. One sees that a region situated between the two large particles is free of such forces, because the small particles cannot penetrate into it. As a consequence, resulting forces appear, the depletion forces indicated by thick arrows, which act as effective attractive forces between the large particles

where $\delta = r - (1/2)(\sigma_{11} + \sigma_{22})$ is the minimum distance between the surfaces of the spheres and k_{B} the Boltzmann constant. The corresponding potential is

$$u_d = -\int_{\sigma_{22}}^{\delta} F_{\mathrm{d}} \mathrm{d}\delta = -\frac{1}{4}\pi\rho k_{\mathrm{B}}T\,(\sigma_{22} - \delta)$$

$$\times \left[\sigma_{22}\left(\sigma_{11} + \frac{2\sigma_{22}}{3}\right) - \left(\sigma_{11} + \frac{\sigma_{22}}{3}\right)\delta - \frac{\delta^2}{3}\right], \quad \delta \leq \sigma_{22}$$

$$0 \qquad\qquad , \quad \delta > \sigma_{22}, \qquad (5.2)$$

The derivation was based on the assumption that the solvent fluid behaves as an ideal gas, which is valid only at low densities. At higher densities, the force acting on a sphere immersed in a fluid of smaller spheres depends on the local density $\rho(\mathbf{r})$ in the form [67, 68]

$$\mathbf{F} = -k_{\mathrm{B}}T \int_S \rho\hat{\mathbf{n}}\mathrm{d}A \qquad (5.3)$$

where S is the surface of a sphere of radius $(\sigma_{11} + \sigma_{22})/2$ centered at the center of the solute sphere, and $\hat{\mathbf{n}}$ is unit vector normal to the surface directed outwards. When the depletion force is spherically symmetric, as in our case, the preceding equation leads to [67, 68]

$$F_{\mathrm{d}} = -\frac{1}{2}\pi k_{\mathrm{B}}T\left(\sigma_{11}+\sigma_{22}\right)^2 \int_0^{\pi} \rho\left(\theta\right)\sin\theta\cos\theta \mathrm{d}\theta, \tag{5.4}$$

where $\rho(\theta)$ is the density of the solvent fluid at contact distance from the large solute sphere and θ is the angle between a reference axis and the line joining the centers of the large sphere and the smaller one when they are in contact with each other. Equation (5.4) represents the local pressure $\rho(\mathbf{r})k_{\mathrm{B}}T$ integrated over the area of the exclusion sphere for spheres of species 2 around a sphere of species 1, and reduces to the Asakura–Oosawa expression (5.1) if the solvent fluid is considered an ideal gas.

In the limit $R \to \infty$ the larger spheres can be considered as planar hard walls. In this case, we can replace the local pressure by the solvation force per unit area $f_{\mathrm{s}}(x)$, that is, the excess pressure of the solvent due to its confinement between two planar hard walls separated a distance x. Then, after some algebra, Eq. (5.4) leads to [69]

$$F_{\mathrm{d}} = \pi\frac{\sigma_{11}+\sigma_{22}}{2}\int_{\delta}^{\infty} f_{\mathrm{s}}\left(x\right)\mathrm{d}x\,. \tag{5.5}$$

which is the Derjaguin approximation [70]. From (5.5), the depletion force between two large spheres immersed in a fluid of smaller ones can be expressed in the form [69]

$$F_{\mathrm{d}} = \pi\frac{\sigma_{11}+\sigma_{22}}{2}\left[p\left(\delta-\sigma_{22}\right)-2\gamma\right]\,, \quad \delta \le \sigma_{22}\,, \tag{5.6}$$

where p is the pressure of the solvent fluid and γ its surface tension at a planar wall. Therefore, the depletion force depends on the pressure and surface tension of a monodisperse fluid of hard spheres with diameter σ_{22}. The former can be obtained from the very simple and accurate Carnahan–Starling [71] equation

$$Z_0^{CS} = \frac{pV}{Nk_{\mathrm{B}}T} = \frac{1+\eta+\eta^2-\eta^3}{\left(1-\eta\right)^3}, \tag{5.7}$$

whereas the surface tension can be obtained from the expression [72]

$$\gamma = -\frac{9}{2\pi}\frac{k_{\mathrm{B}}T}{\sigma_{22}^2}\frac{\eta^2\left(1+\frac{44}{35}\eta-\frac{4}{5}\eta^2\right)}{\left(1-\eta\right)^3}. \tag{5.8}$$

The depletion potential in the Derjaguin approximation is

$$u_{\mathrm{d}}\left(\delta\right) = u_{\mathrm{d}}\left(\sigma_{22}\right) - \pi\frac{\sigma_{11}+\sigma_{22}}{2}\left(\delta-\sigma_{22}\right)\left[\frac{1}{2}p\left(\delta-\sigma_{22}\right)-2\gamma\right], \quad \delta \le \sigma_{22}, \tag{5.9}$$

The Derjaguin approximation is exact in the limit $R \to \infty$. However, for intermediate values of R, we cannot consider the large particles as planar walls. A more appropriate approximation for this case is the so-called wedge approximation [69], in which the depletion force is given by

$$F_{\rm d} = \pi \frac{\sigma_{11} + \sigma_{22}}{2} \left[p\left(\delta - \sigma_{22}\right) - 2\gamma \cos \alpha \right] , \quad \delta \leq \sigma_{22}, \tag{5.10}$$

where $\cos \alpha = (\sigma_{11} + \delta)/(\sigma_{11} + \sigma_{22})$. The depletion potential in this case is

$$\begin{aligned} u_{\rm d}\left(\delta\right) &= u_{\rm d}\left(\sigma_{22}\right) - \pi \frac{\sigma_{11} + \sigma_{22}}{2} \left(\delta - \sigma_{22}\right) \\ &\times \left[\frac{1}{2} p\left(\delta - \sigma_{22}\right) - \gamma \frac{\delta + 2\sigma_{11} + \sigma_{22}}{\sigma_{11} + \sigma_{22}} \right], \quad \delta \leq \sigma_{22}, \end{aligned} \tag{5.11}$$

Finally, we will mention here another expression for the depletion potential, based in its exact expression in the limit $R = 1$, which was proposed in [69]. It reads

$$u_{\rm d}\left(\delta\right) = u_{\rm d}\left(\sigma_{22}\right) - \frac{\sigma_{11} + \sigma_{22}}{2\sigma_{22}} \ln \left[\frac{g_0\left(\sigma_{22} + \delta\right)}{g_0\left(2\sigma_{22}\right)} \right], \quad \delta \leq \sigma_{22}. \tag{5.12}$$

where g_0 is the RDF of a monodisperse HS fluid.

The depletion force between two large spheres immersed in a fluid of smaller spheres has been measured directly by means of computer simulations [68, 73]. These simulations show that the term $u_{\rm d}(\sigma_{22})$, the value of the depletion potential at distance σ_{22}, is negligible. This allows a direct comparison with the precedent theoretical approximations, among others. For diameter ratios $R = 5$ and 10 it was found [68] that the Asakura–Oosawa theory works well only in the low-density limit whereas, even at low densities, it underestimates the depletion force at contact and the deviation between theory and simulation increases with density. In contrast, the Derjaguin approximation was found to be in fairly good agreement with simulations for $R = 10$ and $\eta = 0.229$. However, a different scenario emerged from a further analysis [69], which showed that the depletion potential $u_{\rm d}$ calculated from the Asakura–Oosawa theory is quite accurate for these diameter ratios at least up to a packing fraction $\eta = 0.341$, whereas the Derjaguin and wedge approximations give poorer results, especially for packing fractions $\eta \gtrsim 0.30$. In the same reference it was shown that approximation (5.12) also provides fairly good agreement with simulation data for the same diameter ratios at all densities.

From the fact that in a mixture of large and small spheres a depletion force, or potential, acts on the larger spheres due to the presence of the smaller ones, arises the possibility of treating the larger particles in the actual mixture as an effective monodisperse fluid of spheres with core diameter σ_{11} and an effective potential that, assuming pairwise additivity, would be of the form

$$u\left(r\right) = u_0\left(r\right) + u_d\left(r\right) , \tag{5.13}$$

where $u_0(r)$ is the HS potential for spheres with diameter σ_{11}, and $u_{\rm d}(r)$ the depletion potential acting on them due to the presence of the small spheres.

In the absence of a simple, analytical, and enough accurate expression for the depletion potential, the latter can be determined numerically from computer simulations, as mentioned before. The question that comes up on this subject is whether the depletion potential, determined in the low-density limit, is appropriate for higher densities or, in other words, which is the relative importance of the many-body interactions not accounted for by the depletion potential and, in the end, to what extent the properties of the actual mixture can be obtained from those of the effective fluid.

This question has been addressed in depth by Malherbe and Amokrane [74] who compared the results obtained from simulations of mixtures with $R = 5, 10$, and 20 and those of the effective fluid, with a potential of the form (5.13) determined from computer simulation, under the assumption of pairwise additivity. They found that the contact values $g_{11}(\sigma_{11})$ of the RDF in the effective fluid were too high as compared with those of the actual mixture. The relative deviation showed to be nearly independent on the partial packing fraction $\eta_1 = (\pi/6)\rho_1\sigma_{11}^3$ of the larger spheres but increased with the partial packing fraction η_2 of the solvent spheres and with the diameter ratio R. The fact that the assumption of pairwise additivity is inappropriate for very asymmetric mixtures and low concentrations of the larger spheres was confirmed by the analysis of the mean force. To this end, the mean force for a set of four large spheres in a bath of smaller ones determined directly from simulation was compared with that obtained from the superposition approximation on the basis of simulations for two large spheres in the solvent bath. From the results it was clear that the superposition approximation overestimates the actual force for the set of four spheres.

5.2.3 Phase Diagram

A good deal of the computational effort devoted to binary mixtures of additive hard spheres has been conducted to determine the phase equilibria and phase transitions in these systems. Most of this effort has been performed along two lines: crystallization and demixing. We will examine briefly the main findings on these subjects.

Crystallization

In HS systems, the freezing transition is entropy-driven. It takes place when the decrease in entropy due to increasing order in the solid phase is largely compensated by the increase in entropy due to the increase in the free volume available in the crystalline phase for the particles to move. In a binary mixture of unequal-sized additive hard spheres of species A and B, where A refers to the bigger spheres, the freezing may lead to pure crystalline phases A and B, to substitutionally disordered crystals AB, or to structures of the form AB_n. Which of these crystalline phases will be the most stable one will depend on the diameter ratio $R = \sigma_A/\sigma_B$, the composition x_A, and the packing fraction η.

The maximum packing fraction for a monodisperse face-centered cubic (fcc) or hexagonal close-packed (hcp) crystalline solid of hard spheres with diameter σ is $\eta = \pi\sqrt{2}/6 = 0.7405$. For bidisperse mixtures, structures AB_n having maximum packing fractions higher than this value will be favored with respect to the monodisperse A and B crystals. Theoretical considerations based on the available free space [75] predicted that for diameter ratios $R \lesssim 1.60$ crystalline structures of the form AB_n would be unstable with respect to the monodisperse crystals. For $R > 2.42$ a stable AB phase would occur. This situation corresponds to the placement of one small sphere in each of the octahedral interstitial lattice sites of the crystal formed by the larger spheres, a structure similar to that in the NaCl crystal. A second AB phase, with contacts between spheres A and B, but no contact between spheres of the species A alone, would be possible in the range $2.18 < R < 2.42$. Similar considerations lead to the prediction of an AB_2 phase, with a structure like that in AlB_2, in the range $1.60 < R < 2.07$. A stable AB_{13} lattice, formed by a cubic cell formed by eight A spheres in the vertices, with a B sphere in the center surrounded by twelve B spheres in an icosahedral cluster, like in $NaZn_{13}$, might be possible, according to the above-mentioned criterion, within a small range $1.72 \lesssim R \lesssim 1.86$, provided that some distribution in size of the B spheres is allowed. For higher diameter ratios than those quoted here, other structures of the form AB_n would be possible, formed by filling the interstitial holes of the crystal formed by the bigger spheres with the smaller ones with more than one of the latter in each interstitial site. These limiting diameter ratios were determined [75] on the basis of space-filling considerations for close-packed structures. For lower densities of the solid phase, these structures might be possible for lower diameter ratios than those quoted here.

For diameter ratios R close to 1, a substitutionally disordered fcc solid will form. This solid is stable for $1 \leq R \lesssim 1.14$ [76]. For $1.14 \lesssim R \lesssim 1.18$ two solid phases appear, one rich in the larger spheres and the other rich in the smaller ones [76, 77]. At the critical density $R \approx 1.18$, density functional theory predicts [77] that the larger spheres are absent of the latter phase, whereas in the other, although the larger spheres are predominant, an appreciable proportion of the smaller ones is still present. For higher diameter ratios the substitutionally disordered fcc solid would be unstable [77].

Computer simulations [6, 76, 78, 79, 80, 81, 82, 83, 84] essentially confirm these predictions. In a fcc lattice formed by the larger spheres at close packing, the maximum diameter of the smaller spheres that can fit into the octahedral interstitial sites forming an AB lattice corresponds to a diameter ratio $R = 2.42$, as stated before, but this limiting diameter ratio would reduce up to $R \approx 1.67$ at densities close to melting. However, computer simulations show [78] that the range of metastability of the pure HS fcc lattice $1.35 \leq V/V_0 \leq 1.50$, where $V_0 = \sigma^3/\sqrt{2}$ is the close-packed volume, nearly disappears when smaller particles are dissolved into the lattice, so that the upper limit of the solid phase becomes close to $V/V_0 = 1.35$. For higher densities, when small particles are introduced at constant volume into a fcc lattice formed by the larger ones, the former diffuses in a fluid-like manner throughout the interstitial sites. As the diameter of the smaller particles increases, the diffusion coefficient decreases and, for a given diameter ratio, the small particles finally become trapped into the octahedral interstitial sites and and AB lattice is formed. For $V/V_0 = 1.20$ the critical diameter ratio is $R = 2.5$ [78], as compared with the value $R = 2$ which is the minimum diameter ratio in the AB lattice at that density. Therefore, for lower diameter ratios the whole mixture would be either in the fluid or in a substitutionally disordered solid.

The preceding results are consistent with those reported in [6], in which mixtures of additive hard spheres with diameter ratios $R = 11/10$, $5/3$, $3, 5$, and 20, and with mole fractions $x_A = 0.1019, 0.5$, and 0.8981, were studied by means of MC and MD simulations. The $R = 5/3$ mixture with $x_A = 0.8981$ was found to freeze at a packing fraction $\eta \approx 0.52$, without significative diffusion of none of the two species. For the other two mole fractions no sign of fluid to solid transition was found up to a packing fraction $\eta \approx 0.60$ for that diameter ratio. For $R = 3$ and $x_A = 0.8981$ the system underwent a phase transition at $\eta \approx 0.53$ in which the larger spheres formed a crystal with the smaller ones diffusing throughout the lattice. Again the phase transition was absent for the other mole fractions considered. For the size ratio $R = 5$ the system with $x_A = 0.8981$ showed a fluid to solid transition at a similar packing fraction, and with similar characteristics as the former, and even for $x_A = 0.50$ the system apparently froze, but in this latter case the transition was continuous and somewhat unclear. Finally, for $R = 20$ the transition was present in both, the $x_A = 0.8981$ (at $\eta \approx 0.54$) and the $x_A = 0.5$ (at $\eta \approx 0.52$) mixtures, more sharper in the first case, with the smaller particles diffusing through the lattice formed by the larger ones in both cases. The melting curve of binary mixtures with size ratios $\alpha = 1/R = 0.95$ and 0.90 was determined in [79] as a function of the mole fraction from the conditions of equal chemical potential and pressure in both phases. To this end, the Gibbs free enthalpy and the pressure in both phases were determined from computer simulations. For the second of these diameter ratios the system was found to form an azeotrope at $x_1 \approx 0.22$. The mixtures with $\alpha = 0.95$ and 0.85, in addition to that with $\alpha = 0.73$, were studied also by MD simulations in [84]. For $x_1 = 0.70$, the first of these mixtures was seen to form at high densities a substitutionally disordered crystal. The situation for the mixture with $\alpha = 0.85$ was less clear, although for $x_A = 0.80$ partial crystallization of the two components took place, without signal of demixing within the simulation time. Finally, for $\alpha = 0.73$ a range of pressures was found in which the fluid mixture coexisted with either of the two pure crystals A or B. Eutectic points were found for both the $\alpha = 0.85$ and the $\alpha = 0.73$ size ratios with mole fractions $x_1 \approx 0.26$ and $x_1 \approx 0.15$, respectively.

The thermodynamic stability of AB_n crystals in binary HS systems has been studied in a series of papers [76, 80, 81, 82, 83]. To this end, the free energy of the binary crystal was determined by computer simulations and those for the fluid and the pure component crystal phases were determined from thermodynamic integration of suitable analytical equations of state. The AB_{13} structure was found [76] to be stable within a region $1.62 \lesssim R \lesssim 2.08$ for packing fractions $\eta \gtrsim 0.54$, depending on the diameter ratio, and unstable for lower diameter ratios with respect to the pure A and B crystals. The phase diagram for a mixture with $\alpha = 0.58$ was studied as a function of the pressure and composition in [81]. For different ranges of pressures and compositions the fluid (F), A, B, AB_2, and AB_{13} phases were found to be stable. The region of stability of the AB_2 crystal was established in [80] for $1.62 \lesssim R \lesssim 2.33$ and packing fractions $\eta \gtrsim 0.61$, with a limit $R \approx 1.61$ beyond which this structure would be less stable than the monodisperse A or B crystals, in good agreement with the predictions of [75]. The results reported in [82] for phase diagrams of mixtures with $\alpha = 0.58, 0.60, 0.61$, and 0.625, are in agreement with those from [76, 80, 81]. Finally, the stability of the AB crystal, NaCl structure, was studied in [83]. The phase diagrams for $\alpha = 0.414$ showed the existence of a region of stability of the AB phase, together with others for the A, B, and F phases. In the phase diagram for $\alpha = 0.45$, close to the expected upper limit of stability of the AB phase, an additional AB_2 phase appeared. In none of these two diameter ratios the AB_{13} phase was present. All these results are in agreement with the predictions of [75] based on the maximum packing fraction rule.

Demixing

Another important subject related to the phase transitions is the demixing transition in the fluid, which is predicted to occur by certain theories for high diameter ratios and low mole fractions of the bigger spheres, leading to two fluid phases each of them enriched in one of the components. From the computer simulation side the problem has been tackled by several authors [13, 85, 86]. Thus, in [85] Gibbs ensemble Monte Carlo simulations were performed for mixtures of hard spherocylinders with thickness ratio $R = 10$ and lengths $L/\sigma_{11} = 10, 8, 5$, and 3, where L is the distance between the center of the two hemispheres. In all cases an entropy-driven fluid–fluid demixing transition was found. The extreme value $L = 0$, corresponding to an HS mixture, could not be simulated because the transition is predicted in this case to take place for a very low mole fraction, which requires a huge number of the smaller particles and hinders the larger ones to explore phase space in an efficient manner. To gain insight into the demixing transition in the HS mixtures, the phase coexistence region was determined for the above-mentioned values of L and the the results were extrapolated to $L = 0$. The range of of compositions for which phase coexistence was present decreased with decreasing L, although the location of the critical point remained approximately constant with the length of the spherocylinders. However, the packing fraction of the mixture at the critical point was found to increase with decreasing length, and this increase sped up for low values of L. Therefore, it was concluded that for $L = 0$

no demixing transition occurs for the diameter ratio $R = 10$ at any density or, if any, it would be metastable with respect to the freezing transition.

The existence of a demixing transition in a mixture of hard spheres with diameter ratio $R = 30$ was shown in [86] by means of Monte Carlo simulation. To this end, an efficient algorithm for a system consisting of $N_1 = 32$ large spheres and $N_2 = 864\,000$ small spheres, both of them with the same packing fraction $\eta_1 = \eta_2 = 0.1215$, was used. For the diameter ratio $R = 20$ demixing was not found, but could not be completely discarded because of the presence of important finite size effects. For the diameter ratio $R = 10$ finite size effects were not found, which seems to indicate that demixing does not takes place at this diameter ratio.

HS mixtures with diameter ratio $R = 10$ were also studied in [13] by MC and MD simulations for several compositions in the range $0.0020 \leq x_1 \leq 0.1$ at moderate to high densities. In this range certain theories predict the presence of a demixing transition. However, no sign of such transition was found. For $x_1 = 0.002$ and $\eta = 0.75$, diffusion coefficients revealed that the large particles were trapped in an amorphous solid structure with the small particles still moving.

The Phase Diagram in the Effective One-Fluid Description

The phase diagram of asymmetric binary mixtures of additive hard spheres has been also studied by computer simulations within the scenario of the effective one-fluid description [59, 60, 73, 87, 88]. Monte Carlo simulations with an effective potential determined from MD simulations for $R = 10$ provided some evidence of a tendency to demixing first and to crystallizing of the larger spheres later for partial packing fractions $\eta_1 = 0.1$ and $\eta_2 \approx 0.37$ [73].

The complete phase diagram in the effective one-fluid description for mixtures with diameter ratios ranging from $R = 5$ to $R = 30$ have been determined from computer simulations by Dijkstra et al. in a series of papers [59, 87, 88], using for the depletion potential a theoretical approximation. For $R = 20$ and $R = 30$, stable solid–solid and fluid–solid, and metastable fluid–fluid coexistence were found [59, 88], whereas in the mixture with $R = 10$ the solid–solid transition was found to be metastable with respect to the solid–fluid transition and the mixture with $R = 5$ showed only the second of these transitions. In [59] the coexistence densities for these phase transitions were reported for the same diameter ratios. However, computer simulations performed with a slightly different effective potential for the diameter ratios $R = 10$ and $R = 20$ provided no evidence of a fluid–fluid coexistence, and the stability of the isostructural solid–solid transition for $R = 10$ was uncertain [60].

We may ask whether the depletion potential approach is accurate enough for the study of the phase equilibria of these mixtures. Comparison of the phase diagrams obtained in this way with those resulting from direct simulations show close agreement between the two procedures [59, 88].

5.3 Consistency Conditions

The virial theorem for a mixture of hard spheres with diameters σ_{ii} and mole fractions x_i at number density $\rho = N/V$, or packing fraction $\eta = (\pi/6)\rho\sum_i x_i\sigma_{ii}^3$, can be expressed in the form

$$Z = 1 + \frac{2}{3}\pi\rho\sum_i\sum_j x_i x_j \sigma_{ij}^3 g_{ij}(\sigma_{ij}), \tag{5.14}$$

where $g_{ij}(\sigma_{ij})$ is the contact value of the RDF for particles i and j and σ_{ij} is the distance of closest approach between the centers of spheres i and j which, for additive mixtures, is $\sigma_{ij} = (\sigma_{ii} + \sigma_{jj})/2$. The partial RDFs at contact distance $g_{ij}(\sigma_{ij})$ must satisfy certain consistence conditions. A list of them was compiled in [29] and is summarized here. For simplicity we will restrict ourselves to binary mixtures of species 1 and 2 with mole fractions x_1 and x_2, with $x_2 = 1 - x_1$.

For $x_1 = 1$ or $x_1 = 0$ the mixture reduces to a monodisperse fluid with spheres of the species 1 or 2, respectively. Denoting by σ the diameter of these spheres and by $g_0(\sigma)$ the corresponding contact value of their RDF, we will have two conditions

1)

$$g_{11}(\sigma_{11}) = g_0(\sigma); \quad x_1 = 1, \tag{5.15}$$

and

2)

$$g_{22}(\sigma_{22}) = g_0(\sigma); \quad x_2 = 1. \tag{5.16}$$

A similar situation arises for $\sigma_{22} = \sigma_{11}$ whence three new conditions arise

3)

$$g_{11}(\sigma_{11}) = g_0(\sigma); \quad \sigma_{22} = \sigma_{11}, \tag{5.17}$$

4)

$$g_{22}(\sigma_{22}) = g_0(\sigma); \quad \sigma_{22} = \sigma_{11}, \tag{5.18}$$

and

5)

$$g_{12}(\sigma_{12}) = g_0(\sigma); \quad \sigma_{22} = \sigma_{11}. \tag{5.19}$$

Three more conditions are obtained from the limiting case in which the diameter of the smaller spheres vanishes whereas that of the bigger spheres remains finite, that is, $\sigma_{22}/\sigma_{11} = 0$ or $R \to \infty$. In this case, the equation of state and the contact radial distribution function of the $N_2 = x_2 N$ point particles are like those of an ideal gas in a volume $V - N_1\pi\sigma_{11}^3/6 = V - x_1 N\pi\sigma_{11}^3/6$, that is,

$$Z = \frac{1}{1-\eta}, \tag{5.20}$$

and

$$g(0) = \frac{1}{1-\eta}, \tag{5.21}$$

where in this case $\eta = (\pi/6)N_1\sigma_{11}^3/V$.

Moreover, for this particular mixture the compressibility factor for this system is given exactly by [89]

$$Z = x_1 Z_0 + \frac{x_2}{1-\eta}, \tag{5.22}$$

where Z_0 is the compressibility factor for a pure fluid of hard spheres with packing fraction $\eta = x_1(\pi/6)\rho\sigma_{11}^3$. The virial equation of state (5.14) can be expressed exactly in the form [26]

$$Z = x_1 Z_0 + x_2 \left[1 + \eta g_{12}(\sigma_{11}/2)\right]; \quad \sigma_{22} = 0, \tag{5.23}$$

since $\sigma_{12} = \sigma_{11}/2$, because $\sigma_{22} = 0$.

From Eqs. (5.20), (5.21), and (5.23), together with the virial equation (5.14) we obtain the conditions

6)

$$g_{12}(\sigma_{12}) = \frac{1}{1-\eta}; \quad \sigma_{22} = 0, \tag{5.24}$$

7)

$$g_{11}(\sigma_{11}) = g_0(\sigma); \quad \sigma_{22} = 0, \tag{5.25}$$

where $g_0(\sigma)$ is the contact value of the RDF of a pure fluid of hard spheres of diameter $\sigma = \sigma_{11}$, and

8)

$$g_{22}(\sigma_{22}) = \frac{1}{1-\eta}; \quad \sigma_{22} = 0. \tag{5.26}$$

In the colloidal limit, $x_1 \to 0$ and $\sigma_{11} \to \infty$, we can consider the system as consisting of hard spheres with diameter $\sigma = \sigma_{22}$ and packing fraction $\eta = (\pi/6)\rho\sigma_{22}^3$ in contact with a planar hard wall. In this case, if $\rho \equiv \rho_2 = N_2/V$ is the average density of particles of diameter σ_{22} in the bulk fluid, then $\rho_2 g_{12}(\sigma_{12})$ is the density of particles in contact with the wall. The pressure, which is the moment per unit surface and time transferred by the particles in colliding with the wall, is obtained multiplying the latter density by $k_B T$. This gives rise to the exact result [19, 20]

9)

$$g_{12}(\sigma_{12}) = Z_0(\eta); \quad x_1 \to 0, \sigma_{11} \to \infty. \tag{5.27}$$

10) A condition relating the contact values of the three partial RDFs to each other has been derived in [22] in the form

$$x_1^2\sigma_{11}^2 \frac{\partial g_{11}(\sigma_{11})}{\partial \sigma_{22}} + x_1 x_2 \sigma_{12}^2 \frac{\partial g_{12}(\sigma_{12})}{\partial \sigma_{22}} = x_2^2 \sigma_{22}^2 \frac{\partial g_{22}(\sigma_{22})}{\partial \sigma_{11}} + x_1 x_2 \sigma_{12}^2 \frac{\partial g_{12}(\sigma_{12})}{\partial \sigma_{11}}. \tag{5.28}$$

11) Another useful relationship between the contact values of the three partial RDFs and the Helmholtz free energy F of the mixture has been derived in the same reference. It can be expressed as follows [22, 23]

$$g_{ij}\left(\sigma_{ij}\right)=\frac{1}{2\pi Nk_{\mathrm{B}}T\left(2-\delta_{ij}\right)\rho x_i x_j\sigma_{ij}^2}\left(\frac{\partial F}{\partial\sigma_{ij}}\right)_{\{s_{ab}\}}, \tag{5.29}$$

where δ_{ij} is the Kronecker delta function and $\{s_{ab}\}$ represents the set of parameters other than σ_{ij}.

12) On the basis of the exact expression [21, 22]

$$\left(\frac{\partial lnQ}{\partial\sigma_{22}}\right)_{T,\rho,x_i,\sigma_{11}}=-2\pi N\rho\left[x_2^2\sigma_{22}^2g_{22}\left(\sigma_{22}\right)+x_1x_2\sigma_{12}^2g_{12}\left(\sigma_{12}\right)\right], \tag{5.30}$$

one readily obtains [24]

$$\lim_{\sigma_{22}\to 0}\left[\frac{\partial\left(F/Nk_{\mathrm{B}}T\right)}{\partial\sigma_{22}}\right]_{T,\rho,x_i,\sigma_{11}}=2\pi\rho x_1x_2\lim_{\sigma_{22}\to 0}\left[\sigma_{12}^2g_{12}\left(\sigma_{12}\right)\right], \tag{5.31}$$

which, using Eq. (5.24), can be rewritten in the form [24]

$$\lim_{\sigma_{22}\to 0}\left[\frac{\partial\left(F/Nk_{\mathrm{B}}T\right)}{\partial\sigma_{22}}\right]_{T,\rho,x_i,\sigma_{11}}=\frac{1}{2}\frac{\pi\rho x_1x_2\sigma_{11}^2}{1-\eta}. \tag{5.32}$$

with $\eta=x_1(\pi/6)\rho\sigma_{11}^3$.

Several other conditions are derived from the scaled particle theory [19] and involve the excess chemical potential and its derivatives in the infinite dilution limit of the solute. Thus, for $\sigma_{22}=0$ we have [26]

13)

$$\beta\mu_2^e=-\ln\left(1-\eta\right);\quad\sigma_{22}=0;\quad x_2=0, \tag{5.33}$$

14)

$$\sigma_{11}\beta\frac{\partial\mu_2^e}{\partial\sigma_{22}}=\frac{3\eta}{1-\eta};\quad\sigma_{22}=0;\quad x_2=0. \tag{5.34}$$

In these two expressions $\beta=1/k_{\mathrm{B}}T$ and $\eta=x_1(\pi/6)\rho\sigma_{11}^3$.

15) In the limit $x_1\to 0$ and $\sigma_{11}\to\infty$ it holds [26]

$$\frac{\sigma_{22}^3}{3\sigma_{11}^2}\beta\frac{\partial\mu_1^e}{\partial\sigma_{11}}=\eta Z_0\left(\eta\right);\quad x_1\to 0,\sigma_{11}\to\infty, \tag{5.35}$$

with $\eta=(\pi/6)\rho\sigma_{22}^3$.

16) For $\sigma_{22}=\sigma_{11}$ [26, 27]

$$\sigma_{ii}\beta\frac{\partial\mu_j^e}{\partial\sigma_{jj}}=12\eta g_{ii}\left(\sigma_{ii}\right);\quad\sigma_{jj}=\sigma_{ii},\ x_j=0, \tag{5.36}$$

with $\eta=(\pi/6)\rho\sigma_{ii}^3$.

17) Finally, for either $x_2 \to 0$ or $x_1 \to 0$ [26]

$$\beta \sigma_{ii} \frac{\partial \mu_{jj}^{e}}{\partial \sigma_{jj}} = 3\eta \left(1 + \sigma_{jj}/\sigma_{ii}\right)^2 g_{ij}\left(\sigma_{ij}\right) \quad ; \quad x_j = 0, \tag{5.37}$$

with $\eta = (\pi/6)\rho\sigma_{ii}^3$.

The preceding conditions may be useful to derive new expressions for the partial RDFs of additive HS fluid mixtures or to improve other proposed expressions, as we shall see later.

5.4 Virial Coefficients

Another source of information about the thermodynamic properties of fluids is the virial expansion. For additive HS fluid mixtures, the second and third virial coefficients are known analytically, whereas several of the higher order virial coefficients have been determined numerically with high degree of accuracy and there are available several procedures to obtain approximate, though very accurate, analytical expressions for them. All these points will be addressed in the next subsections.

5.4.1 Exact Results

Exact Analytical Expressions of the Second and Third Virial Coefficients

The virial expansion of the compressibility factor in terms of the number density can be expressed in the form

$$Z = 1 + \sum_{n=2}^{\infty} B^{(n)} \rho^{n-1}, \tag{5.38}$$

where $B^{(n)}$ is the virial coefficient of order n.

The second virial coefficient of a mixture of additive hard spheres is given by

$$B^{(2)} = \sum_i \sum_j x_i x_j B_{ij}^{(2)}, \tag{5.39}$$

where, for a binary mixture, $i, j = 1, 2$, and the contributions $B_{ij}^{(2)}$ can be expressed exactly in the form

$$B_{ii}^{(2)} = 4\left(\frac{\pi}{6}\sigma_{ii}^3\right), \tag{5.40}$$

and

$$B_{ij}^{(2)} = \frac{\pi}{12}\left(\sigma_{ii}^3 + 3\sigma_{ii}^2\sigma_{jj} + 3\sigma_{ii}\sigma_{jj}^2 + \sigma_{jj}^3\right). \tag{5.41}$$

Analogously, the expression of the third virial coefficient of the binary mixture [29] is

$$B^{(3)} = \sum_{i,j,k} x_i x_j x_k B_{ijk}^{(3)}, \tag{5.42}$$

where $i, j, k = 1, 2$ and the partial contributions $B_{ijk}^{(3)}$ can be expressed exactly in the form

$$B_{iii}^{(3)} = 10 \left(\frac{\pi}{6}\sigma_{ii}^3\right)^2, \tag{5.43}$$

and

$$B_{iij}^{(3)} = \left(\frac{\pi}{6}\right)^2 \left(\frac{1}{3}\sigma_{ii}^6 + \frac{8}{3}\sigma_{ii}^3\sigma_{jj}^3 + 5\sigma_{ii}^4\sigma_{jj}^2 + 2\sigma_{ii}^5\sigma_{jj}\right). \tag{5.44}$$

The contributions $B_{ij}^{(2)}$ and $B_{iij}^{(3)}$ to the second and third virial coefficients, respectively, can be written in an alternative way, but equally exact, to Eqs. (5.41) and (5.44), namely [23]

$$B_{ij}^{(2)} = 4\left(\frac{\pi}{6}\right)\sigma_{ij}^3, \tag{5.45}$$

where, for additive hard spheres $\sigma_{ij} = (\sigma_{ii} + \sigma_{jj})/2$, as said before, and

$$B_{iij}^{(3)} = \frac{2}{3}\left(\frac{\pi}{6}\right)^2 \sigma_{ii}^3 \left(\sigma_{ii}^3 - 18\sigma_{ii}\sigma_{ij}^2 + 32\sigma_{ij}^3\right). \tag{5.46}$$

In this form, the crossed contributions to the virial coefficients are expressed in terms of the crossed contact distance σ_{ij}. This is necessary for certain purposes, as we will see later.

Exact Numerical Data for Higher Order Virial Coefficients

Beyond the third virial coefficient, exact analytical expressions for the virial coefficients are not available, and the virial coefficients have to be determined numerically. We denote the virial coefficients thus obtained as "exact numerical data," in spite of the fact that they are affected by some numerical uncertainty. Thus, the partial contributions to the fourth [31, 32, 35], fifth [32, 33, 34, 35], and sixth [34, 35] virial coefficients have been determined numerically for different values of the diameter ratio R, and the seventh virial coefficient has been reported for the size ratio $R = 10$ [35, 36].

5.4.2 Analytical Expressions of the Higher Order Virial Coefficients from the Wheatley's Method

Wheatley [37] assumes the following general form for the virial coefficients

$$B^{(n)} = \sum_{m=0}^{n} \frac{n!}{m!\,(n-m)!} x_1^{n-m} x_2^m B_m^{(n)}, \tag{5.47}$$

where

$$B_m^{(n)} = \sum_{k=0}^{3} a_{nmk}\sigma_{11}^{3n-3m-k}\sigma_{22}^{3m+k-3}, \tag{5.48}$$

and determines the coefficients a_{nmk} from a series of conditions:

1) The virial coefficients must remain unchanged when replacing subscripts 1 and 2 by each other.
2) The virial coefficients must have the correct limiting value for $\sigma_{22}/\sigma_{11} \to 0$.
3) The derivative of the free energy with respect to σ_{22} when $\sigma_{22}/\sigma_{11} \to 0$ must have the correct value [24], given by Eq. (5.32).
4) The virial coefficients must reduce to those for the pure HS fluid when $\sigma_{22} = \sigma_{11}$.
5) When $\sigma_{22} = \sigma_{11}$ the derivative of the excess free energy of mixing with respect to σ_{22} must be zero. To this end, the contributions $B_m^{(n)}$ to the virial coefficients must satisfy the condition [37]

$$\left(\frac{\partial B_m^{(n)}}{\partial \sigma_{22}}\right)_{\sigma_{11}=\sigma_{22}} = 3\frac{m}{n}(n-1)\frac{B_m^{(n)}}{\sigma_{22}}. \tag{5.49}$$

These conditions, allow us to determine the coefficients a_{nmk}. In particular, the expressions for $B^{(2)}$ and $B^{(3)}$ thus obtained coincide with the exact ones (5.39) and (5.42), respectively. However, for higher order virial coefficients the corresponding expressions does not provide, as a particular case, the right values of $b_0^{(n)}$, the virial coefficients for the pure HS fluid. To ensure that this condition is fulfilled, Wheatley [37] adds two more terms of the form $a_{n,1,3/2}\sigma_{11}^{3n-3/2}\sigma_{22}^{3/2}$ and $a_{n,n-1,3/2}\sigma_{11}^{3/2}\sigma_{22}^{3n-3/2}$ respectively, with $a_{n,1,3/2} = a_{n,n-1,3/2}$, to Eq. (5.48). This extra parameter, is determined [37] from the condition that the resulting expression for $B^{(n)}$ must reproduce, for the particular case $\sigma_{11} = \sigma_{22} = \sigma$, any desired value $b_0^{(n)}$, eventually the exact one, for the monodisperse HS fluid. For $n \leq 3$ these new parameters are zero.

In [37] coefficients a_{nmk} were not reported. It seems worthwhile to do that here and therefore they are listed in Table 5.2 for $4 \leq n \leq 7$.

In Tables 5.3–5.6 the partial contributions to the virial coefficients obtained from the Wheatley's procedure are compared with "exact" numerical data of B_4–B_7 for different diameter ratios. We can see that the agreement is excellent. Therefore, the expressions derived by Wheatley [37] will be useful for interpolation purposes in the case of the virial coefficients B_4 and B_5 and even seems to be reliable for estimating the virial coefficients B_6 and B_7, for which the numerical data available are scarce. Of course, the same procedure can be used to estimate higher order virial coefficients, but the lack of numerical data for $n > 7$ prevents any assessment about the reliability of the procedure.

Table 5.2. Coefficients a_{nmk} in the Wheatley's expressions for the virial coefficients $B^{(4)}$ to $B^{(7)}$ of additive hard-sphere fluid mixtures

m	k = 0	1	2	3	3/2
			$n=4$		
0	0	0	0	$b_0^{(4)}$	
1	$\frac{\pi^3}{864}$	$\frac{\pi^3}{96}$	$\frac{3}{2}b_0^{(4)} - \frac{13\pi^3}{144}$	$\frac{5\pi^3}{144}$	$-\frac{1}{2}b_0^{(4)} + \frac{19\pi^3}{432}$
2	$\frac{\pi^3}{108}$	$\frac{1}{2}b_0^{(4)} - \frac{\pi^3}{108}$	$\frac{1}{2}b_0^{(4)} - \frac{\pi^3}{108}$	$\frac{\pi^3}{108}$	
3	$\frac{5\pi^3}{144}$	$\frac{3}{2}b_0^{(4)} - \frac{13\pi^3}{144}$	$\frac{\pi^3}{96}$	$\frac{\pi^3}{864}$	$-\frac{1}{2}b_0^{(4)} + \frac{19\pi^3}{432}$
4	$b_0^{(4)}$	0	0	0	
			$n=5$		
0	0	0	0	$b_0^{(5)}$	
1	$\frac{\pi^4}{6480}$	$\frac{\pi^4}{540}$	$\frac{9}{5}b_0^{(5)} - \frac{2\pi}{5}b_0^{(4)} - \frac{\pi^4}{432}$	$\frac{2\pi}{15}b_0^{(4)}$	$-\frac{4}{5}b_0^{(5)} + \frac{4\pi}{15}b_0^{(4)} - \frac{7\pi^4}{1620}$
2	$\frac{\pi^4}{810}$	$\frac{1}{5}b_0^{(5)} + \frac{7\pi^4}{3240}$	$\frac{4}{5}b_0^{(5)} - \frac{13\pi^4}{1620}$	$\frac{\pi^4}{216}$	
3	$\frac{\pi^4}{216}$	$\frac{4}{5}b_0^{(5)} - \frac{13\pi^4}{1620}$	$\frac{1}{5}b_0^{(5)} + \frac{7\pi^4}{3240}$	$\frac{\pi^4}{810}$	
4	$\frac{2\pi}{15}b_0^{(4)}$	$\frac{9}{5}b_0^{(5)} - \frac{2\pi}{5}b_0^{(4)} - \frac{\pi^4}{432}$	$\frac{\pi^4}{540}$	$\frac{\pi^4}{6480}$	$-\frac{4}{5}b_0^{(5)} + \frac{4\pi}{15}b_0^{(4)} - \frac{7\pi^4}{1620}$
5	$b_0^{(5)}$	0	0	0	

(Continued)

Table 5.2. (Continued)

m	k = 0	1	2	3	3/2
			$n=6$		
0	0	0	0	$b_0^{(6)}$	
1	$\frac{\pi^5}{46656}$	$\frac{5\pi^5}{7776}$	$2b_0^{(6)} - \frac{5\pi}{12}b_0^{(5)} + \frac{11\pi^5}{15552}$	$\frac{5\pi}{36}b_0^{(5)}$	$-b_0^{(6)} + \frac{5\pi}{18}b_0^{(5)} - \frac{\pi^5}{729}$
2	$\frac{\pi^5}{5832}$	$\frac{\pi^2}{54}b_0^{(4)} - \frac{\pi^5}{2916}$	$b_0^{(6)} - \frac{\pi^2}{27}b_0^{(4)} + \frac{\pi^5}{5832}$	$\frac{\pi^2}{54}b_0^{(4)}$	
3	$\frac{5\pi^5}{7776}$	$\frac{1}{2}b_0^{(6)} - \frac{5\pi^5}{7776}$	$\frac{1}{2}b_0^{(6)} - \frac{5\pi^5}{7776}$	$\frac{5\pi^5}{7776}$	
4	$\frac{\pi^2}{54}b_0^{(4)}$	$b_0^{(6)} - \frac{\pi^2}{27}b_0^{(4)} + \frac{\pi^5}{5832}$	$\frac{\pi^2}{54}b_0^{(4)} - \frac{\pi^5}{2916}$	$\frac{\pi^5}{5832}$	
5	$\frac{5\pi}{36}b_0^{(5)}$	$2b_0^{(6)} - \frac{5\pi}{12}b_0^{(5)} + \frac{11\pi^5}{15552}$	$\frac{5\pi^5}{7776}$	$\frac{\pi^5}{46656}$	$-b_0^{(6)} + \frac{5\pi}{18}b_0^{(5)} - \frac{\pi^5}{729}$
6	$b_0^{(6)}$	0	0	0	
			$n=7$		
0	0	0	0	$b_0^{(7)}$	
1	$\frac{\pi^6}{326592}$	$\frac{\pi^6}{18144}$	$\frac{15}{7}b_0^{(7)} - \frac{3\pi}{7}b_0^{(6)} + \frac{\pi^6}{15552}$	$\frac{\pi}{7}b_0^{(6)}$	$-\frac{8}{7}b_0^{(7)} + \frac{2\pi}{7}b_0^{(6)} - \frac{5\pi^6}{40824}$
2	$\frac{\pi^6}{40824}$	$-\frac{1}{7}b_0^{(7)} + \frac{5\pi^2}{252}b_0^{(5)} - \frac{\pi^6}{20412}$	$\frac{8}{7}b_0^{(7)} - \frac{5\pi^2}{126}b_0^{(5)} + \frac{\pi^6}{40824}$	$\frac{5\pi^2}{252}b_0^{(5)}$	
3	$\frac{5\pi^6}{54432}$	$\frac{2}{7}b_0^{(7)} + \frac{\pi^3}{378}b_0^{(4)} - \frac{5\pi^6}{27216}$	$\frac{5}{7}b_0^{(7)} - \frac{\pi^3}{189}b_0^{(4)} + \frac{5\pi^6}{54432}$	$\frac{\pi^3}{378}b_0^{(4)}$	
4	$\frac{\pi^3}{378}b_0^{(4)}$	$\frac{5}{7}b_0^{(7)} - \frac{\pi^3}{189}b_0^{(4)} + \frac{5\pi^6}{54432}$	$\frac{2}{7}b_0^{(7)} + \frac{\pi^3}{378}b_0^{(4)} - \frac{5\pi^6}{27216}$	$\frac{5\pi^6}{54432}$	
5	$\frac{5\pi^2}{252}b_0^{(5)}$	$\frac{8}{7}b_0^{(7)} - \frac{5\pi^2}{126}b_0^{(5)} + \frac{\pi^6}{40824}$	$-\frac{1}{7}b_0^{(7)} + \frac{5\pi^2}{252}b_0^{(5)} - \frac{\pi^6}{20412}$	$\frac{\pi^6}{40824}$	
6	$\frac{\pi}{7}b_0^{(6)}$	$\frac{15}{7}b_0^{(7)} - \frac{3\pi}{7}b_0^{(6)} + \frac{\pi^6}{15552}$	$\frac{\pi^6}{18144}$	$\frac{\pi^6}{326592}$	$-\frac{8}{7}b_0^{(7)} + \frac{2\pi}{7}b_0^{(6)} - \frac{5\pi^6}{40824}$
7	$b_0^{(7)}$	0	0	0	

Table 5.3. Partial contributions $B_m^{(4)}$ to the virial coefficient B_4 for different diameter ratios $\alpha \equiv 1/R$

α	Ref.	$B_1^{(4)}$	$B_2^{(4)}$	$B_3^{(4)}$
0.05	[30, 31]	0.05539(2)	$0.4280(5) \times 10^{-4}$	$0.177483(4) \times 10^{-7}$
	Eq. (5.48)	0.0555682	0.426574×10^{-4}	0.177451×10^{-7}
	Eqs. (5.50–5.51)	0.0552452	0.426574×10^{-4}	0.17702×10^{-7}
0.10	[30, 31]	0.08189(2)	$0.4040(6) \times 10^{-3}$	$0.119704(4) \times 10^{-5}$
	Eq. (5.48)	0.082255	0.400794×10^{-3}	0.119683×10^{-5}
	Eqs. (5.50–5.51)	0.0815397	0.400794×10^{-3}	0.119289×10^{-5}
0.20	[30, 31]	0.1589(4)	$0.4331(5) \times 10^{-2}$	$0.8480(5) \times 10^{-4}$
	Eq. (5.48)	0.15938	0.429468×10^{-2}	0.847949×10^{-4}
	Eqs. (5.50–5.51)	0.158053	0.429468×10^{-2}	0.845044×10^{-4}
0.30	[30, 31]	0.2729(4)	$0.1896(2) \times 10^{-1}$	$0.1064(2) \times 10^{-2}$
	Eq. (5.48)	0.273305	0.188174×10^{-1}	0.106484×10^{-2}
	Eqs. (5.50–5.51)	0.271679	0.188174×10^{-1}	0.106182
0.40	[30, 31]	0.4301(9)	$0.5682(7) \times 10^{-1}$	$0.6570(5) \times 10^{-2}$
	Eq. (5.48)	0.430341	0.565016×10^{-1}	0.65707×10^{-2}
	Eqs. (5.50–5.51)	0.428699	0.565016×10^{-1}	0.655692×10^{-2}
0.50	[30, 31]	0.636(1)	0.1375(2)	$0.2742(2) \times 10^{-1}$
	Eq. (5.48)	0.636861	0.13703	0.274302×10^{-1}
	Eqs. (5.50–5.51)	0.635416	0.13703	0.273913×10^{-1}
0.60	[30, 31]	0.899(1)	0.2898(4)	$0.8933(9) \times 10^{-1}$
	Eq. (5.48)	0.899267	0.289198	0.893421×10^{-1}
	Eqs. (5.50–5.51)	0.898153	0.289198	0.892643×10^{-1}
0.70	[30, 31]	1.224(2)	0.5535(7)	0.2449(3)
	Eq. (5.48)	1.22398	0.553079	0.245002
	Eqs. (5.50–5.51)	1.22325	0.553079	0.244888
0.80	[30, 31]	1.617(2)	0.983(1)	0.5919(7)
	Eq. (5.48)	1.61741	0.982399	0.592035
	Eqs. (5.50–5.51)	1.61705	0.982399	1.64712
0.90	[30, 31]	2.086(3)	1.647(2)	1.298(2)
	Eq. (5.48)	2.08602	1.64712	1.29828
	Eqs. (5.50–5.51)	2.08591	1.64712	1.29822

Table 5.4. Partial contributions $B_m^{(5)}$ to the virial coefficient B_5 for different diameter ratios $\alpha \equiv 1/R$

α	Ref.	$B_1^{(5)}$	$B_2^{(5)}$	$B_3^{(5)}$	$B_4^{(5)}$
0.10	[33]	$4.350(7) \times 10^{-2}$	$1.97(2) \times 10^{-4}$	$5.52(2) \times 10^{-7}$	$1.182(3) \times 10^{-9}$
	Eq. (5.48)	4.43492×10^{-2}	1.93336×10^{-4}	5.48979×10^{-7}	1.18206×10^{-9}
	Eqs. (5.50–5.51)	4.2925×10^{-2}	1.93336×10^{-4}	5.48979×10^{-7}	1.17391×10^{-9}
0.20	[33]	$9.60(1) \times 10^{-2}$	$2.328(6) \times 10^{-3}$	$4.253(7) \times 10^{-5}$	$6.464(8) \times 10^{-7}$
	Eq. (5.48)	9.72868×10^{-2}	2.29943×10^{-3}	4.2266×10^{-5}	6.48156×10^{-7}
	Eqs. (5.50–5.51)	9.46395×10^{-2}	2.29943×10^{-3}	4.2266×10^{-5}	6.43388×10^{-7}
0.30	[33]	0.1781(2)	$1.106(2) \times 10^{-2}$	$5.756(7) \times 10^{-4}$	$2.660(3) \times 10^{-5}$
	Eq. (5.48)	0.179643	1.09415×10^{-2}	5.72973×10^{-4}	2.66731×10^{-5}
	Eqs. (5.50–5.51)	0.176395	1.09415×10^{-2}	5.72973×10^{-4}	2.65069×10^{-5}
0.40	[33]	0.2965(4)	$3.535(2) \times 10^{-2}$	$3.802(4) \times 10^{-3}$	$3.789(5) \times 10^{-4}$
	Eq. (5.48)	0.297749	3.51668×10^{-2}	3.79452×10^{-3}	3.79968×10^{-4}
	Eqs. (5.50–5.51)	0.294465	3.51668×10^{-2}	3.79452×10^{-3}	3.78176×10^{-4}

0.50	[33]	0.4563(6)	$9.09(1) \times 10^{-2}$	$1.695(2) \times 10^{-2}$	$3.020(4) \times 10^{-3}$
	Eq. (5.48)	0.458058	9.03566×10^{-2}	1.69125×10^{-2}	3.02507×10^{-3}
	Eqs. (5.50–5.51)	0.455165	9.03566×10^{-2}	1.69125×10^{-2}	3.01521×10^{-3}
0.60	[33]	0.6662(9)	0.2010(2)	$5.865(6) \times 10^{-2}$	$1.664(2) \times 10^{-2}$
	Eq. (5.48)	0.667082	0.200461	5.85397×10^{-2}	1.66608×10^{-2}
	Eqs. (5.50–5.51)	0.664848	0.200461	5.85397×10^{-2}	1.66268×10^{-2}
0.70	[33]	0.930(1)	0.4011(7)	0.1702(2)	$7.10(1) \times 10^{-2}$
	Eq. (5.48)	0.931361	0.40056	0.16989	7.11295×10^{-2}
	Eqs. (5.50–5.51)	0.929896	0.40056	0.16989	7.10507×10^{-2}
0.80	[33]	1.262(2)	0.739(1)	0.4330(5)	0.2520(4)
	Eq. (5.48)	1.25746	0.739746	0.43283	0.251944
	Eqs. (5.50–5.51)	1.25672	0.739746	0.43283	0.251825
0.90	[33]	1.656(2)	1.284(1)	0.997(1)	0.774(1)
	(5.48)	1.65194	1.28433	0.997334	0.77356
	Eqs. (5.50–5.51)	1.65173	1.28433	0.997334	0.773473

Table 5.5. Partial contributions $B_m^{(6)}$ to the virial coefficient B_6 for the diameter ratio $\alpha \equiv 1/R = 0.10$

Ref.	$B_1^{(6)}$	$B_2^{(6)}$	$B_3^{(6)}$	$B_4^{(6)}$	$B_5^{(6)}$
[34, 35]	$0.2354(3)\times 10^{-1}$	$0.975(3)\times 10^{-4}$	$0.262(2)\times 10^{-6}$	$0.545(3)\times 10^{-9}$	$0.974(2)\times 10^{-12}$
Eq. (5.48)	0.286132×10^{-1}	0.971994×10^{-4}	0.261503×10^{-6}	0.551217×10^{-9}	0.980659×10^{-12}

Table 5.6. Partial contributions $B_m^{(7)}$ to the virial coefficient B_7 for the diameter ratio $\alpha \equiv 1/R = 0.10$

Ref.	$B_1^{(7)}$	$B_2^{(7)}$	$B_3^{(7)}$	$B_4^{(7)}$	$B_5^{(7)}$	$B_6^{(7)}$
[35, 36]	$1.27(2)\times 10^{-2}$	$5.0(1)\times 10^{-5}$	$1.28(6)\times 10^{-7}$	$= 2.6(2)\times 10^{-10}$	$4.6(2)\times 10^{-13}$	$7.2(3)\times 10^{-16}$
Eq. (5.48)	1.3048×10^{-2}	7.85067×10^{-5}	1.28305×10^{-7}	$= 2.63964\times 10^{-10}$	4.62454×10^{-13}	7.35542×10^{-16}

5.4.3 Analytical Expressions of the Higher Order Virial Coefficients from the Barrio–Solana Method

The virial coefficients proposed by Wheatley [37] in the form of Eqs. (5.47) and (5.48) does not depend explicitly on the crossed contact distance σ_{ij}. For certain purposes this explicit dependence may be necessary, as we will see later. Therefore, and alternative analytical approximation for the virial coefficients B_4 and B_5 was proposed by us in [38]. The starting point was the general expression (5.48) and the alternative expressions (5.45) and (5.46) of the crossed contributions to the second and third virial coefficients, respectively, in terms of the crossed contact distance σ_{ij}. Then, on the basis of the form of these contributions, we proposed general expressions for the crossed contributions to the higher order virial coefficients in the form

$$B_m^{(n)} = \left(\frac{\pi}{6}\right)^{n-1} \sigma_{ii}^{3(n-2)} \times \left(A_{nm0}\sigma_{ii}^3 + A_{nm1}\sigma_{ii}\sigma_{ij}^2 + A_{nm2}\sigma_{ij}^3 + A_{nm3}\sigma_{ii}^{3-b_n}\sigma_{ij}^{b_n}\right), \quad (5.50)$$

for $m = 1$, and the same expression for $m = n-1$ but replacing subscripts i and j by each other, and

$$B_m^{(n)} = \left(\frac{\pi}{6}\right)^{n-1} \sigma_{ii}^{3(n-m-1)}\sigma_{jj}^{3(m-1)} \times \left(A_{nm0}\sigma_{ii}^3 + A_{nm1}\sigma_{ii}\sigma_{ij}^2 + A_{nm2}\sigma_{jj}\sigma_{ij}^2 + A_{nm3}\sigma_{jj}^3 + A_{nm4}\sigma_{ij}^3\right) \quad (5.51)$$

for $1 < m \leq n/2$, and the same expression for $n/2 \leq m < n-1$ but again replacing subscripts i and j by each other.

Equation (5.50) retains the form of Eq. (5.46) except for the new term A_{nm3} which has been added in order to impose the condition that the virial coefficients $B^{(n)}$ of the mixture must reproduce, as particular cases, the desired values of virial coefficients $b_0^{(n)}$ of the pure fluid. Of course, for $n \leq 3$ these extra terms are zero. Equation (5.51) has the form of a combination of the equations corresponding to $B_{iij}^{(3)}$ and $B_{jji}^{(3)}$. The explicit expressions of the contributions of the form (5.50) and (5.51) for the fourth and fifth virial coefficients are as follows [38]

$$B_1^{(4)} = \left(\frac{\pi}{6}\right)^3 \sigma_{11}^6 \left(A_{410}\sigma_{11}^3 + A_{411}\sigma_{11}\sigma_{12}^2 + A_{412}\sigma_{12}^3 + A_{413}\sigma_{11}^{3-b_4}\sigma_{12}^{b_4}\right), \tag{5.52}$$

$$B_2^{(4)} = \left(\frac{\pi}{6}\right)^3 \sigma_{11}^3\sigma_{22}^3 \times \left(A_{420}\sigma_{11}^3 + A_{421}\sigma_{11}\sigma_{12}^2 + A_{422}\sigma_{22}\sigma_{12}^2 + A_{423}\sigma_{22}^3 + A_{424}\sigma_{12}^3\right), \tag{5.53}$$

$$B_3^{(4)} = \left(\frac{\pi}{6}\right)^3 \sigma_{22}^6 \left(A_{430}\sigma_{22}^3 + A_{431}\sigma_{22}\sigma_{12}^2 + A_{432}\sigma_{12}^3 + A_{433}\sigma_{22}^{3-b_4}\sigma_{12}^{b_4}\right), \tag{5.54}$$

and

$$B_1^{(5)} = \left(\frac{\pi}{6}\right)^4 \sigma_{11}^9 \left(A_{510}\sigma_{11}^3 + A_{511}\sigma_{11}\sigma_{12}^2 + A_{512}\sigma_{12}^3 + A_{513}\sigma_{11}^{3-b_5}\sigma_{12}^{b_5}\right), \tag{5.55}$$

$$B_2^{(5)} = \left(\frac{\pi}{6}\right)^4 \sigma_{11}^6\sigma_{22}^3 \times \left(A_{520}\sigma_{11}^3 + A_{521}\sigma_{11}\sigma_{12}^2 + A_{522}\sigma_{22}\sigma_{12}^2 + A_{523}\sigma_{22}^3 + A_{524}\sigma_{12}^3\right), \tag{5.56}$$

$$B_3^{(5)} = \left(\frac{\pi}{6}\right)^4 \sigma_{11}^3\sigma_{22}^6 \times \left(A_{530}\sigma_{22}^3 + A_{531}\sigma_{22}\sigma_{12}^2 + A_{532}\sigma_{11}\sigma_{12}^2 + A_{533}\sigma_{11}^3 + A_{534}\sigma_{12}^3\right) \tag{5.57}$$

$$B_4^{(5)} = \left(\frac{\pi}{6}\right)^4 \sigma_{22}^9 \left(A_{540}\sigma_{22}^3 + A_{541}\sigma_{22}\sigma_{12}^2 + A_{542}\sigma_{12}^3 + A_{543}\sigma_{22}^{3-b_5}\sigma_{12}^{b_5}\right), \tag{5.58}$$

respectively.

To determine the coefficients A_{mnk}, the same conditions (1–5) cited in the Sect. 5.4.2 were used, plus an additional condition derived from Eq. (5.29). In that equation, the excess free energy F^e of the mixture can be obtained from the compressibility factor in the form

$$\frac{F^e}{Nk_BT} = \int_0^\rho \frac{Z-1}{\rho}\mathrm{d}\rho. \tag{5.59}$$

For the particular case $\sigma_{11} = \sigma_{22} = \sigma$, Eq. (5.29) should give the contact value of the radial distribution function $g_0(\sigma)$ of the pure HS fluid, which is related to the compressibility factor by

$$g_0(\sigma) = \frac{3}{2} \frac{Z_0 - 1}{\pi \rho}. \tag{5.60}$$

Introducing Eq. (5.60) into Eq. (5.29), after some algebra, we obtain the condition 6)

$$Z_0 = 1 + \lim_{\sigma_{22}=\sigma_{11}=\sigma} \frac{1}{3(2-\delta_{ij}) x_i x_j \sigma_{ij}^2} \frac{\partial (F^{\mathrm{e}}/Nk_{\mathrm{B}}T)}{\partial \sigma_{ij}}. \tag{5.61}$$

When we use for the compressibility factors of the mixture the virial expansion (5.38) and the corresponding virial expansion

$$Z_0 = 1 + \sum_{n=2}^{\infty} b_0^{(n)} \rho^{n-1}, \tag{5.62}$$

for the monodisperse HS fluid, we obtain a condition relating the virial coefficients of the two fluids. It is obvious that the virial coefficients with partial contributions in the form (5.48) cannot fulfill this condition because they do not depend explicitly on $\sigma_{ij}, i \neq j$.

The values of the coefficients A_{nmk}, obtained for $n = 4$ and 5 from conditions 1)–6), taking $b_0^{(4)} = 0.2869495 \times (b_0^{(2)})^3$ and $b_0^{(5)} = 0.110252 \times (b_0^{(2)})^4$ [90],

Table 5.7. Values of the coefficients A_{nmk} in the Barrio and Solana expressions for the virial coefficients $B^{(4)}$ and $B^{(5)}$ of additive hard-sphere fluid mixtures

	k				
m	0	1	2	3	4
			$n = 4$		
0	$b_0^{(4)}$	0	0	0	
1	−0.178022	−60.5171	60	19.0599	
2	−0.394128	−26.3648	−26.3648	−0.394128	71.8826
3	−0.178022	−60.5171	60	19.0599	
4	0	0	0	0	$b_0^{(4)}$
			$n = 5$		
0	$b_0^{(5)}$	0	0	0	
1	−2.50916	−148.734	117.535	61.9335	
2	0.0299348	−67.8286	−52.8898	0.695228	148.218
3	0.0299348	−67.8286	−52.8898	0.695228	148.218
4	−2.50916	−148.734	117.535	61.9335	
5	0	0	0	0	$b_0^{(5)}$

with $b_0^{(2)} = (2/3)\pi\sigma^3$, are listed in Table 5.7. Moreover, $b_4 = 1.24217363$ and $b_5 = 1.29724267$ [38].

The predicted values of the partial contributions to $B^{(4)}$ and $B^{(5)}$ from Eqs. (5.50–5.51), or explicitly from Eqs. (5.52–5.58), are compared in Tables 5.3 and 5.4 with the exact numerical data and the values obtained from the Wheatley's procedure. We can see that the two theoretical calculations agree very well with each other and with numerical data for all diameter ratios.

5.5 Contact Values of the Radial Distribution Functions, Equation of State, and Consistency

In this section, we will review some theoretically based analytical expressions for the contact values of the RDFs and the equation of state. We will emphasize their performance with regard to the consistency conditions quoted in Sect. 5.3. Then, we will review a number of improvements proposed for these functions, derived on the basis of consistency conditions, as well as other expressions for the RDFs and the equation of state derived directly using certain consistency conditions. For convenience we will start from some well-known results for the monocomponent HS fluid.

5.5.1 Monocomponent HS Fluid

A great number of equation of state have been proposed for the monocomponent HS fluid, and most of them have been summarized in Chap. 3. However, we mention here only some results from the scaled particle theory. This theory provides expressions for the contact value $g(\sigma)$ of the RDF and the equation of state of a monodisperse HS fluid in the form [91, 92]

$$g_0^{\mathrm{SPT}}(\sigma) = \frac{1}{1-\eta} + \frac{3}{2}\frac{\eta}{(1-\eta)^2} + \frac{3}{4}\frac{\eta^2}{(1-\eta)^3}, \tag{5.63}$$

and

$$Z_0^{\mathrm{SPT}} = \frac{1+\eta+\eta^2}{(1-\eta)^3}, \tag{5.64}$$

respectively. This equation of state (5.63) is identical to $Z_0^{\mathrm{PY}-c}$, the solution provided by the Percus–Yevick (PY) theory from the compressibility route, derived in an entirely different way [93, 94]. In contrast, the equation of state resulting from the PY theory through the virial theorem is quite different

$$Z_0^{\mathrm{PY}-v} = \frac{1+2\eta+3\eta^2}{(1-\eta)^2}, \tag{5.65}$$

whose corresponding expression for the contact RDF is

$$g_0^{\mathrm{PY}-v}(\sigma) = \frac{1+\eta/2}{(1-\eta)^2}. \tag{5.66}$$

Some other well-known expressions have been summarized in Table 6.1 of Chap. 6

5.5.2 Fluid Mixtures of Additive HS

For mixtures of additive hard spheres, the SPT provides the following expression for the contact values of the RDFs [19]

$$g_{ij}^{\mathrm{SPT}}(\sigma_{ij}) = \frac{1}{1-\xi_3} + \frac{3\xi_2}{(1-\xi_3)^2}\frac{\sigma_{ii}\sigma_{jj}}{\sigma_{ii}+\sigma_{jj}} + \frac{3\xi_2^2}{(1-\xi_3)^3}\left(\frac{\sigma_{ii}\sigma_{jj}}{\sigma_{ii}+\sigma_{jj}}\right)^2, \tag{5.67}$$

which leads to the equation of state [19]

$$Z^{\mathrm{SPT}} = \frac{6}{\pi\rho}\left[\frac{\xi_0}{1-\xi_3} + 3\frac{\xi_1\xi_2}{(1-\xi_3)^2} + 3\frac{\xi_2^3}{(1-\xi_3)^3}\right], \tag{5.68}$$

where

$$\xi_l = \frac{\pi}{6}\rho\sum_l x_i\sigma_i^l, \tag{5.69}$$

so that $\xi_3 = \eta$.

Equation (5.68) is equal to that resulting [39] from the PY theory using the compressibility route, $Z^{\mathrm{PY}-c} = Z^{SPT}$. Using the virial route, the PY theory provides results quite different

$$g_{ij}^{\mathrm{PY}-v}(\sigma_{ij}) = \frac{1}{1-\xi_3} + \frac{3\xi_2}{(1-\xi_3)^2}\frac{\sigma_{ii}\sigma_{jj}}{\sigma_{ii}+\sigma_{jj}}, \tag{5.70}$$

and

$$Z^{\mathrm{PY}-v} = Z^{SPT} - \frac{18}{\pi\rho}\frac{\xi_3\xi_2^3}{(1-\xi_3)^3}. \tag{5.71}$$

5.5.3 Semiempirical Corrections

The values of the equation of state predicted by the PY compressibility for the monodisperse HS fluid are higher, whereas those given by the PY virial equation are lower than the simulation data. Much better agreement with simulation data is achieved by the celebrated Carnahan–Starling (CS) equation [71]

$$Z_0^{\mathrm{CS}} = \frac{1+\eta+\eta^2-\eta^3}{(1-\eta)^3}, \tag{5.72}$$

which can be obtained by adding the PY compressibility and virial equations weighed by 2/3 and 1/3, respectively, although originally it was derived in a different way. The expression for the the contact value of the RDF corresponding to the CS equation of state is

$$g_0^{\mathrm{CS}}(\sigma) = \frac{1}{1-\eta} + \frac{3}{2}\frac{\eta}{(1-\eta)^2} + \frac{1}{2}\frac{\eta^2}{(1-\eta)^3} \tag{5.73}$$

It was observed [40] that the preceding result, and consequently the CS equation of state (5.72), can be obtained from the SPT expression (5.63) by weighting the latter by 2/3. Then, it was proposed to perform the same weighting on the SPT expression (5.67) for HS fluid mixtures. This leads to

$$g_{ij}^{\mathrm{BMCSL}}(\sigma_{ij}) = \frac{1}{1-\xi_3} + \frac{3\xi_2}{(1-\xi_3)^2}\frac{\sigma_{ii}\sigma_{jj}}{\sigma_{ii}+\sigma_{jj}} + \frac{2\xi_2^2}{(1-\xi_3)^3}\left(\frac{\sigma_{ii}\sigma_{jj}}{\sigma_{ii}+\sigma_{jj}}\right)^2 . \tag{5.74}$$

We denote this result as BMCSL because it leads to the equation of state

$$Z^{\mathrm{BMCSL}} = \frac{1}{\xi_0}\left[\frac{\xi_0}{1-\xi_3} + \frac{3\xi_1\xi_2}{(1-\xi_3)^2} + \frac{(3-\xi_3)\,\xi_2^3}{(1-\xi_3)^3}\right], \tag{5.75}$$

derived independently by Boublík [40] in the mentioned way, and by Mansoori, Carnahan, Starling, and Leland [41] by weighting the PY compressibility and virial equations as indicated before for the monodisperse fluid. However, strictly speaking the latter authors did not report the expression for $g_{ij}(\sigma_{ij})$.

5.5.4 Improved Expressions for the RDFs and the Equation of State Based on Consistency Conditions

The BMCSL expression (5.74) for $g_{ij}(\sigma_{ij})$ is remarkably accurate for moderate values of the diameter ratio R and the mole fraction x_1. For extreme values of these quantities, several corrections have been proposed [28, 43, 44, 45] to improve the accuracy. These corrections are often based on consistency conditions.

Thus, on the basis of the exact result (5.27) for $g_{12}(\sigma_{12})$ in the colloidal limit, and an approximate result [95] for $g_{11}(\sigma_{11})$ in that limit, the following expressions were proposed [43]

$$g_{11}^{\mathrm{HC}}(\sigma_{11}) = g_{11}^{\mathrm{BMCSL}}(\sigma_{11}) + \exp\left[\frac{3\,(\xi_2 R\sigma_{22}-\xi_3)}{2\,(1-\xi_3)^2}\right] - 1, \tag{5.76}$$

and

$$g_{12}^{\mathrm{YCH}}(\sigma_{12}) = g_{12}^{\mathrm{BMCSL}}(\sigma_{12}) + \frac{\xi_2^2 (R^2-1)}{(1-\xi_3)^3 (R+1)^2}\sigma_{22}^2 - \frac{\xi_2^3 (R^3-1)}{(1-\xi_3)^3 (R+1)^3}\sigma_{22}^3. \quad (5.77)$$

whereas for $g_{22}(\sigma_{22})$, which corresponds to the smaller spheres, Eq. (5.74) is retained. Subsequently, Eq. (5.76) was slightly modified [44] in the form

$$g_{11}^{\mathrm{HC}}(\sigma_{11}) = g_{11}^{\mathrm{BMCSL}}(\sigma_{11}) + f_1(x), \quad (5.78)$$

where

$$f_1(x) = \exp(x) - 1 - x, \quad (5.79)$$

with

$$x = \left(\frac{\xi_2\sigma_{22}}{\xi_3}\right)^3 \frac{3(\xi_2\sigma_{11} - \xi_3)}{2(1-\xi_3)^2}. \quad (5.80)$$

The equation of state that results from Eqs. (5.74), (5.77), (5.78), and the virial theorem Eq. (5.14), is

$$Z^{\mathrm{YCH}} = Z^{\mathrm{BMCSL}} + \frac{2}{3}\pi\rho\left[x_1^2\sigma_{11}^3 f_1(x) + 2x_1x_2\sigma_{12}^3 f_2(y)\right], \quad (5.81)$$

where

$$f_2(y) = \frac{\xi_2^2}{(1-\xi_3)^3}\frac{R^2-1}{(R+1)^2}\sigma_{22}^2 - \frac{\xi_2^3}{(1-\xi_3)^3}\frac{R^3-1}{(R+1)^3}\sigma_{22}^3. \quad (5.82)$$

Another improvement to the BMCSL results was reported in [28]. It was based on the fact that the BMCSL expression (5.74) for $g_{11}(\sigma_{11})$ predicts values too low, as compared with simulation data, and the deviation increases as R increases and x_1 decreases. In contrast, the BMCSL expressions for $g_{12}(\sigma_{12})$ and $g_{22}(\sigma_{22})$ are more accurate. Moreover, the BMCSL expression for $g_{12}(\sigma_{12})$ does not satisfy condition 9), Eq. (5.27). Therefore, it was proposed to add to Eq. (5.74) and "ad hoc" term fulfilling the following conditions, *i)* the resulting values of $g_{11}(\sigma_{11})$ should be in closer agreement with simulation data with little effect on the values of $g_{12}(\sigma_{12})$ and $g_{22}(\sigma_{22})$; *ii)* conditions 1)–9) from Sect. 5.3 should be satisfied; and *iii)* for the limiting case of a pure fluid, $g_{11}(\sigma_{11})$ will reduce to the CS expression (5.73).

The form proposed for the additional term was [28]

$$\alpha(\xi_3)(1-\delta_{ij}x_i)\frac{f(\xi_l)}{(1-\xi_3)^3}\frac{\sigma_{11}-\sigma_{22}}{\sigma_{12}}\sigma_{11}\sigma_{22}\frac{\sigma_{ii}\sigma_{jj}}{\sigma_{ij}}\left[\delta_{ij} + (1-\delta_{ij})\frac{\sigma_{22}}{\sigma_{11}}\right], \quad (5.83)$$

where δ_{ij} is the Kronecker delta function and $\alpha(\xi_3)$ and $f(\xi_l)$ are functions that were determined from the above-mentioned consistency conditions, with the result $f(\xi_l) = \xi_1\xi_2$ and $\alpha(\xi_3) = (1-\xi_3)/4$. The resulting expressions for the contact RDFs are

$$g_{11}(\sigma_{11}) = g_{11}^{\mathrm{BMCSL}}(\sigma_{11}) + \frac{x_2}{4}\frac{\xi_1\xi_2}{(1-\xi_3)^2}\frac{\sigma_{11}-\sigma_{22}}{\sigma_{12}}\sigma_{11}^2\sigma_{22}, \tag{5.84}$$

$$g_{12}(\sigma_{12}) = g_{21}(\sigma_{21}) = g_{12}^{\mathrm{BMCSL}}(\sigma_{12}) + \frac{1}{4}\frac{\xi_1\xi_2}{(1-\xi_3)^2}\frac{\sigma_{11}-\sigma_{22}}{\sigma_{12}}\frac{\sigma_{11}\sigma_{22}^3}{\sigma_{12}}, \tag{5.85}$$

and

$$g_{22}(\sigma_{22}) = g_{22}^{\mathrm{BMCSL}}(\sigma_{22}) + \frac{x_1}{4}\frac{\xi_1\xi_2}{(1-\xi_3)^2}\frac{\sigma_{11}-\sigma_{22}}{\sigma_{12}}\sigma_{11}\sigma_{22}^2, \tag{5.86}$$

and, using the virial theorem (5.14), the corresponding equation of state is

$$Z^{\mathrm{BS}} = Z^{\mathrm{BMCSL}} + 2\frac{\xi_1\xi_2}{(1-\xi_3)^2}\frac{R-1}{R+1}x_1x_2\sigma_{11}\sigma_{22}\left[\xi_4 + 2\xi_0\sigma_{22}^2\sigma_{12}^2\right]. \tag{5.87}$$

Matyushov and Ladanyi [26], starting from the exact result (5.35) arising from the SPT, proposed for the excess chemical potential of the species 1 in the limit $R \to \infty$ and $x_1 \to 0$ an expression of the form

$$\mu_i^e = \sum_{n=0}^{3} a_n d^n, \tag{5.88}$$

and determined the coefficients a_i by requiring the fulfillment of the consistency conditions 13)–16), given by Eqs. (5.33)–(5.36), respectively, taking $g_{ii}(\sigma_{ii}) = g_0^{CS}$ as given by Eq. (5.73). Then, from condition 17), given by Eq. (5.37), they obtained in the infinite dilution limit of the larger spheres

$$\begin{aligned} g_{12}(\sigma_{12}) = {} & \frac{1}{1-\xi_3}\frac{1}{(d+1)^2} + \frac{(2-\xi_3)(1+\xi_3)}{(1-\xi_3)^2}\frac{d}{(d+1)^2} \\ & + \frac{1+\xi_3+\xi_3^2-\xi_3^3}{(1-\xi_3)^3}\left(\frac{d}{d+1}\right)^2; \quad x_1 \to 0. \end{aligned} \tag{5.89}$$

On the basis of this result, for $x_1 > 0$ Matyushov and Ladanyi [26] proposed the expression

$$\begin{aligned} g_{12}^{\mathrm{ML}}(\sigma_{12}) = {} & \frac{1}{1-\xi_3}\frac{1}{\delta^2(d+1)^2} + \frac{(2-\xi_3)(1+\xi_3)}{(1-\xi_3)^2}\frac{d}{(d+1)^2} \\ & + \frac{1+\xi_3+\xi_3^2-\xi_3^3}{(1-\xi_3)^3}\left(\frac{d}{d+1}\right)^2\left[\delta(1-x_1) + \delta^2 x_1\right], \end{aligned} \tag{5.90}$$

where

$$\delta = \frac{1+(d^2-1)x_1}{1+(d^3-1)x_1}. \tag{5.91}$$

Boublík [46] proposed and improved expression for the equation of state of additive mixtures of hard sphere based on similar grounds as that of Matyushov and Ladanyi. The starting point was the BMCSL equation (5.75) and the consistency conditions 13)–15), Eqs. (5.33)–(5.35). By integrating the equation of state to obtain the chemical potential, it was realized that the BMCSL equation fulfils conditions 13) and 14), but not condition 15). Therefore, there was proposed to correct the BMCSL expression for the chemical potential by adding a suitable correction term, thus

$$\mu_i^e = \left(\mu_i^e\right)^{\mathrm{BMCSL}} + \left(\mu_i^e\right)^{\mathrm{cor}} . \tag{5.92}$$

For the correction term $(\mu_i)^{\mathrm{cor}}$, an expression like that in Eq. (5.88) was assumed. The coefficients a_n were determined by requiring the fulfilment of conditions 14) and 15) plus the condition $(\mu_i)^{\mathrm{cor}} = 0$ for $R = 1$. As a consequence, the resulting correction term is valid only for $x_1 = 0$. Using this result, together with other plausible arguments, the last term in the BMCSL equation (5.75) was modified. This gave rise to an equation of state, self-consistent regarding conditions 13)–15), which reads [46]

$$\begin{aligned} Z^B &= \frac{6}{\pi\rho}\left[\frac{\xi_0}{1-\xi_3} + \frac{3\xi_1\xi_2}{(1-\xi_3)^2} + \frac{\left(3-\xi_3+\xi_3^2/3\right)\xi_2^3}{(1-\xi_3)^3}\right] \\ &= Z^{\mathrm{BMCSL}} + \frac{1}{3}\frac{\xi_3^2\xi_2^3}{\xi_0\,(1-\xi_3)^3} \end{aligned} \tag{5.93}$$

Viduna and Smith [45] start from a general expression of the form of the BMCSL (5.74), namely

$$\begin{aligned} g_{ij}(\sigma_{ij}) = {} & \frac{1}{1-\xi_3} \\ & + f_1(\xi_3)\frac{\xi_2}{\xi_3}\frac{\sigma_{ii}\sigma_{jj}}{\sigma_{ij}} + f_2(\xi_3)\frac{(1-c)\,\xi_2^2 + c\xi_1\xi_3}{\xi_3^2}\left(\frac{\sigma_{ii}\sigma_{jj}}{\sigma_{ij}}\right)^2 , \end{aligned} \tag{5.94}$$

which reduces to the BMCSL form (5.74) for $c = 0$, $f_1(\xi_3) = (3/2)\xi_3/(1-\xi_3)^2$, and $f_2(\xi_3) = (1/2)\xi_3^2/(1-\xi_3)^3$, and require that the general expression (5.94) in the infinite dilution limit of the solute must reproduce the Matyushov and Ladanyi expression (5.89). This gives $f_1(\xi_3) = (3-\xi_3)/2(1-\xi_3)^2$ and $f_2(\xi_3) = (2-\xi_3)\xi_3^2/2(1-\xi_3)^3$. They settle parameter c to 1/3 in order to obtain good agreement with simulation data and introduce small additional modifications into functions f_1 and f_2 on a heuristic way to improve accuracy. The final result is

$$\begin{aligned} g_{ij}^{\mathrm{VS}}(\sigma_{ij}) = {} & \frac{1}{1-\xi_3} + \frac{\left(3-\xi_3+\xi_3^2/2\right)}{2\,(1-\xi_3)^2}\frac{\xi_2}{\xi_3}\frac{\sigma_{ii}\sigma_{jj}}{\sigma_{ij}} \\ & + \frac{\left(2-\xi_3-\xi_3^2/2\right)\left(2\xi_2^2+\xi_1\xi_3\right)}{6\,(1-\xi_3)^3}\left(\frac{\sigma_{ii}\sigma_{jj}}{\sigma_{ij}}\right)^2 , \end{aligned} \tag{5.95}$$

which introduced into the virial theorem (5.14) gives the equation of state [45]

$$Z^{\mathrm{VS}} = \frac{1}{\xi_0}\left[\frac{\xi_0}{1-\xi_3} + \frac{\left(16 - 2\xi_3 + 5\xi_3^3\right)\xi_1\xi_2}{6\left(1-\xi_3\right)^2} + \frac{2\xi_1\xi_2 + \left(18 - 8\xi_3 + \xi_3^2 - 7\xi_3^3\right)\xi_2^3}{6\left(1-\xi_3\right)^3}\right]. \tag{5.96}$$

Conversely, the same authors use consistency conditions to derive expressions for the contact values of the RDFs $g_{ij}(\sigma_{ij})$ from the equation of state [96]. They start from a generalization to multicomponent mixtures of expression (5.88) for the chemical potential. Considering an m-component mixture of hard spheres with diameters σ_{ii}, according to the SPT, the work required to insert an additional sphere with diameter $\sigma_{m+1,m+1}$ in the infinite dilution limit will be

$$\mu_{m+1}^e = \sum_{n=0}^{3} a_n \sigma_{m+1,m+1}^n. \tag{5.97}$$

On the other hand, the generalization of condition 16) (5.36) is [27, 96]

$$\beta\left(\frac{\partial \mu_{m+1}^e}{\partial \sigma_{m+1,m+1}}\right)_{V,T,N_j,\sigma_{jj}\neq m+1,m+1} = 2\pi\rho\sum_{j=1}^{m} x_j \sigma_{j,m+1}^2 g_{j,m+1}\left(\sigma_{j,m+1}\right); \quad x_{m+1} = 0, \tag{5.98}$$

Combining Eqs. (5.97) and (5.98), taking $\sigma_{m+1,m+1} = \sigma_{ii}$ where subscript ii here refers to one of the solvent spheres, and removing the constraint of infinite dilution, Viduna and Smith obtain for the contact values of the RDFs the expression

$$g_{ij}\left(\sigma_{ij}\right) = A + B\frac{\sigma_{ii}\sigma_{jj}}{\sigma_{ij}} + C\left(\frac{\sigma_{ii}\sigma_{jj}}{\sigma_{ij}}\right)^2. \tag{5.99}$$

Coefficient A is determined from this equation and condition (5.98) for $\sigma_{11} = 0$. To determine B and C one can use the same condition for $m = 2$ and any equation of state suitable for a ternary mixture together with Eq. (5.99). Replacing σ_{33} by σ_{11} and σ_{33} by σ_{22} successively, one obtains two equations whose solution yields these unknowns. It is to be noted that when this procedure is used in conjunction with the BMCSL equation of state (5.75) the resulting expressions for the $g_{ij}(\sigma_{ij})$ are different from those given by Eq. (5.74) [96], due to thermodynamic inconsistency. The same is true for other equations of state.

Table 5.8. Comparison of the values of $g_{11}(\sigma_{11})$ obtained from several analytical expressions with the simulation data from [13] for the diameter ratio $R = 10$ and several values of the mole fraction x_1 of the bigger spheres

	η					
	0.35	0.40	0.45	0.50	0.55	0.60
			$x_1 = 0.1$			
Simulation	3.1772	3.9904	5.0432	6.5917	9.1647	
BMCSL	3.1411	3.8993	4.9389	6.4078	8.5604	
YCH	3.1411	3.8993	4.9389	6.4078	8.5604	
BS	3.1628	3.9326	4.9890	6.4827	8.6722	
VS	3.2497	4.0664	5.1878	6.7685	9.0686	
RVE	3.0476	3.9479	5.0017	6.4830	8.6389	
SHS	3.1800	3.9559	5.0195	6.5225	8.7266	
			$x_1 = 0.005$			
Simulation	8.8552	11.627	16.295	23.591	36.875	52.762
BMCSL	6.0242	8.1262	11.167	15.700	22.704	34.018
YCH	6.0246	8.1269	11.169	15.703	22.708	34.027
BS	6.5372	8.9125	12.352	17.469	25.347	37.999
VS	7.8668	10.985	15.514	22.236	32.494	48.726
RVE	3.0613	2.9394	2.2067	0.25619	-4.1246	-13.467
SHS	3.2680	3.3937	3.1648	2.2275	-0.11192	-5.2800

In Tables 5.8–5.11, the results from the preceding expressions for $g_{ij}(\sigma_{ij})$ and/or Z are compared with the simulation data from [13] for the diameter ratio $R = 10$ and several compositions. We can see in these tables that none of the approximations considered provides, on the whole, better performance than the others. One remarkable fact is that the BMCSL expressions (5.74)–(5.75) continue doing an excellent job in most cases. Exceptions are the values of $g_{11}(\sigma_{11})$ predicted by Eq. (5.74) that are too low for extremely low values of the mole fraction x_1. In these cases, expression (5.95) works better, although its accuracy is not completely satisfactory. Regarding the equation of state, the best overall results are those from Eq. (5.87), although the differences between most of the equations of state considered are small.

From the preceding results, we can conclude that the use of consistency conditions can be a useful way to introduce changes in the available expressions of the equation of state or the contact values of the radial distribution functions in order to improve their accuracy, but further developments in the field are still needed, especially for the function $g_{11}(\sigma_{11})$.

Table 5.9. As in Table 5.8 for $g_{12}(\sigma_{12})$

	η					
	0.35	0.40	0.45	0.50	0.55	0.60
			$x_1 = 0.1$			
Simulation	1.7844	1.9916	2.2558	2.5824	3.0013	
BMCSL	1.7911	2.0083	2.2799	2.6278	3.0865	
YCH	1.7952	2.0151	2.2910	2.6460	3.1164	
BS	1.7916	2.0090	2.2810	2.6293	3.0887	
VS	1.7755	1.9853	2.2465	2.5795	3.0162	
RVE	1.7841	1.9969	2.2614	2.5979	3.0377	
SHS	1.7917	2.0130	2.2941	2.6626	3.1640	
			$x_1 = 0.005$			
Simulation	2.1251	2.4724	2.9174	3.5195	4.2953	5.5074
BMCSL	2.1477	2.4985	2.9550	3.5658	4.4107	5.6286
YCH	2.1686	2.5327	3.0106	3.6559	4.5583	5.8755
BS	2.1571	2.5129	2.9767	3.5981	4.4590	5.7014
VS	2.1643	2.5270	3.0025	3.6428	4.5334	5.8227
RVE	2.2151	2.6136	3.1495	3.8942	4.9702	6.6014
SHS	2.2171	2.6177	3.1577	3.9104	5.0024	6.6670

Table 5.10. As in Table 5.8 for $g_{22}(\sigma_{22})$

	η					
	0.35	0.40	0.45	0.50	0.55	0.60
			$x_1 = 0.1$			
Simulation	1.6735	1.8595	2.0642	2.3316	2.6788	
BMCSL	1.6753	1.8510	2.0663	2.3358	2.6817	
YCH	1.6753	1.8510	2.0663	2.3358	2.6817	
BS	1.6755	1.8514	2.0669	2.3366	2.6830	
VS	1.6648	1.8354	2.0436	2.3029	2.6341	
RVE	1.6473	1.8023	1.9826	2.1921	2.4333	
SHS	1.6574	1.8237	2.0260	2.2780	2.6010	
			$x_1 = 0.005$			
Simulation	1.8622	2.1038	2.4107	2.8051	3.3333	4.0636
BMCSL	1.8622	2.1053	2.4125	2.8103	3.3414	4.0776
YCH	1.8622	2.1053	2.4125	2.8103	3.3414	4.0776
BS	1.8625	2.1057	2.4131	2.8112	3.3427	4.0796
VS	1.8538	2.0940	2.3977	2.7915	3.3178	4.0480
RVE	1.8456	2.0771	2.3650	2.7306	3.2062	3.8437
SHS	1.8537	2.0943	2.4003	2.8008	3.3447	4.1177

Table 5.11. As in Table 5.8 for Z

	η 0.35	0.40	0.45	0.50	0.55	0.60
			$x_1 = 0.1$			
Simulation	2.2010	2.6020	3.1330	3.8770	5.0050	
BMCSL	2.1989	2.5958	3.1290	3.8675	4.9292	
YCH	2.2006	2.5991	3.1349	3.8783	4.9488	
BS	2.2021	2.6014	3.1385	3.8833	4.9551	
B	2.2012	2.6009	3.1394	3.8887	4.9718	
VS	2.2074	2.6112	3.1552	3.9098	4.9933	
RVE	2.2006	2.5975	3.1291	3.8624	4.9101	
SHS	2.2044	2.6067	3.1504	3.9100	5.0147	
			$x_1 = 0.005$			
Simulation	2.3040	2.7260	3.2900	4.0670	5.1590	6.7890
BMCSL	2.2960	2.7144	3.2697	4.0284	5.1017	6.6868
YCH	2.3041	2.7295	3.2973	4.0782	5.1915	6.8505
BS	2.3027	2.7261	3.2896	4.0613	5.1558	6.7757
B	2.2977	2.7181	3.2774	4.0439	5.1329	6.7496
VS	2.3112	2.7431	3.3216	4.1192	5.2575	6.9509
RVE	2.3009	2.7232	3.2850	4.0546	5.1464	6.7635
SHS	2.3047	2.7326	3.3068	4.1032	5.2531	6.9976

5.5.5 Consistency of the SPT-Related Expressions for the Contact Values of the Radial Distribution Functions and the Equation of State

It is worth analyzing whether the equations of state cited in the preceding subsections fulfil the consistency conditions 1)–17) given in Sect. 5.3, apart from the fact that several of these equations have been derived on the basis of imposing the fulfilment of a number of these conditions. To this end, we can obtain the excess, or residual, free energy from Eq. (5.59) and then the excess chemical potential for the component i is obtained as $\mu_i^e = (\partial F^e/\partial N_i)$. However, Eq. (5.81) does not lead to an analytical expression for F^e, and therefore conditions 11)–17) cannot be readily checked.

On the other hand, the expressions for the contact values $g_{ij}(\sigma_{ij})$ corresponding to the equation of state (5.93) were not reported. They have been obtained from condition 11). The results are summarized in Table 5.12. From these results it is apparent that the attempts to introduce improvements on existing expressions for the equation of state and/or the RDFs by constraining them to satisfy certain consistency conditions may result in the violation of others. Therefore, while the use of consistency conditions may be useful to improve these expressions, this procedure must be used with caution.

Table 5.12. Summary of the consistency conditions that are satisfied by the expressions for the contact values of the radial distribution functions and the equations of state considered

	EOS						
Condition	SPT	PY	BMCSL	YCH	BS	B	VS
1	√	√	√	√	√	√	√
2	√	√	√	√	√	√	√
3	√	√	√	√	√	×	√
4	√	√	√	√	√	×	√
5	√	√	√	√	√	×	√
6	√	√	√	√	√	×	√
7	√	√	√	√	√	√	√
8	√	√	√	√	√	×	√
9	√	×	×	√	√	×	×
10	√	×	×	×	×	√	×
11	×	×	×		×	√	×
12	√	√	√		×	√	×
13	√	√	√		√	√	×
14	√	√	√		×	√	×
15	×	×	×		√	×	×
16	√	√	√		×	√	√
17	√	√	√		×	×	√

5.6 Deriving Density Expansions of the Contact Values of the Radial Distribution Functions from the Approximate Virial Expansion and Consistency Conditions

The fact that the equation of state can be expressed as a virial expansion of the form (5.38) implies that the contact values of the RDFs $g_{ij}(\sigma_{ij})$ can be expressed in a similar way, namely

$$g_{ij}\left(\sigma_{ij}\right) = 1 + \sum_{n=3}^{\infty} \alpha_{ij}^{(n)} \rho^{n-2} \tag{5.100}$$

where $\alpha_{ij}^{(n)}$ is the contribution of the virial coefficient of order n to $g_{ij}(\sigma_{ij})$. These contributions can be determined [23, 47] from the consistency conditions 1)–17) given in Sect. 5.3, not necessarily all of them, provided that appropriate analytical expressions of the virial coefficients are available. To this end, the virial coefficients must be expressed explicitly in terms of the three collision distances σ_{11}, σ_{22}, and σ_{12}. For this reason, expression (5.46) for the contribution iij to the third virial coefficient is appropriate whereas expression (5.44) is not. For the same reason, the Wheatley approximate general expression (5.48) for the partial contributions to the higher order virial coefficients is not appropriate for this purpose whereas the Barrio and Solana approximate general expressions (5.50) and (5.51) are appropriate.

From these analytical expressions (5.46) and (5.50)–(5.51) for the partial contributions to the third and higher order virial coefficients, respectively, the corresponding expressions for the contributions $\alpha_{ij}^{(n)}$, have the following forms [47]. For $n = 3$

$$\begin{aligned}\alpha_{11}^{(3)} &= \frac{3}{2}\left\{x_1 \alpha_{11}^{(3,1)} \sigma_{11}^3 + 3x_2\left[\alpha_{11}^{(3,2)} \sigma_{11}^3 + \alpha_{11}^{(3,3)} \sigma_{11} \sigma_{12}^2 + \alpha_{11}^{(3,4)} \sigma_{12}^3\right]\right\} \\ &+ \frac{3}{2}\left\{3\frac{x_2^2}{x_1}\left[\alpha_{11}^{(3,5)} \frac{\sigma_{22}^6}{\sigma_{11}^3} + \alpha_{11}^{(3,6)} \frac{\sigma_{22}^4}{\sigma_{11}^3} \sigma_{12}^2 + \alpha_{11}^{(3,7)} \frac{\sigma_{22}^3}{\sigma_{11}^3} \sigma_{12}^3\right] + \frac{x_2^3}{x_1^2} \alpha_{11}^{(3,8)} \frac{\sigma_{22}^6}{\sigma_{11}^3}\right\},\end{aligned} \tag{5.101}$$

$$\begin{aligned}\alpha_{22}^{(3)} &= \frac{3}{2}\left\{\frac{x_1^3}{x_2^2} \alpha_{22}^{(3,1)} \frac{\sigma_{11}^6}{\sigma_{22}^3} + 3\frac{x_1^2}{x_2}\left[\alpha_{22}^{(3,2)} \frac{\sigma_{11}^6}{\sigma_{22}^3} + \alpha_{22}^{(3,3)} \frac{\sigma_{11}^4}{\sigma_{22}^3} \sigma_{12}^2 + \alpha_{22}^{(3,4)} \frac{\sigma_{11}^3}{\sigma_{22}^3} \sigma_{12}^3\right]\right\} \\ &+ \frac{3}{2}\left\{3x_1\left[\alpha_{22}^{(3,5)} \sigma_{22}^3 + \alpha_{22}^{(3,6)} \sigma_{22} \sigma_{12}^2 + \alpha_{22}^{(3,7)} \sigma_{12}^3\right] + x_2 \alpha_{22}^{(3,8)} \sigma_{22}^3\right\},\end{aligned} \tag{5.102}$$

$$\begin{aligned}\alpha_{12}^{(3)} &= \frac{3}{4}\left\{\frac{x_1^2}{x_2} \alpha_{12}^{(3,1)} \frac{\sigma_{11}^6}{\sigma_{12}^3} + 3x_1\left[\alpha_{12}^{(3,2)} \frac{\sigma_{11}^6}{\sigma_{12}^3} + \alpha_{12}^{(3,3)} \frac{\sigma_{11}^4}{\sigma_{12}} + \alpha_{12}^{(3,4)} \sigma_{11}^3\right]\right\} \\ &+ \frac{3}{4}\left\{3x_2\left[\alpha_{12}^{(3,5)} \frac{\sigma_{22}^6}{\sigma_{12}^3} + \alpha_{12}^{(3,6)} \frac{\sigma_{22}^4}{\sigma_{12}} + \alpha_{12}^{(3,7)} \sigma_{22}^3\right] + \frac{x_2^2}{x_1} \alpha_{12}^{(3,8)} \frac{\sigma_{22}^6}{\sigma_{12}^3}\right\}.\end{aligned} \tag{5.103}$$

For $n = 4$

$$\alpha_{11}^{(4)} = \frac{3}{2}\left\{x_1^2\alpha_{11}^{(4,1)}\sigma_{11}^6 + 4x_1x_2\left[\alpha_{11}^{(4,2)}\sigma_{11}^6 + \alpha_{11}^{(4,3)}\sigma_{11}^4\sigma_{12}^2 + \alpha_{11}^{(4,4)}\sigma_{11}^3\sigma_{12}^3\right.\right.$$
$$\left.+\alpha_{11}^{(4,5)}\sigma_{11}^{6-b_4}\sigma_{12}^{b_4}\right] + 6x_2^2\left[\alpha_{11}^{(4,6)}\sigma_{11}^3\sigma_{22}^3 + \alpha_{11}^{(4,7)}\sigma_{11}\sigma_{22}^3\sigma_{12}^2\right.$$
$$\left.+\alpha_{11}^{(4,8)}\sigma_{22}^4\sigma_{12}^2 + \alpha_{11}^{(4,9)}\sigma_{22}^6 + \alpha_{11}^{(4,10)}\sigma_{22}^3\sigma_{12}^3\right] + 4\frac{x_2^3}{x_1}\left[\alpha_{11}^{(4,11)}\frac{\sigma_{22}^9}{\sigma_{11}^3}\right.$$
$$\left.\left.+\alpha_{11}^{(4,12)}\frac{\sigma_{22}^7\sigma_{12}^2}{\sigma_{11}^3} + \alpha_{11}^{(4,13)}\frac{\sigma_{22}^6\sigma_{12}^3}{\sigma_{11}^3} + \alpha_{11}^{(4,14)}\frac{\sigma_{22}^{9-b_4}\sigma_{12}^{b_4}}{\sigma_{11}^3}\right] + \frac{x_2^4}{x_1^2}\alpha_{11}^{(4,15)}\frac{\sigma_{22}^9}{\sigma_{11}^3}\right\}, \tag{5.104}$$

$$\alpha_{22}^{(4)} = \frac{3}{2}\left\{\frac{x_1^4}{x_2^2}\alpha_{22}^{(4,1)}\frac{\sigma_{11}^9}{\sigma_{22}^3} + 4\frac{x_1^3}{x_2}\left[\alpha_{22}^{(4,2)}\frac{\sigma_{11}^9}{\sigma_{22}^3} + \alpha_{22}^{(4,3)}\frac{\sigma_{11}^7\sigma_{12}^2}{\sigma_{22}^3} + \alpha_{22}^{(4,4)}\frac{\sigma_{11}^6\sigma_{12}^3}{\sigma_{22}^3}\right.\right.$$
$$\left.+\alpha_{22}^{(4,5)}\frac{\sigma_{11}^{9-b_4}\sigma_{12}^{b_4}}{\sigma_{22}^3}\right] + 6x_1^2\left[\alpha_{22}^{(4,6)}\sigma_{11}^6 + \alpha_{22}^{(4,7)}\sigma_{11}^4\sigma_{12}^2 + \alpha_{22}^{(4,8)}\sigma_{22}\sigma_{11}^3\sigma_{12}^2\right.$$
$$\left.+\alpha_{22}^{(4,9)}\sigma_{11}^3\sigma_{22}^3 + \alpha_{22}^{(4,10)}\sigma_{11}^3\sigma_{12}^3\right] + 4x_1x_2\left[\alpha_{22}^{(4,11)}\sigma_{22}^6 + \alpha_{22}^{(4,12)}\sigma_{22}^4\sigma_{12}^2\right.$$
$$\left.\left.+\alpha_{22}^{(4,13)}\sigma_{22}^3\sigma_{12}^3 + \alpha_{22}^{(4,14)}\sigma_{22}^{6-b_4}\sigma_{12}^{b_4}\right] + x_2^2\alpha_{22}^{(4,15)}\sigma_{22}^6\right\}, \tag{5.105}$$

and

$$\alpha_{12}^{(4)} = \frac{3}{4}\left\{\frac{x_1^3}{x_2}\alpha_{12}^{(4,1)}\frac{\sigma_{11}^9}{\sigma_{12}^3} + 4x_1^2\left[\alpha_{12}^{(4,2)}\frac{\sigma_{11}^9}{\sigma_{12}^3} + \alpha_{12}^{(4,3)}\frac{\sigma_{11}^7}{\sigma_{12}} + \alpha_{12}^{(4,4)}\sigma_{11}^6\right.\right.$$
$$\left.+\alpha_{12}^{(4,5)}\sigma_{11}^{9-b_4}\sigma_{12}^{b_4-3}\right] + 6x_1x_2\left[\alpha_{12}^{(4,6)}\frac{\sigma_{11}^6\sigma_{22}^3}{\sigma_{12}^3} + \alpha_{12}^{(4,7)}\frac{\sigma_{11}^4\sigma_{22}^3}{\sigma_{12}}\right.$$
$$\left.+\alpha_{12}^{(4,8)}\frac{\sigma_{22}^4\sigma_{11}^3}{\sigma_{12}} + \alpha_{12}^{(4,9)}\frac{\sigma_{22}^6\sigma_{11}^3}{\sigma_{12}^3} + \alpha_{12}^{(4,10)}\sigma_{11}^3\sigma_{22}^3\right] + 4x_2^2\left[\alpha_{12}^{(4,11)}\frac{\sigma_{22}^9}{\sigma_{12}^3}\right.$$
$$\left.\left.+\alpha_{12}^{(4,12)}\frac{\sigma_{22}^7}{\sigma_{12}} + \alpha_{12}^{(4,13)}\sigma_{22}^6 + \alpha_{12}^{(4,14)}\sigma_{22}^{9-b_4}\sigma_{12}^{b_4-3}\right] + \frac{x_2^3}{x_1}\alpha_{12}^{(4,15)}\frac{\sigma_{22}^9}{\sigma_{12}^3}\right\} \tag{5.106}$$

For $n = 5$

$$
\begin{aligned}
\alpha_{11}^{(5)} = \frac{3}{2}\Bigg\{ & x_1^3\alpha_{11}^{(5,1)}\sigma_{11}^9 + 5x_1^2x_2\Big[\alpha_{11}^{(5,2)}\sigma_{11}^9 + \alpha_{11}^{(5,3)}\sigma_{11}^7\sigma_{12}^2 + \alpha_{11}^{(5,4)}\sigma_{11}^6\sigma_{12}^3 \\
& +\alpha_{11}^{(5,5)}\sigma_{11}^{9-b_5}\sigma_{12}^{b_5}\Big] + 10x_1x_2^2\Big[\alpha_{11}^{(5,6)}\sigma_{11}^6\sigma_{22}^3 + \alpha_{11}^{(5,7)}\sigma_{11}^4\sigma_{12}^2\sigma_{22}^3 + \alpha_{11}^{(5,8)}\sigma_{11}^3\sigma_{12}^2\sigma_{22}^4 \\
& +\alpha_{11}^{(5,9)}\sigma_{11}^3\sigma_{22}^6 + \alpha_{11}^{(5,10)}\sigma_{11}^3\sigma_{12}^3\sigma_{22}^3\Big] + 10x_2^3\Big[\alpha_{11}^{(5,11)}\sigma_{22}^9 + \alpha_{11}^{(5,12)}\sigma_{12}^2\sigma_{22}^7 \\
& +\alpha_{11}^{(5,13)}\sigma_{11}\sigma_{12}^2\sigma_{22}^6 + \alpha_{11}^{(5,14)}\sigma_{11}^3\sigma_{22}^6 + \alpha_{11}^{(5,15)}\sigma_{12}^3\sigma_{22}^6\Big] + 5\frac{x_2^4}{x_1}\Bigg[\alpha_{11}^{(5,16)}\frac{\sigma_{22}^{12}}{\sigma_{11}^3} \\
& +\alpha_{11}^{(5,17)}\frac{\sigma_{12}^2\sigma_{22}^{10}}{\sigma_{11}^3} + \alpha_{11}^{(5,18)}\frac{\sigma_{12}^3\sigma_{22}^9}{\sigma_{11}^3} + \alpha_{11}^{(5,19)}\frac{\sigma_{12}^{b_5}\sigma_{22}^{12-b_5}}{\sigma_{11}^3}\Bigg] + \frac{x_2^5}{x_1^2}\alpha_{11}^{(5,20)}\frac{\sigma_{22}^{12}}{\sigma_{11}^3}\Bigg\},
\end{aligned}
\tag{5.107}
$$

$$
\begin{aligned}
\alpha_{22}^{(5)} = \frac{3}{2}\Bigg\{ & \frac{x_1^5}{x_2^2}\alpha_{22}^{(5,1)}\frac{\sigma_{11}^{12}}{\sigma_{22}^3} + 5\frac{x_1^4}{x_2}\Bigg[\alpha_{22}^{(5,2)}\frac{\sigma_{11}^{12}}{\sigma_{22}^3} + \alpha_{22}^{(5,3)}\frac{\sigma_{11}^{10}\sigma_{12}^2}{\sigma_{22}^3} + \alpha_{22}^{(5,4)}\frac{\sigma_{11}^9\sigma_{12}^3}{\sigma_{22}^3} \\
& +\alpha_{22}^{(5,5)}\frac{\sigma_{11}^{12-b_5}\sigma_{12}^{b_5}}{\sigma_{22}^3}\Bigg] + 10x_1^3\Big[\alpha_{22}^{(5,6)}\sigma_{11}^9 + \alpha_{22}^{(5,7)}\sigma_{11}^7\sigma_{12}^2 + \alpha_{22}^{(5,8)}\sigma_{11}^6\sigma_{12}^2\sigma_{22} \\
& +\alpha_{22}^{(5,9)}\sigma_{11}^6\sigma_{22}^3 + \alpha_{22}^{(5,10)}\sigma_{11}^6\sigma_{12}^3\Big] + 10x_1^2x_2\Big[\alpha_{22}^{(5,11)}\sigma_{11}^3\sigma_{22}^6 + \alpha_{22}^{(5,12)}\sigma_{11}^3\sigma_{12}^2\sigma_{22}^4 \\
& +\alpha_{22}^{(5,13)}\sigma_{11}^4\sigma_{12}^2\sigma_{22}^3 + \alpha_{22}^{(5,14)}\sigma_{11}^6\sigma_{22}^3 + \alpha_{22}^{(5,15)}\sigma_{11}^3\sigma_{12}^3\sigma_{22}^3\Big] + 5x_1x_2^2\Big[\alpha_{22}^{(5,16)}\sigma_{22}^9 \\
& +\alpha_{22}^{(5,17)}\sigma_{22}^7\sigma_{12}^2 + \alpha_{22}^{(5,18)}\sigma_{22}^6\sigma_{12}^3 + \alpha_{22}^{(5,19)}\sigma_{22}^{9-b_5}\sigma_{12}^{b_5}\Big] + x_2^3\alpha_{22}^{(5,20)}\sigma_{22}^9\Bigg\},
\end{aligned}
\tag{5.108}
$$

$$
\begin{aligned}
\alpha_{12}^{(5)} = \frac{3}{4}\Bigg\{ & \frac{x_1^4}{x_2}\alpha_{12}^{(5,1)}\frac{\sigma_{11}^{12}}{\sigma_{12}^3} + 5x_1^3\Bigg[\alpha_{12}^{(5,2)}\frac{\sigma_{11}^{12}}{\sigma_{12}^3} + \alpha_{12}^{(5,3)}\frac{\sigma_{11}^{10}}{\sigma_{12}} + \alpha_{12}^{(5,4)}\sigma_{11}^9 \\
& +\alpha_{12}^{(5,5)}\frac{\sigma_{11}^{12-b_5}}{\sigma_{12}^{3-b_5}}\Bigg] + 10x_1^2x_2\Bigg[\alpha_{12}^{(5,6)}\frac{\sigma_{11}^9\sigma_{22}^3}{\sigma_{12}^3} + \alpha_{12}^{(5,7)}\frac{\sigma_{11}^7\sigma_{22}^3}{\sigma_{12}} + \alpha_{12}^{(5,8)}\frac{\sigma_{11}^6\sigma_{22}^4}{\sigma_{12}} \\
& +\alpha_{12}^{(5,9)}\frac{\sigma_{11}^6\sigma_{22}^6}{\sigma_{12}^3} + \alpha_{12}^{(5,10)}\sigma_{11}^6\sigma_{22}^3\Bigg] + 10x_1x_2^2\Bigg[\alpha_{12}^{(5,11)}\frac{\sigma_{11}^3\sigma_{22}^9}{\sigma_{12}^3} + \alpha_{12}^{(5,12)}\frac{\sigma_{11}^3\sigma_{22}^7}{\sigma_{12}} \\
& +\alpha_{12}^{(5,13)}\frac{\sigma_{11}^4\sigma_{22}^6}{\sigma_{12}} + \alpha_{12}^{(5,14)}\frac{\sigma_{11}^6\sigma_{22}^6}{\sigma_{12}^3} + \alpha_{12}^{(5,15)}\sigma_{11}^3\sigma_{22}^6\Bigg] + 5x_2^3\Bigg[\alpha_{12}^{(5,16)}\frac{\sigma_{22}^{12}}{\sigma_{12}^3} \\
& +\alpha_{12}^{(5,17)}\frac{\sigma_{22}^{10}}{\sigma_{12}} + \alpha_{12}^{(5,18)}\sigma_{22}^9 + \alpha_{12}^{(5,19)}\frac{\sigma_{22}^{12-b_5}}{\sigma_{12}^{3-b_5}}\Bigg] + \frac{x_2^4}{x_1}\alpha_{12}^{(5,20)}\frac{\sigma_{22}^{12}}{\sigma_{12}^3}\Bigg\}
\end{aligned}
\tag{5.109}
$$

To determine the coefficients $\alpha_{ij}^{(n,k)}$ in the preceding expressions for $\alpha_{ij}^{(n)}$, we first introduce them into the density expansion (5.100), then we introduce the expressions of the $g_{ij}(\sigma_{ij})$ thus obtained into the virial equation (5.14) and finally, we apply conditions 1)–17) and solve the resulting equations. The

Table 5.13. Parameters $\alpha_{ij}^{(n,k)}$ in Eqs. (5.101)–(5.109)

k	$\alpha_{11}^{(n,k)}$	$\alpha_{22}^{(n,k)}$	$\alpha_{12}^{(n,k)}$
		$n=3$	
1	$5\pi/18$	0	0
2	$\pi/54$	0	0
3	$-2\pi/9$	0	$-\pi/9$
4	$8\pi/27$	0	$8\pi/27$
5	0	$\pi/54$	0
6	0	$-2\pi/9$	$-\pi/9$
7	0	$8\pi/27$	$8\pi/27$
8	0	$5\pi/18$	0
		$n=4$	
1	$b_0^{(4)}/\pi$	0	0
2	−0.00813427	0	0
3	−2.15070	0	−0.614486
4	$5\pi^2/27$	0	$5\pi^2/54$
5	0.750697	0	0.120201
6	−0.0120058	−0.00600291	0
7	−0.535411	−0.401558	−0.267706
8	−0.401558	−0.535411	−0.267706
9	−0.00600291	−0.0120058	0
10	1.09483	1.09483	1.09483
11	0	−0.008134278	0
12	0	−2.15070	−0.614486
13	0	$5\pi^2/27$	$5\pi^2/54$
14	0	0.750697	0.120201
15	0	$b_0^{(4)}/\pi$	0
		$n=5$	
1	$b_0^{(5)}/\pi$	0	0
2	−0.0600306	0	0
3	−2.96534	0	−0.593068
4	2.10897	0	0.702991
5	1.32155	0	0.160181
6	0.000537133	0.000179044	0
7	−0.0601047	−1.29221	−0.270462
8	0.253829	−1.3083	−0.210895
9	0.00831653	0.00831653	0
10	0	2.65954	0.886513
11	0.000179014	0.000537133	0
12	−1.29221	−0.0601047	−0.270462
13	−1.3083	0.253829	−0.210895

(Continued)

Table 5.13. (Continued)

k	$\alpha_{11}^{(n,k)}$	$\alpha_{22}^{(n,k)}$	$\alpha_{12}^{(n,k)}$
14	0.00831653	0.00831653	0
15	2.65954	0	0.886513
16	0	−0.0600306	0
17	0	−2.96534	−0.593068
18	0	2.10897	0.702991
19	0	1.32155	0.160181
20	0	$b_0^{(5)}/\pi$	0

values of these coefficients were reported in [47] and are listed in Table 5.13 for convenience. It is worth to write here the final expressions for the contributions $\alpha_{ij}^{(3)}$ arising from the third virial coefficient. They are given exactly by [23, 47]

$$\alpha_{11}^{(3)} = \frac{\pi}{12}\left\{5x_1\sigma_{11}^3 + x_2\left[\sigma_{11}^3 - 12\sigma_{11}\sigma_{12}^2 + 16\sigma_{12}^3\right]\right\}, \tag{5.110}$$

$$\alpha_{22}^{(3)} = \frac{\pi}{12}\left\{x_1\left[\sigma_{22}^3 - 12\sigma_{22}\sigma_{12}^2 + 16\sigma_{12}^3\right] + 5x_2\sigma_{22}^3\right\}, \tag{5.111}$$

and

$$\alpha_{12}^{(3)} = \frac{\pi}{12}\left\{x_1\left[8\sigma_{11}^3 - 3\frac{\sigma_{11}^4}{\sigma_{12}}\right] + x_2\left[8\sigma_{22}^3 - 3\frac{\sigma_{22}^4}{\sigma_{12}}\right]\right\}. \tag{5.112}$$

The procedure might be continued obtaining higher order contributions to the density expansion of the RDFs $g_{ij}(\sigma_{ij})$, provided that suitable expressions for the virial coefficients for $n > 5$ were available. However, the increasing number of equations involved in obtaining the parameters $\alpha_{ij}^{(n,k)}$ turns this procedure impractical.

It is to be noted that, using expressions (5.45), (5.46), (5.50), and (5.51) for the crossed contributions to the virial coefficients $B^{(2)}$ to $B^{(5)}$ in the virial expansion (5.38) of the compressibility factor Z truncated at $n = 5$, and using expressions (5.101)–(5.109) in the density expansion (5.100) of the RDFs, again truncated at the level of $n = 5$, these density expansions fulfil, by construction, all the consistency conditions 1)–17), expressed as a power series of the density when necessary.

5.7 Rescaling the Density Expansions

While the series expansions (5.38) and (5.100), truncated at $n = 5$, may be useful at low densities, they do not provide enough accuracy at moderate to high densities and, as just stated, obtaining accurate analytical expressions for higher order coefficients in these expansions become increasingly difficult.

In these circumstances, the use of certain procedures to accelerate the convergence of these series may be a solution to the problem. We will mention here two of these procedures namely the *rescaled virial expansion* (RVE) [51] and the *scaled hard sphere* (SHS) [52]. In the former, the compressibility factor is expressed in the form

$$Z^{\mathrm{RVE}} = \frac{1 + \sum_{n=2}^{\infty} c_{n-1} \eta^{n-1}}{(1-\eta)^3}, \tag{5.113}$$

where the coefficients c_n are determined from the condition that the series expansion of Eq. (5.113) must reproduce the known virial coefficients. This gives

$$c_{n-1} = B^{*(n)} - 3B^{*(n-1)} + 3B^{*(n-2)} - B^{*(n-3)}, \tag{5.114}$$

where $B^{*(n)} = B^{(n)}/v_{mix}^{n-1}$, in which $v_{mix} = (\pi/6)\sigma_{mix}^3$, and it is considered that $B^{(n)} = 0$ for $n \leq 0$ and $B^{(1)} = 1$. In applying Eq. (5.113), to determine the coefficients c_n we do not need analytical expressions for them, numerical values are sufficient.

We propose to rescale the contact values of the RDFs in a similar way, namely

$$g_{ij}^{\mathrm{RVE}}(\sigma_{ij}) = \frac{1 + \sum_{n=3}^{\infty} c_{n-2}^{(ij)} \eta^{n-2}}{(1-\eta)^3}, \tag{5.115}$$

where

$$c_{n-2}^{(ij)} = \gamma_n^{(ij)} - 3\gamma_{n-1}^{(ij)} + 3\gamma_{n-2}^{(ij)} - \gamma_{n-3}^{(ij)}, \tag{5.116}$$

with $\gamma_n^{(ij)} = \alpha_{ij}^{(n)}/v_{mix}^n$ and the condition that $\gamma_2^{(ij)} = 1$ and $\gamma_n^{(ij)} = 0$ for $n \leq 1$. As seen before, to determine coefficients $\alpha_{ij}^{(n)}$ we need suitable analytical expressions for the virial coefficients, which at present is limited to the first five virial coefficients.

In the SHS procedure [52], the excess compressibility factor Z of the mixture is determined from scaling of the excess compressibility factor Z_0 of a monodisperse HS fluid, with the same packing fraction as the mixture, in the form

$$Z^{\mathrm{SHS}} = 1 + (Z_0 - 1) \sum_{n=2}^{\infty} a_{n-2} \rho^{n-2}, \tag{5.117}$$

where coefficients a_{n-2} are again determined from the condition that the series expansion of this equation in terms of the number density $\rho = N/V$ must reproduce the known virial coefficients. This leads to

$$a_{n-2} = \frac{B^{(n)} - \sum_{i=0}^{n-3} a_i b_0^{(n-i)}}{b_0^{(2)}}, \quad n \geq 3, \tag{5.118}$$

with $a_0 = B^{(2)}/b_0^{(2)}$, where $B^{(n)}$ is the virial coefficient of order n in the virial expansion of the compressibility factor Z of the mixture in terms of the number density ρ and $b_0^{(n)}$ that corresponds to the monodisperse fluid.

For Z_0 we can use the popular Carnahan–Starling equation (5.7). However, Eq. (5.117) is quite sensitive to the accuracy of Z_0. Therefore, it may be necessary to use a more accurate equation of state for the monodisperse fluid. To this end, we have chosen a RVE reproducing up to the tenth virial coefficient, that is

$$Z_0 = \frac{1 + \sum_{n=2}^{10} c_{n-1} \eta^{n-1}}{(1-\eta)^3}, \tag{5.119}$$

with

$$\begin{gathered} c_1 = 1; \quad c_2 = 1; \quad c_3 = -0.635232; \\ c_4 = -0.869792; \quad c_5 = 0.23591552; \quad c_6 = 0.20774528; \\ c_7 = -0.27478016; \quad c_8 = 0.418304; \quad c_6 = 0.60481536. \end{gathered} \tag{5.120}$$

In a similar way, the SHS expression for the contact values of the RDFs is [11]

$$g_{ij}^{\mathrm{SHS}}(\sigma_{ij}) = g_0(\sigma_{mix}) \sum_{n=2}^{\infty} a_{n-2}^{(ij)} \rho^{n-2}, \tag{5.121}$$

where $g_0(\sigma_{\mathrm{mix}})$ can be obtained from Z_0 using the virial theorem for a monodisperse HS fluid

$$Z_0 = 1 + \frac{2}{3}\pi\rho\sigma^3 g_0(\sigma), \tag{5.122}$$

taking $\sigma = \sigma_{\mathrm{mix}}$.

Coefficients $a_{n-3}^{(ij)}$ in Eq. (5.121) are given by

$$a_{n-2}^{(ij)} = \alpha_n^{(ij)} - \frac{\sum_{l=0}^{n-3} a_l^{(ij)} b_0^{(n-l)}}{b_0^{(2)}}, \quad n \geq 3, \tag{5.123}$$

with $a_0^{(ij)} = 1$.

On the other hand, the relationship between the two sets of coefficients a_{n-2} and $a_{n-2}^{(ij)}$ can be readily obtained by introducing $g_0(\sigma_{\mathrm{mix}})$ into Eq. (5.121) and the result into the expression of the virial theorem (5.14) for a binary mixture of hard spheres and comparing the result with Eq. (5.117). This yields

$$a_{n-2} = \sum_i \sum_j x_i x_j \sigma_{ij}^3 a_{n-2}^{(ij)}. \tag{5.124}$$

Note that this relationship would allow us to obtain the coefficients a_{n-2} from the knowledge of the corresponding coefficients $a_{n-2}^{(ij)}$, but the converse is not true.

On the other hand, in practice, the infinite series in Eqs. (5.117) and (5.121) must be truncated due to the limited number of virial coefficients $B^{(n)}$, in the first case, and of coefficients $\alpha_{ij}^{(n)}$ in the second case, that are known.

In Tables 5.8–5.11, we have included too the results for $g_{ij}(\sigma_{ij})$ and Z obtained from the RVE and SHS scalings. To this end, we have used up to the contributions corresponding to the fifth virial coefficient in both kinds of quantities, as given by the analytical formulas reported in Sect. 5.4.3 and in Sect. 5.6. We can see in these tables that, for extreme values of the diameter ratio and mole fractions, none of the two scalings provide reliable results, on the whole, for $g_{11}(\sigma_{11})$ and $g_{12}(\sigma_{12})$, although they work fairly well for $g_{22}(\sigma_{22})$ and Z. The failure is surely due to the fact that higher order terms in the expansion (5.100), and their subsequent scalings (5.115) and (5.121), need to be used in order to achieve complete accuracy.

5.8 Final Remarks

We will end this review with a brief summary of the main conclusions we can extract from the analysis performed in the preceding Sections.

In Sect. 5.2 we have seen that the computer simulation of very asymmetric binary mixtures of hard spheres with low concentrations of the larger spheres present considerable difficulties at high densities, due to the decay in the mobility of the larger spheres. In these mixtures appear depletion forces which act on the larger spheres due to the presence of the smaller ones. In these circumstances, one way of tackling the problem of simulating these systems is to replace the true mixture by an effective monodisperse fluid with particles interacting by means of an effective potential consisting in a HS potential, for spheres with diameter equal to that of the larger spheres, plus an attractive tail corresponding to the depletion potential. The depletion potential can be determined theoretically in an approximate way or more accurately by means of computer simulation. However, the assumption of pairwise additivity of the potential seems to be inappropriate for very asymmetric mixtures with low concentration of the larger spheres.

On the other hand, the phase diagram of binary asymmetric mixtures of hard spheres of species A and B, A being the larger spheres, is rather complex, as found in computer simulations. Depending on the diameter ratio, relative concentrations, and mole fraction, they can present different fluid and solid phases. The solid phases may be pure A and B crystal phases, substitutionally disordered AB phases or AB_n crystals. In certain situations, the larger spheres freeze into a crystal with the smaller spheres diffusing through it. Regarding the fluid, there is some evidence from computer simulations that a demixing transition occurs for diameter ratios $R \gtrsim 20$. However, simulations performed under the assumption of the effective one-fluid approximation are controversial. Therefore, further simulations are needed to completely clarify this point.

Another source of "exact" information on the thermodynamic properties of asymmetric mixtures of hard spheres is the virial expansion. As seen in Sect. 5.4, the second and third virial coefficients of these mixtures are known exactly in analytical form. The partial contributions to the fourth, fifth, sixth, and seventh virial coefficients have been determined numerically with great accuracy for a number of diameter ratios. In addition, there are available two procedures to obtain approximate, though very accurate, analytical expressions for higher order virial coefficients as seen in Sects. 5.4.2 and 5.4.3.

A third source of exact information on the thermodynamic and structural properties of HS mixtures is the set of consistency conditions cited in Sect. 5.3. Consistency conditions may be useful for certain applications. Thus, in Sect. 5.5, we have illustrated the way certain consistency conditions have been used in the literature to improve existing expressions for the contact values of the radial distribution functions and/or the equation of state (Sect. 5.5.4). However, this procedure may lead to the undesired violation of other consistency conditions, as shown in Sect. 5.5.5. Another application of the consistency conditions has been shown in Sect. 5.6, where the contributions of the first five virial coefficients to the density expansions of the RDFs were determined on this basis, using the exact analytical expression of the second and third virial coefficients and the approximate expressions analyzed in Sect. 5.4.3. The coefficients obtained in this way for the density expansion of the RDFs fulfil, by construction, the whole set of consistency conditions listed in Sect. 5.3.

Finally, as shown in Sect. 5.7, the truncated density expansions of the compressibility factor Z and the RDFs $g_{ij}(\sigma_{ij})$, can be rescaled in several ways, in order to increase the range of densities for which these expansions are accurate. However, although the number of the virial coefficients known exactly or numerically at present seems to be sufficient to achieve satisfactory accuracy for the compressibility factor for diameter ratios $R \leq 10$ and mole fractions of the larger spheres $x_1 \geq 0.005$, the situation is not the same for the coefficients in the expansion of the RDFs.

References

1. E. B. Smith, K. R. Lea: Trans. Faraday Soc. **59**, 1535 (1963)
2. B. J. Alder: J. Chem. Phys. **40**, 2724 (1964)
3. A. Rotenberg: J. Chem. Phys. **43**, 4377 (1965)
4. L. L. Lee, D. Levesque: Mol. Phys. **26**, 1351 (1973)
5. H. P. Fries, J. P. Hansen: Mol. Phys. **48**, 891 (1983)
6. G. Jackson, J. S. Rowlinson, F. van Swol: J. Phys. Chem. **91**, 4907 (1987)
7. W. G. T. Kranendonk, D. Frenkel: Mol. Phys. **72**, 715 (1991)
8. D. H. L. Yau, K. Y. Chan, D. Henderson: Mol. Phys. **88**, 1237 (1996)
9. M. Barosova, A. Malijevsky, S. Labík, W. R. Smith: Mol. Phys. **87**, 423 (1996)
10. C. Barrio, J. R. Solana: Physica A, **351**, 387 (2005)
11. C. Barrrio, J. R. Solana: Proceedings of the VII Iberoamerican Conference on Phase Equilibria and Fluid Properties for Process Design, p. 606 (2006).
12. D. Mathyushov, D. Henderson, K.-Y. Chan: Mol. Phys. **96**, 1813 (1999)
13. L. Lue, L. V. Woodcock: Mol. Phys. **96**, 1435 (1999)

14. L. Lue, L. V. Woodcock: Int. J. Thermophys. **23**, 937 (2002)
15. D. Henderson, A. Trokhymchuk, L. V. Woodcock, K.-Y. Chan: Mol. Phys. **103**, 667 (2005)
16. B. J. Yoon, K. K. Baeck, S. I. Jeon: Chem. Phys. Lett. **301**, 481 (1999)
17. D. Cao, K.-L. Chan, D. Henderson, W. Wang: Mol. Phys. **98**, 619 (2000)
18. A. Malijevsky, M. Barosova, W. R. Smith: Mol. Phys. **91**, 65 (1997)
19. J. L. Lebowitz, E. Helfand, E. Praestegaard: J. Chem. Phys. **43**, 774 (1965)
20. E. Waisman, D. Henderson, J. L. Lebowitz: Molec. Phys. **32**, 1373 (1976)
21. E. Z. Hamad, A. G. Mansoori: Fluid Phase Equilib. **51**, 13 (1989)
22. E. Z. Hamad: J. Chem. Phys. **101**, 10195 (1994)
23. E. Z. Hamad: J. Chem. Phys. **105**, 3222 (1996)
24. N. M. Tukur, E. Z. Hamad, G. A. Mansoori: J. Chem. Phys. **110**, 3463 (1999)
25. C. Vega: J. Chem. Phys. **108**, 3074 (1998)
26. D. V. Matyushov, B. M. Ladanyi: J. Chem. Phys. **107**, 5815 (1997).
27. W. R. Smith, S. Labik: Mol. Phys. **80**, 1561 (1993).
28. C. Barrio, J. R. Solana: J. Chem. Phys. **113**, 10180 (2000)
29. T. Kihara, K. Miyoshi: J. Stat. Phys. **13**, 337 (1975)
30. F. Saija, G. Fiumara, P. V. Giaquinta: Mol. Phys. **87**, 991 (1996)
31. F. Saija, G. Fiumara, P. V. Giaquinta: Mol. Phys. **92**, 1089 (1997)
32. E. Enciso, N. G. Almarza, M. A. González, F. J. Bermejo: Phys. Rev. E **57**, 4486 (1998)
33. R. J. Wheatley, F. Saija, P. V. Giaquinta: Mol. Phys. **94**, 877 (1998)
34. A. Y. Vlasov, X.-M. You, A. J. Masters: Mol. Phys. **100**, 3313 (2002)
35. A. Y. Vlasov, A. J. Masters: Fluid Phase Equilib. **212**, 183 (2003)
36. A. Y. Vlasov, A. J. Masters: Private communication.
37. R. J. Wheatley, *J. Chem. Phys.* **111**, 5455 (1999)
38. C. Barrio, J. R. Solana, Mol. Phys. **101**, 1545 (2003)
39. J. L. Lebowitz: Phys. Rev. A **133**, 895 (1964)
40. T. Boublík: J. Chem. Phys. **53**, 471 (1970)
41. G. A. Mansoori, N. F. Carnahan, K. E. Starling, T. Leland: J. Chem. Phys. **54**, 1523 (1971)
42. E. Z. Hamad: J. Chem. Phys. **100**, 3733 (1995)
43. D. H. L. Yau, K.-Y. Chan, D. Henderson: Mol. Phys. **91**, 1137 (1997)
44. D. Henderson, K.-Y. Chan: Mol. Phys. **98**, 1005 (2000)
45. D. Viduna, W. R. Smith: Mol. Phys. **100**, 2903 (2002)
46. T. Boublík: *Mol. Phys.* **100**, 3443 (2002)
47. C. Barrio, J. R. Solana: J. Chem. Phys. **119**, 3826 (2003)
48. A. Santos, S. B. Yuste, M. López de Haro: Mol. Phys. **96**, 1 (1999)
49. C. Barrio, J. R. Solana: Mol. Phys. **97**, 797 (1999)
50. A. Santos, S. B. Yuste, M. López de Haro: Mol. Phys. **99**, 1959 (2001)
51. M. Baus, J. L. Colot: Phys. Rev. A **36**, 3912 (1987)
52. C. Barrio, J. R. Solana: Physica A **366**, 79 (2006)
53. A. D. Dinsmore, A. G. Yodh, D. J. Pine: Phys. Rev. E **52**, 4045 (1995)
54. T. Coussaert, M. Baus: Phys. Rev. Lett. **79**, 1881 (1997)
55. T. Coussaert, M. Baus: Phys. Rev. Lett. **80**, 4832 (1998a)
56. T. Coussaert, M. Baus: J. Chem. Phys. **109**, 6012 (1998b)
57. M. Dijkstra, van R. Roig, R. Evans: Phys. Rev. Lett. **81**, 2268 (1998)
58. M. Dijkstra, van R. Roig, R. Evans: Phys. Rev. Lett. **82**, 117 (1998)
59. M. Dijkstra, van R. Roig, R. Evans, Phys. Rev. E **59**, 5744 (1999)

60. N. G. Almarza, E. Enciso: Phys. Rev. E **59**, 4426 (1999)
61. M. López de Haro, C. Tejero: J. Chem. Phys. **121**, 6918 (2004)
62. P. D. Kaplan, J. L. Rouke, A. G. Yodh, D. J. Pine: Phys. Rev. Lett. **72**, 582 (1994)
63. U. Steiner, A. Meller, J. Stavans: Phys. Rev. Lett. **74**, 4750 (1995)
64. A. Imhof, J. K. G. Dhont: Phys. Rev. Lett. **75**, 1662 (1995)
65. S. Asakura, F. Oosawa: J. Chem. Phys. **22**, 1255 (1955)
66. S. Asakura, F. Oosawa: J. Polym. Sci. **33**, 183 (1958)
67. P. Attard: J. Chem. Phys. **91**, 3083 (1989)
68. R. Dickman, P. Attard, V. Simonian: J. Chem. Phys. **107**, 205 (1997)
69. B. Götzelmann, R. Evans, S. Dietrich: Phys. Rev. E **57**, 6785 (1998)
70. B. Derjaguin: Kolloid-Z **69**, 155 (1934)
71. N. F. Carnahan, K. E. Starling: J. Chem. Phys. **51**, 635 (1969)
72. D. Henderson, M. Plischke: Proc. R. Soc. London, Ser. A **410**, 409 (1987)
73. T. Biben, P. Bladon, D. Frenkel: J. Phys.: Condens. Matter **85**, 10799 (1996)
74. J. G. Malherbe, S. Amokrane: Mol. Phys. **99**, 355 (2001)
75. M. J. Murray, J. V. Sanders: Phil. Mag. **42**, 721 (1980)
76. M. D. Elridge, P. A. Madden, D. Frenkel: Mol. Phys. **79**, 105 (1993)
77. J. L. Barrat, M. Baus, J. P. Hansen: Phys. Rev. Lett. **56**, 1063 (1986)
78. D. L. Ermak, B. J. Alder, L. R. Pratt: J. Phys. Chem. **85**, 3221 (1981)
79. W. G. T. Kranendonk, D. Frenkel: J. Phys.: Condens. Matter **1**, 7735 (1989)
80. M. D. Elridge, P. A. Madden, D. Frenkel: Mol. Phys. **80**, 987 (1993)
81. M. D. Elridge, P. A. Madden, D. Frenkel: Nature **365**, 35 (1993)
82. M. D. Elridge, P. A. Madden, P. N. Pusey, P. Bartlett: Mol. Phys. **84**, 395 (1995)
83. E. Trizac, M. D. Elridge, P. A. Madden: Mol. Phys. **90**, 675 (1997)
84. T. Gruhn, P. A. Monson: Phys. Rev. E **64**, 061703 (2001)
85. M. Dijkstra, R. van Roij: Phys. Rev. E **56**, 5594 (1997)
86. A. Buhot, W. Krauth: Phys. Rev. Lett. **80**, 3787 (1998)
87. M. Dijkstra, R. van Roij, R. Evans: Phys. Rev. Lett. **81**, 2268 (1998)
88. M. Dijkstra, R. van Roij, R. Evans: Phys. Rev. Lett. **82**, 117 (1999)
89. A. G. McLelland, B. J. Alder: J. Chem. Phys. **24**, 115 (1956)
90. N. Clisby, B.M. McCoy, arXiv:cond-mat/0503525 (2005), available at the URL http://arxiv.org/archive/cond-mat.
91. H. Reiss, H. L. Frisch, J. L. Lebowitz: J. Chem. Phys. **31** 369, (1959)
92. E. Helfand, H. L. Frisch, J. L. Lebowitz: J. Chem. Phys. **34** 1037, (1961)
93. M. S. Wertheim: Phys. Rev. Lett. **10**, 321 (1963)
94. E. Thiele: J. Chem. Phys. **39**, 474 (1963)
95. D. Henderson, M. Lozada-Cassou: J. Colloid Interface Sci. **114**, 180 (1986)
96. D. Viduna, W. R. Smith: J. Chem. Phys. **117**, 1214 (2002).

6

Alternative Approaches to the Equilibrium Properties of Hard-Sphere Liquids

M. López de Haro[1], S.B. Yuste[2] and A. Santos[3]

[1] Centro de Investigación en Energía, Universidad Nacional Autónoma de México (U.N.A.M.), Temixco, Morelos 62580, Mexico
malopez@servidor.unam.mx

[2] Departamento de Física, Universidad de Extremadura, E-06071 Badajoz, Spain
santos@unex.es

[3] Departamento de Física, Universidad de Extremadura, E-06071 Badajoz, Spain
andres@unex.es

An overview of some analytical approaches to the computation of the structural and thermodynamic properties of single component and multicomponent hard-sphere fluids is provided. For the structural properties, they yield a thermodynamically consistent formulation, thus improving and extending the known analytical results of the Percus–Yevick theory. Approximate expressions for the contact values of the radial distribution functions and the corresponding analytical equations of state are also discussed. Extensions of this methodology to related systems, such as sticky hard spheres and square-well fluids, as well as its use in connection with the perturbation theory of fluids are briefly addressed.

6.1 Introduction

In the statistical thermodynamic approach to the theory of simple liquids, there is a close connection between the thermodynamic and the structural properties [1, 2, 3, 4]. These properties depend on the intermolecular potential of the system, which is generally assumed to be well represented by pair interactions. The simplest model pair potential is that of a hard-core fluid (rods, disks, spheres, hyperspheres) in which attractive forces are completely neglected. In fact, it is a model that has been most studied and has rendered some analytical results, although up to this day no general (exact) explicit expression for the equation of state is available, except for the one-dimensional case. Something similar applies to the structural properties. An interesting feature concerning the thermodynamic properties is that in hard-core systems the equation of state depends only on the contact values of the radial

López de Haro, M., et al.: *Alternative Approaches to the Equilibrium Properties of Hard-Sphere Liquids*. Lect. Notes Phys. **753**, 183–245 (2008)
DOI 10.1007/978-3-540-78767-9_6

distribution functions. In the absence of a completely analytical approach, the most popular methods to deal with both kinds of properties of these systems are integral equation theories and computer simulations.

It is well known that in real gases and liquids at high temperatures, the state and thermodynamic properties are determined almost entirely by the repulsive forces among molecules. At lower temperatures, attractive forces become significant, but even in this case they affect very little the configuration of the system at moderate and high densities. These facts are taken into account in the application of the perturbation theory of fluids, where hard-core fluids are used as the reference systems in the computation of the thermodynamic and structural properties of real fluids. However, successful results using perturbation theory are rather limited due to the fact that, as mentioned above, there are in general no exact (analytical) expressions for the thermodynamic and structural properties of the reference systems which are in principle required in the calculations. On the other hand, in the realm of soft condensed matter the use of the hard-sphere model in connection, for instance, with sterically stabilized colloidal systems is quite common. This is due to the fact that nowadays it is possible to prepare (almost) monodisperse spherical colloidal particles with short-ranged harshly repulsive interparticle forces that may be well described theoretically with the hard-sphere potential.

This chapter presents an overview of the efforts we have made over the last few years to compute the thermodynamic and structural properties of hard-core systems using relatively simple (approximate) analytical methods. It is structured as follows. In Sect. 6.2, we describe our proposals to derive the contact values of the radial distribution functions of a multicomponent mixture (with an arbitrary size distribution, either discrete or continuous) of d-dimensional hard spheres (HS) from the use of some consistency conditions and the knowledge of the contact value of the radial distribution function of the corresponding single component system. In turn, these contact values lead to equations of state both for additive and for non-additive HS. Some consequences of such equations of state, in particular the demixing transition, are briefly analyzed. This is followed in Sect. 6.3 by the description of the Rational Function Approximation method to obtain analytical expressions for the structural quantities of three-dimensional single component and multicomponent fluids. The only required inputs in this approach are the contact values of the radial distribution functions and so the connection with the work of the previous section follows naturally. Structural properties of related systems, like sticky HS or square-well fluids, that may also be tackled with the same philosophy are also discussed in Sect. 6.4. Section 6.5 provides an account of the reformulation of the perturbation theory of liquids using the results of the Rational Function Approximation method for a single component hard-sphere fluid and its illustration in the case of the Lennard–Jones fluid. In the final section, we provide some perspectives of the achievements obtained so far and of the challenges that remain ahead.

6.2 Contact Values and Equations of State for Mixtures

As stated in the Introduction, a nice feature of hard-core fluids is that the expressions of all their thermodynamic properties in terms of the radial distribution functions (RDF) are particularly simple. In fact, for these systems the internal energy reduces to that of the ideal gas and in the pressure equation it is only the contact values rather than the full RDF which appear explicitly. In this section, we present our approach to the derivation of the contact values of hard-core fluid mixtures in d dimensions.

6.2.1 Additive Systems in d Dimensions

As defined in Chap. 5, if σ_{ij} denotes the distance of separation at contact between the centers of two interacting fluid particles, one of species i and the other of species j, the mixture is said to be *additive* if σ_{ij} is just the arithmetic mean of the hard-core diameters of each species. Otherwise, the system is *non-additive*. The system of additive HS has been described in detail in Chap. 5. In this section and in Sect. 6.2.2 we deal again with such a system using a somewhat different perspective, while non-additive hard-core mixtures will be treated in Sect. 6.2.3.

Definitions

Let us consider an additive mixture of HS in d dimensions with an arbitrary number N of components. In fact, our discussion will remain valid for $N \to \infty$, i.e., for polydisperse mixtures with a continuous distribution of sizes.

The additive hard core of the interaction between a sphere of species i and a sphere of species j is $\sigma_{ij} = \frac{1}{2}(\sigma_i + \sigma_j)$, where the diameter of a sphere of species i is $\sigma_{ii} = \sigma_i$. Let the number density of the mixture be ρ and the mole fraction of species i be $x_i = \rho_i/\rho$, where ρ_i is the number density of species i. From these quantities one can define the packing fraction $\eta = v_d \rho M_d$, where $v_d = (\pi/4)^{d/2}/\Gamma(1 + d/2)$ is the volume of a d-dimensional sphere of unit diameter and

$$M_n \equiv \langle \sigma^n \rangle = \sum_{i=1}^{N} x_i \sigma_i^n \tag{6.1}$$

denotes the nth moment of the diameter distribution.

In an HS mixture, the knowledge of the contact values $g_{ij}(\sigma_{ij})$ of the RDF $g_{ij}(r)$, where r is the distance, is important for a number of reasons. For example, the availability of $g_{ij}(\sigma_{ij})$ is sufficient to get the equation of state (EOS) of the mixture via the virial expression

$$Z(\eta) = 1 + \frac{2^{d-1}}{M_d} \eta \sum_{i,j=1}^{N} x_i x_j \sigma_{ij}^d g_{ij}(\sigma_{ij}) , \tag{6.2}$$

where $Z = p/\rho k_B T$ is the compressibility factor of the mixture, p being the pressure, k_B the Boltzmann constant, and T the absolute temperature.

The exact form of $g_{ij}(\sigma_{ij})$ as functions of the packing fraction η, the set of diameters $\{\sigma_k\}$, and the set of mole fractions $\{x_k\}$ is only known in the one-dimensional case, where one simply has [5]

$$g_{ij}(\sigma_{ij}) = \frac{1}{1-\eta}, \quad (d = 1) . \tag{6.3}$$

Consequently, for $d \geq 2$, one has to resort to approximate theories or empirical expressions. For hard-disk mixtures, an accurate expression is provided by Jenkins and Mancini's (JM) approximation [6, 7, 8],

$$g_{ij}^{\mathrm{JM}}(\sigma_{ij}) = \frac{1}{1-\eta} + \frac{9}{16}\frac{\eta}{(1-\eta)^2}\frac{\sigma_i\sigma_j M_1}{\sigma_{ij} M_2}, \quad (d = 2) . \tag{6.4}$$

The associated compressibility factor is

$$Z_{\mathrm{JM}}(\eta) = \frac{1}{1-\eta} + \frac{M_1^2}{M_2}\eta\frac{1+\eta/8}{(1-\eta)^2}, \quad (d = 2) . \tag{6.5}$$

In the case of three-dimensional systems, some important analytical expressions for the contact values and the corresponding compressibility factor also exist. For instance, the expressions which follow from the solution of the Percus–Yevick (PY) equation of additive HS mixtures by Lebowitz [9] are

$$g_{ij}^{\mathrm{PY}}(\sigma_{ij}) = \frac{1}{1-\eta} + \frac{3}{2}\frac{\eta}{(1-\eta)^2}\frac{\sigma_i\sigma_j M_2}{\sigma_{ij} M_3}, \quad (d = 3) , \tag{6.6}$$

$$Z_{\mathrm{PY}}(\eta) = \frac{1}{1-\eta} + \frac{M_1 M_2}{M_3}\frac{3\eta}{(1-\eta)^2} + \frac{M_2^3}{M_3^2}\frac{3\eta^2}{(1-\eta)^2}, \quad (d = 3) . \tag{6.7}$$

Also analytical are the results obtained from the Scaled Particle Theory (SPT) [10, 11, 12, 13, 14, 15],

$$g_{ij}^{\mathrm{SPT}}(\sigma_{ij}) = \frac{1}{1-\eta} + \frac{3}{2}\frac{\eta}{(1-\eta)^2}\frac{\sigma_i\sigma_j M_2}{\sigma_{ij} M_3} + \frac{3}{4}\frac{\eta^2}{(1-\eta)^3}\left(\frac{\sigma_i\sigma_j M_2}{\sigma_{ij} M_3}\right)^2, \quad (d = 3) , \tag{6.8}$$

$$Z_{\mathrm{SPT}}(\eta) = \frac{1}{1-\eta} + \frac{M_1 M_2}{M_3}\frac{3\eta}{(1-\eta)^2} + \frac{M_2^3}{M_3^2}\frac{3\eta^2}{(1-\eta)^3}, \quad (d = 3) . \tag{6.9}$$

Neither the PY nor the SPT lead to particularly accurate values and so Boublík [16] and, independently, Grundke and Henderson [17] and Lee and Levesque [18] proposed an interpolation between the PY and the SPT contact values that we will refer to as the BGHLL values:

$$g_{ij}^{\mathrm{BGHLL}}(\sigma_{ij}) = \frac{1}{1-\eta} + \frac{3}{2}\frac{\eta}{(1-\eta)^2}\frac{\sigma_i\sigma_j M_2}{\sigma_{ij} M_3} + \frac{1}{2}\frac{\eta^2}{(1-\eta)^3}\left(\frac{\sigma_i\sigma_j M_2}{\sigma_{ij} M_3}\right)^2, \quad (d = 3) . \tag{6.10}$$

This leads through Eq. (6.2) to the widely used and rather accurate Boublík–Mansoori–Carnahan–Starling–Leland (BMCSL) EOS [16, 19] for HS mixtures:

$$Z_{\mathrm{BMCSL}}(\eta) = \frac{1}{1-\eta} + \frac{M_1 M_2}{M_3}\frac{3\eta}{(1-\eta)^2} + \frac{M_2^3}{M_3^2}\frac{\eta^2(3-\eta)}{(1-\eta)^3}, \quad (d=3). \tag{6.11}$$

Refinements of the BGHLL values have been subsequently introduced, among others, by Henderson et al. [20, 21, 22, 23, 24, 25, 26, 27, 28], Matyushov and Ladanyi [29], and Barrio and Solana [30] (for this latter see also Chap. 5) to eliminate some drawbacks of the BMCSL EOS in the so-called colloidal limit of binary HS mixtures. On a different path, but also having to do with the colloidal limit, Viduna and Smith [31, 32] have proposed a method to obtain contact values of the RDF of HS mixtures from a given EOS. However, none of these proposals may be easily generalized so as to be valid for any dimensionality and any number of components. Therefore, if one wants to have a more general framework able to deal with arbitrary d and N an alternative strategy is called for.

Universality Ansatz

In order to follow our alternative strategy, it is useful to make use of exact limit results that can help one in the construction of approximate expressions for $g_{ij}(\sigma_{ij})$. Let us consider first the limit in which one of the species, say i, is made of point particles, i.e., $\sigma_i \to 0$. In that case, $g_{ii}(\sigma_i)$ takes the ideal gas value, except that one has to take into account that the available volume fraction is $1-\eta$. Thus,

$$\lim_{\sigma_i \to 0} g_{ii}(\sigma_i) = \frac{1}{1-\eta} . \tag{6.12}$$

An even simpler situation occurs when all the species have the same size, $\{\sigma_k\} \to \sigma$, so that the system becomes equivalent to a single component system. Therefore,

$$\lim_{\{\sigma_k\} \to \sigma} g_{ij}(\sigma_{ij}) = g_{\mathrm{s}} , \tag{6.13}$$

where g_{s} is the contact value of the RDF of the single component fluid at the same packing fraction η as that of the mixture. Table 6.1 lists some of the most widely used proposals for the contact value g_{s} and the associated compressibility factor

$$Z_{\mathrm{s}} = 1 + 2^{d-1}\eta g_{\mathrm{s}} \tag{6.14}$$

in the case of the single component HS fluid. A more comprehensive list of expressions for the compressibility factor was provided in Chap. 3.

Equations (6.12) and (6.13) represent the simplest and most basic conditions that $g_{ij}(\sigma_{ij})$ must satisfy. As already pointed out in Chap. 5, there are a number of other less trivial consistency conditions [14, 20, 23, 24, 25, 26, 29, 30, 33, 34, 35], some of which will be used later on.

Table 6.1. Some expressions of g_s and Z_s for the single component HS fluid. In the SHY proposal, $\eta_{cp} = (\sqrt{3}/6)\pi$ is the crystalline close-packing fraction for hard disks. In the LM proposal, b_3 and b_4 are the (reduced) third and fourth virial coefficients, $\zeta(\eta) = 1.2973(59) - 0.062(13)\eta/\eta_{cp}$ for $d = 4$, and $\zeta(\eta) = 1.074(16) + 0.163(45)\eta/\eta_{cp}$ for $d = 5$, where the values of the close-packing fractions are $\eta_{cp} = \pi^2/16 \simeq 0.617$ and $\eta_{cp} = \pi^2\sqrt{2}/30 \simeq 0.465$ for $d = 4$ and $d = 5$, respectively

d	g_s	Z_s	Label	Ref.
2	$\dfrac{1-7\eta/16}{(1-\eta)^2}$	$\dfrac{1+\eta^2/8}{(1-\eta)^2}$	H	[36]
2	$\dfrac{1-\eta(2\eta_{cp}-1)/2\eta_{cp}^2}{1-2\eta+\eta^2(\eta_{cp}-1)/2\eta_{cp}^2}$	$\dfrac{1}{1-2\eta+\eta^2(\eta_{cp}-1)/2\eta_{cp}^2}$	SHY	[37, 38]
2	$g_s^{H} - \dfrac{\eta^3}{2^7(1-\eta)^4}$	$Z_s^{H} - \dfrac{\eta^4}{2^6(1-\eta)^4}$	L	[39, 40, 41]
3	$\dfrac{1+\eta/2}{(1-\eta)^2}$	$\dfrac{1+2\eta+3\eta^2}{(1-\eta)^2}$	PY	[42, 43]
3	$\dfrac{1-\eta/2+\eta^2/4}{(1-\eta)^3}$	$\dfrac{1+\eta+\eta^2}{(1-\eta)^3}$	SPT	[10, 11, 12]
3	$\dfrac{1-\eta/2}{(1-\eta)^3}$	$\dfrac{1+\eta+\eta^2-\eta^3}{(1-\eta)^3}$	CS	[44]
4, 5	$\dfrac{1+[2^{1-d}b_3-\zeta(\eta)b_4/b_3]\eta}{1-\zeta(\eta)(b_4/b_3)\eta+[\zeta(\eta)-1]\,2^{1-d}b_4\eta^2}$	$1+2^{d-1}\eta g_s^{LM}$	LM	[45]

In order to proceed, in line with a property shared by earlier proposals [see, in particular, Eqs. (6.4), (6.6), (6.8), and (6.10)], we assume that, at a given packing fraction η, the dependence of $g_{ij}(\sigma_{ij})$ on the parameters $\{\sigma_k\}$ and $\{x_k\}$ takes place *only* through the scaled quantity

$$z_{ij} \equiv \frac{\sigma_i\sigma_j}{\sigma_{ij}}\frac{M_{d-1}}{M_d} . \tag{6.15}$$

More specifically, we assume

$$g_{ij}(\sigma_{ij}) = \mathcal{G}(\eta, z_{ij}) , \tag{6.16}$$

where the function $\mathcal{G}(\eta, z)$ is *universal* in the sense that it is a common function for all the pairs (i, j), regardless of the composition and number of components of the mixture. Of course, the function $\mathcal{G}(\eta, z)$ is in principle different for each dimensionality d. To clarify the implications of this universality ansatz, let us imagine two mixtures $\mathcal{M}$ and $\mathcal{M}'$ having the same packing fraction η but strongly differing in the set of mole fractions, the sizes of the particles, and even the number of components. Suppose now that there exists a pair (i, j) in mixture $\mathcal{M}$ and another pair (i', j') in mixture $\mathcal{M}'$ such that $z_{ij} = z_{i'j'}$. Then, according to Eq. (6.16), the contact value of the RDF for the pair (i, j) in mixture $\mathcal{M}$ is the same as that for the pair (i', j') in mixture $\mathcal{M}'$, i.e., $g_{ij}(\sigma_{ij}) = g_{i'j'}(\sigma_{i'j'})$. In order to ascribe a physical meaning to the parameter z_{ij}, note that the ratio M_{d-1}/M_d can be understood as a "typical" inverse diameter (or curvature) of the particles of the mixture. Thus,

$z_{ij}^{-1} = \frac{1}{2}(\sigma_i^{-1} + \sigma_j^{-1})/(M_{d-1}/M_d)$ represents the arithmetic mean curvature, in units of M_{d-1}/M_d, of a particle of species i and a particle of species j.

Once the ansatz (6.16) is adopted, one may use the limits in (6.12) and (6.13) to get $\mathcal{G}(\eta, z)$ at $z = 0$ and $z = 1$, respectively. Since $z_{ii} \to 0$ in the limit $\sigma_i \to 0$, insertion of Eq. (6.12) into (6.16) yields

$$\mathcal{G}(\eta, 0) = \frac{1}{1-\eta} \equiv \mathcal{G}_0(\eta) \ . \tag{6.17}$$

Next, if all the diameters are equal, $z_{ij} \to 1$, so that Eq. (6.13) implies that

$$\mathcal{G}(\eta, 1) = g_\mathrm{s} \ . \tag{6.18}$$

Linear Approximation

As the simplest approximation [46], one may assume a linear dependence of $\mathcal{G}$ on z that satisfies the basic requirements (6.17) and (6.18), namely

$$\mathcal{G}(\eta, z) = \frac{1}{1-\eta} + \left(g_\mathrm{s} - \frac{1}{1-\eta}\right) z \ . \tag{6.19}$$

Inserting this into Eq. (6.16), one has

$$g_{ij}^\mathrm{e1}(\sigma_{ij}) = \frac{1}{1-\eta} + \left(g_\mathrm{s} - \frac{1}{1-\eta}\right) \frac{M_{d-1}}{M_d} \frac{\sigma_i \sigma_j}{\sigma_{ij}} \ . \tag{6.20}$$

Here, the label "e1" is meant to indicate that (i) the contact values used are an *extension* of the single component contact value g_s and that (ii) $\mathcal{G}(\eta, z)$ is a *linear* polynomial in z. This notation will become handy below. Although the proposal (6.20) is rather crude and does not produce especially accurate results for $g_{ij}(\sigma_{ij})$ when $d \geq 3$, it nevertheless leads to an EOS that exhibits an excellent agreement with simulations in 2, 3, 4, and 5 dimensions, provided that an accurate g_s is used as input [46, 47, 48, 49, 50]. This EOS may be written as

$$Z_\mathrm{e1}(\eta) = 1 + \frac{\eta}{1-\eta} 2^{d-1}(\Omega_0 - \Omega_1) + \left[Z_\mathrm{s}(\eta) - 1\right] \Omega_1 \ , \tag{6.21}$$

where the coefficients Ω_m depend only on the composition of the mixture and are defined by

$$\Omega_m = 2^{-(d-m)} \frac{M_{d-1}^m}{M_d^{m+1}} \sum_{n=0}^{d-m} \binom{d-m}{n} M_{n+m} M_{d-n} \ . \tag{6.22}$$

In particular, for $d = 2$ and $d = 3$,

$$Z_\mathrm{e1}(\eta) = \frac{1}{1-\eta} + \frac{M_1^2}{M_2}\left[Z_\mathrm{s}(\eta) - \frac{1}{1-\eta}\right], \quad (d = 2) \ , \tag{6.23}$$

$$Z_{\mathrm{e1}}(\eta) = \frac{1}{1-\eta} + \frac{M_1 M_2}{2M_3}\left\{\left[Z_{\mathrm{s}}(\eta) - \frac{1}{1-\eta}\right]\left(1+\frac{M_2^2}{M_1 M_3}\right) + \frac{3\eta}{1-\eta}\left(1-\frac{M_2^2}{M_1 M_3}\right)\right\}, \quad (d=3). \tag{6.24}$$

As an extra asset, from Eq. (6.21) one may write the virial coefficients of the mixture B_n, defined by

$$Z = 1 + \sum_{n=1}^{\infty} B_{n+1}\rho^n \,, \tag{6.25}$$

in terms of the (reduced) virial coefficients of the single component fluid b_n defined by

$$Z_{\mathrm{s}} = 1 + \sum_{n=1}^{\infty} b_{n+1}\eta^n \,. \tag{6.26}$$

The result is

$$B_n = v_d^{n-1} M_d^{n-1}\left[\Omega_1 b_n + 2^{d-1}(\Omega_0 - \Omega_1)\right] \,. \tag{6.27}$$

In the case of binary mixtures, these coefficients are in very good agreement with the available exact and simulation results [46, 48], except when the mixture involves components of very disparate sizes, especially for high dimensionalities. One may perform a slight modification such that this deficiency is avoided and thus get a modified EOS [48, 51, 52]. For $d = 2$ and $d = 3$ it reads

$$Z(\eta) = Z_{\mathrm{s}}(\eta) + x_1\left[\frac{1}{1-\eta_2} Z_{\mathrm{s}}\left(\frac{\eta_1}{1-\eta_2}\right) - Z_{\mathrm{s}}(\eta)\right]\left(\frac{\sigma_2-\sigma_1}{\sigma_2}\right)^{d-1} + x_2\left[\frac{1}{1-\eta_1} Z_{\mathrm{s}}\left(\frac{\eta_2}{1-\eta_1}\right) - Z_{\mathrm{s}}(\eta)\right]\left(\frac{\sigma_1-\sigma_2}{\sigma_1}\right)^{d-1}, \quad (d=2,3), \tag{6.28}$$

where $\eta_i = v_d \rho_i \sigma_i^d$ is the *partial* volume packing fraction due to species i. In contrast to most of the approaches (PY, SPT, BMCSL, e1, ...), the proposal (6.28) expresses $Z(\eta)$ in terms not only of $Z_{\mathrm{s}}(\eta)$ but also involves $Z_{\mathrm{s}}\left(\frac{\eta_1}{1-\eta_2}\right)$ and $Z_{\mathrm{s}}\left(\frac{\eta_2}{1-\eta_1}\right)$. Equation (6.28) should in principle be useful in particular for binary mixtures involving components of very disparate sizes. However, it is slightly less accurate than the one given in Eq. (6.21) for ordinary mixtures [48].

Quadratic Approximation

In order to improve the proposal contained in Eq. (6.20), in addition to the consistency requirements (6.12) and (6.13), one may consider the condition stemming from a binary mixture in which one of the species (say $i = 1$) is much larger than the other one (i.e., $\sigma_1/\sigma_2 \to \infty$), but occupies a negligible volume (i.e., $x_1(\sigma_1/\sigma_2)^d \to 0$). In that case, a sphere of species 1 is felt as a wall by particles of species 2, so that [20, 26, 53, 54]

$$\lim_{\substack{\sigma_1/\sigma_2 \to \infty \\ x_1(\sigma_1/\sigma_2)^d \to 0}} \left[g_{12}(\sigma_{12}) - 2^{d-1}\eta g_{22}(\sigma_2)\right] = 1 \; . \tag{6.29}$$

Hence, in the limit considered in Eq. (6.29), we have $z_{22} \to 1$, $z_{12} \to 2$. Consequently, under the universality ansatz (6.16), one may rewrite Eq. (6.29) as

$$\mathcal{G}(\eta, 2) = 1 + 2^{d-1}\eta\mathcal{G}(\eta, 1) \; . \tag{6.30}$$

Thus, Eqs. (6.17), (6.18), and (6.30) provide complete information on the function $\mathcal{G}$ at $z = 0$, $z = 1$, and $z = 2$, respectively, in terms of the contact value g_s of the single component RDF.

The simplest functional form of $\mathcal{G}$ that complies with the above consistency conditions is a quadratic function of z [55]:

$$\mathcal{G}(\eta, z) = \mathcal{G}_0(\eta) + \mathcal{G}_1(\eta)z + \mathcal{G}_2(\eta)z^2 \; , \tag{6.31}$$

where the coefficients $\mathcal{G}_1(\eta)$ and $\mathcal{G}_2(\eta)$ are explicitly given by

$$\mathcal{G}_1(\eta) = (2 - 2^{d-2}\eta)g_\mathrm{s} - \frac{2 - \eta/2}{1 - \eta} \; , \tag{6.32}$$

$$\mathcal{G}_2(\eta) = \frac{1 - \eta/2}{1 - \eta} - (1 - 2^{d-2}\eta)g_\mathrm{s} \; . \tag{6.33}$$

Therefore, the explicit expression for the contact values is

$$\begin{aligned} g_{ij}^{\mathrm{e2}}(\sigma_{ij}) = {} & \frac{1}{1-\eta} + \left[(2 - 2^{d-2}\eta)g_\mathrm{s} - \frac{2 - \eta/2}{1 - \eta}\right] \frac{M_{d-1}}{M_d} \frac{\sigma_i\sigma_j}{\sigma_{ij}} \\ & + \left[\frac{1 - \eta/2}{1 - \eta} - (1 - 2^{d-2}\eta)g_\mathrm{s}\right] \left(\frac{M_{d-1}}{M_d} \frac{\sigma_i\sigma_j}{\sigma_{ij}}\right)^2 \; . \end{aligned} \tag{6.34}$$

Following the same criterion as the one used in connection with Eq. (6.20), the label "e2" is meant to indicate that (i) the resulting contact values represent an *extension* of the single component contact value g_s and that (ii) $\mathcal{G}(\eta, z)$ is a *quadratic* polynomial in z. Of course, the quadratic form (6.31) is not the only choice compatible with conditions (6.17), (6.18), and (6.30). For instance, a rational function was also considered in [55]. However, although it is rather accurate, it does not lead to a closed form for the EOS. In contrast, when Eq. (6.34) is inserted into Eq. (6.2), one gets a closed expression for the compressibility factor in terms of the packing fraction η and the first few moments M_n, $n \le d$. The result is

$$\begin{aligned} Z_{\mathrm{e2}}(\eta) = {} & 1 + 2^{d-2}\frac{\eta}{1-\eta}\left[2(\Omega_0 - 2\Omega_1 + \Omega_2) + (\Omega_1 - \Omega_2)\eta\right] \\ & + \left[Z_\mathrm{s}(\eta) - 1\right]\left[2\Omega_1 - \Omega_2 + 2^{d-2}(\Omega_2 - \Omega_1)\eta\right] \; , \end{aligned} \tag{6.35}$$

where the quantities Ω_m are defined in Eq. (6.22). Quite interestingly, in the two-dimensional case Eq. (6.35) reduces to Eq. (6.23), i.e.,

$$Z_{\mathrm{e}1}(\eta) = Z_{\mathrm{e}2}(\eta), \quad (d=2)\ . \tag{6.36}$$

This illustrates the fact that two different proposals for the contact values $g_{ij}(\sigma_{ij})$ can yield the same EOS when inserted into Eq. (6.2). On the other hand, for three-dimensional mixtures Eq. (6.35) becomes

$$Z_{\mathrm{e}2}(\eta) = \frac{1}{1-\eta} + \frac{M_1 M_2}{M_3}\left(1-\eta+\frac{M_2^2}{M_1 M_3}\eta\right)\left[Z_{\mathrm{s}}(\eta) - \frac{1}{1-\eta}\right], \quad (d=3), \tag{6.37}$$

which differs from Eq. (6.24). In fact,

$$Z_{\mathrm{e}1}(\eta) - Z_{\mathrm{e}2}(\eta) = \frac{M_1 M_2}{2M_3}\left(1-\frac{M_2^2}{M_1 M_3}\right)\left[\frac{1+\eta}{1-\eta} - (1-2\eta)Z_{\mathrm{s}}(\eta)\right], \quad (d=3)\ . \tag{6.38}$$

Specific Examples

In this subsection, rather than carrying out an exhaustive comparison with the wealth of results available in the literature, we will consider only a few representative examples. In particular, for $d=3$, we will restrict ourselves to a comparison with classical proposals (say BGHLL, PY, and SPT for the contact values). The comparison with more recent ones may be found in [46, 55, 56, 57].

Thus far the development has been rather general since g_{s} remains free in Eqs. (6.20) and (6.34). In order to get specific results, it is necessary to fix g_{s} [cf. Table 6.1]. In the one-dimensional case, one has $g_{\mathrm{s}} = 1/(1-\eta)$ and so one gets the exact result (6.3) after substitution into Eq. (6.20). Similarly, Eqs. (6.32) and (6.33) lead to $\mathcal{G}_1 = \mathcal{G}_2 = 0$ and so we recover again the exact result.

If in the two-dimensional case we take Henderson's value [36] $g_{\mathrm{s}} = g_{\mathrm{s}}^{\mathrm{H}}$, then the linear approximation (6.20) reduces to the JM approximation, Eq. (6.4). This equivalence can be symbolically represented as $g_{ij}^{\mathrm{eH1}} = g_{ij}^{\mathrm{JM}}$, where the label "eH1" refers to the extension of Henderson's *single component* value in the linear approximation. While g_{ij}^{JM} is very accurate, even better results are provided by the quadratic form (6.34), especially if Luding's value [39, 40, 41] $g_{\mathrm{s}} = g_{\mathrm{s}}^{\mathrm{L}}$ is used [58].

In the three-dimensional case, Eq. (6.20) is of the form of the solution of the PY equation [9]. In fact, insertion of $g_{\mathrm{s}} = g_{\mathrm{s}}^{\mathrm{PY}}$ leads to Eq. (6.6), i.e., $g_{ij}^{\mathrm{ePY1}} = g_{ij}^{\mathrm{PY}}$. Similarly, if the SPT expression [10, 11, 12] $g_{\mathrm{s}} = g_{\mathrm{s}}^{\mathrm{SPT}}$ is used for the single component contact value in the quadratic approximation (6.34), we reobtain the SPT expression for the mixture, Eq. (6.8). In other words, $g_{ij}^{\mathrm{eSPT2}} = g_{ij}^{\mathrm{SPT}}$. On the other hand, if the much more accurate CS [44] expression $g_{\mathrm{s}} = g_{\mathrm{s}}^{\mathrm{CS}}$ is used as input, we arrive at the following expression:

$$g_{ij}^{\mathrm{eCS2}} = \frac{1}{1-\eta} + \frac{3}{2}\frac{\eta(1-\eta/3)}{(1-\eta)^2}\frac{\sigma_i\sigma_j M_2}{\sigma_{ij}M_3} + \frac{\eta^2(1-\eta/2)}{(1-\eta)^3}\left(\frac{\sigma_i\sigma_j M_2}{\sigma_{ij}M_3}\right)^2, \quad (d=3), \tag{6.39}$$

which is different from the BGHLL one, Eq. (6.10), improves the latter for $z_{ij} > 1$, and leads to similar results for $z_{ij} < 1$, as comparison with computer simulations shows [55]. The four approximations (6.6), (6.8), (6.10), and (6.39) are consistent with conditions (6.12) and (6.13), but only the SPT and eCS2 are also consistent with condition (6.29). It should also be noted that if one considers a binary mixture in the infinite solute dilution limit, namely $x_1 \to 0$, so that $z_{12} \to 2/(1+\sigma_2/\sigma_1)$, Eq. (6.39) yields the same result for $g_{12}(\sigma_{12})$ as the one proposed by Matyushov and Ladanyi [29] for this quantity on the basis of exact geometrical relations. However, the extension that the same authors propose when there is a non-vanishing solute concentration, i.e., for $x_1 \neq 0$, is different from Eq. (6.39).

Equation (6.34) can also be used in the case of hyperspheres ($d \geq 4$) [55]. In particular, a very good agreement with available computer simulations [49] is obtained for $d = 4$ and $d = 5$ by using Luban and Michels [45] value $g_\mathrm{s} = g_\mathrm{s}^{\mathrm{LM}}$.

Now we turn to the compressibility factors (6.21) and (6.35), which are obtained from the contact values (6.20) and (6.34), respectively. Since they depend on the details of the composition through the d first moments, they are meaningful even for continuous polydisperse mixtures.

As said above, in the two-dimensional case both Eqs. (6.21) and (6.35) reduce to Eq. (6.23), which yields very accurate results when a good Z_s is used as input [50, 55, 58]. For three-dimensional mixtures, insertion of $Z_\mathrm{s} = Z_\mathrm{s}^{\mathrm{CS}}$ in Eqs. (6.24) and (6.37) yields

$$Z_{\mathrm{eCS1}}(\eta) = Z_{\mathrm{BMCSL}}(\eta) + \frac{\eta^3 M_2}{(1-\eta)^3 M_3^2}\left(M_1M_3 - M_2^2\right), \quad (d=3), \tag{6.40}$$

$$Z_{\mathrm{eCS2}}(\eta) = Z_{\mathrm{BMCSL}}(\eta) - \frac{\eta^3 M_2}{(1-\eta)^2 M_3^2}\left(M_1M_3 - M_2^2\right), \quad (d=3), \tag{6.41}$$

where $Z_{\mathrm{BMCSL}}(\eta)$ is given by Eq. (6.11). Note that $Z_{\mathrm{eCS1}}(\eta) > Z_{\mathrm{BMCSL}}(\eta) > Z_{\mathrm{eCS2}}(\eta)$. Since simulation data indicate that the BMCSL EOS tends to underestimate the compressibility factor, it turns out that, as illustrated in Fig. 6.1 for an equimolar binary mixture with $\sigma_2/\sigma_1 = 0.6$, the performance of Z_{eCS1} is, paradoxically, better than that of Z_{eCS2} [55], despite the fact that the underlying linear approximation for the contact values is much less accurate than the quadratic approximation. This shows that a rather crude approximation such as Eq. (6.20) may lead to an extremely good EOS [46, 48, 49, 50], which, as clearly seen in Fig. 6.1, represents a substantial improvement over the classical proposals. Interestingly, the EOS corresponding to Z_{eCS1} has recently been independently derived as the second-order approximation of the Fundamental Measure Theory for the HS fluid by Hansen–Goos and Roth [60].

In the case of $d = 4$ and $d = 5$, use of $Z_\mathrm{s}(\eta) = Z_\mathrm{s}^{\mathrm{LM}}(\eta)$ in Eq. (6.21) produces a simple extended EOS of a mixture of hard additive hyperspheres in

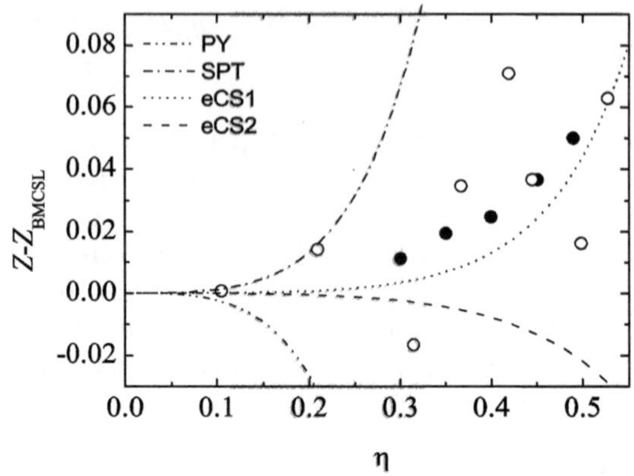

Fig. 6.1. Deviation of the compressibility factor from the BMCSL value, as a function of the packing fraction η for an equimolar three-dimensional binary mixture with $\sigma_2/\sigma_1 = 0.6$. The open [21, 22] and closed [59] *circles* are simulation data. The *lines* are the PY EOS (– · · –), the SPT EOS (– · – ·), the eCS1 EOS (· · ·), and the eCS2 EOS (– – –)

these dimensionalities. The accuracy of these two EOS for hard hypersphere mixtures in the fluid region has been confirmed by simulation data [49] for a wide range of compositions and size ratios. In Fig. 6.2, this accuracy is explicitly exhibited in the case of three equimolar mixtures, two in 4D and one in 5D.

6.2.2 A More Consistent Approximation for Three-Dimensional Additive Mixtures

Up to this point, we have considered an arbitrary dimensionality d and have constructed, under the universality assumption (6.16), the accurate quadratic

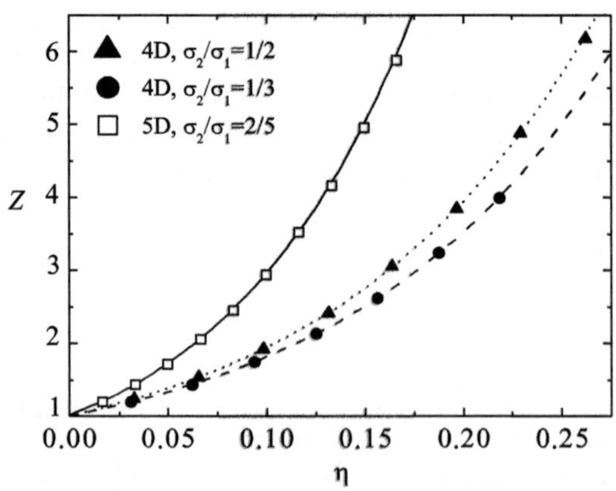

Fig. 6.2. Compressibility factor for three equimolar mixtures in 4D and 5D systems. *Lines* are the eLM1 predictions, while *symbols* are simulation data [49]

approximation (6.34), which fulfills the consistency conditions (6.12), (6.13), and (6.29). However, there exist extra consistency conditions that are not necessarily satisfied by (6.34). In particular, when the mixture is in contact with a hard wall, the state of equilibrium imposes that the pressure evaluated near the wall by considering the impacts with the wall must be the same as the pressure in the bulk evaluated from the particle–particle collisions. This consistency condition is especially important if one is interested in deriving accurate expressions for the contact values of the particle–wall correlation functions.

Since a hard wall can be seen as a sphere of infinite diameter, the contact value g_{wj} of the correlation function of a sphere of diameter σ_j with the wall can be obtained from $g_{ij}(\sigma_{ij})$ as

$$g_{wj} = \lim_{\substack{\sigma_i \to \infty \\ x_i \sigma_i^d \to 0}} g_{ij}(\sigma_{ij}) \, . \tag{6.42}$$

Note that g_{wj} provides the ratio between the density of particles of species j adjacent to the wall and the density of those particles far away from the wall. The sum rule connecting the pressure of the fluid and the above contact values is [61]

$$Z_w(\eta) = \sum_{j=1}^{N} x_j g_{wj} \, , \tag{6.43}$$

where the subscript w in Z_w has been used to emphasize that Eq. (6.43) represents a route alternative to the virial one, Eq. (6.2), to get the EOS of the HS mixture. The condition $Z = Z_w$ is equivalent to (6.29) in the special case where one has a *single* fluid in the presence of the wall. However, in the general case of a mixture plus a wall, the condition $Z = Z_w$ is stronger than Eq. (6.29). In the two-dimensional case, it turns out that the quadratic approximation (6.34) already satisfies the requirement $Z = Z_w$, regardless of the density and composition of the mixture [58]. However, this is not the case for $d \geq 3$.

Our problem now consists of computing $g_{ij}(\sigma_{ij})$ and the associated g_{wj} for the HS mixture in the presence of a hard wall, so that the condition $Z = Z_w$ is satisfied for an arbitrary mixture [56, 57]. Due to the mathematical complexity of the problem, here we will restrict ourselves to three-dimensional systems ($d = 3$). Similarly to what we did in the preceding subsection, we consider a class of approximations of the universal type (6.16), so that conditions (6.12) and (6.13) lead again to Eqs. (6.17) and (6.18), respectively. Notice that Eq. (6.16) implies in particular that

$$g_{wj} = \mathcal{G}(\eta, z_{wj}), \quad z_{wj} = 2\sigma_j \frac{M_2}{M_3}. \tag{6.44}$$

Assuming that $z = 0$ is a regular point and taking into account condition (6.17), $\mathcal{G}(\eta, z)$ can be expanded in a power series in z:

$$\mathcal{G}(\eta, z) = \mathcal{G}_0(\eta) + \sum_{n=1}^{\infty} \mathcal{G}_n(\eta) z^n \, . \tag{6.45}$$

After simple algebra, using the ansatz (6.16) and Eq. (6.45) in Eqs. (6.2) (with $d = 3$) and (6.43) one gets

$$Z = \mathcal{G}_0 + 3\eta \frac{M_1 M_2}{M_3} \mathcal{G}_0 + 4\eta \sum_{n=1}^{\infty} \mathcal{G}_n \frac{M_2^n}{M_3^{n+1}} \sum_{i,j=1}^{N} x_i x_j \sigma_i^n \sigma_j^n \sigma_{ij}^{3-n} , \qquad (6.46)$$

$$Z_w = \mathcal{G}_0 + \sum_{n=1}^{\infty} 2^n \mathcal{G}_n \frac{M_2^n}{M_3^n} M_n . \qquad (6.47)$$

Notice that if the series (6.45) is truncated after a given order $n \geq 3$, Z_w is given by the first n moments of the size distribution only. On the other hand, Z still involves an infinite number of moments if the truncation is made after $n \geq 4$ due to the presence of terms like $\sum_{i,j} x_i x_j \sigma_i^4 \sigma_j{}^4/\sigma_{ij}$, $\sum_{i,j} x_i x_j \sigma_i^5 \sigma_j{}^5/\sigma_{ij}^2$, ... Therefore, if we want the consistency condition $Z = Z_w$ to be satisfied for *any* discrete or continuous polydisperse mixture, either the whole infinite series (6.45) needs to be considered or it must be truncated after $n = 3$. The latter is of course the simplest possibility and thus we make the approximation

$$\mathcal{G}(\eta, z) = \mathcal{G}_0(\eta) + \mathcal{G}_1(\eta) z + \mathcal{G}_2(\eta) z^2 + \mathcal{G}_3(\eta) z^3 . \qquad (6.48)$$

As a consequence, Z and Z_w depend functionally on the size distribution of the mixture only through the first three moments (which is in the spirit of Rosenfeld's Fundamental Measure Theory [62]).

Using the approximation (6.48) in Eqs. (6.46) and (6.47) we are led to

$$Z = \mathcal{G}_0 + \eta \left[\frac{M_1 M_2}{M_3} \left(3\mathcal{G}_0 + 2\mathcal{G}_1\right) + 2\frac{M_2^3}{M_3^2} \left(\mathcal{G}_1 + 2\mathcal{G}_2 + 2\mathcal{G}_3\right) \right] , \qquad (6.49)$$

$$Z_w = \mathcal{G}_0 + 2\frac{M_1 M_2}{M_3} \mathcal{G}_1 + 4\frac{M_2^3}{M_3^2} \left(\mathcal{G}_2 + 2\mathcal{G}_3\right) . \qquad (6.50)$$

Thus far, the dependence of both Z and Z_w on the moments M_1, M_2, and M_3 is explicit and we only lack the packing-fraction dependence of $\mathcal{G}_1$, $\mathcal{G}_2$, and $\mathcal{G}_3$. From Eqs. (6.49) and (6.50) it follows that the difference between Z and Z_w is given by

$$Z - Z_w = \frac{M_1 M_2}{M_3} \left[3\eta \mathcal{G}_0 - 2(1-\eta)\mathcal{G}_1\right] + 2\frac{M_2^3}{M_3^2} \left[\eta \mathcal{G}_1 - 2(1-\eta)\mathcal{G}_2 - 2(2-\eta)\mathcal{G}_3\right] . \qquad (6.51)$$

Therefore, $Z = Z_w$ for *any* dispersity provided that

$$\mathcal{G}_1(\eta) = \frac{3\eta}{2\left(1-\eta\right)^2} , \qquad (6.52)$$

$$\mathcal{G}_2(\eta) = \frac{3\eta^2}{4\left(1-\eta\right)^3} - \frac{2-\eta}{1-\eta} \mathcal{G}_3(\eta) , \qquad (6.53)$$

where use has been made of the definition of $\mathcal{G}_0$, Eq. (6.17). To close the problem, we use the equal size limit given in Eq. (6.18), which yields $\mathcal{G}_0 + \mathcal{G}_1 + \mathcal{G}_2 + \mathcal{G}_3 = g_\mathrm{s}$. After a little algebra we are led to

$$\mathcal{G}_2(\eta) = (2-\eta)g_\mathrm{s} - \frac{2+\eta^2/4}{(1-\eta)^2}\,, \tag{6.54}$$

$$\mathcal{G}_3(\eta) = (1-\eta)\left(g_\mathrm{s}^\mathrm{SPT} - g_\mathrm{s}\right)\,. \tag{6.55}$$

This completes the derivation of our improved approximation, which we will call "e3," following the same criterion as the one used to call "e1" and "e2" to the approximations (6.20) and (6.34), respectively. In Eq. (6.55), $g_\mathrm{s}^\mathrm{SPT}$ is the SPT contact value for a single fluid, whose expression appears in Table 6.1. From Eq. (6.55) it is obvious that the choice $g_\mathrm{s} = g_\mathrm{s}^\mathrm{SPT}$ makes our e3 approximation to become the e2 approximation, both reducing to the SPT for mixtures, Eq. (6.8). This means that the SPT is fully internally consistent with the requirement $Z = Z_w$, although it has the shortcoming of not being too accurate in the single component case. The e3 proposal, on the other hand, satisfies the condition $Z = Z_w$ and has the flexibility of accommodating any desired g_s.

For the sake of concreteness, let us write explicitly the contact values in the e3 aproximation:

$$g_{ij}^\mathrm{e3}(\sigma_{ij}) = \frac{1}{1-\eta} + \frac{3\eta}{2(1-\eta)^2}\frac{M_2}{M_3}\frac{\sigma_i\sigma_j}{\sigma_{ij}} + \left[(2-\eta)g_\mathrm{s} - \frac{2+\eta^2/4}{(1-\eta)^2}\right] \times\left(\frac{M_2}{M_3}\frac{\sigma_i\sigma_j}{\sigma_{ij}}\right)^2 + (1-\eta)\left(g_\mathrm{s}^\mathrm{SPT} - g_\mathrm{s}\right)\left(\frac{M_2}{M_3}\frac{\sigma_i\sigma_j}{\sigma_{ij}}\right)^3\,, \tag{6.56}$$

$$g_{wj}^\mathrm{e3} = \frac{1}{1-\eta} + \frac{3\eta}{(1-\eta)^2}\frac{M_2}{M_3}\sigma_j + 4\left[(2-\eta)g_\mathrm{s} - \frac{2+\eta^2/4}{(1-\eta)^2}\right]\left(\frac{M_2}{M_3}\sigma_j\right)^2 + 8(1-\eta)\left(g_\mathrm{s}^\mathrm{SPT} - g_\mathrm{s}\right)\left(\frac{M_2}{M_3}\sigma_j\right)^3\,. \tag{6.57}$$

With the above results the compressibility factor may be finally written in terms of Z_s as

$$Z_\mathrm{e3}(\eta) = \frac{1}{(1-\eta)} + \left(\frac{M_1M_2}{M_3} - \frac{M_2^3}{M_3^2}\right)\frac{3\eta}{(1-\eta)^2} + \frac{M_2^3}{M_3^2}\left[Z_\mathrm{s}(\eta) - \frac{1}{1-\eta}\right]\,. \tag{6.58}$$

A few comments are in order at this stage. First, from Eq. (6.49) we can observe that, for the class of approximations (6.48), the compressibility factor Z does not depend on the individual values of the coefficients $\mathcal{G}_2$ and $\mathcal{G}_3$, but only on their sum. As a consequence, two different approximations of the form (6.48) sharing the same density dependence of $\mathcal{G}_1$ and $\mathcal{G}_2+\mathcal{G}_3$ also share the same virial EOS. For instance, if one makes the choice $g_\mathrm{s}=g_\mathrm{s}^\mathrm{PY}$, then $Z_\mathrm{ePY3}=Z_\mathrm{PY}$, even though $g_{ij}^\mathrm{ePY3}(\sigma_{ij})\neq g_{ij}^\mathrm{PY}(\sigma_{ij})$. Furthermore, if one makes the more accurate choice $g_\mathrm{s}=g_\mathrm{s}^\mathrm{CS}$, then $Z_\mathrm{eCS3}=Z_\mathrm{BMCSL}$, but again $g_{ij}^\mathrm{eCS3}(\sigma_{ij})\neq g_{ij}^\mathrm{BGHLL}(\sigma_{ij})$. The eCS3 contact values are

$$g_{ij}^\mathrm{eCS3}(\sigma_{ij})=\frac{1}{1-\eta}+\frac{3\eta}{2\,(1-\eta)^2}\frac{M_2}{M_3}\frac{\sigma_i\sigma_j}{\sigma_{ij}}+\frac{\eta^2(1+\eta)}{4(1-\eta)^3}\left(\frac{M_2}{M_3}\frac{\sigma_i\sigma_j}{\sigma_{ij}}\right)^2 +\frac{\eta^2}{4(1-\eta)^2}\left(\frac{M_2}{M_3}\frac{\sigma_i\sigma_j}{\sigma_{ij}}\right)^3, \tag{6.59}$$

$$g_{wj}^\mathrm{eCS3}=\frac{1}{1-\eta}+\frac{3\eta}{(1-\eta)^2}\frac{M_2}{M_3}\sigma_j+\frac{\eta^2(1+\eta)}{(1-\eta)^3}\left(\frac{M_2}{M_3}\sigma_j\right)^2 +\frac{2\eta^2}{(1-\eta)^2}\left(\frac{M_2}{M_3}\sigma_j\right)^3 . \tag{6.60}$$

In Figs. 6.3 and 6.4 we display the performance of the contact values as given by Eqs. (6.59) and (6.60), respectively, by comparison with results of computer simulations for both discrete and polydisperse mixtures. In both figures, we have also included the results that follow from the classical proposals as well as those of the eCS1 and eCS2 approximations. It is clear that for the wall–particle contact values the eCS3 approximation yields the best

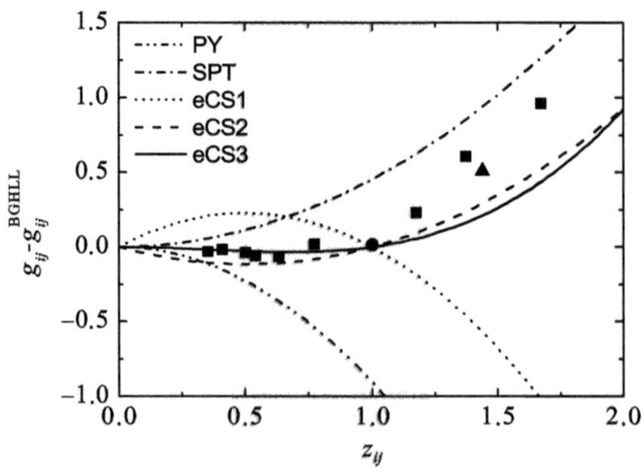

Fig. 6.3. Plot of the difference $g_{ij}(\sigma_{ij})-g_{ij}^\mathrm{BGHLL}(\sigma_{ij})$ as a function of the parameter $z_{ij}=(\sigma_i\sigma_j/\sigma_{ij})M_2/M_3$ for hard spheres ($d=3$) at a packing fraction $\eta=0.49$. The symbols are simulation data for the single fluid (*circle*, [47]), three binary mixtures (*squares*, [63]) with $\sigma_2/\sigma_1=0.3$ and $x_1=0.0625$, 0.125, and 0.25, and a ternary mixture (*triangles*, [64]) with $\sigma_2/\sigma_1=2/3$, $\sigma_3/\sigma_1=1/3$, and $x_1=0.1$, $x_2=0.2$. The *lines* are the PY approximation (– · · · –), the SPT approximation (– · – ·), the eCS1 approximation (· · ·), the eCS2 approximation (– – –), and the eCS3 approximation (—)

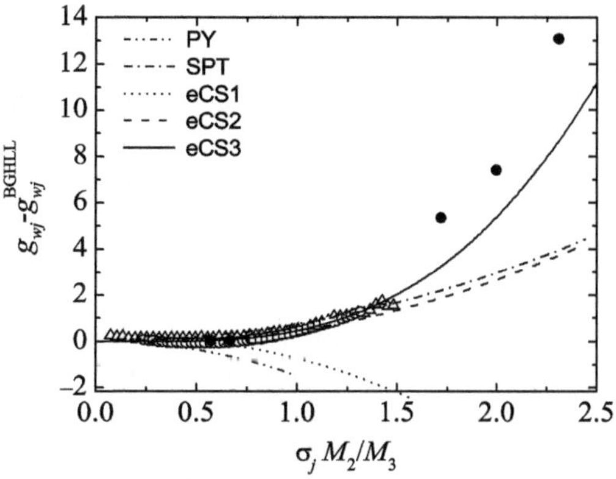

Fig. 6.4. Plot of the difference $g_{wj} - g_{wj}^{\rm BGHLL}$ as a function of the parameter $z_{wj}/2 = \sigma_j M_2/M_3$ for hard spheres ($d = 3$) at a packing fraction $\eta = 0.4$. The symbols are simulation data for a polydisperse mixture with a narrow top-hat distribution (*open squares*, [65]), a polydisperse mixture with a wide top-hat distribution (*open circles*, [65]), a polydisperse mixture with a Schulz distribution (*open triangles*, [65]), and a binary mixture (*closed circles*, [66]). The *lines* are the PY approximation (– · · · –), the SPT approximation (– · – ·), the eCS1 approximation (· · ·), the eCS2 approximation (– – –), and the eCS3 approximation (—)

performance, while for the particle–particle contact values both the eCS2 and the eCS3 are of comparable accuracy. A further feature to be pointed out is that the practical collapse on a common curve of the simulation data in Figs. 6.3 and 6.4 provides *a posteriori* support for the universality ansatz made in Eq. (6.16).

As mentioned earlier, there exist extra consistency conditions (see for instance [15]) that one might use as well within our approach. Assuming that the ansatz (6.16) still holds, some of these conditions are related to the derivatives of $\mathcal{G}$ with respect to z, namely

$$\left.\frac{\partial \mathcal{G}(\eta, z)}{\partial z}\right|_{z=0} = \frac{3\eta}{2(1-\eta)^2}\,, \tag{6.61}$$

$$\left.\frac{\partial^2 \mathcal{G}(\eta, z)}{\partial z^2}\right|_{z=0} = \frac{3\eta}{1-\eta}\left(g_{\rm s}^{\rm PY} - \frac{1}{2}g_{\rm s}\right)\,, \tag{6.62}$$

$$\left.\frac{\partial^3 \mathcal{G}(\eta, z)}{\partial z^3}\right|_{z=2} = 0\,. \tag{6.63}$$

Interestingly enough, as shown by Eq. (6.52), condition (6.61) is already satisfied by our e3 approximation without having to be imposed. On the other hand, condition (6.63) implies $\mathcal{G}_3 = 0$ in the e3 scheme and thus it is only satisfied if $g_{\rm s} = g_{\rm s}^{\rm SPT}$, in which case we recover the SPT. Condition (6.62) is not fulfilled either by the SPT or by the e3 approximation (except for a particular expression of $g_{\rm s}$ which is otherwise not very accurate). Thus, fulfilling the extra conditions (6.62) and (6.63) with a free g_s requires either considering a higher order polynomial in z (in which case the consistency condition $Z = Z_w$ cannot be satisfied for arbitrary mixtures, as discussed before) or not using the universality ansatz at all. In the first case, we have checked that a quartic or even a quintic polynomial does not improve matters, whereas giving up the universality assumption increases significantly the number of parameters to be determined and seems not to be adequate in view of the behavior observed in the simulation data.

An additional comment has to do with the restriction to $d = 3$ in this subsection. As noted before, the approximation e1 reduces to the exact result (6.3) for $d = 1$. For $d = 2$, the approximation e2 already fulfills the condition $Z = Z_w$ and so there is no real need to go further in that case. Since we have needed the approximation e3 to satisfy $Z = Z_w$ for $d = 3$, it is tempting to speculate that a polynomial form for $\mathcal{G}(z)$ of degree d could be found to be consistent with the condition $Z = Z_w$ for $d \geq 4$. However, a detailed analysis shows that this is not the case for an *arbitrary* mixture, since the number of conditions exceeds the number of unknowns, unless the universality assumption is partially relaxed.

As a final comment, let us stress that, although the discussion in this section has referred, for the sake of simplicity, to *discrete* mixtures, all the dependence on the details of the composition occurs through a finite number of moments, so that the results remain meaningful even for continuous *polydisperse* mixtures [67]. In that case, instead of a set of mole fractions $\{x_i\}$ and a set of diameters $\{\sigma_i\}$, one has to deal with a distribution function $w(\sigma)$ such that $w(\sigma)\mathrm{d}\sigma$ is the fraction of particles with a diameter comprising between σ and $\sigma + \mathrm{d}\sigma$. Therefore, the moments (6.1) are now defined as

$$M_n = \int_0^\infty \mathrm{d}\sigma\, \sigma^n w(\sigma) , \tag{6.64}$$

and with such a change the results we have derived for discrete mixtures also hold for polydisperse systems.

6.2.3 Non-Additive Systems

Non-additive hard-core mixtures, where the distance of closest approach between particles of different species is no longer the arithmetic mean of the diameters of both particles, have received much less attention than additive mixtures, in spite of their in principle more versatility to deal with interesting aspects occurring in real systems (such as fluid–fluid phase separation) and of their potential use as reference systems in perturbation calculations on the thermodynamic and structural properties of, say, Lennard–Jones mixtures. Nevertheless, the study of non-additive systems goes back 50 years [68, 69, 70, 71, 72] and is still a rapidly developing and challenging problem. Approaches to deal with this problem based on density functional theories will be presented in Chap. 7.

As mentioned in the paper by Ballone et al. [73], where the relevant references may be found, experimental work on alloys, aqueous electrolyte solutions, and molten salts suggests that hetero-coordination and homo-coordination may be interpreted in terms of excluded volume effects due to non-additivity of the repulsive part of the intermolecular potential. In particular, positive non-additivity leads naturally to demixing in HS mixtures, so that some of the experimental findings of phase separation in the above-mentioned (real) systems may be accounted for by using a model of a binary mixture of (positive) non-additive HS. On the other hand, negative non-additivity seems to account well for chemical short-range order in amorphous and liquid binary mixtures with preferred hetero-coordination [74, 75].

Some Preliminary Definitions

Let us consider an N-component mixture of non-additive HS in d dimensions. In this case, $\sigma_{ij} = \frac{1}{2}(\sigma_i + \sigma_j)(1 + \Delta_{ij})$, where $\Delta_{ij} \geq -1$ is a symmetric matrix with zero diagonal elements ($\Delta_{ii} = 0$) that characterizes the degree of non-additivity of the interactions. If $\Delta_{ij} > 0$, the non-additivity character of the ij interaction is said to be *positive*, while it is *negative* if $\Delta_{ij} < 0$. In the case of a binary mixture ($N = 2$), the only non-additivity parameter is $\Delta \equiv \Delta_{12} = \Delta_{21}$. The virial EOS (6.2) remains being valid in the non-additive case.

The contact values $g_{ij}(\sigma_{ij})$ can be expanded in a power series in density as

$$g_{ij}(\sigma_{ij}) = 1 + v_d\rho \sum_{k=1}^{N} x_k c_{k;ij} + (v_d\rho)^2 \sum_{k,\ell=1}^{N} x_k x_\ell c_{k\ell;ij} + \mathcal{O}(\rho^3) \,. \tag{6.65}$$

The coefficients $c_{k;ij}$, $c_{k\ell;ij}$, ... are independent of the composition of the mixture, but they are in general complicated nonlinear functions of the diameters σ_{ij}, σ_{ik}, σ_{jk}, $\sigma_{k\ell}$, ... Insertion of the expansion (6.65) into Eq. (6.2) yields the virial expansion of Z, namely

$$\begin{aligned} Z(\rho) &= 1 + \sum_{n=2}^{\infty} \overline{B}_n (v_d\rho)^{n-1} \\ &= 1 + v_d\rho \sum_{i,j=1}^{N} \overline{B}_{ij} x_i x_j + (v_d\rho)^2 \sum_{i,j,k=1}^{N} \overline{B}_{ijk} x_i x_j x_k \\ &\quad + (v_d\rho)^3 \sum_{i,j,k,\ell=1}^{N} \overline{B}_{ijk\ell} x_i x_j x_k x_\ell + \mathcal{O}(\rho^4) \,. \end{aligned} \tag{6.66}$$

Note that, for further convenience, we have introduced the coefficients $\overline{B}_n \equiv v_d^{-(n-1)} B_n$, where B_n are the usual virial coefficients [cf. Eq. (6.25)]. The composition-independent second, third, and fourth (barred) virial coefficients are given by

$$\overline{B}_{ij} = 2^{d-1} \sigma_{ij}^d \,, \tag{6.67}$$

$$\overline{B}_{ijk} = \frac{2^{d-1}}{3} \left(c_{k;ij}\sigma_{ij}^d + c_{j;ik}\sigma_{ik}^d + c_{i;jk}\sigma_{jk}^d \right) , \tag{6.68}$$

$$\overline{B}_{ijk\ell} = \frac{2^{d-1}}{6} \left(c_{k\ell;ij}\sigma_{ij}^d + c_{j\ell;ik}\sigma_{ik}^d + c_{i\ell;jk}\sigma_{jk}^d + c_{jk,i\ell}\sigma_{i\ell}^d + c_{ik,j\ell}\sigma_{j\ell}^d \right. \\ \left. + c_{ij;k\ell}\sigma_{k\ell}^d \right) . \tag{6.69}$$

A Simple Proposal for the Equation of State of *d*-Dimensional Non-Additive Mixtures

Our goal now is to generalize the e1 proposal given by Eq. (6.20) to the non-additive case [76]. We will not try to extend the e2 and e3 proposals, Eqs. (6.34) and (6.56), because of two reasons. First, given the inherent complexity of non-additive systems, we want to keep the approach as simple as possible. Second, we are more interested in the EOS than in the contact values themselves and, as mentioned earlier, the e1 proposal provides excellent EOS, at least in the additive case, despite the simplicity of the corresponding contact values.

As the simplest possible extension, we impose again the point particle and equal size consistency conditions, Eqs. (6.12) and (6.13), and thus keep in this case also the ansatz (6.16) and the linear structure of Eq. (6.19). However, instead of using Eq. (6.15), we determine the parameters z_{ij} as to reproduce Eq. (6.65) to first order in the density. The result is readily found to be [76]

$$z_{ij} = \left(\frac{b_3}{b_2} - 1 \right)^{-1} \left(\frac{\sum_k x_k c_{k;ij}}{M_d} - 1 \right) . \tag{6.70}$$

Here $b_2 = 2^{d-1}$ and b_3 are the second and third virial coefficients for the single component fluid, as defined by Eq. (6.26). The proposal of Eq. (6.19) supplemented by Eq. (6.70) is, by construction, accurate for densities low enough as to justify the truncated approximation $g_{ij}(\sigma_{ij}) \approx 1 + v_d \rho \sum_k x_k c_{k;ij}$. On the other hand, the limitations of this truncated expansion for moderate and large densities may be compensated by the use of g_s. When Eqs. (6.16), (6.19), and (6.70) are inserted into Eq. (6.2) one gets

$$Z(\eta) = 1 + \frac{\eta}{1-\eta} \frac{b_3 M_d \overline{B}_2 - b_2 \overline{B}_3}{(b_3 - b_2) M_d^2} + [Z_\mathrm{s}(\eta) - 1] \frac{\overline{B}_3 - M_d \overline{B}_2}{(b_3 - b_2) M_d^2} . \tag{6.71}$$

Equation (6.71) is the sought generalization of Eq. (6.21) to non-additive hard-core systems. As in the additive case, the density dependence in the EOS of the mixture is rather simple: $Z(\eta) - 1$ is expressed as a linear combination of $\eta/(1-\eta)$ and $Z_\mathrm{s}(\eta) - 1$, with coefficients such that the second and third virial coefficients are reproduced. Again, Eq. (6.71) is bound to be accurate for sufficiently low densities, while the limitations of the truncated expansion for moderate and large densities are compensated by the use of the EOS of the pure fluid.

The exact second virial coefficient $\overline{B}_2$ is known from Eq. (6.67). In principle, one should use the exact coefficients $c_{k;ij}$ to compute $\overline{B}_3$. However, to the best of our knowledge they are only known for $d \leq 3$. Since our objective is to have a proposal which is explicit for any d, we can make use of a reasonable approximation for them [76], as described below.

An Approximate Proposal for $c_{k;ij}$

The values of the coefficients $c_{k;ij}$ are exactly known for $d = 1$ and $d = 3$ and from these results one may approximate them in d dimensions as [76]

$$c_{k;ij} = \sigma_{k;ij}^d + \left(\frac{b_3}{b_2} - 1\right) \frac{\sigma_{k;ij}^{d-1}}{\sigma_{ij}} \sigma_{i;jk}\sigma_{j;ik} , \tag{6.72}$$

where we have called

$$\sigma_{k;ij} \equiv \sigma_{ik} + \sigma_{jk} - \sigma_{ij} \tag{6.73}$$

and it is understood that $\sigma_{k;ij} \geq 0$ for all sets ijk. Clearly, $\sigma_{i;ij} = \sigma_i$. For a binary mixture Eq. (6.72) yields

$$\begin{aligned} c_{1;11} &= (b_3/b_2)\sigma_1^d , \\ c_{2;11} &= (2\sigma_{12} - \sigma_1)^d + (b_3/b_2 - 1)\,\sigma_1(2\sigma_{12} - \sigma_1)^{d-1} , \\ c_{1;12} &= \sigma_1^d + (b_3/b_2 - 1)\,(2\sigma_{12} - \sigma_1)\sigma_1^d/\sigma_{12} . \end{aligned} \tag{6.74}$$

Of course, Eqs. (6.72) and (6.74) reduce to the exact results for $d = 1$ ($b_2 = b_3 = 1$) and for $d = 3$ ($b_2 = 4$, $b_3 = 10$).

The quantities $\sigma_{k;ij}$ may be given a simple geometrical interpretation. Assume that we have three spheres of species i, j, and k aligned in the sequence ikj. In such a case, the distance of closest approach between the centers of spheres i and j is $\sigma_{ik} + \sigma_{jk}$. If the sphere of species k were not there, that distance would of course be σ_{ij}. Therefore, $\sigma_{k;ij}$ as given by Eq. (6.73) represents a kind of effective diameter of sphere k, as seen from the point of view of the interaction between spheres i and j.

Inserting Eq. (6.72) into Eq. (6.70), one gets

$$z_{ij} = \left(\frac{b_3}{b_2} - 1\right)^{-1} \left(\frac{\sum_k x_k \sigma_{k;ij}^d}{M_d} - 1\right) + \frac{\sum_k x_k \sigma_{k;ij}^{d-1} \sigma_{i;jk}\sigma_{j;ik}}{M_d \sigma_{ij}} . \tag{6.75}$$

It can be easily checked that in the additive case ($\sigma_{k;ij} \to \sigma_k$), Eq. (6.75) reduces to Eq. (6.15).

Equations (6.72) and (6.74) are restricted to the situation $\sigma_{k;ij} \geq 0$ for any choice of i, j, and k, i.e., $2\sigma_{12} \geq \max(\sigma_1, \sigma_2)$ in the binary case. This excludes the possibility of dealing with mixtures with extremely high negative nonadditivity in which one sphere of species k might "fit in" between two spheres of species i and j in contact. Since for $d = 3$ and $N = 2$ the coefficients $c_{k;ij}$ are also known for such mixtures [77], we may extend our proposal to deal with these cases:

$$\begin{aligned} c_{1;11} &= (b_3/b_2)\sigma_1^d , \\ c_{2;11} &= \widehat{\sigma}_2^d + (b_3/b_2 - 1)\,\sigma_1 \widehat{\sigma}_2^{d-1} , \\ c_{1;12} &= (2\sigma_{12} - \widehat{\sigma}_2)^d + (b_3/b_2 - 1)\,\widehat{\sigma}_2 \sigma_1^d / \sigma_{12} , \end{aligned} \tag{6.76}$$

where we have defined

$$\widehat{\sigma}_2 = \max(2\sigma_{12} - \sigma_1, 0) . \tag{6.77}$$

With such an extension, we recover the exact values of $c_{k;ij}$ for a binary mixture of hard spheres ($d = 3$), even if $\sigma_1 > 2\sigma_{12}$ or $\sigma_2 > 2\sigma_{12}$.

The EOS (6.71) becomes explicit when $\overline{B}_3$ is obtained from Eq. (6.68) by using the approximation (6.72). The resulting virial coefficient is the exact one for $d = 1$ and $d = 3$. For hard disks ($d = 2$), it turns out that the approximate third virial coefficient is practically indistinguishable from the exact one [76]. When the approximate $\overline{B}_3$ is used, Eq. (6.71) reduces to Eq. (6.21) in the additive case.

From the comparison with simulation results, both for the compressibility factor and higher order virial coefficients, we find that the EOS (6.71) does a good job for non-additive mixtures, thus representing a reasonable compromise between simplicity and accuracy, provided that Z_s is accurate enough. This is illustrated in Fig. 6.5, where the proposal (6.71) with $Z_\mathrm{s} = Z_\mathrm{s}^\mathrm{CS}$ and a similar proposal by Hamad [78, 79, 80] are compared with simulation

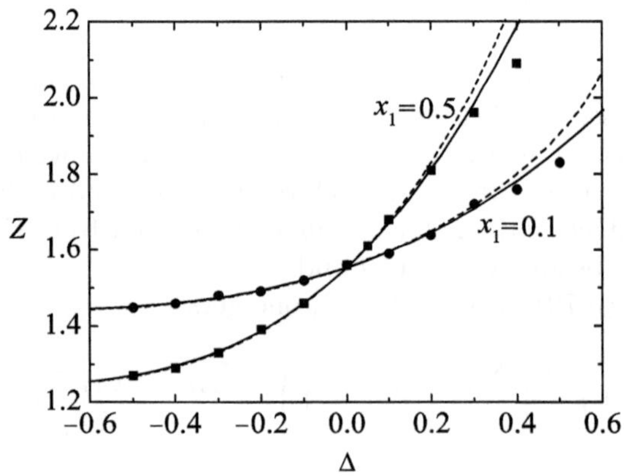

Fig. 6.5. Plot of the compressibility factor versus the non-additivity parameter Δ for a symmetric binary mixture of non-additive hard spheres ($d = 3$) at $\eta = \pi/30$ and two different compositions. The *solid lines* are our proposal, Eq. (6.71), with $Z_\mathrm{s} = Z_\mathrm{s}^\mathrm{CS}$, while the *dashed lines* are Hamad's proposal [78, 79, 80]. The *symbols* are results from Monte Carlo simulations [81, 82]

data [81, 82] for some three-dimensional symmetric mixtures. A more extensive comparison [76] shows that Eq. (6.71) seems to work better (especially as the density is increased) in the case of positive non-additivities, at least for $d = 1$, $d = 2$, and $d = 3$, but its performance is also reasonably good in highly asymmetric mixtures, even for negative Δ. Of course, the full assessment of this proposal is still pending since it involves many facets (non-additivity parameters, size ratios, density, and composition). Without this full assessment and given its rather satisfactory performance so far, going beyond the approximation given by Eq. (6.19) (taking similar steps to the ones described in Sects. 6.2.1 and 6.2.2 for additive systems) does not seem to be necessary at this stage, although it is in principle feasible.

6.2.4 Demixing

Demixing is a common phase transition in fluid mixtures usually originated on the asymmetry of the interactions (e.g., their strength and/or range) between the different components in the mixture. In the case of athermal systems such as HS mixtures in d dimensions, if fluid–fluid separation occurs, it would represent a neat example of an entropy-driven phase transition, i.e., a phase separation based only on the size asymmetry of the components. The existence of demixing in binary additive three-dimensional HS mixtures has been studied theoretically since decades, and the issue is still controversial. In this subsection, we will present our results following different but related routes that attempt to clarify some aspects of this problem.

Binary Mixtures of Additive d-Dimensional Spheres ($d = 3$, $d = 4$, and $d = 5$)

Now we look at the possible instability of a binary fluid mixture of HS of diameters σ_1 and σ_2 ($\sigma_1 > \sigma_2$) in d dimensions by looking at the Helmholtz free energy per unit volume, f, which is given by

$$\frac{f}{\rho k_{\mathrm{B}} T} = -1 + \sum_{i=1}^{2} x_i \ln\left(x_i \rho \lambda_i^d\right) + \int_0^{\eta} \mathrm{d}\eta' \frac{Z(\eta') - 1}{\eta'} , \tag{6.78}$$

where λ_i is the thermal de Broglie wavelength of species i. We locate the spinodals through the condition $f_{11} f_{22} - f_{12}^2 = 0$, with $f_{ij} \equiv \partial^2 f / \partial\rho_i \partial\rho_j$. Due to the spinodal instability, the mixture separates into two phases of different composition. The coexistence conditions are determined through the equality of the pressure p and the two chemical potentials μ_1 and μ_2 in both phases ($\mu_i = \partial f / \partial\rho_i$), leading to binodal (or coexistence) curves.

We begin with the case $d = 3$. It is well known that the BMCSL EOS, Eq. (6.11), does not lead to demixing. However, other EOS for HS mixtures have been shown to predict demixing [53, 54, 83], including the EOS that is obtained by truncating the virial series after a certain number of terms [84, 85].

In particular, it turns out that both $Z = Z_{\mathrm{eCS1}}$, Eq. (6.40), and $Z = Z_{\mathrm{eCS2}}$, Eq. (6.41), lead to demixing for certain values of the parameter $\gamma \equiv \sigma_2/\sigma_1$ that measures the size asymmetry. The critical values of the pressure, the composition, and the packing fraction are presented in Table 6.2 for a few values of γ.

As discussed earlier, the eCS1 EOS and, to a lesser extent, the eCS2 EOS are both in reasonably good agreement with the available simulation results for the compressibility factor [21, 22, 47, 59] and lead to the exact second and third virial coefficients but differ in the predictions for B_n with $n \geq 4$. The scatter in the values for the critical constants shown in Table 6.2 is evident and so there is no indication as to whether one should prefer one equation over the other in connection with this problem. Notice, for instance, that the eCS2 does not predict demixing for $\gamma \geq 0.2$, while both the values of the critical pressures and packing fractions for which it occurs according to the eCS1 EOS suggest that the transition might be metastable with respect to a fluid–solid transition.

Now we turn to the cases $d = 4$ and $d = 5$. Here, we use the extended Luban–Michels equation (eLM1) described in Sect. 6.2.1 [see Eq. (6.21) and Table 6.1]. As seen in Fig. 6.6, the location of the critical point tends to go down and to the right in the η_2 versus η_1 plane as γ decreases for $d = 4$ [86]. On the other hand, while it also tends to go down as γ decreases if $d = 5$, its behavior in the η_2 versus η_1 plane is rather more erratic in this case.

Also, the value of the critical pressure p_c (in units of $k_B T/\sigma_1^d$) is not a monotonic function of γ; its minimum value lies between $\gamma = 1/3$ and $\gamma = 1/2$ when $d = 4$, and it is around $\gamma = 3/5$ for $d = 5$. This non-monotonic behavior is also observed for three-dimensional HS [83, 85].

It is conceivable that the demixing transition in binary mixtures of hard hyperspheres in four and five dimensions described above may be metastable with respect to a fluid–solid transition, as it may also be the case of 3D HS. In fact, the value of the pressure at the freezing transition for the single component fluid is [45] $p_f\sigma^d/k_B T \simeq 12.7$ ($d = 3$), 11.5 ($d = 4$), and 12.2 ($d = 5$), i.e., $p_f\sigma^d/k_B T$ does not change appreciably with the dimensionality

Table 6.2. Critical constants $p_c\sigma_1^3/k_B T$, x_{1c}, and η_c for different γ-values as obtained from the two extended CS Eqs. (6.40) and (6.41)

	eCS1			eCS2		
γ	$p_c\sigma_1^3/k_B T$	x_{1c}	η_c	$p_c\sigma_1^3/k_B T$	x_{1c}	η_c
0.05	3599	0.0093	0.822	1096	0.0004	0.204
0.1	1307	0.0203	0.757	832.0	0.0008	0.290
0.2	653.4	0.0537	0.725	—	—	—
0.3	581.9	0.0998	0.738	—	—	—
0.4	663.4	0.1532	0.766	—	—	—

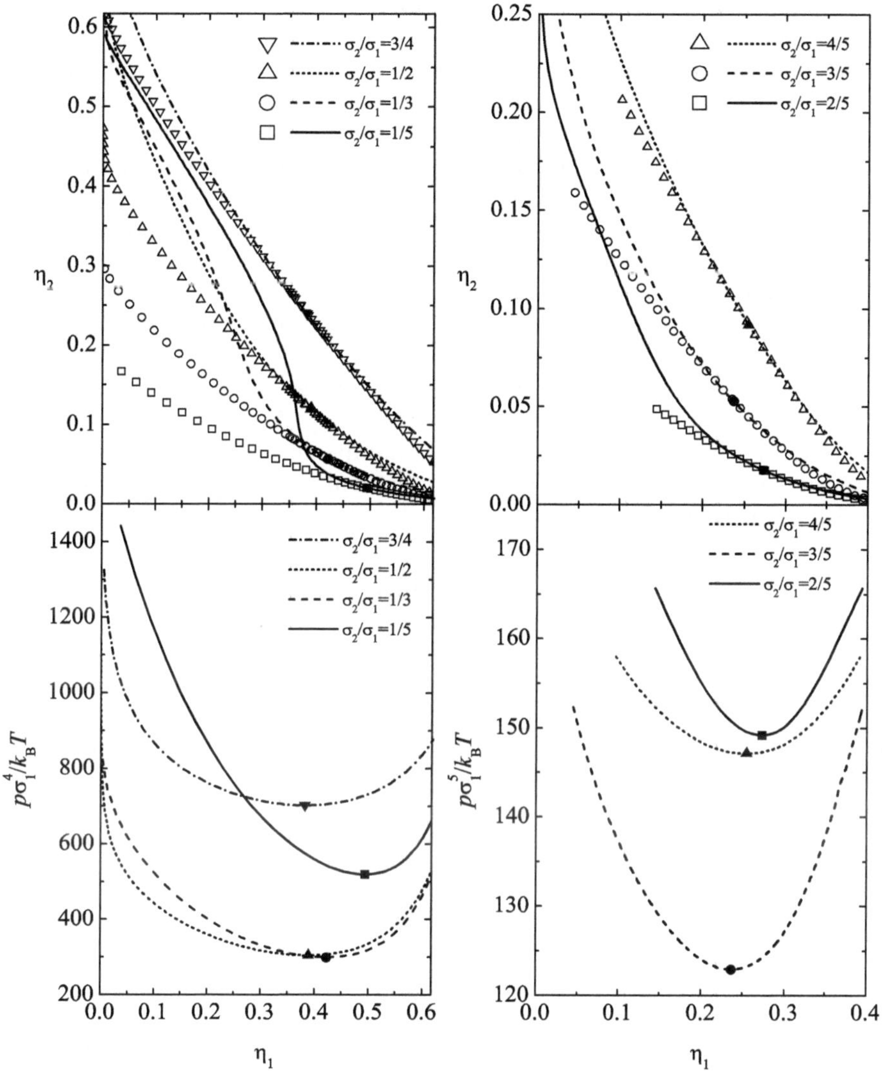

Fig. 6.6. Spinodal curves (*upper panels*: lines) and binodal curves (*upper panels*: open symbols; *lower panels*: lines) in a 4D system (*left panels*) and in a 5D system (*right panels*). The *closed symbols* are the critical consolute points

but is clearly very small in comparison with the critical pressures $p_c\sigma_1^d/k_BT$ we obtain for the mixture; for instance, $p_c\sigma_1^d/k_BT \simeq 600$ ($d = 3$, $\gamma = 3/10$), 300 ($d = 4$, $\gamma = 1/3$), and 123 ($d = 5$, $\gamma = 3/5$). However, one should also bear in mind that if the concentration x_1 of the bigger spheres decreases, the value of the pressure at which the solid–fluid transition in the mixture occurs in three dimension is also considerably increased with respect to p_f

[cf. Fig. 6.6 of [83]]. Thus, for concentrations $x_1 \simeq 0.01$ corresponding to the critical point of the fluid–fluid transition, the maximum pressure of the fluid phase greatly exceeds $p_{\rm f}$. If a similar trend with composition also holds in four and five dimensions, and given that the critical pressures become smaller as the dimensionality d is increased, it is not clear whether the competition between the fluid–solid and the fluid–fluid transitions in these dimensionalities will always be won by the former. The point clearly deserves further investigation.

An interesting feature must be mentioned. There is a remarkable similarity between the binodal curves represented in the $p\sigma_i^d$–η_1 and in the μ_i–η_1 planes [86]. By eliminating η_1 as if it were a parameter, one can represent the binodal curves in a μ_i versus $p\sigma_i^d$ plane. Provided the origin of the chemical potentials is such as to make $\lambda_i = \sigma_i$, the binodals in the μ_i–$p\sigma_i^d$ plane practically collapse into a single curve (which is in fact almost a straight line) for each dimensionality ($d = 3$, $d = 4$, and $d = 5$) [86]. A closer analysis of this phenomenon shows, however, that it is mainly due to the influence on μ_i of terms which are quantitatively dominant but otherwise irrelevant to the coexistence conditions.

Binary Mixtures of Non-Additive Hard Hyperspheres in the Limit of High Dimensionality

Let us now consider a binary mixture of non-additive HS of diameters σ_1 and σ_2 in d dimensions. Thus, in this case $\sigma_{12} \equiv 1/2(\sigma_1 + \sigma_2)(1 + \Delta)$ where as before Δ may be either positive or negative. Furthermore, assume (something that will become exact in the limit $d \to \infty$ [87]) that the EOS of the mixture is described by the second virial coefficient only, namely

$$p = \rho k_{\rm B} T \left[1 + B_2(x_1)\rho\right] , \tag{6.79}$$

where, according to Eq. (6.67),

$$B_2(x_1) = v_d 2^{d-1} \left(x_1^2\sigma_1^d + x_2^2\sigma_2^d + 2x_1x_2\sigma_{12}^d\right) . \tag{6.80}$$

The Helmholtz free energy per unit volume is given by $f/\rho k_{\rm B}T = -1 + \sum_{i=1}^2 x_i \ln\left(x_i\rho\lambda_i^d\right) + B_2\rho$, where Eq. (6.78) has been used. The Gibbs free energy *per particle* is

$$g = (f + p)/\rho = \sum_{i=1}^{2} x_i \ln\left(x_i\rho\lambda_i^d\right) + 2B_2(x_1)\rho , \tag{6.81}$$

where without loss of generality we have set $k_{\rm B}T = 1$. Given a size ratio γ, a value of Δ, and a dimensionality d, the consolute critical point (x_{1c}, p_c) is the solution to $\left(\partial^2 g/\partial x_1^2\right)_p = \left(\partial^3 g/\partial x_1^3\right)_p = 0$, provided of course it exists. Then, one can get the critical density ρ_c from Eq. (6.79).

We now introduce the scaled quantities [88]

$$\widetilde{p} \equiv 2^{d-1} v_d d^{-2} p \sigma_1^d / k_B T, \quad u \equiv d^{-1} B_2 \rho \,. \tag{6.82}$$

Consequently, Eqs. (6.79) and (6.81) can be rewritten as

$$\widetilde{p} = u \left(u + d^{-1} \right) / \widetilde{B}_2 \,, \tag{6.83}$$

$$g = \sum_{i=1}^{2} x_i \ln \left(x_i \Lambda_i \right) + \ln \left(A_d u / \widetilde{B}_2 \right) + 2du \,, \tag{6.84}$$

where $\widetilde{B}_2 \equiv B_2 / 2^{d-1} v_d \sigma_1^d$, $\Lambda_i \equiv (\lambda_i/\sigma_1)^d$, and $A_d \equiv d/2^{d-1} v_d$. Next we take the limit $d \to \infty$ and assume that the volume ratio $\widetilde{\gamma} \equiv \gamma^d$ is kept fixed and that there is a (slight) non-additivity $\Delta = d^{-2}\widetilde{\Delta}$ such that the scaled non-additivity parameter $\widetilde{\Delta}$ is also kept fixed in this limit. Thus, the second virial coefficient can be approximated by

$$\widetilde{B}_2 = \widetilde{B}_2^{(0)} + \widetilde{B}_2^{(1)} d^{-1} + \mathcal{O}(d^{-2}), \; \widetilde{B}_2^{(0)} = \left(x_1 + x_2 \widetilde{\gamma}^{1/2} \right)^2, \; \widetilde{B}_2^{(1)} = x_1 x_2 \widetilde{\gamma}^{1/2} J \,, \tag{6.85}$$

with

$$J \equiv \frac{1}{4} \left(\ln \widetilde{\gamma} \right)^2 + 2\widetilde{\Delta}. \tag{6.86}$$

Let us remark that, in order to find a consolute critical point, it is essential to keep the term of order d^{-1} if $\widetilde{\Delta} \leq 0$. The EOS (6.83) can then be inverted to yield

$$u = u^{(0)} + u^{(1)} d^{-1} + \mathcal{O}(d^{-2}), \quad u^{(0)} = \sqrt{\widetilde{p} \widetilde{B}_2^{(0)}}, \quad u^{(1)} = -\frac{1}{2} \left(1 - u^{(0)} \frac{\widetilde{B}_2^{(1)}}{\widetilde{B}_2^{(0)}} \right) . \tag{6.87}$$

In turn, the Gibbs free energy (6.84) becomes

$$g = g^{(0)} d + g^{(1)} + \mathcal{O}(d^{-1}) \,,$$
$$g^{(0)} = 2u^{(0)}, \quad g^{(1)} = \sum_{i=1}^{2} x_i \ln \left(x_i \Lambda_i \right) + \ln \left(A_d u^{(0)} / \widetilde{B}_2^{(0)} \right) + 2u^{(1)} \,, \tag{6.88}$$

while the chemical potentials $\mu_1 = g + x_2 \left(\partial g / \partial x_1 \right)_p$ and $\mu_2 = g - x_1 \left(\partial g / \partial x_1 \right)_p$ are given by

$$\mu_i = \mu_i^{(0)} d + \mu_i^{(1)} + \mathcal{O}(d^{-1}), \quad \mu_1^{(0)} = 2\widetilde{p}^{1/2} \,,$$
$$\mu_1^{(1)} = \ln \left(A_d x_1 \Lambda_1 \sqrt{\widetilde{p}/\widetilde{B}_2^{(0)}} \right) - 1/\sqrt{\widetilde{B}_2^{(0)}} + (x_2/x_1)(\widetilde{\gamma}\widetilde{p})^{1/2} \widetilde{B}_2^{(1)} / \widetilde{B}_2^{(0)} \,, \tag{6.89}$$

where μ_2 is obtained from μ_1 by the changes $x_1 \leftrightarrow x_2$, $\Lambda_1 \to \Lambda_2/\widetilde{\gamma}$, $\widetilde{\gamma} \to 1/\widetilde{\gamma}$, $\widetilde{p} \to \widetilde{p}\widetilde{\gamma}$, $\widetilde{B}_2 \to \widetilde{B}_2/\widetilde{\gamma}$.

The coordinates of the critical point are readily found to be

$$x_{1c} = \frac{\widetilde{\gamma}^{3/4}}{1+\widetilde{\gamma}^{3/4}}, \quad \widetilde{p}_c = \frac{\left(1+\widetilde{\gamma}^{1/4}\right)^4}{4\widetilde{\gamma}J^2}. \tag{6.90}$$

Note that x_{1c} is independent of $\widetilde{\Delta}$. The coexistence curve, which has to be obtained numerically, follows from the conditions $\mu_i^{(1)}(x_A,\widetilde{p}) = \mu_i^{(1)}(x_B,\widetilde{p})$ $(i = 1,2)$ where $x_1 = x_A$ and $x_1 = x_B$ are the mole fractions of the coexisting phases. Once the critical consolute point has been identified in the pressure/concentration plane, we can obtain the critical density. The dominant behaviors of $\widetilde{B}_2$ and u at the critical point are

$$\widetilde{B}_2^{(0)}(x_{1c}) = \frac{\widetilde{\gamma}}{\left(1-\widetilde{\gamma}^{1/4}+\widetilde{\gamma}^{1/2}\right)^2}, \quad u_c^{(0)} = \frac{\left(1+\widetilde{\gamma}^{1/4}\right)^2}{2\left(1-\widetilde{\gamma}^{1/4}+\widetilde{\gamma}^{1/2}\right)J}. \tag{6.91}$$

Hence, the critical density readily follows after substitution in the scaling relation given in Eq. (6.82). It is also convenient to consider the scaled version $\widetilde{\eta} \equiv d^{-1}2^d\eta$ of the packing fraction $\eta = v_d\rho\sigma_1^d\,(x_1 + x_2\widetilde{\gamma})$. At the critical point, it takes the nice expression

$$\widetilde{\eta}_c = \frac{\left(\widetilde{\gamma}^{1/8}+\widetilde{\gamma}^{-1/8}\right)^2}{J}. \tag{6.92}$$

The previous results clearly indicate that a demixing transition is possible not only for additive or positively non-additive mixtures but even for negative non-additivities. The only requirement is $J > 0$, i.e., $\widetilde{\Delta} > -1/8\,(\ln\widetilde{\gamma})^2$ or, equivalently, $\Delta > -1/8\,(\ln\gamma)^2$. Figure 6.7 shows the binodal curves corresponding to $\widetilde{\gamma} = 0.01$ and $\widetilde{\Delta} = -0.1$ (negative non-additivity), $\widetilde{\Delta} = 0$ (additivity), and $\widetilde{\Delta} = 0.1$ (positive non-additivity).

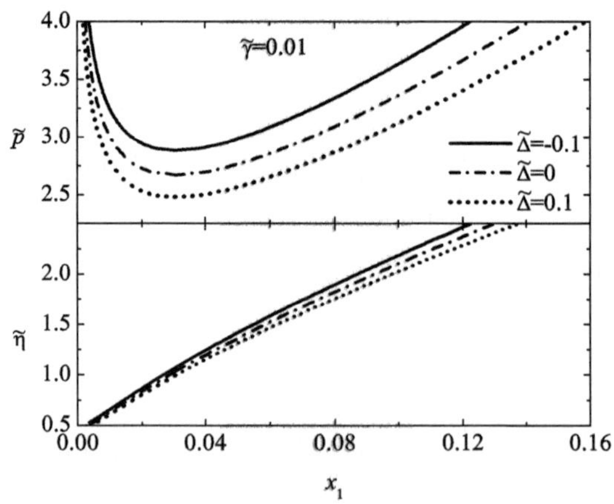

Fig. 6.7. Binodal curves in the planes $\widetilde{p}$ versus x_1 and $\widetilde{\eta}$ versus x_1 corresponding to $\widetilde{\gamma} = 0.01$ and $\widetilde{\Delta} = -0.1$, $\widetilde{\Delta} = 0$, and $\widetilde{\Delta} = 0.1$

While the high-dimensionality limit has allowed us to address the problem in a mathematically simple and clear-cut way, the possibility of demixing with negative non-additivity is not an artifact of that limit. As said before, demixing is known to occur for positive non-additive binary mixtures of HS in three dimensions, and there is compelling evidence on the existence of this phenomenon in the additive case, at least in the metastable fluid region. Even though in a three-dimensional mixture the EOS is certainly more complicated than Eq. (6.79) and the demixing transition that we have just discussed for negative non-additivity is possibly metastable with respect to the freezing transition, the main effects at work (namely the competition between depletion due to size asymmetry and hetero-coordination due to negative non-additivity) are also present. In fact, it is interesting to point out that Roth et al. [89], using the approximation of an effective single component fluid with pair interactions to describe a binary mixture of non-additive three-dimensional HS and employing an empirical rule based on the effective second virial coefficient, have also suggested that demixing is possible for small negative non-additivity and high size asymmetry. Our exact results lend support to this suggestion and confirm that, in some cases, the limit $d \to \infty$ highlights features already present in real systems.

6.3 The Rational Function Approximation (RFA) Method for the Structure of Hard-Sphere Fluids

The RDF $g(r)$ and its close relative the (static) structure factor $S(q)$ are the basic quantities used to discuss the structure of a single component fluid [1, 2, 3, 4]. The latter quantity is defined as

$$S(q) = 1 + \rho\widetilde{h}(q) , \tag{6.93}$$

where

$$\widetilde{h}(q) = \int \mathrm{d}\mathbf{r}\, \mathrm{e}^{\mathrm{i}\mathbf{q}\cdot\mathbf{r}} h(r) \tag{6.94}$$

is the Fourier transform of the total correlation function $h(r) \equiv g(r) - 1$, i being the imaginary unit. An important related quantity is the direct correlation function $c(r)$, which is defined in Fourier space through the Ornstein–Zernike (OZ) relation [1, 2, 3, 4]

$$\widetilde{c}(q) = \frac{\widetilde{h}(q)}{1 + \rho\widetilde{h}(q)} , \tag{6.95}$$

where $\widetilde{c}(q)$ is the Fourier transform of $c(r)$.

As pointed out in Chap. 1, the usual approach to obtain $g(r)$ is through one of the integral equation theories, where the OZ equation is complemented by a closure relation between $c(r)$ and $h(r)$ [1]. However, apart from requiring in general hard numerical labor, a disappointing aspect is that the substitution of the (necessarily) approximate values of $g(r)$ obtained from them in the (exact) statistical mechanical formulae may lead to the thermodynamic inconsistency problem.

The two basic routes to obtain the EOS of a single component fluid of HS are the virial route, Eq. (6.14), and the compressibility route

$$\chi_\mathrm{s} \equiv k_\mathrm{B}T\left(\frac{\partial\rho}{\partial p}\right)_T = [1-\rho\widetilde{c}(0)]^{-1} = S(0)$$
$$= 1 + 2^d d\eta\sigma^{-d}\int_0^\infty \mathrm{d}r\, r^{d-1}h(r)\,. \tag{6.96}$$

where χ_s is the (reduced) isothermal compressibility. Thermodynamic consistency implies that

$$\chi_\mathrm{s}^{-1}(\eta) = \frac{\mathrm{d}}{\mathrm{d}\eta}[\eta Z_\mathrm{s}(\eta)]\,, \tag{6.97}$$

but, in general, this condition is not satisfied by an approximate RDF. In the case of an HS mixture, the virial route is given by Eq. (6.2), while the compressibility route is indicated below [cf. Eq. (6.145)].

In this section we describe the RFA method, which is an alternative to the integral equation approach and in particular leads by construction to thermodynamic consistency.

6.3.1 The Single Component HS Fluid

We begin with the case of a single component fluid of HS of diameter σ. The following presentation is equivalent to the one given in [90, 91], where all details can be found, but more suitable than the former for direct generalization to the case of mixtures.

The starting point will be the Laplace transform

$$G(s) = \int_0^\infty \mathrm{d}r\, \mathrm{e}^{-sr} r g(r) \tag{6.98}$$

and the auxiliary function $\Psi(s)$ defined through

$$G(s) = \frac{s}{2\pi}\left[\rho + \mathrm{e}^{s\sigma}\Psi(s)\right]^{-1}. \tag{6.99}$$

The choice of $G(s)$ as the Laplace transform of $rg(r)$ and the definition of $\Psi(s)$ from Eq. (6.99) are suggested by the exact form of $g(r)$ to first order in density [90].

Since $g(r) = 0$ for $r < \sigma$ while $g(\sigma^+) =$ finite, one has

$$g(r) = \Theta(r-\sigma)\left[g(\sigma^+) + g'(\sigma^+)(r-\sigma) + \cdots\right]\,, \tag{6.100}$$

where $g'(r) \equiv \mathrm{d}g(r)/\mathrm{d}r$. This property imposes a constraint on the large s behavior of $G(s)$, namely

$$e^{\sigma s} sG(s) = \sigma g(\sigma^+) + \left[g(\sigma^+) + \sigma g'(\sigma^+)\right] s^{-1} + \mathcal{O}(s^{-2}). \tag{6.101}$$

Therefore, $\lim_{s\to\infty} e^{s\sigma} sG(s) = \sigma g(\sigma^+) =$ finite or, equivalently,

$$\lim_{s\to\infty} s^{-2}\Psi(s) = \frac{1}{2\pi\sigma g(\sigma^+)} = \text{finite}. \tag{6.102}$$

On the other hand, according to Eq. (6.96) with $d = 3$,

$$\begin{aligned}\chi_\mathrm{s} &= 1 - 24\eta\sigma^{-3} \lim_{s\to 0} \frac{\mathrm{d}}{\mathrm{d}s} \int_0^\infty \mathrm{d}r\, e^{-sr} r\,[g(r) - 1] \\ &= 1 - 24\eta\sigma^{-3} \lim_{s\to 0} \frac{\mathrm{d}}{\mathrm{d}s} \left[G(s) - s^{-2}\right] .\end{aligned} \tag{6.103}$$

Since the (reduced) isothermal compressibility χ_s is also finite, one has $\int_0^\infty \mathrm{d}r\, r^2\,[g(r) - 1] =$ finite, so that the weaker condition $\int_0^\infty \mathrm{d}r\, r\,[g(r) - 1] = \lim_{s\to 0}[G(s) - s^{-2}] =$ finite must hold. This in turn implies

$$\Psi(s) = -\rho + \rho\sigma s - \frac{1}{2}\rho\sigma^2 s^2 + \left(\frac{1}{6}\rho\sigma^3 + \frac{1}{2\pi}\right) s^3 - \left(\frac{1}{24}\rho\sigma^3 + \frac{1}{2\pi}\right) \sigma s^4 + \mathcal{O}(s^5). \tag{6.104}$$

First-Order Approximation (PY Solution)

An interesting aspect to be remarked is that the minimal input we have just described on the physical requirements related to the structure and thermodynamics of the system is enough to determine the small and large s limits of $\Psi(s)$, Eqs. (6.102) and (6.104), respectively. While infinite choices for $\Psi(s)$ would comply with such limits, a particularly simple form is a *rational function.* In particular, the rational function having the least number of coefficients to be determined is

$$\Psi(s) = \frac{E^{(0)} + E^{(1)}s + E^{(2)}s^2 + E^{(3)}s^3}{L^{(0)} + L^{(1)}s} , \tag{6.105}$$

where one of the coefficients can be given an arbitrary non-zero value. We choose $E^{(3)} = 1$. With such a choice and in view of Eq. (6.104), one finds $E^{(0)} = -\rho L^{(0)}$, $E^{(1)} = -\rho(L^{(1)} - \sigma L^{(0)})$, $E^{(2)} = \rho(\sigma L^{(1)} - \frac{1}{2}\sigma^2 L^{(0)})$, and

$$L^{(0)} = 2\pi \frac{1 + 2\eta}{(1 - \eta)^2} , \tag{6.106}$$

$$L^{(1)} = 2\pi\sigma \frac{1 + \eta/2}{(1 - \eta)^2}. \tag{6.107}$$

Upon substitution of these results into Eqs. (6.99) and (6.105), we get

$$G(s)=\frac{\mathrm{e}^{-\sigma s}}{2\pi s^2}\frac{L^{(0)}+L^{(1)}s}{1-\rho\left[\varphi_2(\sigma s)\sigma^3L^{(0)}+\varphi_1(\sigma s)\sigma^2L^{(1)}\right]}\,, \tag{6.108}$$

where

$$\varphi_n(x)\equiv x^{-(n+1)}\left(\sum_{m=0}^{n}\frac{(-x)^m}{m!}-\mathrm{e}^{-x}\right). \tag{6.109}$$

In particular,

$$\varphi_0(x)=\frac{1-\mathrm{e}^{-x}}{x},\quad \varphi_1(x)=\frac{1-x-\mathrm{e}^{-x}}{x^2},\quad \varphi_2(x)=\frac{1-x+x^2/2-\mathrm{e}^{-x}}{x^3}. \tag{6.110}$$

Note that $\lim_{x\to 0}\varphi_n(x)=(-1)^n/(n+1)!$

It is remarkable that Eq. (6.108), which has been derived here as the simplest rational form for $\Psi(s)$ complying with the requirements (6.102) and (6.104), coincides with the solution to the PY closure, $c(r)=0$ for $r>\sigma$, of the OZ equation [42, 43]. Application of Eq. (6.102) yields the PY contact value $g_{\mathrm{s}}^{\mathrm{PY}}$ and compressibility factor $Z_{\mathrm{s}}^{\mathrm{PY}}$ shown in Table 6.1. Analogously, Eq. (6.103) yields

$$\chi_{\mathrm{s}}^{\mathrm{PY}}=\frac{(1-\eta)^4}{(1+2\eta)^2}\,. \tag{6.111}$$

It can be easily checked that the thermodynamic relation (6.97) is not satisfied by the PY theory.

Second-Order Approximation

In the spirit of the RFA, the simplest extension of the rational approximation (6.105) involves two new terms, namely αs^4 in the numerator and $L^{(2)}s^2$ in the denominator, both of them necessary in order to satisfy Eq. (6.102). Such an addition leads to

$$\Psi(s)=\frac{E^{(0)}+E^{(1)}s+E^{(2)}s^2+E^{(3)}s^3+\alpha s^4}{L^{(0)}+L^{(1)}s+L^{(2)}s^2}. \tag{6.112}$$

Applying Eq. (6.104), it is possible to express $E^{(0)}$, $E^{(1)}$, $E^{(2)}$, $E^{(3)}$, $L^{(0)}$, and $L^{(1)}$ in terms of α and $L^{(2)}$. This leads to

$$G(s)=\frac{\mathrm{e}^{-\sigma s}}{2\pi s^2}\frac{L^{(0)}+L^{(1)}s+L^{(2)}s^2}{1+\alpha s-\rho\left[\varphi_2(\sigma s)\sigma^3L^{(0)}+\varphi_1(\sigma s)\sigma^2L^{(1)}+\varphi_0(\sigma s)\sigma L^{(2)}\right]}\,, \tag{6.113}$$

where

$$L^{(0)}=2\pi\frac{1+2\eta}{(1-\eta)^2}+\frac{12\eta}{1-\eta}\left(\frac{\pi}{1-\eta}\frac{\alpha}{\sigma}-\frac{L^{(2)}}{\sigma^2}\right)\,, \tag{6.114}$$

$$L^{(1)} = 2\pi\sigma\frac{1+\frac{1}{2}\eta}{(1-\eta)^2} + \frac{2}{1-\eta}\left(\pi\frac{1+2\eta}{1-\eta}\alpha - 3\eta\frac{L^{(2)}}{\sigma}\right). \tag{6.115}$$

Thus far, irrespective of the values of the coefficients $L^{(2)}$ and α, the conditions $\lim_{s\to\infty} e^{s\sigma} sG(s) =$ finite and $\lim_{s\to 0}[G(s) - s^{-2}] =$ finite are satisfied. Of course, if $L^{(2)} = \alpha = 0$, one recovers the PY approximation. More generally, we may determine these coefficients by prescribing the compressibility factor Z_{s} (or equivalently the contact value g_{s}) and then, in order to ensure thermodynamic consistency, compute from it the isothermal compressibility χ_{s} by means of Eq. (6.97). From Eqs. (6.102) and (6.103) one gets

$$L^{(2)} = 2\pi\alpha\sigma g_{\mathrm{s}} , \tag{6.116}$$

$$\chi_{\mathrm{s}} = \left(\frac{2\pi}{L^{(0)}}\right)^2 \left[1 - \frac{12\eta}{1-\eta}\frac{\alpha}{\sigma}\left(1 + 2\frac{\alpha}{\sigma}\right) + \frac{12\eta}{\pi}\frac{\alpha L^{(2)}}{\sigma^3}\right]. \tag{6.117}$$

Clearly, upon substitution of Eqs. (6.114) and (6.116) into Eq. (6.117) a quadratic algebraic equation for α is obtained. The physical root is

$$\alpha = -\frac{12\eta(1+2\eta)E_4}{(1-\eta)^2 + 36\eta\left[1+\eta - Z_{\mathrm{s}}(1-\eta)\right]E_4} , \tag{6.118}$$

where

$$E_4 = \frac{1-\eta}{36\eta\left(Z_{\mathrm{s}} - \frac{1}{3}\right)}\left\{1 - \left[1 + \frac{Z_{\mathrm{s}} - \frac{1}{3}}{Z_{\mathrm{s}} - Z_{\mathrm{s}}^{\mathrm{PY}}}\left(\frac{\chi_{\mathrm{s}}}{\chi_{\mathrm{s}}^{\mathrm{PY}}} - 1\right)\right]^{1/2}\right\} . \tag{6.119}$$

The other root must be discarded because it corresponds to a negative value of α, which, according to Eq. (6.116), yields a negative value of $L^{(2)}$. This would imply the existence of a positive real value of s at which $G(s) = 0$ [90, 91], which is not compatible with a positive definite RDF. However, according to the form of Eq. (6.119) it may well happen that, once Z_{s} has been chosen, there exists a certain packing fraction η_{g} above which α is no longer positive. For such a packing fraction, the associated χ_{s} becomes equal to $\chi_{\mathrm{s}}^{\mathrm{PY}}$. This condition may be interpreted as an indication that, at the packing fraction η_{g} where α vanishes, the system ceases to be a fluid and a glass transition in the HS fluid occurs [91, 92, 93].

Expanding (6.113) in powers of s and using Eq. (6.101), one can obtain the derivatives of the RDF at $r = \sigma^+$ [94]. In particular, the first derivative is

$$g'(\sigma^+) = \frac{1}{2\pi\alpha\sigma}\left[L^{(1)} - L^{(2)}\left(\frac{1}{\alpha} + \frac{1}{\sigma}\right)\right] , \tag{6.120}$$

which may have some use in connection with perturbation theory [18].

It is worthwhile to point out that the structure implied by Eq. (6.113) coincides in this single component case with the solution of the Generalized Mean Spherical Approximation (GMSA) [95, 96, 97], where the OZ relation is solved under the ansatz that the direct correlation function has a Yukawa form outside the core.

For a given Z_s, once $G(s)$ has been determined, inverse Laplace transformation yields $rg(r)$. First, note that Eq. (6.99) can be formally rewritten as

$$G(s) = -\frac{s}{2\pi} \sum_{n=1}^{\infty} \rho^{n-1} \left[-\Psi(s)\right]^{-n} \mathrm{e}^{-ns\sigma}. \tag{6.121}$$

Thus, the RDF is then given by

$$g\left(r\right) = \frac{1}{2\pi r} \sum_{n=1}^{\infty} \rho^{n-1} \psi_n \left(r - n\sigma\right) \Theta\left(r - n\sigma\right) \ , \tag{6.122}$$

with $\Theta\left(x\right)$ denoting the Heaviside step function and

$$\psi_n\left(r\right) = -\mathcal{L}^{-1}\left\{s\left[-\Psi\left(s\right)\right]^{-n}\right\} \ , \tag{6.123}$$

$\mathcal{L}^{-1}$ denoting the inverse Laplace transform. Explicitly, using the residue theorem,

$$\psi_n\left(r\right) = -\sum_{i=1}^{4} \mathrm{e}^{s_i r} \sum_{m=1}^{n} \frac{a_{mn}^{(i)}}{(n-m)!(m-1)!} r^{n-m} \ , \tag{6.124}$$

where

$$a_{mn}^{(i)} = \lim_{s \to s_i} \left(\frac{\mathrm{d}}{\mathrm{d}s}\right)^{m-1} s\left[-\Psi\left(s\right)/\left(s - s_i\right)\right]^{-n} \ , \tag{6.125}$$

s_i $(i = 1, \ldots, 4)$ being the poles of $1/\Psi(s)$, i.e., the roots of $E^{(0)} + E^{(1)}s + E^{(2)}s^2 + E^{(3)}s^3 + \alpha s^4 = 0$. Explicit expressions of $g(r)$ up to the second coordination shell $\sigma \leq r \leq 3\sigma$ can be found in [98].

On the other hand, the static structure factor $S(q)$ [cf. Eq. (6.93)] and the Fourier transform $\widetilde{h}(q)$ may be related to $G(s)$ by noting that

$$\widetilde{h}(q) = \frac{4\pi}{q} \int_0^{\infty} \mathrm{d}r\, r \sin(qr) h(r) = -2\pi \left. \frac{G(s) - G(-s)}{s} \right|_{s=\mathrm{i}q} . \tag{6.126}$$

Therefore, the basic structural quantities of the single component HS fluid, namely the RDF and the static structure factor, may be analytically determined within the RFA method once the compressibility factor Z_s, or equivalently the contact value g_s, is specified. In Fig. 6.8, we compare simulation data of $g(r)$ for a density $\rho\sigma^3 = 0.9$ [99] with the RFA prediction and a recent approach by Trokhymchuk et al. [100], where $Z_\mathrm{s} = Z_\mathrm{s}^{\mathrm{CS}}$ [cf. Table 6.1] and the associated compressibility

$$\chi_\mathrm{s}^{\mathrm{CS}} = \frac{(1-\eta)^4}{1 + 4\eta + 4\eta^2 - 4\eta^3 + \eta^4} \tag{6.127}$$

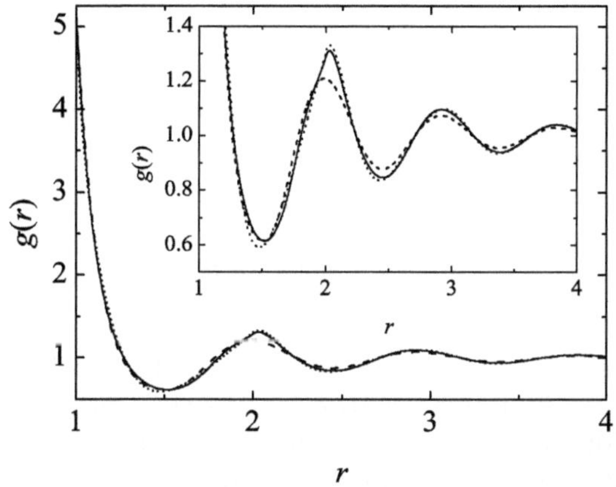

Fig. 6.8. Radial distribution function of a single component HS fluid for $\rho\sigma^3 = 0.9$. The *solid lines* represent simulation data [99]. The *dashed lines* represent the results of the approach of [100], while the *dotted lines* refer to those of the RFA method. The inset shows the oscillations of $g(r)$ in more detail

are taken in both cases. Both theories are rather accurate, but the RFA captures better the maxima and minima of $g(r)$ [101].

It is also possible to obtain within the RFA method the direct correlation function $c(r)$. Using Eqs. (6.95) and (6.126), and applying the residue theorem, one gets, after some algebra,

$$c(r) = \left(K_+ \frac{e^{\kappa r}}{r} + K_- \frac{e^{-\kappa r}}{r} + \frac{K_{-1}}{r} + K_0 + K_1 r + K_3 r^3\right) \Theta(1-r) + K \frac{e^{-\kappa r}}{r} , \tag{6.128}$$

where

$$\kappa = \frac{1}{\alpha}\sqrt{12\alpha\eta L^{(2)}/\pi + 1 - 12\alpha(1+2\alpha)\eta/(1-\eta)} , \tag{6.129}$$

$$\begin{aligned} K_\pm = & \frac{e^{\mp\kappa}}{4\alpha^2(1-\eta)^4\kappa^6}\Big\{2\left[1+2(1+3\alpha)\eta\right] \pm \left[2+\eta+2\alpha(1+2\eta)\right]\kappa \\ & + (1-\eta)\left[\kappa^2 - \eta\left(12+(\kappa\pm 6)\kappa\right)\right] L^{(2)}/\pi\Big\} \Big\{12\eta\left[1+2(1+3\alpha)\eta\right] \\ & \pm 6\eta\left[3\eta - 2\alpha(1-4\eta)\right]\kappa - 6\eta(1+2\alpha)(1-\eta)\kappa^2 - (1-\eta)^2\kappa^3(\alpha\kappa\mp 1) \\ & + 6\eta(1-\eta)\left[\kappa^2 - \eta\left(12+(\kappa\pm 6)\kappa\right)\right] L^{(2)}/\pi\Big\} , \end{aligned} \tag{6.130}$$

$$K_{-1} = -\left(\frac{L^{(2)}}{2\pi\alpha} + K_+ e^{\kappa} + K_- e^{-\kappa} + K_0 + K_1 + K_3\right) , \tag{6.131}$$

$$K_0 = -\left[\frac{1+2(1+3\alpha)\eta - 6\eta(1-\eta)L^{(2)}/\pi}{\alpha\kappa(1-\eta)^2}\right]^2 , \tag{6.132}$$

$$K_1 = \frac{6\eta}{\kappa^2} K_0 + \frac{3\eta}{2\alpha^2\kappa^2(1-\eta)^4} \left\{ [2+\eta+2\alpha(1+2\eta)]^2 - 4(1-\eta)[1+\eta \right.$$
$$\left. \times(7+\eta+6\alpha(2+\eta))] L^{(2)}/\pi + 12\eta(2+\eta)(1-\eta)^2 {L^{(2)}}^2/\pi^2 \right\}, \quad (6.133)$$

$$K_3 = \frac{\eta}{2} K_0 , \quad (6.134)$$

$$K = -(K_+ + K_- + K_{-1}) . \quad (6.135)$$

In Eqs. (6.129)–(6.135) we have taken $\sigma = 1$ as the length unit. Note that Eq. (6.135) guarantees that $c(0) =$ finite, while Eq. (6.131) yields $c(\sigma^+) - c(\sigma^-) = L^{(2)}/2\pi\alpha = g(\sigma^+)$. The latter equation proves the continuity of the indirect correlation function $\gamma(r) \equiv h(r) - c(r)$ at $r = \sigma$. With the above results, Eqs. (6.122) and (6.128), one may immediately write the function $\gamma(r)$. Finally, we note that the bridge function $B(r)$ is linked to $\gamma(r)$ and to the cavity (or background) function $y(r) \equiv \mathrm{e}^{\phi(r)/k_\mathrm{B}T} g(r)$, where $\phi(r)$ is the interaction potential, through

$$B(r) = \ln y(r) - \gamma(r) , \quad (6.136)$$

and so, within the RFA method, the bridge function is also completely specified analytically for $r > \sigma$ once Z_s is prescribed.

If one wants to have $B(r)$ also for $0 \leq r \leq \sigma$, then an expression for the cavity function is required in that region. Here we propose such an expression using a limited number of constraints. First, since the cavity function and its first derivative are continuous at $r = \sigma$, we have

$$y(1) = g_\mathrm{s}, \quad \frac{y'(1)}{y(1)} = \frac{L^{(1)}}{L^{(2)}} - \frac{1}{\alpha} - 1 , \quad (6.137)$$

where Eqs. (6.116) and (6.120) have been used and again $\sigma = 1$ has been taken. Next, we consider the following exact zero-separation theorems [102, 103, 104]:

$$\ln y(0) = Z_\mathrm{s}(\eta) - 1 + \int_0^\eta \mathrm{d}\eta' \frac{Z_\mathrm{s}(\eta') - 1}{\eta'} , \quad (6.138)$$

$$\frac{y'(0)}{y(0)} = -6\eta y(1) . \quad (6.139)$$

The four conditions (6.137)–(6.139) can be enforced by assuming a cubic polynomial form for $\ln y(r)$ inside the core, namely

$$y(r) = \exp\left(Y_0 + Y_1 r + Y_2 r^2 + Y_3 r^3\right), \quad (0 \leq r \leq 1) , \quad (6.140)$$

where

$$Y_0 = Z_\mathrm{s}(\eta) - 1 + \int_0^\eta \mathrm{d}\eta' \frac{Z_\mathrm{s}(\eta') - 1}{\eta'} , \quad (6.141)$$

$$Y_1 = -6\eta y(1) , \quad (6.142)$$

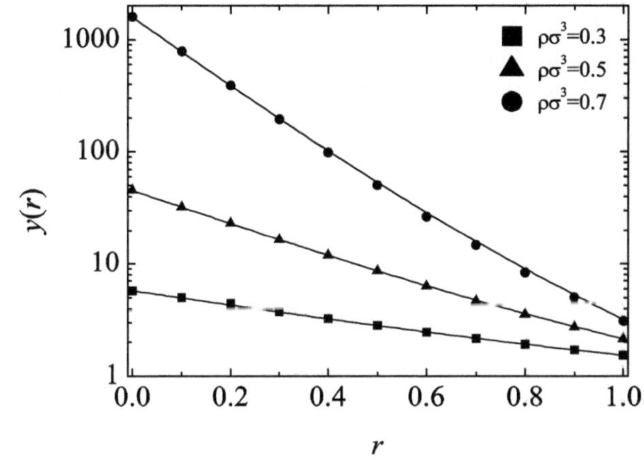

Fig. 6.9. Cavity function of a single component HS fluid in the overlap region for $\rho\sigma^3 = 0.3$, 0.5, and 0.7. The *solid lines* represent our proposal (6.140) with $Z_s = Z_s^{CS}$, while the *symbols* represent Monte Carlo simulation results [105]

$$Y_2 = 3 \ln y(1) - \frac{y'(1)}{y(1)} - 3Y_0 - 2Y_1 \,, \tag{6.143}$$

$$Y_3 = -2 \ln y(1) + \frac{y'(1)}{y(1)} + 2Y_0 + Y_1 . \tag{6.144}$$

The proposal (6.140) is compared with available Monte Carlo data [105] in Fig. 6.9, where an excellent agreement can be observed.

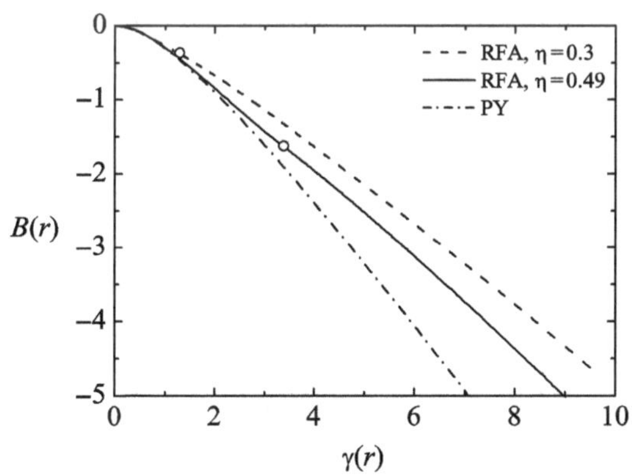

Fig. 6.10. Parametric plot of the bridge function $B(r)$ versus the indirect correlation function $\gamma(r)$. The *dashed line* refers to the RFA for $\eta = 0.3$, while the *solid line* refers to the RFA for $\eta = 0.49$. In each case, the branch of the curve to the *right* of the circle corresponds to $r \leq 1$, while that to the *left* corresponds to $r \geq 1$. For comparison, the PY closure $B(r) = \ln[1 + \gamma(r)] - \gamma(r)$ is also plotted (*dash-dotted line*)

Once the cavity function $y(r)$ provided by the RFA method is complemented by (6.140), the bridge function $B(r)$ can be obtained at any distance. Figure 6.10 presents a parametric plot of the bridge function versus the indirect correlation function as given by the RFA method for two different packing fractions, as well as the result associated with the PY closure. The fact that one gets a smooth curve means that within the RFA the oscillations in $\gamma(r)$ are highly correlated to those of $B(r)$. Further, the effective closure relation in the RFA turns out to be density dependent, in contrast with what occurs for the PY theory. Note that the absolute value $|B(r)|$ for a given value of $\gamma(r)$ is smaller in the RFA than the PY value and that the RFA and PY curves become paradoxically closer for larger densities. Since the PY theory is known to yield rather poor values of the cavity function inside the core [106, 107], it seems likely that the present differences may represent yet another manifestation of the superiority of the RFA method, a point that certainly deserves to be further explored.

6.3.2 The Multicomponent HS Fluid

The method outlined in the preceding subsection will be now extended to an N-component mixture of additive HS. Note that in a multicomponent system the isothermal compressibility χ is given by

$$\begin{aligned}\chi^{-1} &= \frac{1}{k_B T}\left(\frac{\partial p}{\partial \rho}\right)_{T,\{x_j\}} = \frac{1}{k_B T}\sum_{i=1}^{N} x_i \left(\frac{\partial p}{\partial \rho_i}\right)_{T,\{x_j\}} \\ &= 1-\rho\sum_{i,j=1}^{N} x_i x_j \widetilde{c}_{ij}(0)\ ,\end{aligned} \tag{6.145}$$

where $\widetilde{c}_{ij}(q)$ is the Fourier transform of the direct correlation function $c_{ij}(r)$, which is defined by the OZ equation

$$\widetilde{h}_{ij}(q) = \widetilde{c}_{ij}(q) + \sum_{k=1}^{N} \rho_k \widetilde{h}_{ik}(q)\widetilde{c}_{kj}(q)\ , \tag{6.146}$$

where $h_{ij}(r) \equiv g_{ij}(r) - 1$. Equations (6.145) and (6.146) are the multicomponent extensions of Eqs. (6.96) and (6.95), respectively. Introducing the quantities $\widehat{h}_{ij}(q) \equiv \sqrt{\rho_i \rho_j}\widetilde{h}_{ij}(q)$ and $\widehat{c}_{ij}(q) \equiv \sqrt{\rho_i \rho_j}\,\widetilde{c}_{ij}(q)$, the OZ relation (6.146) becomes, in matrix notation,

$$\widehat{\mathsf{c}}(q) = \widehat{\mathsf{h}}(q)\cdot[\mathsf{I}+\widehat{\mathsf{h}}(q)]^{-1}\ , \tag{6.147}$$

where I is the $N \times N$ identity matrix. Thus, Eq. (6.145) can be rewritten as

$$\chi^{-1} = \sum_{i,j=1}^{N} \sqrt{x_i x_j}\,[\delta_{ij} - \widehat{c}_{ij}(0)] = \sum_{i,j=1}^{N} \sqrt{x_i x_j}\left[\mathsf{I}+\widehat{\mathsf{h}}(0)\right]^{-1}_{ij} . \tag{6.148}$$

Similarly to what we did in the single component case, we introduce the Laplace transforms of $rg_{ij}(r)$:

$$G_{ij}(s) = \int_0^\infty \mathrm{d}r\, \mathrm{e}^{-sr} r g_{ij}(r). \tag{6.149}$$

The counterparts of Eqs. (6.100) and (6.101) are

$$g_{ij}(r) = \Theta(r-\sigma_{ij})\left[g_{ij}(\sigma_{ij}^+) + g'_{ij}(\sigma_{ij}^+)(r-\sigma_{ij}) + \cdots\right] , \tag{6.150}$$

$$\mathrm{e}^{\sigma_{ij}s} sG_{ij}(s) = \sigma_{ij} g_{ij}(\sigma_{ij}^+) + \left[g_{ij}(\sigma_{ij}^+) + \sigma_{ij} g'_{ij}(\sigma_{ij}^+)\right] s^{-1} + \mathcal{O}(s^{-2}). \tag{6.151}$$

Moreover, the condition of a finite compressibility implies that $\widetilde{h}_{ij}(0) =$ finite. As a consequence, for small s,

$$s^2 G_{ij}(s) = 1 + H_{ij}^{(0)} s^2 + H_{ij}^{(1)} s^3 + \cdots \tag{6.152}$$

with $H_{ij}^{(0)} =$ finite and $H_{ij}^{(1)} = -\widetilde{h}_{ij}(0)/4\pi =$ finite, where

$$H_{ij}^{(n)} \equiv \frac{1}{n!}\int_0^\infty \mathrm{d}r\, (-r)^n r h_{ij}(r). \tag{6.153}$$

We are now in the position to generalize the approximation (6.113) to the N-component case [108]. While such a generalization may be approached in a variety of ways, two motivations are apparent. On the one hand, we want to recover the PY result as a particular case in much the same fashion as in the single component system. On the other hand, we want to maintain the development as simple as possible. Taking all of this into account, we propose

$$G_{ij}(s) = \frac{\mathrm{e}^{-\sigma_{ij}s}}{2\pi s^2}\left(\mathsf{L}(s)\cdot\left[(1+\alpha s)\mathsf{I} - \mathsf{A}(s)\right]^{-1}\right)_{ij} , \tag{6.154}$$

where $\mathsf{L}(s)$ and $\mathsf{A}(s)$ are the matrices

$$L_{ij}(s) = L_{ij}^{(0)} + L_{ij}^{(1)} s + L_{ij}^{(2)} s^2 , \tag{6.155}$$

$$A_{ij}(s) = \rho_i\left[\varphi_2(\sigma_i s)\sigma_i^3 L_{ij}^{(0)} + \varphi_1(\sigma_i s)\sigma_i^2 L_{ij}^{(1)} + \varphi_0(\sigma_i s)\sigma_i L_{ij}^{(2)}\right] , \tag{6.156}$$

the functions $\varphi_n(x)$ being defined by Eq. (6.109). We note that, by construction, Eq. (6.154) complies with the requirement $\lim_{s\to\infty} \mathrm{e}^{\sigma_{ij}s} sG_{ij}(s) =$ finite. Further, in view of Eq. (6.152), the coefficients of s^0 and s in the power series expansion of $s^2 G_{ij}(s)$ must be 1 and 0, respectively. This yields $2N^2$ conditions that allow us to express $\mathsf{L}^{(0)}$ and $\mathsf{L}^{(1)}$ in terms of $\mathsf{L}^{(2)}$ and α. The solution is [108]

$$L_{ij}^{(0)} = \vartheta_1 + \vartheta_2\sigma_j + 2\vartheta_2\alpha - \vartheta_1\sum_{k=1}^{N}\rho_k\sigma_k L_{kj}^{(2)} , \tag{6.157}$$

$$L_{ij}^{(1)} = \vartheta_1\sigma_{ij} + \frac{1}{2}\vartheta_2\sigma_i\sigma_j + (\vartheta_1 + \vartheta_2\sigma_i)\alpha - \frac{1}{2}\vartheta_1\sigma_i \sum_{k=1}^{N} \rho_k\sigma_k L_{kj}^{(2)} , \tag{6.158}$$

where $\vartheta_1 \equiv 2\pi/(1-\eta)$ and $\vartheta_2 \equiv 6\pi(M_2/M_3)\eta/(1-\eta)^2$.

In parallel with the development of the single component case, $\mathsf{L}^{(2)}$ and α can be chosen arbitrarily. Again, the choice $L_{ij}^{(2)} = \alpha = 0$ gives the PY solution [9, 109]. Since we want to go beyond this approximation, we will determine those coefficients by taking prescribed values for $g_{ij}(\sigma_{ij})$, which in turn, via Eq. (6.2), give the EOS of the mixture. This also leads to the required value of $\chi^{-1} = \partial(\rho Z)/\partial\rho$, thus making the theory thermodynamically consistent. In particular, according to Eq. (6.151),

$$L_{ij}^{(2)} = 2\pi\alpha\sigma_{ij}g_{ij}(\sigma_{ij}^+). \tag{6.159}$$

The condition related to χ is more involved. Making use of Eq. (6.152), one can get $\widetilde{h}_{ij}(0) = -4\pi H_{ij}^{(1)}$ in terms of $\mathsf{L}^{(2)}$ and α and then insert it into Eq. (6.148). Finally, elimination of $L_{ij}^{(2)}$ in favor of α from Eq. (6.159) produces an algebraic equation of degree $2N$, whose physical root is determined by the requirement that $G_{ij}(s)$ is positive definite for positive real s. It turns out that the physical solution corresponds to the smallest of the real roots. Once α is known, upon substitution into Eqs. (6.154), (6.157), (6.158), and (6.159), the scheme is complete. Also, using Eq. (6.151), one can easily derive the result

$$g_{ij}'(\sigma_{ij}^+) = \frac{1}{2\pi\alpha\sigma_{ij}}\left[L_{ij}^{(1)} - L_{ij}^{(2)}\left(\frac{1}{\alpha} + \frac{1}{\sigma_{ij}}\right)\right]. \tag{6.160}$$

It is straightforward to check that the results of the preceding subsection are recovered by setting $\sigma_i = \sigma$, regardless of the values of the mole factions.

Once $G_{ij}(s)$ has been determined, inverse Laplace transformation directly yields $rg_{ij}(r)$. Although in principle this can be done analytically, it is more practical to use one of the efficient methods discussed by Abate and Whitt [110] to numerically invert Laplace transforms.[1]

In Fig. 6.11, we present a comparison between the results of the RFA method with the PY theory and simulation data [64] for the RDF of a ternary mixture. In the case of the RFA, we have used the eCS2 contact values and the corresponding isothermal compressibility. The improvement of the RFA over the PY prediction, particularly in the region near contact, is noticeable. Although the RFA accounts nicely for the observed oscillations, it seems to somewhat overestimate the depth of the first minimum.

Explicit knowledge of $G_{ij}(s)$ also allows us to determine the Fourier transform $\widetilde{h}_{ij}(q)$ through the relation

[1] A code using the Mathematica computer algebra system to obtain $G_{ij}(s)$ and $g_{ij}(r)$ with the present method is available from the web page http://www.unex.es/eweb/fisteor/santos/filesRFA.html

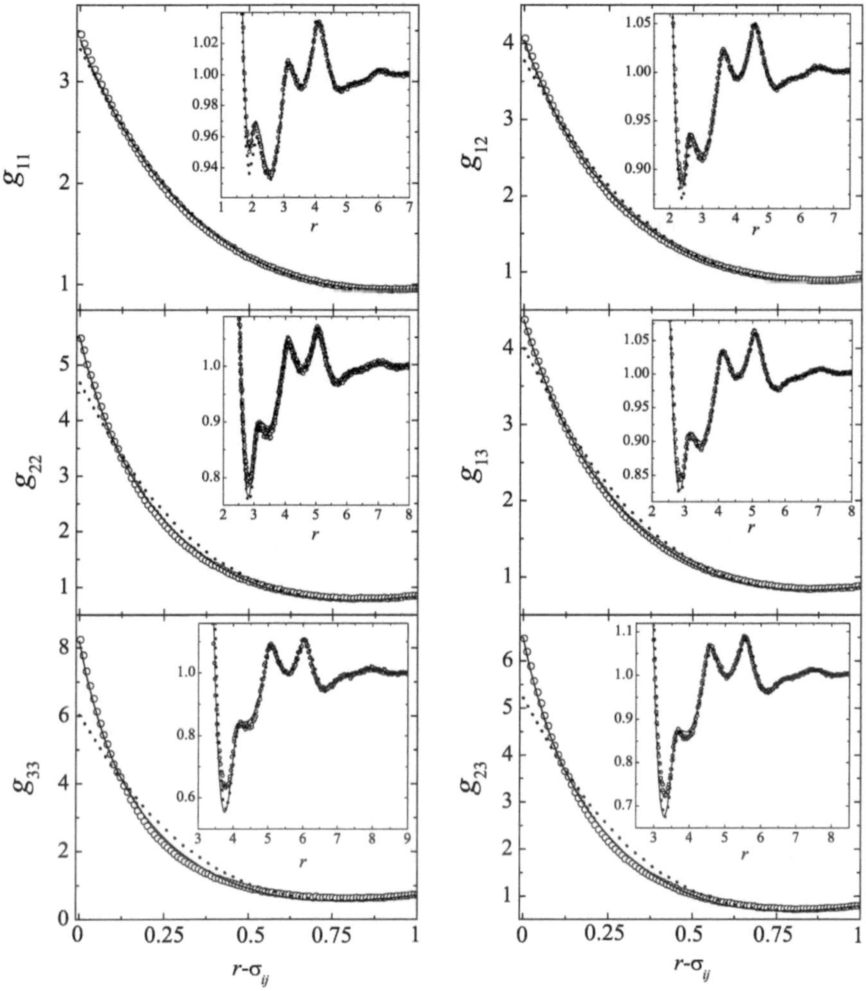

Fig. 6.11. Radial distribution functions $g_{ij}(r)$ for a ternary mixture with diameters $\sigma_1 = 1$, $\sigma_2 = 2$, and $\sigma_3 = 3$ at a packing fraction $\eta = 0.49$ with mole fractions $x_1 = 0.7$, $x_2 = 0.2$, and $x_3 = 0.1$. The *circles* are simulation results [64], the *solid lines* are the RFA predictions, and the *dotted lines* are the PY predictions

$$\widetilde{h}_{ij}(q) = -2\pi \left. \frac{G_{ij}(s) - G_{ij}(-s)}{s} \right|_{s=\mathrm{i}q} . \tag{6.161}$$

The structure factor $S_{ij}(q)$ may be expressed in terms of $\widetilde{h}_{ij}(q)$ as [4]

$$S_{ij}(q) = x_i\delta_{ij} + \rho x_i x_j \widetilde{h}_{ij}(q). \tag{6.162}$$

In the particular case of a binary mixture, rather than the individual structure factors $S_{ij}(q)$, it is some combination of them which may be easily associated with fluctuations of the thermodynamic variables [111, 112]. Specifically, the quantities [4]

$$S_{nn}(q) = S_{11}(q) + S_{22}(q) + 2S_{12}(q) , \tag{6.163}$$

$$S_{nc}(q) = x_2 S_{11}(q) - x_1 S_{22}(q) + (x_2 - x_1) S_{12}(q) , \tag{6.164}$$

$$S_{cc}(q) = x_2^2 S_{11}(q) + x_1^2 S_{22}(q) - 2x_1 x_2 S_{12}(q) \tag{6.165}$$

are sometimes required.

After replacement of $\widehat{h}_{ij}(q) = \sqrt{\rho_i \rho_j}\widetilde{h}_{ij}(q)$ in Eq. (6.147), one easily gets $\widetilde{c}_{ij}(q)$. Subsequent inverse Fourier transformation yields $c_{ij}(r)$. The result gives $c_{ij}(r)$ for $r > \sigma_{ij}$ as the superposition of N Yukawas [113], namely

$$c_{ij}(r) = \sum_{\ell=1}^{N} K_{ij}^{(\ell)} \frac{\mathrm{e}^{-\kappa_\ell r}}{r} , \tag{6.166}$$

where $q = \pm \mathrm{i}\kappa_\ell$ with $\ell = 1, \ldots, N$ are the zeros of $\det\left[\mathsf{I} + \widehat{\mathsf{h}}(q)\right]$ and the amplitudes $K_{ij}^{(\ell)}$ are obtained by applying the residue theorem as

$$K_{ij}^{(\ell)} = \frac{\mathrm{i}\kappa_\ell}{2\pi} \lim_{q \to \mathrm{i}\kappa_\ell} \widetilde{c}_{ij}(q)(q - \mathrm{i}\kappa_\ell) . \tag{6.167}$$

The indirect correlation functions $\gamma_{ij}(r) \equiv h_{ij}(r) - c_{ij}(r)$ readily follow from the previous results for the RDF and direct correlation functions. Finally, in this case the bridge functions $B_{ij}(r)$ for $r > \sigma_{ij}$ are linked to $g_{ij}(r)$ and $c_{ij}(r)$ through

$$B_{ij}(r) = \ln g_{ij}(r) - \gamma_{ij}(r) \tag{6.168}$$

and so once more we have a full set of analytical results for the structural properties of a multicomponent fluid mixture of HS once the contact values $g_{ij}(\sigma_{ij})$ are specified.

6.4 Other Related Systems

The philosophy behind the RFA method to derive the structural properties of three-dimensional HS systems can be adapted to deal with other related systems. The main common features of the RFA can be summarized as follows. First, one chooses to represent the RDF in Laplace space. Next, using as a guide the low-density form of the Laplace transform, an auxiliary function is defined which is approximated by a rational or a rational-like form. Finally, the coefficients are determined by imposing some basic consistency conditions. In this section we consider the cases of sticky-hard-sphere (SHS), square-well, and hard-disk fluids. In the two former cases the RFA program is followed quite literally, while in the latter case it is done more indirectly through the RFA method as applied to hard rods ($d = 1$) and hard spheres ($d = 3$).

6.4.1 Sticky Hard Spheres

The SHS fluid model has received a lot of attention since it was first introduced by Baxter in 1968 [114] and later extended to multicomponent mixtures by Perram and Smith [115] and, independently, by Barboy [116, 117]. In this model, the molecular interaction may be defined via square-well (SW) potentials of infinite depth and vanishing width, thus embodying the two essential characteristics of real molecular interactions, namely a harsh repulsion and an attractive part. In spite of their known shortcomings [118, 119], an important feature of SHS systems is that they allow for an exact solution of the OZ equation in the PY approximation [114, 115]. Furthermore, they are thought to be appropriate for describing structural properties of colloidal systems, micelles, and microemulsions, as well as some aspects of gas–liquid equilibrium, ionic fluids and mixtures, solvent-mediated forces, adsorption phenomena, polydisperse systems, and fluids containing chainlike molecules [120, 121, 122, 123, 124, 125, 126, 127, 128, 129, 130, 131, 132, 133, 134, 135, 136, 137, 138, 139, 140, 141, 142, 143, 144, 145, 146, 147, 148, 149, 150, 151, 152, 153, 154, 155].

Let us consider an N-component mixture of spherical particles interacting according to the SW potential

$$\phi_{ij}(r)=\begin{cases}\infty, & r<\sigma_{ij}\ ,\\ -\epsilon_{ij}, & \sigma_{ij}<r<R_{ij}\ ,\\ 0, & r>R_{ij}\ .\end{cases} \tag{6.169}$$

As in the case of additive HS, $\sigma_{ij}=(\sigma_i+\sigma_j)/2$ is the distance between the centers of a sphere of species i and a sphere of species j at contact. In addition, ϵ_{ij} is the well depth and $R_{ij}-\sigma_{ij}$ indicates the well width. We now take the SHS limit [114], namely

$$R_{ij}\to\sigma_{ij},\quad \epsilon_{ij}\to\infty,\quad \tau_{ij}\equiv\frac{1}{12}\frac{\sigma_{ij}}{R_{ij}-\sigma_{ij}}\mathrm{e}^{-\epsilon_{ij}/k_\mathrm{B}T}=\ \text{finite}\ , \tag{6.170}$$

where the τ_{ij} are monotonically increasing functions of the temperature T and their inverses measure the degree of "adhesiveness" of the interacting spheres i and j. Even without strictly taking the mathematical limits (6.170), short-range SW fluids can be well described in practice by the SHS model [156].

The virial EOS for the SHS mixture is given by

$$\begin{aligned}Z&=1+\frac{1}{6}\rho\sum_{i,j=1}^{N}x_ix_j\int \mathrm{d}\mathbf{r}\, r y_{ij}(r)\frac{\mathrm{d}}{\mathrm{d}r}\mathrm{e}^{-\phi_{ij}(r)/k_\mathrm{B}T}\\ &=1+\frac{2\pi}{3}\rho\sum_{i,j=1}^{N}x_ix_j\sigma_{ij}^3y_{ij}(\sigma_{ij})\left[1-\frac{1}{12\tau_{ij}}\left(3+\frac{y'_{ij}(\sigma_{ij})}{y_{ij}(\sigma_{ij})}\right)\right]\ ,\end{aligned} \tag{6.171}$$

where $y_{ij}(r)\equiv g_{ij}(r)\mathrm{e}^{\phi_{ij}(r)/k_\mathrm{B}T}$ is the cavity function and $y'_{ij}(r)=\mathrm{d}y_{ij}(r)/\mathrm{d}r$. Since $y_{ij}(r)$ must be continuous, it follows that

$$g_{ij}(r) = y_{ij}(r)\left[\Theta(r-\sigma_{ij}) + \frac{\sigma_{ij}}{12\tau_{ij}}\delta(r-\sigma_{ij})\right]. \tag{6.172}$$

The case of an HS system is recovered by taking the limit of vanishing adhesiveness $\tau_{ij}^{-1} \to 0$, in which case Eq. (6.171) reduces to the three-dimensional version of Eq. (6.2). On the other hand, the compressibility EOS, Eq. (6.145), is valid for any interaction potential, including SHS.

As in the case of HS, it is convenient to define the Laplace transform (6.149). The condition $y_{ij}(\sigma_{ij})$ = finite translates into the following large s behavior of $G_{ij}(s)$:

$$\mathrm{e}^{\sigma_{ij}s}G_{ij}(s) = \sigma_{ij}^2 y_{ij}(\sigma_{ij})\left(\frac{1}{12\tau_{ij}} + \sigma_{ij}^{-1}s^{-1}\right) + \mathcal{O}(s^{-2}), \tag{6.173}$$

which differs from (6.151): $\mathrm{e}^{\sigma_{ij}s}G_{ij}(s) \sim s^{-1}$ for HS and $\mathrm{e}^{\sigma_{ij}s}G_{ij}(s) \sim s^0$ for SHS. However, the small s behavior is still given by Eq. (6.152), as a consequence of the condition χ^{-1} = finite.

The RFA proposal for SHS mixtures [157] keeps the form (6.154), except that now

$$L_{ij}(s) = L_{ij}^{(0)} + L_{ij}^{(1)}s + L_{ij}^{(2)}s^2 + L_{ij}^{(3)}s^3, \tag{6.174}$$

$$A_{ij}(s) = \rho_i\left[\varphi_2(\sigma_i s)\sigma_i^3 L_{ij}^{(0)} + \varphi_1(\sigma_i s)\sigma_i^2 L_{ij}^{(1)} + \varphi_0(\sigma_i s)\sigma_i L_{ij}^{(2)} - \mathrm{e}^{-\sigma_i s}L_{ij}^{(3)}\right], \tag{6.175}$$

instead of Eqs. (6.155) and (6.156). By construction, Eqs. (6.154), (6.174), and (6.175) comply with the requirement $\lim_{s\to\infty}\mathrm{e}^{\sigma_{ij}s}G_{ij}(s)$ = finite. Further, in view of Eq. (6.152), the coefficients of s^0 and s in the power series expansion of $s^2G_{ij}(s)$ must be 1 and 0, respectively. This yields $2N^2$ conditions that allow us to express $\mathsf{L}^{(0)}$ and $\mathsf{L}^{(1)}$ in terms of $\mathsf{L}^{(2)}$, $\mathsf{L}^{(3)}$, and α as [157]

$$L_{ij}^{(0)} = \vartheta_1 + \vartheta_2\sigma_j + 2\vartheta_2\alpha - \vartheta_1\sum_{k=1}^N \rho_k\left(\sigma_k L_{kj}^{(2)} - L_{kj}^{(3)}\right) - \vartheta_2\sum_{k=1}^N \rho_k\sigma_k L_{kj}^{(3)}, \tag{6.176}$$

$$L_{ij}^{(1)} = \vartheta_1\sigma_{ij} + \frac{1}{2}\vartheta_2\sigma_i\sigma_j + (\vartheta_1 + \vartheta_2\sigma_i)\alpha - \frac{1}{2}\vartheta_1\sigma_i\sum_{k=1}^N \rho_k\left(\sigma_k L_{kj}^{(2)} - L_{kj}^{(3)}\right)$$
$$-\frac{1}{2}(\vartheta_1 + \vartheta_2\sigma_i)\sum_{k=1}^N \rho_k\sigma_k L_{kj}^{(3)}, \tag{6.177}$$

where ϑ_1 and ϑ_2 are defined below Eq. (6.158). We have the freedom to choose $\mathsf{L}^{(3)}$ and α, but $\mathsf{L}^{(2)}$ is constrained by the condition (6.173), i.e., the ratio between the first and second terms in the expansion of $\mathrm{e}^{\sigma_{ij}s}G_{ij}(s)$ for large s must be exactly equal to $\sigma_{ij}/12\tau_{ij}$.

First-Order Approximation (PY Solution)

The simplest approximation consists of making $\alpha = 0$. In view of the condition $e^{\sigma_{ij}s}G_{ij}(s) \sim s^0$ for large s, this implies that $L_{ij}^{(3)} = 0$. In that case, the large s behavior that follows from Eq. (6.154) is

$$2\pi e^{\sigma_{ij}s}G_{ij}(s) = L_{ij}^{(2)} + \left[L_{ij}^{(1)} + \left(\mathsf{L}^{(2)} \cdot \mathsf{D}\right)_{ij}\right] s^{-1} + \mathcal{O}(s^{-2}) , \qquad (6.178)$$

where

$$D_{ij} \equiv \rho_i \left(\frac{1}{2}\sigma_i^2 L_{ij}^{(0)} - \sigma_i L_{ij}^{(1)} + L_{ij}^{(2)}\right) . \qquad (6.179)$$

Comparison with Eq. (6.173) yields

$$y_{ij}(\sigma_{ij}) = \frac{6\tau_{ij}}{\pi\sigma_{ij}^2} L_{ij}^{(2)} , \qquad (6.180)$$

$$\frac{12\tau_{ij}L_{ij}^{(2)}}{\sigma_{ij}} = L_{ij}^{(1)} + \sum_{k=1}^{N} L_{ik}^{(2)} D_{kj} . \qquad (6.181)$$

Taking into account Eqs. (6.176) and (6.177) (with $L_{ij}^{(2)} = L_{ji}^{(2)}$ and of course also with $\alpha = 0$ and $\mathsf{L}^{(3)} = 0$), Eq. (6.181) becomes a closed equation for $\mathsf{L}^{(2)}$:

$$\frac{12\tau_{ij}L_{ij}^{(2)}}{\sigma_{ij}} = \vartheta_1\sigma_{ij} + \frac{1}{2}\vartheta_2\sigma_i\sigma_j - \frac{1}{2}\vartheta_1 \sum_{k=1}^{N} \rho_k\sigma_k \left(L_{ki}^{(2)}\sigma_j + L_{kj}^{(2)}\sigma_i\right) + \sum_{k=1}^{N} \rho_k L_{ki}^{(2)} L_{kj}^{(2)} . \qquad (6.182)$$

The physical root $\mathsf{L}^{(2)}$ of Eq. (6.182) is the one vanishing in the HS limit $\tau_{ij} \to \infty$. Once known, Eq. (6.180) gives the contact values.

This first-order approximation obtained from the RFA method turns out to coincide with the exact solution of the PY theory for SHS [115].

Second-Order Approximation

As in the case of HS mixtures, a more flexible proposal is obtained by keeping α (and, consequently, $L_{ij}^{(3)}$) different from zero. In that case, instead of Eq. (6.178), one has

$$2\pi e^{\sigma_{ij}s}G_{ij}(s) = \frac{L_{ij}^{(3)}}{\alpha}\left[1 + \left(\frac{L_{ij}^{(2)}}{L_{ij}^{(3)}} - \frac{1}{\alpha}\right) s^{-1}\right] + \mathcal{O}(s^{-2}) . \qquad (6.183)$$

This implies

$$L_{ij}^{(3)} = \frac{\pi\sigma_{ij}^2}{6\tau_{ij}}\alpha y_{ij}(\sigma_{ij}) , \tag{6.184}$$

$$\frac{12\tau_{ij}L_{ij}^{(3)}}{\sigma_{ij}} = L_{ij}^{(2)} - \frac{L_{ij}^{(3)}}{\alpha} . \tag{6.185}$$

If we fix $y_{ij}(\sigma_{ij})$, Eqs. (6.176), (6.177), (6.184), and (6.185) allow one to express $\mathsf{L}^{(0)}$, $\mathsf{L}^{(1)}$, $\mathsf{L}^{(2)}$, and $\mathsf{L}^{(3)}$ as *linear* functions of α. Thus, only the scalar parameter α remains to be fixed, analogously to what happens in the HS case. As done in the latter case, one possibility is to choose α in order to reproduce the isothermal compressibility χ given by Eq. (6.148). To do so, one needs to find the coefficients $H_{ij}^{(1)}$ appearing in Eq. (6.152). The result is [157]

$$\mathsf{H}^{(0)} = \mathsf{C}^{(0)} \cdot \left(\mathsf{I} - \mathsf{A}^{(0)}\right)^{-1} , \tag{6.186}$$

$$\mathsf{H}^{(1)} = \mathsf{C}^{(1)} \cdot \left(\mathsf{I} - \mathsf{A}^{(0)}\right)^{-1} , \tag{6.187}$$

where

$$C_{ij}^{(0)} = \frac{1}{2\pi}L_{ij}^{(2)} + \sum_{k=1}^{N} A_{kj}^{(2)} - \sum_{k=1}^{N} \sigma_{ik}\left(\alpha\delta_{kj} - A_{kj}^{(1)}\right) - \sum_{k=1}^{N} \frac{1}{2}\sigma_{ik}^2\left(\delta_{kj} - A_{kj}^{(0)}\right) , \tag{6.188}$$

$$C_{ij}^{(1)} = \frac{1}{2\pi}L_{ij}^{(3)} + \sum_{k=1}^{N} A_{kj}^{(3)} + \sum_{k=1}^{N} \sigma_{ik}A_{kj}^{(2)} - \sum_{k=1}^{N}\left(\frac{1}{2}\sigma_{ik}^2 + H_{ik}^{(0)}\right)\left(\alpha\delta_{kj} - A_{kj}^{(1)}\right)$$
$$-\sum_{k=1}^{N}\left(\frac{1}{6}\sigma_{ik}^3 + \sigma_{ik}H_{ik}^{(0)}\right)\left(\delta_{kj} - A_{kj}^{(0)}\right) , \tag{6.189}$$

$$A_{ij}^{(n)} = (-1)^n\rho_i\left[\frac{\sigma_i^{n+3}}{(n+3)!}L_{ij}^{(0)} - \frac{\sigma_i^{n+2}}{(n+2)!}L_{ij}^{(1)} + \frac{\sigma_i^{n+1}}{(n+1)!}L_{ij}^{(2)} - \frac{\sigma_i^n}{n!}L_{ij}^{(3)}\right] . \tag{6.190}$$

Equation (6.187) gives $\mathsf{H}^{(1)}$ in terms of α: $H_{ij}^{(1)} = P_{ij}(\alpha)/[Q(\alpha)]^2$, where $P_{ij}(\alpha)$ denotes a polynomial in α of degree $2N$ and $Q(\alpha)$ denotes a polynomial of degree N. It turns out then that, seen as a function of α, χ is the ratio of two polynomials of degree $2N$. Given a value of χ, one may solve for α. The physical solution, which has to fulfill the requirement that $G_{ij}(s)$ is positive definite for positive real s, corresponds to the smallest positive real root.

Once α is known, the scheme is complete: Eq. (6.184) gives $\mathsf{L}^{(3)}$, then $\mathsf{L}^{(2)}$ is obtained from Eq. (6.185), and finally $\mathsf{L}^{(1)}$ and $\mathsf{L}^{(0)}$ are given by Eqs. (6.176) and (6.177), respectively. Explicit knowledge of $G_{ij}(s)$ through Eqs. (6.154), (6.174), and (6.175) allows one to determine the Fourier transform $\widetilde{h}_{ij}(q)$

and the structure factor $S_{ij}(q)$ through Eqs. (6.161) and (6.162), respectively. Finally, inverse Laplace transformation of $G_{ij}(s)$ yields $g_{ij}(r)$.[2]

Single-Component SHS Fluids

The special case of single-component SHS fluids [158, 159] can be obtained from the multicomponent one by taking $\sigma_{ij} = \sigma$ and $\tau_{ij} = \tau$. Thus, the Laplace transform of $rg(r)$ in the RFA is

$$G(s) = \frac{e^{-s}}{2\pi s^2} \frac{L^{(0)} + L^{(1)}s + L^{(2)}s^2 + L^{(3)}s^3}{1 + \alpha s - \rho\left[\varphi_2(s)L^{(0)} + \varphi_1(s)L^{(1)} + \varphi_0(s)L^{(2)} - e^{-s}L^{(3)}\right]}, \tag{6.191}$$

where we have taken $\sigma = 1$. Equations (6.176) and (6.177) become

$$L^{(0)} = 2\pi\frac{1+2\eta}{(1-\eta)^2} + \frac{12\eta}{1-\eta}\left(\frac{\pi\alpha}{1-\eta} - L^{(2)}\right) + \frac{12\eta}{(1-\eta)^2}(1-4\eta)L^{(3)}, \tag{6.192}$$

$$L^{(1)} = 2\pi\frac{1+\frac{1}{2}\eta}{(1-\eta)^2} + \frac{2}{1-\eta}\left(\pi\frac{1+2\eta}{1-\eta}\alpha - 3\eta L^{(2)}\right) - \frac{18\eta^2}{(1-\eta)^2}L^{(3)}. \tag{6.193}$$

The choice $\alpha = L^{(3)} = 0$ makes Eq. (6.191) coincide with the exact solution to the PY approximation for SHS [114], where $L^{(2)}$ is the physical root (i.e., the one vanishing in the limit $\tau \to \infty$) of the quadratic equation [see Eq. (6.182)]

$$12\tau L^{(2)} = 2\pi\frac{1+2\eta}{(1-\eta)^2} - \frac{12\eta}{1-\eta}L^{(2)} + \frac{6}{\pi}\eta {L^{(2)}}^2. \tag{6.194}$$

We can go beyond the PY approximation by prescribing a contact value $y(1)$, so that, according to Eqs. (6.184) and (6.185),

$$L^{(3)} = \frac{\pi}{6}\frac{\alpha}{\tau}y(1), \tag{6.195}$$

$$L^{(2)} = \left(12\tau + \frac{1}{\alpha}\right)L^{(3)}. \tag{6.196}$$

By prescribing the isothermal compressibility χ, the parameter α can be obtained as the physical solution (namely, the one remaining finite in the limit $\tau \to \infty$) of a quadratic equation [159]. Thus, given an EOS for the SHS fluid, one can get the thermodynamically consistent values of $y(1)$ and χ and determine from them all the coefficients appearing in Eq. (6.191).

Figure 6.12 shows the cavity function for $\eta = 0.164$ and $\tau = 0.13$ as obtained from Monte Carlo simulations [154] and as predicted by the PY and RFA theories, the latter making use of the EOS recently proposed by Miller and Frenkel [155]. It can be observed that the RFA is not only more accurate than the PY approximation near $r = 1$ but also near $r = 2$. On the other hand, none of these two approximations account for the singularities (delta peaks and/or discontinuities) of $y(r)$ at $r = \sqrt{8/3}, 5/3, \sqrt{3}, 2, \ldots$ [150, 151, 152, 153, 154].

[2] See Footnote one.

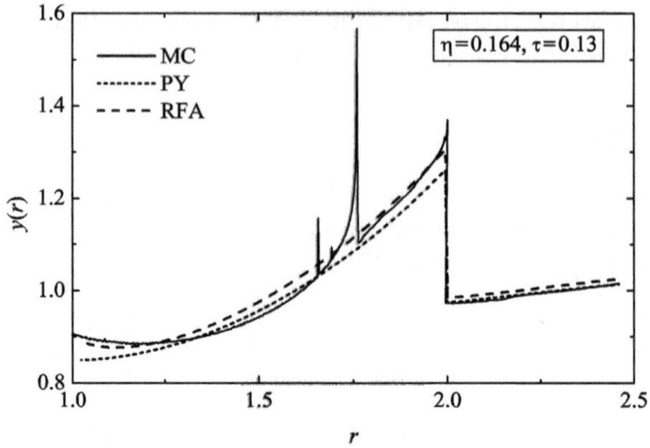

Fig. 6.12. Cavity function of a single-component SHS fluid for $\eta = 0.164$ and $\tau = 0.13$. The *solid line* represents simulation data [154]. The *dotted* and *dashed lines* represent the PY and RFA approaches, respectively

6.4.2 Single-Component Square-Well Fluids

Now, we consider again the SW interaction potential (6.169) but for a single fluid, i.e., $\sigma_{ij} = \sigma$, $\epsilon_{ij} = \epsilon$, $R_{ij} = R$. Since no exact solution of the PY theory for the SW potential is known, the application of the RFA method is more challenging in this case than for HS and SHS fluids.

As in the cases of HS and SHS, the key quantity is the Laplace transform of $rg(r)$ defined by Eq. (6.98). It is again convenient to introduce the auxiliary function $\Psi(s)$ through Eq. (6.99). As before, the conditions $g(r) =$ finite and $\chi =$ finite imply Eqs. (6.102) and (6.104), respectively. However, the important difference between HS and SHS fluids is that in the latter case $G(s)$ must reflect the fact that $g(r)$ is discontinuous at $r = R$ as a consequence of the discontinuity of the potential $\phi(r)$ and the continuity of the cavity function $y(r)$. This implies that $G(s)$, and hence $\Psi(s)$, must contain the exponential term $\mathrm{e}^{-(R-\sigma)s}$. This manifests itself in the low-density limit, where the condition $\lim_{\rho\to 0} y(r) = 1$ yields

$$\lim_{\rho\to 0} \Psi(s) = \frac{1}{2\pi} \frac{s^3}{\mathrm{e}^{1/T^*}(1+s) - \mathrm{e}^{-(R-1)s}(\mathrm{e}^{1/T^*} - 1)(1+Rs)}, \tag{6.197}$$

where $T^* \equiv k_\mathrm{B}T/\epsilon$ and we have taken $\sigma = 1$.

In the spirit of the RFA method, the simplest form that complies with Eq. (6.102) and is consistent with Eq. (6.197) is [160]

$$\Psi(s) = \frac{1}{2\pi} \frac{-12\eta + E_1 s + E_2 s^2 + E_3 s^3}{1 + Q_0 + Q_1 s - \mathrm{e}^{-(R-1)s}\,(Q_0 + Q_2 s)}, \tag{6.198}$$

where the coefficients Q_0, Q_1, Q_2, E_1, E_2, and E_3 are functions of η, T^*, and R. The condition (6.104) allows one to express the parameters Q_1, E_1, E_2, and E_3 as linear functions of Q_0 and Q_2 [160, 161]:

$$Q_1 = \frac{1}{1+2\eta}\left[1+\frac{\eta}{2}+2\eta(R^3-1)Q_2-\frac{\eta}{2}(R-1)^2(R^2+2R+3)Q_0\right] + Q_2-(R-1)Q_0 \,, \tag{6.199}$$

$$E_1 = \frac{6\eta^2}{1+2\eta}\left[3-4(R^3-1)Q_2+(R-1)^2(R^2+2R+3)Q_0\right] \,, \tag{6.200}$$

$$E_2 = \frac{6\eta}{1+2\eta}\{1-\eta-2(R-1)\left[1-2\eta R(R+1)\right]Q_2 +(R-1)^2\left[(1-\eta(R+1)^2\right]Q_0\} \,, \tag{6.201}$$

$$E_3 = \frac{1}{1+2\eta}\{(1-\eta)^2+6\eta(R-1)\left(R+1-2\eta R^2\right)Q_2 -\eta(R-1)^2[4+2R-\eta(3R^2+2R+1)]Q_0\} \,. \tag{6.202}$$

From Eq. (6.102), we have

$$g(1^+) = \frac{Q_1}{E_3} \,. \tag{6.203}$$

The complete RDF is given by Eq. (6.122), where now Eq. (6.198) must be used in Eq. (6.123). In particular, $\psi_1(r)$ and $\psi_2(r)$ are

$$\psi_1(r) = \psi_{10}(r)\Theta(r)+\psi_{11}(r+1-R)\Theta(r+1-R) \,, \tag{6.204}$$

$$\psi_2(r) = \psi_{20}(r)\Theta(r)+\psi_{21}(r+1-R)\Theta(r+1-R)+\psi_{22}(r+2-2R)\Theta(r+2-2R) \,, \tag{6.205}$$

where

$$\psi_{1k}(r) = 2\pi\sum_{i=1}^{3}\frac{W_{1k}(s_i)}{E'(s_i)}s_i\mathrm{e}^{s_i x} \,, \tag{6.206}$$

$$\psi_{2k}(r) = -4\pi^2\sum_{i=1}^{3}\left[rW_{2k}(s_i)+W'_{2k}(s_i)-W_{2k}(s_i)\frac{E''(s_i)}{E'(s_i)}\right]\frac{\mathrm{e}^{s_i r}}{[E'(s_i)]^2} \,. \tag{6.207}$$

Here, s_i are the three distinct roots of $E(s) \equiv -12\eta+E_1 s+E_2 s^2+E_3 s^3$ and

$$W_{10}(s) \equiv 1+Q_0+Q_1 s, \quad W_{11}(s) \equiv -(Q_0+Q_2 s) \tag{6.208}$$

$$W_{20}(s) \equiv s[W_{10}(s)]^2, \quad W_{21}(s) \equiv 2sW_{10}(s)W_{11}(s), \quad W_{22}(s) \equiv s[W_{11}(s)]^2 \,. \tag{6.209}$$

To close the proposal, we need to determine the parameters Q_0 and Q_2 by imposing two new conditions. An obvious condition is the continuity of the cavity function at $r = R$, that implies

$$g(R^+) = \mathrm{e}^{1/T^*}g(R^-) \,. \tag{6.210}$$

This yields

$$\left(1-e^{-1/T^*}\right)\psi_{10}(R-1)=-\psi_{11}(0)=2\pi\frac{Q_2}{E_3}\,. \qquad (6.211)$$

As an extra condition, we could enforce the continuity of the first derivative $y'(r)$ at $r=R$ [162]. However, this complicates the problem too much without any relevant gain in accuracy. In principle, it might be possible to impose consistency with a given EOS, via either the virial route, the compressibility route, or the energy route. But this is not practical since no simple EOS for SW fluids is at our disposal for wide values of density, temperature, and range. As a compromise between simplicity and accuracy, we fix the parameter Q_0 at its exact zero-density limit value, namely $Q_0=e^{1/T^*}-1$ [160]. Therefore, Eq. (6.211) becomes a transcendental equation for Q_2 that needs to be solved numerically. For narrow SW potentials, however, it is possible to replace the exact condition (6.210) by a simpler one allowing Q_2 to be obtained analytically [161], which is especially useful for determining the thermodynamic properties [161, 163].

It can be proven that the RFA proposal (6.198) reduces to the exact solutions of the PY equation [42, 43, 114] in the HS limit, i.e., $\epsilon\to 0$ or $R\to 1$, and in the SHS limit, i.e., $\epsilon\to\infty$ and $R\to 1$ with $(R-1)e^{1/T^*}=$ finite [160, 161].

Comparison with computer simulations [160, 161, 163, 164] shows that the RFA for SW fluids is rather accurate at any fluid density if the potential well

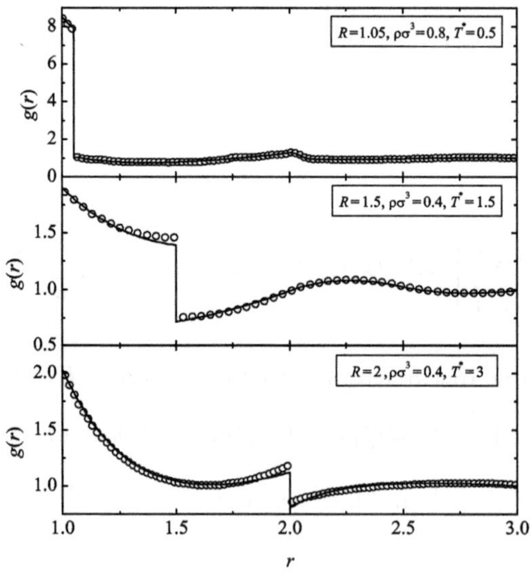

Fig. 6.13. Radial distribution function of a single-component SW fluid for $R=1.05$, $\rho\sigma^3=0.8$, and $T^*=0.5$ (*top panel*); for $R=1.5$, $\rho\sigma^3=0.4$, and $T^*=1.5$ (*middle panel*); and for $R=2.0$, $\rho\sigma^3=0.4$, and $T^*=3.0$ (*bottom panel*). The *circles* represent simulation data [164] and the *solid lines* refer to the results obtained from the RFA method

is sufficiently narrow (say $R \leq 1.2$), as well as for any width if the density is small enough (say $\rho\sigma^3 \leq 0.4$). However, as the width and/or the density increases, the RFA predictions worsen, especially at low temperatures. As an illustration, Fig. 6.13 compares the RDF provided by the RFA with Monte Carlo data [164] for three representative cases.

6.4.3 Hard Disks

As is well known, the PY theory is exactly solvable for HS fluids with an odd number of dimensions [165, 166, 167, 168, 169]. In particular, in the case of hard rods ($d = 1$), the PY theory provides the exact RDF $g(r)$ or, equivalently, the exact cavity function $y(r)$ outside the hard core (i.e., for $r > \sigma$). However, it does not reproduce the exact $y(r)$ in the overlapping region (i.e., for $r < \sigma$) [106]. The full exact one-dimensional cavity function is [106]

$$y_{\mathrm{HR}}(r|\eta) = \frac{\mathrm{e}^{-(r-1)\eta/(1-\eta)}}{1-\eta} + \sum_{n=2}^{\infty} \frac{\eta^{n-1}\mathrm{e}^{-(r-n)\eta/(1-\eta)}}{(1-\eta)^n(n-1)!}(r-n)^{n-1}\Theta(r-n) , \tag{6.212}$$

where the subscript HR stands for hard rods and, as usual, $\sigma = 1$ has been taken. Consequently, one has

$$g_{\mathrm{HR}}(1^+|\eta) = \frac{1}{1-\eta}, \quad \int_0^\infty \mathrm{d}r\, r h_{\mathrm{HR}}(r|\eta) \equiv H_{\mathrm{HR}}^{(0)}(\eta) = -\frac{1}{2} + \frac{2}{3}\eta - \frac{1}{4}\eta^2 . \tag{6.213}$$

When d is even, the PY equation is not analytically solvable for the HS interaction. In particular, in the important case of hard disks ($d = 2$), one must resort to numerical solutions of the PY equation [1, 170]. Alternatively, a simple heuristic approach has proven to yield reasonably good results [171]. Such an approach is based on the naïve assumption that the structure and spatial correlations of a hard-disk fluid share some features with those of a hard-rod and a HS fluid. This fuzzy idea becomes a more specific one by means of the following simple model [171]:

$$g_{\mathrm{HD}}(r|\eta) = \nu(\eta) g_{\mathrm{HR}}(r|\omega_1(\eta)\eta) + [1-\nu(\eta)] g_{\mathrm{HS}}(r|\omega_3(\eta)\eta) . \tag{6.214}$$

Here, the subscript HD stands for hard disks ($d = 2$) and the subscript HS stands for hard spheres ($d = 3$). The parameter $\nu(\eta)$ is a density-dependent mixing parameter, while $\omega_1(\eta)\eta$ and $\omega_3(\eta)\eta$ are the packing fractions in one and three dimensions, respectively, which are "equivalent" to the packing fraction η in two dimensions. In Eq. (6.214), it is natural to take for $g_{\mathrm{HR}}(r|\eta)$ the exact solution, Eq. (6.212). As for $g_{\mathrm{HR}}(r|\eta)$, one might use the RFA recipe described in Sect. 6.3. However, in order to keep the model (6.214) as simple as possible, it is sufficient for practical purposes to take the PY solution, Eq. (6.108). In the latter approximation,

$$g_{\mathrm{HS}}(1^+|\eta) = \frac{1+\eta/2}{(1-\eta)^2}, \quad \int_0^\infty \mathrm{d}r\, r h_{\mathrm{HS}}(r|\eta) \equiv H_{\mathrm{HS}}^{(0)}(\eta) = -\frac{10-2\eta+\eta^2}{20(1+2\eta)} . \tag{6.215}$$

In order to close the model (6.214), we still need to determine the parameters $\nu(\eta)$, $\omega_1(\eta)$, and $\omega_3(\eta)$. To that end, we first impose the condition that Eq. (6.214) must be consistent with a prescribed contact value $g_{\mathrm{HD}}(1^+|\eta)$ or, equivalently, with a prescribed compressibility factor $Z_{\mathrm{HD}}(\eta) = 1 + 2\eta g_{\mathrm{HD}}(1^+|\eta)$, with independence of the choice of the mixing parameter $\nu(\eta)$. In other words,

$$g_{\mathrm{HD}}(1^+|\eta) = g_{\mathrm{HR}}(1^+|\omega_1(\eta)\eta) = g_{\mathrm{HS}}(1^+|\omega_3(\eta)\eta) . \tag{6.216}$$

Making use of Eqs. (6.213) and (6.215), this yields

$$\omega_1(\eta) = \frac{g_{\mathrm{HD}}(1^+|\eta) - 1}{\eta g_{\mathrm{HD}}(1^+|\eta)}, \quad \omega_3(\eta) = \frac{4g_{\mathrm{HD}}(1^+|\eta) + 1 - \sqrt{24 g_{\mathrm{HD}}(1^+|\eta) + 1}}{4\eta g_{\mathrm{HD}}(1^+|\eta)} . \tag{6.217}$$

Once $\omega_1(\eta)$ and $\omega_3(\eta)$ are known, we can determine $\nu(\eta)$ by imposing that the model (6.214) reproduces the isothermal compressibility $\chi_{\mathrm{HD}}(\eta)$ thermodynamically consistent with the prescribed $Z_{\mathrm{HD}}(\eta)$ [cf. Eq. (6.97)]. From Eqs. (6.96) and (6.214), one has

$$\chi_{\mathrm{HD}}(\eta) = 1 + 8\eta \int_0^\infty \mathrm{d}r\, r \left\{\nu(\eta) h_{\mathrm{HR}}(r|\omega_1(\eta)\eta) + [1-\nu(\eta)]\, h_{\mathrm{HS}}(r|\omega_3(\eta)\eta)\right\} , \tag{6.218}$$

so that

$$\nu(\eta) = \frac{\left[\chi_{\mathrm{HD}}(\eta) - 1\right]/8\eta - H_{\mathrm{HS}}^{(0)}(\omega_3(\eta)\eta)}{H_{\mathrm{HR}}^{(0)}(\omega_1(\eta)\eta) - H_{\mathrm{HS}}^{(0)}(\omega_3(\eta)\eta)} , \tag{6.219}$$

where $H_{\mathrm{HR}}^{(0)}(\eta)$ and $H_{\mathrm{HS}}^{(0)}(\eta)$ are given by Eqs. (6.213) and (6.215), respectively.

Once a sensible EOS for hard disks is chosen (see, for instance, Table 6.1 and Chap. 3), Eqs. (6.217) and (6.219) provide the parameters of the model (6.214). The results show that the scaling factor $\omega_1(\eta)$ is a decreasing function, while $\omega_3(\eta)$ is an increasing function [171]. As for the mixing parameter $\nu(\eta)$, it is hardly dependent on density and takes values around $\nu(\eta) \simeq 0.35$–0.40.

Comparison of the interpolation model (6.214) with computer simulation results shows a surprisingly good agreement despite the crudeness of the model and the absence of empirical fitting parameters, especially at low and moderate densities [171]. The discrepancies become important only for distances beyond the location of the second peak and for densities close to the stability threshold.

6.5 Perturbation Theory

When one wants to deal with realistic intermolecular interactions, the problem of deriving the thermodynamic and structural properties of the system becomes rather formidable. Thus, perturbation theories of liquids have been devised since the mid-twentieth century. In the case of single-component fluids, the use of an accurate and well-characterized RDF for the HS fluid in a perturbation theory opens up the possibility of deriving a closed theoretical scheme for the determination of the thermodynamic and structural properties of more realistic models, such as the Lennard–Jones (LJ) fluid. In this section, we will consider this model system, which captures the basic physical properties of real non-polar fluids, to illustrate the procedure when the RFA method is used.

In the application of the perturbation theory of liquids, the stepping stone has been the use of the HS RDF obtained from the solution to the PY equation. Unfortunately, the absence of thermodynamic consistency present in the PY approximation (as well as in other integral equation theories) may clearly contaminate the results derived from its use within a perturbative treatment. In what follows, we will reanalyze the different theoretical schemes for the thermodynamics of LJ fluids that have been constructed with perturbation theory, taking as the reference system the HS fluid. This includes the consideration of the RDF as obtained with the RFA method, which embodies thermodynamic consistency, as well as the proposal of a unifying framework in which all schemes fit in. With our development, we will be able to present a formulation which lends itself to relatively easy numerical calculations while retaining the merits that analytical results provide, namely a detailed knowledge and control of all the approximations involved.

Let us consider a three-dimensional fluid system defined by a pair interaction potential $\phi(r)$. The virial and energy EOS express the compressibility factor Z and the excess part of the Helmholtz free energy per unit volume f^{ex}, respectively, in terms of the RDF of the system as

$$Z = 1 - \frac{2}{3}\pi\rho\beta \int_0^\infty \mathrm{d}r\, \frac{\partial\phi(r)}{\partial r} g(r) r^3 , \tag{6.220}$$

$$\frac{f^{\text{ex}}}{\rho k_{\text{B}}T} = 2\pi\rho\beta \int_0^\infty \mathrm{d}r\, \phi(r) g(r) r^2 , \tag{6.221}$$

where $\beta \equiv 1/k_{\text{B}}T$. Let us now assume that $\phi(r)$ is split into a known (reference) part $\phi_0(r)$ and a perturbation part $\phi_1(r)$. The usual perturbative expansion for the Helmholtz free energy to first order in β leads to [2]

$$\frac{f}{\rho k_{\text{B}}T} = \frac{f_0}{\rho k_{\text{B}}T} + 2\pi\rho\beta \int_0^\infty \mathrm{d}r\, \phi_1(r) g_0(r) r^2 + \mathcal{O}\left(\beta^2\right) , \tag{6.222}$$

where f_0 and $g_0(r)$ are the free energy and the RDF of the reference system, respectively.

The LJ potential is

$$\phi_{\rm LJ}(r) = 4\epsilon\left(r^{-12} - r^{-6}\right) , \tag{6.223}$$

where ϵ is the depth of the well and, for simplicity, we have taken the distance at which the potential vanishes as the length unit, i.e., $\phi_{\rm LJ}(r=1)=0$. For this potential, the reference system may be forced to be a HS system, i.e., one can set

$$\phi_0(r) = \phi_{\rm HS}(r) = \begin{cases} \infty, & r \le \sigma_0 , \\ 0, & r > \sigma_0 , \end{cases} \tag{6.224}$$

where σ_0 is a conveniently chosen effective HS diameter. In this case, the Helmholtz free energy to this order is approximated by

$$\frac{f_{\rm LJ}}{\rho k_{\rm B}T} \approx \frac{f_{\rm HS}}{\rho k_{\rm B}T} + 2\pi\rho\beta \int_{\sigma_0}^{\infty} {\rm d}r\, \phi_{\rm LJ}(r) g_{\rm HS}(r/\sigma_0) r^2 . \tag{6.225}$$

Note that Eq. (6.225) may be rewritten in terms of the Laplace transform $G(s)$ of $(r/\sigma_0)g_{\rm HS}(r/\sigma_0)$ as

$$\frac{f_{\rm LJ}}{\rho k_{\rm B}T} \approx \frac{f_{\rm HS}}{\rho k_{\rm B}T} + 2\pi\rho\beta\sigma_0^3 \int_0^{\infty} {\rm d}s\, \Phi_{\rm LJ}(s) G(s) , \tag{6.226}$$

where $\Phi_{\rm LJ}(s)$ satisfies

$$r\phi_{\rm LJ}(r) = \sigma_0 \int_0^{\infty} {\rm d}s\, {\rm e}^{-rs/\sigma_0} \Phi_{\rm LJ}(s) , \tag{6.227}$$

so that

$$\Phi_{\rm LJ}(s) = 4\epsilon\sigma_0^{-2} \left[\frac{(s/\sigma_0)^{10}}{10!} - \frac{(s/\sigma_0)^4}{4!}\right] . \tag{6.228}$$

Irrespective of the value of the diameter σ_0 of the reference system, the right-hand side of Eq. (6.226) represents *always* an upper bound for the value of the free energy of the real system. Therefore, it is natural to determine σ_0 so as to provide the least upper bound. This is precisely the variational scheme of Mansoori and Canfield [172, 173] and Rasaiah and Stell [174], usually referred to as MC/RS, and originally implemented with the PY theory for $G(s)$, Eq. (6.108). In our case, however, we will consider $G(s)$ as given by the RFA method, Eq. (6.113). Therefore, at fixed ρ and β, the effective diameter σ_0 in the MC/RS scheme is obtained from the conditions

$$\frac{\partial}{\partial\sigma_0}\left\{\int_0^{\eta_0} {\rm d}\eta\, \frac{Z_{\rm HS}(\eta)-1}{\eta} + 48\beta\epsilon\sigma_0^{-2}\int_0^{\infty} {\rm d}s\, G(s|\eta_0) \times \left[\frac{(s/\sigma_0)^{10}}{10!} - \frac{(s/\sigma_0)^4}{4!}\right]\right\} = 0 , \tag{6.229}$$

$$\frac{\partial^2}{\partial\sigma_0^2}\left\{\int_0^{\eta_0} d\eta\,\frac{Z_{\rm HS}(\eta)-1}{\eta} + 48\beta\epsilon\sigma_0^{-2}\int_0^\infty ds\,G(s|\eta_0) \times\left[\frac{(s/\sigma_0)^{10}}{10!}-\frac{(s/\sigma_0)^4}{4!}\right]\right\} > 0. \quad (6.230)$$

In these equations, use has been made of the thermodynamic relationship between the free energy and the compressibility factor, Eq. (6.78). Moreover, we have called $\eta_0 \equiv (\pi/6)\rho\sigma_0^3$ and have made explicit with the notation $G(s|\eta_0)$ the fact that the HS RDF depends on the packing fraction η_0.

Even if the reference system is not forced to be an HS fluid, one can still use Eq. (6.226) provided an adequate choice for σ_0 is made such that the expansion involved in the right-hand side of Eq. (6.222) yields the right-hand side of Eq. (6.226) to order β^2. This is the idea of the Barker and Henderson [175] first-order perturbation scheme (BH_1), where the effective HS diameter is

$$\sigma_0 = \int_0^\infty dr\,\left[1-e^{-\beta\phi_{\rm LJ}(r)}\right] . \quad (6.231)$$

The same ideas may be carried out to higher order in the perturbation expansion. The inclusion of the second-order term in the expansion yields the so-called macroscopic compressibility approximation [2] for the free energy, namely

$$\frac{f_{\rm LJ}}{\rho k_{\rm B}T} = \frac{f_0}{\rho k_{\rm B}T} + 2\pi\rho\beta\int_0^\infty dr\,\phi_1(r)g_0(r)r^2 -\pi\rho\beta^2\chi_0\int_0^\infty dr\,\phi_1^2(r)g_0(r)r^2 + \mathcal{O}\left(\beta^3\right) , \quad (6.232)$$

where χ_0 is the (reduced) isothermal compressibility of the reference system.[3]

To implement a particular perturbation scheme in this approximation under a unifying framework that eventually leads to easy numerical evaluation, two further assumptions may prove convenient. First, the perturbation potential $\phi_1(r) \equiv \phi_{\rm LJ}(r) - \phi_0(r)$ may be split into two parts using some "molecular size" parameter $\xi \geq \sigma_0$ such that

$$\phi_1(r) = \begin{cases} \phi_{1a}(r)\ , \ 0 \le r \le \xi\ , \\ \phi_{1b}(r)\ , \ r > \xi\ . \end{cases} \quad (6.233)$$

Next, a choice for the RDF for the reference system is done in the form

$$g_0\,(r) \approx \theta(r) y_{\rm HS}(r/\sigma_0)\ , \quad (6.234)$$

[3] The macroscopic compressibility approach is only one of the possibilities of approximation to the second order Barker–Henderson perturbation theory term. Another successful approach is the local-compressibility approximation (see [2], p 308). This expresses the free energy in terms of $\phi_1(r)$ and HS quantities

where y_{HS} is the cavity (background) correlation function of the HS system and $\theta(r)$ is a step function defined by

$$\theta(r) = \begin{cases} \theta_a(r) \, , 0 \le r \le \xi \, , \\ \theta_b(r) \, , r > \xi \, , \end{cases} \tag{6.235}$$

in which the functions $\theta_a(r)$ and $\theta_b(r)$ depend on the scheme.

With these assumptions, the integrals involved in Eq. (6.232) may be rewritten as

$$\begin{aligned} I_n &\equiv \int_0^\infty \mathrm{d}r \, \phi_1^n(r) g_0(r) r^2 \\ &= \int_0^{\sigma_0} \mathrm{d}r \, \phi_{1a}^n(r)\theta_a(r) y_{\mathrm{HS}}(r/\sigma_0) r^2 + \int_{\sigma_0}^{\xi} \mathrm{d}r \, \phi_{1a}^n(r)\theta_a(r) g_{\mathrm{HS}}(r/\sigma_0) r^2 \\ &\quad + \int_\xi^\infty \mathrm{d}r \, \phi_{1b}^n(r)\theta_b(r) g_{\mathrm{HS}}(r/\sigma_0) r^2 \, , \end{aligned} \tag{6.236}$$

with $n = 1, 2$ and where the fact that $y_{\mathrm{HS}}(r/\sigma_0) = g_{\mathrm{HS}}(r/\sigma_0)$ when $r > \sigma_0$ has been used. Decomposing the last integral as $\int_\xi^\infty = \int_{\sigma_0}^\infty - \int_{\sigma_0}^\xi$ and applying the same step as in Eq. (6.226), Eq. (6.236) becomes

$$\begin{aligned} I_n &= \sigma_0^3 \int_0^\infty \mathrm{d}s \, \Phi_{nb}(s) G(s) + \int_0^{\sigma_0} \mathrm{d}r \, \phi_{1a}^n(r)\theta_a(r) y_{\mathrm{HS}}(r/\sigma_0) r^2 \\ &\quad + \int_{\sigma_0}^{\xi} \mathrm{d}r \, \left[\phi_{1a}^n(r)\theta_a(r) - \phi_{1b}^n(r)\theta_b(r)\right] g_{\mathrm{HS}}(r/\sigma_0) r^2 \, , \end{aligned} \tag{6.237}$$

where the functions $\Phi_{1b}(s)$ and $\Phi_{2b}(s)$ are defined by the relation

$$r\phi_{1b}^n(r)\theta_b(r) = \sigma_0 \int_0^\infty \mathrm{d}s \, \mathrm{e}^{-rs/\sigma_0} \Phi_{nb}(s) \, . \tag{6.238}$$

In the Barker–Henderson second-order perturbation scheme (BH_2), one takes

$$\theta_a(r) = 0, \quad \theta_b(r) = 1, \quad \xi = \sigma_0, \quad \phi_{1a}(r) = 0, \quad \phi_{1b}(r) = 4\epsilon \left(r^{-12} - r^{-6}\right) \, , \tag{6.239}$$

and σ_0 is computed according to Eq. (6.231). This choice ensures that

$$\begin{aligned} \frac{f_{\mathrm{LJ}}}{\rho k_{\mathrm{B}} T} &= \frac{f_{\mathrm{HS}}}{\rho k_{\mathrm{B}} T} + 2\pi\rho\beta \int_{\sigma_0}^\infty \mathrm{d}r \, \phi_1(r) g_{\mathrm{HS}}(r/\sigma_0) r^2 \\ &\quad - \pi\rho\beta^2 \chi_{\mathrm{HS}} \int_{\sigma_0}^\infty \mathrm{d}r \, \phi_1^2(r) g_{\mathrm{HS}}(r/\sigma_0) r^2 + O\left(\beta^3\right) . \end{aligned} \tag{6.240}$$

On the other hand, if one chooses

$$\theta_a(r) = \exp\left[-\beta\left(\phi_{\mathrm{LJ}}(r) + \epsilon\right)\right], \quad \theta_b(r) = 1, \quad \xi = 2^{1/6} \, , \tag{6.241}$$

$$\phi_{1a}(r) = -\epsilon, \quad \phi_{1b}(r) = 4\epsilon\left(r^{-12} - r^{-6}\right) , \tag{6.242}$$

the scheme leads to the Weeks–Chandler–Andersen (WCA) theory [176] if one determines the HS diameter through the condition $\chi_0 = \chi_{\text{HS}}$ [177], which in turn implies

$$\int_0^{\sigma_0} \mathrm{d}r\, r^2 \mathrm{e}^{-\beta\phi_0(r)} y_{\text{HS}}(r/\sigma_0) = \int_{\sigma_0}^{2^{1/6}} \mathrm{d}r\, r^2 g_{\text{HS}}(r/\sigma_0)\left[1 - \mathrm{e}^{-\beta\phi_0(r)}\right] . \tag{6.243}$$

To close the scheme, the HS cavity function has to be provided in the range $0 \leq r \leq \sigma_0$. Fortunately, relatively simple expressions for $y_{\text{HS}}(r/\sigma_0)$ are available in the literature [178, 179, 180], apart from our own proposal, Eq. (6.140).

Note that $\theta_b(r)$ and $\phi_{1b}(r)$, and thus also $\Phi_{nb}(s)$, are the same functions in the BH_2 and WCA schemes. It is convenient, in order to have all the quantities needed to evaluate f_{LJ} in these schemes, to provide explicit expressions for $\Phi_{1b}(s)$ and $\Phi_{2b}(s)$. These are given by [cf. Eq. (6.228)]

$$\Phi_{1b}(s) = \Phi_{\text{LJ}}(s) , \tag{6.244}$$

$$\Phi_{2b}(s) = 16\epsilon^2\sigma_0^{-2}\left[\frac{(s/\sigma_0)^{22}}{22!} - 2\frac{(s/\sigma_0)^{16}}{16!} + \frac{(s/\sigma_0)^{10}}{10!}\right] . \tag{6.245}$$

Up to this point, we have embodied the most popular perturbation schemes within a unified framework that requires as input *only* the EOS of the HS fluid in order to compute the Helmholtz free energy of the LJ system and leads to relatively easy numerical computations. It should be clear that a variety of other possible schemes, requiring the same little input, fit in our unified framework, which is based on the RFA method for $g_{\text{HS}}(r/\sigma_0)$ and $G(s)$. Once f_{LJ} has been determined, the compressibility factor of the LJ fluid at a given order of the perturbation expansion readily follows from Eqs. (6.222) or (6.232) through the thermodynamic relation

$$Z_{\text{LJ}} = \rho\left(\frac{\partial}{\partial\rho}\frac{f_{\text{LJ}}}{\rho k_{\text{B}}T}\right)_T . \tag{6.246}$$

Taking into account that the HS fluid presents a fluid–solid transition at a freezing packing fraction $\eta_{\text{f}} \simeq 0.494$ [181] and a solid–fluid transition at a melting packing fraction $\eta_{\text{m}} \simeq 0.54$ [181], the fluid–solid and solid–fluid coexistence lines for the LJ system may be computed from the values (ρ, T) determined from the conditions $(\pi/6)\rho\sigma_0^3(\rho,T) = \eta_{\text{f}}$ and $(\pi/6)\rho\sigma_0^3(\rho,T) = \eta_{\text{m}}$, respectively, with the effective diameter $\sigma_0(\rho,T)$ obtained using any of the perturbative schemes. Similarly, admitting that there is a glass transition in the HS fluid at the packing fraction $\eta_{\text{g}} \simeq 0.56$ [182], one can now determine the location of the liquid–glass transition line for the LJ fluid in the (ρ, T) plane from the simple relationship $(\pi/6)\rho\sigma_0^3(\rho,T) = \eta_{\text{g}}$. With a proper choice for Z_{HS}, it has been shown [93, 183, 184] that the critical point, the structure, and the phase diagram (including a glass transition) of the LJ fluid may be adequately described with this approach.

6.6 Perspectives

In this chapter, we have given a self-contained account of a simple (mostly analytical) framework for the study of the thermodynamic and structural properties of hard-core systems. Whenever possible, the developments have attempted to cater for mixtures with an arbitrary number of components (including polydisperse systems) and arbitrary dimensionality. We started considering the contact values of the RDF because they enter directly into the EOS and are required as input in the RFA method to compute the structural properties. With the aid of consistency conditions, we were able to devise various approximate proposals which, when used in conjunction with a sensible choice for the contact value of the RDF of the single-component fluid (required in the formulation but otherwise chosen at will), have been shown to be in reasonably good agreement with simulation results and lead to accurate EOS both for additive and for non-additive mixtures. Some aspects of the results that follow from the use of these EOS were illustrated by looking at demixing problems in these mixtures, including the far from intuitive case of a binary mixture of non-additive hard spheres in infinite dimensionality.

After that, restricting ourselves to three-dimensional systems, we described the RFA method as applied to a single-component HS fluid and to a multicomponent mixture of HS. Using this approach, we have been able to obtain explicit analytical results for the RDF, the direct correlation function, the static structure factor, and the bridge function, in the end requiring as input *only* the contact value of the RDF of the single-component HS fluid (or equivalently its compressibility factor). One of the nice assets of the RFA approach is that it eliminates the thermodynamic consistency problem which is present in most of the integral equation formulations for the computation of structural quantities. Once again, when a sensible choice for the single-component EOS is made, we have shown, through the comparison between the results of the RFA approach and simulation data for some illustrative cases, the very good performance of our development. Also, the use of the RFA approach in connection with some other related systems (sticky hard spheres, square-well fluids, and hard disks) has been addressed.

The final part of the chapter concerns the use of HS results for more realistic intermolecular potentials in the perturbation theory of liquids. In this instance, we have been able to provide a unifying scheme in which the most popular perturbation theory formulations may be expressed and which was devised to allow for easy computations. We illustrated this for a LJ fluid, but it should be clear that a similar approach might be followed for other fluids, and in fact, it has recently been done in connection with the glass transition of hard-core Yukawa fluids [185].

Finally, it should be clear that there are many facets of the equilibrium and structural properties of hard-core systems that may be studied with a similar approach but that up to now have not been considered. For instance, the generalizations of the RFA approach for systems such as hard hyperspheres, non-additive hard spheres, square-well mixtures, penetrable spheres [186], or the Jagla potential [187], appear as interesting challenges. Similarly, the extension of the perturbation theory scheme to the case of LJ mixtures seems a worthwhile task. We hope to address some of these problems in the future and would be very much rewarded if some others were taken up by researchers who might find these developments also a valuable tool for their work.

References

1. J.A. Barker, D. Henderson: Rev. Mod. Phys. **48**, 587 (1976)
2. D.A. McQuarrie: *Statistical Mechanics* (Harper & Row, New York 1976)
3. H.L. Friedman: *A Course in Statistical Mechanics* (Prentice Hall, Englewood Cliffs 1985)
4. J.-P. Hansen, I.R. McDonald: *Theory of Simple Liquids, Third Edition* (Academic Press, London 2006)
5. J.L. Lebowitz, D. Zomick: J. Chem. Phys. **54**, 3335 (1971)
6. J.T. Jenkins, F. Mancini: J. Appl. Mech. **54**, 27 (1987)
7. C. Barrio, J.R. Solana: J. Chem. Phys. **115**, 7123 (2001)
8. C. Barrio, J.R. Solana: J. Chem. Phys. **117**, 2451(E) (2002)
9. J.L. Lebowitz: Phys. Rev. A **133**, 895 (1964)
10. H. Reiss, H.L. Frisch, J.L. Lebowitz: J. Chem. Phys. **31**, 369 (1959)
11. E. Helfand, H.L. Frisch, J.L. Lebowitz: J. Chem. Phys. **34**, 1037 (1961)
12. J.L. Lebowitz, E. Helfand, E. Praestgaard: J. Chem. Phys. **43**, 774 (1965)
13. M.J. Mandell, H. Reiss: J. Stat. Phys. **13**, 113 (1975)
14. Y. Rosenfeld: J. Chem. Phys. **89**, 4272 (1988)
15. M. Heying, D.S. Corti: J. Phys. Chem. B **108**, 19756 (2004)
16. T. Boublík: J. Chem. Phys. **53**, 471 (1970)
17. E.W. Grundke, D. Henderson: Mol. Phys. **24**, 269 (1972)
18. L.L. Lee, D. Levesque: Mol. Phys. **26**, 1351 (1973)
19. G.A. Mansoori, N.F. Carnahan, K.E. Starling, J.T.W. Leland: J. Chem. Phys. **54**, 1523 (1971)
20. D. Henderson, A. Malijevský, S. Labík, K.-Y. Chan: Mol. Phys. **87**, 273 (1996)
21. D.H.L. Yau, K.-Y. Chan, D. Henderson: Mol. Phys. **88**, 1237 (1996)
22. D.H.L. Yau, K.-Y. Chan, D. Henderson: Mol. Phys. **91**, 1137 (1997)
23. D. Henderson, K.-Y. Chan: J. Chem. Phys. **108**, 9946 (1998)
24. D. Henderson, K.-Y. Chan: Mol. Phys. **94**, 253 (1998)
25. D. Henderson, K.-Y. Chan: Mol. Phys. **98**, 1005 (2000)
26. D. Henderson, D. Boda, K.-Y. Chan, D.T. Wasan: Mol. Phys. **95**, 131 (1998)
27. D. Matyushov, D. Henderson, K.-Y. Chan: Mol. Phys. **96**, 1813 (1999)
28. D. Cao, K.-Y. Chan, D. Henderson, W. Wang: Mol. Phys. **98**, 619 (2000)
29. D.V. Matyushov, B.M. Ladanyi: J. Chem. Phys. **107**, 5815 (1997)
30. C. Barrio, J.R. Solana: J. Chem. Phys. **113**, 10180 (2000)
31. D. Viduna, W.R. Smith: Mol. Phys. **100**, 2903 (2002)
32. D. Viduna, W.R. Smith: J. Chem. Phys. **117**, 1214 (2002)
33. E. Hamad: J. Chem. Phys. **101**, 10195 (1994)
34. C. Vega: J. Chem. Phys. **108**, 3074 (1998)
35. N.M. Tukur, E.Z. Hamad, G.A. Mansoori: J. Chem. Phys. **110**, 3463 (1999)
36. D. Henderson: Mol. Phys. **30**, 971 (1975)
37. A. Santos, M. López de Haro, S.B. Yuste: J. Chem. Phys. **103**, 4622 (1995)
38. M. López de Haro, A. Santos, S.B. Yuste: Eur. J. Phys. **19**, 281 (1998)

39. S. Luding: Phys. Rev. E **63**, 042201 (2001)
40. S. Luding: Adv. Compl. Syst. **4**, 379 (2002)
41. Luding S., Strauß O.: The equation of state for almost elastic, smooth, polydisperse granular gases for arbitrary density. In: Pöschel T., Luding S. (eds.): *Granular Gases*, Lect. Notes Phys. **564**, 389–409. Springer, Berlin (2001)
42. M.S. Wertheim: Phys. Rev. Lett. **10**, 321 (1963)
43. E. Thiele: J. Chem. Phys. **39**, 474 (1963)
44. N.F. Carnahan, K.E. Starling: J. Chem. Phys. **51**, 635 (1969)
45. M. Luban, J.P.J. Michels: Phys. Rev. A **41**, 6796 (1990)
46. A. Santos, S.B. Yuste, M. López de Haro: Mol. Phys. **96**, 1 (1999)
47. A. Malijevský, J. Veverka: Phys. Chem. Chem. Phys. **1**, 4267 (1999)
48. A. Santos, S.B. Yuste, M. López de Haro: Mol. Phys. **99**, 1959 (2001)
49. M. González-Melchor, J. Alejandre, M. López de Haro: J. Chem. Phys. **114**, 4905 (2001)
50. M. López de Haro, S.B. Yuste, A. Santos: Phys. Rev. E **66**, 031202 (2002)
51. A. Santos: Mol. Phys. **96**, 1185 (1999)
52. A. Santos: Mol. Phys. **99**, 617(E) (2001)
53. C. Regnaut, A. Dyan, S. Amokrane: Mol. Phys. **99**, 2055 (2001)
54. C. Regnaut, A. Dyan, S. Amokrane: Mol. Phys. **100**, 2907(E) (2002)
55. A. Santos, S.B. Yuste, M. López de Haro: J. Chem. Phys. **117**, 5785 (2002)
56. A. Santos, S.B. Yuste, M. López de Haro: J. Chem. Phys. **123**, 234512 (2005)
57. M. López de Haro, S.B. Yuste, A. Santos: Mol. Phys. **104**, 3461 (2006)
58. S. Luding, A. Santos: J. Chem. Phys. **121**, 8458 (2004)
59. M. Barošová, A. Malijevský, S. Labík, W.R. Smith: Mol. Phys. **87**, 423 (1996)
60. H. Hansen-Goos, R. Roth: J. Chem. Phys. **124**, 154506 (2006)
61. R. Evans: Microscopic theories of simple fluids and their interfaces. In: *Liquids at Interfaces*, ed by J. Charvolin, J.F. Joanny, J. Zinn-Justin (Elsevier, Amsterdam 1990) pp 3–98
62. Y. Rosenfeld: Phys. Rev. Lett. **63**, 980 (1989)
63. A. Malijevský, M. Barošová, W.R. Smith: Mol. Phys. **91**, 65 (1997)
64. Al. Malijevský, A. Malijevský, S.B. Yuste, A. Santos, M. López de Haro: Phys. Rev. E **66**, 061203 (2002)
65. M. Buzzacchi, I. Pagonabarraga, N.B. Wilding: J. Chem. Phys. **121**, 11362 (2004)
66. Al. Malijevský, S.B. Yuste, A. Santos, M. López de Haro: Phys. Rev. E **75**, 061201 (2007)
67. F. Lado: Phys. Rev. E **54**, 4411 (1996)
68. I. Prigogine, S. Lafleur: Bull. Classe Sci. Acad. Roy. Belg. **40**, 484 (1954)
69. I. Prigogine, S. Lafleur: Bull. Classe Sci. Acad. Roy. Belg. **40**, 497 (1954)
70. S. Asakura, F. Oosawa: J. Chem. Phys. **22**, 1255 (1954)
71. S. Asakura, F. Oosawa: J. Polym. Sci. **33**, 183 (1958)
72. R. Kikuchi: J. Chem. Phys. **23**, 2327 (1955)
73. P. Ballone, G. Pastore, G. Galli, D. Gazzillo: Mol. Phys. **59**, 275 (1986)
74. D. Gazzillo, G. Pastore, S. Enzo: J. Phys.: Condens. Matter **1**, 3469 (1989)
75. D. Gazzillo, G. Pastore, R. Frattini: J. Phys.: Condens. Matter **2**, 8465 (1990)
76. A. Santos, M. López de Haro, S.B. Yuste: J. Chem. Phys. **122**, 024514 (2005)
77. E.Z. Hamad: J. Chem. Phys. **105**, 3229 (1996)
78. E.Z. Hamad: J. Chem. Phys. **105**, 3222 (1996)
79. H. Hammawa, E.Z. Hamad: J. Chem. Soc. Faraday Trans. **92**, 4943 (1996)

80. M. Al-Naafa, J.B. El-Yakubu, E.Z. Hamad: Fluid Phase Equilibria **154**, 33 (1999)
81. J. Jung, M.S. Jhon, F.H. Ree: J. Chem. Phys. **100**, 528 (1994)
82. J. Jung, M.S. Jhon, F.H. Ree: J. Chem. Phys. **100**, 9064 (1994)
83. T. Coussaert, M. Baus: J. Chem. Phys. **109**, 6012 (1998)
84. A.Y. Vlasov, A.J. Masters: Fluid Phase Equilibria **212**, 183 (2003)
85. M. López de Haro, C.F. Tejero: J. Chem. Phys. **121**, 6918 (2004)
86. S.B. Yuste, A. Santos, M. López de Haro: Europhys. Lett. **52**, 158 (2000)
87. H.-O. Carmesin, H.L. Frisch, J.K. Percus: J. Stat. Phys. **63**, 791 (1991)
88. A. Santos, M. López de Haro: Phys. Rev. E **72**, 010501(R) (2005)
89. R. Roth, R. Evans, A.A. Louis: Phys. Rev. E **64**, 051202 (2001)
90. S.B. Yuste, A. Santos: Phys. Rev. A **43**, 5418 (1991)
91. S.B. Yuste, M. López de Haro, A. Santos: Phys. Rev. E **53**, 4820 (1996)
92. M. Robles, M. López de Haro, A. Santos, S.B. Yuste: J. Chem. Phys. **108**, 1290 (1998)
93. M. Robles, M. López de Haro: Europhys. Lett. **62**, 56 (2003)
94. M. Robles, M. López de Haro, J. Chem. Phys. **107**, 4648 (1997)
95. E. Waisman: Mol. Phys. **25**, 45 (1973)
96. D. Henderson, L. Blum: Mol. Phys. **32**, 1627 (1976)
97. J.S. Høye, L. Blum: J. Stat. Phys. **16**, 399 (1977)
98. A. Díez, J. Largo, J.R. Solana: J. Chem. Phys. **125**, 074509 (2006)
99. J. Kolafa, S. Labík, A. Malijevský: Phys. Chem. Chem. Phys. **6**, 2335 (2004) See also http://www.vscht.cz/fch/software/hsmd/for molecular dynamics results of $g(r)$
100. A. Trokhymchuk, I. Nezbeda, J. Jirsák, D. Henderson: J. Chem. Phys. **123**, 024501 (2005)
101. M. López de Haro, A. Santos, S.B. Yuste: J. Chem. Phys. **124**, 236102 (2006)
102. L.L. Lee: J. Chem. Phys. **103**, 9388 (1995)
103. L.L. Lee, D. Ghonasgi, E. Lomba: J. Chem. Phys. **104**, 8058 (1996)
104. L.L. Lee, A. Malijevský: J. Chem. Phys. **114**, 7109 (2001)
105. S. Labík, A. Malijevský: Mol. Phys. **53**, 381 (1984)
106. Al. Malijevský, A. Santos: J. Chem. Phys. **124**, 074508 (2006)
107. A. Santos, Al. Malijevský: Phys. Rev. E **75**, 021201 (2007)
108. S.B. Yuste, A. Santos, M. López de Haro: J. Chem. Phys. **108**, 3683 (1998)
109. L. Blum, J.S. Høye: J. Phys. Chem. **81**, 1311 (1977)
110. J. Abate, W. Whitt: Queuing Systems **10**, 5 (1992)
111. N.W. Ashcroft, D.C. Langreth: Phys. Rev. **156**, 685 (1967)
112. A.B. Bathia, D.E. Thornton: Phys. Rev. B **8**, 3004 (1970)
113. S.B. Yuste, A. Santos, M. López de Haro: Mol. Phys. **98**, 439 (2000)
114. R.J. Baxter: J. Chem. Phys. **49**, 2270 (1968)
115. J.W. Perram, E.R. Smith: Chem. Phys. Lett. **35**, 138 (1975)
116. B. Barboy: Chem. Phys. **11**, 357 (1975)
117. B. Barboy, R. Tenne: Chem. Phys. **38**, 369 (1979)
118. G. Stell: J. Stat. Phys. **63**, 1203 (1991)
119. B. Borštnik, C. G. Jesudason, G. Stell: J. Chem. Phys. **106**, 9762 (1997)
120. B. Barboy: J. Chem. Phys. **61**, 3194 (1974)
121. J.W. Perram, E.R. Smith: Chem. Phys. Lett. **39**, 328 (1975)
122. P.T. Cummings, J.W. Perram, E.R. Smith: Mol. Phys. **31**, 535 (1976)
123. E.R. Smith, J.W. Perram: J. Stat. Phys. **17**, 47 (1977)

124. J.W. Perram, E.R. Smith: Proc. R. Soc. London A **353**, 193 (1977)
125. W.G.T. Kranendonk, D. Frenkel: Mol. Phys. **64**, 403 (1988)
126. C. Regnaut, J.C. Ravey: J. Chem. Phys. **91**, 1211 (1989)
127. G. Stell, Y. Zhou: J. Chem. Phys. **91**, 3618 (1989)
128. J.N. Herrera, L. Blum: J. Chem. Phys. **94**, 6190 (1991)
129. A. Jamnik, D. Bratko, D.J. Henderson: J. Chem. Phys. **94**, 8210 (1991)
130. S.V.G. Menon, C. Manohar, K.S. Rao: J. Chem. Phys. **95**, 9186 (1991)
131. Y. Zhou, G. Stell: J. Chem. Phys. **96**, 1504 (1992)
132. E. Dickinson: J. Chem. Soc. Faraday Trans. **88**, 3561 (1992)
133. C.F. Tejero, M. Baus: Phys. Rev. E, **48**, 3793 (1993)
134. K. Shukla, R. Rajagopalan: Mol. Phys. **81**, 1093 (1994)
135. C. Regnaut, S. Amokrane, Y. Heno: J. Chem. Phys. **102**, 6230 (1995)
136. C. Regnaut, S. Amokrane, P. Bobola: Prog. Colloid Polym. Sci. **98**, 151 (1995)
137. Y. Zhou, C.K. Hall, G. Stell: Mol. Phys. **86**, 1485 (1995)
138. J.N. Herrera-Pacheco, J.F. Rojas-Rodríguez: Mol. Phys. **86**, 837 (1995)
139. Y. Hu, H. Liu, J.M. Prausnitz: J. Chem. Phys. **104**, 396 (1996)
140. O. Bernard, L. Blum: J. Chem. Phys. **104**, 4746 (1996)
141. L. Blum, M.F. Holovko, I.A. Protsykevych: J. Stat. Phys. **84**, 191 (1996)
142. S. Amokrane, P. Bobola, C. Regnaut: Prog. Colloid Polym. Sci. **100**, 186 (1996)
143. S. Amokrane, C. Regnaut: J. Chem. Phys. **106**, 376 (1997)
144. C. Tutschka, G. Kahl, E. Riegler: Mol. Phys. **100**, 1025 (2002)
145. D. Gazzillo, A. Giacometti: Mol. Phys. **100**, 3307 (2002)
146. M.A. Miller, D. Frenkel: Phys. Rev. Lett. **90**, 135702 (2003)
147. D. Gazzillo, A. Giacometti: J. Chem. Phys. **120**, 4742 (2004)
148. R. Fantoni, D. Gazzillo, A. Giacometti: Phys. Rev. E **72**, 011503 (2005)
149. A. Jamnik: Chem. Phys. Lett. **423**, 23 (2006)
150. A.J. Post, E. D. Glandt: J. Chem. Phys. **84**, 4585 (1986)
151. N.A. Seaton, E.D. Glandt: J. Chem. Phys. **84**, 4595 (1986)
152. N.A. Seaton, E.D. Glandt: J. Chem. Phys. **86**, 4668 (1986)
153. N.A. Seaton, E.D. Glandt: J. Chem. Phys. **87**, 1785 (1987)
154. M.A. Miller, D. Frenkel: J. Phys.: Condens. Matter **16**, S4901 (2004)
155. M.A. Miller, D. Frenkel: J. Chem. Phys. **121**, 535 (2004)
156. Al. Malijevský, S.B. Yuste, A. Santos: J. Chem. Phys. **125**, 074507 (2006)
157. A. Santos, S.B. Yuste, M. López de Haro: J. Chem. Phys. **109**, 6814 (1998)
158. S.B. Yuste, A. Santos: J. Stat. Phys. **72**, 703 (1993)
159. S.B. Yuste, A. Santos: Phys. Rev. E **48**, 4599 (1993)
160. S.B. Yuste, A. Santos: J. Chem. Phys. **101**, 2355 (1994)
161. L. Acedo, A. Santos: J. Chem. Phys. **115**, 2805 (2001)
162. L. Acedo: J. Stat. Phys. **99**, 707 (2000)
163. J. Largo, J.R. Solana, L. Acedo, A. Santos: Mol. Phys. **101**, 2981 (2003)
164. J. Largo, J.R. Solana, S.B. Yuste, A. Santos: J. Chem. Phys. **122**, 084510 (2005)
165. C. Freasier, D.J. Isbister: Mol. Phys. **42**, 927 (1981)
166. E. Leutheusser: Physica A **127**, 667 (1984)
167. M. Robles, M. López de Haro, A. Santos: J. Chem. Phys. **120**, 9113 (2004)
168. M. Robles, M. López de Haro, A. Santos: J. Chem. Phys. **125**, 219903(E) (2006)
169. M. Robles, M. López de Haro, A. Santos: J. Chem. Phys. **126**, 016101 (2007)

170. D.G. Chae, F.H. Ree, T. Ree: J. Chem. Phys. **50**, 1581 (1976)
171. S.B. Yuste, A. Santos: J. Chem. Phys. **99**, 2020 (1993)
172. G.A. Mansoori, F.B. Canfield: J. Chem. Phys. **51**, 4958 (1969)
173. G.A. Mansoori, J.A. Provine, F.B. Canfield: J. Chem. Phys. **51**, 5295 (1969)
174. J. Rasaiah, G. Stell: Mol. Phys. **18**, 249 (1970)
175. J.A. Barker, D. Henderson: J. Chem. Phys. **47**, 2856 (1967)
176. J.D. Weeks, D. Chandler, H.C. Andersen: J. Chem. Phys. **53**, 149 (1971)
177. L. Verlet, J.J. Weis: Phys. Rev. A **5**, 939 (1972)
178. D. Henderson, E.W. Grundke: J. Chem. Phys. **63**, 601 (1975)
179. J.A. Ballance, R.J. Speedy: Mol. Phys. **54**, 1035 (1985)
180. Y. Zhou, G. Stell: J. Stat. Phys. **52**, 1389 (1988)
181. J.-P. Hansen, L. Verlet: Phys. Rev. **184**, 151 (1969)
182. R.J. Speedy: J. Chem. Phys. **100**, 6684 (1994)
183. M. Robles, M. López de Haro,: Phys. Chem. Chem. Phys. **3**, 5528 (2001)
184. M. López de Haro, M. Robles: J. Phys.: Condens. Matt. **16**, S2089 (2004)
185. M. López de Haro, M. Robles: Physica A **372**, 307 (2006)
186. C.N. Likos: Phys. Rep. **348**, 267 (2001)
187. E.A. Jagla: J. Chem. Phys. **111**, 8980 (1999)

7

Density Functional Theories of Hard Particle Systems

P. Tarazona[1], J.A. Cuesta[2], and Y. Martínez-Ratón[2]

[1] Departamento de Física Teórica de la Materia Condensada, Universidad Autónoma de Madrid, E-28049 Madrid, Spain
pedro.tarazona@uam.es

[2] Grupo Interdisciplinar de Sistemas Complejos (GISC), Departamento de Matemáticas, Universidad Carlos III, Avda. de la Universidad 30, E-28911 Leganés, Spain
cuesta@math.uc3m.es, yuri@math.uc3m.es

To the memory of our friend Yasha Rosenfeld,
who discovered the Fundamental Measure Theory,
making this chapter grow into a thick one.

This chapter deals with the applications of the density functional (DF) formalism to the study of inhomogeneous systems with hard core interactions. It includes a brief tutorial on the fundamentals of the method, and the exact free energy DF for one-dimensional hard rods obtained by Percus. The development of DF approximations for the free energy of hard spheres (HS) is presented through its milestones in the weighted density approximation (WDA) and the fundamental measure theory (FMT). The extensions of these approaches to HS mixtures include the FMT treatment of polydisperse systems and the approximations for mixtures with non-additive core radii. The DF treatment of non-spherical hard core systems is presented within the generic context of the study of liquid crystals phases. The chapter is directed to the potential users of these theoretical techniques, with clear explanations of the practical implementation details of the most successful approximations.

7.1 Introduction

The density functional (DF) formalism for classical particles [1] was developed to find out the equilibrium density distribution $\rho(\boldsymbol{r})$ of inhomogeneous systems at interfaces or in the presence of an external potential $V(\boldsymbol{r})$. In most cases, like the layering of fluids against walls or liquids confined in nano-capillaries, the sharpest level of structure in $\rho(\boldsymbol{r})$ comes from the effects of molecular packing, and hence the development of DF theories for hard-core models has been a

Tarazona P., et al.: *Density Functional Theories of Hard Particle Systems*. Lect. Notes Phys. **753**, 247–341 (2008)
DOI 10.1007/978-3-540-78767-9_7

main objective in the field [2, 3]. The free energy DF for one-dimensional (1D) hard rods (HR), presented by Percus [4] in 1976, provided both an exact case to test the internal relations of the DF formalism and a hint on how to approximate the free energy of three-dimensional (3D) hard spheres (HS) and two-dimensional (2D) hard disk (HD) systems. Over the last decades, there has been an impressive improvement in the quality of these approximations, with milestones in the weighted density approximation (WDA) in the middle 1980s and the fundamental measure theory (FMT) over the 1990s.

After a brief tutorial to the DF formalism, to be skipped by the expert reader, this chapter will first deal with the most successful DF schemes for HS, presenting them in an easy-to-use fashion and comparing their relative advantages and difficulties. We do not intend an exhaustive presentation of the many approaches and variants, which have been developed to describe the DF free energy of these systems, but rather to provide the reader with an easy to follow introduction to the basis and the practical use of some extensively used approaches.

The second part of the chapter will be dedicated to the free energy DF for HS mixtures, for which the FMT appears to be the most natural approximate scheme. Different versions of FMT will be presented and their relative merits compared, including their use for polydisperse systems and the extensions to non-additive HS.

The third part of the chapter will deal with non-spherical hard core systems, describing again the most successful schemes which have been developed to study rigid cores.

7.2 Brief Tutorial to the Density Functional Formalism

The DF formalism for systems with classical statistics [1] establishes that for any given temperature ($\beta = 1/k_{\mathrm{B}}T$) and form of the pair molecular interaction potential energy, $u(\boldsymbol{r}_i - \boldsymbol{r}_j)$, there is a unique intrinsic free energy $\mathcal{F}[\rho]$, which is a functional of the density distribution $\rho(\boldsymbol{r})$ and not of the external potential. The grand canonical equilibrium density distribution for the system in presence of any external potential $V(\boldsymbol{r})$ and in contact with a reservoir of particles at chemical potential μ is that which minimizes the grand potential energy DF,

$$\Omega[\rho] \equiv \mathcal{F}[\rho] + \int \mathrm{d}\boldsymbol{r} \rho(\boldsymbol{r})(V(\boldsymbol{r}) - \mu) , \tag{7.1}$$

with respect to all possible functions $\rho(\boldsymbol{r})$. That minimum condition on Ω may be expressed through the Euler–Lagrange equation

$$\frac{\delta \Omega[\rho]}{\delta \rho(\boldsymbol{r})} \equiv \frac{\delta \mathcal{F}[\rho]}{\delta \rho(\boldsymbol{r})} + V(\boldsymbol{r}) - \mu = 0 , \tag{7.2}$$

in terms of the functional derivative of $\mathcal{F}[\rho]$.

The explicit knowledge of the (exact or approximate) free energy DF would reduce the equilibrium statistical mechanics of inhomogeneous systems to a problem of functional minimization with respect to the one-particle distribution $\rho(\boldsymbol{r})$, and this would represent a huge simplification with respect to the direct evaluation of the grand partition function,

$$\Xi = \mathrm{e}^{-\beta\Omega_0} = \sum_N \frac{\mathrm{e}^{\beta\mu N}}{N!\Lambda^{3N}} \int \prod_{i=1}^{N} \mathrm{d}\boldsymbol{r}_i \mathrm{e}^{-\beta U_N} , \tag{7.3}$$

and its use to get the equilibrium density distribution

$$\rho(\boldsymbol{r}) = \left\langle \sum_i \delta(\boldsymbol{r}-\boldsymbol{r}_i) \right\rangle = \frac{1}{\Xi} \sum_N \frac{\mathrm{e}^{\beta\mu N}}{(N-1)!\Lambda^{3N}} \int \prod_{i=1}^{N} \mathrm{d}\boldsymbol{r}_i \mathrm{e}^{-\beta U_N} \delta(\boldsymbol{r}-\boldsymbol{r}_i) , \tag{7.4}$$

where $U_N(\boldsymbol{r}_1, \boldsymbol{r}_2, \ldots, \boldsymbol{r}_N) = \sum_{ij} u(\boldsymbol{r}_{ij}) + \sum_i V(\boldsymbol{r}_i)$ is the total potential energy of N particles, including both the molecular interactions $u(\boldsymbol{r})$ and any external potential $V(\boldsymbol{r})$. The thermal wavelength Λ includes in (7.3) and (7.4) the contribution from the momentum integrals, which for classical particles is fully factorized from those over the positions. Since changing the value of Λ (i.e. changing the particle mass) produces only a trivial shift of the chemical potential and it does not affect any thermodynamic or structural property, we follow here the usual choice $\Lambda = 1$ to waive it out.

7.2.1 The Ideal Gas and the Excess Free Energy Density Functional

The classical ideal (non-interacting) gas, $u(\boldsymbol{r}_{ij}) = 0$, provides the simplest pedagogical example of the DF formalism, since the exact sampling of the atomic configurations in (7.3) may be readily computed to get

$$\Xi_{\mathrm{id}} = \sum_{N=0}^{\infty} \frac{(\mathrm{e}^{\beta\mu} Z_1)^N}{N!} = \exp(Z_1 e^{\beta\mu}) , \tag{7.5}$$

in terms of the one-particle partition function

$$Z_1 = \int \mathrm{d}\boldsymbol{r} e^{-\beta V(\boldsymbol{r})} . \tag{7.6}$$

Therefore, the equilibrium grand potential energy is

$$\Omega_{\mathrm{id}}[V] \equiv -\frac{1}{\beta} \ln(\Xi_{\mathrm{id}}) = -\frac{\mathrm{e}^{\beta\mu}}{\beta} \int \mathrm{d}\boldsymbol{r} \mathrm{e}^{-\beta V(\boldsymbol{r})} , \tag{7.7}$$

and the equilibrium density distribution follows the simple form

$$\rho_{\mathrm{id}}(\boldsymbol{r}) = \mathrm{e}^{\beta(\mu - V(\boldsymbol{r}))} . \tag{7.8}$$

These results may be inverted to get, via equation (7.1), the explicit functional form of the ideal gas intrinsic free energy,

$$\beta \mathcal{F}_{\rm id}[\rho] = \int d\boldsymbol{r} \rho(\boldsymbol{r}) \big[\ln \rho(\boldsymbol{r}) - 1 \big] \ . \tag{7.9}$$

The functional derivative of $\mathcal{F}_{\rm id}[\rho]$ is

$$\frac{\delta \mathcal{F}_{\rm id}[\rho]}{\delta \rho(\boldsymbol{r})} = \frac{1}{\beta} \ln \rho(\boldsymbol{r}) \ , \tag{7.10}$$

so that the Euler–Lagrange equation (7.2) leads directly to the equilibrium density (7.8) and its substitution in (7.9) and (7.1) gives back the equilibrium grand potential energy (7.7), which reduces to $\beta\Omega_{\rm id} = -\int d\boldsymbol{r}\rho(\boldsymbol{r}) = -\langle N \rangle$.

The free energy DF of hard core systems is purely entropic, and it may be split in the ideal gas contribution plus an *excess* free energy, $\mathcal{F}_{\rm ex}[\rho]$, to account for the entropy reduction due to the non-overlap of the molecular cores. Hence, it is usually written in $k_{\rm B}T$ units as

$$\beta \mathcal{F}[\rho] = \beta \mathcal{F}_{\rm id}[\rho] + \beta \mathcal{F}_{\rm ex}[\rho] \equiv \int d\boldsymbol{r} \left\{ \Phi_{\rm id}\big(\rho(\boldsymbol{r})\big) + \Phi\big([\rho];\boldsymbol{r}\big) \right\} \ , \tag{7.11}$$

with the ideal free energy density as a function, $\Phi_{\rm id}(\rho) = \rho(\ln \rho - 1)$, evaluated at the local density, while the excess contribution is expressed as the volume integral of an excess free energy density $\Phi\big([\rho];\boldsymbol{r}\big)$, which is a function of $\boldsymbol{r}$ and a functional of $\rho(\boldsymbol{r})$. Notice that there could be multiple choices of $\Phi\big([\rho];\boldsymbol{r}\big)$ leading to the same $\mathcal{F}_{\rm ex}[\rho]$, since there are many ways to separate the total free energy excess of the system in terms of local contributions, and we will see that different approaches to $\mathcal{F}_{\rm ex}[\rho]$ may give similar results with very different $\Phi\big([\rho];\boldsymbol{r}\big)$.

The development and use of approximations for $\Phi\big([\rho];\boldsymbol{r}\big)$ is the goal of the DF formalism [2, 3]. The main ingredients to build DF approximations are the results for uniform systems of HS and other hard-core fluids presented in the accompanying chapters of this book. Within the DF formalism, the properties of bulk uniform fluids, with homogeneous density distributions $\rho(\boldsymbol{r}) = \rho_0$, are associated to the case of null external potential, $V(\boldsymbol{r}) = 0$, so that the spatial symmetry is recovered in (7.4) from the translational invariance of U_N. From (7.2), we get that the functional derivative of $\mathcal{F}[\rho]$, evaluated at $\rho(\boldsymbol{r}) = \rho_0$, has to be constant and equal to the equilibrium chemical potential of the system, while $\Omega[\rho_0]$ and $\mathcal{F}[\rho_0]$ in (7.1) become the extensive thermodynamic potentials Ω_0 and F_0, proportional to the total volume of the system.

7.2.2 The Correlation Structure of Uniform Systems

A most interesting result of the DF formalism [1] is the relation between $\mathcal{F}_{\rm ex}[\rho]$ and the direct correlation function, $c(r, \rho_0)$, of a bulk fluid. This function was originally defined through the Ornstein–Zernike equation,

$$h(r,\rho_0) = c(r,\rho_0) + \rho_0 \int \mathrm{d}\boldsymbol{r}' c(r',\rho_0) h(|\boldsymbol{r}-\boldsymbol{r}'|,\rho_0) \ , \tag{7.12}$$

in terms of the total correlation function $h(r,\rho_0) \equiv g(r,\rho_0) - 1$. For any system with classical statistics, there is an exact DF relation,

$$\left. \frac{\delta^2 \beta \mathcal{F}_{\mathrm{ex}}[\rho]}{\delta\rho(\boldsymbol{r}_1)\delta\rho(\boldsymbol{r}_2)} \right|_{\rho_0} = -c(r_{12},\rho_0) \ , \tag{7.13}$$

with the second functional derivative of $\mathcal{F}_{\mathrm{ex}}[\rho]$ evaluated at $\rho(\boldsymbol{r}) = \rho_0$.

The functional Taylor expansion of $\mathcal{F}_{\mathrm{ex}}[\rho]$ in terms of the density difference, $\Delta\rho(\boldsymbol{r}) = \rho(\boldsymbol{r}) - \rho_0$, with respect to a uniform bulk reference system, takes the generic form

$$\begin{aligned} \beta\mathcal{F}_{\mathrm{ex}}[\rho] =& \beta F_{\mathrm{ex}}(\rho_0) + \beta\mu_{\mathrm{ex}}(\rho_0) \int \mathrm{d}\boldsymbol{r} \Delta\rho(\boldsymbol{r}) - \\ & - \frac{1}{2} \int \mathrm{d}\boldsymbol{r}\mathrm{d}\boldsymbol{r}' c(|\boldsymbol{r}-\boldsymbol{r}'|,\rho_0) \Delta\rho(\boldsymbol{r}) \Delta\rho(\boldsymbol{r}') + \mathcal{O}^3(\Delta\rho(\boldsymbol{r})), \end{aligned} \tag{7.14}$$

where the excess free energy, $F_{\mathrm{ex}}(\rho_0)$, and chemical potential $\mu_{\mathrm{ex}}(\rho_0)$, of the bulk liquid are directly obtained from its equation of state, while $c(r,\rho_0)$ may be obtained from the bulk correlation structure. The excellent approximations for these bulk quantities, reviewed in the accompanying chapters of this book, may then give a good approximation to $\beta\mathcal{F}_{\mathrm{ex}}[\rho]$ for systems with weak variations of the density. The drawback of the truncated functional Taylor expansion is that for density distributions having a large constant plateau at a value different from ρ_0, the local thermodynamics at the plateau is poorly represented beyond the quadratic dependence of $F_{\mathrm{ex}}(\rho)$ around $\rho = \rho_0$.

From the opposite point of view, we may take any given DF approximation for $\mathcal{F}_{\mathrm{ex}}[\rho]$ and use it to predict in two different ways the correlation structure of any bulk liquid. The *OZ route* gets $c(r,\rho_0)$ from (7.13) and then $h(r,\rho_0)$ from (7.12), while for systems with pairwise interactions, the *test-particle route* uses the pair potential between two molecules $u(r)$ as an "external" potential acting on a bulk fluid from a particle pinned at the origin. The solution of the Euler–Lagrange equation (7.2) would then give the density distribution of the remaining molecules, which is interpreted as $\rho(r) \equiv \rho_0 [1 + h(r,\rho)]$ in terms of the total correlation in the bulk fluid. The exact DF form of $\mathcal{F}[\rho]$, for a given molecular interaction, should give the same result for $h(r,\rho_0)$ along the two routes, but the use of any DF approximation would lead to inconsistent predictions. In particular, for hard-core interactions, the test-particle route would always fulfill the exact requirement $h(r,\rho_0) = -1$ inside the core, while the compressibility route would (in general) fail to satisfy that condition. On the other hand, the sum rule on the total correlation function in the grand-canonical ensemble [5]

$$\rho_0 \int \mathrm{d}\boldsymbol{r} h(r,\rho_0) = -1 + \rho_0 k_{\mathrm{B}} T \chi_T \ , \tag{7.15}$$

in terms of the bulk liquid isothermal compressibility $\chi_T \equiv \rho_0^{-1} (\partial \rho_0/\partial p)_T$ (where p stands for pressure), is always satisfied by the compressibility route to $h(r, \rho_0)$, but not by the test-particle one with DF approximations for $\mathcal{F}[\rho]$. In general, the degree of inconsistency between the compressibility and the test-particle routes to $h(r, \rho_0)$ gives a quality test for any DF approximation.

7.2.3 Density Functional Virial Expansion of the Excess Free Energy

The usual virial expansion for the thermodynamics, and correlation structure, of bulk fluids may be extended to get the generic density expansion of $\mathcal{F}_{\text{ex}}[\rho]$ in systems with pairwise interactions,

$$\begin{aligned}\beta\mathcal{F}_{\text{ex}}[\rho] =& \frac{1}{2}\int \mathrm{d}\boldsymbol{r}_1\rho(\boldsymbol{r}_1)\int \mathrm{d}\boldsymbol{r}_2\rho(\boldsymbol{r}_2)f(r_{12})+\\ &+\frac{1}{6}\int \mathrm{d}\boldsymbol{r}_1\rho(\boldsymbol{r}_1)\int \mathrm{d}\boldsymbol{r}_2\rho(\boldsymbol{r}_2)\int \mathrm{d}\boldsymbol{r}_3\rho(\boldsymbol{r}_3)f(r_{12})f(r_{23})f(r_{31})+\cdots,\end{aligned} \tag{7.16}$$

where $f(\boldsymbol{r}) = 1 - \exp(-\beta u(\boldsymbol{r}))$ is the Mayer function,[1] with the pairwise potential $u(\boldsymbol{r})$. The expansion for the direct correlation function in a system with uniform density follows from (7.13) and (7.16),

$$-c(r_{12}, \rho_0) = f(r_{12}) + \rho_0 f(r_{12})\int \mathrm{d}\boldsymbol{r}_3 f(r_{23})f(r_{31}) + \cdots . \tag{7.17}$$

As expected, the lowest order of the excess free energy DF produced by the interactions between the particles is quadratic in $\rho(\boldsymbol{r})$, while its second functional derivative has a zero density limit $c(r, 0) = -f(r)$. These exact low-density results for $\beta\mathcal{F}_{\text{ex}}[\rho]$ provide useful hints for the development of DF approximations.

7.2.4 The Local Density Approximation and the Gradient Expansion

The local density approximation (LDA) is the simplest scheme to build a DF approximation for $\beta\mathcal{F}_{\text{ex}}[\rho]$ from the restricted knowledge of the thermodynamics of uniform systems. This DF form approximates the local excess free energy density in (7.11) as a function of the local density, $\Phi_{\text{LDA}}([\rho]; \boldsymbol{r}) = \Phi(\rho(\boldsymbol{r}))$, where $\Phi(\rho_0)$ is the excess free energy per unit volume in a system with uniform density ρ_0, directly accessible from its equation of state. The ideal gas free energy DF does have exactly the local density form (7.9), but the generic

[1] Notice the different sign of this definition with respect to the standard one.

form of the excess $\Phi([\rho]; \boldsymbol{r})$ depends on the density distribution at different positions, as already observed in the first term of the virial expansion (7.16), so that the LDA can only give acceptable results if $\rho(\boldsymbol{r})$ changes very smoothly over molecular size distances. The LDA description of the HS excess free energy may also be regarded as a *coarse grained* description of $\rho(\boldsymbol{r})$, when we are interested in its long-ranged aspects rather than in the molecular layering structure [6].

For systems with attractive interactions, there is a systematic, although limited, way to improve the LDA in terms of density gradient expansions [1], assuming that the generic functional $\Phi([\rho]; \boldsymbol{r})$ is not just a local function of $\rho(\boldsymbol{r})$ but also of its local gradient $\nabla\rho(\boldsymbol{r})$. van der Waals theory for the structure of the liquid surface may be considered as the precursor of that DF approximation [7], well before the DF formalism was established. However, the approach fails from the beginning when applied to systems with hard-core interactions. The non-local dependence of $\Phi([\rho]; \boldsymbol{r})$ implies a sharp finite range of interference between $\rho(\boldsymbol{r})$ and $\rho(\boldsymbol{r}')$, associated to the geometry of the molecule, and this cannot be represented as a simple expansion in terms of the local density gradient.

7.2.5 The Mean Field Approximation

For systems with very soft molecular potentials, the particles may be assumed to be fully uncorrelated, as in the ideal gas, so that $\mathcal{F}_{\mathrm{ex}}[\rho]$ has got no entropic contribution and is given by the interaction energy

$$\mathcal{F}_{\mathrm{ex}}[\rho] = \frac{1}{2} \int \mathrm{d}\boldsymbol{r} \mathrm{d}\boldsymbol{r}' \rho(\boldsymbol{r}) \rho(\boldsymbol{r}') u(\boldsymbol{r} - \boldsymbol{r}') . \tag{7.18}$$

This is the mean field approximation (MFA) DF [2], since the functional derivative in (7.2) takes the form

$$\frac{\delta \mathcal{F}_{\mathrm{ex}}[\rho]}{\delta \rho(\boldsymbol{r})} = \int \mathrm{d}\boldsymbol{r}' \rho(\boldsymbol{r}') u(\boldsymbol{r} - \boldsymbol{r}') , \tag{7.19}$$

and it is interpreted as the potential created on a particle at $\boldsymbol{r}$ by the (uncorrelated) mean distribution of particles over the whole system. The MFA may give an accurate description of the full excess free energy for ultra-soft interactions, like those between polymer chains in bad solvents [8], but it is obviously out of question for the description of hard-core interactions with an infinite $u(r)$ inside the molecular cores. The relevance of the MFA in this book comes from the usual treatment of simple liquids to split the interaction potential in a repulsive core, $u_{\mathrm{rep}}(r)$, often described as a reference HS with temperature-dependent diameter and the soft attractive part $u_{\mathrm{att}}(r)$, which may be included through the MFA [2]. Such a simple addition of the molecular packing and the attractive interaction effects is sometimes referred to as a generalized van der Waals approximation, and it is the simplest and very successful approach to the DF of realistic model interactions.

7.3 The Exact Density Functional for Hard Rods

In 1976, Percus [4] presented the exact intrinsic free energy DF for a 1D system of HR. Within the 1D version of (7.11), the excess free energy density for rods of length σ is given by

$$\Phi^{(1\mathrm{D})}([\rho];x) = -\rho(x)\ln\left(1-\int_{x}^{x+\sigma}\mathrm{d}x'\rho(x')\right) . \tag{7.20}$$

The asymmetry of this expression, with the integral running only for values of x' larger than the local position variable, x, comes from the transfer matrix procedure employed to compute it, in which the entropy associated to the insertion of each particle is calculated by fixing the previously inserted particles on its right. Exactly equivalent expressions may be obtained with the opposite asymmetry and with the symmetric combination

$$\Phi^{(1\mathrm{D})}([\rho];x) = -\frac{\rho(x+\sigma/2)+\rho(x-\sigma/2)}{2}\ln\left[1-\eta(x)\right] , \tag{7.21}$$

where the *local packing fraction*

$$\eta(x) = \int_{-\frac{\sigma}{2}}^{\frac{\sigma}{2}}\mathrm{d}x'\rho(x+x') , \tag{7.22}$$

represents the probability that the point x is covered by a HR.

The explicit DF form of $\beta\mathcal{F}_{\mathrm{ex}}[\rho]$ for HR provides an excellent example to analyze the practical use of the DF formalism and also a most valuable hint to guess the free energy DF approximations for hard bodies in higher dimensions. The minimum principle (7.2), in the presence of any external potential $V(x)$, gives

$$\ln\rho(x) + \frac{\delta\beta\mathcal{F}_{\mathrm{ex}}[\rho]}{\delta\rho(x)} + \beta V(x) - \beta\mu = 0 , \tag{7.23}$$

with *non-local* dependence of the excess free energy functional derivative

$$\frac{\delta\beta\mathcal{F}_{\mathrm{ex}}[\rho]}{\delta\rho(x)} = -\ln\left[1-\eta\left(x+\frac{\sigma}{2}\right)\right] - \int_{x-\sigma}^{x}\mathrm{d}x'\frac{\rho(x')}{1-\eta\left(x'+\frac{\sigma}{2}\right)} , \tag{7.24}$$

so that $\rho(x)$ cannot be obtained by solving (7.23) independently for each x, as in the ideal gas case (7.8). The interactions couple the values of $\rho(x)$ at positions within a rod length σ, and (7.23) becomes an integral equation to be simultaneously solved for the whole $\rho(x)$, rather than separately for each x. Qualitatively, this integral equation is similar to those described below for DF approximations for HS and HD, and several numerical techniques have been used either to solve it or equivalently to perform the numerical minimization of $\Omega[\rho]$, as described in the next sections.

The exact direct correlation function of a uniform distribution of HR of length σ and homogeneous 1D density ρ_0 is directly obtained from the second functional derivative of (7.20) or (7.21),

$$c(|x-x'|,\rho_0) = -\left.\frac{\delta^2\beta\mathcal{F}_{\rm ex}[\rho]}{\delta\rho(x)\delta\rho(x')}\right|_{\rho(x)=\rho_0} = -\Theta(\sigma-|x-x'|)\frac{1-\rho_0|x-x'|}{(1-\eta_0)^2}\,, \quad (7.25)$$

where $\Theta(y)$ is the Heaviside step function ($=1$ if $y>0$ and $=0$ if $y<0$), which makes $c(|x-x|,\rho_0)=0$ beyond the contact distance $|x-x'|=\sigma$, and $\eta_0=\sigma\rho_0$ is the packing fraction of the fluid. Notice that the Mayer function for 1D HR is precisely $f(x)=\Theta(\sigma-|x|)$, so that, as it should be, (7.25) is consistent with the generic density expansion (7.17).

The exact total correlation function $h(x,\rho_0)$ in a system with uniform density ρ_0 may be consistently obtained from $c(x,\rho_0)$, through the 1D version of the Ornstein–Zernike equation (7.12), or by the test-particle route, solving (7.24) with $V(x)=0$ for $|x|>\sigma$ and $V(x)=\infty$ for $|x|<\sigma$. The result may be expressed as the infinite sum [9]

$$\begin{aligned} h(x,\rho_0) =& 1+\sum_{k=0}^{\infty}\Theta(|x|-(k+1)\sigma)\frac{\rho_0^k(|x|-(k+1)\sigma)^k}{k!(1-\eta_0)^{k+1}} \\ &\times\exp\left\{-\frac{\rho_0(|x|-(k+1)\sigma)}{1-\eta_0}\right\}, \end{aligned} \quad (7.26)$$

with each term in the sum representing the contribution from the successive neighbors of the particle centered at $x=0$ (which in a 1D system with hard core interactions are forced to keep the order along the line). As in other problems in statistical physics, this 1D ordering is the key element to get the exact functional form of $\beta\mathcal{F}_{\rm ex}[\rho]$, and it is clearly absent in higher dimensions. Early attempts [10, 11] to extend the DF form (7.21) to approximate the free energy DF of 3D HS were not very successful, and the hints offered by 1D system were of little use for the design of DF approximations in two and three dimensions, until the advent of the most recent family of DF approximations based on Rosenfeld's FMT [12, 13] (that will be introduced in Sect. 7.5).

7.4 The Weighted Density Approximation

In 1980, Nordholm et al. [14] presented a theory for the free energy of an inhomogeneous HS fluid which inspired a whole family of DF approximations. The most successful members of that family had come to share the name of

WDA, originally applied in the context of DF theory for the exchange and correlation energy of electronic systems[2] [15].

The common feature of the WDAs for classical fluids is to approximate the local free energy DF by $\Phi_{\rm WDA}[\rho, \boldsymbol{r}] = \rho(\boldsymbol{r})\psi(\bar{\rho}(\boldsymbol{r}))$, where $\psi(\rho_0) \equiv \Phi(\rho_0)/\rho_0$ is the excess free energy per particle of a bulk system with uniform density ρ_0 and $\bar{\rho}(\boldsymbol{r})$ is a *weighted density* to represent a sampling of the density around the point $\boldsymbol{r}$ through the convolution

$$\bar{\rho}(\boldsymbol{r}) = \int \mathrm{d}\boldsymbol{r}' \rho(\boldsymbol{r} + \boldsymbol{r}') w(r', \bar{\rho}(\boldsymbol{r})) \,, \tag{7.27}$$

with a normalized *weight function*, $w(r, \rho)$, which may be assumed to depend on the local value of $\bar{\rho}(\boldsymbol{r})$. The choice of that weight function is the key element of the WDA because it determines the non-local dependence of $\mathcal{F}_{\rm ex}[\rho]$. Notice that the LDA is recovered with a delta-function weight, $w(r) = \delta(\boldsymbol{r})$. The MFA may also be cast into the WDA form by taking the weight function $w(r) = u(r)/u_0$, with the pair potential normalized by its total volume integral, u_0, and the linear function $\psi(\rho_0) = u_0\rho_0/2$ for the excess free energy per particle. The use of non-local weights and non-linear functions $\psi(\rho_0)$ made the WDA a very successful DF scheme for HS and other hard-core particles.

Nordholm et al. [14] proposed to use the HS Mayer function to account for that non-local dependence, together with the simplest excluded volume approximation for $\psi(\rho) = -\ln(1 - 2\pi\sigma^3\rho/3)$. The use of more accurate equations of state and the exploration of other analytical forms for $w(r)$ paved the road [16] to the crucial element of the WDA: the (density dependent) weight function $w(r, \rho_0)$ may be tailored, through the relationship (7.13), to reproduce any given approximation for the direct correlation function of the bulk liquid, so that DF approximations for $\mathcal{F}_{\rm ex}[\rho]$ may be systematically built, using as ingredients the equation of state and the correlation structure of the HS liquid. The resulting functionals are much more effective than the plain functional Taylor expansion (7.16) built with the same ingredients. The price to pay with respect to the earlier density-independent weight functions is that (7.27) becomes an equation for $\bar{\rho}(\boldsymbol{r})$ which has to be solved at every point $\boldsymbol{r}$ and for every density distribution $\rho(\boldsymbol{r})$ arising along the numerical procedure to minimize $\Omega[\rho]$. In practice, this difficulty may be temperated assuming a simplified polynomial dependence [17]

$$w(r, \rho) = w_0(r) + \rho w_1(r) + \rho^2 w_2(r) + \cdots \,, \tag{7.28}$$

so that the evaluation of $\bar{\rho}(\boldsymbol{r})$ for a given $\rho(\boldsymbol{r})$ requires a few plain convolutions to get the weighted density components

[2] Notice that two different non-local DF schemes were proposed in this work, one under the name of averaged density approximation (ADA) and the other under the name of WDA; but what came to be called as WDA in the context of classical fluids corresponds to the ADA in electronic DF theory. The name ADA was used during some time to refer to the DF approximation in [17].

$$\bar{\rho}_\nu(\boldsymbol{r}) = \int \mathrm{d}\boldsymbol{r}' \rho(\boldsymbol{r}+\boldsymbol{r}') w_\nu(r'), \qquad \nu = 0, 1, 2, \dots , \tag{7.29}$$

and the solution of the algebraic equation

$$\bar{\rho}(\boldsymbol{r}) = \bar{\rho}_0(\boldsymbol{r}) + \bar{\rho}_1(\boldsymbol{r})\bar{\rho}(\boldsymbol{r}) + \bar{\rho}_2(\boldsymbol{r})\bar{\rho}(\boldsymbol{r})^2 + \cdots . \tag{7.30}$$

The normalization of $w(r,\rho)$, imposed so that $\bar{\rho}$ equals the bulk density in a homogeneous system, implies

$$\int \mathrm{d}\boldsymbol{r}\, w_0(\boldsymbol{r}) = 1, \qquad \int \mathrm{d}\boldsymbol{r}\, w_\nu(\boldsymbol{r}) = 0, \quad \nu > 0. \tag{7.31}$$

The form of the first elementary weight functions, $w_\nu(r)$, may be obtained to reproduce the exact first two terms of the exact density expansions (7.16) and (7.17), which are also exactly given by the Percus–Yevick (PY) approximation for the bulk liquid $c_{\mathrm{PY}}(r,\rho_0)$. The exact zero-order weight function is the normalized step function

$$w_0(r) = \frac{3}{4\pi\sigma^3}\Theta(\sigma - r) \tag{7.32}$$

employed in the preliminary explorations of the approach [14, 16], while the exact $w_1(r)$ is the solution to the integral equation

$$\begin{aligned} &\frac{10\pi}{3}\sigma^3 w_0(r) + 8w_1(r) + \frac{5\pi}{3}\sigma^3 \int \mathrm{d}\boldsymbol{r}' w_0(r') w_0(|\boldsymbol{r}+\boldsymbol{r}'|) \\ &\qquad + 8\int \mathrm{d}\boldsymbol{r}' w_0(r') w_1(|\boldsymbol{r}+\boldsymbol{r}'|) = \left[8 - 6\frac{r}{\sigma} + \frac{1}{2}\left(\frac{r}{\sigma}\right)^3\right]\Theta(\sigma - r). \end{aligned} \tag{7.33}$$

This equation is solved in Fourier space as

$$\hat{w}_1(k) = \frac{\pi}{6}\sigma^3 \frac{\hat{f}(k) - 20\hat{w}_0(k) - 10\hat{w}_0(k)^2}{8\left[1+\hat{w}_0(k)\right]} , \tag{7.34}$$

where, using the notation $q = k\sigma$, we have

$$\hat{w}_0(k) = \frac{3}{q^3}(\sin q - q\cos q), \tag{7.35}$$

$$\hat{f}(k) = \frac{288}{q^6}\left[1 + q^2 - \left(1 + \frac{q^2}{2} + \frac{5q^4}{24}\right)\cos q - q\left(1 + \frac{q^2}{6}\right)\sin q\right] . \tag{7.36}$$

Hence, the function $w_1(r)$ is obtained by the inverse Fourier transform,

$$w_1(r) = \frac{\sigma^3}{12\pi r}\int_0^\infty \mathrm{d}k\; k\sin(kr)\frac{\hat{f}(k) - 20\hat{w}_0(k) - 10\hat{w}_0(k)^2}{8\left[1+\hat{w}_0(k)\right]} , \tag{7.37}$$

which may be integrated numerically to get the result represented in Fig. 7.1. The weight function $w_1(r)$ has a core region, for $r \le \sigma$, with negative radial

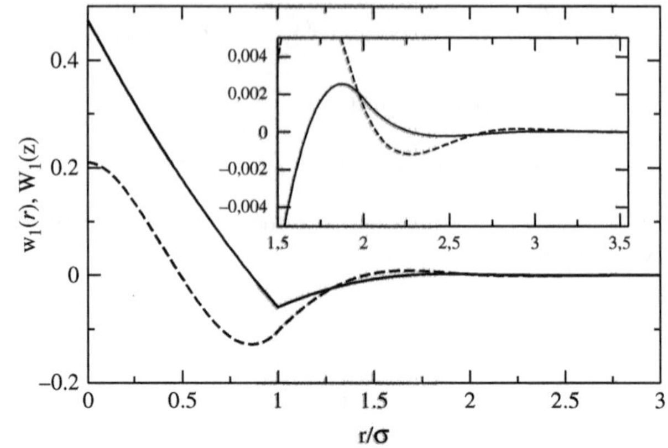

Fig. 7.1. The *full line* gives the radial weight function $w_1(r)$ and the *broken line* the function $W_1(z)$, defined in Eq. (7.42), to be used in problems with planar symmetry. The inset shows the oscillating tails of both functions

slope and an oscillating tail for $r > \sigma$. The shape of $w_1(r)$ already indicates that, in order to reproduce the direct correlation function of a dense HS liquid, the WDA representation of $\mathcal{F}_{\mathrm{ex}}[\rho]$ requires a delicate choice of the weighting function.

The WDA proposed under this name by Curtin and Ashcroft in 1985 [18] makes formally consistent use of the PY approximation both for the thermodynamics, $\psi_{\mathrm{PY}}(\rho)$, and for the direct correlation function $c_{\mathrm{PY}}(r,\rho)$; here we will refer to it as PY-WDA. The slightly earlier version, proposed by Tarazona [17], uses (7.29) with the exact forms for $w_0(r)$ and $w_1(r)$ but truncates the expansion at second order with a purely empirical weight [19]

$$w_2(r) = \frac{5\pi\sigma^3}{24}\left[1 - 2\frac{r}{\sigma} + \frac{5}{6}\left(\frac{r}{\sigma}\right)^2\right]\Theta(\sigma - r) \tag{7.38}$$

to obtain a global good fit to $c_{\mathrm{PY}}(r,\rho_0)$ over the entire range of liquid densities. This approximation is in the same spirit as the one that leads to the semi-empirical Carnahan–Starling (CS) equation of state, which does not improve on the PY result for the virial expansion but produces *quasi-exact* results for the pressure of the HS fluid at any density (see Chap. 3). The CS equation of state was used by Tarazona to fix

$$\psi_{\mathrm{CS}}(\rho) = \frac{4\eta - 3\eta^2}{(1-\eta)^2}\,, \tag{7.39}$$

with the packing fraction $\eta = \pi\sigma^3\rho/6$, so that we will refer to that DF approach as CS-WDA. From a practical point of view, the CS-WDA truncation of (7.29) at second order provides a simple analytical solution for (7.27) in terms of the three direct convolutions of the density distribution, $\bar{\rho}_0(r)$, $\bar{\rho}_1(r)$

and $\bar{\rho}_2(r)$. The applications of the PY-WDA may also be carried out with some simplified prescription for $w(r, \rho)$, similar to (7.28) and (7.29). Both versions require, in practice, a truncation of the range of $w(r, \rho)$ to a few molecular diameters.

Notice that in any WDA version there are density distributions for which the non-linear equation (7.27) has none or more than one solution. In the CS-WDA, the lack of solutions corresponds to having a negative discriminant, $(1 - \rho_1(\boldsymbol{r}))^2 - 4\rho_0(\boldsymbol{r})\rho_2(\boldsymbol{r}) < 0$, while for a positive discriminant we have to select the appropriate root. In the PY-WDA, the multiplicity or lack of solutions cannot be controlled in an analytical way, but they can still appear along the numerical minimization. It was implicitly assumed that density profiles leading to no positive solution for (7.27) were *unphysical*, and they may be discarded within the minimization of $\Omega[\rho]$. Whenever (7.27) had more than one positive solution, the *physical* one is that closest to $\bar{\rho}_0(\boldsymbol{r})$. In the CS-WDA, this corresponds to

$$\bar{\rho}(r) = \frac{2\bar{\rho}_0(r)}{1 - \bar{\rho}_1(r) + \sqrt{\left[1 - \bar{\rho}_1(r)\right]^2 - 4\bar{\rho}_0(r)\bar{\rho}_2(r)}} . \tag{7.40}$$

Both the PY-WDA and the CS-WDA have been extensively applied in the last twenty years to many problems concerning inhomogeneous distributions of HS. Several variants, like splitting $\psi(\bar{\rho})$ in several components, each with a different elementary weight function, have been explored, some of them with similarly good results but with no significant improvement over the whole range of analyzed problems. Therefore, the two 1985 versions have become the standard forms of the WDA. The remaining part of this section presents the application of the CS-WDA to two problems of interest: the density profile $\rho(z)$ of a HS fluid against a flat wall potential $V(z)$ and the description of the HS crystal as a self-stabilized inhomogeneity of the HS fluid. In general, the results with the PY-WDA are fairly close to those of the CS-WDA, so we will only point out their more important differences.

7.4.1 Hard Sphere Fluid Against a Planar Wall

In the absence of any kind of symmetry breaking effect, the density distribution of a HS fluid near a planar wall potential $V(z)$ should depend only on the distance to the wall, which we take as the z coordinate. The DF minimization of $\Omega[\rho]$ in (7.1) is then restricted to functions $\rho(\boldsymbol{r}) = \rho(z)$, so that the numerical problem of minimization is hugely reduced with respect to that of a 3D dependence of $\rho(x, y, z)$. The weighted density inherits the same symmetry, and the elementary weighted densities may be calculated as 1D convolutions

$$\bar{\rho}_\nu(z) = \int \mathrm{d}z' \rho(z + z') W_\nu(z') , \tag{7.41}$$

where the functions

$$W_\nu(z) \equiv \int\limits_{-\infty}^{\infty}\int\limits_{-\infty}^{\infty} \mathrm{d}x\mathrm{d}y\; w_\nu\left(\sqrt{x^2+y^2+z^2}\right) = 2\pi \int\limits_{|z|}^{\infty} \mathrm{d}r\; r\; w_\nu(r) \qquad (7.42)$$

can be analytically obtained for $\nu = 0$,

$$W_0(z) = \frac{3}{4\sigma^3}(\sigma^2 - z^2)\Theta(\sigma - |z|)\;, \qquad (7.43)$$

and $\nu = 2$ in (7.38),

$$W_2(z) = \frac{5\pi^2\sigma^5}{288}\left[1 - 12\left(\frac{z}{\sigma}\right)^2 + 16\left(\frac{|z|}{\sigma}\right)^3 - 5\left(\frac{z}{\sigma}\right)^4\right]\Theta(\sigma - |z|), \qquad (7.44)$$

while $W_1(z)$ may be integrated numerically from (7.37) to get the function represented in Fig. 7.1. For practical purposes, $W_1(z)$ is tabulated over the discrete mesh of z used to describe the density profile $\rho(z)$ and usually truncated at $|z| = 3.55\sigma$. The total weighted density $\bar{\rho}(z)$ in the CS-WDA version is then readily obtained from (7.40), and the free energy density $\Phi_{\mathrm{WDA}}([\rho]; z) = \rho(z)\psi(\bar{\rho}(z))$ plus the ideal gas contribution is integrated along the z axis to get the grand potential energy per unit area

$$\frac{\beta\Omega[\rho]}{A} = \int \mathrm{d}z\,\rho(z)\left[\ln\rho(z) - 1 + \psi(\bar{\rho}(z)) + \beta V(z) - \beta\mu\right]\;. \qquad (7.45)$$

This integral should cover all the region of relevance, where $\rho(z)$ is non-zero, up to the bulk region where $\Phi[\rho; z] = \Phi(\rho_0)$ is constant. For dense HS fluids near sharp walls, the region of variable $\rho(z)$ to be used in the minimization of (7.45) has to cover $10 - 20\sigma$ beyond the point where $V(z)$ vanishes. In determining an appropriate integration interval for (7.45), one has to bear in mind that the region where $\bar{\rho}(z) \neq \rho_0$ exceeds the one where $\rho(z) \neq \rho_0$ by the maximum distance for which $W_1(z) \neq 0$. The bulk fluid contribution to $\Omega[\rho]$ is $\Omega_0 = -pV$, in terms of the bulk pressure $p(\rho_0)$. A βp term may be added to the integral in (7.45) so that the integrand vanishes away from the wall and the integral directly gives $\Delta\Omega[\rho]/A = (\Omega[\rho] + pV)/A = \gamma$, i.e. the surface tension of the wall–fluid interface.

The Euler–Lagrange equation (7.2) takes, in the WDA, the form

$$\ln\rho(z) + \psi(\bar{\rho}(z)) + \int \mathrm{d}z'\rho(z')\psi'(\bar{\rho}(z'))\frac{\delta\bar{\rho}(z')}{\delta\rho(z)} + \beta V(z) = \beta\mu\;, \qquad (7.46)$$

where $\psi'(\rho) \equiv d\psi(\rho)/d\rho$, and from (7.27),

$$\frac{\delta\bar{\rho}(z')}{\delta\rho(z)} = \frac{W(|z'-z|, \bar{\rho}(z'))}{1 - \int \mathrm{d}z''\rho(z'')W'(|z'-z''|, \bar{\rho}(z'))}\;, \qquad (7.47)$$

and $W'(z,\rho) = \partial W(r,\rho)/\partial\rho$. Within the CS-WDA, we get a simple expression for this functional derivative

$$\frac{\delta\bar{\rho}(z')}{\delta\rho(z)} = \frac{W_0(|z'-z|) + \bar{\rho}(z')W_1(|z'-z|) + \bar{\rho}(z')^2 W_2(|z'-z|)}{1 - \bar{\rho}_1(z') - 2\bar{\rho}(z')\bar{\rho}_2(z')} , \qquad (7.48)$$

in terms of the same elementary weighted densities which define $\bar{\rho}(z)$. The bulk chemical potential is $\beta\mu = \ln\rho_0 + \psi(\rho_0) + \rho_0\psi'(\rho_0)$, so (7.46) is trivially solved within the bulk liquid with $\rho(z) = \rho_0$.

The numerical solution of (7.46) is usually obtained with a discretized description of the density profile, $\rho_i = \rho(z_i)$, along a regular mesh $\{z_i\}$ over the z axis. If there are N_1 mesh points in a length σ, the numerical convolutions to obtain $\bar{\rho}_\nu(z)$ (computed with a fast Fourier algorithm [20]) requires a computational cost proportional to $N_1 \ln N_1$, and the evaluation of the integral in (7.46) requires similar convolutions. This should be compared with the computational cost to evaluate the exact expression (7.16) for $\mathcal{F}_{\text{ex}}[\rho]$. The first term of this expression may be exactly cast into the WDA form, $\Phi_{\text{WDA}}[\rho] = \rho(\boldsymbol{r})\psi(\bar{\rho}(\boldsymbol{r}))$, with $\psi(\rho) = 4\pi\rho\sigma^3/3 + \cdots$ and $w(r,\rho) = w_0(r) + \cdots$, so that the required number of operations goes also like $N_1 \ln N_1$; but the second term contains a triple integral, which cannot be factorized in terms of a one-center convolution, to get $\bar{\rho}(\boldsymbol{r})$, and an external integral over $\rho(\boldsymbol{r})\psi(\bar{\rho}(\boldsymbol{r}))$. Therefore, the numerical effort to get the second term in (7.16) is proportional to $N_1^2 \ln N_1$. The exponent of N_1 systematically grows for higher order terms.

This is most relevant to understand the practical limitations in the design of DF approximations, since any application would require many evaluations of $\mathcal{F}_{\text{ex}}[\rho]$ and of its functional derivative to minimize $\Omega[\rho]$. It would be useless to have a prescription for $\mathcal{F}_{\text{ex}}[\rho]$, which requires an unfeasible computational cost, like e.g. the expansion (7.16) extended up to a higher order in $\rho(\boldsymbol{r})$. The successful forms of $\mathcal{F}_{\text{ex}}[\rho]$ are good compromises between the accuracy of the approximation and its computational cost. The free energy for the 1D HR system may be obtained from the computation of the infinite series (7.16), but the exact result (7.20) or (7.21) has a computational cost equivalent to just the first term of the series. Therefore, these DF forms for $\mathcal{F}_{\text{ex}}[\rho]$ are extremely useful rearrangements of (7.16), which achieve an enormous reduction of its computational cost. It is only because of the 1D character that this can be done exactly; the WDA for 3D HS represents an attempt to estimate the second and higher order terms in (7.16) by means of a one-center convolution with a weight function $w(r,\rho)$. The requirement that $\mathcal{F}_{\text{ex}}^{\text{WDA}}[\rho]$ reproduces the direct correlation function of a bulk liquid implies that $w(r,\rho)$ must be tailored to resum the infinite series (7.16) up to its second functional derivative evaluated at any $\rho(\boldsymbol{r}) = \rho_0$ and that rather stringent requirement can only be achieved at the price of using a careful tuning of the oscillating tail of $w(r,\rho)$.

The minimization of $\Omega[\rho]$ requires many times the computational cost of its evaluation for a single-density profile. Several numerical methods have been devised to minimize $\Omega[\rho]$ with the discretized description of $\rho(z)$, over the

$\{z_i, \rho_i\}$ mesh. A simple iterative scheme was used in the earliest applications, rewriting (7.46) as

$$\rho(z) = \rho_0 \exp\left(-\beta\left[\frac{\delta\mathcal{F}_{\mathrm{ex}}}{\delta\rho(z)} + V(z) - \mu_{\mathrm{ex}}\right]\right) \tag{7.49}$$

and using a fixed point iteration, i.e. inserting an initial guess for $\rho(z)$ to evaluate the right-hand side and getting the new estimation for the density profile from the left-hand side. Unless $\rho(z)$ is low everywhere, the convergence of that procedure is very slow. The reason is that, as such, the algorithm seldom converges. In order to make it convergent, the new estimate has to be weight-averaged with the old one, with a stronger weight on the latter, and this severely slows down the iteration. A more efficient method is the use of conjugate gradient techniques, exploring the minimum of $\Omega[\rho]$ along the functional direction set by $\delta\Omega[\rho]/\delta\rho(z)$ and projecting on the subspace which has not been previously explored [20]. In any case, the density profiles like those shown in Fig. 7.2, for HS fluids against a hard wall, present an increasingly strong layering structure for growing ρ_0, which reflects the ordered packing of the spheres induced by the planar wall. The functional dependence of $\mathcal{F}_{\mathrm{ex}}^{\mathrm{WDA}}[\rho]$ is sharp enough to reproduce those layered structures as the minimum of $\Omega[\rho]$. The rather costly numerical minimization process is intrinsic to the strongly non-local dependence of the free energy with $\rho(z)$.

Compared with computer simulations [21], the qualitative results obtained either with the CS-WDA or the PY-WDA are similarly good; only for very high-bulk densities a difference in favor of the CS-WDA is observed for the contact value of the density at the wall, directly related to the bulk pressure

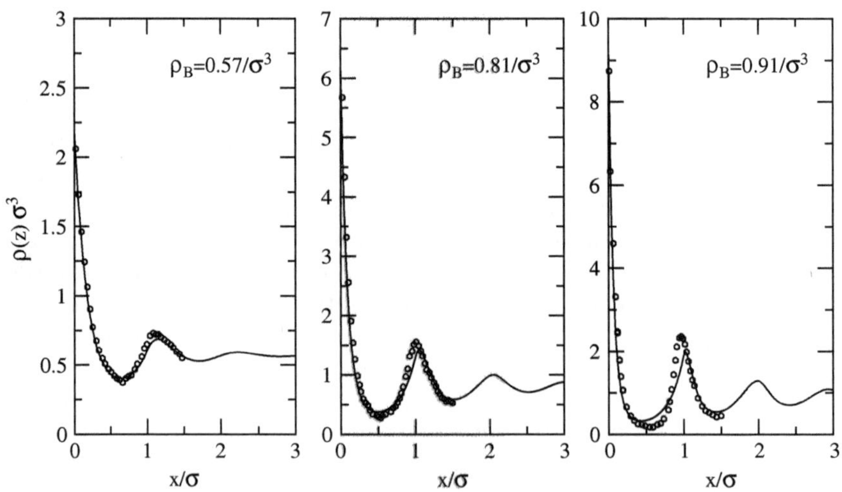

Fig. 7.2. Density profiles of the HS fluid at a hard wall, for three values of the bulk density. The *lines* are the results of CS-WDA, and the *circles* are the computer simulations by Snook and Henderson [21]

through the mechanical equilibrium balance $\rho(0) = \beta p$, which is consistently given by any DF approximation beyond the LDA [2]. Therefore, the use of the CS bulk equation of state gives quasi-exact values for $\rho(0)$, while the PY-WDA version tends to overestimate that contact value. Within the oscillatory part of $\rho(z)$, the two DF approximations give similar results, in reasonably good agreement with the computer simulations, over the full range of fluid densities (notice that the highest ρ_0 in Fig. 7.2 is only slightly below the freezing value). For $\rho_0 > 0.8\sigma^{-3}$, the shape and position of the second peak in the layering structure shows some discrepancies with the computer simulation results, but the overall quality of the WDA results is quite satisfactory and generally better than those obtained from the application of integral equation approximations with the wall–particle representation. Moreover, the WDA (as any DF approximation) provides consistent results for the density profiles and for the free energy of the system, which is not accessible from the integral-equation approach to $\rho(z)$. The results for the wall–fluid surface tension are presented in Fig. 7.3, and they have a very good agreement with the best computer simulations [22].

The consistent access to the density structure and the thermodynamics is a major advantage of the DF formalism for the study of phase transitions in surface or confined systems, like the problems of wetting and capillary condensation. These problems have been extensively explored [2, 19] for models with reference HS cores plus attractive tails, within the generalized van der Waals DF treatment described in Sect. 7.2.5.

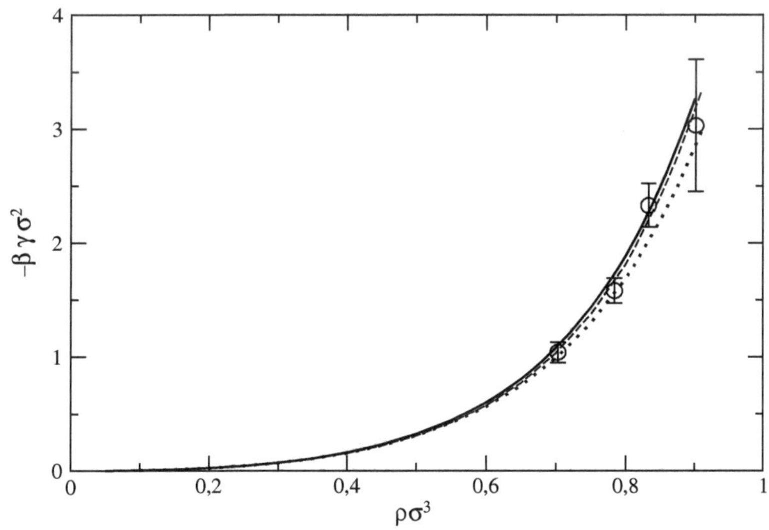

Fig. 7.3. Surface tension (with the sign changed) in kT/σ^2 units for the HS fluid at a hard wall. The *circles* with error bars are the results from the computer simulations by Henderson and van Swol [22]. The *dotted line* is the result of the CS-WDA; the *broken line* is the SPT in [23, 24]; the *full line* is the result from the DI-FMT

The WDA has also been successfully applied to systems with non-planar geometries, like cylindrical pores [19, 25] or spherical [26] external potentials, including the test-particle approach to get the pair distribution function $g(r)$ as the (normalized) density profile around a fixed molecule [27]. In those cases, the density distribution still depends on a single variable (the distance to the cylinder axis or to the center of the sphere), so that the computational effort is similar to that of planar walls, although the convolutions to get $\bar{\rho}(r)$ involve a kernel with two variables, $W(r, r')$, rather than $W(z - z')$ in (7.42). Some applications to systems with non-planar symmetry, like structured surfaces [28] or slit capillaries [29], have also been explored, with the obvious increase of the computational cost. The present computational capabilities, compared to those in 1985 when the WDA was proposed, certainly play in favor of any DF approach to solve the many interesting problems posed by systems with complex density dependence.

7.4.2 Weighted Density Approximation for the Hard Sphere Crystal

The problem of HS crystallization was a main target for the original development of the WDA. The earliest molecular dynamic simulations by Alder and Wainwright [30] in 1960 showed the crystallization of HS at high density. The most accurate estimation for the coexisting mean densities in this first-order phase transition is $\rho_c = 1.05\sigma^{-3}$ for the FCC crystal and $\rho_f = 0.94\sigma^{-3}$ for the fluid [31, 32, 33] (see also [34] for a comprehensive review of the subject). [3]

Within the scope of this chapter, the HS crystallization was probably the most important problem for the development of DF approximations. The DF approach to this problem sets that the crystal phase should be regarded as a *self-structured* fluid, for which, even in the absence of any external potential, the minimum of $\Omega[\rho]$ is not achieved by a homogeneous density $\rho(\boldsymbol{r}) = \rho_0$, but rather by a strongly modulated density $\rho(\boldsymbol{r})$, with the symmetry of the crystal lattice. After some early attempts [35, 36, 37] to describe crystals with the functional expansion (7.16) around a reference fluid, the first non-perturbative result for the coexisting densities was obtained with the zero-order WDA by Tarazona [16], using a Gaussian parametrization for the density,

$$\rho(\boldsymbol{r}) = \left(\frac{\alpha}{\pi}\right)^{\frac{3}{2}} \sum_{\mathbf{R}} \mathrm{e}^{-\alpha|\mathbf{r}-\mathbf{R}|^2} , \tag{7.50}$$

with normalized peaks at the sites $\mathbf{R}$ of a crystal lattice. This parametrization became standard with the more elaborated CS-WDA and PY-WDA.

The only free parameters to minimize $\Omega[\rho]$ within this restricted functional family are the inverse square Gaussian width α and the crystal lattice parameters. Taking the FCC lattice as the most obvious candidate, we only have

[3] The main branches of the phase diagram of a HS fluid are shown in Fig. 3.5 of Chap. 3

to fix the lattice length parameter, which is directly given by equating the mean density ρ_0 to the inverse volume of the unit crystal cell. Therefore, for a fixed mean density, we have to minimize the free energy per unit volume, $f(\alpha, \rho_0) \equiv \rho_0 \mathcal{F}[\rho]/N$, with respect to the parameter α. The results plotted in Fig. 7.4 were obtained with the CS-WDA for three values of ρ_0, and they show the three qualitatively different regimes which may be found. At low density, the free energy density is minimum at $\alpha = 0$, which corresponds to the sum of infinitely flat Gaussians, so that the contributions from all the lattice sites add to a constant homogeneous density $\rho(\boldsymbol{r}) = \rho_0$, i.e. the parametrization (7.50) contains the uniform fluid phase as the $\alpha = 0$ limit, and that is the only stable phase at low density, with a free energy density $f_f(\rho_0)$ given by the CS bulk equation of state. For densities slightly above $\rho_0 = 0.9\sigma^{-3}$, the global minimum of $f(\alpha, \rho_0)$ is still at $\alpha = 0$, but there is a local minimum at a large value $\alpha \geq 50\sigma^{-2}$, which corresponds to a strongly structured density distribution and has to be interpreted as a metastable crystal phase, with free energy $f_c(\rho_0)$ per unit volume. Increasing ρ_0 pushes $f_c(\rho_0)$ below $f_f(\rho_0)$, so that the crystal phase becomes the most stable phase for $\rho_0 > 1.\sigma^{-3}$, as shown in Fig. 7.5. The minimization of the grand potential energy per unit volume, $\omega(\mu) \equiv f(\rho_0) - \mu\rho_0$, at fixed μ, leads to the usual double tangent construction for the mean densities of the coexisting fluid ($\rho_f \approx 0.94\sigma^{-3}$) and crystal ($\rho_c \approx 1.05\sigma^{-3}$) phases,

$$f'_f(\rho_f) = f'_c(\rho_c) \equiv \mu_{\text{coex}} \tag{7.51}$$

and

$$\omega_f = \omega_c \equiv -p_{\text{coex}} \,, \tag{7.52}$$

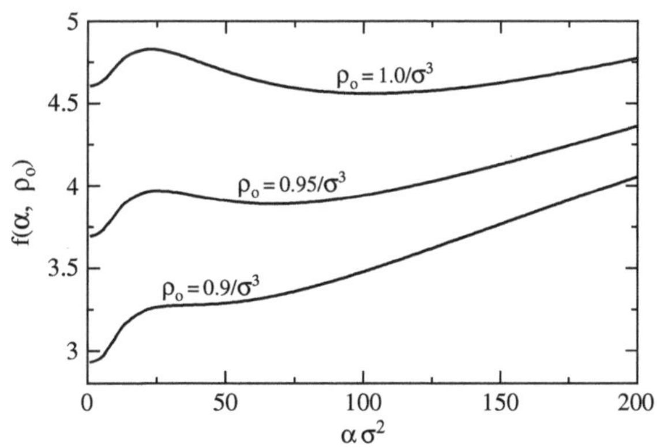

Fig. 7.4. The free energy per unit volume $f(\alpha, \rho_0)$ for the HS crystal, described as a FCC lattice with Gaussian peaks, as a function of the inverse squared width parameter α. The minimum at $\alpha = 0$ gives the free energy of the fluid phase while at the higher values of the mean density ρ_0, the crystal phase appears as the (relative or absolute) minimum at $\alpha > 50/\sigma^2$

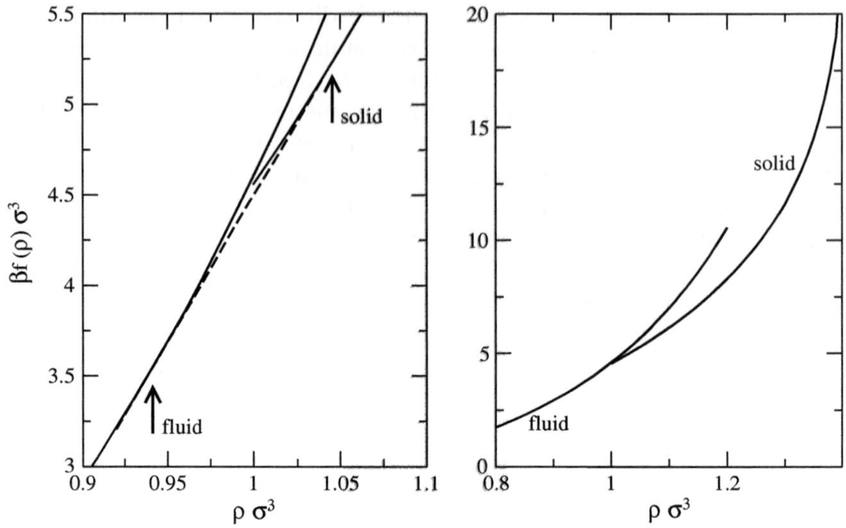

Fig. 7.5. The free energy per unit volume $f(\rho_0)$ for the HS fluid and solid phases, as functions of their mean density, within the CS-WDA with Gaussian peaks on the FCC lattice for the crystal. The *left panel* gives a detailed view of the transition region, and the broken straight line is the double tangent construction to get the densities of the coexisting phases. The *right panel* presents the results over the entire range for the solid phase density

with their common pressure, $\beta p_{\mathrm{coex}}\sigma^3 \approx 11.66$ and chemical potential $\beta\mu_{\mathrm{coex}} \approx 16.16$.

The numerical evaluation of $f(\alpha, \rho_0)$ within the CS-WDA is simplified with the use of bispherical coordinates for the convolution of a single, normalized, Gaussian peak (set at the origin) with the spherical weight functions centered at a distance $r = |\boldsymbol{r}|$,

$$\begin{aligned}\bar{\rho}_\nu^{\mathrm{gauss}}(r,\alpha) &\equiv \left(\frac{\alpha}{\pi}\right)^{\frac{3}{2}} \int \mathrm{d}\boldsymbol{r}' w_\nu(|\boldsymbol{r}-\boldsymbol{r}'|)\mathrm{e}^{-\alpha r'^2} \\ &= \frac{2\pi}{r}\left(\frac{\alpha}{\pi}\right)^{\frac{3}{2}} \int_0^\infty \mathrm{d}r' r' w_\nu(r') \int_{|r-r'|}^{r+r'} \mathrm{d}r'' r'' \mathrm{e}^{-\alpha r''^2} \\ &= \frac{1}{r}\sqrt{\frac{\alpha}{\pi}} \int_0^\infty \mathrm{d}r' r' w_\nu(r') \left(\mathrm{e}^{-\alpha|r-r'|^2} - \mathrm{e}^{-\alpha(r+r')^2}\right).\end{aligned} \tag{7.53}$$

The $\nu = 0$ step weight function gives rise to the analytical expression

$$\begin{aligned}\bar{\rho}_0^{\mathrm{gauss}}(r,\alpha) = \frac{3}{8\pi\sigma^3}\Big[&\mathrm{erf}\left(\sqrt{\alpha}(\sigma - r)\right) + \mathrm{erf}\left(\sqrt{\alpha}(\sigma + r)\right) \\ &- \frac{\mathrm{e}^{-\alpha(\sigma-r)^2} - \mathrm{e}^{-\alpha(\sigma+r)^2}}{r\sqrt{\pi\alpha}}\Big].\end{aligned} \tag{7.54}$$

with the error function

$$\mathrm{erf}(x) = \frac{2}{\sqrt{\pi}} \int_0^x \mathrm{e}^{-t^2}\, \mathrm{d}t. \tag{7.55}$$

More cumbersome, but still analytical, expressions may be obtained for the empirical $w_2(r)$ of the CS-WDA (7.38) and for piecewise polynomial approximants to $w_1(r)$. Therefore, the evaluation of the elementary weighted densities $\bar{\rho}_\nu(\boldsymbol{r})$ at any position of the crystal is performed via the sum of analytical expressions over the lattice sites $\{\boldsymbol{R}\}$,

$$\bar{\rho}_\nu(\boldsymbol{r}, \alpha) = \sum_{\{\boldsymbol{R}\}} \bar{\rho}_\nu^{\mathrm{gauss}}(|\boldsymbol{r} - \boldsymbol{R}|, \alpha) \;. \tag{7.56}$$

For small values of α (near the limit of the uniform fluid), the lattice sum would require many values of $\boldsymbol{R}$, since $\bar{\rho}_\nu^{\mathrm{gauss}}(r, \alpha)$ would decay very slowly with r. However, the minimum value of α at which the crystal phase appears (even as a metastable phase) is above $50/\sigma^2$, and the contributions to (7.56) decay very fast with $|\boldsymbol{r} - \boldsymbol{R}|$. The numerical calculation is thus very effective on the right-hand side of Fig. 7.4, which is the only relevant part to obtain $f_c(\rho_0)$. The evaluation of the full $\bar{\rho}(\boldsymbol{r})$ follows directly from (7.40), and the WDA excess free energy $\Phi\big([\rho]; \boldsymbol{r}\big) = \rho(\boldsymbol{r})\psi(\bar{\rho}(\boldsymbol{r}))$ may be obtained at any point of the space. The integrals of this free energy density and of the ideal gas contribution have to be done over the lattice unit cell, with any quadrature which takes advantage of the point symmetries shared by $\rho(\boldsymbol{r})$ and $\bar{\rho}(\boldsymbol{r})$. Alternatively, we can place a lattice site at the origin and integrate the contribution of that single Gaussian peak,

$$\beta f(\alpha, \rho_0) = \rho_0 \left(\frac{\alpha}{\pi}\right)^{\frac{3}{2}} \int \mathrm{d}\boldsymbol{r} \mathrm{e}^{-\alpha r^2} \left[\ln \rho(\boldsymbol{r}) - 1 + \psi(\bar{\rho}(\boldsymbol{r}))\right] \;, \tag{7.57}$$

extended to the whole volume. Notice that both $\rho(\boldsymbol{r})$, in the ideal gas logarithmic term, and $\bar{\rho}(\boldsymbol{r})$, in the excess free energy Φ, have to be evaluated with the full functions, including the contribution of all lattice sites, so that the integrand of (7.57) depends on the direction of $\boldsymbol{r}$. The radial integral may be very effectively done with Gauss–Hermite quadratures, to take advantage of its Gaussian decay, while the double angular integration has to be performed with any standard numerical method. The computation of $f_c(\alpha, \rho_0)$ was perfectly feasible twenty years ago, and it is trivially done nowadays.

In Table 7.1 we compare the theoretical and computer simulation results for the coexisting fluid and crystal densities. We also compare the Lindemann ratio of the crystal, which gives the relative mean square displacement of the particles from their lattice positions and which within the Gaussian parametrization (7.50) is given by $L = (\rho_c/2)^{1/3}(3/\alpha)^{1/2}$. The results of the CS-WDA, and to a lesser extent those of the PY-WDA, gave a clear improvement over the existing theories [35, 36, 37], based on the DF

Table 7.1. The results for the coexisting fluid and crystal densities, in HS diameter units. The Lindemann ratio of the crystal at coexistence is given by $L = (\rho_c/2)^{1/3}(3/\alpha)^{1/2}$ in terms of the Gaussian parametrization of the crystal density distribution

	ρ_f	ρ_c	$\rho_c - \rho_f$	L
Computer simulations [31]	0.94	1.04	0.10	0.126
CS-WDA	0.943	1.061	0.118	0.109
PY-WDA	0.905	1.025	0.120	0.104
Zero-order WDA [16]	0.892	0.966	0.074	
Early perturbative theories [35, 36, 37]	0.976	1.035	0.059	

expansion (7.14) around a uniform reference fluid. The density gap between the coexisting phases, which was grossly underestimated by the DF expansion, became slightly overestimated by both WDA versions. The actual values of the coexisting densities are very sensitive to changes in the theory, because as shown in Fig. 7.5, the free energies of the fluid and crystal phases have a nearly tangent intersection. More relevant is the systematic underestimation, by 15–20% in the Lindemann parameter of the crystal at any given density, which was obtained with any theoretical approach until the advent of the recent FMT versions. L measures the mean squared departures of a HS from its lattice position in the crystal, and a good description of the correlation between the neighbor particles is needed to get the correct amplitudes for the collective excursions of the HS within the crystal. Notice in this respect that the HS crystal is very different from the usual harmonic-crystal model analyzed in solid-state text books; the movement of each HS particle does not follow from the sum of the harmonic normal (i.e. uncorrelated) modes associated to the quadratic expansion of the energy, but it is made of purely free flights between collisions, and all the thermodynamic and elastic properties of the HS crystal follows from purely entropic effects.

The extension of the variational form (7.50) to more flexible shapes was suggested by the observation of a small anisotropy of the particle deviations from their lattice position in computer simulations of the HS crystal. The free minimization over the Fourier space [38] did not change significantly the results of the simplest Gaussian description for the coexisting densities but gave small anisotropies with a sign opposite to that of the simulations. However, the CS-WDA evaluation of the elastic constants of the HS crystal, which also required a deformation of the unit cell density distribution [39], gave good results compared with computer simulations despite the qualitative failure of other theories [40] for the HS crystal.

The most important failure of the WDA description of the HS crystal appears when the density distribution (either within the Gaussian parametrization or the full Fourier space description) is allowed to change the lattice space independently of the mean density. The restriction to have exactly one particle per unit cell, i.e. no vacancies or defect interstitial particles, is a very natural simplification of the variational space for $\rho(\boldsymbol{r})$, and it corresponds to what is observed in computer simulations. Nevertheless, we might expect that a good DF approximation should be able to find that unit occupancy of the crystal unit cell from the minimization of $\Omega[\rho]$, but this is not the case for the WDA. The free variation of the normalization in (7.50) increases the mean density of the coexisting crystal by 20% and destroys the good prediction for the coexistence itself. This problem was not peculiar of the WDA: other DF treatments of the HS crystal, based on refined versions of the functional expansion around a uniform fluid (7.14), shared the same problem. Another qualitative failure of the WDA appears when this DF approximation is used to describe a more open crystal structure, like the BCC lattice. Although that crystal structure (with only eight nearest neighbors) is not expected to be the equilibrium one for HS at any mean density, it is useful to have theoretical results for the HS BCC crystal as a reference system to describe systems with softer repulsions. However, the WDA results for any non-compact crystal structures are unphysical, with the Gaussian parameter α shrinking (instead of growing) when the mean density approaches the complete packing.

These failures of the WDA are not surprising because the full DF approximation is built on the information of the bulk HS fluid. The local correlation structure in the dense HS fluid is not very different from that in an expanded crystal with close-packed structure, with twelve neighbors around each particle, but the local structure in a BCC crystal is completely different from any one likely to appear in a bulk HS fluid, and therefore the WDA has no input to estimate the excess free entropy of such configurations.

7.4.3 Other Applications and Variants of the WDA

The generic DF structure of the WDA is not specific for HS; in fact, it was first proposed (under the name of averaged density approximation) for the exchange and correlation energy of the electron liquid [15]. However, its applications to HS systems have probably been the most successful. The WDA concept has also been used to get DF approximations for 2D HD fluids and to study their crystallization [41, 42], although without reaching the specific features of 2D crystallization. The role of dislocations and disclinations of the perfect crystal phase and the possible presence of an intermediate hexatic phase would be very difficult targets for any DF treatment of the crystal, described as a self-structured fluid at the level of the one-particle density distribution, and they are certainly beyond the WDA. Whether or not the effects of those topological defects may be included with a more accurate representation of $\mathcal{F}_{\text{ex}}[\boldsymbol{\rho}]$ is still an open question. A further reduction of the dimension leads to the exact results for 1D HR (7.20) and (7.21), which includes the disruption of the long-ranged 1D order by point defects, so that there is no phase transition and the fluid phase is stable at any mean density $\rho_0\sigma < 1$. The fact that a relatively simple analytic form of $\mathcal{F}_{\text{ex}}[\boldsymbol{\rho}]$ is able to describe, at the $\rho(x)$ level, an effect arising from global correlations extended to the whole 1D system gives hope to the future DF description of the complex 2D crystallization. On the opposite sense, the accuracy of the WDA when applied to systems of hyper-spheres, in $D > 3$ dimensions, should increase with D, because its DF structure contains the basic elements of the low-density expansion (7.16), and the maximum density allowed by the packing constraints decreases as D increases.

Some early attempts were made to apply the WDA to simple fluid models, with both attractive and repulsive regions of their interatomic potential, but it was soon clear that the non-local aspects associated to the entropy loss by the core repulsions cannot be directly added to those associated to the energy of the attractive interactions. The best DF approximations for these systems are based on the separated treatment of a reference HS system and the MFA [2, 19], or similar DF schemes, for the description of the attractive interactions [43]. For purely repulsive, but soft, pair interaction potentials, the WDA has been applied with moderate success, allowing for temperature-dependent weight functions [44, 45, 46]. However, the quality of the approximation rapidly decreases as the repulsive potential $u(r)$ becomes softer, down to a rather discrete accuracy for the soft repulsion of the one-component plasma.

Several variants of the WDA scheme for HS have been developed and explored, among them that of Meister and Kroll [2, 47, 48], who explored a variational method to determine $\bar{\rho}(\boldsymbol{r}, \rho)$ within a DF expansion from a WDA scheme (see relevant comments in [2]). Other variants of the WDA pointed to a simplification of its practical use for the HS crystal, for its own sake or as reference system for the solid phase of other simple fluid models. The modified weighted density approximation (MWDA) of Denton and Ashcroft [49, 50, 51] takes a constant weighted density over the entire system, rather than being a function of the local position. That provides a strong simplification in the numerical use of the theory, which becomes equivalent to the effective liquid approximation (ELA) and its variants (GELA, MELA) [52, 53], in which the crystal phase is described in a perturbative scheme around a uniform fluid phase. The obvious drawback of these approaches, with respect to any DF approximation, is that they cannot be applied to generic inhomogeneous distributions $\rho(\mathbf{r})$.

7.5 The Fundamental Measure Theory

In 1989, a new DF approximation for the HS fluid was proposed by Rosenfeld [13] under the name of FMT, and it represented a breaking point in the theory of inhomogeneous hard body systems. That original version will be referred here as the oFMT to distinguish it from its subsequent variants, proposed and tested by different authors. By the end of the century, it was clear that the generic FMT scheme is qualitatively superior to any WDA although it requires a much more careful design of the functional form and it also has some technical complexities, like the use of vector and tensor weighting functions, which have probably scared some potential users. This section gives a brief account of the main concepts of the FMT and the most advanced version, based on the idea of *dimensional interpolation*. We present detailed recipes for the practical application of the FMT to the most usual symmetries and comments on the FMT results for the most important problems.

7.5.1 Rosenfeld's Original Fundamental Measure Theory

Fundamental and subtle clues led Rosenfeld to the original development of the FMT [12, 13], which was directly set as a DF theory for HS mixtures, as we review in Sect. 7.8. Here, we present first the mono-component case, with some changes in the original notation, to give a simpler account of both the original FMT version and the most recent developments. The first point to set up the FMT is to include the *packing fraction* as an essential non-local measure for the free energy of HS. The best approximations for the equation of state and correlation structure of the HS fluid cast the density dependence in terms of the dimensionless packing fraction $\eta = \pi\rho\sigma^3/6$, i.e. the ratio of the volume occupied by the HS with respect to the total system volume. The natural extension to inhomogeneous systems is the local packing fraction,

$$\eta(\boldsymbol{r}) \equiv \int \mathrm{d}\boldsymbol{r}' \; \rho(\boldsymbol{r}+\boldsymbol{r}') \; \Theta\left(\frac{\sigma}{2} - |\boldsymbol{r}'|\right) \; , \tag{7.58}$$

which is a crucial piece of the exact free energy DF for 1D HR (7.21) and (7.22). Notice that $\eta(\boldsymbol{r})$ has the appealing interpretation of being the probability that for random molecular configurations over the equilibrium statistical ensemble, the point $\boldsymbol{r}$ happens to be inside a HS core. A density distribution producing anywhere a value $\eta(\boldsymbol{r}) > 1$ is therefore forbidden for the HS fluid, so that the use of this non-local measure gives a simple tool to locate the DF borders of the accessible $\rho(\boldsymbol{r})$, much more accurately than the lack of solutions for $\bar{\rho}(\boldsymbol{r})$ in the WDA, although still not in a complete form.[4]

Notice that (7.58) is very different from the zero-order weighted function in the WDA (7.32), since $w_0(r)$ is a (normalized) Mayer function, i.e. a radial step function with radius equal to the HS diameter σ while the radius of the step function defining $\eta(\boldsymbol{r})$ is the molecular radius $R \equiv \sigma/2$. A main concept of the FMT is precisely that the *fundamental measure* to describe the non-local dependence of $\Phi([\rho];\boldsymbol{r})$ is the shape of a single molecule rather than the excluded volume between two molecules; therefore the free energy DF is

[4] The restriction to $\eta(\boldsymbol{r}) \leq 1$ is a necessary but not sufficient condition to have a density distribution $\rho(\boldsymbol{r})$ compatible with the HS core repulsion. Obviously, $\eta(\boldsymbol{r})$ cannot take, over extensive regions, values larger than the close packing of spheres $\eta = 0.74$ although it may approach 1 over small regions of the size of a sphere. See the discussion in Sect. 7.5.2 and in [54] for the smallest cases in which the accessible DF boundary moves below $\eta = 1$.

represented as a function of $\eta(\boldsymbol{r})$ and other weighted densities, which are all the convolutions of $\rho(\boldsymbol{r})$ within the range of the HS radius.

The exact free energy for 1D HR given in (7.21) has precisely the form

$$\Phi^{(1D)}([\rho];x) = \phi_1(\eta(x), n(x)) \equiv -n(x)\ln\left[1-\eta(x)\right] , \qquad (7.59)$$

with the local packing fraction (7.22) and the combination $n(x) \equiv \left[\rho(x+R)+\rho(x-R)\right]/2$, which may be regarded as the 1D version of a generic *surface weighted density*,

$$n(\boldsymbol{r}) \equiv \int \mathrm{d}\boldsymbol{r}' \rho(\boldsymbol{r}+\boldsymbol{r}') w(\boldsymbol{r}') , \qquad (7.60)$$

with the normalized *molecular surface* weight function,

$$w(\boldsymbol{r}) \equiv \frac{\delta(|\boldsymbol{r}|-R)}{s_{\mathrm{D}}} , \qquad (7.61)$$

s_{D} being the total *molecular surface* in D dimensions ($s_1 = 2$, $s_2 = 2\pi R$, $s_3 = 4\pi R^2$). Notice that we are not following Rosenfeld's original notation: the scalar density $n(\boldsymbol{r})$ defined in (7.60) and (7.61) equals the homogeneous density for any distribution $\rho(\boldsymbol{r}) = \rho_0$.

There is an interesting difference between the two alternative forms (7.20) and (7.21) which can be used to describe the exact excess free energy of 1D HR systems. In the first one, $\Phi^{(1D)}([\rho];x)$ is explicitly proportional to the local density $\rho(x)$, so that (like in the WDA) the free energy excess is *locally assigned* to the position of the particle centers, with a value per particle which depends on the sampling of the density up to distances σ around the point $\boldsymbol{r}$.

In contrast, the expression (7.21) for $\Phi^{(1D)}([\rho];x)$ is not proportional to $\rho(x)$, and this functional form may assign a local density of excess free energy to regions with $\rho(x) = 0$ but $\eta(x) > 0$. The proportionality with $n(x)$ in (7.59) has to be interpreted as the assignment of the free energy excess to the positions of the *molecular surface*, i.e. of the two ends of the 1D HR. Obviously, there is nothing wrong in having different exact expressions for $\Phi^{(1D)}[\rho;x]$, as far as they all integrate to the same $\mathcal{F}_{\mathrm{ex}}[\rho]$ for any density distribution. The no-overlap restriction imposed by the hard cores produces a global entropy decrease, which may be *locally assigned* in different ways, to get the free energy density $\Phi^{(1D)}([\rho];x)$.

For 1D HR, the exact $\rho(x) \to 0$ limit of (minus) the direct correlation function (7.17) is recovered from (7.21) as

$$f(x) \equiv \Theta(\sigma-|x|) = \frac{1}{2}\int \mathrm{d}x' \delta\left(\frac{\sigma}{2}-|x+x'|\right)\Theta\left(\frac{\sigma}{2}-|x'|\right) , \qquad (7.62)$$

where the Mayer function range, $-\sigma \leq x \leq \sigma$, arises from the convolution of the two fundamental measures, each with half that range. The 3D geometry of the HS makes it impossible to represent the spherical step of the Mayer function, $f(\boldsymbol{r}) = \Theta(\sigma-|\boldsymbol{r}|)$, in terms of the half-ranged ($R = \sigma/2$) step

$w_0(\boldsymbol{r}) = \Theta(R - |\boldsymbol{r}|)$ in (7.58) and the spherical shell (7.61). That explains the relatively poor performance of the direct extension of the 1D free energy DF form to describe 3D HS, attempted by Robledo et al. [10, 11]. To solve this problem, Rosenfeld made the crucial proposal to enlarge the set of fundamental measure functions with a *vector weight*,

$$\boldsymbol{w}(\boldsymbol{r}) \equiv \frac{\boldsymbol{r}}{R} \frac{\delta(R - |\boldsymbol{r}|)}{s_{\mathrm{D}}} , \tag{7.63}$$

to define a vector-weighted density,

$$\boldsymbol{v}(\boldsymbol{r}) \equiv \int \mathrm{d}\boldsymbol{r}' \; \rho(\boldsymbol{r} + \boldsymbol{r}') \; \boldsymbol{w}(\boldsymbol{r}') \, . \tag{7.64}$$

The packing fraction (7.58), the scalar-weighted density (7.60) and the vector weighted density (7.64) are used as the three local variables to describe $\Phi_{\mathrm{FMT}}([\rho]; \boldsymbol{r}) = \phi(\eta(\boldsymbol{r}), n(\boldsymbol{r}), \boldsymbol{v}(\boldsymbol{r}))$. The lowest (quadratic) order of this function must depend on its three variables in the form $\phi(\eta, n, \boldsymbol{v}) = \eta n + 4\pi R^3 (n^2 - \boldsymbol{v} \cdot \boldsymbol{v}) + \mathcal{O}^3(\eta, n, \boldsymbol{v})$. This is the unique combination which recovers the exact low-density expansion of the direct correlation function (7.17), since the unique geometrical construction to extend (7.62) to 3D is

$$\begin{aligned} f(r) \equiv \Theta(\sigma - |\boldsymbol{r}|) = \frac{1}{2\pi R^2} \int \mathrm{d}\boldsymbol{r}' \delta\left(R - |\boldsymbol{r}'|\right) \Big[\Theta\left(R - |\boldsymbol{r} + \boldsymbol{r}'|\right) \\ + R\,\delta\left(R - |\boldsymbol{r} + \boldsymbol{r}'|\right) \left(1 - \frac{\boldsymbol{r}' \cdot (\boldsymbol{r} + \boldsymbol{r}')}{R^2}\right) \Big] . \end{aligned} \tag{7.65}$$

The final clue followed by Rosenfeld was that the structure of the spherical shell weight functions, $w(\boldsymbol{r})$ and $\boldsymbol{w}(\boldsymbol{r})$, is directly related to the derivatives of $\Theta(R - |\boldsymbol{r}|)$ with respect to the molecular size, suggesting a connection with the scaled particle theory (SPT) for the thermodynamic properties of hard bodies [23, 24, 55]. This theory reproduces the exact equation of state in 1D and the PY compressibility result for 3D HS.

The search for a function $\phi(\eta, n, \boldsymbol{v})$ with the exact low-density limit and reproducing the PY direct correlation function led to the oFMT form

$$\Phi_{\mathrm{oFMT}}([\rho]; \boldsymbol{r}) = \sum_{i=1}^{3} \phi_i^{(3\mathrm{D})}(\eta(\boldsymbol{r}), n(\boldsymbol{r}), \boldsymbol{v}(\boldsymbol{r})) \, , \tag{7.66}$$

with the first term $\phi_1^{(3\mathrm{D})}(\eta, n) = -n \ln(1 - \eta)$ identical to the full 1D result (7.59), and two more terms given by

$$\phi_2^{(3\mathrm{D})}(\eta, n, \boldsymbol{v}) = 4\pi R^3 \, \frac{n^2 - \boldsymbol{v} \cdot \boldsymbol{v}}{1 - \eta} \tag{7.67}$$

and

$$\phi_3^{(3\mathrm{D})}(\eta, n, \boldsymbol{v}) = 8\pi^2 R^6 \; n \, \frac{n^2/3 - \boldsymbol{v} \cdot \boldsymbol{v}}{(1 - \eta)^2} \, . \tag{7.68}$$

The results of this DF approximation were a puzzling mixture of successes and failures: the density profiles for HS against a hard wall are clearly better than those obtained with the PY-WDA version, particularly in reproducing the strong oscillatory structure for a high-density bulk fluid, as will be shown in Fig. 7.9. The equation of state plugged into the CS-WDA provides a better value of the contact density, but it still does not match the excellent description of $\rho(z)$ given by the oFMT for the interlayer spacing, reflecting a sharper representation of the correlation structure in inhomogeneous systems. Also, the second functional derivative of the oFMT excess free energy is restricted to the PY range, $|\boldsymbol{r} - \boldsymbol{r}'| \le \sigma \equiv 2R$, not only for bulk liquids but for any density distribution. In contrast, the PY-WDA built on the same results for the bulk liquid direct correlation has a second functional derivative which goes beyond that range for inhomogenous density distributions, since the precise tailoring of the oscillating tail in $w(r, \rho)$ may only restrict the range for the homogeneous fluid direct correlation. Another very appealing feature of the oFMT was its natural extension to HS mixtures, in contrast with the conceptual difficulties of the WDA for that task. We will elaborate more on this point later in Sect. 7.6.

However, the oFMT has some qualitative pitfalls related to the use of delta-function shells in $w(\boldsymbol{r})$ and $\boldsymbol{w}(\boldsymbol{r})$. The overlap of these shells produces strong divergences, which have to be tamed by the careful choice of the combinations (7.67), (7.68) of $n(\boldsymbol{r})$ and $\boldsymbol{v}(\boldsymbol{r})$. It is a generic feature of the FMT that slight variants of a good approximation may be qualitatively wrong, in contrast with the mild dependence of the WDA on the details of each particular version. Still, the combination of n and $\boldsymbol{v}$ used in (7.68) can only avoid the unphysical divergences of $\Phi([\rho]; \boldsymbol{r})$ for homogeneous systems, and its application to highly inhomogeneous systems, like the DF description of the HS crystal, leads to a dramatic failure: as the Gaussian peaks shrink to delta functions, the free energy of the crystal phase decreases without bound. It took ten years to develop new versions of the FMT with this problem corrected and able to provide an overall improvement over the WDA in all aspects, as we review below.

This notwithstanding, the oFMT remains an excellent, and fairly used, DF choice to solve problems of highly packed HS (and their mixtures) with planar geometry, where the homogeneity over the transverse directions is enough to avoid the divergences of the radial shells overlap. The practical implementation of those applications is discussed below, together with that of the more recent FMT versions. The vector weighted density, $\boldsymbol{v}(\boldsymbol{r})$, gave the oFMT a less intuitive character than the WDA, which has probably restrained some potential users. The FMT version proposed by Kierlik and Rosinberg [56, 57] substituted $\boldsymbol{v}(\boldsymbol{r})$ with the gradient of $\eta(\boldsymbol{r})$. The resulting DF is perhaps more intuitive, but with little (or none) advantage for the practical use of the theory. In fact, the same authors proved that both versions of the FMT (with and without vector density) are identical because they both yield the same free energy DF [58].

7.5.2 Dimensional Crossover and the Cavity Theory

The concept of *dimensional crossover* is a strong test for DF approximations which was first applied [19] to the CS-WDA for HS. The idea is that any DF approximation for the 3D excess free energy of HS, $\mathcal{F}_{\rm ex}[\rho]$, contains also predictions for any 2D distribution of HD, $\rho_{\rm 2D}(x,y)$, because the latter should be fully equivalent to a 3D distribution with a delta function along the Z axis: $\rho_{\rm 3D}(x,y,z) = \rho_{\rm 2D}(x,y)\delta(z)$; see Fig. 7.6. The extremely strong inhomogeneity of that 3D density distribution provides a severe test for any DF approximation to $\mathcal{F}_{\rm ex}[\rho]$, and the WDA showed a rather limited success in this respect. The dimensional crossover may be extended to represent 1D distributions of HR as 3D distributions, with $\rho_{\rm 3D}(x,y,z) = \rho_{\rm 1D}(x)\delta(y)\delta(z)$, and the equivalent reduction from 2D to 1D can also be defined. The original FMT version for 3D HS and its 2D counterpart were subjected to those tests [59] in order to extract the predictions for the excess free energy of uniform 2D and 1D systems from their higher dimensional representations. The results were again a mixture of success and pitfalls: the DF reduction of the oFMT from 3D to 2D is very accurate; however, the reduction from 3D to 1D was a complete failure, since the third term $\phi_3^{(3D)}$ in (7.66) diverges. Nevertheless, it was pointed out that if that term were eliminated, the sum $\phi_1^{(3D)} + \phi_2^{(3D)}$ would give the exact result for $\Phi^{(1D)}$.

The DF dimensional crossover was pushed forward with the concept of the zero-dimensional (0D) limit, leading to the development of the *cavity*

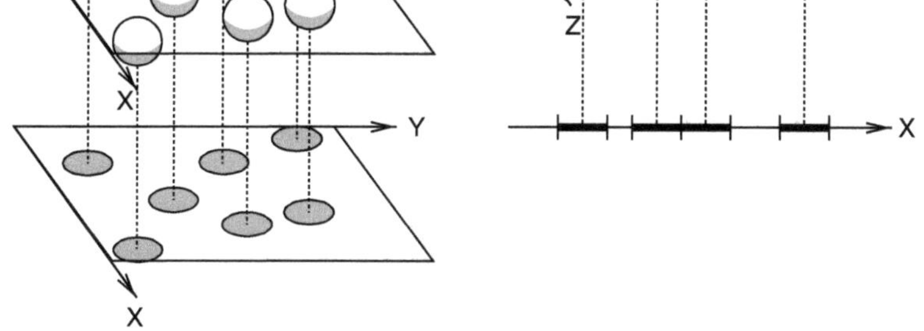

Fig. 7.6. Sketches of dimensional crossover within the DF formalism. On the *left*, a 3D distribution with the hard sphere centers located on the $z = 0$ plane $\rho_{\rm 3D}(x,y,z) = \delta(z)\rho_{\rm 2D}(x,y)$ is fully equivalent to a 2D distribution of hard disks $\rho_{\rm 3D}(x,y,z) = \delta(y)\delta(z)\rho_{\rm 1D}(x)$. On the *right*, a 3D distribution with all the HS centers along the X axis is fully equivalent to a system of 1D hard rods

theory by Tarazona and Rosenfeld [54], which gave a fresh view to the FMT. Consider a small cavity which cannot hold more than one HS and which is connected to a particle reservoir at chemical potential μ. We use $\eta_0 \leq 1$ as the mean number of particles in the cavity, i.e. the probability that the cavity is occupied. Unlike the 2D and 1D cases, the exact 0D limit does not only cover the strict reduction of the geometrical dimension, $\rho(\boldsymbol{r}) = \eta_0 \delta(x)\delta(y)\delta(z)$, but also a wide range of smoother 3D density distributions, with the only restriction that the distance between any two points with $\rho(\boldsymbol{r}) \neq 0$ is less than a HS diameter $\sigma = 2R$. The exact excess free energy $\mathcal{F}_{ex}[\rho]$ may be obtained for any density distribution $\rho(\boldsymbol{r})$ allowed within the cavity, because its grand partition function (7.3) takes exactly the form $\Xi_{0\mathrm{D}} = 1 + Z_1 \exp(\beta\mu)$, with the two first terms of the ideal gas expression (7.5) and the total integral of the density distribution, i.e. the total occupancy of the cavity, is exactly $\eta_0 = Z_1 \exp(\beta\mu)/\Xi_{0\mathrm{D}}$. Using this expression, eliminate the dependence of $\Xi_{0\mathrm{D}}$ with the chemical potential and the confining external potential in Z_1. The exact generic form of the 0D excess free energy is obtained as $\beta\mathcal{F}_{\mathrm{ex}} = \phi_0(\eta_0) \equiv (1-\eta_0)\ln(1-\eta_0)+\eta_0$, which depends on $\rho(\boldsymbol{r})$ only through its total integral η_0, independently of the shape of the density distribution within the cavity.

That 0D excess free energy has the low-density expansion $\phi_0(\eta_0) = \eta_0^2/2 + \eta_0^3/6 + \cdots$, which could have been extracted from (7.16), since the 0D character of the cavity implies that $f(r_{ij}) = 1$ for any pair of points with $\rho(\boldsymbol{r}_i) > 0$ and $\rho(\boldsymbol{r}_j) > 0$. In the opposite limit, for cavities which are nearly always occupied, the free energy excess goes to $\phi_0(1) = 1$ with diverging slope, as it corresponds to the physical condition that the full occupancy of the cavity implies the lack of empty states in the total partition function, which can only be achieved with a divergently high chemical potential in the reservoir.

The requirement that any $\rho(\boldsymbol{r})$ within a 0D cavity has an excess free energy $\phi_0(\eta_0)$ is a very severe condition which imposes strong restrictions on the non-local dependence of $\mathcal{F}_{\mathrm{ex}}[\rho]$. The cavity theory [54] shows that a full $\mathcal{F}_{\mathrm{ex}}[\rho]$ for 3D HS (and also for 2D HD) may be built *using only* the requirement that the exact 0D limit is recovered over a *partial set* of cavity shapes. That free energy DF has the essential elements of the FMT: the use of $\eta(\boldsymbol{r})$ and density convolutions with spherical delta-function shells or radius R. The geometrical combinations of those convolutions are fixed in such way that the divergences produced when they overlap are fully eliminated. Within this 0D FMT approach, the thermodynamics of the uniform bulk HS fluid is a particular output of the DF approximation rather than an input to be used in the construction of $\mathcal{F}_{\mathrm{ex}}[\rho]$. The 0D FMT excess free energy density for the bulk HS fluid is directly given in terms of the successive derivatives of the 0D free energy excess, $\phi_0(\eta)$, with respect to η: $\phi_0'(\eta) = -\ln(1-\eta)$, $\phi_0''(\eta) = 1/(1-\eta)$ and $\phi_0'''(\eta) = 1/(1-\eta)^2$, something that already appears in the result of the SPT and that was directly plugged into the oFMT (7.66) and (7.68). Rosenfeld had coined the term *ideal liquid* [60] to refer to the (unreachable) limit $\eta \to 1$ of the HS fluid, since the inverse powers of $1-\eta$ appear in any good representation of the equation of state for the HS bulk fluid despite the much lower geometrical close packing of spheres (with maximum mean value $\eta_{\max} = 0.740$). Within the cavity theory, that *ideal liquid* limit of the bulk fluid appears naturally as a reminiscence of the physical singularity in the 0D excess free energy at $\eta_0 = 1$.

The DF extracted from the 0D limit confirms that the FMT elements are the natural building blocks of $\mathcal{F}_{\text{ex}}[\rho]$, but it also unveils the limitations of any DF approximation based on one-center convolutions. The structure of the 0D FMT free energy DF follows closely that set in the oFMT: the free energy density for isotropic hard core bodies in D dimensions has precisely $\nu = 1, \ldots, D$ terms, with the dependence on the local packing fraction given by the νth-derivative of $\phi_0(\eta)$ and with an order ν convolution of the density with spherical delta-function shells, plus geometrical factors which eliminate all the spurious divergences. The first-order term in any dimension has the exact form of the 1D free energy, $\phi_1^{(1D)}(\eta, n) = n\phi_0'(\eta)$, identical to Rosenfeld's choice in (7.66). Also the second term for the 3D case becomes exactly that in the oFMT form, $\phi_2^{(3D)}(\eta, n, \boldsymbol{v}) = 4\pi R^3(n^2 - \boldsymbol{v}\cdot\boldsymbol{v})\phi_0''(\eta)$. However, the third term for HS (and the second term for 2D HD) cannot be separated in terms of one-center convolutions, like $n(\boldsymbol{r})$ and $\boldsymbol{v}(\boldsymbol{r})$. The appearance of non-separable convolutions was already discussed in the comparison between the WDA and the exact density expansion (7.16), and there is a strong practical bias toward one-center DF forms. Moreover, even if the non-separable kernels were accepted in $\phi_3^{(3D)}$, the resulting DF approximations still leave out a class of 0D cavities, which were called *the lost cases* of the FMT [54], for which the local packing fraction $\eta(\boldsymbol{r})$ does not reach anywhere the total integral η_0 of the density distribution $\rho(\boldsymbol{r})$. The simplest example of a *lost-case* cavity is made of three point-like subcavities, at the vertex of an equilateral triangle, separated by a distance between $\sqrt{3}R$ and $2R$, so that any two of them cannot be occupied at the same time, but there is no point closer than R to the three subcavities, where $\eta(\boldsymbol{r})$ would be the sum of the three partial occupancies. Forbidden density distributions, with $1 < \eta_0 \le 3/2$, are compatible with the restriction $\eta(\boldsymbol{r}) \le 1$, and they cannot be excluded within the FMT scheme of one-center convolutions.

The importance of these *lost cases* increases with the dimension of the supporting space. They are absent in cavities with 1D support, and the exact $\Phi^{(1D)}([\rho]; \boldsymbol{r})$ of Percus may be built from the 0D FMT. In 2D, there are some 0D *lost cases*, and the 0D FMT does not lead to the exact result, but still it leads to the (fairly accurate) SPT for the equation of state of homogeneous systems of HD. In 3D, the increasing role of the *lost cases* is reflected in a rather poor approximation for the bulk HS equation of state given by the 0D FMT free energy DF. The compromise between a good dimensional reduction, a good description of the HS bulk fluid and the practical handling of the DF approximation leads to the recent DF approach presented in the next section.

7.5.3 The Dimensional Interpolation Fundamental Measure Theory

The dimensional interpolation (DI) version of the FMT for HS was introduced [61] as a practical approximation which balances the advantages of the previous proposals. It is built to recover the PY direct correlation function (and hence the PY-compressibility equation of state) for bulk homogeneous HS fluids, as in the original FMT version of Rosenfeld, but using the 0D FMT basic elements to systematically eliminate the spurious divergences in the 1D and 0D reductions. As the original FMT version, the DI-FMT excess free energy $\Phi^{(3D)}([\rho];\boldsymbol{r})$ has three additive terms;

$$\Phi^{(3D)}([\rho];\boldsymbol{r}) = \sum_{\nu=1}^{3} \phi_\nu^{(3D)}(\eta(\boldsymbol{r}), n(\boldsymbol{r}), \dots) \ , \tag{7.69}$$

and the first two have exactly the same form proposed by Rosenfeld, in terms of the local packing fraction, the scalar and the vector density convolutions, namely

$$\phi_1^{(3D)}(\eta, n) = -n \ln(1-\eta) \ , \qquad \phi_2^{(3D)}(\eta, n, \boldsymbol{v}) = 4\pi R^3 \frac{n^2 - \boldsymbol{v}\cdot\boldsymbol{v}}{1-\eta} \ . \tag{7.70}$$

The third term requires a new tensor density, $\mathcal{T}(\boldsymbol{r})$, with cartesian components ($\alpha, \beta = x, y, z$) defined by the convolution

$$\mathcal{T}_{\alpha\beta}(\boldsymbol{r}) \equiv \int \mathrm{d}\boldsymbol{r}' \rho(\boldsymbol{r}+\boldsymbol{r}') \frac{r'_\alpha r'_\beta}{R^2} \frac{\delta(R-|\boldsymbol{r}'|)}{4\pi R^2} \ . \tag{7.71}$$

There is a unique combination of this tensor with $n(\boldsymbol{r})$, $\boldsymbol{v}(\boldsymbol{r})$ and $\eta(\boldsymbol{r})$ which added to $\Phi_1^{(3D)} + \Phi_2^{(3D)}$ recovers the bulk PY direct correlation for $\rho(\boldsymbol{r}) = \rho_0$ and the exact free energy for the strict 0D limit, $\rho(\boldsymbol{r}) = \eta_0 \delta(\boldsymbol{r})$. That combination defines the DI-FMT version with

$$\Phi_3^{(3D)}(\eta, n, \boldsymbol{v}, \mathcal{T}) = 12\pi^2 R^6 \frac{\boldsymbol{v}\cdot\mathcal{T}\cdot\boldsymbol{v} - n\,\boldsymbol{v}\cdot\boldsymbol{v} - \mathrm{Tr}[\mathcal{T}^3] + n\,\mathrm{Tr}[\mathcal{T}^2]}{(1-\eta)^2} \ , \tag{7.72}$$

in terms of the rotational invariants formed by the index contractions of $\mathcal{T}$ and $\boldsymbol{v}$, as well as the traces of $\mathcal{T}$, $\mathcal{T}^2$ and $\mathcal{T}^3$. Notice that the scalar density is precisely $n(\boldsymbol{r}) = \mathrm{Tr}[\mathcal{T}]$.

The original Rosenfeld's proposal (7.68) for $\Phi_3^{(3D)}(\eta, n, \boldsymbol{v})$ may be recovered from (7.72) with the extra assumption that the tensor $\mathcal{T}$ is approximated as $n/3$ times the unit 3D tensor, i.e. neglecting its anisotropy but keeping its trace. That explains the good results of the oFMT for planar density profiles, $\rho(\boldsymbol{r}) = \rho(z)$, since the homogeneity over the XY directions renders the anisotropy of $\mathcal{T}(\boldsymbol{r})$ nearly irrelevant because, by symmetry, this tensor is diagonal and has $\mathcal{T}_{xx} = \mathcal{T}_{yy} = (n - \mathcal{T}_{zz})/2$. In contrast, the strong inhomogeneity of the 3D representation for a 1D system of HR, $\rho(x,y,z) = \rho_{1D}(x)\delta(y)\delta(z)$, leaves a crucial role to the anisotropy of $\mathcal{T}(\boldsymbol{r})$ as the geometrical tool to obtain the *complete cancellation* of this third term for any 1D density distribution, so that the exact 1D result is recovered from the sum of the first two terms in $\Phi^{(3D)}([\rho];\boldsymbol{r})$. Neglecting the anisotropy of $\mathcal{T}(\boldsymbol{r})$ in the oFMT produces the spurious divergence of $\phi_3^{(3D)}(\eta,n,\boldsymbol{v})$ which invalidates the DF approximation for that dimensional reduction.

7.5.4 Comparative Application of the Fundamental Measure Theory and the Weighted Density Approximation to a Spherical 0D Cavity

The application of different DF approximations to a simple 0D cavity offers an interesting perspective of their essential aspects. Let us consider a spherical density distribution, with total integral η_0, homogeneously distributed inside a small sphere of radius ϵ. For any $\epsilon < R$ and $\eta_0 \leq 1$, this density distribution represents a 0D cavity, with total excess free energy $\phi_0(\eta_0) = (1-\eta_0)\ln(1-\eta_0) + \eta_0$. This value should be obtained if the exact 3D excess free energy, $\beta\mathcal{F}_{\rm ex}[\rho]$, were evaluated for the density distribution

$$\rho(\boldsymbol{r}) = \frac{3\eta_0}{4\pi\epsilon^3}\Theta(\epsilon - |\boldsymbol{r}|)\ , \tag{7.73}$$

with a radial step function of radius ϵ. This is a very severe test for any DF approximation, particularly with respect to the singular behavior of $\phi_0(\eta_0)$, which has to reach the value $\phi_0(1) = 1$ with infinite derivative, to represent the boundary between accessible and unaccessible density distributions of the HS within the cavity. The WDA represents the local density of excess free energy as $\Phi_{\rm WDA} = \rho(\boldsymbol{r})\psi(\bar{\rho}(\boldsymbol{r}))$.

As presented in the left panel of Fig. 7.7 for a cavity of $\epsilon/\sigma = 0.1$, the weighted density $\bar{\rho}(\boldsymbol{r})$ exceeds the range of $r = \sigma$, but only the values within $r \leq \epsilon$ are relevant for the evaluation of the free energy, through $\psi_{\rm CS}(\bar{\rho})$. In the left panel of Fig. 7.8, the free energy excess for low occupation is shown to be fairly independent of the cavity radius ϵ and recovers the exact quadratic dependence $\mathcal{F}_{\rm ex}^{\rm WDA}[\rho] = \eta_0^2/2 + \mathcal{O}(\eta_0^3)$. For larger η_0, the CS-WDA results show some dependence on ϵ, deviating from the exact $\phi_0(\eta_0)$. The representation of the boundary to unaccessible ($\eta_0 > 1$) distributions in the WDA may only come from the lack of solutions for $\bar{\rho}(\boldsymbol{r})$, which in the CS-WDA occurs when the discriminant $(1-\bar{\rho}_1(\boldsymbol{r}))^2 - 4\bar{\rho}_0(\boldsymbol{r})\bar{\rho}_2(\boldsymbol{r})$ in (7.40) vanishes. This is found to happen first at the center of the cavity, $r = 0$, when the cavity occupation reaches a value $\eta_0 = \eta_{\rm max}^{\rm WDA}$, which is marked by the full circles in Fig. 7.8. The WDA is not able to reproduce the exact result $\eta_{\rm max} = 1$ for any $\epsilon < 0.5\sigma$, but taking into account that this DF approximation was designed using only information of the bulk liquid correlation structure, it is rewarding to observe that it gives a reasonable estimate for the accessible 0D states, ranging from $\eta_{\rm max}^{\rm WDA} \approx 0.8$ at $\epsilon = 0$ to $\eta_{\rm max}^{\rm WDA} \approx 1.3$ at $\epsilon = R$. Moreover, the values of $\beta\mathcal{F}_{\rm ex}^{\rm WDA}[\rho]$ are reasonably close to the exact result, shown by the thick full lines in Fig. 7.8. The performance of the WDA for the description of the HS crystal is directly

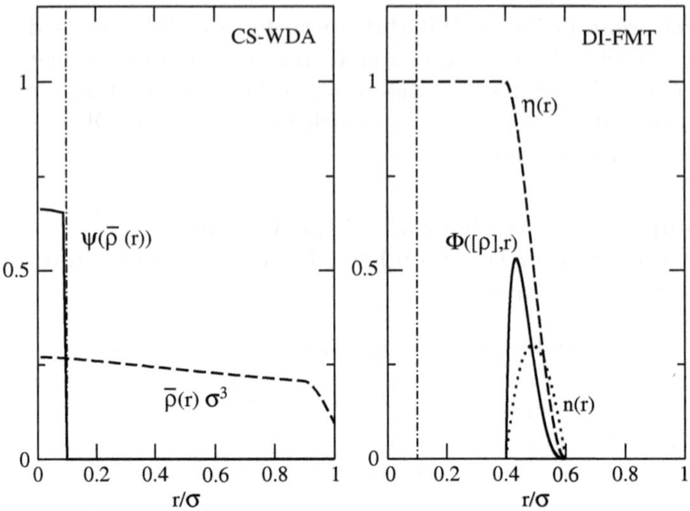

Fig. 7.7. Excess free energy density for a 0D cavity, with uniform density within a sphere of radius $\epsilon = 0.1\sigma$, marked by the vertical *dash-dotted lines*. The *left panel* shows the CS-WDA weighted density (*dashed line*) and the free energy excess per particle (*full line*) evaluated at $\bar{\rho}$, within the spherical density distribution, because that DF approximation assigns $\Phi([\rho], r)$ to the positions of the HS center; η_0 is taken close to its maximum allowed value. The *right panel* shows the main elements of the FMT for $\eta_0 = 1$, the local packing fraction (*dashed line*) is a smoothed radial step around $r = \sigma/2$, with the inner value set by η_0. The scalar density $n(r)$ (*dotted line*) is restricted to the region $\sigma/2 - \epsilon < r < \sigma/2 + \epsilon$, and the same applies to the (not shown) vectorial and tensorial densities. Therefore, the excess free energy density (*full line*) is locally assigned to the positions of the HS surface

related to that for 0D cavities; when the unit cell occupancy is set to unity, the WDA estimation for the free energy is fairly good, because for $\eta_0 = 1$ the values of $\beta\mathcal{F}_{\rm ex}^{\rm WDA}[\rho]$ are not far from the exact $\phi_0(1) = 1$. However, when the unit cell occupancy is used as a free variational parameter, the WDA result for the HS crystal is spoiled, because it is not able to locate the precise DF boundary at $\eta_0 = 1$.

In contrast, the FMT appears to have the precise non-local elements to extract the 0D limit, and it does it in a very different manner than the WDA, as shown in the right panel of Fig. 7.7. The local packing fraction $\eta(\boldsymbol{r})$ obtained from (7.73) reaches the value set by the total cavity occupancy η_0 for any $|\boldsymbol{r}| \leq R - \epsilon$, while the scalar, vector and tensor densities have non-zero values only over the range $R - \epsilon \leq |\boldsymbol{r}| \leq R + \epsilon$, so that the free energy excess is locally assigned to that region. The singularity of $\Phi^{\rm FMT}([\rho]; \boldsymbol{r})$ at $\eta_0 = 1$ comes from the integration of the local free energy around $|\boldsymbol{r}| = R - \epsilon$, with vanishing combinations of $n(\boldsymbol{r})$, $\boldsymbol{v}(\boldsymbol{r})$ and $\mathcal{T}(\boldsymbol{r})$, divided by powers of $1 - \eta(\boldsymbol{r})$. Both for the FMT based only on the OD limit and for the DI-FMT, the integration by parts of $\Phi^{\rm FMT}([\rho]; \boldsymbol{r})$ recovers the singular behavior of the exact $\phi_0(\eta_0)$,

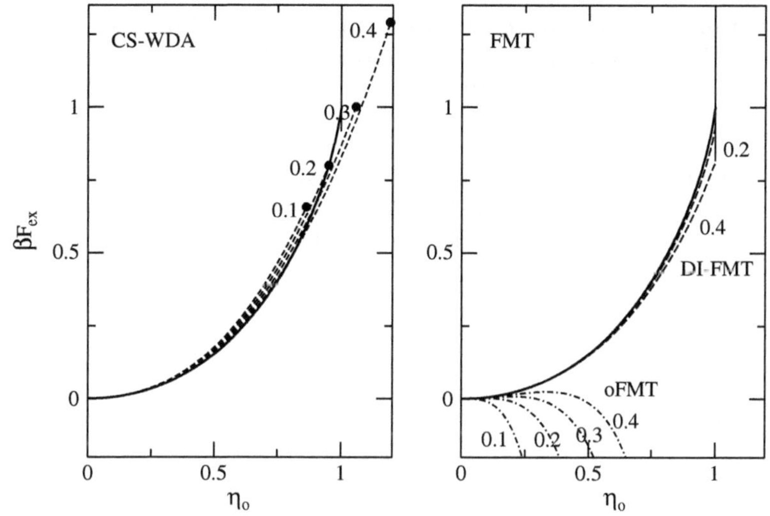

Fig. 7.8. Excess free energy, in kT units, for 0D spherical cavities with $\rho(r) = 3\eta_0/(4\pi\epsilon^3)$ for $r \leq \epsilon < \sigma/2$ and $\rho(r) = 0$ otherwise. The exact result (*full line* in both panels) is independent of ϵ, and it is restricted to $\eta_0 \leq 1$, as marked by the vertical lines. The *dashed lines* in the *left panel* show the CS-WDA results for $\epsilon/\sigma = 0.1, 0.2, 0.3$ and 0.4 as labeled. The *full circles* mark the upper limit for η_0 set by the existence of real roots for equation (7.30). The *right panel* shows the results for the original FMT (*dash-dotted lines*) and the DI-FMT (*dashed lines*). The latter is very close to the exact results, so that only the results for $\epsilon/\sigma = 0.2$ and 0.4 are presented. The oFMT presents qualitatively wrong negative values, diverging for $\eta_0 \to 1$ at any ϵ and for $\epsilon \to 0$ at any η_0

with diverging slope at $\eta_0 = 1$, so that for any finite $\epsilon < R$, the boundary between accessible and unaccessible density distributions is exactly located. Nevertheless, there are small shifts, of order $\eta_0^3\epsilon^2$, which generate a weak dependence of the free energy excess with the cavity radius ϵ. For the DI-FMT results presented in the right panel of Fig. 7.8, that deviation is hardly visible at $\epsilon < 0.2\sigma$. The FMT version based only on the 0D limit would reduce this small discrepancy with the exact $\phi_0(\eta_0)$ but at the cost of a qualitative degradation of the results for the bulk HS fluid.

The oFTM proposed by Rosenfeld shows here a qualitative difference, because the contribution $\phi_3^{(3D)}(\eta, n, \boldsymbol{v})$ in (7.68) has a negative divergence, proportional to $-[\epsilon\ (1-\eta_0)]^{-2}$, so that it ruins the good behavior of $\phi_1^{(3D)} + \phi_2^{(3D)}$. As shown in Fig. 7.8, the results become even worse as ϵ decreases. From the results for the simple 0D cavity analyzed here, it becomes clear that the DF description of the HS crystal with the oFMT is bound to failure: the narrow Gaussian peaks describing the fluctuations of each particle around its lattice site would produce the same pathology in the evaluation of $\Phi^{\mathrm{FMT}}([\rho]; \boldsymbol{r})$, and the minimization with respect to the Gaussian width would lead to a spurious negative divergence for delta-function peaks.

7.5.5 Application of the Fundamental Measure Theory to Planar Density Profiles: Practical Handling of Vector and Tensor Weights

The application of the FMT to describe the density profiles of a HS fluid against a planar wall, or any other system with $\rho(\boldsymbol{r}) = \rho(z)$, is technically similar in either the oFMT or the DI-FMT, and they also produce very similar results. The convolutions of the density profile adopt the simple forms

$$\eta(z) = \pi \int_{-R}^{R} \mathrm{d}z' \rho(z+z')(R^2 - z'^2) , \tag{7.74}$$

$$n(z) = \int_{-R}^{R} \mathrm{d}z' \frac{\rho(z+z')}{2R} . \tag{7.75}$$

The only non-zero component of the vector density is

$$v_z(z) = \int_{-R}^{R} \mathrm{d}z' \; \rho(z+z') \; \frac{z'}{2R^2} , \tag{7.76}$$

while the tensor weighted density becomes diagonal, with

$$\mathcal{T}_{zz}(z) = \int_{-R} \mathrm{d}z' \; \rho(z+z') \frac{z'^2}{2R^3} \equiv n(z) - \frac{\eta(z)}{2\pi R^3} \tag{7.77}$$

and $\mathcal{T}_{xx} = \mathcal{T}_{yy} = (n - \mathcal{T}_{zz})/2 \equiv \eta(z)/(4\pi R^3)$.

There are only three independent convolutions (7.74)–(7.76), carried over the restricted range $-R \leq z' \leq R$, to get the components of the excess free energy density, $\phi_1^{(3\mathrm{D})} = -n(z) \ln[1 - \eta(z)]$, $\phi_2^{(3\mathrm{D})} = 4\pi R^3 [n(z)^2 - v_z(z)^2]/[1 - \eta(z)]$ and any of the two versions for the third term, either in the oFMT version (7.68)

$$\phi_3^{(3\mathrm{D})} = 8\pi^2 R^6 n(z) \frac{n(z)^2/3 - v_z(z)^2}{[1-\eta(z)]^2} , \tag{7.78}$$

or in the DI-FMT (7.72)

$$\phi_3^{(3\mathrm{D})} = 3\pi^2 R^6 \frac{[n(z) - \mathcal{T}_{zz}(z)][n(z)^2 - 4v_z(z)^2 + 3\mathcal{T}_{zz}(z)^2]}{[1-\eta(z)]^2} . \tag{7.79}$$

Therefore, the formal use of vector and tensor weighted densities for these planar profiles adds no extra practical difficulty to the FMT with respect to the WDA. As commented above, the original expression (7.78) is recovered from the DI-FMT (7.79) if we neglect the anisotropy of the tensor $\mathcal{T}$ while keeping its trace, $\mathcal{T}_{xx} = \mathcal{T}_{yy} \approx \mathcal{T}_{zz} \approx n/3$.

The integration of $\Phi([\rho]; z) = \sum_{i=1}^{3} \phi_i^{(3D)}$ is similar to that in (7.45), with the only difference that it has to be extended to the full region with $n(z) \neq 0$, which goes up to a distance $R = \sigma/2$ beyond that with $\rho(z) \neq 0$. Thus, to get the excess free energy of a HS fluid against a hard wall, with $\rho(z) = 0$ for $z < 0$, the integral over the excess free energy has to be extended from $z = -R$ to cover all the region accessible to a HS surface, rather than accessible to its center. As in the WDA, we have to extend the integration beyond the maximum distance allowed for variable $\rho(z)$ in order to include the full range with $n(z) \neq \rho_0$. Altogether, the reduced range and the simpler analytical forms of the required density convolutions make the practical use of any of the FMT versions even easier than that of the WDA for planar density profiles. The subtleties of the delta-function convolutions disappear when applied to density distributions which only depend on z, and the two versions give similarly good results. In Fig. 7.9, we present a detailed view of the density profiles $\rho(z)$ for a HS fluid at bulk density $\rho_B = 0.81/\sigma^3$ against a hard wall. The computer simulation results by Snook and Henderson [21] (circles) are compared with the CS-WDA (dotted line) and several FMT versions. It is hard to distinguish the original formulation (oFMT, dashed line) from the dimensional interpolation version (DI-FMT, full line), and they both clearly improve the CS-WDA results in the interstitial region between the first two HS layers although the expanded view presented in this figure allows the observation of some deviations from the simulation data in the detailed shape the layer at $z \lesssim \sigma$, which would be hard to observe in the scale of Fig. 7.2. The DI-FMT results for the surface tension at a hard wall, as a function of the bulk HS density, are shown in Fig. 7.3. They are very similar to those of the

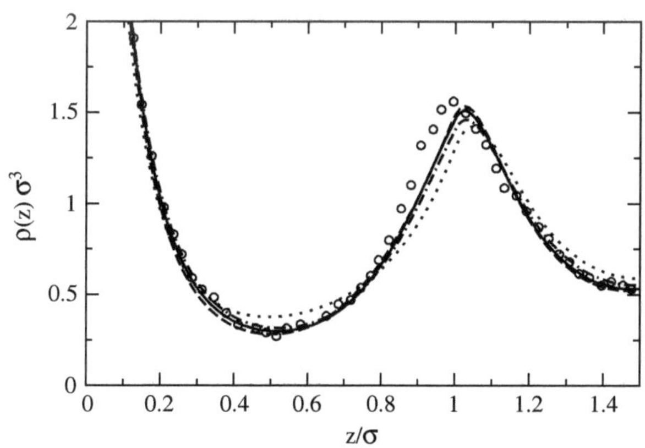

Fig. 7.9. Density profile for the HS fluid, with bulk density $\rho_B = 0.81/\sigma^3$ against a hard wall. The figure presents a detailed view of the central panel in Fig. 7.2. The computer simulations results [21] are compared with the CS-WDA (*dotted line*) and three FMT versions: oFMT (*dashed line*), DI-FMT (*full line*) and CS(WB)-FMT (*dash-dotted line*)

oFMT, slightly above the SPT predictions, and within the error bars of the computer simulations.

Dimensional Reduction to the 2D Hard Disks Fluid

The most extreme inhomogeneity within the category of planar density profiles is the representation of the 2D homogeneous fluid of HD as 3D spheres with $\rho(z) = \rho_{2D}\delta(z)$. All the FMT weighted densities have the simple analytical forms derived from (7.74)–(7.77), and even the integrals of $\Phi_i^{(3D)}([\rho]; z)$ can be done analytically although their expressions are rather cumbersome [62]. All these DF approaches recover the first term in the generic DF virial expansion (7.16), and hence they give the exact value $\bar{B}_2 = 2$ in the virial expansion $\beta p/\rho_{2D} = 1 + \bar{B}_2\eta_{2D} + \bar{B}_3\eta_{2D}^2 + \cdots$, with $\eta_{2D} = \pi\rho_{2D}\sigma^2/4$. For the next term, the DI-FMT gives $\bar{B}_3 = 328/105 = 3.1238$, in very good agreement with the exact value $\bar{B}_3 = 3.128$, slightly above the SPT theory result $\bar{B}_3 = 3$. The CS-WDA ($\bar{B}_3 = 3.54$) and Rosenfeld's oFMT ($\bar{B}_3 = 8/3 = 2.666$) have larger deviations, with opposite signs.

However, the general aspect of $\beta p/\rho_{2D}$ at large ρ_{2D}, presented in Fig. 7.10, gives better overall results for the oFMT than for the DI-FMT. The quasi-exact semi-empirical equation of state of Colot and Baus [63] appears between the SPT and the oFMT, while the DI-FMT deviates upward, but is still much better than the CS-WDA. It was shown in [62] that all the FMT versions produce a dependence $\beta p/\rho_{2D} \sim (1-\eta_{2D})^{-5/2}$, which fits worse than the SPT form, $\beta p/\rho_{2D} \sim (1-\eta_{2D})^{-2}$, to the density dependence of $\beta p/\rho_{2D}$ at large 2D fluid densities. Therefore, the excellent result of the oFMT comes from the fortunate cancellation of errors, between the underestimation of $\bar{B}_3$ for the

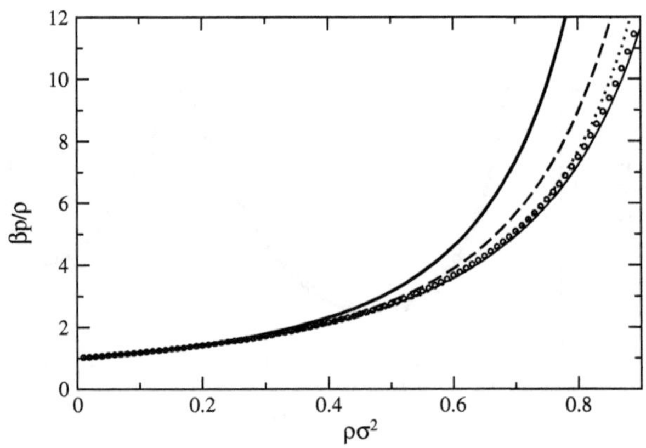

Fig. 7.10. The equations of state for the 2D hard disks fluid given by the SPT (*thin full line*) and the quasi-exact semi-empirical fit of Colot and Baus [63] (*circles*) are compared to those derived from the dimensional crossover of the free energy DF for 3D hard spheres: CS-WDA (*thick full line*), oFMT (*dotted line*) and DI-FMT (*dashed line*)

low-density expansion and the overestimation of the *ideal liquid* divergence in the power of $1-\eta_{2D}$. The DI-FMT, built to interpolate between the 0D cavity and the 3D bulk fluid, goes exactly through the 1D free energy and gives an excellent result for B_3 in 2D, but it cannot keep the same quality at higher 2D densities. A different FMT version, with excellent interpolation between 0D, 1D and 2D, was also found [62], but it would give a rather poor estimation for the 3D equation of state. Altogether, we may say that the dimensional reduction from the DF for 3D HS to the equation of state of the 2D HD fluid is a good test for the DF dependence of $\Phi^{(3D)}([\rho], r)$ and that the different FMT versions perform better than any WDA. Nevertheless, the best FMT versions are reaching the limits within that DF class, so that any consistent improvement would probably require the inclusion of a qualitatively different DF structure.

7.5.6 Dimensional Interpolation Fundamental Measure Theory Applied to the Hard Sphere Crystal

The application of the oFMT to describe the HS crystal was precluded by the spurious negative divergence of the $\Phi_3^{(3D)}$ term. It was suggested [13] that this failure represented a fundamental flaw in the DF description of a crystal as a self-structured fluid, through the minimization of a density distribution like (7.50). The analysis of the 0D limit showed that the problem was directly associated to the negative divergence of $\Phi_3^{(3D)}$, and an empirical modification of the oFMT was proposed [64, 65] to eliminate it and to recover a good description of the HS crystal within the FMT. However, it was only with the DI-FMT, and through the use of the tensor weight function, that the problem could be eliminated from its root, with the use of delta-function shell convolutions vanishing in the strict 0D limit. The practical application of the DI-FMT to density distributions made of Gaussian peaks is computationally more demanding than with the WDA, since the local distribution of excess free energy is assigned to the interstitial spaces, rather than being proportional to the narrow peaks in $\rho(\boldsymbol{r})$. Within the density parametrization (7.50), we may obtain analytical expressions for the convolutions with a single, normalized, Gaussian peak centered at the origin. The scalar function is

$$n^{\text{gauss}}(\boldsymbol{r}) = \sqrt{\frac{\alpha}{\pi}} \, \frac{\mathrm{e}^{-\alpha(|\boldsymbol{r}|-R)^2} - \mathrm{e}^{-\alpha(|\boldsymbol{r}|+R)^2}}{4\pi R|\boldsymbol{r}|} \approx \sqrt{\frac{\alpha}{\pi}} \, \frac{\mathrm{e}^{-\alpha(|\boldsymbol{r}|-R)^2}}{4\pi R|\boldsymbol{r}|} , \tag{7.80}$$

where the second exponential term may be obviously neglected for any relevant value of $r \approx R$ and $\alpha\sigma^2 \gg 1$. Within the same restrictions, the vector function is

$$\boldsymbol{v}^{\text{gauss}}(\boldsymbol{r}) = n^{\text{gauss}}(\boldsymbol{r}) \left(1 - \frac{1}{2\alpha R|\boldsymbol{r}|}\right) \frac{\boldsymbol{r}}{|\boldsymbol{r}|} , \tag{7.81}$$

parallel to $\boldsymbol{r}$. The tensor components $\mathcal{T}_{\mu\nu}$, with $\mu, \nu = x, y, z$, take the form

$$\mathcal{T}^{\text{gauss}}_{\beta\gamma}(\boldsymbol{r}) = n^{\text{gauss}}(\boldsymbol{r}) \left[\frac{1-\xi(|\boldsymbol{r}|)}{2}\,\delta_{\beta\gamma} + \frac{3\xi(|\boldsymbol{r}|)-1}{2}\,\frac{r_\beta r_\gamma}{|\boldsymbol{r}|^2} \right], \tag{7.82}$$

with a diagonal term, and the direct product of the radial directions, given in terms of the function

$$\xi(r) = 1 - \frac{1}{\alpha R r} + \frac{1}{2(\alpha R r)^2}. \tag{7.83}$$

Notice that $\boldsymbol{v}^{\text{gauss}}(\boldsymbol{r})$ and $\mathcal{T}^{\text{gauss}}(\boldsymbol{r})$ have also exact (but more cumbersome) expressions to be used away from the practical restriction to $r \approx R$ and $\alpha\sigma^2 \gg 1$. The local packing fraction takes the form

$$\begin{aligned}\eta^{\text{gauss}}(r,\alpha) = \frac{3}{2}\Big[&\operatorname{erf}\left(\sqrt{\alpha}(R-|\boldsymbol{r}|)\right) + \operatorname{erf}\left(\sqrt{\alpha}(R+|\boldsymbol{r}|)\right) \\ &- \frac{\mathrm{e}^{-\alpha(R-|\boldsymbol{r}|)^2} - \mathrm{e}^{-\alpha(R+|\boldsymbol{r}|)^2}}{|\boldsymbol{r}|\sqrt{\pi\alpha}} \Big],\end{aligned} \tag{7.84}$$

The contributions of all the Gaussian peaks on the crystal lattice have to be added, with fast convergence in terms of the distance to the point $\boldsymbol{r}$ to the lattice position. After adding all the relevant contributions to the vector and tensor cartesian components, the rotational invariant combinations in (7.72) are evaluated. The final integral of $\Phi([\rho];\boldsymbol{r})$ has to be done much more carefully than for the WDA, since it extends to a narrow region on the lattice interstitial space and it contains inverse powers of $1-\eta(\boldsymbol{r})$ with (integrable) divergences in the strict limit of full unit cell occupancy. See [66] for a discussion of the present best numerical results.

The larger computational cost necessary to obtain results for the HS crystal with this DF approximation is clearly worthwhile because the equation of state of the HS crystal and its structure (represented by the optimal parameter α in (7.50)) are in excellent agreement with computer simulation results [61], as shown in Fig. 7.11, and certainly improve on the WDA results. Moreover, the extension of the DF minimization beyond the normalized Gaussian representation provides an even stronger support for the DI-FMT. The wider parametrization

$$\rho(\boldsymbol{r}) = \eta_0 \left(\frac{\alpha}{\pi}\right)^{\frac{3}{2}} \sum_{\{\boldsymbol{R}\}} \mathrm{e}^{-\alpha|\boldsymbol{r}-\boldsymbol{R}|^2} \left[1 + \tau\xi(\boldsymbol{r}-\boldsymbol{R})\right], \tag{7.85}$$

where $\xi(\boldsymbol{r}) = x^4 + y^4 + z^4 - 3|\boldsymbol{r}|^4/5$ is the leading term in the anisotropy of the cubic lattices, was used to minimize $\Omega[\rho]$ with respect to η_0, α and τ. The restriction to unit cell occupancy, $\eta_0 = 1$, which was essential for the accuracy of the WDA, is removed and the minimization of the DI-FMT free energy produces the self-adjusted normalization to $1-\eta_0 \approx 10^{-8}$, i.e. a crystal with a negligibly small number of vacancies, well below the limit of observation in computer simulations (see the remarks on [67]). The anisotropy

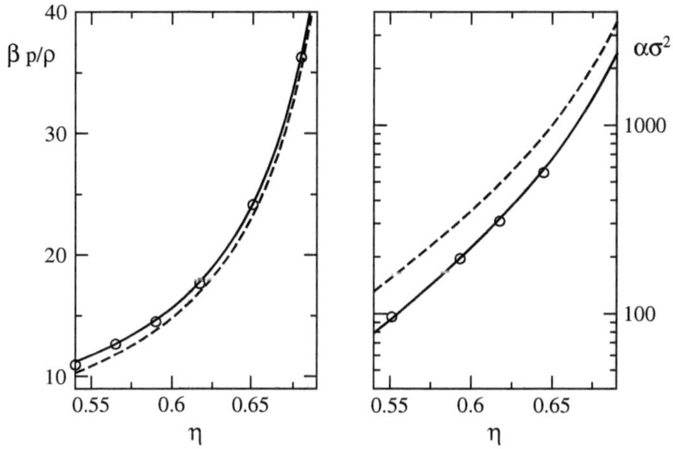

Fig. 7.11. The equation of state (*left*) and the square inverse Gaussian width parameter (*right*) for the HS FCC crystal, in terms of the mean packing fraction $\eta = \rho_0 \pi \sigma^3/6$. The *full lines* are the results of the DI-FMT, the *dashed lines* those of the CS-WDA and the circles present the computer simulation data from [32, 33].

of the density distribution was characterized with $\tau \approx 0.02$ for a crystal with mean density $\rho_0 = 1.03/\sigma^3$, which is consistent with the computer simulation observation of a maximum spreading between 10 and 15% of the density along the interstitial $[1,1,1]$ direction, with respect to that of the nearest neighbor $[1,1,0]$ direction, of the FCC lattice. This contrasts with the opposite sign for the anisotropy ($\tau < 0$) predicted by the WDA, which would correspond to a small, but unphysical, accumulation of the density along the nearest neighbor directions [38].

As with the WDA, the DI-FMT does not discern between the free energy of the different closed packed structures. The FCC and HCP lattices have identical free energies, up to the present best numerical accuracy. However, the application to other (metastable) crystal structures marks a qualitative difference, again in favor of the DI-FMT. A HS crystal with BCC structure is unstable with respect to shears which would transform it into the FCC, but its DF characterization is still useful for the description of systems with soft repulsions from a HS reference system [68]. However, the WDA with normalized Gaussian peaks on a BCC lattice gives fully unphysical predictions, with the Gaussian peaks becoming wider as the density approaches its maximum compatible with the BCC structure. The use of the DI-FMT solves the problem, and it gives very reasonable predictions for the metastable BCC and even for the simple cubic, crystal structures [66].

Altogether, the representation of the HS crystal within the DI-FMT is excellent and qualitatively superior to the WDA or any other DF theory. In fact, the main practical problem of the DI-FMT to study the crystallization of HS comes from its unprecedented success in the representation of the HS crystal, since its equation of state is much more accurate than that of the compressibility PY approximation for the bulk HS fluid, which is taken as an input to design the DF approximation. The predicted densities of the coexisting fluid and crystal phases are rather poor, $\rho_f = 0.892/\sigma^3$ and $\rho_c = 0.985/\sigma^3$, far from the accuracy of the CS-WDA presented in Table 7.1, which makes more paired errors in the thermodynamics of the two phases. In the following subsection, we comment on an attempt to solve this puzzling situation, having a DF approximation which is too good in the description of the structured phase, compared with the (input) thermodynamics of the homogeneous liquid. However, we may anticipate that the question is still open, as a challenge for the future, in the amazing development of the FMT.

7.5.7 Other Applications and Versions of the Fundamental Measure Theory

Beside those which have already been commented, DF approximations within the FMT scheme have been applied to other interesting problems, e.g. the use of the test-particle route for the correlation structure [69] or fluids confined in narrow cylindrical or spherical cavities [70]. Also, we have to comment on several variants of the FMT which have been explored and used in different contexts, like that of Gonzalez et al. [71], who build a FMT version using only the two simplest elements: $\eta(\boldsymbol{r})$ and the scalar delta-function shell convolution $n(\boldsymbol{r})$. Their construction is based on the idea that $\phi_0(\eta_0) - 1$ is the ideal free energy of a gas of holes (of density $1 - \eta_0$) and it may be used as a generating functional. It is clear that such a DF approximation will have stronger unphysical divergences than the original FMT, since none of the delta-function shell overlaps are eliminated; however, all these divergences have positive sign, and the DF minimization of $\Omega[\rho]$ always avoids them. The quality of the results for systems with planar symmetry was clearly worse than with the oFMT, since avoiding the spurious divergences $\rho(\boldsymbol{r})$ develops unphysical kinks, but at least the results represent true minima of $\Omega[\rho]$, since $\mathcal{F}_{\mathrm{ex}}[\rho]$ does not have the spurious bottomless pits for 0D cavities. An interesting feature of that simple FMT version was its flexibility to plug any bulk thermodynamics, like the CS equation of state which could improve on the PY approximation.

The idea of plugging the CS equation of state within the FMT is most appealing with the DI-FMT, since the excellent quality of the results for the crystal phase made (comparatively) bad the PY approximation for the bulk fluid. Along the lines of Gonzalez et al. [71], a CS version of the FMT was proposed and tested by Tarazona [62] maintaining the DI-FMT structure, but changing the η-dependence of $\phi_3^{(3D)}$ in (7.72) to

$$\begin{aligned}\phi_3^{(3D)}(\eta, n, \boldsymbol{v}, \mathcal{T}) = {} & 12\pi^2 R^6 \varphi_{\mathrm{CS}}(\eta(\boldsymbol{r})) \\ & \times \left(\boldsymbol{v}\cdot\mathcal{T}\cdot\boldsymbol{v} - n\,\boldsymbol{v}\cdot\boldsymbol{v} - \mathrm{Tr}[\mathcal{T}^3] + n\,\mathrm{Tr}[\mathcal{T}^2]\right),\end{aligned} \tag{7.86}$$

with

$$\varphi_{\mathrm{CS}}(\eta) = \frac{2}{3\eta^2}\left[\frac{\eta}{(1-\eta)^2} + \ln(1-\eta)\right] = 1 + \frac{16}{9}\eta + \frac{5}{2}\eta^2 + \cdots \qquad (7.87)$$

instead of $\varphi_{\mathrm{PY}}(\eta) = (1-\eta)^{-2} = 1 + 2\eta + 3\eta^2 + \cdots$ used in (7.72). This DI-FMT version, which we will refer to as the CS-FMT, was used to study the HS crystal and its dimensional reduction to the 2D fluid of HD [62]. The extension of this approximation to HS mixtures was later proposed by Roth et al. [72], under the name of the White Bear (WB) FMT. That name appears sometimes associated also to the mono-component case, which was used in the same paper as a check of the results in [62] for the HS crystal and to explore other problems of interest. The results of that CS(WB)-FMT are reasonably good, keeping all the qualitative advantages of the DI-FMT with respect to the WDA and improving the value of the coexisting fluid and crystal densities: $\rho_f = 0.934/\sigma^3$ and $\rho_c = 1.023/\sigma^3$. However, it is clear that the quality of the description for the HS crystal is degraded with respect to the (quasi-exact) results of the DI-FMT. Also in the application to planar density profiles, like the HS fluid against a hard wall shown in Fig. 7.9, the accuracy of the CS bulk equation of state is reflected in a better value for the contact density $\rho(0) = \beta p(\rho_B)$ given by the CS(WB)-FMT, but the effect is restricted to a very narrow range of distances to the wall, and for $z \approx \sigma$ the CS(WB)-FMT results are marginally worse than those of the DI-FMT or the oFTM.

Therefore, the adjustment of the free energy for the bulk fluid in (7.87) is done somehow against the natural structure of the FMT, changing the $1/(1-\eta)^2$ dependence in (7.72) which emanates from the third derivative of the exact 0D free energy $\phi_0(\eta)$. It was also noticed by Tarazona [62] that the low-density expansion of the PY approximation is exact, up to order ρ_0^2 (included), for any $|\boldsymbol{r}| \leq \sigma$. The first discrepancy between the PY and the exact $c(r, \rho_0)$ arises at order ρ_0^3 inside the core, but it already appears at order ρ_0^2 outside the core, where the tail for $\sigma < r \leq \sqrt{2}\sigma$ in the exact expansion is neglected by the PY assumption $c(r, \rho_0) = 0$ for $r \geq \sigma$. The volume integral of this tail produces the difference between the fourth virial coefficients in the PY and the CS equations of state (bracketing the exact result), but within the CS(WB)-FMT that difference is assigned to the core of $c(r, \rho_0)$ rather than to create a tail for $r > \sigma$.

Therefore, the empirical corrections leading to (7.86), or to the generic WB-FMT for HS mixtures, are giving the wrong non-local dependence to the thermodynamic improvement, beyond the PY approximation. The FMT appears to be the natural extension of the PY approximation for inhomogeneous HS fluids, with the assumption $\delta^2\mathcal{F}_{\mathrm{ex}}[\rho]/\delta\rho(\boldsymbol{r})\delta\rho(\boldsymbol{r}') = 0$ for $|\boldsymbol{r} - \boldsymbol{r}'| \geq \sigma$ emanating from the use of the R-ranged *fundamental measures*. Hence, it would be difficult to achieve a really consistent improvement of the FMT, with a better thermodynamics for the bulk fluid, without going beyond the use of one-center convolutions of the density. The analysis of HS mixtures presented in the next sections also supports the FMT as the most satisfactory framework to get very good DF approximations for hard core particles, but still with its intrinsic limitations, which seem difficult to overcome without a drastic transformation of the generic description of the non-local dependence in $\mathcal{F}_{\text{ex}}[\rho]$.

7.6 Density Functionals for Hard Sphere Mixtures

There are two obvious extensions of any DF approach to the HS model: including soft, and maybe attractive, potentials and studying mixtures of HS. For the former, it has already been commented that one way to deal with such models is to use a MFA for the attractive part of the interaction potential over a reference HS system (with possibly a temperature-dependent diameter) that accounts for the repulsive part. The latter, however, turns out to be more problematic, because we need special-purpose DFs to deal with mixtures of HS, and the DFs we have already studied for a mono-component fluid of HS are only half-way the final result. We need further insights to study mixtures.

When facing this problem, the situation is very different if one considers WDA-like or FMT-like DFs. While the WDA is the 'easiest' non-local DF theory for HS, extending it for mixtures becomes a headache. FMT, however, is a DF theory whose formulation for mixtures is rather straightforward, and (except for some subtleties that will be explained later) its construction adds no further complications to the theory. We will review both theories in what follows, but first let us discuss the modifications we have to make in the general scheme of DF theory in order to account for mixtures.

First of all, every component or species of an m-component inhomogeneous mixture will have its own density profile, $\rho_\nu(\boldsymbol{r})$, $\nu = 1, 2, \ldots, m$. To simplify notation, we introduce the vector

$$\boldsymbol{\rho}(\boldsymbol{r}) = \Big(\rho_1(\boldsymbol{r}), \ldots, \rho_m(\boldsymbol{r})\Big) \tag{7.88}$$

to refer to all density profiles collectively. As for one-component systems, there exist a unique intrinsic free energy $\mathcal{F}[\boldsymbol{\rho}]$ which is a functional of the vector density $\boldsymbol{\rho}(\boldsymbol{r})$. If on particles of species ν acts an external potential $V_\nu(\boldsymbol{r})$ and there is a reservoir of particles with chemical potential μ_ν for that species, the grand potential

$$\Omega[\boldsymbol{\rho}] = \mathcal{F}[\boldsymbol{\rho}] + \sum_{\nu=1}^{m} \int \mathrm{d}\boldsymbol{r}\, \rho_\nu(\boldsymbol{r})[V_\nu(\boldsymbol{r}) - \mu_\nu] \tag{7.89}$$

reaches its absolute minimum for the equilibrium density profile vector $\boldsymbol{\rho}(\boldsymbol{r})$. This minimum principle leads to the set of Euler–Lagrange equations

$$\frac{\delta\Omega[\boldsymbol{\rho}]}{\delta\rho_\nu(\boldsymbol{r})} = \frac{\delta\mathcal{F}[\boldsymbol{\rho}]}{\delta\rho_\nu(\boldsymbol{r})} + V_\nu(\boldsymbol{r}) - \mu_\nu = 0, \qquad \nu = 1, \ldots, m\,. \tag{7.90}$$

A similar reasoning to that of Sect. 7.2.1 leads to the DF for an ideal gas mixture

$$\beta\mathcal{F}_{\text{id}}[\boldsymbol{\rho}] = \sum_{\nu=1}^{m} \int \mathrm{d}\boldsymbol{r}\, \rho_\nu(\boldsymbol{r})[\ln \rho_\nu(\boldsymbol{r}) - 1]. \tag{7.91}$$

A remark is on purpose here. As in the one-component case, there is a thermal wavelength Λ_ν which is different for each component. When dealing with 'quenched' mixtures, the most common kind of mixture, for which the global composition of the system is fixed beforehand, we can ignore this wavelength, as we have done so far, and set it to 1. This is what we will do henceforth. However, maintaining this factor is relevant when species are actually aggregates of particles whose composition is determined by chemical equilibrium. When describing this kind of systems, the thermal wavelength (or rather the thermal volume) is replaced by the internal partition function of an aggregate of species ν, and its specific shape is most relevant for the equilibrium configuration of the system [73] (for an illustration of the dramatic effects this may have on the system, see [74, 75, 76, 77]).

The ideal term contains the entropy of mixing, and the remaining contribution to a DF will be the excess free energy,

$$\beta\mathcal{F}[\boldsymbol{\rho}] = \beta\mathcal{F}_{\text{id}}[\boldsymbol{\rho}] + \beta\mathcal{F}_{\text{ex}}[\boldsymbol{\rho}] \equiv \int \mathrm{d}\boldsymbol{r}\, \left\{\Phi_{\text{id}}(\boldsymbol{\rho}(\boldsymbol{r})) + \Phi([\boldsymbol{\rho}];\boldsymbol{r})\right\} , \tag{7.92}$$

where $\Phi_{\text{id}}(\boldsymbol{\rho}) = \sum_\nu \rho_\nu(\ln\rho_\nu - 1)$. The direct correlation function will now be a matrix, which can be obtained trough

$$\left.\frac{\delta^2\beta\mathcal{F}_{\text{ex}}[\boldsymbol{\rho}]}{\delta\rho_\mu(\boldsymbol{r}_1)\delta\rho_\nu(\boldsymbol{r}_2)}\right|_{\boldsymbol{\rho}(\boldsymbol{r})=\boldsymbol{\rho}_0} = -c_{\mu\nu}(r_{12}, \boldsymbol{\rho}_0), \qquad \mu,\nu = 1,\ldots,m , \tag{7.93}$$

and which is related to the total correlation function $h_{\mu\nu}(r, \boldsymbol{\rho}_0)$ via the (matrix) Ornstein–Zernike equation

$$h_{\mu\nu}(r, \boldsymbol{\rho}_0) = c_{\mu\nu}(r, \boldsymbol{\rho}_0) + \sum_{\lambda=1}^{m} \rho_{0,\lambda} \int \mathrm{d}\boldsymbol{r}' c_{\mu\lambda}(r', \boldsymbol{\rho}_0) h_{\lambda\nu}(|\boldsymbol{r} - \boldsymbol{r}'|, \boldsymbol{\rho}_0). \tag{7.94}$$

One final remark concerns polydisperse systems. In all expressions above, we have considered, for simplicity, multicomponent mixtures. Polydisperse mixtures are a special kind of mixtures in which there are virtually an infinity of species, labeled by a continuous parameter $s \in (s_0, s_1)$ (s may refer to size, shape, charge or whatever parameter or set of parameters that renders the mixture polydisperse). For these systems, $\rho_\nu(\boldsymbol{r})$ becomes a function $\rho(s;\boldsymbol{r})$ and dependencies on $\boldsymbol{\rho}(\boldsymbol{r})$ become functional dependencies on that function. Hence, all expressions change accordingly and all sums on ν must be replaced by integrals on s. The resulting equations are formally the same, so we will only refer to this kind of mixtures when there are relevant issues to mention.

7.7 Weighted Density Approximation for Mixtures of Hard Spheres

In 1990, Denton and Ashcroft proposed an extension of the WDA to binary mixtures [50] and applied it to determine freezing. The theory was one of the many conceivable generalization of the WDA for HS, and in its construction several simplifications were assumed. Although some of them may be questionable, the theory represents a reasonable trade off between flexibility and numerical tractability. The form we are presenting it here would be its formulation for an m-component mixture although, as it will be clear, applying the resulting DF to more than a few components (most of its applications are for just 2 components) is only for the braves.

Denton and Ashcroft's proposal for the excess free energy density is

$$\Phi_{\mathrm{WDA}}([\boldsymbol{\rho}],\boldsymbol{r}) = \sum_{\mu=1}^{m} \rho_\mu(\boldsymbol{r})\psi(\bar{\rho}^{(\mu)},\boldsymbol{x}) , \tag{7.95}$$

where $\boldsymbol{x} = (x_1,\dots,x_m)$ is the *composition* vector whose components, $x_\nu = \rho_\nu/\rho$, $\rho = \sum_\nu \rho_\nu$, are the global values of the molar fractions of the species over the entire system of volume V, and $\psi(\rho,\boldsymbol{x})$ is the free energy per particle of a mixture of total density ρ and composition vector $\boldsymbol{x}$. The effective *total densities* $\bar{\rho}^{(\mu)}$, $\mu = 1,\dots,m$, are defined through weighted convolutions of the density profiles

$$\bar{\rho}^{(\mu)}(\boldsymbol{r}) = \sum_{\nu=1}^{m} \int \mathrm{d}\boldsymbol{r}' \rho_\nu(\boldsymbol{r}+\boldsymbol{r}') w_{\mu\nu}(r';\bar{\rho}^{(\mu)}(\boldsymbol{r}),\boldsymbol{x}). \tag{7.96}$$

Weights are assumed to be normalized to 1 and symmetric ($w_{\mu\nu} = w_{\nu\mu}$).

Now, as in the WDA for HS, determining the weights amounts to relate the second derivative of $\beta\mathcal{F}_{\mathrm{ex}}[\boldsymbol{\rho}]$ of the uniform mixture with its direct correlation function, i.e.

$$\left.\frac{\delta^2\beta\mathcal{F}_{\mathrm{ex}}[\boldsymbol{\rho}]}{\delta\rho_\mu(\boldsymbol{r})\delta\rho_\nu(\boldsymbol{r}')}\right|_{\boldsymbol{\rho}(\boldsymbol{r})=\boldsymbol{\rho}_0} = -c_{\mu\nu}(|\boldsymbol{r}-\boldsymbol{r}'|,\boldsymbol{\rho}_0) . \tag{7.97}$$

Before going any further, let us comment on the assumptions leading to this proposal. There are two main simplifications: one is the dependence on $\boldsymbol{x}$, rather than assuming a dependence on the whole set of effective densities $\bar{\rho}^{(\mu)}(\boldsymbol{r})$; the other one is to assume the same form $\psi(\rho,\boldsymbol{x})$ for all partial free energy per particle functions in the ansatz (7.95). The aim of both simplifications is to keep the complexity of the theory at a reasonable level. However, there is a rationale behind them: the approximation implies that each particle of the non-uniform mixture contributes to the excess free energy as if it interacted with an effective locally uniform mixture whose total density is different for each species.

It turns out that, of the two assumptions made by Denton and Ashcroft, the most questionable is the constant composition hypothesis. A decade later, Davidchack and Laird made a version of this WDA without this assumption [78]. They replaced equation (7.95) by

$$\Phi_{\mathrm{WDA}}([\boldsymbol{\rho}],\boldsymbol{r}) = \sum_{\mu=1}^{m} \rho_\mu(\boldsymbol{r})\psi(\bar{\boldsymbol{\rho}}^{(\mu)}(\boldsymbol{r})) \,, \tag{7.98}$$

where now $\psi(\boldsymbol{\rho})$ is the free energy per particle of a mixture with partial densities $\boldsymbol{\rho} = (\rho_1,\dots,\rho_m)$, and the effective densities $\bar{\boldsymbol{\rho}}^{(\mu)}(\boldsymbol{r}) = \left(\bar{\rho}_1^{(\mu)},\dots,\bar{\rho}_m^{(\mu)}\right)$, $\mu = 1,\dots,m$, are defined by the averages

$$\bar{\rho}_\nu^{(\mu)}(\boldsymbol{r}) = \int \mathrm{d}\boldsymbol{r}'\rho_\nu(\boldsymbol{r}+\boldsymbol{r}')w_{\mu\nu}(r';\bar{\boldsymbol{\rho}}^{(\mu)}(\boldsymbol{r})) \,. \tag{7.99}$$

Not only is this version of the WDA for mixtures more symmetric, but it also leads to simpler equations for the weights. Because of the constraint on the composition, in the version of Denton and Ashcroft there appears a term proportional to the inverse volume of the system, V^{-1}, which is absent in the Davidchack and Laird formulation.

When equations (7.98) and (7.99) are used in (7.97), the resulting equations for the weights are (in Fourier space)

$$\begin{aligned} -\hat{c}_{\mu\nu}(k) = {} & \left(\frac{\partial\psi}{\partial\rho_\mu} + \frac{\partial\psi}{\partial\rho_\nu}\right)\hat{w}_{\mu\nu}(k) + \frac{\partial^2\psi}{\partial\rho_\mu\partial\rho_\nu}\sum_{\lambda=1}^{m}\rho_\lambda\hat{w}_{\mu\lambda}(k)w_{\lambda\nu}(k) \\ & + \frac{\partial\psi}{\partial\rho_\mu}\sum_{\lambda=1}^{m}\rho_\lambda\frac{\partial\hat{w}_{\mu\lambda}}{\partial\rho_\nu}(k)\hat{w}_{\lambda\nu}(k) + \frac{\partial\psi}{\partial\rho_\nu}\sum_{\lambda=1}^{m}\rho_\lambda\hat{w}_{\mu\lambda}(k)\frac{\partial\hat{w}_{\lambda\nu}}{\partial\rho_\nu}(k), \end{aligned} \tag{7.100}$$

where the dependence on $\boldsymbol{\rho}$ is implicitly assumed. Equation (7.100) represents a computational challenge: they form a system of partial differential equations very hard to solve even numerically. One could try the CS-WDA strategy, namely expanding $w_{\mu\nu}(r,\boldsymbol{\rho})$ in powers of the densities ρ_λ,

$$w_{\mu\nu}(r,\boldsymbol{\rho}) = w_{\mu\nu}^{(0)}(r) + \sum_{\lambda=1}^{m}\rho_\lambda w_{\mu\nu\lambda}^{(1)}(r) + \sum_{\lambda,\gamma=1}^{m}\rho_\lambda\rho_\gamma w_{\mu\nu\lambda\gamma}^{(2)}(r) + \cdots \,, \tag{7.101}$$

and truncate at second order. The weights $w_{\mu\nu}^{(0)}(r)$, $w_{\mu\nu\lambda}^{(1)}(r)$ and $w_{\mu\nu\lambda\gamma}^{(2)}(r)$ are determined by inserting the expansion (7.101) in equation (7.100) and replacing $\hat{c}_{\mu\nu}(k)$ by its density expansion to second order. This eliminates the problem of the partial derivatives with respect to the densities. If once the weights are determined, we evaluate the convolutions

$$\bar{\rho}^{(0)}_{\mu\nu}(\boldsymbol{r}) = \int \mathrm{d}\boldsymbol{r}' \rho_\nu(\boldsymbol{r}+\boldsymbol{r}') w^{(0)}_{\mu\nu}(r'), \tag{7.102}$$

$$\bar{\rho}^{(1)}_{\mu\nu\lambda}(\boldsymbol{r}) = \int \mathrm{d}\boldsymbol{r}' \rho_\nu(\boldsymbol{r}+\boldsymbol{r}') w^{(1)}_{\mu\nu\lambda}(r'), \tag{7.103}$$

$$\bar{\rho}^{(2)}_{\mu\nu\lambda\gamma}(\boldsymbol{r}) = \int \mathrm{d}\boldsymbol{r}' \rho_\nu(\boldsymbol{r}+\boldsymbol{r}') w^{(1)}_{\mu\nu\lambda\gamma}(r') , \tag{7.104}$$

then the weighted densities are the solution to the (approximate) system of algebraic equations

$$\bar{\rho}^{(\mu)}_\nu(\boldsymbol{r}) = \bar{\rho}^{(0)}_{\mu\nu}(\boldsymbol{r}) + \sum_{\lambda=1}^{m} \bar{\rho}^{(1)}_{\mu\nu\lambda}(\boldsymbol{r})\bar{\rho}^{(\mu)}_\lambda(\boldsymbol{r}) + \sum_{\lambda,\gamma=1}^{m} \bar{\rho}^{(2)}_{\mu\nu\lambda\gamma}(\boldsymbol{r})\bar{\rho}^{(\mu)}_\lambda(\boldsymbol{r})\bar{\rho}^{(\mu)}_\gamma(\boldsymbol{r}). \tag{7.105}$$

This problem is definitely simpler than the one posed by equation (7.100), but still not computationally trivial. We have a system of m^2 algebraic equations rather than a simple quadratic equation, and obtaining it requires the determination of $m(m+1)/2$ weights $w^{(0)}_{\mu\nu}$ (they are symmetric), $m^2(m+1)/2$ weights $w^{(1)}_{\mu\nu\lambda}$ and $m^2(m+1)^2/2$ weights $w^{(2)}_{\mu\nu\lambda\gamma}$ (they are symmetric in μ, ν and in λ, γ), that makes a total of $m(m+1)^2(m+2)/4$ weights. Just for a binary mixture, $m = 2$, this amounts to computing 18 weights; for a ternary mixture this number rises to 60 weights! And then we need to compute convolutions of them all with the densities and solve the equations...

This is probably the reason why this theory has never been applied as such. Denton and Ashcroft explicitly mention the numerical difficulty of this extension of the WDA and switch to the so-called modified WDA (MWDA), an effective liquid theory, to study freezing in a binary mixture of HS [50]. Their results are quite good compared to simulations, specially if one takes into account that effective liquid theories entirely refuse to account for the fluid structure appropriately. A year later [79], they changed the theory and introduced the WDA at the level of the first-order direct correlation functions $c^{(1)}_\mu(\boldsymbol{r}, [\boldsymbol{\rho}]) = -\delta\beta\mathcal{F}_{\mathrm{ex}}/\delta\rho_\mu(\boldsymbol{r})$. This simplifies the equations for the weights to the extent of making them proportional to the direct correlation functions, so the resulting theory is much easier to apply. The results for the layering of a binary HS mixture near a hard wall that they obtained with this new version were reasonable although there were important discrepancies with respect to simulations. Davidchack and Laird did compute the weights of the WDA as presented here by numerically solving the system (7.100), but they did it just for illustration purposes because, in order to apply it to the freezing problem, they also resorted to the MWDA version of their theory. There are hybrid approaches [80] in which local effective densities are used in (7.98), but on the right-hand side of (7.99), these local densities are replaced by their averages. This leads to simpler equations for the weights and to slightly improved results for the density profiles of a binary HS mixture near a hard wall.

We will not pursue this approach any further. By now its computational complexity should be clear, even for few component mixtures; if the number of components is large it is hopeless and its applicability to study polydisperse mixtures is out of question. On the other hand, at present it is not worthwhile to invest more effort in this or similar approaches because FMT provides a much better alternative to study, not only mixtures with an arbitrary number of species, but even polydisperse mixtures. So let us see how FMT deals with mixtures.

7.8 Fundamental Measure Theory for Mixtures of Hard Spheres

Two key pieces in Rosenfeld's inspiration to create FMT were the 1D DF and SPT. Let us see what they tell us about mixtures. We shall begin with the 1D DF. The exact functional for an additive mixture of HR in a segment was obtained by Vanderlick et al. [81]. Appropriately written, its form is most revealing. Formally, it looks the same as the one derived by Percus for a one-component fluid,

$$\Phi^{(\mathrm{1D})}\big([\rho];x\big) = -n(x)\ln\big[1-\eta(x)\big]; \tag{7.106}$$

the difference lies in the definition of the weighted densities, which now is

$$n(x) = \frac{1}{2}\sum_{\nu=1}^{m}\big[\rho_\nu(x+\sigma_\nu/2)+\rho_\nu(x-\sigma_\nu/2)\big], \tag{7.107}$$

$$\eta(x) = \sum_{\nu=1}^{m}\int_{-\sigma_\nu/2}^{\sigma_\nu/2}\rho_\nu(x+x')\,\mathrm{d}x'. \tag{7.108}$$

In other words, the mixture is accounted for by introducing a weight function for each species and adding up all the resulting weighted densities.

Let us see now what we can learn from SPT. The excess free energy density (in $k_{\mathrm{B}}T$ units) provided by SPT can be written in the form [12, 23, 24, 55]

$$\Phi = -\xi_0\ln(1-\xi_3) + \frac{\xi_1\xi_2}{1-\xi_3} + \frac{1}{24\pi}\frac{\xi_3^3}{(1-\xi_3)^2}, \tag{7.109}$$

where the variables ξ_i, $i = 0,1,2,3$ are geometrically weighted density averages,

$$\xi_i = \sum_{\nu=1}^{m}\rho_\nu R_\nu^{(i)}, \tag{7.110}$$

with $R_\nu^{(0)} = 1$, $R_\nu^{(1)} = R_\nu$, $R_\nu^{(2)} = 4\pi R_\nu^2$ and $R_\nu^{(3)} = 4\pi R_\nu^3/3$ being 1, the curvature radius, the surface and the volume of a sphere of diameter R_ν, respectively. This suggests that a free energy functional should depend on weighted densities whose weights have locally the same geometrical meaning.

Thus, Rosenfeld's proposal [13] for the weight functions was to define, for each species $\nu = 1, \ldots, m$, the same normalized weighted densities as for HS, i.e.

$$\begin{aligned} \eta_\nu(\boldsymbol{r}) &= \int d\boldsymbol{r}' \rho_\nu(\boldsymbol{r}+\boldsymbol{r}')\Theta(R_\nu - |\boldsymbol{r}'|), \\ \boldsymbol{v}_\nu(\boldsymbol{r}) &= \int d\boldsymbol{r}' \rho_\nu(\boldsymbol{r}+\boldsymbol{r}') \frac{\boldsymbol{r}'}{R_\nu} \frac{\delta(R_\nu - |\boldsymbol{r}'|)}{4\pi R_\nu^2}, \\ n_\nu(\boldsymbol{r}) &= \int d\boldsymbol{r}' \rho_\nu(\boldsymbol{r}+\boldsymbol{r}') \frac{\delta(R_\nu - |\boldsymbol{r}'|)}{4\pi R_\nu^2}. \end{aligned} \tag{7.111}$$

With them, one can define six *collective densities*

$$\begin{aligned} \eta(\boldsymbol{r}) &= \sum_{\nu=1}^{m} \eta_\nu(\boldsymbol{r}), & n^{(2)}(\boldsymbol{r}) &= \sum_{\nu=1}^{m} 4\pi R_\nu^2 \, n_\nu(\boldsymbol{r}), \\ \boldsymbol{v}^{(2)}(\boldsymbol{r}) &= \sum_{\nu=1}^{m} 4\pi R_\nu^2 \, \boldsymbol{v}_\nu(\boldsymbol{r}), & n^{(1)}(\boldsymbol{r}) &= \sum_{\nu=1}^{m} R_\nu \, n_\nu(\boldsymbol{r}), \\ \boldsymbol{v}^{(1)}(\boldsymbol{r}) &= \sum_{\nu=1}^{m} R_\nu \, \boldsymbol{v}_\nu(\boldsymbol{r}), & n^{(0)}(\boldsymbol{r}) &= \sum_{\nu=1}^{m} n_\nu(\boldsymbol{r}), \end{aligned} \tag{7.112}$$

where the scaling factors are chosen such that in the limit of uniform densities we have $\eta(\boldsymbol{r}) \to \xi_3$, $n^{(2)}(\boldsymbol{r}) \to \xi_2$, $n^{(1)}(\boldsymbol{r}) \to \xi_1$ and $n^{(0)}(\boldsymbol{r}) \to \xi_0$. By repeating the same procedure that led to the functional (7.66)–(7.68), one can obtain Rosenfeld's original excess free DF as $\Phi_{\text{oFMT}}([\rho], \boldsymbol{r}) = \Phi_1([\rho], \boldsymbol{r}) + \Phi_2([\rho], \boldsymbol{r}) + \Phi_3([\rho], \boldsymbol{r})$, where [13]

$$\Phi_1([\rho], \boldsymbol{r}) = -n^{(0)} \ln(1-\eta) \,, \tag{7.113}$$

$$\Phi_2([\rho], \boldsymbol{r}) = \frac{n^{(1)} n^{(2)} - \boldsymbol{v}^{(1)} \cdot \boldsymbol{v}^{(2)}}{1-\eta} \,, \tag{7.114}$$

$$\Phi_3([\rho], \boldsymbol{r}) = \frac{1}{24\pi} n^{(2)} \frac{n^{(2)} n^{(2)} - 3\boldsymbol{v}^{(2)} \cdot \boldsymbol{v}^{(2)}}{(1-\eta)^2} \,. \tag{7.115}$$

Actually, Rosenfeld found his functional directly for mixtures in the form just presented. For later convenience and for computational simplicity, we will use a form involving only the normalized weighted densities (7.111),

$$\Phi_1([\rho], \boldsymbol{r}) = -\sum_{\nu=1}^{m} n_\nu \ln(1-\eta), \tag{7.116}$$

$$\Phi_2([\rho], \boldsymbol{r}) = 2\pi \sum_{\nu,\mu=1}^{m} R_\nu R_\mu (R_\nu + R_\mu) \frac{n_\nu n_\mu - \boldsymbol{v}_\nu \cdot \boldsymbol{v}_\mu}{1-\eta} \,, \tag{7.117}$$

$$\Phi_3([\rho], \boldsymbol{r}) = 8\pi^2 \sum_{\nu,\mu,\lambda=1}^{m} R_\nu^2 R_\mu^2 R_\lambda^2 n_\nu \frac{\frac{1}{3} n_\mu n_\lambda - \boldsymbol{v}_\mu \cdot \boldsymbol{v}_\lambda}{(1-\eta)^2} \,. \tag{7.118}$$

As its one-component counterpart, this DF can give very accurate density profiles of HS mixtures near a hard wall. As a matter of fact, because of its expressibility in terms of collective densities, it can be as easily applied to polydisperse mixtures. As such, it has been successfully used by Pagonabarraga et al. [82] to study size segregation at the adsorption profiles of a polydisperse mixture of HS near a hard wall, something inconceivable with a WDA. But as it comes to study freezing, this functional has the same divergences that cause the breakdown of the solid minimization for the one-component HS fluid.

With an exception that we will comment later on, fixing these divergences requires the same strategy as the one followed for the one-component FMT. In fact, when expressed in terms of the collective densities, the resulting functionals are formally the same. Thus, apart from the empirical corrections introduced to avoid these singularities [64, 65] (and which have been applied with success to study entropic selectivity in microporous materials [83]), we can formulate an analog of the DI-FMT by introducing the new tensor weighted density

$$(\mathcal{T}_\nu)_{\alpha\beta}(\boldsymbol{r}) = \int \mathrm{d}\boldsymbol{r}'\rho_\nu(\boldsymbol{r}+\boldsymbol{r}')\frac{r'_\alpha r'_\beta}{R_\nu^2}\frac{\delta(R-|\boldsymbol{r}'|)}{4\pi R_\nu^2} \tag{7.119}$$

and replacing Φ_3 of the oFMT by

$$\Phi_3([\boldsymbol{\rho}],\boldsymbol{r}) = 12\pi^2\sum_{\nu,\mu,\lambda=1}^{m} R_\nu^2 R_\mu^2 R_\lambda^2 \frac{\varphi_{\nu\mu\lambda}(\boldsymbol{r})}{\left[1-\eta(\boldsymbol{r})\right]^2}\,, \tag{7.120}$$

$$\varphi_{\nu\mu\lambda}(\boldsymbol{r}) = \boldsymbol{v}_\nu\cdot\mathcal{T}_\mu\cdot\boldsymbol{v}_\lambda - n_\mu\boldsymbol{v}_\nu\cdot\boldsymbol{v}_\lambda - \mathrm{Tr}(\mathcal{T}_\nu\mathcal{T}_\mu\mathcal{T}_\lambda) + n_\mu\,\mathrm{Tr}(\mathcal{T}_\nu\mathcal{T}_\lambda). \tag{7.121}$$

The fact that the coefficient in (7.120) is a polynomial in the radii allows for an alternative form in terms of collective densities. Thus, if

$$\mathcal{T}(\boldsymbol{r}) = \sum_{\nu=1}^{m} 4\pi R_\nu^2\,\mathcal{T}_\nu(\boldsymbol{r})\,, \tag{7.122}$$

then (7.120) and (7.121) can be rewritten as

$$\Phi_3([\boldsymbol{\rho}],\boldsymbol{r}) = \frac{3}{16\pi}\frac{\boldsymbol{v}^{(2)}\cdot\mathcal{T}\cdot\boldsymbol{v}^{(2)} - n^{(2)}\,\boldsymbol{v}^{(2)}\cdot\boldsymbol{v}^{(2)} - \mathrm{Tr}(\mathcal{T}^3) + n^{(2)}\,\mathrm{Tr}(\mathcal{T}^2)}{(1-\eta)^2}\,. \tag{7.123}$$

But if one checks dimensional crossover with this new functional, one encounters the surprise that it does not even recover the exact 1D limit! It is surprising because its one-component counterpart does. The reason is that the third term is not zero in this dimensional reduction. In order to understand what is going on, we should imagine situations in which the centers of three spheres are aligned and their three surfaces intersect at a common circle. For spheres of the same size, this can only occur if at least two spheres are at the same position. This means that their surfaces fully overlap, and the correction introduced by the DI-FMT DF is such that the third term vanishes when this happens. However, if the spheres are different, such a triple intersection can occur and the three spheres can sit at different positions. The DI-FMT DF does not account for these situations and so they do contribute to the third term. In [84], we considered this problem and showed that it can be fixed by adding a rank-3 tensor weighted density

$$(\mathcal{M}_\nu)_{\alpha,\beta,\gamma}(\boldsymbol{r}) = \int \mathrm{d}\boldsymbol{r}' \rho_\nu(\boldsymbol{r}+\boldsymbol{r}') \frac{r'_\alpha r'_\beta r'_\gamma}{R_\nu^3} \frac{\delta(R_\nu - |\boldsymbol{r}'|)}{4\pi R_\nu^2} , \tag{7.124}$$

and correcting $\varphi_{\nu\mu\lambda}(\boldsymbol{r})$ in (7.121) with the addition of

$$\Delta\varphi_{\nu\mu\lambda}(\boldsymbol{r}) = \frac{2R_\mu^2(R_\nu - R_\lambda)}{R_\nu(R_\nu + R_\mu)(R_\mu + R_\lambda)} (\boldsymbol{v}_\nu \cdot \mathcal{M}_\mu : \mathcal{T}_\lambda - \boldsymbol{v}_\nu \cdot \mathcal{T}_\lambda \cdot \boldsymbol{v}_\mu) \tag{7.125}$$

(the symbol ':' denotes the contraction of two indices). The vanishing of this term for a one-component fluid is patent from the factor $R_\nu - R_\lambda$. Also, it gives no contribution for uniform fluids because then $v_\nu(\boldsymbol{r}) = 0$ and $\mathcal{M}_{\nu\mu\lambda}(\boldsymbol{r}) = 0$. One striking feature is that the coefficient dependent on the radii is no longer a polynomial, but a rational function. The dramatic consequence of this is that the new functional cannot be written in terms of a finite number of collective densities, as the previous ones, thus spoiling one very nice feature of FMT, specially when applied to polydisperse mixtures. The difficulty in applying this functional with respect to the DI-FMT one is considerable, so it can only be justified if the gain in accuracy is enough.

In [84], density profiles of a binary HS mixture near a hard wall and within a narrow slit were obtained with both the corrected functional and the DI-FMT one. The profiles are shown in Figs. 7.12 and 7.13. As it can be seen, the differences between the two DFs are negligible. Also the comparison between the dimensional reduction to 1D of the DI-FMT and the corrected one (which is exact) shows, again for a binary mixture, that the differences are smaller than 2% for all packing fractions up to $\eta = 0.8$ and is not larger than 10% for such a high packing fraction as $\eta = 0.95$. The conclusion so far is that the higher complexity of the corrected FMT does not justify using it instead of the DI-FMT one, which is not only much simpler in that it contains no rank-3 tensor, but also in that it can be written in terms of collective variables, which gives a considerable advantage when studying polydisperse mixtures. It remains to study what we could call the worst case scenario: namely, inhomogeneous 1D-like situations like cylindrical pores. As far as we know, nobody has tested these two functionals in this situation yet.

Irrespective of this analysis, no matter whether we consider the DI-FMT or the corrected one, the problem of the *lost cases* commented on in Sect. 7.5.2 is even worse for mixtures than it is for a one-component fluid [84]. The reason is the inability of these FMT functionals to recover the lowest order in a density expansion of the third-order direct correlation function. As already mentioned, circumventing this problem would amount to modifying the structure of FMT

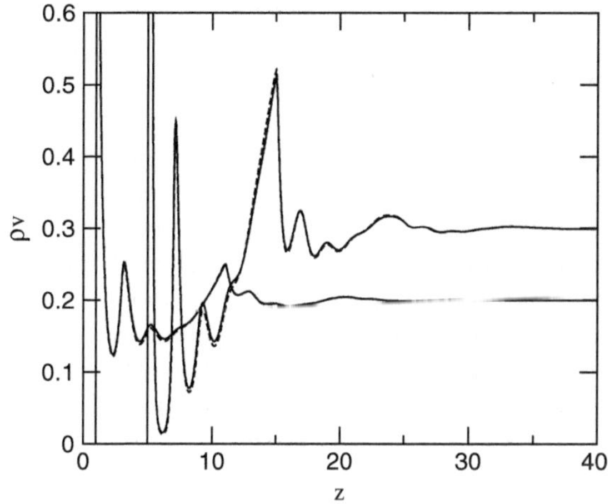

Fig. 7.12. Density profiles of the two components of a binary mixture of radii $R_1 = 1$, $R_2 = 5$ near a hard wall (the density is scaled with the volume of each sphere, denoted v in the axis label). With a *dashed line* we plot the results of the DI-FMT, with a *full line* we plot the results of the corrected FMT functional

functionals dramatically, and it is not clear at all whether such modifications, even if they corrected the problem, would be computationally amenable. So this seems to be as far as FMT can go.

Nevertheless, empirical modifications of FMT have proposed that, even though they necessarily spoil some of their nice features, they can still produce very good results when applied to specific problems. As already mentioned in Sect. 7.4.3, one of the obvious modifications is to replace the SPT excess free

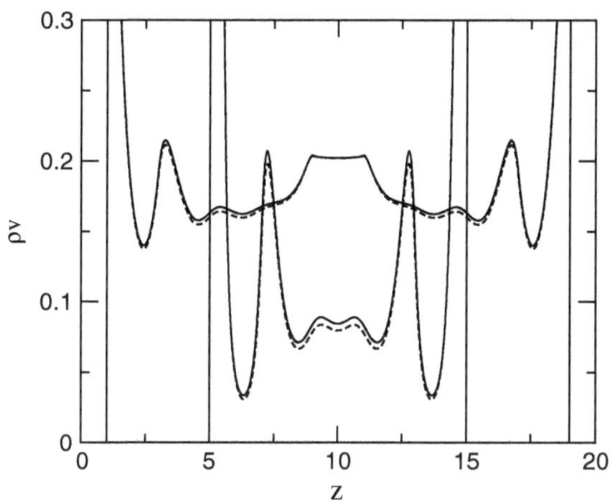

Fig. 7.13. Same as Fig. 7.12 for a slit of width $20R_1$

energy density by the empirical BMCSL (from Boublík–Mansoori–Carnahan–Starling–Leland). This has been done for the oFMT version [72, 85] as well as for the DI-FMT version [72]. The results for the density profiles of binary mixtures near a hard wall are in excellent agreement with simulations. When applied to polydisperse mixtures [86] they even improve those obtained with the original Rosenfeld's functional [82]. Finally, another empirical modification based on the CS free energy density [87] has been applied with excellent results to the calculation of the surface tension and bending rigidities of HS near a spherical surface (although the fluid is mono-disperse in this case, the analytical expressions derived for these magnitudes require the DF for a binary mixture, where one of the components becomes the spherical surface). Thus we see that, as it comes to concrete applications, FMT can still tolerate enough flexibility to gain in accuracy.

7.9 Non-Additive Mixtures of Hard Spheres

So far we have only considered ordinary, or *additive*, HS mixtures, but this is not the only interesting model of HS mixtures, there is a more general class of them which are *non-additive*. This means that, although the interaction between particles of either the same or different species is hard core, the effective diameters are related through the relationship

$$\sigma_{\nu\mu} = \frac{1}{2}(\sigma_{\nu\nu} + \sigma_{\mu\mu})(1 + \Delta_{\nu\mu}). \tag{7.126}$$

The case $\Delta_{\nu\mu} = 0$ for $\nu \neq \mu$ describes the situation for additive HS and the cases $\Delta_{\nu\mu} > 0$ or $\Delta_{\nu\mu} < 0$ describe non-additive HS. Such a model describes well certain mixtures of chemical compounds or mixtures of colloidal particles and polymers.

In spite of the apparent minor modification we have made, the resulting model is extraordinarily more difficult than the additive one, so much that there is no known functional for a single non-additive HS mixture even in 1D. Just to illustrate the problem, one must bear in mind that, while additive HR in 1D remain mixed for all densities up to close packing, if the mixture is non-additive with $\varDelta_{\nu\mu} > 0$, the mixed state is entropically disfavored with respect to the mixed one, so that close packing is only attainable for a fully demixed state. This fact forces the system to undergo a glass transition because, upon increasing density, the system gets trapped in smaller regions of the phase space as soon as some mixed configuration stops being reachable for that given density [88]. Thus, non-additive HS mixtures bring about a higher complexity than additive ones.

Because of this, there are fewer DFs for non-additive HS mixtures than there are for additive ones. Nevertheless we should mention a few exceptions in which functionals in the FMT spirit have been proposed, although to obtain them some of the nice features of this theory have had to be sacrificed. We will mostly describe two of them, plus an attempt at generalizing FMT to non-additive HS mixtures with arbitrary $\Delta_{\nu\mu}$.

7.9.1 The Asakura–Oosawa–Vrij Model

Let us consider the following simplified model of a colloid–polymer mixture introduced first by Asakura and Oosawa [89] and later by Vrij [90]. Colloidal particles have a typical radius R_c and polymers R_p. Interaction between colloidal particles is HS-like with diameter $\sigma_{cc} = 2R_c$, between colloidal particles and polymers is also HS-like with diameter $\sigma_{cp} = R_c + R_p$, but polymers are ideal particles to themselves (i.e. $\sigma_{pp} = 0$). This is a non-additive HS mixture with $\Delta_{cp} = R_p/R_c > 0$. For this model, Schmidt et al. [91] proposed the following functional. They kept the general FMT scheme, so their functional is formally the same as (7.116), (7.117), (7.120), (7.121),

$$\Phi_1\big([\boldsymbol{\rho}],\boldsymbol{r}\big) = \sum_{\nu=c,p} n_\nu \phi_\nu(\eta_c,\eta_p), \tag{7.127}$$

$$\Phi_2\big([\boldsymbol{\rho}],\boldsymbol{r}\big) = 2\pi \sum_{\nu,\mu=c,p} R_\nu R_\mu (R_\nu + R_\mu)\big[n_\nu n_\mu - \boldsymbol{v}_\nu\cdot\boldsymbol{v}_\mu\big]\phi_{\nu\mu}(\eta_c,\eta_p), \tag{7.128}$$

$$\begin{aligned}\Phi_3\big([\boldsymbol{\rho}],\boldsymbol{r}\big) = 8\pi^2 \sum_{\nu,\mu,\lambda=c,p} & R_\nu^2 R_\mu^2 R_\lambda^2 \big[\boldsymbol{v}_\nu\cdot\mathcal{T}_\mu\cdot\boldsymbol{v}_\lambda - n_\mu \boldsymbol{v}_\nu\cdot\boldsymbol{v}_\lambda \\ & - \mathrm{Tr}(\mathcal{T}_\nu\mathcal{T}_\mu\mathcal{T}_\lambda) + n_\mu \mathrm{Tr}(\mathcal{T}_\nu\mathcal{T}_\lambda)\big]\phi_{\nu\mu\lambda}(\eta_c,\eta_p)\,,\end{aligned} \tag{7.129}$$

where the weighted densities $\eta_\nu(\boldsymbol{r})$, $n_\nu(\boldsymbol{r})$, $\boldsymbol{v}_\nu(\boldsymbol{r})$ and $\mathcal{T}_\nu(\boldsymbol{r})$, $\nu = c,p$, are defined as in (7.111), (7.119) and

$$\phi_{\nu_1\cdots\nu_l}(\eta_c,\eta_p) = \frac{\partial^l \phi_0(\eta_c,\eta_p)}{\partial\eta_{\nu_1}\cdots\partial\eta_{\nu_l}}\,, \tag{7.130}$$

$\phi_0(\eta_c,\eta_p)$ being the excess free energy of a 0D cavity.

If the colloidal particles and polymers were just additive HS, then the grand partition function of a 0D cavity would be

$$\Xi_0 = 1 + z_c + z_p\,, \tag{7.131}$$

with z_c, z_p their respective fugacities. Then $\eta_\nu = z_\nu/\Xi_0$, from which $\Xi_0^{-1} = 1 - \eta_c - \eta_p$ and therefore

$$\begin{aligned}\phi_0(\eta_c,\eta_p) &\equiv \eta_c + \eta_p - \ln\Xi_0 + \eta_c\ln(z_c/\eta_c) + \eta_p\ln(z_p/\eta_p) \\ &= \eta_c + \eta_p + (1-\eta_c-\eta_p)\ln(1-\eta_c-\eta_p).\end{aligned} \tag{7.132}$$

Substituting this function back into (7.127)–(7.129), one recovers the DI-FMT for an additive HS mixture (7.116), (7.117), (7.120), (7.121).

But if particles are like in the Asakura–Oosawa–Vrij (AOV) model, then

$$\Xi_0 = z_c + e^{z_p} \tag{7.133}$$

because in a 0D cavity there may be one colloidal particle or an arbitrary number of polymers. Therefore,

$$\eta_c = \frac{z_c}{\Xi_0}, \qquad \eta_p = \frac{z_p e^{z_p}}{\Xi_0}, \tag{7.134}$$

so $z_c = \eta_c \Xi_0$, from (7.133) $e^{z_p} = \Xi_0(1-\eta_c)$ and thus

$$z_p = \frac{\eta_p}{1-\eta_c}, \qquad \Xi_0 = \frac{\exp\left(\frac{\eta_p}{1-\eta_c}\right)}{1-\eta_c}, \qquad z_c = \frac{\eta_c \exp\left(\frac{\eta_p}{1-\eta_c}\right)}{1-\eta_c}. \tag{7.135}$$

The excess free energy that follows from these expressions is

$$\phi_0(\eta_c, \eta_p) = \eta_c + (1-\eta_c-\eta_p)\ln(1-\eta_c). \tag{7.136}$$

The functional (7.127)–(7.129) possesses a set of good properties: (i) by construction (it has the same structure as the DI-FMT), it recovers the correct 0D limit for 0D cavities with no lost cases; (ii) the first two terms in the low-density limit of the direct correlation function are exact; (iii) it generates the correct depletion potential between two colloidal particles in a sea of ideal polymers at any density [92], and (iv) it is linear in $\eta_p(\boldsymbol{r})$. The latter has some important implications: on the one hand, the direct correlation function $c_{pp}(r, \rho_c, \rho_p) = 0$ as in the PY approximation; on the other hand, the Euler–Lagrange equation for $\rho_p(\boldsymbol{r})$ permits to obtain explicitly this density as a functional of only $\rho_c(\boldsymbol{r})$, something that greatly simplifies the minimization process. As a matter of fact, this nice feature can be used to obtain explicitly a DF for the effective fluid of colloidal particles interacting with the depletion potential induced by the polymers. This can be achieved through the procedure described in [93].

As for the results one can get from this functional, for uniform densities [91, 94] the resulting free energy coincides with that of the free-volume approximation of the AOV model [95], which predicts a stable gas–liquid coexistence for $\Delta_{cp} \gtrsim 0.32$. For $\Delta_{cp} < 0.1547$, the model can be exactly mapped onto a one-component colloidal fluid with pairwise interaction [96]. This allows to perform simulations for such an asymmetric mixture that would otherwise be hindered by severe equilibration problems. The comparison of the structure factors is quite good [91].

A variant of this model in which a third component, point-like HS, mimics a poor solvent can also be dealt with in the same manner [97]. The model exhibits a richer phenomenology, with a reentrant demixing transition for a sufficiently high solvent concentration.

The procedure sketched here is a very ingenious twist of the standard FMT for additive HS. Its results are very good, considering the simplicity of the proposal and the difficulty of the model. This notwithstanding, there is an important and unavoidable flaw in this extension of FMT to non-additive models: the 1D limit cannot be correctly recovered. Having in mind the dimensional crossover behavior of the DI-FMT, it is obvious that, for $\Delta_{\nu\mu} = 1$, the

1D limit of this AOV DF will be just Φ_1 in (7.127), a result that is only a crude approximation to the exact excess free energy of the 1D AOV model [88]. Even worse, if $\Delta_{\nu\mu} \neq 1$, then the term Φ_3 in (7.129) also makes a contribution for the same reasons that it does for an additive HS mixture (see Sect. 7.8) [94]. Actually, this problem can be fixed by patching out the DF in a way similar to that employed in Sect. 7.8, i.e. adding a contribution like (7.125) to the term (7.129). Notice that this will change neither the excess free energy nor the direct correlation function of the uniform fluid, so all features of the AOV FMT described above remain intact; only the 1D limit will change and will then be given by (7.127) for any particle radii.

7.9.2 The Widom–Rowlinson Model

Another model whose DF can be obtained with the same procedure is the Widom–Rowlinson model [98]. This is an m-component non-additive HS mixture of radii R_ν, $\nu = 1, \ldots, m$ such that

$$\sigma_{\nu\mu} = R_\nu + R_\mu, \qquad \sigma_{\nu\nu} = 0. \tag{7.137}$$

This extreme non-additivity, in which there is interaction only between different species, favors demixing. The scheme to construct a FMT DF for this model is the same as for the AOV model [99]; the only thing that needs to be changed is $\phi_0(\eta_1, \ldots, \eta_m)$. This function is determined from the grand partition function of a 0D cavity,

$$\Xi_0 = 1 - m + \sum_{\nu=1}^{m} e^{z_\nu}. \tag{7.138}$$

Unfortunately, an analytic expression for ϕ_0 cannot be obtained explicitly, so it has to be solved numerically using the relationships

$$\eta_\nu = \frac{z_\nu e^{z_\nu}}{\Xi_0}, \qquad \nu = 1, \ldots, m\,. \tag{7.139}$$

The dimensional crossover behavior of this DF is similar to that of the AOV DF described above, so it shares the same fundamental flaw in 1D. On the other hand, the bulk fluid has no other thermodynamic stable phases than uniform ones (either mixed or demixed): because there is no excluded volume between particles of the same species, no solid phase is stable at high densities. Only when the number of components is large can the solid be more stable than the demixed fluid phases (notice that increasing the number of species increases the stability of the mixed phase, as the probability for a particle to have a neighbor of its same species decreases). There is evidence of this fact, when the number of components $m > 31$, coming from a different model: the parallel hard hypercubic model [100].

7.9.3 General Non-additive Mixtures

Schmidt has recently proposed a FMT for rather general HS binary mixtures [101]. The shape of his DF is

$$\beta\mathcal{F}_{\text{ex}}[\rho_1,\rho_2] = \int \mathrm{d}\boldsymbol{r}\mathrm{d}\boldsymbol{r}' \sum_{ij=0}^{3} K_{ij}(|\boldsymbol{r}-\boldsymbol{r}'|)\Phi_{ij}\left(\{n_\nu^{(k)}\},\{n_\mu^{(l)}\}\right) , \qquad (7.140)$$

where the Φ_{ij} are functions of weighted densities, as usual in FMT. Two novelties render this functional peculiar: first of all, the weights are all scalar, as in Kierlik and Rosinberg's version of FMT [56, 57] (using derivatives of delta functions instead of vector and tensor weights); secondly, there is a kernel $K_{ij}(|\boldsymbol{r}-\boldsymbol{r}'|)$ which accounts for the non-additivity. The components of this kernel are also weights *á la* Kierlik-Rosinberg (with higher order derivatives), but their radius is R_{12} instead of the radii of the spheres. Both the kernel and the weights are tailored so as to recover the lowest order of the density expansion of $\beta\mathcal{F}_{\text{ex}}$.

At low densities of both components, the structure of the fluid is satisfactorily reproduced, something to be expected because it is in the very construction of FMT. The prediction of the demixing transition is also quite accurate compared to simulations. However, this functional has two serious drawbacks. The first one is that in the limit $\Delta_{12} \to -1$ (so that $R_{12} = 0$ and the two species do not see each other), the functional is not the sum of two DF, one for each species, as it should. The second one is that the choice of weights carries the same flaw for the solid phase as the original Rosenfeld's functional, so it can only be used to study density profiles close to a wall and similar problems. But in spite of these problems, the idea of introducing a kernel depending on R_{12} is worth exploring and might be the clue to construct the definitive FMT able to deal with such complex models.

7.10 Density Functionals for Anisotropic Bodies

This section is devoted to give a view on different approximations commonly used to construct DFs of fluids composed of anisotropic particles. These particles, which can be molecules or colloidal particles, have, apart from positional, orientational degrees of freedom. They can have prolate (rod-like) or oblate (disk-like) shapes and, because of that, form a set of new phases. The isotropic fluid is the most disordered phase that can be conceived, without positional and orientational order. A direct consequence of the particle anisotropy is the appearance of orientational order where at least one of the principal axes of particles are on average aligned along a preferential direction while the positional order is absent. This phase was discovered in 1888 by the Austrian botanic Friedrich Reinitzer and later classified by Friedel as the nematic phase [102]. The coupling between positional and orientational degrees of freedoms makes these fluids to exhibit also, under certain conditions, a phase transition to the so-called smectic phase [102]. The latter is a periodically layered phase composed of prolate particles aligned perpendicular to the layers. In the layer planes, the centers of mass are randomly positioned, which constitutes the main reason why one can think of smectics as a 1D periodic fluids. Some fluids composed of particles with disk-like symmetries exhibit a transition to a columnar phase, in which the centers of mass are randomly positioned in liquid-like columns. The columns are located at the nodes of a 2D grid, while the particle axes are oriented parallel to the columns. One of the crystalline phases present in liquid crystals is the plastic solid, characterized by long-ranged 3D positional order while the orientational order is absent. Although we will not mention other liquid crystalline phases, the number of them found experimentally is much greater than the few examples presented here.

As was already pointed out at the beginning of this chapter, the DF formalism describes any fluid in terms of its density distribution function. The complete set of variables necessary to describe the degrees of freedom of the anisotropic particle is now $\boldsymbol{x} \equiv \{\boldsymbol{r}, \hat{\boldsymbol{\Omega}}\}$, where $\hat{\boldsymbol{\Omega}} \equiv \{\phi, \theta, \chi\}$ is the set of Euler angles that fix the particle orientation. Thus, the density distribution function $\rho(\boldsymbol{x})$ depends, in general, on six variables that makes the study of these systems much more difficult.

All DFs used in the study of liquid crystals belong to one of the two following sets: (i) those that make emphasis on the attractive contribution of the anisotropic pair potential and (ii) those that describe the fluids in terms of their purely repulsive (and usually hard core) interactions. The use of a simple hard core potential to model a real liquid crystal molecule is, of course, a crude simplification, but as it was shown firstly by Onsager [103, 104] and further by Frenkel [105, 106, 107], this simple interaction is enough to stabilize the nematic, smectic, columnar and different crystalline phases. These works showed that the main liquid crystal symmetries can be explained by the entropic nature of the hard core potentials.

As we will see below, many of these functionals have as an important ingredient the thermodynamic (the equation of state) and structural (the direct correlation function) functions taken from the HS fluid. The reason behind this is the lack of good approximations for these functions for fluids composed of anisotropic particles. In the following section, we will describe the DFs mainly based on the attractive anisotropic part of the pair interaction and further, in Sect. 7.10.2, we will present DFs based on purely hard core interactions.

7.10.1 Mean Field and Related Density Functionals

The bulk and interfacial phase behavior of liquid crystals has been crucial on determining the origin of several DF approximations. In this section, we will describe their evolution for the particular case of fluids interacting through anisotropic attractive pair potentials. At the same time, the main features of liquid crystal phase behavior will be briefly introduced.

One of the first approaches used to describe the liquid crystal isotropic-nematic phase transition is due to Maier and Saupe [108, 109, 110]. Their approximation is equivalent to using the most simple DF which includes, apart from the ideal gas term, $\beta\mathcal{F}_{\rm id}[\rho] = \int d\boldsymbol{x}\rho(\boldsymbol{x})\left[\ln\rho(\boldsymbol{x})-1\right]$ (note that $\boldsymbol{x}$ represents the set of spatial and angular variables), an excess part which is approximated in a mean field way with the following attractive pair potential between particles with axial symmetry: $V(r,\theta_{12}) = -V_0(r)P_2(\cos\theta_{12})$, where θ_{12} is the angle between the principal axes of the molecules and $P_2(x)$ is the second-order Legendre polynomial [111]. The factor $V_0(r) > 0$ constitutes the spatial dependence of the pair potential, which is assumed to be decoupled from the angular one. This potential, for a fixed r, reaches its minimum value for parallel configuration of particles, favoring in such way the nematic order, while is positive for the mutual perpendicular configuration. The choice of this potential, apart from its simplicity, is motivated by the fact that the adequate order parameter that describes the degree of nematic order is just $S \equiv \langle P_2(\cos\theta)\rangle$, where θ is the angle between the particle axes and the so-called nematic director (the direction of preferential alignment), and the average is taken over the orientational distribution function $h(\theta) = h(\pi-\theta)$, with head–tail symmetry.

The excess part of the free energy functional has the standard mean field form

$$\mathcal{F}_{\rm ex}[\rho] = -\frac{1}{2}\int {\rm d}\boldsymbol{x}_1 \int {\rm d}\boldsymbol{x}_2 \rho(\boldsymbol{x}_1)\rho(\boldsymbol{x}_2)V_0(r_{12})P_2(\cos\theta_{12}) \,. \qquad (7.141)$$

Due to the absence of positional ordering in the nematic phase, the density distribution function can be expressed as $\rho(\boldsymbol{x}) = \rho h(\theta)/(4\pi)$, where ρ is the fluid density and the angular density function is normalized as $\int_{-1}^{1} {\rm d}(\cos\theta)\, h(\theta) = 1$. The equation (7.141) can be drastically simplified by taking into account that $\hat{\boldsymbol{\Omega}}_i = \{\theta_i,\phi_i\}$ (the polar and azimuthal angles of the principal axes are sufficient to describe the orientation of axially symmetric bodies) and carrying out the integration over the angles ϕ_i with the use of the identity [111] $\int {\rm d}\phi_1 \int {\rm d}\phi_2 P_2(\cos\theta_{12}) = (2\pi)^2 P_2(\cos\theta_1)P_2(\cos\theta_2)$ (θ_i being the angle between the axis of particle i and the nematic director). The ideal part of the free energy density, $\beta\mathcal{F}_{\rm id}[\rho]/V \equiv \Phi_{\rm id}[h(\theta)]$, is a functional of $h(\theta)$ and can be written as $\Phi_{\rm id}[h(\theta)] = \rho\left[\ln\rho - 1 + \int_{-1}^{1} d(\cos\theta)h(\theta)\ln h(\theta)\right]$, where the second term is the orientational entropy.

The constrained functional minimization of $\Phi[h(\theta)] = \beta\mathcal{F}[\rho]/V$ with respect to $h(\theta)$ results in

$$h(\theta) = \frac{\exp\left[aSP_2(\cos\theta)\right]}{\int_{-1}^{1} {\rm d}(\cos\theta)\exp\left[aSP_2(\cos\theta)\right]} \,, \qquad (7.142)$$

with $a = \beta v_0 \rho$ and $v_0 = 4\pi \int\limits_0^\infty \mathrm{d}r r^2 V_0(r)$. Multiplying (7.142) by $P_2(\cos\theta)$ and integrating over $\cos\theta$ between -1 and 1, we obtain a self-consistent integral equation to calculate the equilibrium value of S for a fixed temperature β and density ρ. The insertion of the found equilibrium orientational distribution $h_{\mathrm{eq}}(\theta)$ in the free energy density and the application of the usual double tangent construction allow us to find the coexisting densities for the isotropic and nematic phases at fixed temperature and to calculate the phase diagram of the Maier–Saupe model.

In order to consider within the same model the smectic phase, we need to impose that $\rho(\boldsymbol{x},\theta) = \rho(z,\theta)$, i.e. the spatial symmetry is broken in the direction of preferential alignment and the density profile is a periodic function with period d. In principle, its value should be found through the DF minimization, but as the Maier–Saupe functional does not include a repulsive contribution from the particle core, its absolute minimum is reached at $d = 0$, i.e. the density of aligned molecules builds up in and infinitely small region. Thus, the parameter d should be fixed at some value comparable with the molecule length. The constrained functional minimization (with the constraint $d^{-1} \int\limits_0^d \mathrm{d}z \int\limits_{-1}^1 \mathrm{d}(\cos\theta)\rho(z,\theta) = \rho$, ρ being the mean density) leads to the following self-consistent integral equation for the density profile

$$\rho(z,\theta) = \frac{\rho \exp\left[-V_{\mathrm{eff}}(z,\theta)\right]}{d^{-1} \int\limits_0^d \mathrm{d}z \int\limits_{-1}^1 \mathrm{d}(\cos\theta) \exp\left[-V_{\mathrm{eff}}(z,\theta)\right]}, \tag{7.143}$$

where the effective potential in reduced thermal units is a functional of the density profile

$$V_{\mathrm{eff}}(z,\theta) = -\beta \left[\int\limits_{-\infty}^{\infty} \mathrm{d}z' \int\limits_{-1}^{1} \mathrm{d}(\cos\theta) V_0(z-z') P_2(\cos\theta')\rho(z',\theta') \right] P_2(\cos\theta)\,, \tag{7.144}$$

with $V_0(z) = 2\pi \int\limits_{|z|}^\infty \mathrm{d}r r V_0(r)$. In principle, this equation can be solved iteratively for a periodic function $\rho(z+d,\theta) = \rho(z,\theta)$ at each point $(z_i, \cos\theta_i)$ of the constructed rectangular grid $[0,d] \times [0,1]$.

The phase diagram in the coordinates T and d (conveniently scaled with the attractive potential parameters) qualitatively coincides with the experimental results. For small molecule lengths, the system goes from the isotropic to the nematic and then to the smectic phases as the temperature decreases; for higher values of d, there is a direct isotropic to smectic transition [112, 113].

Selinger and Nelson applied the same model to study the vapor–fluid interface. For this purpose, they modeled the interface through an external potential of the form $V_{\text{surf}}(z, \cos\theta) = -\epsilon_s \exp\left[-(z/r_0)^2\right] P_2(\cos\theta)$ favoring perpendicular alignment of molecules at the interface. They found complete wetting by the nematic phase as the temperature approaches the isotropic–nematic coexistence temperature from the isotropic side. They also found incomplete wetting by the smectic phase with one and two layering transitions at temperatures slightly higher than the isotropic–smectic coexistence temperature [113].

As it was mentioned above, the absence of repulsive interactions makes the model non-predictive with respect to the equilibrium value of the smectic period. For the same reason, the model does not exhibit a vapor–liquid phase transition, thus the necessity to model the interface by an external potential. The first DF which included repulsive interactions between particles was developed by Telo da Gama [114, 115]. The first proposed version was

$$\mathcal{F}[\rho] = \mathcal{F}_{\text{id}}[\rho] + \mathcal{F}_{\text{HS}}[\rho] + \frac{1}{2}\int \mathrm{d}\boldsymbol{x}_1 \mathrm{d}\boldsymbol{x}_2 \rho(\boldsymbol{x}_1)\rho(\boldsymbol{x}_2)V_{\text{atr}}(\boldsymbol{x}_1, \boldsymbol{x}_2) \,, \quad (7.145)$$

where the repulsive part was approximated by the HS potential and its contribution was calculated using the LDA (see Sect. 2.4), i.e.

$$\mathcal{F}_{\text{HS}}[\rho] = \int \mathrm{d}\boldsymbol{r} \Phi_{\text{HS}}(\rho(\boldsymbol{r})) \,, \quad (7.146)$$

where $\Phi_{\text{HS}}(\rho)$ is the HS free energy density derived from the compressibility equation of state. With respect to the attractive part, the following approximation was used in [114, 115]

$$V_{\text{atr}}(\boldsymbol{x}_1, \boldsymbol{x}_2) = V_0(r_{12}) + V_1(r_{12})P_2(\theta_{12}) \,, \quad (7.147)$$

with $V_i(r) = 0$ for $r < \sigma$ (the HS diameter) while $V_i(r) = -\epsilon_i(r_0/r)^6$ for $r > \sigma$, $i = 0, 1$. This potential can be regarded as the first terms in a general spherical harmonic expansion of the full anisotropic potential [116]. The first one, $V_0(r_{12})$, is the isotropic part responsible, together with the repulsive core, of the vapor–liquid phase separation, while the second term stabilizes the nematic phase if $\epsilon_1 > 0$. It should be noted that (7.145) is nothing but a DF constructed from a perturbative scheme with the HS fluid as the reference system and the attractive contribution as the perturbation. In this scheme, the radial distribution function of the reference fluid is approximated by its low-density limit $g_{\text{ref}}(r;\eta) = \Theta(r - \sigma)$.

The temperature–density bulk phase diagram predicted by the present model includes a vapor (V)–isotropic liquid (I) coexistence ending in a critical point and a vapor–nematic (N) or I–N coexistence for temperatures below or above the V–I–N triple point; within this functional different interfaces, as the V–I, V–N and I–N ones can be calculated self-consistently in the neighborhood of the triple point. The numerical calculations of the density and order parameter profiles at the V–N interface for some fixed value of ϵ_1 (selected so as to model the real liquid crystal anisotropy) give complete wetting by the

I phase (i.e. a surface-induced disordering transition) at the triple point. As explained in Sect. 7.4.1, the surface tension is calculated through

$$\gamma = \frac{\Omega + pV}{A}, \tag{7.148}$$

with $\Omega[\rho]$ the equilibrium value of the grand potential, obtained from its minimization with respect to the density $\rho(z)$ and the angular distribution function $h(z,\theta)$ profiles $[\rho(z,\theta) = \rho(z)h(z,\theta)]$, p the bulk pressure at coexistence, V the total system volume and A the surface area of the planar interface. The functional minimization with respect to $h(z,\theta)$ is a constrained minimization with a Lagrange multiplier required to fulfill the condition $\int \mathrm{d}(\cos\theta)h(z,\theta) = 1$. The nematic order parameter profile, calculated as $S(z) = \int \mathrm{d}(\cos\theta)h(z,\theta)P_2(\cos\theta)$, is a function of the interface coordinate. To implement the numerical minimization, the density and order parameter are fixed at the left and right sides of the minimization box and are set to the bulk coexisting values. The box length should be large enough for the full interface to fit in. A pair of coupled integral equations for $\{\rho(z), S(z)\}$ is obtained after the exponentiation of the functional minimization of $\Omega[\rho]$ with respect to $\rho(z)$ and $h(z,\theta)$, the latter multiplied by $P_2(\cos\theta)$ and integrated over $\cos\theta$. These equations can be solved iteratively selecting adequate initial guesses for these functions.

The structure of the V–N and I–N interfaces resulting from this model has monotonic density and order parameter profiles. Also, there is no orientational order at the V–I interface (the order parameter is always zero along the interface). Finally, the surface tension does not depend on the orientation of the nematic director through the interface. All these results are in contradiction with experiments on liquid crystals which usually find an excess surface ordering near the interface with particles aligned in preferential directions with respect to it. Besides, the density and order parameter profiles have an oscillating behavior near the interface. All these features can be obtained by improving the model with (i) the use of a non-local DF approximation to evaluate the free energy of the HS reference fluid, which makes the density profile to be an oscillating function around the interface position and (ii) the inclusion of higher order terms in the spherical harmonic expansion of the attractive potential; specially, those terms which couple the positional and orientational degree of freedoms of particles (those proportional to $P_2(\cos\theta_i')$, with θ_i' being the angle between the axis of particle $i = 1,2$ and the radius vector joining the centers of mass of particles 1 and 2). The inclusion of these terms can favor a perpendicular alignment of particles at the interface. They also contribute to generate an excess of nematic ordering in regions of the space where the density gradient is large. Moreover, as we will see shortly, the presence of these terms are necessary to stabilize the smectic phase.

All these requirements were included in the DF version for smectic liquid crystals developed by Mederos and Sullivan [117]. As the reference part of the free energy functional, they took a parallel hard ellipsoid (PHE) fluid whose functional form is exactly the same as that of a HS fluid with the appropriate scaling along the z direction to transform the HS into ellipsoids of whatever axis ratio, i.e.

$$\Phi_{\mathrm{PHE}}[\rho(\mathbf{r})] = \rho(\boldsymbol{r})\psi_{\mathrm{HS}}(\bar{\rho}(\boldsymbol{r})) , \tag{7.149}$$

where the CS-WDA was used to evaluate the free energy per particle. The weighted density is calculated through

$$\bar{\rho}(\boldsymbol{r}) = \int \mathrm{d}\boldsymbol{r}' w(\boldsymbol{r}'; \bar{\rho}(\boldsymbol{r}))\rho(\boldsymbol{r} + \hat{\sigma}\boldsymbol{r}') , \tag{7.150}$$

with $w(\boldsymbol{r}'; \bar{\rho}(\boldsymbol{r}))$ given by (7.28), (7.32), (7.34) and (7.38). The scaling is accounted for the diagonal tensor $\hat{\sigma}$, with components $\sigma_\perp$ along the x and y directions and $\sigma_\parallel$ along the z direction (the direction of alignment). Sufficiently large particle aspect ratio $\sigma_\parallel/\sigma_\perp$ and the use of attractive pair potential,

$$\begin{aligned} V(\boldsymbol{x}_1, \boldsymbol{x}_2) = V_0(r_{12}) + V_1(r_{12})P_2(\cos\theta_{12}) + V_2(r_{12})\left[P_2(\cos\theta_1')\right. \\ \left. + P_2(\cos\theta_2')\right] , \end{aligned} \tag{7.151}$$

are necessary to stabilize the smectic phase. Note the presence of the new terms proportional to $P_2(\cos\theta_i')$, which couple the positions and orientations of particles. Taking the proper signs, the third term in (7.151) reaches its minimum value for parallel alignment of particles with their axes perpendicular to the joining vector, i.e. the usual configuration for smectics.

In order to numerically minimize the functional (7.145) for the smectic density profiles, it should be noted that the orientational part of the ideal gas contribution,

$$S_{\mathrm{or}}[h] = -\int_{-1}^{1} \mathrm{d}\cos\theta \, h(z,\theta) \ln\left[h(z,\theta)\right] , \tag{7.152}$$

is an universal function of the nematic order parameter $S(z)$. Thus, this function can be calculated for a uniform nematic phase and tabulated for a regular set of points $S_i \in [-1/2, 1]$. Taking into account that the mean field contribution of the free energy is a functional only of $\rho(z)$ and $S(z)$, the grand potential can be minimized with respect to those functions evaluated at the points of the discretized minimization box which contains at least one smectic period. The conjugate gradient algorithm [20] can be used to achieve this minimization. The values of the orientational entropy and its derivatives with respect to $S(z_i)$ necessary to calculate the energy and the gradients can be evaluated using a cubic spline interpolation from the already tabulated set of points.

The authors of [117] calculated the phase diagrams for different attractive potentials. They found apart from the usual V, I and N phases an stable region of the smectic phase for high densities. The smectic can coexist with the nematic or with the isotropic liquid, for small or large values of ϵ_2/ϵ_0 as compared to ϵ_1/ϵ_0, ϵ_i being the strength of $V_i(r)$.

This model has also been applied to the study of the surface ordering at the wall–isotropic fluid interface, where the external potential of type $V_s(z) = -\epsilon_s P_2(\cos\theta)/(z_0 + z)^3$ was used to model a flat surface which favors the homeotropic anchoring (perpendicular to the wall–particle alignment). Complete and incomplete wetting by the smectic phase was found together with layering transitions as the strength of the external potential is increased [118, 119]. Finally, the thinning transitions in freely suspended smectic films was also studied with this model [120, 121].

The surface ordering and layering transitions was also found as a result of applying this model to the calculation of the V–I interface. To illustrate the strong ordering generated at the interface, we have plotted in Fig. 7.14(a)–(b) the density and order parameter profiles resulting from the minimization of the grand potential at the equilibrium V–I interface for a fixed set of parameters $\{\epsilon_i\}$ corresponding to a layering phase transition between two and three smectic layers.

We end this subsection with a very successful version of liquid crystal interparticle potential that has been included in the DF formalism via a perturbation theory and has also been applied in MC simulations of liquid crystal molecules. The origin of this potential can be traced back to the original work of Berne and Pechukas [122], who proposed the so-called Gaussian overlap model. In this work, the authors calculated the effective potential between two ellipsoids with a Gaussian repulsive core. The amplitude and the range of the repulsive potential was calculated through the overlap region between two Gaussians representing the mass distribution of two particles separated by a distance $\boldsymbol{r}$ with their uniaxial axes pointing along $\hat{\boldsymbol{\Omega}}_i$ ($i = 1, 2$). They

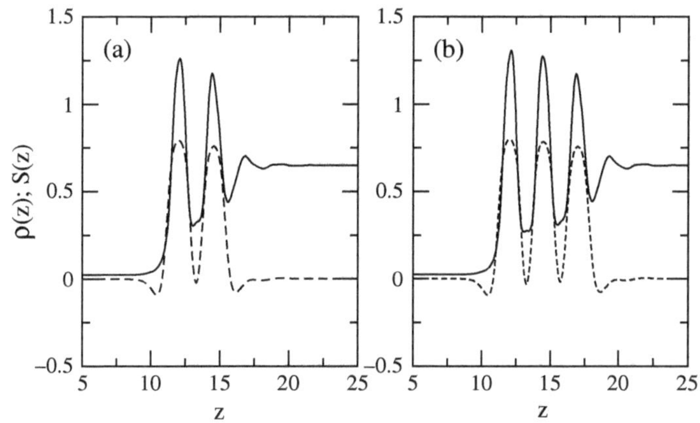

Fig. 7.14. Density $\rho(z)$ (*solid lines*) and order parameter $S(z)$ (*dashed lines*) profiles corresponding to the V–I interface containing two (**a**) and three (**b**) coexisting smectic peaks as a result of the density functional minimization

proposed a quite general form for interparticle potentials by the simple rescaling of the interparticle distance by the found range parameter $\sigma(\hat{\mathbf{\Omega}}, \hat{\mathbf{\Omega}}_1, \hat{\mathbf{\Omega}}_2)$ given by

$$\frac{\sigma_0^2}{\sigma(\hat{\mathbf{\Omega}}, \hat{\mathbf{\Omega}}_1, \hat{\mathbf{\Omega}}_2)^2} = 1 - \frac{\chi}{2} \left[\frac{\left(\hat{\mathbf{\Omega}} \cdot \hat{\mathbf{\Omega}}_1 + \hat{\mathbf{\Omega}} \cdot \hat{\mathbf{\Omega}}_2\right)^2}{1 + \chi \hat{\mathbf{\Omega}}_1 \cdot \hat{\mathbf{\Omega}}_2} + \frac{\left(\hat{\mathbf{\Omega}} \cdot \hat{\mathbf{\Omega}}_1 - \hat{\mathbf{\Omega}} \cdot \hat{\mathbf{\Omega}}_2\right)^2}{1 - \chi \hat{\mathbf{\Omega}}_1 \cdot \hat{\mathbf{\Omega}}_2} \right] , \tag{7.153}$$

which explicitly depends on their particle orientations and on the unit vector joining their center of mass $\hat{\mathbf{\Omega}}$. The anisotropy parameter χ is determined by the aspect ratio of the ellipsoidal Gaussians $\kappa = \sigma_{\|}/\sigma_{\perp}$ through $\chi = (\kappa^2 - 1)/(\kappa^2 + 1)$. The potential amplitude was later modified from its original form so as to take into account appropriately the anisotropy by including some fitting parameters in its functional form [123]. These parameters were selected in such a way as to quantitatively reproduce the well depths and well widths for the side-by-side and end-to-end particle configurations of the Lennard-Jones (LJ) site potentials (where each particle is formed by n LJ centers positioned along a straight line). The resulting pair potential, known as the Gay–Berne potential [123], was later included in a perturbative scheme of DF formalism [124, 125] to calculate the phase diagram of a liquid crystal model. The resulting temperature–density phase diagram, which includes vapor, isotropic liquid and nematic phases, compares reasonably well with the simulation results [124, 125]. A further extension of the model was made to include the smectic symmetry. A comparison of the results with the available computer simulations indicates that the theory leads to qualitatively correct predictions [124, 125].

7.10.2 Density Functionals for Hard Anisotropic Particles

Among hard anisotropic particles, the hard spherocylinder (HSC) is the paradigmatic hard body on which most theoretical and simulation studies have been done. It consists of a cylinder of length L and diameter D capped by two hemispheres of the same diameter. The reason for this particular choice is the simplicity to implement the HSC pair overlap criterion in simulations and also the simple analytic expression that the HSC excluded volume has. The excluded volume between two hard bodies is defined as the spatial region excluded to the center of mass of a second particle, for fixed orientations of both particles, and has for HSCs the following form

$$V_{\text{excl}}(\mathbf{\Omega}_1, \mathbf{\Omega}_2) = 2L^2 D |\sin\gamma| + 2\pi D^2 L + \frac{4}{3}\pi D^3 , \tag{7.154}$$

with γ the relative angle between the long axes of both particles.

Density Functionals for Uniform Fluid Phases

For uniform fluids (as the isotropic or nematic phases), the virial expansion of the excess part of the free energy density up to third order reads

$$\Phi_{\rm ex}[h(\hat{\boldsymbol{\Omega}})] = B_2[h(\hat{\boldsymbol{\Omega}})]\rho^2 + \frac{1}{2}B_3[h(\hat{\boldsymbol{\Omega}})]\rho^3 + \cdots , \tag{7.155}$$

$$B_2[h(\hat{\boldsymbol{\Omega}})] = \frac{1}{2}\int \mathrm{d}\hat{\Omega}_1 \int \mathrm{d}\hat{\Omega}_2 h(\hat{\boldsymbol{\Omega}}_1)h(\hat{\boldsymbol{\Omega}}_2)V_{\rm excl}\,\hat{\boldsymbol{\Omega}}_1, \hat{\boldsymbol{\Omega}}_2 , \tag{7.156}$$

$$B_3\left[h(\hat{\boldsymbol{\Omega}})\right] = \frac{1}{3V}\left[\prod_{i=1}^{3}\int d\boldsymbol{x}_i h(\hat{\boldsymbol{\Omega}}_i)\right] f(\boldsymbol{r}_{12}, \hat{\boldsymbol{\Omega}}_1, \hat{\boldsymbol{\Omega}}_2)$$
$$\times f(\boldsymbol{r}_{23}, \hat{\boldsymbol{\Omega}}_2, \hat{\boldsymbol{\Omega}}_3) f(\boldsymbol{r}_{13}, \hat{\boldsymbol{\Omega}}_1, \hat{\boldsymbol{\Omega}}_3) , \tag{7.157}$$

where V is the system volume and $f(\boldsymbol{r}_{ij}, \hat{\boldsymbol{\Omega}}_i, \hat{\boldsymbol{\Omega}}_j)$ is the Mayer function between two particles with fixed orientations $\hat{\boldsymbol{\Omega}}_i$ and $\hat{\boldsymbol{\Omega}}_j$. Onsager showed that for the isotropic fluid $[h(\hat{\boldsymbol{\Omega}}) = 1/(4\pi)]$, the following asymptotic behavior is fulfilled for large aspect ratios $\kappa \equiv (L+D)/D$: $B_3^{\rm iso}/\left(B_2^{\rm iso}\right)^2 \sim \kappa^{-1}\ln\kappa$ [103, 104]. Then, after scaling the density ρ with $B_2^{\rm iso} = \frac{\pi}{4}DL^2$ to define the dimensionless density $\rho^* = \rho B_2^{\rm iso}$, the excess free energy in the limit of infinite elongation $\kappa \to \infty$ retains only the second virial contribution. This approach, extended also to the study of HSC with a finite aspect ratio κ, is known as the Onsager approximation.

The constrained functional minimization of the total free energy with respect to $h(\hat{\boldsymbol{\Omega}})$ within the Onsager approach results in an integral equation of the form

$$h(\theta) = \frac{\exp\left[-\frac{8\rho^*}{\pi}\int \mathrm{d}(\cos\theta')K(\theta,\theta')h(\theta)\right]}{\int_{-1}^{1}\mathrm{d}(\cos\theta)\exp\left[-\frac{8\rho^*}{\pi}\int \mathrm{d}(\cos\theta')K(\theta,\theta')h(\theta')\right]}, \tag{7.158}$$

$$K(\theta,\theta') = \int_0^{2\pi} \mathrm{d}\phi\sqrt{1-(\cos\theta\cos\theta'\cos\phi + \sin\theta\sin\theta')^2}, \tag{7.159}$$

where the axial symmetry of the HSC has been used $[h(\hat{\boldsymbol{\Omega}}) = h(\theta)]$, θ is the angle between the HSC axis and the nematic director. Equation (7.158) can be solved iteratively for the equilibrium orientational distribution function $h(\theta)$ at a fixed ρ^*. Then, after the insertion of this function into the free-energy density and the use of the usual double-tangent construction to find the coexistence densities of the isotropic and nematic phases, the following values are obtained: $\rho_I^* = 3.290$ and $\rho_N^* = 4.191$ while the nematic order parameter at ρ_N^* is $S = 0.7922$ [126].

The Parsons–Lee Approach

For finite values of κ, further virial coefficients should be included if we want to accurately reproduce the I–N coexistence densities. The inclusion of the third virial coefficient $B_3[h(\theta)]$ in the virial expansion involves the numerical calculation of the integral (7.157), which is not an analytic function of the relative orientations $\hat{\mathbf{\Omega}}_i$ and of the particle characteristic lengths. Thus, a multidimensional integral must be evaluated, making the functional minimization a numerically difficult task. Parsons and later Lee conceived a DF which includes all the virial coefficients after the third approximately [127, 128, 129]. They proposed as the excess free energy density of the HSC

$$\Phi_{\text{ex}}[h(\theta)] = \rho\psi_{\text{HS}}(\eta)\frac{B_2[h(\theta)]}{B_2^{(\text{HS})}}\,, \tag{7.160}$$

where $B_2^{(\text{HS})} = 4v$ is the HS second virial coefficient, with v the hard rod volume selected equal to the HS volume and $\psi_{\text{HS}}(\eta) = \eta(4-3\eta)/(1-\eta)^2$ is the CS excess free energy per particle of the HS fluid, with $\eta = \rho v$ its packing fraction. This approximation, known as the Parsons–Lee (PL) approximation, recovers the second virial low-density limit while it approximates the remain virial coefficients by the expressions

$$B_n[h(\theta)] = (n-1)(n+2)v^{n-1}\frac{B_2[h(\theta)]}{B_2^{(HS)}}\,. \tag{7.161}$$

This simple approach gives quite reasonable values for the coexisting densities as compared to the simulation results. For example, for $\kappa = 5$, the PL approach gives $\eta_I = 0.3995$ and $\eta_N = 0.4172$, while simulations show an isotropic–nematic phase transition at packing fraction equal to 0.4.

The Scaled Particle Theory

The SPT, originally developed by Reiss [23, 24] for HS and later extended to anisotropic particles by several workers [130, 131, 132, 133, 134], has been successfully applied to the study of liquid crystal models composed of hard convex anisotropic particles. This is a systematically constructed theory which approximates the thermodynamic work to insert a scaled particle with a fixed orientation in a fluid. The approximation amounts to interpolating that work between its two known limits: those corresponding to small and large values of the scaling parameters.

For the scaled HSC with length and width equal to $\lambda_1 L$ and $\lambda_2 D$, respectively (λ_i are the scaling parameters), in the limit $\lambda_i \to 0$ the probability that the inserted particle does not overlap with any of the fluid particles is $\exp[-\beta W(\hat{\mathbf{\Omega}}; \lambda_1, \lambda_2)]$, ($W(\hat{\mathbf{\Omega}}; \lambda_1, \lambda_2)$ being the thermodynamic work necessary to insert the scaled particle with orientation $\hat{\mathbf{\Omega}}$. Taking only into account overlaps of two particles, one finds

$$\exp[-\beta W(\hat{\mathbf{\Omega}};\lambda_1,\lambda_2)] = 1-\rho\int \mathrm{d}\hat{\mathbf{\Omega}}' h(\hat{\mathbf{\Omega}}')V_{\mathrm{excl}}(\hat{\mathbf{\Omega}},\hat{\mathbf{\Omega}}';\lambda_1,\lambda_2)\,, \quad (7.162)$$

where $V_{\mathrm{excl}}(\hat{\mathbf{\Omega}},\hat{\mathbf{\Omega}}';\lambda_1,\lambda_2)$ is the excluded volume between the scaled particle and a fluid particle with orientation $\hat{\mathbf{\Omega}}'$. For large values of λ_i, the work required to insert the particle is just a work required to open a macroscopic cavity of volume equal to that of the scaled particle against the fluid pressure, i.e $W(\hat{\mathbf{\Omega}};\lambda_1,\lambda_2) = pV(\lambda_1,\lambda_2)$. In the scaled particle treatment, it is assumed that the work to add a particle with arbitrary values of λ_i can be calculated from the expansion of $W(\hat{\mathbf{\Omega}};\lambda_1,\lambda_2)$ obtained from Eq. (7.162) by the Taylor series around $(0,0)$ up to second order and adding $pV(\lambda_1,\lambda_2)$ as the third-order term. The excess chemical potential of HSPCs with length L and diameter D can be obtained by setting $\lambda_i = 1$ and integrating over all possible orientations with the distribution function, i.e. $\mu_{\mathrm{ex}} = \int \mathrm{d}\hat{\mathbf{\Omega}} h(\hat{\mathbf{\Omega}})W(\hat{\mathbf{\Omega}};1,1)$. The pressure can then be obtained through the Gibbs–Duhem equation $\partial p/\partial\rho = 1+\rho\partial\mu_{\mathrm{ex}}/\partial\rho$. This yields an excess free energy per particle $\psi[h] = \mu_{\mathrm{ex}}+1-p/\rho$, given by

$$\psi[h] = -\ln(1-\eta)+A[h]y+\frac{1}{2}B[h]y^2\,, \quad (7.163)$$

$$A[h] = 3\left(1+\frac{4(\kappa-1)^2}{\pi(3\kappa-1)}\langle\langle|\sin\gamma|\rangle\rangle\right)\,, \quad (7.164)$$

$$B[h] = \frac{12\kappa}{(3\kappa-1)^2}\left(2\kappa-1+\frac{4}{\pi}(\kappa-1)^2\langle\langle|\sin\gamma|\rangle\rangle\right)\,, \quad (7.165)$$

with $y = \eta/(1-\eta)$ and where $\langle\langle u\rangle\rangle = \int \mathrm{d}\hat{\mathbf{\Omega}}\int \mathrm{d}\hat{\mathbf{\Omega}}' h(\hat{\mathbf{\Omega}})h(\hat{\mathbf{\Omega}}')u(\hat{\mathbf{\Omega}},\hat{\mathbf{\Omega}}')$ defines the double angular average of function u with respect to $h(\hat{\mathbf{\Omega}})$. The values that the SPT gives for the I–N coexistence packing fractions for $\kappa = 5$ are $\eta_I = 0.36$ and $\eta_N = 0.39$, less accurate that the Parson's approach. However, the estimation that the SPT makes for the third virial coefficient is better than Parsons result.

Density Functionals for Non-uniform Liquid Crystal Phases

All the DFs extended to the study of non-uniform liquid crystal phases (as the smectic phases), except the last versions based on the FMT which will be described in next subsection, include as basic ingredient in their constructions the thermodynamic and structural functions of the HS fluid. See the review of Vroege et al. [135] for a more detailed discussion. Practically, all of them approximate the exact second-order expansion of the excess free energy of the non-uniform fluid around some reference uniform fluid of density ρ_0

$$\mathcal{F}_{\text{ex}}[\rho(\boldsymbol{x})] = \mathcal{F}_{\text{ex}}[\rho_0(\hat{\Omega})] + \int \mathrm{d}\boldsymbol{x}_1 \mu_{\text{ex}}[\rho_0(\hat{\boldsymbol{\Omega}}_1)]\Delta\rho(\boldsymbol{x}_1)$$

$$- kT \int \mathrm{d}\boldsymbol{x}_1 \int \mathrm{d}\boldsymbol{x}_2 \int_0^1 \mathrm{d}\lambda \int_0^\lambda \mathrm{d}\lambda' c(\boldsymbol{x}_1, \boldsymbol{x}_2; [\rho_0(\hat{\boldsymbol{\Omega}})$$

$$+ \lambda'\Delta\rho(\boldsymbol{x})])\Delta\rho(\boldsymbol{x}_1)\Delta\rho(\boldsymbol{x}_2) , \quad (7.166)$$

where $\Delta\rho(\boldsymbol{x}) = \rho(\boldsymbol{x}) - \rho_0(\hat{\boldsymbol{\Omega}})$ and $\rho_0(\hat{\boldsymbol{\Omega}}) + \lambda'\Delta\rho(\boldsymbol{x})$ represents a parameterized integration path from the uniform isotropic or nematic state to the non-uniform final state.

Selecting a reference fluid of zero density and taking the low-density limit of the direct correlation function equal to $f(\boldsymbol{x}_1, \boldsymbol{x}_2)$, the Mayer function, we obtain the Onsager approximation extended to the non-uniform fluid

$$\mathcal{F}_{\text{ex}}[\rho(\boldsymbol{x})] = \frac{1}{2} \int \mathrm{d}\boldsymbol{x}_1 \int \mathrm{d}\boldsymbol{x}_2 \rho(\boldsymbol{x}_1)\rho(\boldsymbol{x}_2) f(\boldsymbol{x}_1, \boldsymbol{x}_2) . \quad (7.167)$$

This approximation has been successfully applied to study the isotropic-nematic interface by Shundyak and van Roij, in particular the analysis of particle orientations along the interface [136]. The trivial extension of (7.167) to binary mixtures allowed also the study of the isotropic–nematic and nematic–nematic interfaces near an I–N–N triple point that can exhibit certain mixtures of HSCs [137, 138].

The Decoupling Approximation

Selecting as the reference fluid an isotropic fluid and carrying out a density expansion of the direct correlation function around this reference fluid (7.166) leads to the approximation

$$\mathcal{F}_{\text{ex}}[\rho(\boldsymbol{x})] = \mathcal{F}_{\text{ex}}[\rho_0] + \mu_{\text{ex}}[\rho_0] \int \mathrm{d}\boldsymbol{x}_1 \Delta\rho(\boldsymbol{x}_1)$$

$$- \frac{1}{2} kT \int \mathrm{d}\boldsymbol{x}_1 \int \mathrm{d}\boldsymbol{x}_2 c(|\boldsymbol{r}_1 - \boldsymbol{r}_2|, \hat{\boldsymbol{\Omega}}, \hat{\boldsymbol{\Omega}}_1, \hat{\boldsymbol{\Omega}}_2; \rho_0)\Delta\rho(\boldsymbol{x}_1)\Delta\rho(\boldsymbol{x}_2) , \quad (7.168)$$

which has been used as the starting point for studies of the smectic [139] and crystalline [140, 141, 142] phases. However, the function $c(|\boldsymbol{r}_1 - \boldsymbol{r}_2|, \hat{\boldsymbol{\Omega}}, \hat{\boldsymbol{\Omega}}_1, \hat{\boldsymbol{\Omega}}_2; \rho_0)$, the direct correlation function for the isotropic fluid of density ρ_0, is in general unknown (except for the case of HS where we have the PY approximation). Different attempts to approximate this function are all based in the known PY analytic form of the HS direct correlation function. For example, the first recipe was proposed by Pynn, who approximated the direct correlation function of the isotropic HR fluid by the HS PY correlation function of the scaled distance [143, 144], taking the contact distance between two HR $\sigma(\hat{\boldsymbol{\Omega}}, \hat{\boldsymbol{\Omega}}_1, \hat{\boldsymbol{\Omega}}_2)$ as the scaling factor, i.e.

$$c(r, \hat{\boldsymbol{\Omega}}, \hat{\boldsymbol{\Omega}}_1, \hat{\boldsymbol{\Omega}}_2; \eta) = c_{\mathrm{PY}} \left(\frac{r}{\sigma(\hat{\boldsymbol{\Omega}}, \hat{\boldsymbol{\Omega}}_1, \hat{\boldsymbol{\Omega}}_2)}; \eta \right) . \tag{7.169}$$

This single anisotropic dependence is known as the decoupling approximation. However, in the limit $r \to 0$, the correlation function becomes unphysically isotropic. To circumvent this problem, Baus and coworkers [145, 146] introduced the following approximation for the hard ellipsoid correlation function

$$c(r, \hat{\boldsymbol{\Omega}}, \hat{\boldsymbol{\Omega}}_1, \hat{\boldsymbol{\Omega}}_2, \eta) = \frac{V_{\mathrm{excl}}(\hat{\boldsymbol{\Omega}}_1, \hat{\boldsymbol{\Omega}}_2)}{v_0} c_{\mathrm{PY}} \left(\frac{r}{\sigma_0}; \eta \right) , \tag{7.170}$$

with $v_0 = 4\pi\sigma_0^3/3$, the excluded HS volume selected to be equal to the ellipsoid volume ($\sigma_0^3 = a^2 b$). Again, the angular and translational dependence are decoupled.

Weighted Density Approximations

Another route to extend the DF theory to the study of non-uniform phases was developed by Somoza and Tarazona [147, 148, 149] and Poniewierski and Hołyst [150, 151]. Both approximations are based on the WDA. Both theories introduce the anisotropic correlations through a new weight $w(\boldsymbol{r}, \hat{\boldsymbol{\Omega}}_1, \hat{\boldsymbol{\Omega}}_2)$ which takes care of the orientations of the interacting particles. Somoza and Tarazona selected for the weight the Mayer function $f(\boldsymbol{r}, \hat{\boldsymbol{\Omega}}, \hat{\boldsymbol{\Omega}}')$ and imposed the requirement the functional recovered the Onsager limit at low densities. Thus, if we define the number of HR which interact with a given particle placed at $\boldsymbol{r}$ and oriented along $\hat{\boldsymbol{\Omega}}$ as

$$N(\boldsymbol{r}, \hat{\boldsymbol{\Omega}}) = \int \mathrm{d}\boldsymbol{r}' \int \mathrm{d}\hat{\boldsymbol{\Omega}}' \rho(\boldsymbol{r}', \hat{\boldsymbol{\Omega}}') f(\boldsymbol{r} - \boldsymbol{r}', \hat{\boldsymbol{\Omega}}, \hat{\boldsymbol{\Omega}}') \tag{7.171}$$

and the number of interacting rods in the parallel particle approximation (considering that all of them have the symmetry of an ellipsoid of revolution with the same volume)

$$N_{\mathrm{PHE}}(\boldsymbol{r}) = \int \mathrm{d}\boldsymbol{r}' \rho(\boldsymbol{r}) f_{\mathrm{PHE}}(\boldsymbol{r} - \boldsymbol{r}') , \tag{7.172}$$

with $\rho(\boldsymbol{r}) = \int \mathrm{d}\hat{\boldsymbol{\Omega}} \rho(\boldsymbol{r}, \hat{\boldsymbol{\Omega}})$, the proposed functional is defined by

$$\mathcal{F}_{\mathrm{ex}}[\rho(\boldsymbol{r}, \hat{\boldsymbol{\Omega}})] = \int \mathrm{d}\boldsymbol{r} \int \mathrm{d}\hat{\boldsymbol{\Omega}} \rho(\boldsymbol{r}, \hat{\boldsymbol{\Omega}}) \psi_{\mathrm{CS}}[\bar{\rho}(\boldsymbol{r})] \frac{N(\boldsymbol{r}, \hat{\boldsymbol{\Omega}})}{N_{\mathrm{PHE}}(\boldsymbol{r})} , \tag{7.173}$$

where the CS-WDA for a fluid of PHEs was selected to calculate the free energy per particle. Thus, the angular correlations are taken through the scaling factor $N(\boldsymbol{r},\hat{\boldsymbol{\Omega}})/N_{\mathrm{PHE}}(\boldsymbol{r})$. It should be noted that this approximation recovers Parson's approach for uniform fluids. The aspect ratio of the reference hard ellipsoid was calculated by also requiring that its average tensor of inertia be proportional to that of the HSC.

Hołyst and Poniewierski chose the weight $w(\boldsymbol{r},\hat{\boldsymbol{\Omega}},\hat{\boldsymbol{\Omega}}') = f(\boldsymbol{r},\hat{\boldsymbol{\Omega}},\hat{\boldsymbol{\Omega}}')/(2B_2^{\mathrm{iso}})$ which satisfies the normalization condition $\int \mathrm{d}\boldsymbol{r}\langle\langle\omega(\boldsymbol{r},\hat{\boldsymbol{\Omega}},\hat{\boldsymbol{\Omega}}')\rangle\rangle = 1$, with the angular brackets meaning the double angular average with respect to the uniform isotropic distribution function $h(\hat{\boldsymbol{\Omega}}) = 1/(4\pi)$ [150, 151]. Thus, the weighted density is calculated through

$$\bar{\rho}(\boldsymbol{r}) = \int \mathrm{d}\boldsymbol{r}'\rho(\boldsymbol{r}')\left\langle\left\langle w(\boldsymbol{r}-\boldsymbol{r}',\hat{\boldsymbol{\Omega}},\hat{\boldsymbol{\Omega}}')\right\rangle\right\rangle , \tag{7.174}$$

where the double angular average is now with respect to the orientational distribution functions $h(\boldsymbol{r},\hat{\boldsymbol{\Omega}})$ and $h(\boldsymbol{r}',\hat{\boldsymbol{\Omega}}')$, respectively. The free energy functional in this approximation is calculated as

$$\beta\mathcal{F}_{\mathrm{ex}}[\rho(\boldsymbol{r},\hat{\boldsymbol{\Omega}})] = \int \mathrm{d}\boldsymbol{r}\rho(\boldsymbol{r})\psi(\bar{\rho}(\boldsymbol{r})) , \tag{7.175}$$

with the excess free energy per particle

$$\psi(\rho) = \rho B_2^{\mathrm{iso}} + \psi_{\mathrm{CS}}(\rho) - 4\eta , \tag{7.176}$$

which recovers the exact second virial form for the isotropic fluid.

Phase diagrams of HSC's including isotropic, nematic and smectic phases were calculated using both theories. The nematic–smectic transition was calculated through functional minimization of the Somoza–Tarazona functional and through bifurcation analysis of the Poniewierski–Hołyst functional. The transition densities and order parameters of the smectic phase compare reasonably well with simulation results for $\kappa = 5$. The Somoza–Tarazona theory predicts a N–Sm tricritical point located at $L/D = 50$ and a triple I–N–Sm coexistence at $L/D = 3$ in agreement with simulations. This model was also successfully implemented in the study of the isotropic–nematic interface of HSCs by Velasco et al. [152].

7.11 Fundamental Measure Theory for Anisotropic Particles

The first attempt to extend the FMT to hard convex anisotropic bodies was made by Rosenfeld [153]. He realized that the exact decomposition of the Mayer function between two HS of different radii in terms of the one-particle weights $w_\mu^{(\alpha)}(\boldsymbol{r})$ can be regarded as the direct application of the Gauss–Bonnet theorem of differential geometry [154]. This theorem constitutes one of the fundamental results in the theory of surfaces and relates the average over a surface of its Gaussian curvature to a topological property of the surface called its Euler characteristic. If a convex body is formed by the intersection volume between two particles μ and ν, then the Gauss–Bonnet theorem says that Euler characteristics of the surfaces of this intersection volume can be obtained as

$$4\pi f_{\mu\nu}(\boldsymbol{r}) = \iint_{\partial\mu\cap\nu} K_\mu \mathrm{d}S_\mu + \iint_{\partial\nu\cap\mu} K_\nu \mathrm{d}S_\nu + \int_{\partial\mu\cap\partial\nu} \left[k_g^{(\mu)} + k_g^{(\nu)}\right] \mathrm{d}l \,, \tag{7.177}$$

where $\partial\mu\cap\nu$ is the surface of that part of body μ which is inside body ν (the same definition works for $\partial\nu\cap\mu$), while $\partial\mu\cap\partial\nu$ is the closed curve obtained by the intersection of both surfaces. K_μ and K_ν are the Gaussian curvatures of surfaces of μ and ν, while $k_g^{(\mu)}$ and $k_g^{(\nu)}$ are their geodesic curvatures along their common intersection curve. The first two integrals are surface integrals, while the third one is a line integral. For spheres, it can be shown that

$$k_g^{(\mu)} = \frac{H_\mu\left[1 - \boldsymbol{u}_\mu \cdot \boldsymbol{u}_\nu\right]}{|\boldsymbol{u}_\mu \times \boldsymbol{u}_\nu|} \,, \tag{7.178}$$

with H_μ the mean curvature of the surface of body μ, and $\boldsymbol{u}_\mu$ its unit external normal.

If we insert this expression in (7.177), we end up with the decomposition of the Mayer function in terms of a sum of convolutions between one-body weights that defines Rosenfeld's original FMT [13], cf. Eq. (7.65).

The use of (7.178) for any convex body is of course an approximation firstly proposed by Rosenfeld to construct a fundamental measure functional for anisotropic hard bodies. Accordingly, the structure of the functional is exactly the same as his first proposal for HS mixtures (7.113)–(7.115) but with the extended weights

$$w_\mu^{(3)}(\boldsymbol{r}) = \Theta(r - |\boldsymbol{R}_\mu(\hat{\boldsymbol{\Omega}})|), \qquad w_\mu^{(2)}(\boldsymbol{r}) = \delta(r - |\boldsymbol{R}_\mu(\hat{\boldsymbol{\Omega}})|), \tag{7.179}$$

$$\boldsymbol{w}_\mu^{(2)}(\boldsymbol{r}) = \boldsymbol{u}_\mu \delta(r - |\boldsymbol{R}_\mu(\hat{\boldsymbol{\Omega}})|), \qquad w_\mu^{(0)}(\boldsymbol{r}) = \frac{K_\mu}{4\pi} w_\mu^{(2)}(\boldsymbol{r}) \,, \tag{7.180}$$

$$w_\mu^{(1)}(\boldsymbol{r}) = \frac{H_\mu}{4\pi} w^{(2)}(\boldsymbol{r}), \qquad \boldsymbol{w}_\mu^{(1)}(\boldsymbol{r}) = \frac{H_\mu}{4\pi} \boldsymbol{w}_\mu^{(2)}(\boldsymbol{r}) \,, \tag{7.181}$$

where $\boldsymbol{R}_\mu(\hat{\boldsymbol{\Omega}})$ is the radius vector from the center of body μ with orientation $\hat{\boldsymbol{\Omega}}$ to its surface. Once the parametrization of the surface is given, we can calculate the Gaussian K_μ and mean curvature H_μ of body μ using the standard formulas from differential geometry [154]. The weighted densities are obtained by convoluting the density profiles with (7.179)–(7.181) and integrating the result over $\hat{\boldsymbol{\Omega}}$

$$n^{(\alpha)}(\boldsymbol{r}) = \sum_\mu \int \mathrm{d}\hat{\boldsymbol{\Omega}} \rho_\mu * w^{(\alpha)}(\boldsymbol{r}, \hat{\boldsymbol{\Omega}}), \quad \alpha = 0, 1, 2 \tag{7.182}$$

$$\boldsymbol{v}^{(\alpha)}(\boldsymbol{r}) = \sum_\mu \int \mathrm{d}\hat{\boldsymbol{\Omega}} \rho_\mu * \boldsymbol{w}^{(\alpha)}(\boldsymbol{r}, \hat{\boldsymbol{\Omega}}), \quad \alpha = 1, 2 \tag{7.183}$$

$$\eta(\boldsymbol{r}) = \sum_\mu \int \mathrm{d}\hat{\boldsymbol{\Omega}} \rho_\mu * w^{(3)}(\boldsymbol{r}, \hat{\boldsymbol{\Omega}}). \tag{7.184}$$

If we impose to the functional that it recovers the DI-FMT form (7.113), (7.114) and (7.120)–(7.121) for the HS limit, we also need to introduce the second-rank tensors $\mathcal{T}_\mu(\boldsymbol{r})$ with components $(\mathcal{T}_\mu)_{ij}(\boldsymbol{r}) = u_i^{(\mu)} u_i^{(\mu)} \delta(r - |\boldsymbol{R}_\mu(\hat{\boldsymbol{\Omega}})|)$, with $u_i^{(\mu)}$ $(i = x, y, z)$ the external unit normals components. With this definition and the Eqs. (7.179)–(7.181), the free energy density obtained from the DI-FMT can in principle be used to obtain a fundamental measure functional for general convex bodies.

This approximation gives the exact second virial coefficient for the isotropic fluid

$$B_{\mu\nu}^{\mathrm{iso}} = \frac{1}{2}\left[V_\mu + A_\mu \bar{R}_\nu + \bar{R}_\mu A_\nu + V_\nu\right] , \tag{7.185}$$

with V_μ, A_μ and $\bar{R}_\mu = 1/(4\pi)\int_{\partial\mu} H_\mu \mathrm{d}S_\mu$, the volume, surface area and mean curvature radius of body μ, respectively. The calculation of the direct correlation function from this generalized functional gives an expression with different terms proportional to $\Delta V_{\mu\nu}$, $\Delta A_{\mu\nu}$ and $\Delta R_{\mu\nu}$, the volume, the surface area and the mean curvature radius of the intersection of the two bodies, which is consistent with the PY approximation. However, the low density term gives only an approximation for the Mayer function $f_{\mu\nu}$. One can think that this is a minor fault of the present theory, unfortunately, the substitution of the exact Mayer function by the approximated one in terms of a sum of convolutions of single body weights has as a consequence that the Fourier transform of the direct correlation function at the zero wave vector $\boldsymbol{q} = \boldsymbol{0}$ is always isotropic [155]. This in turn means that the isotropic fluid is always stable with respect to an orientationally ordered phase (as the nematic phase), which is obviously unphysical.

To solve this problem, different approximations for the direct correlation function of anisotropic fluids were worked out by Chamoux and Perera [155]. One of them is obtained from the second functional derivative of the extended Rosenfeld functional replacing its low-density term by another one proportional to the exact Mayer function $f_{\mu\nu}$ [155]. This approximation gives, from the compressibility route $1 - \sum_{ij} \rho_i \rho_j \hat{c}_{ij}(\boldsymbol{q} = \boldsymbol{0}) = \partial(\beta p)/\partial\rho$, the scaled particle equation of state and also exhibits an spinodal isotropic–nematic instability at $\boldsymbol{q} = \boldsymbol{0}$.

Another framework to construct DFs for anisotropic particles is based on the deconvolution of the exact Mayer function in terms of weights that can also depend on the orientations of both particles. If we renounce to have single body weights, then the decomposition can be exactly obtained, but for some special limits of the particle anisotropy. This approximation will be described later.

7.11.1 FMT for Parallel Hard Cubes

Beside HS, there is another particle geometry for which a FMT has been consistently developed, namely the case of a fluid of parallel hard cubes [156, 157, 158]. We briefly describe the theory here. The Mayer function between two parallel cubes of edge lengths σ_μ and σ_ν is

$$f_{\mu\nu}(\boldsymbol{r}) = \Theta(\sigma_{\mu\nu} - |x|)\Theta(\sigma_{\mu\nu} - |y|)\Theta(\sigma_{\mu\nu} - |z|) , \tag{7.186}$$

with $\sigma_{\mu\nu} = (\sigma_\mu + \sigma_\nu)/2$. This Mayer function can be decomposed exactly in terms of a finite sum of convolutions between one-particle weights as

$$\begin{aligned} f_{\mu\nu}(\boldsymbol{r}) = w_\mu^{(3)} * w_\nu^{(0)}(\boldsymbol{r}) + w_\mu^{(0)} * w_\nu^{(3)}(\boldsymbol{r}) + \boldsymbol{w}_\mu^{(2)} * \boldsymbol{w}_\nu^{(1)}(\boldsymbol{r}) \\ + \boldsymbol{w}_\mu^{(1)} * \boldsymbol{w}_\nu^{(2)}(\boldsymbol{r}) , \end{aligned} \tag{7.187}$$

with the weights defined as

$$w_\mu^{(0)}(\boldsymbol{r}) \equiv \zeta_\mu(x)\zeta_\mu(y)\zeta_\mu(z), \tag{7.188}$$

$$\begin{aligned} \boldsymbol{w}_\mu^{(1)}(\boldsymbol{r}) \equiv (\tau_\mu(x)\zeta_\mu(y)\zeta_\mu(z), \zeta_\mu(x)\tau_\mu(y)\zeta_\mu(z), \\ \zeta_\mu(x)\zeta_\mu(y)\tau_\mu(z)) , \end{aligned} \tag{7.189}$$

$$\begin{aligned} \boldsymbol{w}_\mu^{(2)}(\boldsymbol{r}) \equiv (\zeta_\mu(x)\tau_\mu(y)\tau_\mu(z), \tau_\mu(x)\zeta_\mu(y)\tau_\mu(z), \\ \tau_\mu(x)\tau_\mu(y)\zeta_\mu(z)) , \end{aligned} \tag{7.190}$$

$$w_\mu^{(3)}(\boldsymbol{r}) \equiv \tau_\mu(x)\tau_\mu(y)\tau_\mu(z) , \tag{7.191}$$

where $\zeta_\mu(x) = \frac{1}{2}\delta\left(\frac{\sigma_\mu}{2} - |x|\right)$ and $\tau_\mu(x) = \Theta\left(\frac{\sigma_\mu}{2} - |x|\right)$. Introducing two more weights as $\omega_\mu^{(1)}(\boldsymbol{r}) \equiv \boldsymbol{u} \cdot \boldsymbol{\omega}_\mu^{(1)}(\boldsymbol{r})$ and $\omega_\mu^{(2)}(\boldsymbol{r}) = \boldsymbol{u} \cdot \boldsymbol{\omega}_\mu^{(2)}(\boldsymbol{r})$ [$\boldsymbol{u} = (1, 1, 1)$], the most general form of the excess part of free energy density according to a dimensional analysis is

$$\begin{aligned} \Phi = a_0 n^{(0)} + a_1 n^{(1)} n^{(2)} + b_1 \boldsymbol{v}^{(1)} \cdot \boldsymbol{v}^{(2)} + a_2 \left(n^{(2)}\right)^3 \\ + b_2 n^{(2)} \boldsymbol{v}^{(2)} \cdot \boldsymbol{v}^{(2)} + c \boldsymbol{v}^{(2)} \cdot \boldsymbol{v}^{(2)} \cdot \boldsymbol{v}^{(2)} , \end{aligned} \tag{7.192}$$

where the weighted densities are introduced as usual, $n^{(\alpha)}(\boldsymbol{r}) = \sum_\mu \rho_\mu * w_\mu^{(\alpha)}$, $\boldsymbol{v}^{(\alpha)}(\boldsymbol{r}) = \sum_\mu \rho_\mu * \boldsymbol{w}_\mu^{(\alpha)}$ and $\boldsymbol{v} \cdot \boldsymbol{v} \cdot \boldsymbol{v} \equiv v_x^3 + v_y^3 + v_z^3$. The coefficients $a_i(\eta)$, $b_i(\eta)$ and $c(\eta)$ are all functions of the dimensionless local packing fraction $\eta = \sum_\mu \rho_\mu * w_\mu^{(3)}$. Introducing Eq. (7.192) in the SPT differential equation [156, 157, 158]

$$-\Phi + \sum_{\alpha} m^{(\alpha)} \frac{\partial \Phi}{\partial m^{(\alpha)}} + n^{(0)} = (1-\eta) \frac{\partial \Phi}{\partial \eta} , \qquad (7.193)$$

with $m^{(\alpha)} \equiv \{n^{(\alpha)}, \boldsymbol{v}^{(\alpha)}\}$, we obtain a set of first-order differential equations to determine $\{a_i, b_i, c\}$ up to the integration constants. Note that Eq. (7.193) can be easily obtained from the usual definition of the fluid pressure together with the identity $\beta p = \partial \Phi / \partial \eta$ valid for the SPT. The density expansion of the direct correlation function $c_{ij}(\boldsymbol{r})$ obtained from (7.192) together with the exact knowledge of the Mayer and the triangle diagrams, obtained as sums of convolutions between single particle weights, allow us to fix all the integration constants but one. Finally, imposing the exact dimensional crossover to zero dimension, this constant can be determined, resulting in the functional

$$\Phi^{(3)} = -n^{(0)} \ln(1-\eta) + \frac{\boldsymbol{v}^{(1)} \cdot \boldsymbol{v}^{(2)}}{1-\eta} + \frac{v^{(2x)} v^{(2y)} v^{(2z)}}{(1-\eta)^2} . \qquad (7.194)$$

It should be noted that the free energy evaluated at density profiles corresponding to arbitrary 0D cavities exactly recovers the 0D limit, i.e. the parallel hard cubes fluid is free from *lost cases*. The fundamental reason behind this is that there are no configurations of three particles having pairwise overlap but no triple overlap.

Based on this property, FMT functionals for hard core particles on a lattice were constructed by the projections of the parallel hard hypercube lattice model on different planes. The FMT for lattice model can then be reformulated based on the exact 0D crossover, and functionals for arbitrary hard particles on a lattice can be obtained that are free from lost cases [159, 160, 161, 162, 163, 164]. The same procedure of projecting the hard hypercube geometry on a particular plane was recently implemented to obtain a FMT for hard hexagons in the continuum [165]. It was shown that the number of weighted densities necessary to obtain a functional free from lost cases is infinite. Note that the hexagon geometry allows to have configurations of three particles with pairwise overlap but no triple overlap.

Following the same procedure, we can obtain the 2D functional for a fluid of parallel hard squares as

$$\Phi^{(2)} = -n^{(0)} \ln(1-\eta) + \frac{v^{(1x)} v^{(1y)}}{1-\eta} , \qquad (7.195)$$

where the 2D weights are

$$w_\mu^{(0)}(\boldsymbol{r}) \equiv \zeta_\mu(x) \zeta_\nu(y) , \qquad (7.196)$$

$$\boldsymbol{w}_\mu^{(1)}(\boldsymbol{r}) \equiv (\tau_\mu(x) \zeta_\mu(y), \zeta_\mu(x) \tau_\mu(y)) , \qquad (7.197)$$

$$w_\mu^{(2)}(\boldsymbol{r}) \equiv \tau_\mu(x) \tau_\mu(y). \qquad (7.198)$$

One-component Parallel Hard Cube Model

It is known that the equation of state of the uniform fluid obtained from the FMT is the SPT applied to the parallel hard cubes fluid; the other only stable phase for this model is a cubic crystal, which was calculated in [166]. For this model, the following parametrization of density profile was used

$$\rho(\boldsymbol{r}) = \eta_0 \left(\frac{\alpha}{\pi}\right)^{\frac{3}{2}} \sum_{\boldsymbol{k}} \exp\left[-\alpha(\boldsymbol{r}-\boldsymbol{R_k})^2\right] = \Psi_0(x)\Psi_0(y)\Psi_0(z)\,, \tag{7.199}$$

$$\Psi_0(x) = \eta_0^{\frac{1}{3}} \left(\frac{\alpha}{\pi}\right)^{\frac{1}{2}} \sum_{k=-\infty}^{\infty} \exp\left[-\alpha(x-kd)^2\right]\,, \tag{7.200}$$

where $\boldsymbol{R_k} = \boldsymbol{k}d$ $[\boldsymbol{k} = (k_1, k_2, k_3)]$ are the vector positions of the nodes of the simple cubic lattice with period d and η_0 representing the particle occupancy probability per unit cell. With this parametrization, the excess free energy density (7.194) for the one-component parallel hard cube fluid becomes

$$\Phi^{(3)} = n^{(0)} \left[-\ln(1-\eta) + \frac{3\eta}{1-\eta} + \frac{\eta^2}{(1-\eta)^2}\right]\,, \tag{7.201}$$

with $n^{(0)}(\boldsymbol{r}) = n^{(0)}(x)n^{(0)}(y)n^{(0)}(z)$, $\eta(\boldsymbol{r}) = \eta(x)\eta(y)\eta(z)$, where

$$n^{(0)}(x) = \frac{1}{2}\left[\Psi_0\left(x-\frac{\sigma}{2}\right) + \Psi_0\left(x+\frac{\sigma}{2}\right)\right]\,, \tag{7.202}$$

$$\eta(x) = \frac{1}{2}\left[\Psi_1\left(x+\frac{\sigma}{2}\right) - \Psi_1\left(x-\frac{\sigma}{2}\right)\right]\,, \tag{7.203}$$

$$\Psi_1(x) = \eta_0^{\frac{1}{3}} \sum_k \operatorname{erf}\left[\sqrt{\alpha}\,(x+kd)\right]\,. \tag{7.204}$$

Thus, the total free energy per unit volume $\beta\mathcal{F}/V$ is minimized with respect to α and η_0 to find the equilibrium density profile. The mean density ρ, the cell period d and the occupancy probability η_0 are related through $\rho = \eta_0 d^{-3}$. The fluid–solid transition is continuous and occurs at $\eta = 0.314$, with a fraction of vacancies of 31% [166]. The equation of state for the solid phase compares very well with simulation results for mean packing fractions $\eta > 0.5$ as can be seen in Fig. 7.15. However, the simulations predict a fluid–solid continuous transition at $\eta \approx 0.48$, higher than the FMT-predicted value of 0.314 [168]. As for HS, the fluid phase is to blame on this discrepancy in the FMT prediction for crystallization. It is known that the SPT equation of state for the fluid of parallel hard cubes deviates from simulation results at intermediate densities ~ 0.4. In this sense, the situation is even worse than for the HS fluid because the exact virial expansion has negative coefficients while the expansion of the SPT equation of state has only positive coefficients.

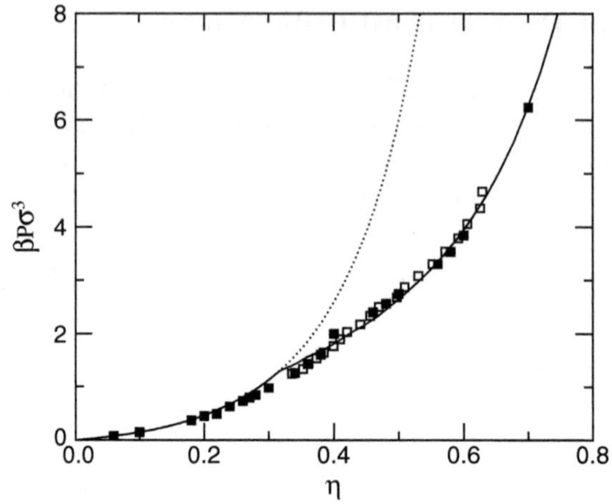

Fig. 7.15. Equation of state of the parallel hard cube system. *Solid lines* correspond to the pressure of the stable phase (fluid or solid) at a given packing fraction η. The *dotted line* is the unstable fluid branch beyond freezing. *Full squares* are the simulations of [167] and *empty squares* those of [168]

Binary Mixtures of Parallel Hard Cubes

The fundamental measure functional defined trough the free energy density (7.194) was used in the study of the demixing transitions in mixtures of parallel hard cubes [166]. Note that again the uniform limit of that functional coincides with the SPT result, and it was shown that for an asymmetry ratio $r \equiv \sigma_1/\sigma_2 > 9.98$, the parallel hard cube mixture phase separates in two phases, each one richer in one of the components [166]. In Fig. 7.16(a), we show the fluid–fluid coexistence for different mixtures. The same figure also shows the spinodal curves of the fluid–solid transition. It can be seen in the figure that the fluid–fluid demixing is always preempted by a fluid–solid phase separation. The same analysis for parallel hard square mixtures, using the uniform limit of (7.195), shows that the demixing is forbidden in 2D. However, using the SPT for a mixture of freely rotating hard squares, it can be shown that demixing occurs between an isotropic fluid and a tetratic nematic phase for $r > 4$ [169, 170, 171]. The last one, also observed in simulations [172], is an orientationally ordered phase invariant under rotations of $\pi/2$ (the tetratic symmetry), i.e. in terms of the orientational distribution function, we have $h(\theta) = h(\theta + \pi/2)$ [169]. In Fig. 7.16 (b), we show the demixing scenario of a mixture of freely rotating hard squares with asymmetry ratio $r = 10$ [173]. We can observe the presence of a lower and upper critical points, apart from a tricritical point below which the isotropic–tetratic nematic transition is of second order.

These results show that certain cautions should be taken in the extrapolation of results given from any DF theory of parallel anisotropic particles

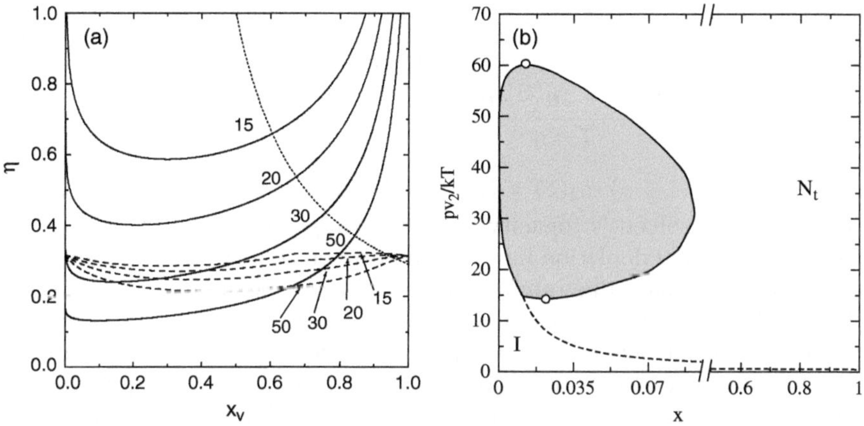

Fig. 7.16. (**a**): Phase diagram of a binary mixture of parallel hard cubes. Packing fraction η versus fraction of volume of big species $x_v \equiv \eta_1/\eta$. The numbers label the asymmetry ratio r. *Solid lines* show the binodals, while *dashed lines* represent the fluid–solid spinodals. The positions of critical points are shown with a *dotted line.* (**b**): Phase diagram of freely rotating hard squares for $r = 10$. Pressure in reduced units versus the molar fraction of big species x. *Dashed line* shows the continuous isotropic–tetratic nematic transition. *Open circles* show the critical points

to predict the phase behavior of freely rotating particles. The restriction of orientations allows us to obtain a consistent FMT for parallel hard cubes but paying the price of losing information about the orientational ordering phenomena inherent to the particle anisotropy.

The Adhesive Parallel Hard Cube Fluid

The parallel hard cube model is far from being a realistic model for particle interactions in colloids or in molecular fluids. However, it has been successfully applied to clarify the entropic mechanism behind the fluid–solid demixing that the highly asymmetric binary mixture of hard core particles usually exhibit [174, 175]. For this purpose, a DF for an adhesive hard cube fluid was obtained taking the limit of infinite asymmetry $\sigma_2/\sigma_1 \to 0$ of the functional $\Upsilon(\mu_2, [\rho_1]) = \mathcal{F}[\rho_1, \rho_2] - \mu_2 \int \mathrm{d}\boldsymbol{r} \rho_2(\boldsymbol{r})$ (the thermodynamic potential for the semi-grand ensemble), where 1 and 2 label the solute and solvent particles, respectively. The chemical potential of the smaller component μ_2 is fixed to a constant value. As a result, an effective one-component functional for the solute particles is obtained with the following expressions for excess free energy density in dimensions 2 and 3

$$\Phi^{(D)}_{\mathrm{eff}} = \Phi^{(D)} + \Phi^{(D)}_{\mathrm{ad}}\,, \tag{7.205}$$

$$\Phi_{\mathrm{ad}}^{(2)} = \frac{z}{8} \frac{\left[g_1(z)|\nabla\eta|^2 - 4g_2(z)\boldsymbol{v}^{(1)} \cdot \boldsymbol{v}^{(1)} + 8g_3(z)v^{(1x)}v^{(1y)}\right]}{1-\eta}, \quad (7.206)$$

$$\Phi_{\mathrm{ad}}^{(3)} = \frac{z}{8} \frac{\left[|\nabla\eta|^2 - 4\boldsymbol{v}^{(2)} \cdot \boldsymbol{v}^{(2)}\right]}{1-\eta}, \quad (7.207)$$

where $g_1(z) = 1/(1+z)$, $g_2(z) = (1+z)/(1+2z)$ and $g_3(z) = z/(1+2z)$ are functions of the solvent's fugacity $z = \exp(\beta\mu_2)$. The terms $\Phi_{\mathrm{ad}}^{(D)}$ take into account the residual depletion interaction between solute particles induced by the solvent. The phase diagram of this 3D adhesive hard cube fluid is shown in Fig. 7.17. The adhesiveness makes the fluid metastable with respect to a phase separation between a close-packed crystal and an infinitely diluted fluid. Different metastable or unstable phases are separated by spinodal lines plotted in the phase diagram (see Fig. 7.17). If a small degree of polydispersity $\Delta_\sigma \equiv \sqrt{\langle\sigma^2\rangle/\langle\sigma\rangle^2 - 1}$ is included to avoid the presence of the close-packed solid, then the phase diagram exhibits a fluid–solid or solid–solid coexistence. This result is qualitatively similar to that found in the asymmetric HS binary mixture [176, 177, 178, 179].

7.11.2 Fundamental Measure Theory for the Zwanzig Model

It can be shown that the free energy functional for parallel hard hypercube systems in dimension D can be obtained from the 0D free energy density

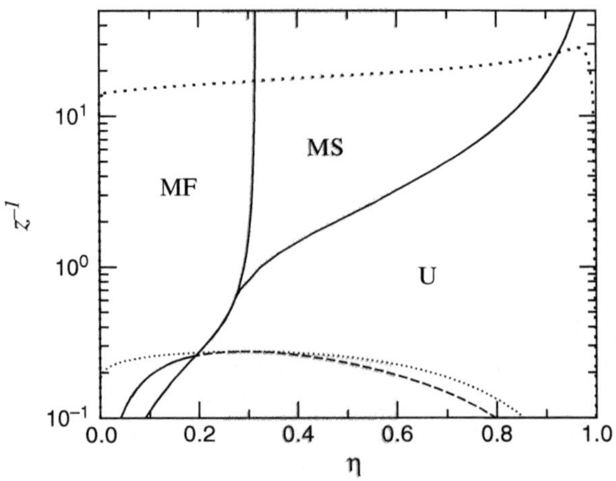

Fig. 7.17. Inverse of the solvent fugacity z^{-1} versus solute packing fraction η of the infinitely asymmetric binary mixture of parallel hard cubes. The *thick solid line* separates the unstable region (U) from the metastable one; the thin one marks the (continuous) transition from a metastable fluid (MF) to a metastable solid (MS); the dashed one is the fluid–fluid spinodal. The *dotted lines* mark the fluid–solid or solid–solid transition and the fluid–fluid metastable coexistence for a polydisperse fluid with polydispersity $\Delta_\sigma = 4.5\%$

$\Phi^{(0)}$ by applying a differential operator [156, 157, 158]. Moreover, the same expression can be used to obtain the functional corresponding to a ternary mixture of hard parallelepipeds with restricted orientations (the orientations are restricted to each of the cartesian axes). This result reads

$$\Phi^{(D)} = \frac{\partial^D}{\partial \boldsymbol{\sigma}^{(D)} \cdots \partial \boldsymbol{\sigma}^{(1)}} \Phi^{(0)}(\eta) , \tag{7.208}$$

where the D-dimensional local packing fraction is defined as

$$\eta(\boldsymbol{r}) = \sum_{\mu} \prod_{i=1}^{D} \left(\int_{x_i - \sigma_\mu^{(i)}/2}^{x_i + \sigma_\mu^{(i)}/2} \mathrm{d}x_i' \right) \rho_\mu(x_1', \ldots, x_D') , \tag{7.209}$$

and we have introduced the short-hand notation

$$\frac{\partial}{\partial \boldsymbol{\sigma}^{(i)}} \equiv \sum_{\mu} \frac{\partial}{\partial \sigma_\mu^{(i)}} , \tag{7.210}$$

$\sigma_\mu^{(i)}$ being the edge length parallel to the i axis of the species μ (each species has a different orientation). This result means that we have found a FMT for the model introduced by Zwanzig to study the isotropic–nematic phase transition by restricting the orientations of parallelepipeds to the three cartesian axes [180, 181]. Through this approximation, the virial coefficients can be calculated up to seventh order. In this way, Zwanzig proved the robust character of the isotropic–nematic phase transition, i.e. that the theoretically predicted transition is not a mere effect of the low-order virial expansion.

Bulk Phase Diagram

The complete phase diagram of the Zwanzig model, including not only isotropic and nematic phases but also non-uniform phases as smectic, columnar, plastic and oriented solid, has been calculated using FMT [182]. Both, rod and plate symmetries were included in this study. For this purpose, it is better to use a different parametrization of the density profiles which includes all possible symmetries. The truncated Fourier series of the density profile reads

$$\rho_\mu(\boldsymbol{r}) = \rho \gamma_\mu \sum_{\boldsymbol{k}}^{\boldsymbol{N}} \alpha_{\boldsymbol{k}}^{(\mu)} \prod_{i=1}^{3} \cos\left(\frac{2\pi}{d_i} k_i x_i \right) , \tag{7.211}$$

where γ_μ ($\mu = x, y, z$) is the average occupancy probability of the unit cell of species μ, $\alpha_{\boldsymbol{k}}^{(\mu)}$ are the Fourier amplitudes of the same species and d_i ($i = 1, 2, 3$) are the simple parallelepipedic lattice periods in the x, y and z spatial directions. The cut-off vector $\boldsymbol{N} \equiv (N_1, N_2, N_3)$ is selected in such a way as to guarantee $\alpha_{\boldsymbol{N}}^{(\mu)} < 10^{-7}$ for all μ. For uniaxial parallelepipeds, the nematic order parameter S is enough to define all the coefficients γ_μ, because the symmetry of the problem requires that $\gamma_\perp = \gamma_x = \gamma_y = (1-S)/3$ and $\gamma_\parallel \equiv \gamma_z = (1+2S)/3$. The nematic director is set parallel to the z axis. Also, the Fourier amplitude $\alpha_{0,0,0}$ is set to unity. These density profiles should be included in the definitions of the weighted densities $n^{(\alpha)}(\boldsymbol{r}) = \sum_{\mu=x,y,z} \rho_\mu * \omega_\mu^{(\alpha)}(\boldsymbol{r})$, $\boldsymbol{v}^{(\alpha)}(\boldsymbol{r}) = \sum_{\mu=x,y,z} \rho_\mu * \boldsymbol{w}_\mu^{(\alpha)}(\boldsymbol{r})$ and $\eta(\boldsymbol{r}) = \sum_\mu \rho_\mu * w_\mu^{(3)}(\boldsymbol{r})$. The weights are the same functions (7.188)–(7.191) with the functions ζ_μ and τ_μ replaced by

$$\zeta_\mu(x_i) = \frac{1}{2}\delta\left(\frac{\sigma_\mu^i}{2} - |x_i|\right), \quad \tau_\mu(x_i) = \Theta\left(\frac{\sigma_\mu^i}{2} - |x_i|\right), \tag{7.212}$$

with $\sigma_\mu^i = \sigma + (L-\sigma)\delta_{\mu i}$ (L and σ being the length and width of the parallelepiped with $\delta_{\mu i}$ the Kronecker delta). All convolutions defining $n^{(\alpha)}(\boldsymbol{r})$, $\boldsymbol{v}^{(\alpha)}(\boldsymbol{r})$ and $\eta(\boldsymbol{r})$ are trivially computed so they are analytic functions of the minimization variables $S, d_\perp, d_\parallel$ and $\alpha_{\boldsymbol{k}}^{(\perp,\parallel)}$. Note that the symmetries of all possible phases are now conveniently included. The smectic, columnar and solid symmetries have vectors $\boldsymbol{k} = (0,0,k_3)$, $\boldsymbol{k} = (k_1,k_2,0)$ and $\boldsymbol{k} = (k_1,k_2,k_3)$, respectively.

In Figs. 7.18(a) and 7.18(b), we plot, respectively, the phase diagrams for prolate and oblate particles calculated with FMT. Note the presence of the discotic smectic (DS) phase, where the uniaxial particle axes are randomly oriented in the plane of smectic layers. The stability of this peculiar phase is due to the restriction of orientations imposed on the particle axes as was shown by Casey and Harrowell in their MC simulation study of the Zwanzig

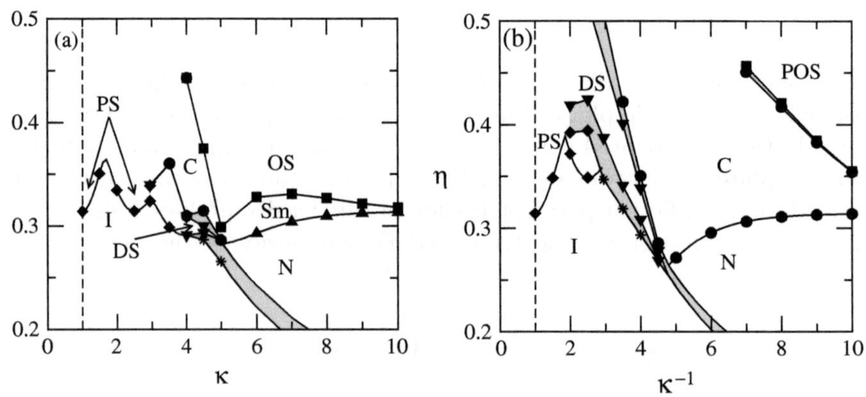

Fig. 7.18. Phase diagrams of prolate (**a**) and oblate (**b**) parallelepipeds. Several phases are labeled as I: isotropic, N: nematic, Sm: smectic, DS: discotic smectic, C: columnar, PS: plastic solid, OS: orientationally ordered solid and POS: perfectly orientationally ordered solid. The transition densities are labeled with different symbols. The shaded areas limit the regions of two-phase coexistence

model [183]. We should say that although the restriction of orientations seems to be a crude simplification to treat the orientational degrees of freedoms, all the liquid crystal phases (smectic, columnar, plastic and oriented solid) found in the theoretical and simulation studies of freely rotating hard particles are present in the Zwanzig phase diagram obtained from FMT. Also, the characteristic aspect ratios at which the smectic and columnar phases of rods and plates become stable ($\kappa_r = \kappa_p^{-1} \sim 5$) are similar to those found in simulations of HSCs ($\kappa \sim 4.5$) [184] and cut spheres $\kappa \sim 0.2$ [185].

Rod-plate Polydisperse Mixture

The second virial approximation of the free energy of the Zwanzig model has been employed to investigate the phase diagram of symmetric mixtures of rods and plates [186]. Stimulated by theoretical calculations made in the early 1970s [187] which show that a binary mixture of rods and plates can stabilize a biaxial nematic phase (in which the symmetry axes of particles of different types point along mutually perpendicular directions), van Roij and Mulder studied the relative stability of this phase against a N–N phase separation [186]. The FMT of the Zwanzig model has been applied to study the effect of polydispersity on the stability of the biaxial nematic phase in a mixture of plates and rods [188, 189]. The polydispersity was introduced in the particle aspect ratio $\kappa = L/\sigma$ (with L and σ the length and width of the parallelepiped) around two values $\kappa_p < 1$ and $\kappa_r > 1$ symmetrically positioned with respect to $\kappa = 1$. It was shown that enough amount of polydispersity can stabilize the biaxial nematic phase even for values of κ_p and κ_r, for which the bidisperse mixture does not exhibit this phase [188, 189]. In Fig. 7.19(a), we show the phase diagram corresponding to a binary mixture of rods and plates with the same particle volumes and with $(\kappa_r, \kappa_p) = (5, 0.2)$, while Fig. 7.19(b) represents the phase diagram of a polydisperse binary mixture with length and breath polydispersities equal to $\Delta_L = 0.610$ and $\Delta_\sigma = 0.302$, respectively. As we can see, enough polydispersity can stabilize the biaxial nematic phase [the shaded region in Fig. 7.19(b)], which is not stable in the binary mixture.

Onsager's theory applied to the mixture of freely rotating plates and rods also confirms the presence of the biaxial nematic phase [190, 191]. However, the inclusion of end-effects in the expressions for the excluded volumes destabilizes the biaxial phase with respect to I–N or N–N phase transitions [192]. There are also several studies employing the extension of the Parsons–Lee theory to a mixture of freely rotating rods and plates [193, 194, 195]. The aim of these works was to shed some light on the very rich phase behavior of the polydisperse rod–plate colloidal mixtures observed in experiments, in which the system can exhibit triphasic equilibrium between phases with different symmetries [196]. The theoretical works predicted different demixing scenarios, some of them in qualitative agreement with experiments [193, 194, 195]. The Parsons–Lee approach for a binary mixture of rods was later extended by Cinacchi et al. to take into account also smectic inhomogeneities

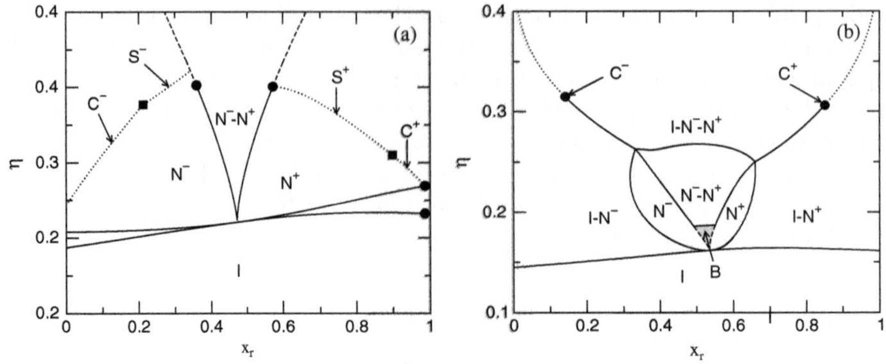

Fig. 7.19. (**a**): Phase diagram of a binary mixture of rods and plates with $\kappa_r = \kappa_p^{-1} = 5$; x_r is the fraction of rods. Phases are labeled I: isotropic, N^-: nematic phase of plates, N^+: nematic phase of rods, $C^\pm$ and $S^\pm$: columnar and smectic phases of rods (+) and plates (−). N^-–N^+ represents the two-phase coexistence region. (**b**): Phase diagram of polydisperse mixture of rods and plates with the same aspect ratio and length and breath polydispersities $\Delta_L = 0.610$ and $\Delta_\sigma = 0.302$. I–$N^\pm$ and I–N^-–N^+ represent the two-phase and three-phase coexistence regions. The *shaded* area shows the region of stability of the biaxial phase (B)

[197, 198, 199]. The main results of their numerical minimization of the proposed functional provided phase diagrams including demixing phase transitions between different smectics and also between smectic and isotropic or nematic phases. For mixtures with species having very different lengths, they found mesophases in which the shortest component is microsegregated at the interstitials between the layers mainly formed by the large component [197, 198, 199].

Interfacial Phase Behavior

The treatment of the spatial degrees of freedoms given by the FMT of the Zwanzig model is very accurate, as it was shown by Bier et al. in their theoretical study of the isotropic–nematic interface of a binary mixture of oblate parallelepipeds [200]. Van Roij et al. studied the interface of the one-component HR fluid confined in a slit using the Zwanzig model in the limit of infinite elongation, i.e. with the excess part of the free energy density equal to $\Phi_{\rm ex} = \boldsymbol{n}_1 \cdot \boldsymbol{n}_2$ [201, 202]. Despite the simplicity of the model, the obtained results are in qualitative agreement with MC simulations where the confined fluid exhibits a biaxial nematic phase close to the walls whose length increases with the bulk chemical potential up to a value at which a capillary nematization of the pore occurs. This surface phase transition ends in a critical point (a critical value for the pore width) [201, 202]. Later, Harnau and Dietrich extended this study to a binary mixture of HR by applying the Zwanzig model of hard parallelepipeds confined in a slit geometry or in geometrically structured surfaces. See their recent review [203] for a compilation of works on the study of liquid crystal interfaces, specifically those using the Zwanzig approach, for particles with prolate and oblate symmetries.

The 2D wall–fluid interface of the HR fluid was recently studied by the 2D version of the FMT for hard rectangles with restricted orientations, Eq. (7.195) and [204]. It was found that for a particular case of hard rectangles having aspect ratio $\kappa = 3$, complete wetting by the columnar phase of the wall–isotropic interface occurs. The isotropic fluid confined by two hard lines also exhibits capillary ordering and layering transitions in analogy with the phenomenology found in the study of the confined 3D smectic phase [205, 206]. We show in Fig. 7.20 the interfacial phase diagram (the pore width H versus chemical potential μ) obtained from the DF minimization. The figure shows that for chemical potentials well bellow its bulk coexistence value, the confined fluid exhibits a first-order transition to an interfacial phase with columnar symmetry. The 2D columnar phase is a layered phase with the long particle axes parallel to the layers. In Fig. 7.21, we plot the coexisting isotropic phase and the 25 layered columnar phase confined in a pore with $H/\sigma = 30$.

For a fixed value of the pore width H and increasing the chemical potential above its bulk coexisting value, an $n-1 \rightarrow n$ layering transition occurs. The capillary ordering and layering transitions coalesce in a set of triple points, two of them shown in Fig. 7.20. The same phase diagram topology was found in MC simulations of a confined hard-sphere fluid [207].

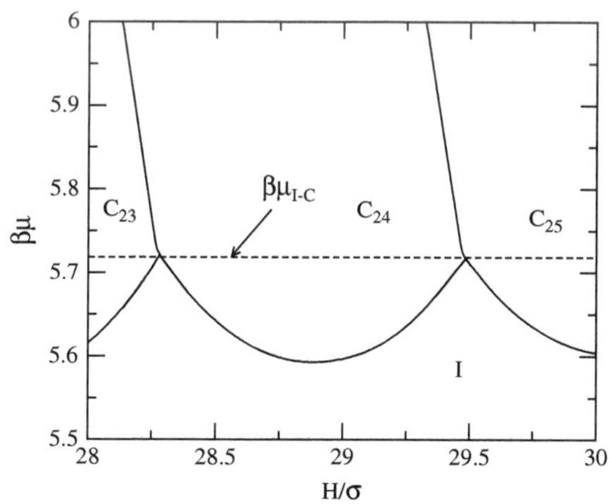

Fig. 7.20. Surface phase diagram, chemical potential μ versus pore width H, of a confined hard rectangle fluid. The *solid lines* show the capillary isotropic (I)–columnar (C) and $n-1 \rightarrow n$ layering transitions bellow and above the bulk chemical potential μ_{IC}, respectively. The regions of stability of the isotropic and columnar interfacial phases with 23, 24 and 25 layers are shown

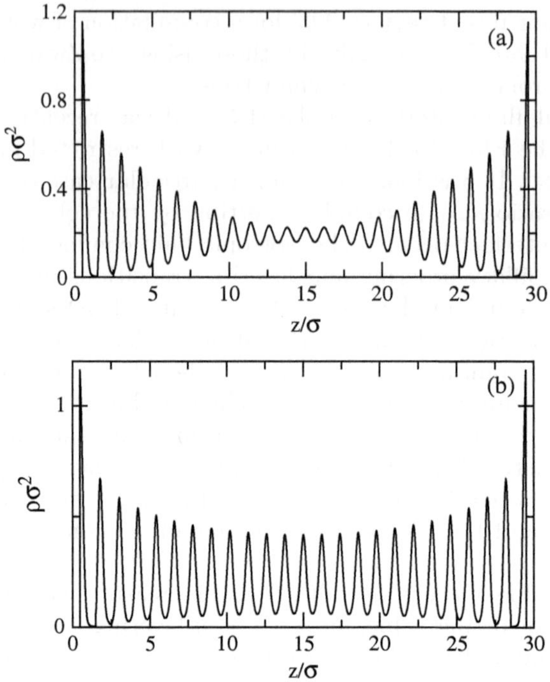

Fig. 7.21. Density profiles of the coexisting isotropic (**a**) and columnar (**b**) interfacial phases

7.11.3 Fundamental Measure Theory for Freely Rotating Hard Anisotropic Particles

This subsection is devoted to present the recent advances of the extension of the FMT to freely rotating particles. Cinacchi and Schmid proposed a DF approximation for general anisotropic particles interpolating between Rosenfeld's DF for HS and Onsager's DF for HR [208]. However, the first attempt to construct a DF with the use of the exact deconvolution of the Mayer function was made by Schmidt for a mixture of HS and infinitely thin HR [209]. In this model, the interaction between the spheres and needles is also hard, i.e. the pair interaction potential is zero if they do not overlap and is infinite otherwise. Finally, the needles do not interact with each other. The DF is constructed in such a way to recover the exact 0D limit for this model, which has the form $\Phi^{(0)} = (1-\eta-\eta')\ln(1-\eta)+\eta$, with η and η' the packing fractions of the hard (spheres) and ideal (needles) particles, respectively [209]. Also, the low-density limit of the corresponding functional (the second virial term) has the correct form. Schmidt showed that the Mayer function between a needle and a sphere can be calculated exactly as

$$f_{\mathrm{sn}}(\boldsymbol{r},\hat{\boldsymbol{\Omega}}) = w_{\mathrm{s}}^{(3)} * w_{\mathrm{n}}^{(0)}(\boldsymbol{r},\hat{\boldsymbol{\Omega}}) + w_{\mathrm{sn}}^{(2)} * w_{\mathrm{n}}^{(1)}(\boldsymbol{r},\hat{\boldsymbol{\Omega}}) , \tag{7.213}$$

where the subindices s and n stand for spheres and needles, respectively. The weights corresponding to HS are the same as those of the original FMT (7.179)–(7.181) with the spherical symmetry while those for the needles correspond to the Rosenfeld extension (7.179)–(7.181)

$$w_{\rm n}^{(1)}(\boldsymbol{r},\hat{\boldsymbol{\Omega}}) = \frac{1}{4}\int_{-L/2}^{L/2} {\rm d}l\delta\left(\boldsymbol{r}+\hat{\boldsymbol{\Omega}}l\right) , \tag{7.214}$$

$$w_{\rm n}^{(0)}(\boldsymbol{r},\hat{\boldsymbol{\Omega}}) = \frac{1}{2}\left[\delta\left(\boldsymbol{r}+\hat{\boldsymbol{\Omega}}L/2\right)+\delta\left(\boldsymbol{r}-\hat{\boldsymbol{\Omega}}L/2\right)\right] , \tag{7.215}$$

except the new one

$$w_{\rm sn}^{(2)}(\boldsymbol{r},\hat{\boldsymbol{\Omega}}) = 2|\boldsymbol{v}_{\rm s}^{(2)}(\boldsymbol{r})\cdot\hat{\boldsymbol{\Omega}}| , \tag{7.216}$$

which depends on the position of the sphere and on the orientation of the needle. In this sense, this is not strictly a one-particle weight, but its definition is necessary to recover the Mayer function decomposition exactly. The functional for the sphere–needle fluid proposed in [209] has the form $\Phi = \Phi_{\rm s} + \Phi_{\rm n}$, with $\Phi_{\rm s}$ the one-component HS free energy density (7.69)–(7.72), while $\Phi_{\rm s}$ includes the interaction between spheres and needles and it has the form

$$\Phi_{\rm sn} = -n_{\rm n}^{(0)}\ln(1-\eta_{\rm s}) + \frac{n_{\rm n}^{(1)}n_{\rm sn}^{(2)}}{1-\eta_{\rm s}} , \tag{7.217}$$

where $n_{\mu}^{(\alpha)}(\boldsymbol{r},\hat{\boldsymbol{\Omega}}) = \rho_\mu * w_\mu^{(\alpha)}(\boldsymbol{r},\hat{\boldsymbol{\Omega}})$ for $\mu = {\rm n},{\rm sn}$, while $\eta_{\rm s}(\boldsymbol{r})$ is the HS local packing fraction. Thus, the excess free energy is calculated as $\beta\mathcal{F}_{\rm ex}[\rho_{\rm s},\rho_{\rm n}] = \int {\rm d}\boldsymbol{x}\Phi\left(\{n_\alpha^{(\beta)}\}\right)$, where the integration is taken over $\boldsymbol{x} = \{\boldsymbol{r},\hat{\boldsymbol{\Omega}}\}$, the spatial and angular variables. This functional was applied to the calculation of the bulk phase diagram of the isotropic rod–sphere mixture for different values of the aspect ratio L/σ. The phase diagrams are identical to those obtained from the free-volume theory, which predicts a demixing transition ending in critical points. The direct sphere–sphere correlation function obtained from the DF is in good agreement with simulations.

To include the rod–rod interaction in the theory, Brader et al. expressed the Mayer function between two needles as the following convolution

$$f_{\rm nn}(\boldsymbol{r},\hat{\boldsymbol{\Omega}},\hat{\boldsymbol{\Omega}}') = w_{\rm nn}^{(2)}(\boldsymbol{r},\hat{\boldsymbol{\Omega}};\hat{\boldsymbol{\Omega}}') * w_{\rm n}^{(1)}(\boldsymbol{r},\hat{\boldsymbol{\Omega}}') , \tag{7.218}$$

where the new weight

$$w_{\rm nn}^{(2)}(\boldsymbol{r},\hat{\boldsymbol{\Omega}};\hat{\boldsymbol{\Omega}}') = 16D\sqrt{1-\left(\hat{\boldsymbol{\Omega}}\cdot\hat{\boldsymbol{\Omega}}'\right)^2}\,w_{\rm n}^{(1)}(\boldsymbol{r},\hat{\boldsymbol{\Omega}}) \tag{7.219}$$

was introduced to describe the residual rod surface in the limit of large aspect ratios [210]. Note that the new weight $w_{\rm nn}^{(2)}$ depends on the orientations of both needles. Thus, the new weighted density is calculated as

$$n_{\rm nn}^{(2)}(\boldsymbol{r},\hat{\boldsymbol{\Omega}}) = \int \mathrm{d}\hat{\boldsymbol{\Omega}}' \rho_{\rm n}(\boldsymbol{r},\hat{\boldsymbol{\Omega}}') * w_{\rm nn}^{(2)}(\boldsymbol{r},\hat{\boldsymbol{\Omega}}';\hat{\boldsymbol{\Omega}}) \,, \tag{7.220}$$

where, apart from the spatial convolution, the angular average with respect to the angles defining the unit vector $\hat{\boldsymbol{\Omega}}'$ is taken. This fact complicates the calculations involved in the excess free energy evaluation. The new form of the excess free energy density now becomes $\Phi = \Phi_{\rm s} + \Phi_{\rm sn} + \Phi_{\rm snn}$, with

$$\Phi_{\rm snn} = \frac{n_{\rm n}^{(1)} n_{\rm nn}^{(2)}}{1-\eta_{\rm s}}. \tag{7.221}$$

This functional was applied to study the planar fluid–fluid interface of the HS HR mixture. The sphere and needle density profiles show either monotonic or damped oscillatory behavior on the sphere-rich side of the interface, depending on which side of the Fisher–Widom line is the bulk fluid in the phase diagram for pure HS. On the needle-rich side, both density profiles decay monotonically toward their respective bulk densities. Needles have biaxial order (they lie preferentially parallel to the interface) on the needle-rich side, while on the sphere-rich side they are oriented normal to the interface [210].

A further extension of the sphere–needle functional was made by Esztermann and Schmidt to account for the first correction (of order D/L) to the Mayer function decomposition between two HSCs [211]. This is accomplished by introducing four new geometric weight functions into the framework. Their corresponding weighted densities, $\tilde{n}_i$, combine in the new contribution $\Delta\Phi = f(\{\tilde{n}_i\})/(1-\eta_{\rm s})$ to the excess free energy density, $\Phi = \Phi_{\rm s}+\Phi_{\rm sn}+\Phi_{\rm snn}+\Delta\Phi$, with $\{\tilde{n}_i\}$, the set of new introduced weighted densities. See the details in [211].

The same formalism was extended to propose a FMT for a ternary mixture of HS, hard platelets and HR, with both needles and platelets of vanishing thickness [212]. The geometric weights were constructed to ensure the exact decomposition of the Mayer functions between different species, except that corresponding to the sphere–platelet interaction, for which the same level of approximation proposed by Rosenfeld for the HD fluid was used. The functional in the one-component platelet limit was tested to consider the isotropic–nematic bulk transition, which was found to be weakly first order, with values for the coexistence densities and the nematic order parameter that compare well with simulation results [212].

7.12 Summary

In this chapter, we have presented a review of DF theory, one of the most important developments of the theory of liquids, whose history is strongly entangled to that of the HS model. Beginning with a general description of the conceptual framework of the theory, we have presented the most important approximations developed to describe the thermodynamics and structure of the HS fluid. Rosenfeld's FMT has been given a special treatment both because of its more involved structure and because it nowadays stands as the most successful DF approximation. The perspective of the advances made during the past two decades is certainly impressive: in the early 1980s, the LDA was nearly the only available choice for the treatment of the HS packing effects within the DF formalism, and now we have DF approximations which may extract from $\rho(\boldsymbol{r})$ the discreteness of the particles, to give the exact 0D limit in a narrow cavity. Despite such impressive progress, there are challenges ahead, which are probably beyond the scope of the FMT. The consistent improvement over the PY equation of state for the bulk fluid, but keeping all the advantages of the DI-FMT for the crystal phase, would be a most important goal for the future.

Beyond the simplest mono-component HS model, we have considered two extensions: mixtures of HS and anisotropic hard convex bodies. The reason is that most approximations initially developed for HS have been later generalized, with more or less success, in these two directions. Again, FMT plays a crucial role in the extension to mixtures, and it has been applied not only to ordinary additive mixtures but also to non-additive ones, of particular interest in the theory of colloids and polymers. The superiority of the FMT with respect to any other DF approximation is absolute for these systems. Before its publication, only binary mixtures were amenable to the DF formalism and even then with great difficulty. FMT has permitted to study multicomponent mixtures, as well as polydisperse systems, with only a manageable level of difficulty. But it is also for mixtures that the limits of the theory become more evident. Future workers in the field have plenty of opportunities to develop better DF approximations for HS mixtures of very different size and to deal with the effects of non-additivity of their excluded cores.

Finally, we have described two alternative approaches for a problem which is qualitatively more difficult than the HS packing: the construction of density functionals for hard anisotropic bodies, beyond the generic low-density expansion pioneered by Onsager. The first route is to try the approximate assembly of the excellent DF for isotropic hard cores, with those magnitudes characterizing the anisotropic pair interactions (such as the Mayer function or the contact distance). The second approach is based on the extension of the FMT to these systems, as it was already attempted by Rosenfeld, so that the crucial ingredient for the non-local dependence of the free energy DF would be the shape of a single molecule rather than the excluded volume between two molecules. The approach has been very successful for systems of parallel anisotropic bodies, in which the orientational degrees of freedom are frozen. A partial extension to systems with orientational disorder has been done within Zwanzig model for liquid crystals: hard parallelepipeds with restricted orientations, which may be treated as a mixture of perfectly oriented bodies. At this point, an interesting question arises: What kind of particle geometry is suitable for first principles derivation of the fundamental measure functional? A useful criterion requires that the pair overlap volume between particles should have the same symmetry of constituent particles. This requirement is fulfilled for HS and hard parallelepipeds, but it does not apply to freely rotating anisotropic bodies. This appears to be the fundamental reason why the recent extensions of FMT to freely rotating anisotropic particles are forced to make strong simplifications, like an expansion in the limit of large particle anisotropy. Our only certainty here is that the field is open and game goes on!

References

1. R. Evans: Avd. Phys. A **28**, 143 (1979)
2. R. Evans: Density functionals on the theory of nonuniform fluids. In *Fundamentals of Inhomogeneous Fluids*, ed by D. Henderson (Dekker, New York 1992) pp 85.
3. H. Löwen: J. Phys.: Cond. Matter **14**, 11897 (2002)
4. J. K. Percus: J. Stat. Phys. **15**, 505 (1976)
5. J.-P. Hansen and I. R. McDonald: *Theory of simple liquids*, 2nd edn (Academic Press, London 1990) p 39
6. P. Tarazona and R. Evans: Mol. Phys. **47**, 1033 (1982)
7. J. S. Rowlinson and B. Widom: *Molecular Theory of Capillarity* (Oxford University Press, Oxford 1989) (reprinted by Dover, New York 2002)
8. A. A. Louis, P. G. Bolhuiss, and J.-P. Hansen: Phys. Rev. E **62**, 7961 (2000)
9. S. Torquato: *Random Heterogenous Materials* (Springer, New York 2001) p 81
10. A. Robledo: J. Chem. Phys. **72**, 1701 (1980)
11. A. Robledo and C. Varea: J. Stat. Phys. **28**, 513 (1980)
12. Y. Rosenfeld: J. Chem. Phys. **89**, 4272 (1988)
13. Y. Rosenfeld: Phys. Rev. Lett. **63**, 980 (1989)
14. S. Nordholm, M. Johnson and B. C. Freasier: Aust. J. Chem. **33**, 2139 (1980)
15. O. Gunnarsson, M. Johnson, and B.I. Lundqvist: Phys. Rev. B **20**, 3136 (1979).
16. P. Tarazona: Mol. Phys. **52**, 81 (1984)
17. P. Tarazona: Phys. Rev. A **31**, 2672 (1985)
18. W. A. Curtin and N. W. Ashcroft: Phys. Rev. A **32**, 2909 (1985)
19. P. Tarazona, U. Marini Bettolo Marconi and R. Evans: Molec. Phys. **60**, 573 (1987)
20. W. H. Press, B. P. Flannery, S. A. Teukolsky, W. T. Vetterling: *Numerical Recipes in FORTRAN: The Art of Scientific Computing* (Cambridge University Press, Cambridge 1992)
21. I. K. Snook, and D. Henderson: J. Chem. Phys. **68**, 2134 (1978)
22. J. R. Henderson and F. van Swol: Mol. Phys. **51**, 991 (1984)
23. H. Reis, H. Frisch and J. L. Lebowitz: J. Chem. Phys. **31**, 369 (1959)
24. H. Reis, H. Frisch, E. Helfand and J. L. Lebowitz: J. Chem. Phys. **32**, 119 (1960)
25. A. Papadopoulou, F. van Swol and U. Marini Bettolo Marconi: J. Chem. Phys. **97** 6942 (1992)
26. S.-C. Kim, J.-K. Suh and S.-H. Suh: Mol. Phys. **79**, 1369 (1993)
27. G. P. Brenan and R. Evans: Mol Phys. **73**, 789 (1991)
28. E. Velasco and P. Tarazona: J. Chem. Phys. **91**, 7916 (1989)

29. P. Röcken, A. Somoza, P. Tarazona and G. Findenegg: J. Chem. Phys. **108**, 8689 (1998)
30. B. J. Alder, T. E. Wainwright: J. Chem. Phys. **33**, 1439 (1960)
31. W. G. Hoover and F. H. Ree: J. Chem. Phys. **49**, 3609 (1968)
32. B. J. Alder, W. G. Hoover and D.A. Young: J. Chem. Phys. **49**, 3688 (1968)
33. D. A. Young and B. J. Alder: J. Chem. Phys. **60**, 1254 (1974)
34. H. Löwen: Phys. Rep. **237** 249 (1994)
35. T. V. Ramakrishnan: Phys. Rev. Lett. **48** 541 (1982)
36. M. Yussouff: Phys. Rev. B **23** 5871 (1981)
37. A. D. J. Haymet: J. Chem. Phys. **78** 4641 (1983)
38. R. Ohnesorge, H. Löwen and H. Wagner: Phys. Rev. E **50**, 4801 (1994)
39. E. Velasco, P. Tarazona: Phys. Rev. A **36**, 979 (1987)
40. M. Jaric, U. Mohanty: Phys. Rev. Lett. **58**, 230 (1987)
41. M. Takamiya and K. Nakanishi: Mol. Phys. **70**, 767 (1990)
42. E. Velasco and L. Mederos: Phys. Rev. E **56**, 2433 (1997)
43. L. Mederos, G. Navascués, P. Tarazona and E. Chacón: Phys. Rev. E **47**, 4284 (1993).
44. A. de Kuijper, W. L. Vos, J.-L. Barrat, J.-P. Hansen and J. A. Schouten: J. Chem. Phys. **93**, 5187 (1990)
45. B. B. Laird and D.M. Kroll: Phys. Rev. A **42**, 4810 (1990)
46. R. Groot: J. Chem. Phys. **95**, 9191 (1990)
47. T.F. Meister and D.M. Kroll: Phys. Rev. A **31**, 4155 (1985)
48. R.D. Groot and J.P. van Eerden, Phys. Rev. A **36**, 4356 (1987)
49. A. R. Denton and N. W. Ashcroft: Phys. Rev. A **39**, 4701 (1989)
50. A. R. Denton and N. W. Ashcroft: Phys. Rev. A **42**, 7312 (1990)
51. A. Khein and N.W. Ashcroft: Phys. Rev. Lett. **78**, 3346 (1997)
52. M. Baus: J. Phys. Conds. Matter. **1**, 3131 (1989)
53. J.F. Lutsko and M. Baus: Phys. Rev. Lett. **64**, 761 (1990)
54. P. Tarazona and Y. Rosenfeld: Phys. Rev. E **55**, R4873 (1997)
55. B. Barboy and W. M. Gelbart: J. Stat. Phys. **22**, 685 (1980)
56. E. Kierlik and M.-L. Rosinberg: Phys. Rev. A **42**, 3382 (1990)
57. E. Kierlik and M.-L. Rosinberg: Phys. Rev. A **33**, 2025 (1986)
58. S. Phan, E. Kierlik, M.-L. Rosinberg, B. Bildstein and G. Kahl: Phys. Rev. E **48**, 618 (1993)
59. Y. Rosenfeld: J. Chem. Phys. **98**, 8126 (1993)
60. Y. Rosenfeld: Phys. Rev. A **33**, 2025 (1986)
61. P. Tarazona: Phys. Rev. Lett. **84**, 694 (2000)
62. P. Tarazona: Physica A **306**, 243 (2002)
63. J. L. Colot and M. Baus: Phys. Rev. A **36**, 3912 (1987)
64. Y. Rosenfeld, M. Schmidt, H. Löwen and P. Tarazona: J. Phys.: Cond. Matter **8**, L577 (1997)
65. Y. Rosenfeld, M. Schmidt, H. Löwen and P. Tarazona: Phys. Rev. E **55**, 4245 (1999)
66. F. J. Lutsko: Phys. Rev. E **74**, 021121 (2006)
67. B. Groh: Phys. Rev. E **61**, 5218 (2000)
68. C. Rascón, E. Velasco, L. Mederos and G. Navascués: J. Chem. Phys. **106**, 6689 (1997)
69. A. González, F. L. Román and J. A. White: J. Phys.: Condens. Matter **11**, 3789 (1999)

70. A. González, F. L. Román, S. Velasco and J. A. White: J. Chem. Phys. **125** 064703 (2006)
71. A. González, J. A. White and R. Evans: J. Phys.: Condens. Matter **9**, 2375 (1997)
72. R. Roth, R. Evans, A. Lang and G. Kahl: J. Phys.: Condens. Matter **14**, 12063 (2002)
73. T. L. Hill: *An introduction to Statistical Thermodynamics* (Dover, New York, 1986), pp 177–188
74. J. N. Israelachvili, D. J. Mitchell and B. W. Ninham: J. Chem. Soc. Faraday Trans. II **72**, 1525 (1976)
75. R. P. Sear and J. A. Cuesta: Europhys. Lett. **55**, 451 (2001)
76. R. Blaak and J. A. Cuesta: J. Chem. Phys. **115**, 963 (2001)
77. J. A. Cuesta and R. P. Sear: Phys. Rev. E **65**, 031406 (2002)
78. R. L. Davidchack and B. B. Laird: Phys. Rev. E **60**, 3417 (1999)
79. A. R. Denton and N. W. Ashcroft: Phys. Rev. A **44**, 8242 (1991)
80. S.-C. Kim, C. H. Lee and B. S. Seong: Phys. Rev. E **60**, 3413 (1999)
81. T. K. Vanderlick, H. T. Davis and J. K. Percus: J. Chem. Phys. **91**, 7136 (1989)
82. I. Pagonabarraga, M. E. Cates and G. J. Ackland: Phys. Rev. Lett. **94**, 911 (2000)
83. D. Goulding, S. Melchionna and J.-P. Hansen: Phys. Chem. Chem. Phys. **3**, 1644 (2001)
84. J. A. Cuesta, Y. Martínez-Ratón and P. Tarazona: J. Phys.: Condens. Matter **14**, 11965 (2002)
85. Y.-X. Yu and J. Wu: J. Chem. Phys. **117**, 10156 (2002)
86. Y.-X. Yu, J. Wu, Y.-X. Xin and G.-H. Gao: J. Chem. Phys. **121**, 1535 (2004)
87. H. Hansen-Goos and R. Roth: J. Phys.: Condens. Matter **18**, 8413 (2006)
88. R. K. Bowles: Physica A **275**, 217 (2000)
89. S. Asakura and F. Oosawa: J. Chem. Phys. **22**, 1255 (1954)
90. A. Vrij: Pure Appl. Chem. **48**, 471 (1976)
91. M. Schmidt, H. Löwen, J. M. Brader and R. Evans: Phys. Rev. Lett. **85**, 1934 (2000)
92. B. Götzelmann, R. Roth, S. Dietrich, M. Dijkstra and R. Evans: Europhys. Lett. **47**, 398 (1999)
93. J. A. Cuesta and Y. Martínez-Ratón: J. Phys.: Condens. Matter **11**, 10107 (1999)
94. M. Schmidt, H. Löwen, J. M. Brader and R. Evans: J. Phys.: Condens. Matter **14**, 9353 (2002)
95. H. N. W. Lekkerkerker, W. C.-K. Poon, P. N. Pusey, A. Stroobants and P. B. Warren: Europhys. Lett. **20**, 559 (1992)
96. M. Dijkstra, J. M. Brader and R. Evans: J. Phys.: Condens. Matter **11**, 10079 (1999)
97. M. Schmidt and A. R. Denton: Phys. Rev. E **65**, 061410 (2002)
98. B. Widom and J. S. Rowlinson: J. Chem. Phys. **52**, 1670 (1970)
99. M. Schmidt: Phys. Rev. E **63**, 010101(R) (2000)
100. R. P. Sear: J. Chem. Phys. **104**, 9948 (1996)
101. M. Schmidt, J. Phys.: Condens. Matter 16, L351 (2004)
102. G. Friedel: Ann. Phys. **18**, 273 (1922)
103. L. Onsager: Phys. Rev. **62**, 558 (1942)

104. L. Onsager: Ann. N. Y. Acad. Sci. **51**, 627 (1949)
105. A. Stroobants, H. N. W. Lekkerkerker and D. Frenkel: Phys. Rev. A **36**, 2929 (1987)
106. D. Frenkel: J. Phys. Chem. **91**, 4912 (1987)
107. D. Frenkel, H. N. W. Lekkerkerker and A. Stroobants: Nature (London) **332**, 882 (1988)
108. W. Maier and A. Saupe: Z. Naturforsch. **13A**, 564 (1958)
109. W. Maier and A. Saupe: Z. Naturforsch. **14A**, 882 (1959)
110. W. Maier and A. Saupe: Z. Naturforsch. **15A**, 287 (1960)
111. M. Abramowitz and I. A. Stegun: *Handbook of mathematical functions* (Dover, New York 1965)
112. W. L. McMillan: Phys. Rev. A **4**, 1238 (1971)
113. J. V. Selinger and D. R. Nelson: Phys. Rev. A **35**, 1736 (1988)
114. M. M. Telo da Gama: Mol. Phys. **52**, 585 (1984)
115. M. M. Telo da Gama: Mol. Phys. **52**, 611 (1984)
116. C. Gray and K. E. Gubbins: *Theory of Molecular Fluids* (Oxford University Press 1984)
117. L. Mederos and D. E. Sullivan: Phys. Rev. A **39**, 854 (1989)
118. A. M. Somoza, L. Mederos and D. E. Sullivan: Phys. Rev. Lett **72**, 3674 (1994)
119. A. M. Somoza, L. Mederos and D. E. Sullivan: Phys. Rev. E **52**, 5017 (1995)
120. Y. Martínez-Ratón, A. M. Somoza, L. Mederos and D. E. Sullivan: Faraday Discuss. **104**, 111 (1996)
121. Y. Martínez-Ratón, A. M. Somoza, L. Mederos and D. E. Sullivan: Phys. Rev. E **55**, 2030 (1997)
122. B. J. Berne and P. Pechukas: J. Chem. Phys. **56**, 4213 (1972)
123. J. G. Gay and B. J. Berne: J. Chem. Phys. **74**, 3316 (1981)
124. E. Velasco, A. M. Somoza and L. Mederos: J. Chem. Phys. **102**, 8107 (1995)
125. E. Velasco and L. Mederos: J. Chem. Phys. **109**, 2361 (1998)
126. H. N. W. Lekkerkerker, P. Coulon, R. van der Haegen and R. Deblieck: J. Chem. Phys. **80**, 3427 (1984)
127. J. D. Parsons: Phys. Rev. A **A19**, 1225 (1979)
128. S. D. Lee: J. Chem. Phys. **87**, 4972 (1987)
129. S. D. Lee: J. Chem. Phys. **89**, 7036 (1988)
130. R. M. Gibbons: Mol. Phys. **17**, 81 (1969)
131. M. A. Cotter and D. E. Martire: J. Chem. Phys. **52**, 1902 (1970)
132. M. A. Cotter and D. E. Martire: J. Chem. Phys. **53**, 4500 (1970)
133. M. A. Cotter: Phys. Rev. A **10**, 625 (1974)
134. M. A. Cotter: J. Chem. Phys. **66**, 1098 (1977)
135. G. J. Vroege and H. N. W. Lekkerkerker: Rep. Prog. Phys. **55**, 1241 (1992)
136. K. Shundyak and R. van Roij: J. Phys.: Condens. Matter **13**, 4789 (2001)
137. K. Shundyak and R. van Roij: Phys. Rev. Lett **88**, 205501 (2002)
138. K. Shundyak and R. van Roij: Phys. Rev. E **68**, 061703 (2003)
139. M. D. Lipkin and D. W. Oxtoby: J. Chem. Phys. **79**, 1939 (1983)
140. U. P. Singh and Y. Singh: Phys. Rev. A **33**, 2725 (1986)
141. J. F. Marko: Phys. Rev. Lett. **60**, 325 (1988)
142. J. F. Marko: Phys. Rev. A **39**, 2050 (1989)
143. R. Pynn: Solid State Commun. **14**, 29 (1974)
144. R. Pynn: J. Chem. Phys. **60**, 4579 (1974)
145. M. Baus: J. Stat. Phys. **48**, 1129 (1987)

146. J. L. Colot, X. G. Wu, H. Xu and M. Baus: Phys. Rev. A **38**, 2022 (1988)
147. A. M. Somoza and P. Tarazona: Phys. Rev. Lett **61**, 2566 (1988)
148. A. M. Somoza and P. Tarazona: J. Chem. Phys. **91**, 517 (1989)
149. A. M. Somoza and P. Tarazona: Phys. Rev. A. **40**, 4161 (1989)
150. A. Poniewierski and R. Holyst: Phys. Rev. Lett. **61**, 2461 (1988)
151. R. Holyst and A. Poniewierski: Mol. Phys. **68**, 381 (1989)
152. E. Velasco, L. Mederos and D. E. Sullivan: Phys. Rev. E **62**, 3708 (2000)
153. Y. Rosenfeld: Phys. Rev. E **50**, R3318 (1994)
154. A. Pressley: *Elementary Differential Geometry* (Springer 2001)
155. A. Chamoux and A. Perera: J. Chem. Phys. **104**, 1493 (1996)
156. J. A. Cuesta: Phys. Rev. Lett. **76**, 3742 (1996)
157. J. A. Cuesta and Y. Martínez-Ratón: Phys. Rev. Lett. **78**, 3681 (1997)
158. J. A. Cuesta, Y. Martínez-Ratón: J. Chem. Phys. **107**, 6379 (1997)
159. L. Lafuente and J. A. Cuesta: Phys. Rev. Lett. **89**, 145701 (2002)
160. L. Lafuente and J. A. Cuesta: J. Phys.: Condens. Matter **14**, 12079 (2002)
161. L. Lafuente and J. A. Cuesta: J. Chem. Phys. **119**, 10832 (2003)
162. L. Lafuente and J. A. Cuesta: Phys. Rev. E **68**, 066120 (2003)
163. L. Lafuente and J. A. Cuesta: Phys. Rev. Lett. **93**, 130603 (2004)
164. L. Lafuente and J. A. Cuesta: J. Phys. A: Math. Gen. **38**, 7461 (2005)
165. J. A. Capitán and J. A. Cuesta: Phys. Rev. E **76**, 011403 (2007)
166. Y. Martínez-Ratón and J. A. Cuesta: J. Chem. Phys. **111**, 317 (1999)
167. F. van Swol and V. Woodcock: Mol. Simul. **1**, 95 (1987)
168. E. A. Jagla: Phys. Rev. E **58**, 4701 (1998)
169. H. Schlacken, H.-J. Moegel and P. Schiller: Mol. Phys. **93**, 777 (1998)
170. Y. Martínez-Ratón, E. Velasco and L. Mederos: Phys. Rev. E **72**, 031703 (2005)
171. Y. Martínez-Ratón, E. Velasco and L. Mederos: J. Chem. Phys. **125**, 014501 (2006)
172. A. Donev, J. Burton, F. H. Stillinger and S. Torquato: Phys. Rev. B **73**, 054109 (2006)
173. D. de las Heras, Y. Martínez-Ratón and E. Velasco: Phys. Rev. E **76**, 031704 (2007)
174. Y. Martínez-Ratón and J. A. Cuesta: Phys. Rev. E **58**, R4080 (1998)
175. J. A. Cuesta and Y. Martínez-Ratón: J. Phys.: Condens. Matt. **12**, A109 (2000)
176. M. Dijkstra, R. Van Roij and R. Evans: Phys. Rev. Lett. **81**, 2268 (1998)
177. M. Dijkstra, R. Van Roij and R. Evans: **82**, 117 (1999)
178. M. Dijkstra, R. Van Roij and R. Evans: Phys. Rev. E **59**, 5744 (1999)
179. N. G. Almarza and E. Enciso: Phys. Rev. E **59**, 4426 (1999)
180. R. W. Zwanzig: J. Chem. Phys. **24**, 855 (1956)
181. R. W. Zwanzig: J. Chem. Phys. **39**, 1714 (1963)
182. Y. Martínez-Ratón: Phys. Rev. E **69**, 061712 (2004)
183. A. Casey and P. Harrowell: J. Chem. Phys. **103**, 6143 (1995)
184. P. Bolhuis and D. Frenkel: J. Chem. Phys. **106**, 666 (1997)
185. J. A. C. Veerman and D. Frenkel: Phys. Rev. A **45**, 5632 (1992)
186. R. van Roij and B. Mulder: J. Phys. II France **4**, 1763 (1994)
187. R. Alben: J. Chem. Phys. **59**, 4299 (1973)
188. Y. Martínez-Ratón and J. A. Cuesta: Phys. Rev. Lett. **89**, 185701 (2001)
189. Y. Martínez-Ratón and J. A. Cuesta: J. Chem. Phys. **118**, 10164 (2003)

190. H. H. Wensink, G. J. Vroege and H. N. Lekkerkerker: Phys. Rev. E **66**, 041704 (2002)
191. S. Varga, A. Galindo and G. Jackson: Phys. Rev. E **66**, 011707 (2002)
192. S. Varga, A. Galindo and G. Jackson: J. Chem. Phys. **117**, 10412 (2002)
193. H. H. Wensink, G. J. Vroege and H. N. Lekkerkerker: J. Chem. Phys. **115**, 7319 (2001)
194. S. Varga, A. Galindo and G. Jackson: J. Chem. Phys. **117**, 7207 (2002)
195. S. Varga, A. Galindo and G. Jackson: Mol. Phys. **101**, 817 (2003)
196. F. M. van der Kooij and H. N. W. Lekkerkerker: Phys. Rev. Lett. **84**, 781 (2000)
197. G. Cinacchi, E. Velasco and L. Mederos: J. Phys. Condens. Matt. **16**, S2003 (2004)
198. G. Cinacchi, E. Velasco and L. Mederos: J. Chem. Phys. **121**, 3854 (2004)
199. G. Cinacchi, Y. Martínez-Ratón, L. Mederos and E. Velasco: J. Chem. Phys. **124**, 234904 (2006)
200. M. Bier, L. Harnau and S. Dietrich: Phys. Rev. E **69**, 021506 (2004)
201. R. van Roij, M. Dijkstra and R. Evans: J. Chem. Phys. **113**, 7689 (2000)
202. M. Dijkstra, R. Van Roij and R. Evans: Phys. Rev. E **63**, 051703 (2001)
203. L. Harnau and S. Dietrich: Soft Matter **3**, 156 (2007)
204. Y. Martínez-Ratón: Phys. Rev. E **75**, 051708 (2007)
205. D. de las Heras, E. Velasco and L. Mederos: Phys. Rev. Lett. **94**, 017801 (2005)
206. D. de las Heras, E. Velasco and L. Mederos: Phys. Rev. E **70**, 011709 (2006)
207. A. Fortini and M. Dijkstra: J. Phys. Condens. Matter **18**, L371 (2006)
208. G. Cinacchi and F. Schmid: J. Phys.: Condens. Matter **14**, 12223 (2002)
209. M. Schmidt: Phys. Rev. E **63**, 050201(R) (2001)
210. J. M. Brader, A. Esztermann and M. Schmidt: Phys. Rev. E **66**, 031401 (2002)
211. A. Esztermann and M. Schmidt: Phys. Rev. E **70**, 022501 (2004)
212. A. Esztermann, H. Reich and M. Schmidt: Phys. Rev. E **73**, 011409 (2006)

100. H. H. Wensink, G. J. Vroege and H. N. W. Lekkerkerker: Phys. Rev. E 66, [illegible] (2002)
101. S. Varga, A. Galindo and G. Jackson: Phys. Rev. E 66, [illegible]
102. S. Varga, A. Galindo and G. Jackson: J. Chem. Phys. 117, [illegible]
103. H. H. Wensink, G. J. Vroege and H. N. W. Lekkerkerker: J. Chem. Phys. [illegible] 7319 (2001)
104. S. Varga, A. Galindo and G. Jackson: J. Chem. Phys. 117, [illegible]
105. S. Varga, A. Galindo and G. Jackson: [illegible]
[illegible]

8

Ensemble Effects in Small Systems

F.L. Román, J.A. White, A. González and S. Velasco

Departamento de Física Aplicada, Universidad de Salamanca, E-37003 Spain.
romanh@usal.es

We present some recent advances in the analysis of finite size effects in small hard particle systems. Two different size effects can be identified: explicit size effects that arise due to the consideration of different statistical mechanics ensembles, and implicit size effects related to the use of periodic boundary conditions in computer simulations. These effects can take place both in kinetic and in structural properties of the system.

Introduction

The study of systems consisting in a reduced number of particles has been the object of great interest in the last few years. The industrial development and control of new materials that involve very small systems reinforce the study of such systems, and the importance of the practical applications justify the increasing scientific activity in the field. This is the case of materials such as zeolites or porous glasses that act as hosts for a fluid that is absorbed in the pores of the confining solid material. The properties of the confined fluid and sometimes those of the host material are altered in this process so that they become significantly different from those in the bulk phase. In situations where the size of the pore is comparable to the fluid particle size the number of particles in the pore can be very small ($N \sim 10$) so that one is dealing with a small system.

Computer simulations have become an essential tool for the theoretical understanding of small systems. Of course, a key aspect in the computer simulation of a small system is the reduced size of the system. In this case finite size effects play a major role that must be carefully determined in order to relate the properties of the small system with those of an infinite system in the thermodynamic limit. Two kinds of finite size effect can be identified in a typical computer simulation of a small system. On the one hand, it is well known that, contrary to what happens in the thermodynamic limit, for a small system the different statistical mechanics ensembles are no longer equivalent

Román, F.L. et al.: *Ensemble Effects in Small Systems*. Lect. Notes Phys. **753**, 343–381 (2008)
DOI 10.1007/978-3-540-78767-9_8

and this leads to the so-called *explicit finite-size effects.* This implies that one must carefully determine the influence of the surroundings in order to choose the most appropriate ensemble for the study of the system. On the other hand, the consideration of periodic boundary conditions in the simulations leads to the emergence of *implicit finite-size effects.* When periodic boundary conditions are considered in a computer simulation, the system can be assumed to consist of an infinite set of replica images of the simulation cell. In this case the pair correlation function becomes anisotropic and the system properties measured through this function can be modified.

From a different point of view, one can establish a close relation between small systems and inhomogeneous systems. It is clear that the former can only be found in nature under confinement constraints and thus they are an example of inhomogeneous system. Density functional theory (DFT) is probably the best-suited theoretical tool for the study of inhomogeneous systems and, consequently, it seems also appropriate to consider the DFT analysis of small, inhomogeneous systems. An important point here is the fact that DFT was initially formulated for inhomogeneous fluids in the grand canonical ensemble (GCE). This ensemble might not be adequate for the study of some small systems, and one must develop new DFT strategies for the analysis of such systems.

In Chap. 7 the application of the DFT to hard particle systems has been extensively studied. This chapter presents some recent advances in the statistical mechanics study of small hard particle systems, paying special attention to the finite size effects that arise both in kinetic and in structural properties of the system.

The single-particle velocity or momentum distribution is perhaps the most common kinetic observable of the system. It is well known that Maxwellian velocity distributions arise in molecular dynamics (MD) simulations of systems at equilibrium with a large number of degrees of freedom. However, very small systems only have a few degrees of freedom and, in this case, non-Maxwellian distributions may come into play. Moreover, the system's degrees of freedom can be further reduced by adding constraints in the mechanical description of the system. These constraints are related to simple considerations such as assuming an isolated system so that the total energy of the system is constant, or considering periodic boundary conditions that lead to the fact that the total momentum of the system is conserved. In Sect. 8.1 we shall address this question by comparing MD data with theoretical results for isolated small systems.

In what respect to structural properties, it is well known that many properties of the system can be obtained from its pair distribution function. The GCE is the framework in which the pair distribution function and the system properties are usually studied. In many cases, however, the simulations are performed in a small system with a fixed finite number of particles. Under these circumstances, the pair distribution function may differ from that of an infinite system and the measured properties may also differ from those of the thermodynamic limit. In Sect. 8.2, we will show how the usually neglected size effects of order $1/N$ in the pair distribution function may lead to important deviations in the density fluctuations and hence in the measured isothermal compressibility of the fluid. In an inhomogeneous system, the one-body density plays the role of the pair distribution for a homogeneous system, and thus, one expects that both functions will show a similar explicit finite-size behavior. In Sect. 8.3 we shall consider some variations on the usual GCE-DFT analysis that will allow to deal with the explicit finite-size effects that arise in the density profiles of highly confined small systems in ensembles other than the GCE.

8.1 Velocity Distributions in Hard Particle Systems

Many MD simulations consist of a system with a finite number of particles N enclosed in a region of volume V and having a constant total energy E. This means that the appropriate framework for the statistical mechanics study of the system is provided by the microcanonical ensemble. However, the consideration of additional constraints in the system must be taken into account in order to determine what kind of microcanonical ensemble should be considered. For example, the conservation of the total linear momentum, or of the total angular momentum of the system can lead to different, special microcanonical ensembles.

In this context, the boundaries of the system play a crucial role. When a particle collides with a hard wall its momentum changes, external forces are exerted on the system, and the total linear momentum $\mathbf{M}$ is not conserved. In this case, where only the energy, the volume, and the number of particles are constant, the system must be treated in the usual microcanonical or EVN ensemble. If the system has periodic boundaries, the particles do not collide with the walls and the total linear momentum is conserved, because no external forces take place in the dynamics of the system. We note, however, that in these systems the total angular momentum is not a constant of the motion. The additional constraint of a fixed total linear momentum gives rise to a special microcanonical ensemble. This ensemble and their differences with the usual microcanonical ensemble was described by Wood [1], who named it *molecular dynamics* ensemble, although the fact that MD generates an ensemble which is a special case of the microcanonical ensemble with constant linear total momentum was previously stressed by Lebowitz et al. [2], and it has been the subject of interest during many years [3, 4, 5, 6, 7]. Some authors have pointed out that in this ensemble there is another conserved quantity related to the position of the center of mass [8, 9]. The consideration of this symmetry does not affect to the calculation of the single particle momentum distribution and will not be taken into account in this work. Nevertheless, it is convenient to point out that this symmetry can be of relevance in the calculation of structural properties since the system can exhibit a lack of homogeneity. In order to distinguish from other molecular dynamics ensembles which have been discussed in the literature [4, 10], here we shall speak of the $EVN\mathbf{M}$ or the microcanonical molecular dynamics ensemble in cases where the system is characterized with a fixed number of particles, volume, energy and total linear momentum.

Another special microcanonical ensemble, the $EVN\mathbf{L}$ ensemble, must be considered in situations where the total angular momentum $\mathbf{L}$ of the system is conserved [11, 12]. These situations take place in systems whose boundaries exhibit spherical symmetry like, e.g., a hard sphere fluid in a spherical cavity but also in self-gravitating systems where a number of classical particles interact via Newtonian gravity [13, 14, 15, 16, 17].

For some simple systems such as the ideal gas and hard-core fluids, EVN and $EVN\mathbf{M}$ analytical expressions for the one-particle velocity and energy distributions have been derived and compared with MD simulations. The calculation of the velocity distribution for an ideal gas in the EVN ensemble was made by Schlüter [18] in 1948. The exact calculation of the phase-space density and momentum distribution function for an ideal gas in the $EVN\mathbf{M}$ ensemble was made by Lado [19] in 1981 by resorting to Fourier transform techniques. This calculation was extended in 1996 by Román et al. [6, 7] for a hard-core fluid. Some exact results are also available in the $EVN\mathbf{L}$ ensemble [11].

The analytical calculation for the single-particle momentum distributions EVN and $EVN\mathbf{M}$ ensembles is based on the fact that velocities and positions are uncorrelated. This allows one to factorize the corresponding microcanonical density of states in configurational and momentum-space components [20]. However, this decomposition is not possible in the $EVN\mathbf{L}$ ensemble for all of the components of the velocity, since in this case positions and velocities are correlated through the total angular momentum. Therefore, analytical results for the one-particle velocity distribution for these systems are not available (not even for the ideal gas!), and MD simulations provide a natural way to obtain information about these distributions.

8.1.1 Microcanonical Ensemble

Let us consider a system of N f-dimensional particles enclosed in a volume V and with total energy E. The classical microcanonical or EVN ensemble is characterized by the equilibrium density function [21]

$$\rho_{\mathrm{m}}(\mathbf{r},\mathbf{p}) = \frac{1}{C_N \Omega(E,V,N)} \delta\left[E - H(\mathbf{r},\mathbf{p})\right] , \tag{8.1}$$

where the label m stands for microcanonical, $(\mathbf{r},\mathbf{p})$ are the phase coordinates of the system, $H(\mathbf{r},\mathbf{p})$ is its Hamiltonian, $\delta(x)$ is the Dirac delta function, and $\Omega(E,V,N)$ is the density of states (or phase-space density) defined by

$$\Omega(E,V,N) = \frac{1}{C_N} \int \mathrm{d}^{fN}\mathbf{r}\mathrm{d}^{fN}\mathbf{p}\, \delta\left[E - H(\mathbf{r},\mathbf{p})\right] = \frac{\partial \Phi(E,V,N)}{\partial E} , \tag{8.2}$$

being $\Phi(E,V,N)$ the phase-space volume,

$$\Phi(E,V,N) = \frac{1}{C_N}\int \mathrm{d}^{fN}\mathbf{r}d^{fN}\mathbf{p}\,\theta\left[E - H(\mathbf{r},\mathbf{p})\right] , \tag{8.3}$$

where $\theta(x)$ is the Heaviside step function and $C_N = Ah^{3N}N!$, being h the Plank's constant and A a constant that makes the phase volume dimensionless.

The density function (8.1) contains all the statistical information of the system. In particular, the probability distribution function associated to the momentum of one particle (single-particle momentum distribution) can be derived by integrating (8.1) over the remaining phase coordinates,

$$\omega_{\mathrm{m}}(\mathbf{p}_1) = \int \mathrm{d}^{fN}\mathbf{r}d^{f(N-1)}\mathbf{p}\,\rho_{\mathrm{m}}(\mathbf{r},\mathbf{p}). \tag{8.4}$$

The classical f-dimensional hard-core fluid is described by the Hamiltonian

$$H(\mathbf{r},\mathbf{p}) = \sum_{i=1}^{N}\frac{\mathbf{p}_i^2}{2m} + U(\mathbf{r}) = \sum_{i=1}^{N}\sum_{\alpha=1}^{f}\frac{p_{i,\alpha}^2}{2m} + U(\mathbf{r}_1,\ldots,\mathbf{r}_N) , \tag{8.5}$$

where m is the mass of a particle, $p_{i,\alpha}$ a cartesian component of the momentum $\mathbf{p}_i$ of the ith particle, and the interaction potential energy $U(\mathbf{r})$ is either ∞ if particles intersect each other or 0 if they do not. From Eq. (8.3) the microcanonical phase-space volume for the hard-core fluid can be written as

$$\begin{aligned}\Phi(E,V,N) &= \frac{1}{C_N}\frac{(2m\pi)^{fN/2}}{\Gamma(fN/2+1)}\int \left[E - U(\mathbf{r})\right]^{fN/2}\theta\left[E - U(\mathbf{r})\right]\mathrm{d}^{fN}\mathbf{r} \\ &= \frac{1}{C_N}\frac{Q_N(2m\pi E)^{fN/2}}{\Gamma(fN/2+1)},\end{aligned} \tag{8.6}$$

being $\Gamma(x)$ the Euler gamma function, and

$$Q_N = \int S(\mathbf{r})\,\mathrm{d}^{fN}\mathbf{r} \;, \tag{8.7}$$

with

$$S(\mathbf{r}) = \begin{cases} 0 & \text{if } U(\mathbf{r}) = \infty \,, \\ 1 & \text{if } U(\mathbf{r}) = 0. \end{cases} \tag{8.8}$$

Differentiating with respect to the energy in (8.6) one can write

$$\Omega(E,V,N) = \frac{1}{C_N}\frac{Q_N(2\pi m)^{fN/2}E^{fN/2-1}}{\Gamma(fN/2)} . \tag{8.9}$$

Therefore, if we use the statistical definition of temperature

$$\frac{1}{k_{\mathrm{B}}T} = \frac{\Omega}{\Phi} , \tag{8.10}$$

from Eqs. (8.6) and (8.9), the temperature of the system in the microcanonical ensemble is given by

$$\frac{1}{k_{\mathrm{B}}T} = \frac{fN}{2E} . \tag{8.11}$$

Two interesting remarks can be pointed out from expression (8.9). First we note that Eq. (8.9) can be written in the form

$$\Omega(E,V,N) = \frac{1}{C_N} Q_N \Omega_p(E,N) , \tag{8.12}$$

where

$$\Omega_p(E,N) = \int \mathrm{d}^{fN}\mathbf{p}\, \delta\left[E - \sum_{i=1}^{N} \frac{\mathbf{p}_i^2}{2m}\right] = \frac{(2m\pi)^{fN/2}}{\Gamma(fN/2)} \tag{8.13}$$

is the density of states of a momentum-space microcanonical ensemble. And second, Eq. (8.9) [or (8.12)] also holds for the ideal gas where $Q_N = V^N$, so that both the ideal gas and the hard-core fluid have the same density of states in the momentum-space.

The microcanonical single-particle momentum distribution is derived from Eqs. (8.1), (8.4), and (8.9) (see e.g. [22]),

$$\omega(\mathbf{p}_1) = \frac{\Gamma(fN/2)(2mE)^{-f/2}}{\Gamma(f(N-1)/2)} \left(1 - \frac{\mathbf{p}_1^2}{2mE}\right)^{f(n-1)/2-1} \theta\left(1 - \frac{\mathbf{p}_1^2}{2mE}\right) \tag{8.14}$$

and the microcanonical single-particle distribution of momentum moduli is given by

$$\begin{aligned}\omega_{\mathrm{m}}(p_1) =& \frac{2\Gamma(fN/2)(2mE)^{-f/2}}{\Gamma(f/2)\Gamma(f(N-1)/2)} p_1^{f-1} \\ & \times \left(1 - \frac{p_1^2}{2mE}\right)^{f(n-1)/2-1} \theta\left(1 - \frac{p_1^2}{2mE}\right) .\end{aligned} \tag{8.15}$$

In the thermodynamic limit (TL)

$$N \to \infty, \quad E \to \infty, \quad \frac{E}{N} = \text{finite} \neq 0 , \tag{8.16}$$

the microcanonical distributions must approach the canonical ensemble result, since the infinite system acts as a thermal bath for the single particle. Taking into account that in this limit the temperature given by (8.11) remains finite, the TL of distributions (8.14) and (8.15) leads, respectively, to:

$$\omega_{\mathrm{c}}(\mathbf{p}_1) = (2\pi m k_{\mathrm{B}}T)^{-f/2} \exp\left(-\frac{\mathbf{p}_1^2}{2mk_{\mathrm{B}}T}\right) , \tag{8.17}$$

$$\omega_{\mathrm{c}}(p_1) = \frac{2(2mk_{\mathrm{B}}T)^{-f/2}}{\Gamma(f/2)} p^{f-1} \exp\left(-\frac{p_1^2}{2mk_{\mathrm{B}}T}\right) , \tag{8.18}$$

where the label c means canonical.

Comparison between Microcanonical and Canonical Ensembles

Figure 8.1 shows a comparison between the canonical (8.18) and microcanonical (8.15) distributions for $N = 2, 3, 4$, and 10 in two dimensions. As one can observe microcanonical distributions approach the (canonical) Maxwell–Boltzmann form as N increases. We note that the differences between distributions are only significant for systems with a very small number of particles. We have used the temperature definition (8.11) for the canonical distributions. These results have been corroborated by molecular dynamics simulations in [6] where the calculations were done by means of the Laplace transform technique [4, 5].

8.1.2 Microcanonical Molecular Dynamics Ensemble

We now consider a microcanonical system where the total linear momentum $\mathbf{M}$ is conserved. This additional constraint must be taken into account, so that the macrostate of the system is now defined by fixed values of E, V, N, and $\mathbf{M}$. The associated Gibbs ensemble is known as the microcanonical molecular dynamics (or $EVN\mathbf{M}$) ensemble. Therefore, we are dealing with a special microcanonical ensemble in which a constraint of constant total momentum $\mathbf{M}$ is included.

At this point it is important to remark that a thermodynamic study of the $EVN\mathbf{M}$ ensemble [4] shows that the pressure obtained using the entropy differs from the pressure obtained from the virial theorem. This apparent paradox was solved in 1999 by Ray and Zhang [8] by realizing that the *correct* microcanonical molecular dynamics ensemble density function must take into account yet another constant of the motion. This constant of the motion,

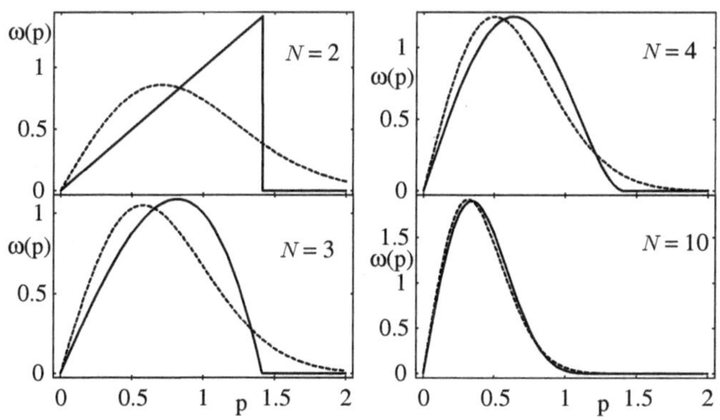

Fig. 8.1. Single particle momentum moduli distributions for a two-dimensional ($f = 2$) hard-core fluid consisting of 2, 3, 4, and 10 particles. *Solid lines* correspond to the microcanonical ensemble distributions whereas the *dashed lines* correspond to the canonical distributions. In all cases we have considered $m = 1$ and $E = 1$ (arbitrary units)

named **G** is associated with Galilean boosts, i.e., transformations between inertial reference frames that have infinitesimally different velocities—**G** is related to the position of the center of mass of the system. The correct molecular dynamics ensemble $EVN\mathbf{MG}$ does not show differences between the results of entropic and virial derivations of the pressure. Nevertheless, the introduction of the new constraint in the density function does not affect averages of quantities that only depend on momentum coordinates [8] and thus the calculation of the single particle momentum distribution, which is the goal of the present subsection, will not be modified by the new constraint. Thus, for clarity, instead of the $EVN\mathbf{MG}$ ensemble, we will continue considering the $EVN\mathbf{M}$ ensemble. More details about the $EVN\mathbf{MG}$ ensemble can be found in [8, 9].

The microcanonical expressions previously presented must be modified in order to incorporate the constraint of constant **M**. In particular, the natural extension for the microcanonical density function (8.1) is given by

$$\rho_{\mathrm{mmd}}(\mathbf{r},\mathbf{p}) = \frac{1}{C_N \Omega(E,V,N,\mathbf{M})} \delta\left(E - H(\mathbf{r},\mathbf{p})\right) \delta\left(\mathbf{M} - \sum_{i=1}^{N} \mathbf{p}_i\right) , \tag{8.19}$$

where the label mmd denotes microcanonical molecular dynamics,

$$\delta\left(\mathbf{M} - \sum_{i=1}^{N} \mathbf{p}_i\right) = \prod_{\alpha=1}^{f} \delta\left(M_\alpha - \sum_{i=1}^{N} p_{i,\alpha}\right) , \tag{8.20}$$

and $\Omega(E,V,N,\mathbf{M})$ is the density of states, defined by

$$\begin{aligned} \Omega(E,V,N,\mathbf{M}) &= \frac{1}{C_N} \int \mathrm{d}^{fN}\mathbf{r}\, d^{fN}\mathbf{p}\, \delta\left(E - H(\mathbf{r},\mathbf{p})\right) \delta\left(\mathbf{M} - \sum_{i=1}^{N} \mathbf{p}_i\right) \\ &= \frac{\partial \Phi(E,V,N,\mathbf{M})}{\partial E}, \end{aligned} \tag{8.21}$$

where $\Phi(E,V,N,\mathbf{M})$ is the phase volume

$$\Phi(E,V,N,\mathbf{M}) = \frac{1}{C_N} \int \mathrm{d}^{fN}\mathbf{r}\, d^{fN}\mathbf{p}\, \theta\left(E - H(\mathbf{r},\mathbf{p})\right) \delta\left(\mathbf{M} - \sum_{i=1}^{N} \mathbf{p}_i\right) . \tag{8.22}$$

The density of states (8.21) was evaluated by Lado [19] for the ideal gas by performing the integral over momentum coordinates by Fourier-expanding the Dirac delta function. This method or the alternative derivation proposed in [7] can be used for the hard-core fluid leading to the following expression for the phase volume

$$\begin{aligned} \Phi(E,V,N,\mathbf{M}) =& \frac{1}{C_N} \frac{Q_N (2\pi m)^{f(N-1)/2}}{N^{f/2} \Gamma(f(N-1)/2+1)} \\ &\times \left(E - \frac{\mathbf{M}^2}{2mN}\right)^{f(N-1)/2} \theta\left(E - \frac{\mathbf{M}^2}{2mN}\right) . \end{aligned} \tag{8.23}$$

According to (8.21), differentiation with respect to the energy in (8.23) gives the density of states,

$$\Omega(E,V,N,\mathbf{M}) = \frac{1}{C_N}\begin{cases} V\delta\left(E-\frac{\mathbf{M}^2}{2m}\right), & \text{for } N=1; \\ \frac{Q_N(2\pi m)^{f(N-1)/2}}{N^{f/2}\Gamma(f(N-1)/2)}\times \\ \times\left(E-\frac{\mathbf{M}^2}{2mN}\right)^{f(N-1)/2-1}\theta\left(E-\frac{\mathbf{M}^2}{2mN}\right), & \text{for } N\geq 2. \end{cases} \tag{8.24}$$

where we have taken into account that the derivative of the step function is the delta function. We note that Eq. (8.24) can be factorized in the form

$$\Omega(E,V,N,\mathbf{M}) = \frac{1}{C_N}Q_N\Omega_p(E,N,\mathbf{M}) \tag{8.25}$$

where

$$\Omega_p(E,N,\mathbf{M}) = \int \mathrm{d}^{fN}\mathbf{p}\,\delta\left[E-\sum_{i=1}^{N}\frac{\mathbf{p}_i^2}{2m}\right]\delta\left[\mathbf{M}-\sum_{i=1}^{N}\mathbf{p}_i\right] \tag{8.26}$$

is the density of states of a momentum-space microcanonical molecular dynamics ensemble. Factorization (8.25) also holds for the ideal gas where $Q_N = V^N$. In particular, expression (8.24) for $N \geq 2$ and $Q_N = V^N$ coincides with the one reported by Lado for the ideal gas [19].

Using (8.23), we get the following expression for the temperature of the system

$$\frac{1}{k_\mathrm{B}T} = \frac{f(N-1)}{2(E-\frac{\mathbf{M}^2}{2mN})}. \tag{8.27}$$

For $\mathbf{M}=0$ (the usual case in molecular dynamics calculations) the phase-space volume (8.23) takes the form

$$\Phi(E,V,N,\mathbf{M}=0) = \frac{1}{C_N}\frac{V^N(2m\pi E)^{f(N-1)/2}}{N^{f/2}\Gamma(f(N-1)/2+1)}, \tag{8.28}$$

whereas the temperature (8.27) becomes

$$\frac{1}{k_\mathrm{B}T} = \frac{f(N-1)}{2E}. \tag{8.29}$$

Comparison between (8.6) and (8.28) shows that the special case of the microcanonical molecular dynamics ensemble with $\mathbf{M}=0$ differs from the microcanonical ensemble, so that the conservation of the total momentum $\mathbf{M}$ implies a reduction in the number of particles from N to $N-1$. This is more clearly shown by a comparison of the temperatures given by expressions (8.11) and (8.29), with a slightly higher temperature in the microcanonical molecular dynamics ensemble than in the microcanonical one.

Single Particle Momentum Distributions for a Hard-Core Fluid in the Microcanonical Molecular Dynamics Ensemble

Similarly to Eq. (8.4), the single-particle momentum distribution in the $EVN\mathbf{M}$ ensemble can be derived from

$$\omega_{\mathrm{mmd}}(\mathbf{p}_1) = \int \mathrm{d}^{fN}\mathbf{r}\, d^{f(N-1)}\mathbf{p}\, \rho_{\mathrm{mmd}}(\mathbf{r},\mathbf{p})\,, \tag{8.30}$$

with $\rho_{\mathrm{mmd}}(\mathbf{r},\mathbf{p})$ given by (8.19). The single-particle momentum distribution for a hard-core fluid in the $EVN\mathbf{M}$ ensemble was first derived by Lado [19] by Fourier-expanding the Dirac delta functions appearing in the integral over momenta. By using the Laplace transform technique [3], it is possible to write the following expression for the single particle momentum distribution [7]

$$\begin{aligned}\omega_{\mathrm{mmd}}(\mathbf{p}_1) =& \frac{\Gamma(f/2)}{(\pi m)^{f/2}\left(E-\frac{\mathbf{M}^2}{4m}\right)^{f/2-1}}\\ &\times\delta\left(E-\frac{\mathbf{p}_1^2}{2m}-\frac{(\mathbf{M}-\mathbf{p}_1)^2}{2m}\right)\quad(\text{for } N=2),\end{aligned} \tag{8.31}$$

$$\begin{aligned}\omega_{\mathrm{mmd}}(\mathbf{p}_1) =& \left(\frac{N}{N-1}\right)^{f/2}\frac{\Gamma(f(N-1)/2)\left(E-\frac{\mathbf{p}_1^2}{2m}-\frac{(\mathbf{M}-\mathbf{p}_1)^2}{2m(N-1)}\right)^{f(N-2)/2-1}}{\Gamma(f(N-2)/2)(2m\pi)^{f/2}\left(E-\frac{\mathbf{M}^2}{2mN}\right)^{f(N-1)/2-1}}\\ &\times\theta\left[E-\frac{\mathbf{p}_1^2}{2m}-\frac{(\mathbf{M}-\mathbf{p}_1)^2}{2m(N-1)}\right]\quad(\text{for } N\geq 3).\end{aligned} \tag{8.32}$$

Result (8.32) coincides with the obtained by Lado [19]. This result is here completed with expression (8.31), which gives the single-particle momentum distribution for $N = 2$ particles.

In order to analyze the TL of the single-particle momentum distribution (8.32), we consider the limit (8.16) giving,

$$\omega_{\mathrm{cmd}}(\mathbf{p}_1) = (2m\pi k_{\mathrm{B}}T)^{-f/2}\exp\left(-\frac{(\mathbf{p}_1-m\mathbf{u})^2}{2mk_{\mathrm{B}}T}\right)\,, \tag{8.33}$$

which gives the single-particle momentum distribution for a hard-core fluid in a generalized canonical ensemble whose mean molecular velocity (momentum) $\mathbf{u}(\mathbf{M} = mN\mathbf{u})$ is nonzero [10, 19]. We refer to this ensemble as the canonical molecular dynamics (cmd) ensemble.

For $\mathbf{M} = 0$ ($\mathbf{u} = 0$), distribution (8.33) coincides with the canonical single-particle momentum distribution (8.17) [we note that, in the TL, temperatures (8.11) and (8.29) coincide], i.e., the special case $\mathbf{M} = 0$ of the canonical molecular dynamics ensemble gives the usual canonical results.

Comparison between Microcanonical Molecular Dynamics and Canonical Ensembles

Figures 8.2 and 8.3 show the single particle momentum moduli distributions for several representative numbers of particles in the microcanonical molecular dynamics ($\mathbf{M} = 0$ and $\mathbf{M} \neq 0$) and in the canonical ensembles.

Figure 8.2 shows a comparison between the canonical and the microcanonical molecular dynamics distributions for $N = 3, 4, 5$, and 10 particles in two dimensions. We can see that, as happened for the microcanonical distribution, the microcanonical molecular dynamics distributions approach the (canonical) Maxwell–Boltzmann form very quickly as N increases. Again, we note that the differences between distributions are only significant for systems with a very small number of particles. We have used the temperature definition (8.27) with $\mathbf{M} = 0$ for the canonical distributions. We note that the microcanonical molecular dynamics distributions with $\mathbf{M} = 0$ behave as the microcanonical ones but considering that the system has one particle less. It is easy to find that the special case of a system consisting in $N = 2$ particles in the microcanonical molecular dynamics ensemble with $\mathbf{M} = 0$ consist of two particles moving in opposite directions with constant velocity, giving rise to a delta distribution.

Finally, Fig. 8.3 shows a comparison between the single-particle momentum moduli distributions for a two-dimensional ($f = 2$) hard-core fluid in the microcanonical molecular dynamics ensemble with $M \neq 0$, and the corresponding results for the canonical molecular dynamics ensemble. We have made this comparison for $N = 2, 3, 4$, and 10, and we have taken $E = 1$, $m = 1$, and $M = 1$. Explicit functional forms for the single particle momentum moduli distributions in the $EVN\mathbf{M}$ ensemble as well as the corresponding canonical ones can be found in [7].

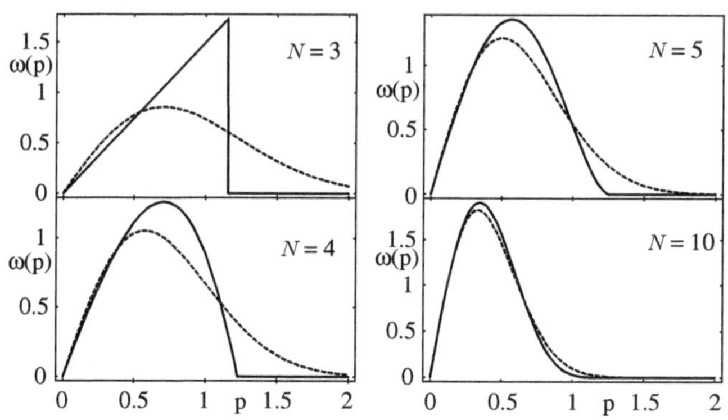

Fig. 8.2. Single particle momentum moduli distributions for a two-dimensional ($f = 2$) hard-core fluid consisting of 3, 4, 5, and 10 particles. *Solid lines* correspond to the microcanonical molecular dynamics ensemble distributions whereas the *dashed lines* correspond to the canonical distributions. In all cases we have considered $m = 1$, $E = 1$, and $\mathbf{M} = 0$ (arbitrary units)

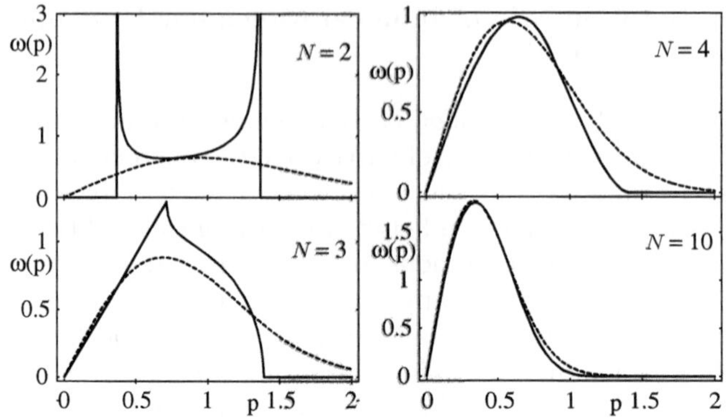

Fig. 8.3. Single particle momentum moduli distributions for a two-dimensional ($f = 2$) hard-core fluid consisting of 3, 4, 5, and 10 particles. *Solid lines* correspond to the microcanonical molecular dynamics ensemble distributions whereas the *dashed lines* correspond to the canonical distributions. In all cases we have considered $m = 1$, $E = 1$, and $\mathbf{M} = 1$ (arbitrary units)

8.1.3 Constant Angular Momentum Microcanonical Ensemble

Here, we restrict ourselves to the calculation of the single-particle velocity distributions of a hard-disk fluid in a microcanonical ensemble where the total angular momentum is conserved. Now the microcanonical expressions (8.1) and (8.2) must be modified in order to incorporate the new constraint. For these ensembles one can write

$$\Omega(E, V, N, \mathbf{L}) \equiv \frac{1}{C_N} \int \mathrm{d}^{fN}\mathbf{r}\, \mathrm{d}^{fN}\mathbf{p}\, \delta[E - H(\mathbf{r}^N, \mathbf{p}^N)]\, \delta(\mathbf{L} - \sum_{i=1}^{N} \mathbf{r}_i \times \mathbf{p}_i) , \tag{8.34}$$

where $\mathbf{L}$ is the total angular momentum of the system. We note that in this case positions and momenta of the particles are correlated and the density of states (8.34) cannot be factorized in the form (8.12) even for the ideal gas. Due to this correlation between positions and momenta, it is not easy to obtain general analytical expressions for the density of states (8.34) and the single particle momentum distributions. Nevertheless, if $\mathbf{L} = \mathbf{0}$, one can obtain the one-particle distribution for the radial component of the velocity, as we will show below.

We now consider a system consisting of N hard disks of mass m and diameter σ confined to a circular cavity of radius R_c. For the particular case $L = 0$, the following analytical expressions for the distribution of the radial component of the velocity can be obtained [11]

$$\omega_{L=0}(v_r) = \frac{\Gamma\left(N - \frac{1}{2}\right)}{\Gamma\left(N - 1\right)} \left(\frac{m}{2\pi E}\right)^{\frac{1}{2}} \left(1 - \frac{m v_r^2}{2E}\right)^{N-2} \Theta\left(1 - \frac{m v_r^2}{2E}\right) , \tag{8.35}$$

where the label denotes the $EVNL$ ensemble with $L = 0$. The procedure for obtaining expression (8.35) is based on the Laplace transform technique [3, 11].

In this case one cannot derive analytic expressions for the tangential velocity distributions. Nevertheless, based on (8.35), approximate expressions have been proposed in [11] for the tangential velocity distribution.

Simulation Results

A basic MD simulation of the system under study reduces to a sequence of collision events of two types: (i) disk–disk collisions and (ii) disk–wall collisions. The interparticle collisions are assumed to be elastic collisions between hard disks where the total energy of the system, its total linear momentum, and the total angular momentum are conserved quantities. The symmetry of the confining potential implies particle–wall elastic collisions that preserve the energy of the particle while its velocity vector is subjected to a specular reflection against a circular wall [see Fig. 8.4(a)], so that the angular momentum of the particle is also conserved. These specular reflections are a consequence of considering a cavity with *smooth* walls (SW). Thus, taking into account the nature of the collisions, the kind of walls considered, and the shape of the cavity, the total energy and the total angular momentum of the system are conserved. Therefore, the MD simulations of a hard-disk system in a SW circular cavity yield results in the $EVNL$ ensemble. The value of L depends on the initial state of the simulation. In this chapter, we shall only consider *non-rotating* systems with $L = 0$.

In order to compare $EVNL$ results with those of the EVN ensemble, it is convenient to change the simulation scheme in order to avoid the fixed L constraint. The simplest way to avoid angular momentum conservation is to consider a cavity with *rough* walls (RW) instead of the above SW cavity. This RW cavity is based on a model suggested by Rapaport [23]. In an RW cavity, for every particle–wall collision event one chooses at random, with equal probability, whether the event is a specular collision or a velocity reversal [12] [see Fig. 8.4]. It is clear that the energy will be conserved in a velocity reversal event whereas the sign of the angular momentum of the colliding disk will

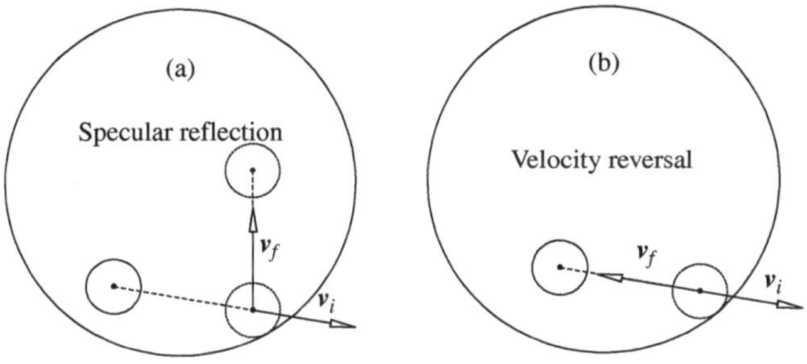

Fig. 8.4. Sketch of the collisions in the cavity walls. (**a**) Specular collisions characteristic of a smooth wall (SW). (**b**) Velocity reversal that may occur in a rough wall (RW)

change. Therefore, during a simulation run L will fluctuate about its mean value $\langle L \rangle = 0$.

As commented above, the SW–MD simulations were done for total angular momentum $L = 0$ (in units of $\sigma\sqrt{mE}$). In the RW simulations, we have checked that the measured value of the mean total angular momentum is equal to zero, within the statistical error of the simulation. More details about the simulation are given in [11].

An essential characteristic of the $EVNL$ ensemble is the existence of correlations between positions and velocities of the particles. In order to analyze this correlation, the cavity is divided in 25 concentric shells of the same thickness $\Delta r = (R_c - \sigma/2)/25$ and the SW–MD distributions associated to the radial component, v_r, and the tangential component, $v_t \equiv \ell/mr = r\dot{\theta}$, of the velocity have been obtained for each shell. More concretely, the one-particle velocity distributions $\hat{\omega}_{L=0}(v_r, r) \equiv \omega_{L=0}(v_r, r)/\rho_{L=0}(r)$ and $\hat{\omega}_{L=0}(v_t, r) \equiv \omega_{L=0}(v_t, r)/\rho_{L=0}(r)$ have been obtained, where $\omega_{L=0}(v_r, r)$ and $\omega_{L=0}(v_t, r)$ have their usual meaning and $\rho_{L=0}(r)$ is the one-particle density, so that one verifies that

$$\int_{-\sqrt{2E/m}}^{\sqrt{2E/m}} \hat{\omega}_{L=0}(v_r, r) \mathrm{d}v_r = \int_{-\sqrt{2E/m}}^{\sqrt{2E/m}} \hat{\omega}_{L=0}(v_t, r) \mathrm{d}v_t = 1 \, . \tag{8.36}$$

Figures 8.5 and 8.6 present the case $N = 3$ with $R_c = 2\sigma$. Figure 8.5 shows the SW–MD radial velocity distribution $\hat{\omega}_{L=0}(v_r, r)$. We only present the result for one shell the SW simulation yields the same radial velocity distribution for different shells, i.e., one verifies that $\hat{\omega}_{L=0}(v_r, r) = \omega_{L=0}(v_r)$, where $\omega_{L=0}(v_r)$ is given by the analytical form (8.35). Therefore, one has $\omega_{L=0}(v_r, r) = \omega_{L=0}(v_r)\rho_{L=0}(r)$, in agreement with the fact that r and v_r are

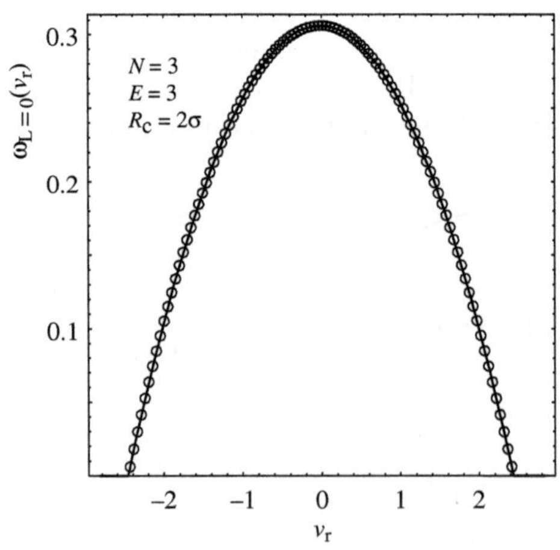

Fig. 8.5. The one-particle distribution for the radial component of the velocity of a system with $N = 3$ hard disks in a circular cavity of radius $R_c = 2\sigma$. The symbols are SW simulation data. The *solid line* is the theoretical distribution (8.35)

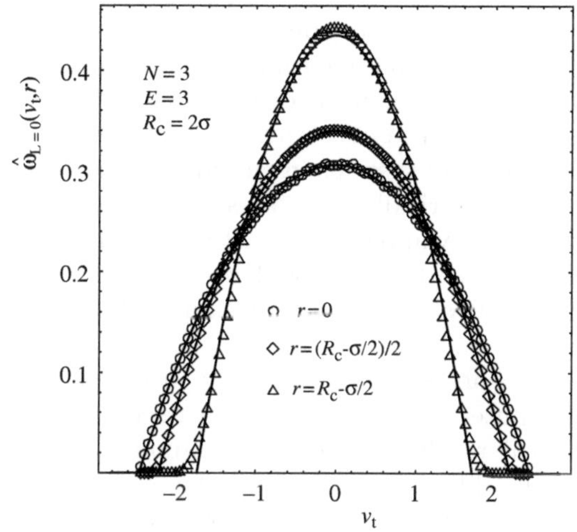

Fig. 8.6. The one-particle distribution $\hat{\omega}_{L=0}(v_t, r)$ for the tangential component of the velocity of a system with $N = 3$ hard disks in a circular cavity of radius $R_c = 2\sigma$. The symbols are results of SW simulations for different values of r. *Solid lines* are an approximate distribution proposed in [11]

not correlated. Figure 8.6 shows the SW–MD tangential velocity distribution $\hat{\omega}_{L=0}(v_t, r)$ for the inner shell ($n = 1$; $r = 0$), the middle shell ($n = 13$; $r = 12\Delta r$), and the outer shell ($n = 25$; $r = 24\Delta r$). One can observe that the more external the shell, the more narrower the distribution $\hat{\omega}_{L=0}(v_t, r)$, that is, the average of the modulus of the tangential component of the velocity of one particle decreases when the particle moves away from the center of the cavity. This result can be explained by means of the following simple physical argument. On the one hand, the angular momentum of a disk is proportional to its distance to the center of the cavity and to the tangential component of its velocity ($\ell = mrv_t$). On the other hand, the total angular momentum is constrained to the fixed value $L = \sum_i \ell_i = 0$. Thus, a particle located close to the border of the cavity (r near its maximum value) has a higher probability to have small values for the tangential component of its velocity. Therefore, since the distribution of the radial component v_r is r-independent, the particles near the border of the cavity are, on the average, slower than the particles located in the center of the cavity. The lines in Fig. 8.6 correspond to an approximate single particle velocity distribution proposed in [11]. The points correspond to simulation data (see also [11])

8.2 Density Fluctuations in Hard-Particle Systems

We have already shown that some of the most important finite-size effects that one has to deal with are those related to the statistical mechanics ensemble used to describe the system under study. These are named as *explicit effects.*

Consideration of periodic boundary conditions in microcanonical systems has revealed the appearance of a modified microcanonical ensemble, and thus, if one considers velocity or momentum distributions, it can be viewed as an explicit effect. Nevertheless the effects of the periodic boundary conditions, when structural properties of systems in ensembles other than microcanonical are studied, have been historically considered as *implicit* or *anomalous effects* [24, 25, 26, 27, 28] since they appear as an artifact of the simulation in order to avoid border effects. This section will be devoted to the study of both explicit and implicit finite size effects in the calculation of the density fluctuations of a hard disk gas.

Explicit finite-size effects arise commonly in computer simulations. For example, many computer simulations in the canonical ensemble are compared with theoretical results in the grand canonical ensemble. If the system is small enough, differences between the results for both ensembles arise. Nevertheless, since both ensembles are equivalent in the thermodynamic limit, one can extrapolate to this limit by using systems of increasing size [29, 30, 31, 32, 33, 34, 35, 36, 37] although there are situations in which care must be taken [38]. A theory of explicit finite-size effects was developed by Lebowitz and Percus [24] by relating the infinite-system pair correlation function to its finite-system counterpart by means of a Taylor series expansion in powers of the inverse number of particles. This theory has been employed by Salacuse et al. [39] for correcting explicit finite-size effects in the calculation of the static structure factor of a uniform bulk fluid from simulation data. Also, Lebowitz et al. [2] have pointed out the ensemble dependence of fluctuations in a quite general context. In some more recent works [7, 38], the Lebowitz and Percus theory has been applied to analyze the corrections due to explicit finite-size effects in two different problems: the fluctuation in the number of hard disks and hard parallel squares deposited on a finite uniform flat surface through a random sequential adsorption process, and the density fluctuations over subvolumes on an uniform hard disk fluid.

Implicit or anomalous finite-size effects appear when periodic boundary conditions are used in computer simulations [24, 26, 27, 28]. The first analysis of implicit effects was reported by Lebowitz and Percus [25] which showed that periodic boundary conditions induce *anisotropy* in the pair distribution function. In 1981 Pratt and Haan [26, 27] reported a theory of implicit finite-size effects. These authors showed that the pair distribution function of a periodic finite system with short-ranged interactions can be approximated by a product of periodically replicated grand canonical pair distribution functions. By using Pratt and Haan's theory, Salacuse et al. [39] have shown that, if $g(r)$ decays very rapidly with increasing r, implicit finite-size corrections can be neglected in the calculation of the static structure factor for sufficiently large system sizes. A study of the implicit finite size effects on the density fluctuations of a hard disk fluid was presented in 1999 by Román et al. [40].

8.2.1 Explicit Finite-Size Effects

Let us consider an equilibrium fluid with temperature T and density ρ. The isothermal compressibility, χ_T, of the fluid is given by [41]

$$\frac{\Delta^2(N)}{\langle N\rangle} = \frac{\langle N^2\rangle - \langle N\rangle^2}{\langle N\rangle} = \rho k_{\mathrm{B}} T \chi_T \,, \tag{8.37}$$

where k_{B} is the Boltzmann constant, and $\Delta^2(N)$ and $\langle N\rangle$ refer, respectively, to the fluctuation in the number of particles and to the mean number of particles in a region of volume V.

Equation (8.37) holds in the thermodynamic limit, ($V \to \infty$, $\langle N\rangle \to \infty$, and $\rho = \langle N\rangle/V =$ finite), although it is also used in computer simulation studies where the extensive variables of the system are finite and usually not very large. In practice, the simulations consider a finite system of N_0 (of the order of a few hundred) particles occupying a volume of size V_0. Then, the system is divided in small subvolumes of size V where the fluctuations in the number of particles can be measured. In this way, one can obtain the particle number distribution, $W(N, V; N_0)$, and the associated moments

$$\langle N^k\rangle_{V,N_0} = \sum_{N=0}^{N_0} N^k W(N, V; N_0) \,. \tag{8.38}$$

In particular, the second moment leads to

$$\frac{\Delta^2(N, V; N_0)}{\langle N\rangle_{V,N_0}} = \frac{\langle N^2\rangle_{V,N_0} - \langle N\rangle^2_{V,N_0}}{\langle N\rangle_{V,N_0}} = \rho k_{\mathrm{B}} T \chi_T^{(V,N_0)}, \tag{8.39}$$

where $\chi_T^{(V,N_0)}$ is the isothermal compressibility of the system of finite size, which can be extrapolated to the thermodynamic limit, yielding the isothermal compressibility of the corresponding infinite system. Usually, this process is handled by choosing periodic boundary conditions for the total system V_0 and then extrapolating for large V (which can become comparable to V_0) [42, 43].

Alternatively, the relative fluctuation (8.37) in the number of particles in V can be obtained from [41, 44]

$$\frac{\Delta^2(N)}{\langle N\rangle} = 1 + \frac{\rho}{V} \int_V \int_V \mathrm{d}\mathbf{r}_1 \mathrm{d}\mathbf{r}_2 \left[g(\mathbf{r}_1, \mathbf{r}_2) - 1\right], \tag{8.40}$$

where $g(\mathbf{r}_1, \mathbf{r}_2) \equiv g(\mathbf{r})$ is the infinite-system pair distribution function, being $\mathbf{r} = \mathbf{r}_2 - \mathbf{r}_1$ the relative position vector between the centers of two particles. In the case of an isotropic fluid one has $g(\mathbf{r}) = g(r)$ with $r = |\mathbf{r}|$. An important feature of the N_0-particle system pair correlation function $g(r; N_0)$ comes from its large-r behavior as 1-const/N_0 instead of tending to unity like its infinite-system counterpart $g(r)$. This fact is usually irrelevant when one is dealing with very large systems ($N_0 \gg 1$). However, if a term of order $1/N_0$ is integrated over a volume of size comparable to the total volume of the system, a finite contribution is obtained.

Fluctuations and Explicit Finite-Size Effects

Consider a uniform fluid consisting of N_0 particles occupying a volume V_0. Like in Eq. (8.40), the relative fluctuation in the number of particles N in a subvolume V is given by [41]

$$\frac{\Delta^2(N,V;N_0)}{\langle N\rangle} = 1 + \frac{\rho}{V}\int_V\int_V \mathrm{d}\mathbf{r}_1\mathrm{d}\mathbf{r}_2\left[g(\mathbf{r}_1,\mathbf{r}_2;N_0) - 1\right], \qquad (8.41)$$

where $\rho = N_0/V_0 = \langle N\rangle/V$ is the number density of the system and $g(\mathbf{r}_1,\mathbf{r}_2;N_0) \equiv g(\mathbf{r};N_0)$ is the pair distribution function of the N_0-particle system. Therefore, in order to connect the finite-system relative fluctuation (8.41) with its infinite-system counterpart (8.40) a relation between $g(\mathbf{r};N_0)$ and $g(\mathbf{r})$ must be used. Lebowitz and Percus [24] proposed a Taylor series expansion of $g(\mathbf{r};N_0)$ in powers of $1/N_0$. Salacuse et al. [39] have extended this series expansion by derivating the coefficient of the $O(1/N_0^2)$ term, and they have shown that for sufficiently large systems this contribution is small enough to be neglected. Basically, the Lebowitz and Percus approach starts from the following relation between the pair distribution function of a system in the grand canonical and in the canonical ensembles:

$$g(\mathbf{r}) = \sum_{N=0}^{\infty} W_0(N)g(\mathbf{r};N), \qquad (8.42)$$

$W_0(N)$ is the probability that the system, in contact with a particle reservoir, has N particles. By expanding $g(\mathbf{r};N)$ around the mean number of particles in the grand canonical system, $\langle N\rangle_0 = \sum_{N=0}^{\infty} NW_0(N) \equiv N_0$, a straightforward calculation leads to [39, 45]

$$g(\mathbf{r}) = g(\mathbf{r};N_0) + \frac{S(\mathbf{0})}{N_0}\left\{1 + \frac{1}{2}\frac{\partial^2}{\partial\eta^2}\left[\eta^2 g(\mathbf{r};N_0)\right]\right\} + O\left(\frac{1}{N_0^2}\right) \qquad (8.43)$$

where $S(\mathbf{k})$ is the structure factor of the fluid [41]

$$S(\mathbf{k}) = 1 + \rho\int \mathrm{d}\mathbf{r}\exp\left(-i\mathbf{k}\cdot\mathbf{r}\right)\left[g(\mathbf{r}) - 1\right], \qquad (8.44)$$

and $\eta = N_0 v_0/V_0$ is the packing fraction, where v_0 is the volume of one particle. Inverting the series (8.43) one obtains

$$g(\mathbf{r};N_0) = g(\mathbf{r}) - \frac{S(\mathbf{0})}{N_0}\left\{1 + \frac{1}{2}\frac{\partial^2}{\partial\eta^2}\left[\eta^2 g(\mathbf{r})\right]\right\} + O\left(\frac{1}{N_0^2}\right). \qquad (8.45)$$

Neglecting terms of $O(1/N_0^2)$ and higher in this series expansion, and substituting Eq. (8.45) into Eq. (8.41), the relative fluctuation is given by

$$\frac{\Delta^2(N,V;N_0)}{\langle N\rangle} = 1 + I[\eta;V] - \frac{VS(\mathbf{0})}{V_0}\left\{1 + \frac{v_0}{2V}\frac{\partial^2}{\partial\eta^2}\left(\eta^2 I[\eta;V]\right)\right\}, \qquad (8.46)$$

where

$$I[\eta; V] = \frac{\eta}{v_0 V} \int_V \int_V \mathrm{d}\mathbf{r}_1 \mathrm{d}\mathbf{r}_2 \left[g(\mathbf{r}) - 1\right] , \tag{8.47}$$

is an integral of the infinite-system pair distribution function in the region delimited by the subvolume V. Expressions (8.46) and (8.47) clearly show that the explicit finite-size corrections come from the term in $1/V_0$, but also that an additional finite-size effect should be taken into account: the edge effect due to consider a finite subvolume V. This effect takes place through the integral $I[\eta; V]$. Furthermore, we note that Eq. (8.46) does not contain any information about the possible edge effects coming from the total system boundaries. However, as mentioned above, the total system periodic boundaries give rise to implicit finite-size effects. In order to minimize these implicit effects, one must choose a total system size with linear dimensions much larger than the range of the interactions. Implicit finite-size effects will be considered later.

Low Density

We first explicitly calculate the leading terms in the relative fluctuation (8.46) at low densities of a fluid consisting of N_0 hard disks in a square *volume* V_0 (of side L_0). For a hard-disk fluid, $g(r)$ can be written as a diagrammatic expansion in the density [44]. Up to the first order, it is given by

$$g(r) = \begin{cases} 0, & \text{if } r < \sigma; \\ 1 + \rho B_2(r), & \text{if } \sigma \le r < 2\sigma; \\ 1, & \text{if } r \ge 2\sigma, \end{cases} \tag{8.48}$$

where σ is the diameter of a hard disk and $B_2(r)$ is the overlapping area between two disks at distance r,

$$B_2(r) = 2\sigma^2 \arccos\left(\frac{r}{2\sigma}\right) - r\sqrt{\sigma^2 - \frac{r^2}{4}} . \tag{8.49}$$

Since the radial distribution function (8.48) has a finite range, the integrals in Eq. (8.47) can be performed by following the method presented in [45]. For sufficiently large systems one has

$$\frac{\Delta^2(N,V)}{\langle N \rangle} = \left(\frac{\epsilon - 1}{\epsilon}\right) \Delta_0(\eta) + \left(\frac{v_0}{V}\right)^{\frac{1}{2}} \Delta_1(\eta) + \left(\frac{v_0}{V}\right) \Delta_2(\eta), \tag{8.50}$$

where $\epsilon \equiv V_0/V$, and

$$\Delta_0(\eta) = 1 - 4\eta + \frac{12\sqrt{3}}{\pi}\eta^2 + O(\eta^3), \tag{8.51}$$

$$\Delta_1(\eta) = \frac{64}{3\pi^{3/2}}\eta + \left[\frac{512}{9\pi^{3/2}} - \frac{1152\sqrt{3}}{5\pi^{5/2}}\right]\eta^2 + O(\eta^3), \tag{8.52}$$

$$\Delta_2(\eta) = -\frac{8}{\pi^2}\eta + \left[\frac{9216\sqrt{3} + 5\lambda}{5\pi^3} - \frac{4096 + 6912\sqrt{3}}{9\pi^2}\right]\eta^2 + O(\eta^3), \tag{8.53}$$

being $\lambda = 2531.7428$. From Eq. (8.50), it is not difficult to realize that the function $\Delta_0(\eta)$ is the relative fluctuation in the number of particles of a system in the thermodynamic limit: $V_0 \to \infty$, $V \to \infty$, and $\epsilon = V_0/V \to \infty$. The functions $\Delta_1(\eta)$ and $\Delta_2(\eta)$ come from edge effects due to consider a finite square subvolume V (see [45]). Approximation (8.50) is equivalent to neglecting the second derivative term of the pair distribution function in Eq. (8.45). The factor $(\epsilon - 1)/\epsilon$ in (8.50) is a consequence of considering a fixed number of particles N_0, i.e., it is due to explicit finite-size effects. This correcting factor indicates that the explicit finite-size effects are relevant when V is comparable to V_0, even if the system contains a large number of particles. Furthermore, this factor is the relative fluctuation in the number of particles in the ideal-gas limit $\eta \to 0$.

The above considerations can be of interest in the calculation of the infinite-system isothermal compressibility χ_T from the extrapolation of simulation data for the isothermal compressibility $\chi_T^{(V)}$ of the finite-size system. In fact, considering square subvolumes of side L and taking into account Eq. (8.39), Eq. (8.50) leads to

$$\chi_T^{(L)} = \left(\frac{\epsilon - 1}{\epsilon}\right) \chi_T + c_1 L^{-1} + c_2 L^{-2} , \qquad (8.54)$$

where c_1 and c_2 are factors that depend on density and temperature. If $\epsilon \gg 1$ ($L \ll L_0$) Eq. (8.54) simplifies to the usual expression

$$\chi_T^{(L)} = \chi_T + c_1 L^{-1} + c_2 L^{-2} . \qquad (8.55)$$

Therefore, an extrapolation process using simulation data for $\chi_T^{(L)}$ must show an asymptotic behavior like in Eq. (8.54) rather than that in Eq. (8.55) [46].

Full Density Range

We will now treat the study of the density dependence of the relative fluctuation (8.46). With this aim we consider a finite-size hard disk fluid at any density below the fluid–solid phase transition. In order to achieve this objective, one requires the knowledge of the infinite-system $g(r)$ for the densities considered. There are many approximations for $g(r)$ of a uniform hard-disk fluid. Here, we shall consider a semiempirical approximation proposed by Baus and Colot [47, 48] for the direct correlation function $c(r)$ of a hard-disk system, which can be related to $g(r)$ by means of the Ornstein–Zernike equation [41]. We consider a subvolume V with undetermined geometry. Equation (8.41) for the fluctuation in the number of particles in V can be written in the form

$$\frac{\Delta^2(N, V; N_0)}{\langle N \rangle} = 1 + \frac{\rho}{V} \int d\mathbf{r}_1 R(\mathbf{r}_1) \int d\mathbf{r}_2 R(\mathbf{r}_2) h(|\mathbf{r}_1 - \mathbf{r}_2|; N_0) , \qquad (8.56)$$

where $h(|\mathbf{r}_1 - \mathbf{r}_2|; N_0) = g(|\mathbf{r}_1 - \mathbf{r}_2|; N_0) - 1$ is the total correlation function [41] and $R(\mathbf{r})$ is a function that represents the geometry of the subvolume [$R(\mathbf{r}) = 1$ for $\mathbf{r}$ inside the subvolume and $R(\mathbf{r}) = 0$ otherwise]. Using $R(\mathbf{r})$, the integrals in (8.56) can be extended to the whole space. Taking into account the Fourier representation of the integrands in (8.56), one obtains

$$\frac{\Delta^2(N,V;N_0)}{\langle N\rangle} = 1 + \frac{\rho}{V(2\pi)^f}\int \mathrm{d}\mathbf{k}\hat{R}^2(\mathbf{k})\hat{h}(\mathbf{k};N_0)\ , \tag{8.57}$$

where $\hat{R}(\mathbf{k})$ and $\hat{h}(\mathbf{k}) \equiv \hat{h}(k)$ are, respectively, the Fourier transform of the geometry function and of the total correlation function, and f is the dimension of the system.

In order to analyze the explicit finite-size effects in Eq. (8.57), an expression similar to (8.45) but relating $\hat{h}(k;N_0)$ to its infinite-system counterpart $\hat{h}(k)$ can be used. Substituting Eq. (8.45) into Eq. (8.57), one obtains Eq. (8.46) with $I[\eta;V]$ given by

$$I[\eta;V] = \frac{\rho}{V(2\pi)^f}\int \mathrm{d}\mathbf{k}\hat{R}^2(\mathbf{k})\hat{h}(k)\ . \tag{8.58}$$

The Fourier transform, $\hat{h}(k)$, of the infinite-system total correlation function is related to the Fourier transform, $\hat{c}(k)$, of the direct correlation function via the Ornstein–Zernike equation [41] which leads to

$$\hat{h}(k) = \frac{\hat{c}(k)}{1-\rho\hat{c}(k)}\ , \tag{8.59}$$

where, for $\hat{c}(k)$ of the hard disk fluid, we consider the approximation of Baus and Colot [47, 48]. This approximation requires the compressibility factor $Z(\eta)$ as an input. Although no exact analytical expressions of $Z(\eta)$ are available for a hard disk fluid there exist several approximate proposals, most of which have been listed in Chap. 3. Many of them have the form [47]

$$Z(\eta) = \frac{1+a\eta^2}{(1-\eta)^2} - \frac{b\eta^{3+c}}{(1-\eta)^{2+d}}\ , \tag{8.60}$$

where a, b, c, and d are constants specific of each equation of state. The choice of the parameter values provides different approximations to the equation of state. For example $a = b = 0$ yields the Scaled Particle Theory (SPT) equation of state, while the values $a = 0.125$ and $b = 0$ correspond to the Henderson equation of state.

In summary, the relative fluctuation in the number of particles in a subvolume of a fluid of N_0 hard disks is given by Eq. (8.46), with $I(\eta, V)$ given by Eq. (8.58) and $\hat{h}(k)$ obtained from Eq. (8.59). We have performed, both for the Henderson and the SPT equation of state, a numerical calculation of the relative fluctuation. In what follows, the obtained results are compared with Monte Carlo (MC) simulation data.

Comparison with Simulation Results

Figure 8.7 shows the relative fluctuation versus the packing fraction η for systems of different size with fixed volume ratio $\epsilon = 64$. The symbols represent

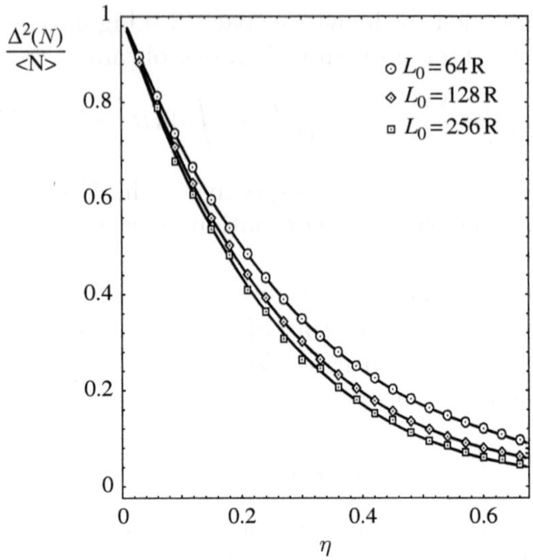

Fig. 8.7. Relative fluctuation versus the packing fraction for a fluid of hard disks. The symbols correspond to MC data for systems with $L_0 = 64R$, $128R$, and $256R$ [38]. The *solid lines* represent theoretical results. Volume ratio $\epsilon = 64$

the results obtained by computer simulation and the solid lines represent the results obtained from Eqs. (8.46), (8.58), and (8.60), and by using the Henderson equation of state [$a = 0.125$ and $b = 0$ in Eq. (8.60)]. Excellent agreement between theoretical and simulation results is obtained even for high values of the packing fraction ($\eta = 0.65$). The use of other equations of state does not give appreciable differences with respect to the results obtained from the Henderson equation of state. This fact seems to indicate that the choice of equation of state is not critical for the study of the relative fluctuations.

Another set of simulations is considered in Fig. 8.8 in order to analyze the dependence of the relative fluctuation with the ratio L/L_0 for a fixed value of the density. In these simulations the total system size is fixed, with $L_0 = 80R$. Figure 8.8 shows results for several packing fractions $\eta = 0.1$, 0.2, 0.4, and 0.6. In every case, one can observe an excellent agreement between simulation and theoretical results [from Eqs. (8.46), (8.58), and (8.60), and by using the Henderson equation of state]. This shows that in the simulated systems explicit finite-size effects are well described by Eq. (8.46) while implicit effects due to periodic boundary conditions may be ignored [38].

8.2.2 Implicit Finite-Size Effects

Border or edge effects are a major disadvantage when performing simulations of small systems in a box. In these systems the number of particles influenced by the walls is very large. In order to avoid this undesired effect, it

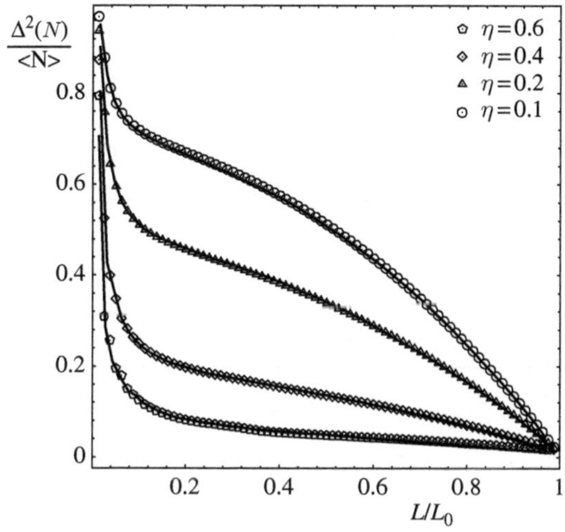

Fig. 8.8. Relative fluctuation versus the ratio L/L_0. The symbols correspond to MC data for systems with packing fractions $\eta = 0.1, 0.2, 0.4$, and 0.6 [38]. The *solid lines* represent theoretical results $L_0 = 80R$

is usual to consider periodic boundary conditions that eliminate to a large extent the inhomogeneities associated to the walls. However, periodic boundary conditions induce an anisotropy in the pair correlation function of the fluid [24, 25, 26, 27, 28]. The effects associated to the introduction of periodic boundary conditions in the system are usually named *implicit finite-size effects* and they can be of some importance, specially in very small systems. When studying the explicit effects we considered systems whose size made the implicit effects negligible. In what follows we shall consider systems whose size makes the implicit finite-size effects relevant.

Like in the calculation of the explicit finite-size effects, we start by considering a two-dimensional fluid consisting of N_0 hard disks enclosed on a square *volume* $V_0 = L_0^2$ with periodic boundary conditions (see Fig. 8.9). We have already shown that Eq. (8.57) gives an appropriate expression for calculating the relative fluctuation in the number of particles inside a subvolume of size $V = L^2$ in terms of the total correlation function $h(\mathbf{r}; N_0)$. Instead of using this size-dependent function, it is convenient to rewrite Eq. (8.57) in terms of the infinite-system $h(\mathbf{r})$ [45] since accurate expressions for $g(\mathbf{r})$ are available in the literature. One has

$$\frac{\Delta^2(N, V; N_0)}{\langle N \rangle} = 1 + \frac{\rho}{(2\pi)^f V} \int \mathrm{d}\mathbf{k} \hat{R}^2(\mathbf{k}) \hat{h}(\mathbf{k}) - \frac{V S(\mathbf{0})}{V_0} . \tag{8.61}$$

We recall that the term $VS(\mathbf{0})/V_0$ in (8.61) is related to explicit finite-size effects. However, the effect of the periodic boundary conditions is not taken into account since the total correlation function considered in (8.61) corresponds to a bulk fluid. In order to include in our analysis these implicit effects, we will consider an approximation for $h(\mathbf{r})$ according to the scheme proposed

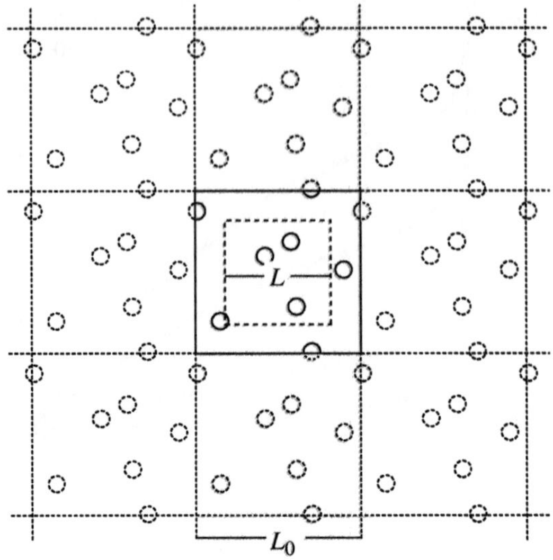

Fig. 8.9. Snapshot of a small system of hard disks including the neighbor images of the central cell

by Pratt and Haan [26, 27]. These authors considered that each particle in a system with periodic boundary conditions can be treated like an oriented supermolecule. This supermolecule consists of the proper particle (or core particle) and their images replicated periodically. Relative to the core particle, the images are located on the sites of a lattice whose unit cell is the system with periodic boundaries itself (see Fig. 8.9). For short-ranged interactions this gives rise to an approximate, superposition-like expression for the pair distribution function which is written as the product of image pair distribution functions [26, 27]. Following these ideas, we assume that the total correlation function of our system can be considered as a periodic function whose period are the characteristic lengths $L_1, \ldots, L_f$ of the system [in our two-dimensional square system these lengths are (L_0, L_0)]. In Fourier space this total correlation function can be written as

$$\hat{h}^{(\mathrm{imp})}(\mathbf{k}) = \sum_{n_1,\ldots,n_f} \mathrm{e}^{-i\mathbf{k}\cdot\mathbf{s}_{n_1,\ldots,n_f}} \hat{h}(\mathbf{k}) \,, \tag{8.62}$$

where the superscript (imp) denotes the inclusion of implicit finite-size effects, $\mathbf{s}_{n_1,\ldots,n_f} = (n_1 L_1, \ldots, n_f L_f)$, and $n_1, \ldots, n_f$ run over all integer values. For our two-dimensional square system Eq. (8.62) leads to the following expression for the relative fluctuation in the number of particles, including both explicit and implicit finite-size effects [40]

$$\frac{\Delta^2(N,V;N_0)}{\langle N\rangle} = 1 - \frac{VS(0)}{V_0} + \frac{\rho}{V(2\pi)^2}\int \mathrm{d}\mathbf{k}\hat{R}^2(\mathbf{k})\hat{h}(\mathbf{k})\Big[1 + 2\cos(k_x L_0)$$
$$+2\cos(k_y L_0) + 4\cos(k_y L_0)\cos(k_y L_0) + \cdots\Big] \,, \tag{8.63}$$

where the terms $2\cos(k_x L_0)$ and $2\cos(k_y L_0)$ appear as a consequence of the interaction of the system with their nearest-neighbor images (up, down, left, and right images), while the term $4\cos(k_y L_0)\cos(k_y L_0)$ is the consequence of the interaction with the images located at the corners of the central cell. Like in the analysis of explicit finite-size effects, we shall consider here an expression for $\hat{h}(\mathbf{k})$ coming from the approximation proposed by Baus and Colot for $\hat{c}(k)$ [47] but using the approximation Z_6 for the compressibility factor. This accurate approximation involves six known virial coefficients of the fluid (see [47] for details). Hereafter we shall refer to this approximation scheme for $\hat{h}(\mathbf{k})$ as BC-Z_6. Other approximations for $\hat{h}(\mathbf{k})$ like the Percus–Yevick result [49], have been considered in [40].

Comparison with CEMC Monte Carlo Simulations

The theoretical results of Eq. (8.63) with BC-Z_6 will be compared with canonical ensemble Monte Carlo (CEMC) simulations of the system [40]. Prior to this, and in order to show the importance of implicit finite-size effects in the present problem, we also compare CEMC data for the relative fluctuation with the results of Eq. (8.61) with BC-Z_6, in which the implicit effects were not taken into account. This comparison is shown in Fig. 8.10 where we plot the relative fluctuation in the number of particles versus the ratio L/L_0 (which represents the relative size of the subvolume) for a system of $N_0 = 128$ hard disks with packing fraction $\eta = 0.59$. As one can see, Eq. (8.61) with BC-Z_6

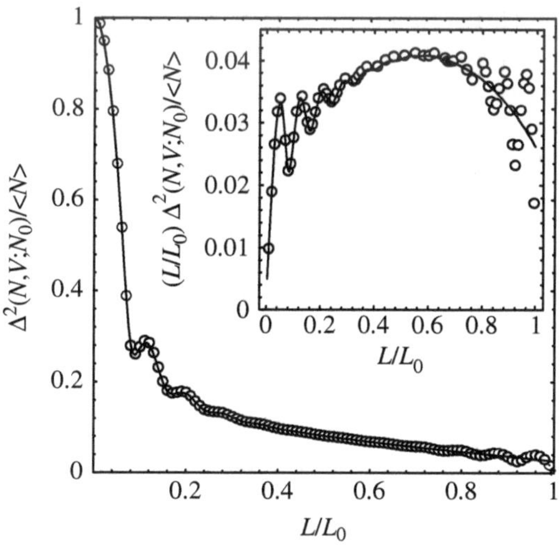

Fig. 8.10. Relative fluctuation versus the ratio between the sides of total system and subvolume. Packing fraction $\eta = 0.59$ and total number of particles $N_0 = 128$. The symbols represent CEMC data [40]. The line corresponds to the result obtained from Eq. (8.61) with BC-Z_6 (see text). In the inset we plot the relative fluctuation times L/L_0

already gives an excellent account of the relative fluctuation for ratios up to $L/L_0 \approx 0.6$, but for larger values of L/L_0 there is an oscillating behavior that is not reproduced with Eq. (8.61). This is illustrated more clearly in the inset of Fig. 8.10 where we plot the relative fluctuation times L/L_0.

In the simulation results presented in Fig. 8.10 one can observe two oscillating zones. For small L/L_0, the oscillations are due to short-range correlations and are already taken into account in Eq. (8.61). However, for large L/L_0, the oscillating behavior is due to short-range correlations with nearest neighbor images of the system. This is a pure implicit finite-size effect not considered in Eq. (8.61). Furthermore, we note that in the central zone of the inset one observes a smooth behavior with a parabolic shape which arises from explicit finite-size effects, i.e., from the term $VS(0)/V_0$ of (8.61) (see [46] for details). As it is shown in Fig. 8.11 this smooth behavior is more apparent for larger systems where the oscillations only arise for small values of L/L_0 (and of $1 - L/L_0$).

In Fig. 8.11, we compare the results of Eq. (8.63) with BC-Z_6 with simulations for systems with the same packing fraction $\eta = 0.59$ but different number of disks ($N_0 = 32$, and 512) [40]. Note that, like in the inset of Fig. 8.10, we represent the relative fluctuation times L/L_0. The Fourier series that appears in Eq. (8.63) has been truncated to the first shell of neighbor images of the system. This approximation seems to be enough to reproduce the main effects of the periodic boundaries.

Finally, we note that Eq. (8.63) can be easily generalized to three dimensions. The only significative difference with the two-dimensional case would be an increase in the computation time of the integral in Eq. (8.63). This study is also relevant for three-dimensional systems such as the hard-sphere fluid or even for more realistic model-fluids such as the Lennard–Jones system. Of course, when studying Lennard–Jones systems (either in three or in two dimensions) one should avoid the critical zone where correlations of long range arise. Further details on the Lennard–Jones system can be found in [46].

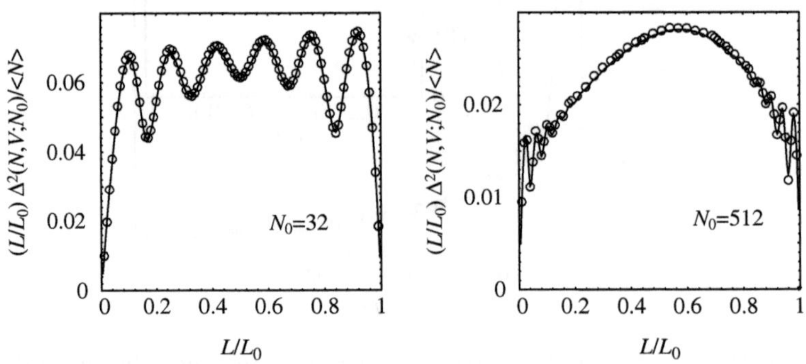

Fig. 8.11. Relative fluctuation times L/L_0 versus the ratio between the sides of total system and subvolume. Packing fraction $\eta = 0.59$ and total numbers of particles $N_0 = 32$, and 512. The symbols represent CEMC data [40]. The line corresponds to the results obtained from Eq. (8.63) with BC-Z_6

8.3 Density Profiles of Inhomogeneous Systems. Structural Properties

The goal of this section is to study the explicit finite-size effects that arise in the structure of a HS fluid confined to a small cavity due to the choice of a particular statistical mechanics ensemble. For simplicity, we shall focus on spherical cavities although our results can be extended to other geometries like narrow cylindrical pores of finite length. In this context one can consider three types of spherical cavities:

- Open spherical cavity. The system can exchange particles with an external reservoir, and the structure of the confined fluid can be conveniently computed by means of grand canonical ensemble Monte Carlo (GCEMC) simulations (see, e.g., [50]) or via density functional theory (DFT). The latter is usually formulated in terms of a minimum grand-potential principle, and thus, it is appropriate for open systems.
- Closed spherical cavity. In this case canonical ensemble Monte Carlo or molecular dynamics simulations are appropriate for the study of the fluid but a conventional (grand canonical) DFT approach at fixed temperature and chemical potential is not adequate in principle and therefore one must consider a different approach. There are two different strategies to perform a DFT study of the system:
  - The canonical density can be expanded in powers of the inverse average number of particles in the corresponding grand canonical system. The expansion makes use of grand canonical results that can be obtained from either DFT or GCEMC.
  - An alternative approach consists of developing a canonical ensemble DFT in which a free-energy variational principle is established instead of the usual minimum grand-potential principle.
- Spherical cavity with fluctuating radius. Now the appropriate framework is the isothermal-isobaric ensemble. The analysis of the system can be done by means of isothermal-isobaric ensemble Monte Carlo or via DFT. In the later case one must reformulate the theory in the isothermal-isobaric ensemble.

8.3.1 Canonical Ensemble

Series Expansion of the Canonical Ensemble Density Profile

A series expansion of the canonical density profile [51, 52] in terms of grand canonical ensemble quantities can be made following the method used in the preceding section for the pair distribution function $g(r)$ [24, 39], which was applied to the study of size effects in the fluctuation in the number of particles [45]. The method is based in the following relation between the grand canonical ensemble inhomogeneous density $\rho(\mathbf{r})$ and the canonical density $\rho_N(\mathbf{r})$,

$$\rho(\mathbf{r}) \equiv \rho^{(1)}(\mathbf{r}) = \sum_{N=0}^{\infty} W(N)\rho_N(\mathbf{r}). \tag{8.64}$$

Like in Eq. (8.42), $W(N)$ is the probability that the system contains N particles [41]. Expanding $\rho_N(\mathbf{r})$ in (8.64) about the average number of particles in the grand canonical ensemble, $\langle N\rangle$, and inverting the series one obtains

$$\rho_{\langle N\rangle}(\mathbf{r}) = \rho(\mathbf{r}) - \frac{1}{2}\langle(N-\langle N\rangle)^2\rangle \frac{\partial^2}{\partial\langle N\rangle^2}\rho(\mathbf{r}) + \cdots , \tag{8.65}$$

In this expression $\rho_{\langle N\rangle}(\mathbf{r})$ denotes the density profile of a system in the canonical ensemble with a number of particles equal to $\langle N\rangle$. Since this number must be an integer, one is forced to consider a corresponding grand canonical ensemble system whose chemical potential leads to an integer value of $\langle N\rangle$. This can be easily done in DFT where the chemical potential of the system can be varied along the process of minimization of the free energy to yield the desired value of $\langle N\rangle$. Furthermore, in this case, the derivatives of $\rho(\mathbf{r})$ with respect to $\langle N\rangle$ can be done numerically. Quantities such as $\langle(N-\langle N\rangle)^2\rangle$ can be expressed in terms of derivatives of $\langle N\rangle$ with respect to $(\beta\mu)$, which can be also performed numerically [52]. Of course, instead of using DFT results, the series expansion can be made on the basis of GCEMC results which is a direct test on the accuracy of the expansion without the limitations of the approximate character of the DFT approach [52].

In order to perform a DFT analysis of an inhomogeneous fluid, one must choose an approximate expression for the free energy functional. Here, we consider the prescriptions of the Fundamental Measure Theory (FMT) of Rosenfeld [53]. As shown in Chap. 7, this theory was shown to be very successful in highly inhomogeneous situations [54, 55, 56, 57, 58, 59] such as the ones that we are going to consider. In addition, this theory can be refined in order to account appropriately for zero-dimensional (0D) situations that may arise due to the packing constraints in the system [60, 61, 62, 63]. As we shall see below, in a spherical cavity one can arrive to quasi-0D situations for which such a theory should yield accurate results.

Next, we will proceed to compare CEMC simulations with the results of the series expansion (8.65) truncated to second order. The system is a fluid of hard spheres of diameter σ confined to a hard spherical cavity of radius R_{cav} so that the fluid is subjected to the following external potential

$$V_{\text{ext}}(r) = \begin{cases} 0, & r < R_{\text{cav}} - \sigma/2 \\ \infty, & r > R_{\text{cav}} - \sigma/2 \end{cases} , \tag{8.66}$$

where r is the distance to the center of the cavity. Not surprisingly, the main differences between canonical and grand canonical results usually appear for cases with few particles. Consequently, the most demanding tests of the expansion should arise in these cases. On the other hand, in certain high packing situations the density profile develops a pronounced peak in the center of the cavity. This peak is related to a zone of strong localization, indicating that a quasi-zero dimensional situation takes place. Accounting properly for such quasi-0D situation is a severe test for any density functional approximation [64]. In addition, the main differences between ensembles arise in the center of the cavity where the peak develops. Therefore, we shall focus mainly on the analysis of these quasi-0D situations.

In Fig. 8.12 (a) we compare DFT results with simulation data (see [51, 52]) for a cavity of radius $R_{\text{cav}} = 1.85\sigma$ with $\langle N \rangle = 6$ particles. As one can observe, canonical and grand canonical results are very different, especially in the center of the cavity where the canonical profile shows a peak less pronounced than that of the grand canonical case. The results of the series expansion both for DFT and for simulation show good agreement with CEMC data, even in the quasi 0D peak. For this central zone, we note that DFT slightly overestimates GCEMC results both for the series expansion and for the grand canonical profile. This behavior is enhanced for the more inhomogeneous situation presented in Fig. 8.12 (b) where we consider a cavity with $\langle N \rangle = 10$ particles. The observed differences between DFT and GCEMC are related to the accuracy of the FMT in quasi-0D situations [64]. Certainly, using a theory specially built for dealing with these situations [62, 63] would improve the obtained results. See Chap. 7 for other approaches and detailed results.

For a larger number of particles, the corrections of the series expansion are negligible except for quasi-0D situations. Fig. 8.13 depicts one of these

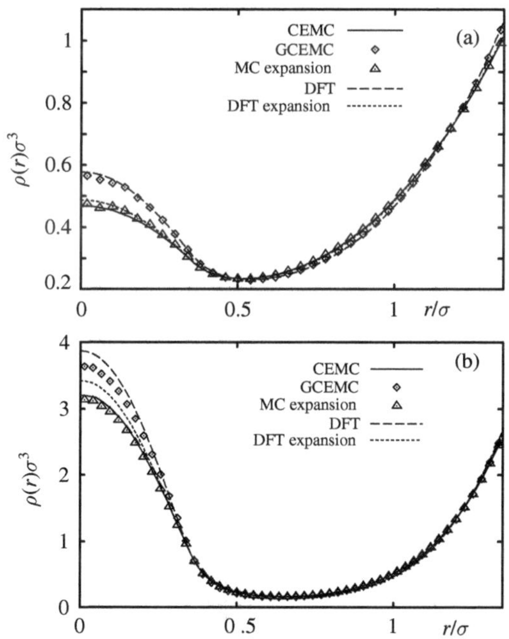

Fig. 8.12. Density profiles of hard spheres in a spherical cavity of radius $R_{\text{cav}} = 1.85\sigma$. (**a**) $\langle N \rangle = 6$ and (**b**) $\langle N \rangle = 10$. The symbols are the results of GCEMC simulations. The solid lines are the results of CEMC simulations. The other lines are the results of DFT (see [51, 52] for simulation and DFT results). The results of the series expansion to second order are also presented (see text)

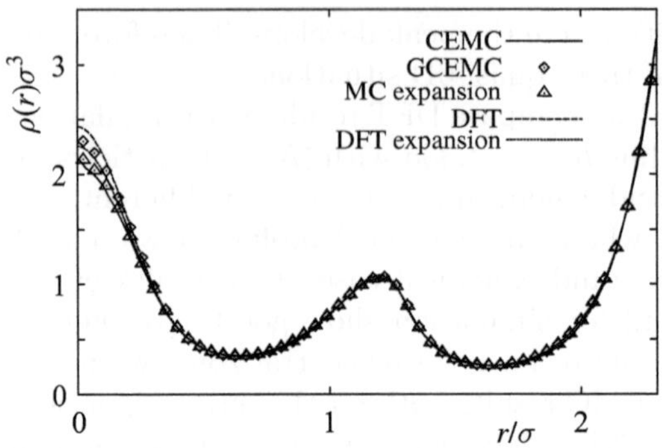

Fig. 8.13. Density profiles of hard spheres in a spherical cavity of radius $R_{\text{cav}} = 2.8\sigma$ for $\langle N \rangle = 45$. The symbols are the results of GCEMC simulations. The *solid line* represents the results of CEMC simulations. The other lines are the results of DFT. The results of the series expansion to second order are also presented (see text)

situations in which we consider a cavity of radius $R_{\text{cav}} = 2.8\sigma$ with $\langle N \rangle = 45$ particles. In this case, canonical and grand canonical results coincide except for the central zone where some differences between both ensembles are observed. These differences are well described by the GCEMC series expansion and less satisfactorily by the DFT series expansion. Like in Fig. 8.12, the DFT results for the quasi-0D peak overestimate those of simulation.

Canonical Ensemble Density Functional Theory

In the density functional theory of the canonical ensemble [65, 66], the equilibrium density profile $\rho_N(\mathbf{r})$ of a closed system with N particles and intrinsic free energy functional $\mathcal{F}_{\text{c}}[\rho]$ is obtained by minimizing the functional

$$\mathcal{F}_{\text{c}}[\rho] + \int \mathrm{d}\mathbf{r}\rho(\mathbf{r})V_{\text{ext}}(\mathbf{r}) \tag{8.67}$$

over all densities that can be associated with the external potential V_{ext} and that are normalized to the integer number of particles N. Assuming that the functional (8.67) is also valid for non-integer values of N and using the Lagrange multiplier technique one has

$$\left.\frac{\delta \mathcal{F}_{\text{c}}[\rho]}{\delta \rho(\mathbf{r})}\right|_{\rho=\rho_N} + V_{\text{ext}}(\mathbf{r}) = \lambda \tag{8.68}$$

where $\rho_N(\mathbf{r})$ denotes the equilibrium density of the fluid in the CE and the Lagrange multiplier λ must be calculated from the constraint

$$\int \rho_N(\mathbf{r})\mathrm{d}\mathbf{r} = N. \tag{8.69}$$

Comparing Eq. (8.68) with the usual Euler–Lagrange equation in the grand canonical ensemble

$$\delta \mathcal{F}_{\rm gc}[\rho]/\delta \rho(\mathbf{r}) + V_{\rm ext}(\mathbf{r}) = \mu \,, \tag{8.70}$$

we note that λ plays the role of the chemical potential μ and the only relevant differences between both equations are due to the fact that the free energy functionals in the CE and the GCE are different. These functionals can be related by means of the following saddle point approximation in terms of the fluctuation in the number of particles [66]:

$$\beta \mathcal{F}_{\rm c}[\rho] - \beta \mathcal{F}_{\rm gc}[\rho] \approx \frac{1}{2} \ln 2\pi \Delta^2(N;[\rho]) \,. \tag{8.71}$$

Equation (8.71) extends the usual saddle point approximation for homogeneous fluids [67] to free energy functionals of fluids with arbitrary inhomogeneous distributions. The chemical potential is adjusted in $\mathcal{F}_{\rm gc}$ so that the average number of particles in the GCE coincides with the number of particles in the CE.

The solution of (8.68) with (8.71) requires expressing the mean square fluctuation $\Delta^2(N;[\rho])$ as a functional of the density [66]. This can be done by noting that Δ^2 is essentially the integral of the two-body density correlation function. Via the Ornstein–Zernike equation, the two-body correlation function is related to the direct correlation function which is by definition a functional of the density and thus the required functional relation can be established [66]. Although we perform a CE-DFT study, we need the GCE free-energy functional $\mathcal{F}_{\rm gc}$ as an input. As in the previous case we use the accurate FMT of Rosenfeld [68, 69, 70]. In particular, for simplicity, we use the Kierlik and Rosinberg version of the FMT [71]. This version is completely equivalent to the original FMT [72] and yields very simple expressions for the direct correlation function, thus simplifying the calculation of Δ^2. In order to compare with the series expansion presented previously, the theory is applied to an HS fluid confined to a hard spherical cavity of radius $R_{\rm cav}$ so that the external potential is given by Eq. (8.66).

In Fig. 8.14 we consider a cavity of radius $R_{\rm cav} = 1.85\sigma$ with 6 and 10 particles. This case was also presented in Fig. 8.12 but using the results of the series expansion of the CE density profile. We compare MC data [51, 52] with DFT results for the density profile both in the CE and in the GCE where, for consistency, we also use the Kierlik and Rosinberg version of the FMT [71]. Figure 8.14 (a) shows the results for six particles. In this case of moderate packing, the CE-DFT yields better agreement with simulation than the series expansion considered in Fig. 8.12 (a). Figure 8.14 (b) ($N_{\rm FMT} = 10$) shows two remarkable features. First, the results of the Kierlik and Rosinberg theory for the GCE largely overestimate the central peak. This is due to the failure of the original FMT to account for the quasi-0D situation that arises in the center of the cavity [64] —compare with Fig. 8.12 (a) where a theory better suited for this situation is used. And, second, the CE-DFT result for this case of high

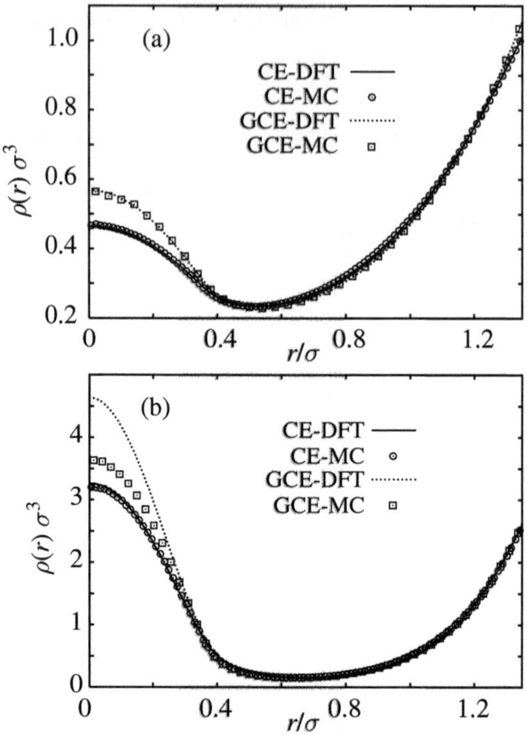

Fig. 8.14. Density profiles of hard spheres in a spherical cavity of radius $R_{cav} = 1.85\sigma$. (**a**) $N = 6$ and (**b**) $N = 10$ particles. The lines represent the results of DFT and the symbols are MC data. Note the different vertical scales

packing remains accurate even in the central peak. This can be ascribed to the fact that the quasi-0D situation is more prominent in the GCE than in the CE so that the theory can yield an accurate result for the CE while it fails in the GCE where contributions from $N > 10$ enhance the inhomogeneity in the center of the cavity.

For extreme cases with $N = 1$ or 2 particles at moderate packing, the theory gives good results in spite of the fact that the CE functional is based on a saddle point approximation scheme [66].

8.3.2 Isothermal-Isobaric Ensemble

Theory

The isothermal-isobaric partition function Δ_N of a fluid with N particles at temperature T and external pressure P can be expressed as

$$\Delta_N = \int_0^\infty \exp(-\beta PV) Q_{N,V}[V_{\text{ext}}] \frac{\mathrm{d}V}{\delta_V} = \int_0^\infty Q_{N,V}[\nu] \frac{\mathrm{d}V}{\delta_V} \tag{8.72}$$

where $\nu(\mathbf{r}, V) \equiv V_{\text{ext}}(\mathbf{r}, V) + PV/N$ is a generalized external potential that includes a pressure term and the usual external potential V_{ext} which, in general, can also depend on the volume V. $Q_{N,V}$ denotes the (volume dependent) CE partition function. A constant volume scale δ_V is included to ensure that Δ_N is a dimensionless quantity.

A generalized volume-dependent density $n(\mathbf{r}, V)$ can be defined as the functional derivative of the logarithm of Δ_N with respect to ν. From Eq. (8.72) we obtain

$$n(\mathbf{r}, V) = -k_{\mathrm{B}}T \frac{\delta \ln \Delta_N}{\delta \nu(\mathbf{r}, V)} = \omega(V)\rho_N(\mathbf{r}, V) \tag{8.73}$$

where ρ_N is the CE density and

$$\omega(V) = \frac{Q_{N,V}[\nu]}{\int_0^\infty Q_{N,V}[\nu]\,\mathrm{d}V} = \frac{Q_{N,V}[V_{\text{ext}}]}{\delta_V \Delta_N} e^{-\beta PV} \tag{8.74}$$

is the volume probability density. From Eq. (8.73) we note that the integral of $n(\mathbf{r}, V)$ over V gives rise to the usual density in the isothermal-isobaric ensemble, namely,

$$\int_0^\infty n(\mathbf{r}, V)\,\mathrm{d}V = \int_0^\infty \omega(V)\rho_N(\mathbf{r}, V)\,\mathrm{d}V = \rho(\mathbf{r}). \tag{8.75}$$

The Legendre transform of $-\ln \Delta_N$ allows us to introduce the following free-energy functional

$$\beta\mathcal{F}[n] = -\ln \Delta_N[\nu[n]] - \beta \int n(\mathbf{r}, V)\nu(\mathbf{r}, V; [n])\,\mathrm{d}\mathbf{r}\,\mathrm{d}V \tag{8.76}$$

where the functional dependence on n is made explicit. From Eq. (8.76), using (8.72) and (8.73), the following expression for the isothermal-isobaric functional $\mathcal{F}$ can be obtained,

$$\beta\mathcal{F}[n] = \int \omega(V) \ln[\delta_V \omega(V)]\,\mathrm{d}V + \int \omega(V)\beta\mathcal{F}_{\mathrm{c}}[\rho_N]\,\mathrm{d}V \tag{8.77}$$

where $\mathcal{F}_{\mathrm{c}}[\rho_N]$ is the free-energy functional in the CE, and $\omega(V)$, ρ_N and N are themselves functionals of n:

$$\omega(V) = \frac{1}{N} \int n(\mathbf{r}, V)\,\mathrm{d}\mathbf{r}, \tag{8.78}$$

$$\rho_N(\mathbf{r}, V) = \frac{Nn(\mathbf{r}, V)}{\int n(\mathbf{r}, V)\,\mathrm{d}\mathbf{r}}, \tag{8.79}$$

and

$$\int n(\mathbf{r}, V)\,\mathrm{d}\mathbf{r}\,\mathrm{d}V = N. \tag{8.80}$$

The free-energy functional $\mathcal{F}$ can be shown to obey the following Euler–Lagrange equation:

$$\frac{\delta \mathcal{F}[n]}{\delta n(\mathbf{r}, V)} + \nu(\mathbf{r}, V) = \lambda \tag{8.81}$$

where the Lagrange multiplier λ arises as a consequence of Eq. (8.80), i.e., it is related to the fixed-N constraint. We note that this Euler–Lagrange equation together with the isothermal-isobaric functional (8.77) and Eqs. (8.78)–(8.80) is fully consistent with Eq. (8.73) that expresses the generalized volume-dependent density in terms of $\omega(V)$ and ρ_N.

Applying Eq. (8.77) to the ideal gas one has [73]:

$$\begin{aligned}\beta \mathcal{F}_{\text{id}}[n] = \frac{1-N}{N} \int n(\mathbf{r}, V) \ln[\delta_V \int n(\mathbf{r}, V)\, \mathrm{d}\mathbf{r}]\, \mathrm{d}\mathbf{r}\, \mathrm{d}V \\ + \int n(\mathbf{r}, V) \ln[\delta_V \Lambda^3 n(\mathbf{r}, V)]\, \mathrm{d}\mathbf{r}\, \mathrm{d}V,\end{aligned} \tag{8.82}$$

and thus, the excess isothermal-isobaric free-energy functional is given by

$$\mathcal{F}_{\text{ex}}[n] \equiv \mathcal{F}[n] - \mathcal{F}_{\text{id}}[n] = \frac{1}{N} \int n(\mathbf{r}, V) \mathcal{F}_{\text{c,ex}}[\rho_N]\, \mathrm{d}\mathbf{r}\, \mathrm{d}V \tag{8.83}$$

where $\mathcal{F}_{\text{c,ex}}$ is the excess contribution to the free energy in the CE. An accurate expression for this excess quantity can be obtained by noting that the free-energy functional in the CE can be approximated by Eq. (8.71) in terms of the GCE free energy. As in previous applications, in the present case we have considered the original FMT of Rosenfeld [53].

Application

As an application of the theory, we consider a model consisting of a fluid of hard spheres of diameter σ confined to a hard, spherical cavity with fluctuating radius. This means that the external potential of the system is given by Eq. (8.66) where R_{cav} is now a fluctuating variable. From Eq. (8.66), the volume available to the center of the hard spheres is given by $V = 4\pi(R_{\text{cav}} - \sigma/2)^3/3$ and thus V_{ext} depends on the volume V through the radius of the cavity and the generalized potential becomes $\nu(\mathbf{r}, V) = V_{\text{ext}}(\mathbf{r}, R_{\text{cav}}(V)) + PV/N$.

The results of the theory have been compared with isothermal-isobaric MC simulations [73]. In the simulations, either a particle displacement or a volume change are selected with probabilities $N/(N+1)$ and $1/(N+1)$, respectively. To perform a volume change, a new value for the radius of the cavity R_{new} is proposed, accepting the change with probability

$$\begin{aligned}p = \min\{1, \exp\Big[-\tfrac{4}{3}\pi\beta P\big((R_{\text{cav}} - \tfrac{\sigma}{2})^3 - (R_{\text{new}} - \tfrac{\sigma}{2})^3\big) \\ +2 \ln \frac{R_{\text{cav}} - \frac{\sigma}{2}}{R_{\text{new}} - \frac{\sigma}{2}}\Big]\}.\end{aligned} \tag{8.84}$$

Any change that leads to a wall–particle overlap is rejected. We note that the present simulation scheme does not involve the use of scaled coordinates.

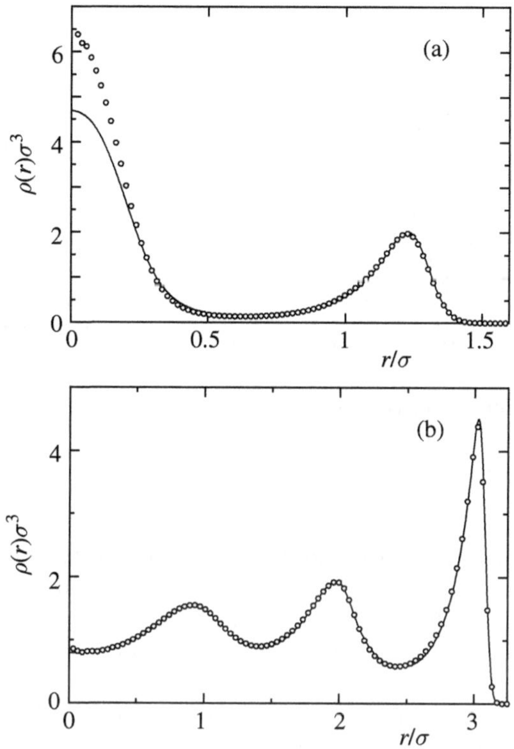

Fig. 8.15. Isothermal-isobaric density profiles of hard spheres in a spherical cavity (see Eq. 8.75). $\beta P = 3.2\sigma^{-3}$ and (**a**) $N = 10$, (**b**) $N = 100$. *Solid lines*, DFT results; *circles*, MC results

In Fig. 8.15 (a) we compare DFT results with MC data for the density profile of a system with $N = 10$ and $\beta P = 3.2\sigma^{-3}$. As in previous situations [see, e.g., Fig. 8.14 (b)], the original version of the FMT theory used in the DFT calculations cannot correctly describe the quasi-0D peak that arises in the center of the cavity [64] but yields excellent agreement with simulation out of this central zone. Figure 8.15 (b) shows the density profile for a system with $N = 100$ and the same pressure. In this case, the packing constraints of the system do not lead to a quasi-0D situation and the agreement between theory and simulation is excellent for the whole profile.

In Fig. 8.15 we have seen that considering the same external pressure but different numbers of particles can lead to very different values for the density in the center of the cavity. It seems therefore natural to analyze the behavior of the density in the center of the cavity for different numbers of particles in order to determine the influence of N in the structure of the confined fluid.

Figure 8.16 shows the results of theory and simulation for the mean density in the center of the cavity as a function of the number of particles [64]. The

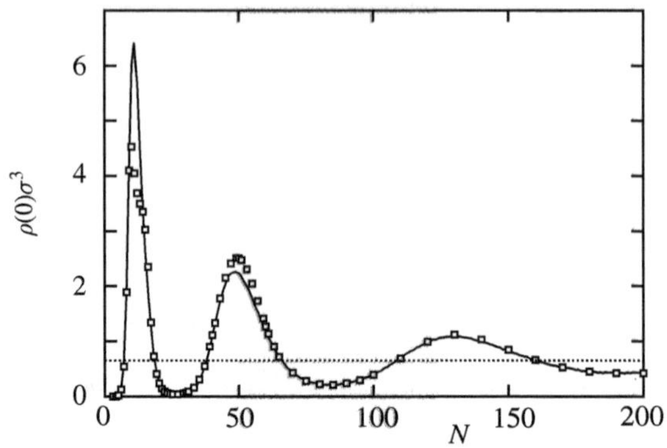

Fig. 8.16. Mean density in the center of the cavity as a function of the number of particles. The *solid lines* are MC results and the symbols are DFT results. $\beta P = 3.2\sigma^{-3}$. The *dashed line* indicates the bulk density

dashed line indicates the Carnahan–Starling value for the bulk density ρ_b of a homogeneous system at a pressure $\beta P = 3.2\sigma^{-3}$. Of course, the mean density $\hat{\rho}(0)$ approaches the bulk density ρ_b for large N. A salient feature of Fig. 8.16 is the appearance of peaks near the magic numbers $N = 13$, 55, and 147 characteristic of cluster structures. The peaks are related to quasi-0D situations, and therefore, they correspond to the larger differences between theory and simulation. The theory slightly overestimates the results of simulation for the second and third peaks while it underestimates MC for the first peak where it is clearly failing. Recent work on similar situations in the GCE [64] shows that the observed differences are related to the use of the original version of the FMT [53] instead of recent refinements of the theory that can properly deal with quasi-0D situations [60, 61, 62, 63].

Summary

This chapter reports on the effects that appear in systems of hard particles in which the number of particles is small. It is shown that the effects appearing in such systems can be twofold. On the one hand, there are explicit finite-size effects that arise due to the consideration of different ensembles —it is shown that the results obtained in these ensembles can be related by an appropriate series expansion. On the other hand, one has to deal with implicit finite-size effects that are specific of some computer simulations techniques. These effects appear as a consequence of considering periodic boundary conditions, and they can be accurately handled by assuming infinite replica images of the real system.

For a better understanding of finite-size effects in hard particle systems, we have presented and analyzed three different problems related to them:

- When considering isolated systems with a small number of particles, the momentum distributions can differ from the usual Maxwell–Boltzmann form. In these isolated systems the conserved quantities modify the appropriate ensemble, providing different results for the momentum distributions. In particular, we have first presented the microcanonical ensemble and later on we have taken into account the conservation of the total linear momentum and the conservation of the total angular momentum.
- Finite-size effects arise in the fluctuations of the density in different ways and require different theoretical approaches. On the one hand, explicit finite-size effects are properly handled by means of a series expansion of the pair distribution function. This series expansion relates the canonical pair distribution function to its grand canonical counterpart introducing the moments of the probability distribution for the number of particles in the grand canonical ensemble. On the other hand, implicit finite-size effects are taken into account by considering the particles in a system with periodic boundary conditions as oriented supermolecules, i.e., each particle is considered together with its infinite replica images.
- The density profiles of small hard-particle systems confined in hard spherical cavities are analyzed. The study is made in the framework of the density functional theory which minimizes the free energy of the system considered as a functional of the density profile. It is shown how the explicit finite-size effects manifest in the inhomogeneous structure of the fluid. Two different approaches are proposed to deal with these effects. First, a series expansion for the canonical profile in terms of the grand canonical one is shown to give accurate results. And second, a version of the density functional theory is proposed for the canonical ensemble as well as for the isothermal isobaric ensemble.

Acknowledgement

This chapter has been supported by *Ministerio de Educación y Ciencia* under grants FIS2005-05081 FEDER and FIS2006-03764 FEDER.

References

1. W. W. Wood, in: *Fundamental Problems in Statistical Mechanics*, Vol 3, ed. by E.D.G. Cohen (North-Holland, Amsterdam, 1975), p. 331.
2. J.L. Lebowitz, J.K. Percus, L. Verlet: Phys. Rev. **153**, 250 (1967).
3. E. M. Pearson, T. Halicioglu, W. A. Tiller: Phys. Rev. A **32**, 3030 (1985).
4. T. Çağin, J. R. Ray: Phys. Rev. A **37**, 247 (1988).
5. J. R. Ray, H. W. Graben: Phys. Rev. A **44**, 6905 (1991).
6. F. L. Román, J. A. White, S. Velasco: Phys. Rev. E **51**, 6271 (1995).
7. F. L. Román, A. González, J. A. White, S. Velasco: Physica A **234**, 53 (1996).
8. J. R. Ray, H. Zhang: Phys. Rev. E **59**, 4781 (1999).

9. W. W. Wood, J. J. Erpenbeck, G. A. Baker, J. D. Johnson: Phys. Rev. E **63**, 011106 (2000).
10. D.C. Wallace, G.K. Straub: Phys. Rev. A **27**, 2201 (1983).
11. F. L. Román, J. A. White, A. González, S. Velasco: Physica A **334**, 312 (2004).
12. F.L. Román, A. González, J.A. White, S. Velasco: J. Chem. Phys. **118**, 7930 (2003).
13. V. Laliena: Phys. Rev. E. **59**, 4786 (1999).
14. D.H.E. Gross, E.V. Votyakov: Eur. Phys. J. B **15**, 115 (2000).
15. O. Fliegans, D.H.E. Gross: Phys. Rev. E **65**, 046143 (2002).
16. E.V. Votyakov, H.I. Hidmi, A. De Martino, D.H.E. Gross: Phys. Rev. Lett. **89**, 031101 (2002).
17. Gross, D. H. E.: *Microcanonical Thermodynamics: Phase Transitions in "Small" Systems* Lect. Notes Phys. **66** World Scientific Publishing Co. Pte. Ltd., Singapore (2001).
18. A. Schlüter: Z. Naturf. A **3**, 350 (1948).
19. F. Lado: J. Chem. Phys. **75**, 5461 (1981).
20. A.I. Khinchin: *Mathematical Foundations of Statistical Mechanics* (Dover, New York, 1949) Chap. II.
21. A. Münster: *Statistical Thermodynamics* (Springer, Berlin, 1969), Chap. 1.
22. S. Velasco, J.A. White, J. Güémez: Eur. J. Phys. **14**, 166 (1993).
23. D.C. Rapaport: *The Art of Molecular Dynamics Simulation* (Cambridge University Press, Cambridge 1997).
24. J. L. Lebowitz, J. K. Percus: Phys. Rev. **122**, 1675 (1961).
25. J. L. Lebowitz, J. K. Percus: Phys. Rev. **124**, 1673 (1961).
26. L. R. Pratt, S. W. Haan: J. Chem. Phys. **74**, 1864 (1981).
27. L. R. Pratt, S. W. Haan: J. Chem. Phys. **74**, 1873 (1981).
28. A. R. Denton, P. A. Egelstaff: Z. Phys. B **103**, 343 (1997).
29. K. Binder, D. W. Hermann: *Monte Carlo Simulation in Statistical Physics: an Introduction* (Springer, Berlin, 1988).
30. K. Binder: Phys. Rev. Lett. **47**, 693 (1981).
31. K. Binder: Z. Phys. B **43**, 119 (1981).
32. M. Rovere, D. W. Hermann, K. Binder: Europhys. Lett. **6**, 585 (1988).
33. M. Rovere, D. W. Hermann, K. Binder: J. Phys.: Condens. Matter **2**, 7009 (1990).
34. M. Rovere, P. Nielaba, K. Binder: Z. Phys. B **90**, 215 (1993).
35. D. Marx, P. Nielaba, K. Binder: Phys. Rev. B **47**, 7788 (1993).
36. H. Weber, D. Marx: Europhys. Lett. **27**, 593 (1994).
37. H. Weber, D. Marx, K. Binder: Phys. Rev. B**51**, 14636 (1995).
38. F. L. Román, J. A. White, S. Velasco: J. Chem. Phys. **106**, 4196 (1997).
39. J. J. Salacuse, A. R. Denton, P. A. Egelstaff: Phys Rev. E **53**, 2382 (1996).
40. F. L. Román, J. A. White, A. González, S. Velasco: J. Chem. Phys. **110**, 9821 (1999).
41. J.P. Hansen, I.R. McDonald: *Theory of Simple Liquids, 2nd Ed.* (Academic Press, London 1986).
42. M. Rovere, D.W. Hermann, K. Binder: Europhys. Lett. **6**, 585 (1988).
43. M. Rovere, D.W. Hermann, K. Binder: J. Phys.: Condens. Matter **2**, 7009 (1990).
44. D. A. McQuarrie: *Statistical Mechanics* (Harper-Row, New York, 1976).
45. F. L. Román, J. A. White, S. Velasco: J. Chem. Phys. **107**, 4635 (1997).

46. F. L. Román, J. A. White, S. Velasco: Europhys. Lett. **42**, 371 (1998).
47. M. Baus, J.L. Colot: Phys. Rev. A **36**, 3912 (1987).
48. M. Baus, J.L. Colot: J. Phys. C **19**, L643 (1986).
49. F. Lado: J. Chem.Phys. **49**, 3092 (1968).
50. M. Schoen, D. J. Diestler, J. H. Cushman: J. Chem. Phys. **87**, 5464 (1987).
51. A. González, J. A. White, F. L. Román, S. Velasco, R. Evans: Phys. Rev. Lett. **79**, 2466 (1997).
52. A. González, J. A. White, F. L. Román, R. Evans: J. Chem. Phys. **109**, 3637 (1998).
53. Y. Rosenfeld: Phys. Rev. Lett. **63**, 980 (1989).
54. Y. Rosenfeld: Phys. Rev. A **42**, 5978 (1990).
55. Y. Rosenfeld, D. Levesque, J. J. Weis: J. Chem. Phys. **92**, 6818 (1990).
56. Y. Rosenfeld: J. Chem. Phys. **98**, 8126 (1993).
57. E. Kierlik, M. L. Rosinberg: Phys. Rev. A **42**, 3382 (1990).
58. E. Kierlik, M. L. Rosinberg: Phys. Rev A **44**, 5025 (1991).
59. S. Phan, E. Kierlik, M. L. Rosinberg, B. Bildstein, G. Kahl: Phys. Rev. E **48**, 618 (1993).
60. Y. Rosenfeld, M. Schmidt, H. Löwen, P. Tarazona: J. Phys.: Condens. Matter **8**, l577 (1996).
61. Y. Rosenfeld, M. Schmidt, H. Löwen, P. Tarazona: Phys Rev. E **55**, 4245 (1997).
62. P. Tarazona: Physica A **306**, 243 (2002).
63. P. Tarazona: Phys. Rev. Lett. **84**, 694 (2000).
64. A. González, J. A. White, F. L. Román, S. Velasco: J. Chem. Phys. **124**, 154708 (2006).
65. R. G. Parr, W. Yang: *Density-Functional Theory of Atoms and Molecules*, (Oxford University Press, Oxford, 1989).
66. J. A. White, A. González, F. L. Román, S. Velasco: Phys. Rev. Lett **84**, 1220 (2000).
67. D. N. Zubarev: *Nonequilibrium Statistical Thermodynamics*, (Plenum, New York, 1974), Chap. 13.
68. Y. Rosenfeld: Phys. Rev. Lett. **63**, 980 (1989).
69. Y. Rosenfeld: Phys. Rev. A **42**, 5978 (1990).
70. Y. Rosenfeld: J. Chem. Phys. **98**, 8126 (1993).
71. E. Kierlik, M. L. Rosinberg: Phys. Rev. A **42**, 3382 (1990).
72. S. Phan, E. Kierlik, M. L. Rosinberg, B. Bildstein, G. Kahl: Phys. Rev. E **48**, 618 (1993).
73. A. González, J. A. White, F. L. Román, S. Velasco: J. Chem. Phys. **120**, 10634 (2004).

66. [illegible]
67. [illegible]
68. [illegible]

9

Modelling of Transport Properties of Hard Sphere Fluids and Related Systems, and its Applications

C.M. Silva[1] and H. Liu[2]

[1] CICECO. Department of Chemistry, University of Aveiro, Campus Universitário de Santiago, 3810-193 Aveiro, Portugal
carlos.manuel@ua.pt

[2] Kronos Canadian Systems Inc., AD OPT Division, 3535 Queen Mary Rd, Suite 650, Montreal, Quebec, H3V 1H8 Canada
hqliu2000@yahoo.com

Fluid transport properties, such as viscosity, diffusion coefficients, and thermal conductivity, are extremely important in both research and industry. Therefore, it is very important to accurately model the behaviour of these kinetic coefficients.

At present, it is not yet possible to give a rigorous theoretical interpretation of the transport properties of dense fluids, since there are many-body interactions involved and because the pair potential energy functions are only known for simple molecules. Nonetheless, for liquids and dense gases, where the repulsive molecular interaction plays a major role in determining their properties, an accurate model for hard spheres can be used as an appealing and tractable first approximation for the major excluded volume and packing effects of real substances. In fact, it has been found that for the liquid viscosity and diffusivity the contribution of the hard sphere term is up to about 70%.

This chapter will address the most important issues involved in the modelling of transport properties of hard sphere and modified systems. Special attention is devoted to the development and analysis of the models, as well as their final applications to real substances over wide ranges of temperature and density, from dilute gases to dense fluids, including supercritical fluids and liquids.

9.1 Introduction

This chapter is devoted to the transport properties of fluids, namely diffusion coefficients, viscosity, and thermal conductivity. Their accurate knowledge is fundamental in both research and industry. For instance, viscosity is the key quantity related to structure relaxation in various dense fluids, stable liquids,

Silva, C.M., Liu, H.: *Modelling of Transport Properties of Hard Sphere Fluids and its Applications*. Lect. Notes Phys. **753**, 383–492 (2008)
DOI 10.1007/978-3-540-78767-9_9

supercooled liquids, and glasses. In various industries and areas of technology, such as chemical, pharmaceutical, metallurgical, and environmental, transport properties are essential for optimum process designs. Particularly emblematic are rate-controlled separations and solid catalysed reactions, which require a mass-transfer and eventually heat-transfer analysis. Typical examples of rate-based separations are membrane processes, molecular distillation, crystallization from solution (crystal size distribution), ion exchange, adsorption, chromatography, and electrophoresis. Concerning catalysed reactions, it is well known that the existence of thermal and concentration gradients within catalyst pellets results in a product distribution different from that which would otherwise be observed.

The self-diffusion coefficient refers to the transport of tracer molecules through a fluid of otherwise identical particles. Starting with an uneven distribution of tracer molecules in a fluid, the diffusion process will homogenize it at microscopic level, and the rate at which the fluid inhomogeneities are smoothed out is determined by the value of the self-diffusion coefficient, D. Dealing with mixtures of different components, the constant with interest is the common diffusion coefficient mostly used in engineering applications.

When different moving parts of a fluid possess unequal velocities, then friction between them tends to bring the fluid to equilibrium, and the average velocity of the molecules—which is in fact the quantity transported throughout the fluid—becomes uniform. The rate at which the average velocity of the particles is transferred is determined by the viscosity coefficient, η.

When a fluid exhibits non-uniform temperature, the molecules in hotter regions will have greater average kinetic energy than those in cooler parts. Such initial temperature distribution will even out and become uniform through heat conduction. Macroscopic motion is not necessary. The only motion is on a molecular level. The rate at which the temperatures in different regions of the fluid are equalized is determined by the thermal conductivity coefficient, λ.

According to the above-mentioned description of the transport properties, these may be defined in terms of the response of a system to a perturbation, e.g. the diffusion coefficient relates the particle flux to a concentration gradient, the shear viscosity is a measure of the shear stress induced by an applied velocity gradient, and the thermal conductivity connects energy flux to a temperature gradient. Such flux-to-gradient relations are the phenomenological linear laws of Fick, Newton, and Fourier, respectively, which may be described mathematically by the general expression [1].

$$\mathbf{J} = -Y \cdot \nabla X \tag{9.1}$$

where Y is the transport coefficient (D, η or λ), and ∇X is the imposed gradient.

At low densities, fluid diffusivity, viscosity, and thermal conductivity may be estimated by some theoretical methods, like those based on the kinetic theory of Enskog and its modifications [2, 3].

For dense fluids, including liquids and supercritical fluids, no theory can provide reliable predictions, because departures from dilute values become significantly larger. Nonetheless, since repulsive molecular interactions play a major role in determining dense gas and liquid properties, accurate models for the hard sphere (HS) fluid can be used as a reasonably realistic description of real fluids. In fact, it has been found that the HS contribution for liquid viscosity and diffusion coefficient is up to about 70%.

For proper application of the HS theory, it is necessary to take into account corrections for correlated motions dominant at intermediate and high densities, which may be assessed by computer simulations, specifically by equilibrium and/or non-equilibrium molecular dynamics. Since the 1970s, numerous computer simulations for various simple fluids have been carried over wide ranges of density/pressure and temperature, particularly for hard sphere and Lennard-Jones fluids. These data have been employed successfully for developing models for these systems and further extensions to real fluids.

9.1.1 Model Fluids: Intermolecular Potential Functions

As has been mentioned above, transport of mass, momentum, and energy through a fluid are microscopic level phenomena, and therefore must result from particle motion and interaction. At low densities, the free molecular motions are largely dominant; collisions are rare events that involve only two molecules. Increasing density, interactions among molecules and groups of molecules dominate the transport process. (Near the critical point the intermolecular interaction becomes irrelevant, because transport properties are determined by the behaviour of clusters and their size.)

The computer simulation of hypothetical fluid models demands certain assumed pair potentials of interaction usually expressed in the form of mathematical functions. In most calculations pairwise additivity is adopted, and the intermolecular potential energy of a system composed of particles is taken to be a sum of isolated pair interactions:

$$U = \sum_{i=1}^{N} \sum_{j>i}^{N} \phi(r_{ij}) \tag{9.2}$$

where $\phi(r_{ij})$ is the pair potential function, and r_{ij} is the interparticle distance.

In principle, the knowledge of the intermolecular potential function enables all fluid properties to be evaluated at any temperature, and we may expect the inverse to be hold if accurate data are available. Such assumption is obeyed especially for dilute gases, while in condensed phases the N-body problem is too complex to be solved exactly.

In the following, we present some pair potentials commonly found in computer simulation studies on transport properties: HS, square-well (SW), soft sphere (SS), Coulomb or one-component plasma (OCP), Repulsive Lennard-Jones or Weeks–Chandler–Andersen (WCA), and Lennard-Jones (LJ). It will be seen in the following sections that these model fluids are the basis for the development and analysis of equations for the dynamic properties of real systems.

Hard sphere potential

It is the first correction to the ideal gas, as one surrounds each molecule with an excluded volume:

$$\phi_{\mathrm{HS}}(r) = \begin{cases} \infty, & r < \sigma \\ 0, & r \geq \sigma \end{cases} \tag{9.3}$$

where σ is the HS diameter and r is the radial coordinate.

Square-well potential

This is the simplest potential possessing the basic characteristics of a real fluid, since, in a crude fashion, it takes into account the ideas of attractive and repulsive forces. It represents rigid spheres of diameter σ surrounded by an attractive shell of strength $\varepsilon_{\mathrm{SW}}$ extending to separations $K\sigma$:

$$\phi_{\mathrm{SW}}(r) = \begin{cases} \infty, & r \leq \sigma \\ -\varepsilon_{\mathrm{SW}}, & \sigma < r < K\sigma \\ 0, & r \geq K\sigma \end{cases} \tag{9.4}$$

Soft sphere and one-component plasma potentials

SS potential contains no attractive part and assumes the generic form

$$\phi_{\mathrm{SS}}(r) = \varepsilon(\sigma/r)^{\nu} \tag{9.5}$$

where σ is the collision diameter, ε is the energy parameter, and ν is usually an integer which characterizes the hardness of molecules. Extreme cases are the HS ($\nu = \infty$) and the Coulomb or OCP ($\nu = 1$) potentials. It is worth noting that some authors consider the SS fluid with $\nu = 12$.

Lennard-Jones potential

It is a more realistic potential function that possesses both the repulsive and the inverse sixth power London forces:

$$\phi_{\mathrm{LJ}}(r) = 4\varepsilon_{\mathrm{LJ}}\left[\left(\frac{\sigma_{\mathrm{LJ}}}{r}\right)^{12} - \left(\frac{\sigma_{\mathrm{LJ}}}{r}\right)^{6}\right] \tag{9.6}$$

where $\varepsilon_{\mathrm{LJ}}$ is the depth of the potential well, which occurs at $r = 2^{1/6}\sigma_{\mathrm{LJ}}$ and marks the change from attraction to repulsion, and σ_{LJ} is the collision diameter for low-energy collisions (i.e. the separation for which $\phi_{\mathrm{LJ}}(r) = 0$). These are the so-called LJ parameters.

Repulsive Lennard-Jones or Weeks–Chandler–Andersen potential

Weeks et al. [4] split up the LJ potential into a reference part containing all repulsive forces and a perturbing part containing all attractive forces. Accordingly, the repulsive LJ fluid is defined as

$$\phi_{\mathrm{WCA}}(r) = \begin{cases} \phi_{\mathrm{LJ}}(r) + \varepsilon_{\mathrm{LJ}}, & r \leq 2^{1/6}\sigma_{\mathrm{LJ}} \\ 0, & r > 2^{1/6}\sigma_{\mathrm{LJ}} \end{cases} \tag{9.7}$$

In particular, the WCA reference system has been already considered in Chap. 4.

9.1.2 Outline of the Chapter

This chapter is divided into seven sections. In this first part, the transport properties that are the subject of our study—diffusion coefficients, viscosity, and thermal conductivity—are briefly introduced, along with the key intermolecular potential functions frequently adopted in literature to devise models.

The fundamentals of the molecular dynamic simulations accomplished to compute kinetic coefficients are given in Sect. 9.2. Only essential concepts and basic mathematical relations are provided, whereas details and specific considerations are left for very good standard texts cited. Nonetheless, it is important to present how the interested coefficients are *measured* computationally.

Before treating each property separately, the main theories of transport in HS and modified systems are discussed in Sect. 9.3, since they provide general tools to analyse and describe the three constants.

The individual study of diffusivity, viscosity, and thermal conductivity of pure fluids and mixtures is focused in Sects. 9.4, 9.5, and 9.6, respectively. A special and comprehensive attention is devoted to the self- and tracer diffusion coefficients. Afterwards, the similarity of modelling allows us to provide smaller sections for the remaining properties. On the whole, the importance that accurate models for hard spheres play in the interpretation and representation of modified systems is emphasized, particularly real fluids.

Some conclusions and final remarks are given in Sect. 9.7, and references compiled at the end.

9.2 Molecular Dynamics Simulations

Research on hypothetical fluid models allows us to understand the effect molecular interactions have upon their macroscopic properties. In addition, comparing results for model fluids and data from experiments on real systems enables us to check the extent to which a hypothetical system approximates a real one.

Two approaches are susceptible for the study of model fluids: the so-called computer experimental or simulation method, and the theoretical prediction method. The first one includes both Monte Carlo calculations of thermodynamic averages and static distribution functions, and equilibrium molecular dynamics (MD) simulations of the time evolution of small samples consisting of a few hundreds or thousands of interacting molecules. Analytical models representing these simulation data are then obtained empirically or semi-empirically. In the second approach, analytical expressions for various properties are obtained by purely theoretical methods, based on some simplifying assumptions. Although simple potential functions of model fluids cannot adequately describe real physical effects, they do serve as reference potentials, while additional effects may be treated as perturbations.

As it will be seen in the following sections, due to the non-existence of good theoretical expressions for transport properties of dense fluids, empirical and semi-empirical correlations based on computer simulation data still hold their relevance. Such fact justifies the importance accurate data have on the development of new precise equations.

An outstanding feature of equilibrium MD is that it allows the calculation of equilibrium and transport properties of an N-body system. The method is designed to generate many microstates consistent with constraints imposed on the system, by solving the Newton's equations of motion of the N-interacting particles. At a given time t, the solution gives an instantaneous picture of the system—i.e. a microstate—comprising its spatial configuration and velocity distribution. Hence, the time evolution of the system is constituted by thousands of microstates, whose instantaneous properties (e.g. energy, pressure, and transport coefficients) may be time averaged subsequently to obtain the statistical averages corresponding to the macroscopic properties of the fluid.

In addition, non-equilibrium methods have also been developed as an alternative simulation technique to compute transport coefficients, namely shear viscosity, bulk viscosity, thermal conductivity, and diffusion coefficients. Following non-equilibrium molecular dynamics (NEMD), an external force is applied to the system to establish the non-equilibrium situation of interest, and the system's response to this force is then determined from the simulation. Hence, according to the NEMD principles, the coefficients are obtained analogously to an experimentalist in the laboratory, i.e. by averaging the non-equilibrium steady-state ratio of the dissipative flux induced by the driving force applied. Presently, it is almost certain that both MD and NEMD generate equivalent results.

Details and specific considerations about MD and NEMD may be found in standard texts, for instance those by Vesely [5], Hoover [6, 7], Evans and Morriss [8], Haile [9], and Allen and Tildesley [10]. In Sect. 9.1.2, the conventional MD approach using Green–Kubo formulas and the associated Einstein relations is briefly described.

9.2.1 Equilibrium Molecular Dynamics

In this section the Green–Kubo formulas and the related Einstein equations for the self-diffusion coefficient, shear viscosity, and thermal conductivity are discussed.

Self-Diffusion Coefficient

In the framework of equilibrium MD, the self-diffusion coefficient is given by the integral of the velocity autocorrelation function as

$$D = \frac{1}{3}\int_0^\infty \langle \mathbf{v}_i(0)\cdot\mathbf{v}_i(t)\rangle \,\mathrm{d}t \tag{9.8}$$

where the velocity autocorrelation function is defined as

$$\langle \mathbf{v}_i(0) \cdot \mathbf{v}_i(t) \rangle = \lim_{\mathcal{T} \to \infty} \frac{1}{\mathcal{T}} \int_0^{\mathcal{T}} \mathbf{v}_i(s) \cdot \mathbf{v}_i(s+t) \, \mathrm{d}s \tag{9.9}$$

It represents the time average of the correlation of the velocities of particle i, $\mathbf{v}_i$, at two different times separated by an interval t. Hence, it is a measure of the memory of a particle of its previous velocity. The Einstein relation, valid at long times, associated with the Green–Kubo formula, Eq. (9.8), is

$$2tD = \lim_{t \to \infty} \frac{1}{3} \left\langle \left| \mathbf{r}_i(t) - \mathbf{r}_i(0) \right|^2 \right\rangle \tag{9.10}$$

where $\mathbf{r}_i(t)$ is the particle vector position, and the angular brackets indicate an average over an equilibrium ensemble, or, on a computer, an average starting from several initial times. In practice, these averages would be calculated for each of the particles in the simulation, the results added together and divided by N to improve statistical accuracy. Frequently, these equations appear in their one-dimensional analog:

$$D = \int_0^{\infty} \langle \dot{x}_i(0) \cdot \dot{x}_i(t) \rangle \, \mathrm{d}t \tag{9.11}$$

$$2tD = \lim_{t \to \infty} \left\langle \left| x_i(t) - x_i(0) \right|^2 \right\rangle \tag{9.12}$$

where $x_i(t)$ and $\dot{x}_i(t)$ are the x-components of the position and velocity of particle i.

Shear Viscosity

With respect to shear viscosity, η, the Green–Kubo relation is an integral over the stress autocorrelation function

$$\eta_{xy} = \frac{1}{V k_{\mathrm{B}} T} \int_0^{\infty} \langle P_{xy}(t) \cdot P_{xy}(0) \rangle \, \mathrm{d}t \tag{9.13}$$

where $P_{xy}(t)$ are defined as the off-diagonal elements of the instantaneous stress tensor $P(t)$ given by

$$P(t) = \begin{pmatrix} P_{xx}(t) & P_{xy}(t) & P_{xz}(t) \\ P_{yx}(t) & P_{yy}(t) & P_{yz}(t) \\ P_{zx}(t) & P_{zy}(t) & P_{zz}(t) \end{pmatrix} \tag{9.14}$$

V is the volume, m is the particle mass, k_{B} is the Boltzmann's constant, and T is the absolute temperature. The instantaneous stress tensor elements are composed of a kinetic and a potential part:

$$P(t) = m \sum_{i=1}^{N} \mathbf{v}_i \cdot \mathbf{v}_i - \frac{1}{2} \sum_{i=1}^{N} \sum_{i \neq j}^{N} \mathbf{r}_{ij} \nabla \phi (r_{ij}) \tag{9.15}$$

where $\mathbf{r}_{ij} = \mathbf{r}_j - \mathbf{r}_i$ is the vector separating particles i and j. The element $P_{xy}(t)$ measures the rate at which y-directed momentum is transported in the x-direction. Equation (9.15) proves that the shear viscosity is a collective function, i.e. involves entire system in calculations, hence the statistical precision cannot be improved by averaging over the N particles of the ensemble. Consequently, η is subjected to much greater imprecision than self-diffusivities. The statistics can only be enhanced by averaging over all six terms that result from the stress tensor:

$$\eta = \frac{1}{6} (\eta_{xy} + \eta_{yx} + \eta_{xz} + \eta_{zx} + \eta_{yz} + \eta_{zy}) \tag{9.16}$$

The Einstein relation corresponding to the Green–Kubo formula given by Eq. (9.13) is

$$\eta = \frac{m^2}{2Vk_{\mathrm{B}}T} \lim_{t\to\infty} \frac{1}{t} \left\langle \left\{ \sum_{i=1}^{N} [\dot{x}_i (t) \, y_i (t) - \dot{x}_i (0) \, y_i (0)] \right\}^2 \right\rangle \tag{9.17}$$

It is worth noting that Green-Kubo expression for viscosity, Eqs. (9.13) and (9.16), is not directly usable for hard spheres because the potential function of hard bodies is discontinuous and the stress tensor, Eq. (9.15), involves $\nabla\phi(r_{ij})$.

Thermal Conductivity

The thermal conductivity, λ, is calculated by the Green–Kubo formula

$$\lambda = \frac{1}{Vk_{\mathrm{B}}T^2} \int_0^{\infty} \langle J_T (t) \, J_T (0) \rangle \, \mathrm{d}t \tag{9.18}$$

where the energy flux, $J_T(t)$, is given by

$$J_T(t) = \frac{\mathrm{d}}{\mathrm{d}t} \sum_{i=1}^{N} x_i E_i \tag{9.19}$$

and E_i is the sum of potential and kinetic energies of particle i. The corresponding Einstein expression is

$$\lambda = \frac{1}{2Vk_{\mathrm{B}}T^2} \lim_{t\to\infty} \frac{1}{t} \left\langle \left\{ \sum_{i=1}^{N} [x_i (t) \, E_i (t) - x_i (0) \, E_i (0)] \right\}^2 \right\rangle \tag{9.20}$$

Once the transport properties have been defined, it is possible to calculate them with the aid of fast computers for systems composed of particles interacting by means of potential functions as those presented in Sect. 9.1.1. Note that such calculations are, in effect, experiments performed on models rather than on real physical systems. The final objective is to compare the simulation results thus obtained with the transport properties data of real fluids, to evaluate if an approximate model provides the most satisfactory basis for their interpretation, correlation, and prediction.

9.3 Theories of Transport in Hard Sphere and Modified Systems

In this section the main theories for transport properties of dense HS and modified systems are presented. It starts with the milestone Enskog theory for the self-diffusivity, viscosity, and thermal conductivity of the HS fluid, and discusses the indispensable corrections that have been assessed by computer simulation since 1970. Special attention is dedicated to the modified Enskog theory, the effective hard sphere diameter method, the free-volume approach, the van der Waals theory of transport, the rough hard sphere theory, and, finally, excess entropy scaling laws. This section is the basis for the remaining chapters, where each property is discussed separately.

9.3.1 Enskog Theory

The coefficients of self-diffusion, viscosity, and thermal conductivity of a pure dilute gas (i.e. in the limit of zero density) may be estimated from the rigorous kinetic theory formulae

$$D_o = \frac{3}{8\rho_o\sigma^2}\left(\frac{k_B T}{\pi m}\right)^{1/2}\frac{1}{\Omega^{(1,1)*}} \tag{9.21}$$

$$\eta_o = \frac{5}{16\sigma^2}\left(\frac{m k_B T}{\pi}\right)^{1/2}\frac{1}{\Omega^{(2,2)*}} \tag{9.22}$$

$$\lambda_o = \frac{25}{32\sigma^2}\left(\frac{k_B T}{\pi m}\right)^{1/2}\frac{C_V}{\Omega^{(2,2)*}} = \frac{75}{64\sigma^2}\left(\frac{k_B^3 T}{\pi m}\right)^{1/2}\frac{1}{\Omega^{(2,2)*}} \tag{9.23}$$

where subscript "o" stands for dilute gas, ρ is the number density, $C_V = (3/2)k_B$ is the specific heat per molecule at constant volume for the translational degrees of freedom, $\Omega^{(s,l)*}$ are dimensionless collision integrals obtained by dividing them by their corresponding HS values, and σ is a characteristic distance parameter between colliding molecules. For the HS system, σ is the sphere diameter and $\Omega^{(s,l)*}$ becomes unitary. These equations are valid for monatomic gases and have to be modified if the molecules have internal structure. Such theory is not applicable to dense gases and liquids since it is based upon the Boltzmann equation for the distribution function, which accounts for two-body collisions only and assumes that molecules have no finite extension in space.

A kinetic theory for transport coefficients of a dense HS system has been developed by Enskog [11]. Taking into account that the diameter of a molecule is no longer negligible compared with the interparticle distance, he was able to graft the first successful theory of dense gases onto the former dilute gas model. The restriction to rigid spheres is made because multibody collisions need not to be considered. Introducing the finite size of molecules, Enskog reckons the collisional transfer and the important effect density has upon the frequency of collisions in dense systems. Accordingly, Enskog equations comprise two distinct contributions: a kinetic, due to translational motion, and a collisional or potential, whereby momentum can be passed to another molecule upon collision. The first one contributes to the three coefficients, whereas the second only affects viscosity and thermal conductivity. Concerning the number of binary collisions that occur in a gas, its frequency was modified by a factor $g(\sigma)$: the radial distribution function at contact, which is unity at low density. The resultant Enskog equations for the transport coefficients of dense HS fluid, subscript E, are

$$\frac{\rho D_{\mathrm{E}}}{\rho_o D_o} = \frac{1}{g(\sigma)} \tag{9.24}$$

$$\frac{\eta_{\mathrm{E}}}{\eta_o} = \frac{1}{g(\sigma)} + 0.8b\rho + 0.761g(\sigma)b^2\rho^2 \tag{9.25}$$

$$\frac{\lambda_{\mathrm{E}}}{\lambda_o} = \frac{1}{g(\sigma)} + 1.2b\rho + 0.755g(\sigma)b^2\rho^2 \tag{9.26}$$

The first term on the right-hand side of Eqs. (9.24)–(9.26) is the kinetic component and just scales in time the solution of the Boltzmann equation for dilute gases; the last term in η_{E} and λ_{E} is purely collisional and dominates at liquid-like densities; their middle term takes account of cross contribution. These two last terms include the second virial coefficient, b—often set equal to the van der Waals excluded volume—which for HS is defined as

$$b = 2\pi\sigma^3/3 \tag{9.27}$$

Thus, as explained in Chaps. 1–3, the value of $g(\sigma)$ can be obtained from the equation of state (EOS) as follows:

$$\frac{P}{\rho k_{\mathrm{B}}T} = 1 + b\rho g(\sigma) \tag{9.28}$$

Thus, by using the HS virial expansion

$$\frac{P}{\rho k_{\mathrm{B}}T} = 1 + b\rho + 0.625\,(b\rho)^2 + 0.2869\,(b\rho)^3 + 0.115\,(b\rho)^4 + \cdots \tag{9.29}$$

one easily finds

$$g(\sigma) = 1 + 0.625b\rho + 0.2869\,(b\rho)^2 + 0.115\,(b\rho)^3 + \cdots \tag{9.30}$$

As mentioned in Chap. 3, the Carnahan–Starling (CS) EOS [12] is frequently used owing to its simplicity and sufficient accuracy to derive thermodynamic properties of the HS system. Its analytical expression is

$$\frac{P}{\rho k_{\mathrm{B}}T} = \frac{1+\varphi+\varphi^2-\varphi^3}{(1-\varphi)^3} \tag{9.31}$$

and then

$$g(\sigma) = \frac{1-\varphi/2}{(1-\varphi)^3} \tag{9.32}$$

where φ is the HS packing fraction, which for N spheres occupying a volume V is given by

$$\varphi = \frac{N\pi\sigma^3}{6V} = \frac{\pi}{6}\rho\sigma^3 \tag{9.33}$$

The close-packed molar volume, V_o, is an alternative parameter, which is also frequently used, and is defined as

$$V_o \equiv \frac{N\sigma^3}{\sqrt{2}} \tag{9.34}$$

From Eqs. (9.27), (9.33), and (9.34), the following interrelations between b, V_o, and φ may be established:

$$\varphi \equiv \frac{\pi\rho^*}{6} = \frac{b\rho}{4} = \frac{\pi\sqrt{2}}{6}\left(\frac{V_o}{V}\right) \tag{9.35}$$

$\rho^* \equiv \rho\sigma^3$ being the reduced number density.

Heyes and Woodcock [13] published another simple accurate EOS of the HS fluid utilized in the scope of thermophysical properties:

$$\frac{P}{\rho k_{\mathrm{B}}T} = 1 + \frac{bv}{(v-v_o^1)^2} \tag{9.36}$$

where $v = V/N$ is the molecular volume, and v_o^1 can be considered to be an effective occupied volume that is used to define effective free volume $v - v_o^1$. From Eqs. (9.28) and (9.36), another analytical expression for $g(\sigma)$ arises:

$$g(\sigma) = \frac{X_1^2}{1-X_1^2} \tag{9.37}$$

Defining $X_1 = v/v_o^1 = V/V_o^1$ and $X = v/v_o = V/V_o$, the relationship $X_1 = fX$ is proposed, with the value $f = 1.1491$ giving good overall agreement for the EOS of the HS fluid [13]. Therefore, taking into account Eq. (9.37) and the interrelation given by Eq. (9.35) it is possible to express Enskog formulae, Eqs. (9.24)–(9.26), explicitly in terms of X_1.

Enskog theory is based on the molecular chaos approximation, and therefore is not applicable over large density range. It is an approximate solution

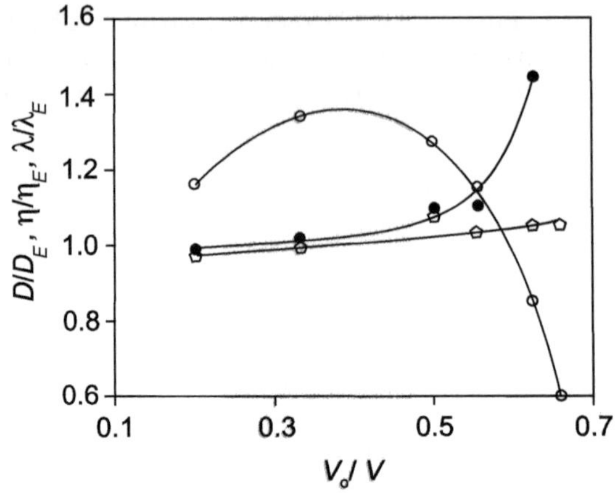

Fig. 9.1. Density dependence of the corrections to the Enskog transport coefficients computed by Alder et al. [14]. (○) D/D_{E}; (●) η/η_{E}; (⬠) $\lambda/\lambda_{\mathrm{E}}$

of Newton's equation of motion, extending the kinetic theory of gases to finite packing fraction and is based on the same random collision assumption. Molecules approach each other from random directions with velocities randomly chosen from the Maxwell–Boltzmann distribution for the appropriate temperature. The errors introduced neglecting the effects of correlated motion between core collisions have been studied in computer simulations, which made possible to appropriatly correcting to Eqs. (9.24)–(9.26). Figure 9.1 shows the corrections determined by Alder et al. [14] and plotted as the ratio between the essentially exact MD transport coefficients of HS and those predicted by Enskog formulae as function of V_0/V.

At high densities, the principal correlation effect is back-scattering, whereby a sphere closely surrounded by a shell of neighbour spheres becomes increasingly locked in and reverses its velocity on collision, which decreases diffusion and increases viscosity till nearly 40%. At intermediate densities, there is a distinct correlation effect associated with an unexpected persistence of velocities, which exceeds Enskog diffusivity by up to 30%; corrections to η_{E} are less than 10%. This was found to be due to a hydrodynamically derived vortex flow that develops around each particle as a result of its velocity at an early time. The deviations from the Enskog theory for the thermal conductivity are barely perceptible within the few per cent accuracy of the data [14].

9.3.2 Modified Enskog Theory

The modified Enskog theory (MET) adapts Enskog expressions, Eqs. (9.24)–(9.26), to represent the dynamic behaviour of real fluids. Its basic assumptions are that the transport mechanisms in real fluids do not essentially differ from those in hard spheres, and that the main discrepancy between them lies in the temperature dependence of the collisions frequency. Hence, similar expressions for transport properties should apply, if appropriate $g(\sigma)$ and b are previously determined from EOS data of the real fluid [15]. This is accomplished by replacing the external pressure, P, in Eq. (9.28) by the thermal pressure,[1] $T(\partial P/\partial T)_\rho$, as follows:

$$\frac{1}{\rho k_{\mathrm{B}}}\left(\frac{\partial P}{\partial T}\right)_\rho = 1 + b\rho g(\sigma) \tag{9.38}$$

Adopting the virial expansion for the real fluid, $\frac{P}{\rho k_{\mathrm{B}} T} = 1 + B\rho + C\rho^2 + \cdots$, Eq. (9.38) gives rise to

$$b\rho g(\sigma) = \left(B + T\frac{\mathrm{d}B}{\mathrm{d}T}\right)\rho + \left(C + T\frac{\mathrm{d}C}{\mathrm{d}T}\right)\rho^2 + \cdots \tag{9.39}$$

Obviously, the values of b cannot be assigned arbitrarily. In the dilute gas limit, since $\lim_{\rho\to 0} g(\sigma) = 1$, it follows from Eq. (9.39) that

$$b = B + T\left(\frac{\mathrm{d}B}{\mathrm{d}T}\right) \tag{9.40}$$

Concluding, EOS data of the real fluid are used in conjunction with Eqs. (9.38) and (9.40) to compute $g(\sigma)$ and b, respectively. The MET can be further modified to allow for real-fluid characteristics by making two additional adjustments. First, experimental dilute gas coefficients (D_o, η_o, and λ_o) may be used in Eqs. (9.24)–(9.26) in spite of the estimated ones by Eqs. (9.21)–(9.23). Second, several modifications are proposed as follows if molecules have internal structure, in particular for thermal conductivity.

Mason and Monchick [16], based on the Wang–Chang–Uhlenbeck–De Boer theory for a dilute polyatomic gas, showed that λ_o can be split into a translational part and an internal part, i.e. $\lambda_o = \lambda_o' + \lambda_o''$, where λ_o' can be calculated from Eq. (9.23). Attending to the exact splitting of the specific heat per molecule at constant volume into translational and internal contributions, $C_V = C_V' + C_V''$, they wrote

$$\lambda_o = \frac{25}{32\sigma^2}\left(\frac{k_{\mathrm{B}}T}{\pi m}\right)^{1/2}\frac{C_V'}{\Omega^{(2,2)*}} + \rho_o D_o C_V'' \tag{9.41}$$

which may be recast as

$$\lambda_0 = \frac{5}{2}\eta_o\frac{C_V'}{m} + \frac{6}{5}\eta_o\frac{C_V''}{m}\frac{\Omega^{(2,2)*}}{\Omega^{(1,1)*}} \tag{9.42}$$

according to the relationships between D_o, η_o, and λ_o, Eqs. (9.21)–(9.23).

[1] The pressure experienced by a molecule embodies two components: the external pressure due to the containing walls of the vessel and the internal pressure that represents the force of cohesion of the molecules existing in real gases. Thermal pressure sums both contributions.

Hanley et al. [15] assumed equivalent λ splitting for dense polyatomic gases and liquids,

$$\lambda = \lambda' + \lambda'' \tag{9.43}$$

where they identified

$$\lambda'' = \rho D C''_V \tag{9.44}$$

by analogy to Eq. (9.41). Taking into account Eqs. (9.24) and (9.26) to calculate $\rho D/\rho_o D_o$ and λ'/λ'_o, respectively, it is possible to establish

$$\lambda'' = \rho D C''_V = \frac{\rho_o D_o}{g(\sigma)} C''_V \equiv \frac{\lambda''_o}{g(\sigma)} \tag{9.45}$$

$$\frac{\lambda'}{\lambda'_o} = \frac{1}{g(\sigma)} + 1.2b\rho + 0.755 g(\sigma)(b\rho)^2 \tag{9.46}$$

The thermal conductivity of the MET is finally established combining Eqs. (9.43), (9.45), and (9.46):

$$\lambda = \lambda'_o \left(\frac{1}{g(\sigma)} + 1.2b\rho + 0.755 g(\sigma)(b\rho)^2 \right) + \frac{\lambda''_o}{g(\sigma)} \tag{9.47}$$

Finally, the shear viscosity remains unchanged, being estimated by Eq. (9.25).

Hanley et al. [15] carried out previous MET calculations for the viscosity and thermal conductivity of argon as a function of density for three isotherms (173, 183, and 348 K), and made a comparison with experimental data. In Figs. 9.2 and 9.3, the results obtained following this procedure are plotted in terms of excess functions over ideal-gas values: $\triangle\eta(\rho,T) = \eta(\rho,T) - \eta_0(T)$ and $\triangle\lambda(\rho,T) = \lambda(\rho,T) - \lambda_0(T)$, respectively. The shading indicates the regions where the densities are greater than the critical density, ρ_c. It is possible to conclude that the MET: (i) can represent data to within 10% for densities up to about 0.3 g/cm^3 ($\approx 0.6\rho_c$); (ii) predicts the same temperature dependence for both $\triangle\eta$ and $\triangle\lambda$ at fixed density; according to the theory both

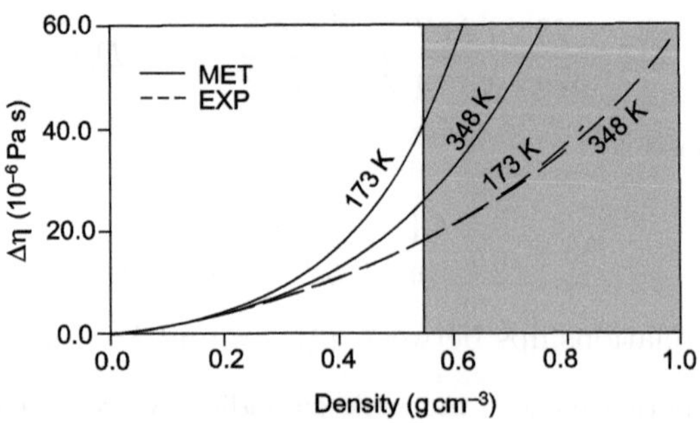

Fig. 9.2. Comparison of the MET and experimental excess viscosity for argon plotted against density, following the Hanley et al. approach [15]. The *shaded region* identifies densities higher than the critical density

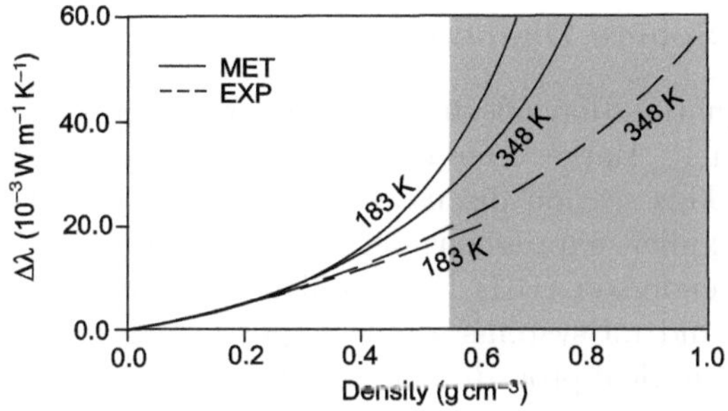

Fig. 9.3. Comparison of the MET and experimental excess thermal conductivity for argon plotted against density, following the Hanley et al. approach [15]. The *shaded region* identifies densities higher than the critical density

$(\partial\triangle\eta/\partial T)_\rho$ and $(\partial\triangle\lambda/\partial T)_\rho$ are negative, while experiment indicates dissimilar dependence [15].

One advantage of the MET is that it can be used to predict transport properties of gases in the intermediate range of densities, even when there are no experimental data available. Nonetheless there are important drawbacks, as that cited above, which implies it should be applied with caution.

9.3.3 Effective Hard Sphere Diameter Method

The effective hard sphere diameter (EHSD) method has been widely used to calculate both equilibrium and transport properties. Kirkwood and Monroe-Boggs [17] have noticed that the form of the radial distribution function is primarily determined by repulsive forces, while attractive interactions play a secondary role. Such results promoted significantly the development of the perturbation approaches for dense fluids, which usually combine hard spheres as an appealing and tractable first approximation, for the major excluded volume and packing effects, with an effective diameter dependent on temperature and possibly on density, to account for the softness of the repulsive potential. Hence, the essential assumption of the EHSD method is that the properties of a fluid can be calculated by the corresponding HS model, if the molecular diameter is replaced by an EHSD, σ_{eff}.

This method has already been used in Chap. 4 to calculate some properties of the WCA reference system. In this chapter, the method is applied to transport properties. Thus, for a general transport coefficient Y, it is expressed as

$$Y(T^*, \rho^*) = Y_{\text{HS}}\left[\rho^*, \sigma_{\text{eff}}\left(T^*, \rho^*\right)\right] \tag{9.48}$$

where $T^* = T/(\varepsilon/k_{\text{B}})$ and ρ^* represent reduced temperature and density, respectively.

Effective Hard Sphere Diameter Criteria

Various EHSD equations have been proposed in the literature according to several criteria, namely, Barker–Henderson, Weeks–Chandler–Andersen, Lado, and Boltzmann. This section discusses them briefly as well as compiles the most important explicit expressions derived from those criteria.

Barker and Henderson (BH) [18] published the first successful analysis, adopting the HS and LJ systems as the unperturbed and perturbing potentials. Their mathematical procedure gave rise to the so-called BH criterion for the EHSD and involves integrating over the repulsive portion of the potential:

$$\sigma_{\mathrm{BH}} = \int_0^{r_o} \left[1 - \exp\left(-\beta\phi_{LJ}\right)\right] \mathrm{d}r \tag{9.49}$$

Here, r is the radial position, r_0 is the value of r for which the Lennard–Jones potential vanishes, i.e. $\phi_{\mathrm{LJ}} = 0$ and $\beta = 1/(k_{\mathrm{B}}T)$. According to this definition, the EHSD is only temperature-dependent. Several analytical expressions appeared from Eq. (9.49), based on numerical calculation results—see Table 9.1.

Another milestone in the perturbation theory is the work of Weeks, Chandler and Andersen [4, 19], who splitted up the LJ potential into a reference part containing all repulsive forces, and a perturbing part containing all forces of attraction (see Eq. (9.7)). This theory asserts that the response given by the Fourier transform of the reference correlation function can be approximated by the response of an HS system with an as yet undefined diameter. Imposing the long-wavelength responses to be exactly the same, a new physically reasonable criterion for the EHSD (henceforth denoted by WCA EHSD) has been arisen. The well-known WCA perturbation theory yields an EHSD-dependent on both temperature and density:

$$\int_0^{\infty} y_{\mathrm{HS}}\left(r; \sigma_{\mathrm{WCA}}(T^*, \rho^*)\right) \left[\exp\left(-\beta\phi_{\mathrm{WCA}}\right) - \exp\left(-\beta\phi_{\mathrm{HS}}\right)\right] r^2 \mathrm{d}r = 0 \tag{9.50}$$

where y_{HS} is the background correlation function of the HS fluid with diameter $\sigma_{\mathrm{WCA}}(T^*, \rho^*)$, and $\rho^* = \rho\sigma_{\mathrm{LJ}}^3$ is the LJ reduced density. In Table 9.2, various analytical expressions for $\sigma_{\mathrm{WCA}}(T^*, \rho^*)$ proposed by some authors are listed.

The WCA theory converges faster than the BH one, although its success is based upon the fact that errors from some rather poor approximations finally almost cancel each other in the end result. Verlet and Weis [20] have developed corrections and analytical expressions for the thermodynamic functions with better results at high densities, which render possible a new formulation to estimate the WCA EHSD. However, their expression is implicit and a trial and error method is needed for the calculations.

Lado [21] proposed another modification to improve consistency with thermodynamic criteria, since the pressure calculated from the virial equation does not confirm the value obtained from the WCA free energy. To overcome this problem, he recommended an alternative temperature- and density-dependent criterion for the EHSD, identified here by LWCA:

$$\int_0^\infty [\exp(-\beta\phi_{\mathrm{WCA}}) - \exp(-\beta\phi_{\mathrm{HS}})] \frac{\partial y_{\mathrm{HS}}}{\partial\sigma} r^2 \,\mathrm{d}r = 0 \qquad (9.51)$$

In particular, Ben-Amotz and Herschbach [22] presented an expression for $\sigma_{\mathrm{LWCA}}(T^*, \rho^*)$ following Eq. (9.51) (see Eq. (9.67) in Table 9.2).

Concerning the Boltzmann criterion, it approximates the EHSD by the distance of closest approach, σ_{B}, of a colliding pair of molecules with the average relative kinetic energy, subjected to soft repulsive interactions. For a central force potential $\phi(\sigma_{\mathrm{B}})$, this corresponds to

$$\phi(\sigma_{\mathrm{B}}) = fkT \qquad (9.52)$$

where f is a constant determined by a suitable thermal average of the kinetic energy. This appealing criterion has been used by several authors working mainly on transport properties [22, 23, 24], and although not rigorous it gives good results for diffusion coefficients over wide ranges of density and temperature [24, 25]. The resultant EHSD is only temperature-dependent (Table 9.1).

Effective Hard Sphere Diameter Expressions

Since the 1980s many authors have proposed explicit equations for the EHSD according to the BH, B, WCA, and LWCA criteria. In Tables 9.1 and 9.2, temperature- and both temperature- and density-dependent expressions are listed according to those criteria. Also given are three additional models: (i) an empirical equation with three parameters published by Hammonds and Heyes [26]; (ii) the previous expression with parameters regressed by Sun et al. [27] using *PVT* data of some real fluids by using the Ross variational approach; and (iii) the work of Liu et al. [25], where the parameters of a general expression by Ben-Amotz and Herschbach [22] were optimized using a self-diffusion HS model to reproduce the MD simulation results available for the WCA LJ fluid (see Sect. 9.4.2, and Eqs. (9.57) and (9.175)).

To facilitate and legitimate the selection between the 15 models compiled, Silva et al. [28] carried out a systematic comparison performed from the point of view of transport phenomena. Figure 9.4 presents the temperature-dependence of the EHSD given by Eqs. (9.53)–(9.62). In Fig. 9.5 the EHSD versus reduced density at fixed temperature is shown for Eqs. (9.63)–(9.67). Furthermore, the temperature dependence at constant density is shown in Fig. 9.6. It is evident from those figures that temperature has a large impact on effective diameter, in spite of density which is almost absent until reduced densities around 0.8.

Following Silva et al. [28], the temperature-dependent B EHSD, and the temperature- and density-dependent LWCA EHSD give the best predictions of the LJ self-diffusivities, when the HS model proposed by Erpenbeck and Wood [29] is utilized (the model of Erpenbeck and Wood is presented in Sect. 9.4.1, Eq. (9.122)). An interesting and valuable observation of Silva et al. [28] is the consistency found between the B EHSD derived by Ben-Amotz and

Table 9.1. Explicit expressions for temperature-dependent EHSD, reduced according to $\sigma^*_{\text{eff}}(T^*) = \sigma_{\text{eff}}(T^*)/\sigma_{\text{LJ}}$

Criterion	Ref.	Model	Eq.
Empirical correlation	[26]	$\sigma^*_{\text{HH}}(T^*) = 1.0217\left(1 - \frac{0.0178}{T^{*1.256}}\right)\frac{1}{T^{*\frac{1}{12}}}$	(9.53)
Ross variational perturbation theory	[27]	$\sigma^*_{\text{SBT}}(T^*) = 1.2566\left(1 - \frac{0.2075}{T^{*0.177}}\right)\frac{1}{T^{*\frac{1}{12}}}$	(9.54)
Boltzmann	[24]	$\sigma^*_{\text{BS}}(T^*) = 2^{\frac{1}{6}}\left[1 + (2T^*)^{\frac{1}{2}}\right]^{-\frac{1}{6}}$	(9.55)
	[22]	$\sigma^*_{\text{BAH}}(T^*) = 1.1532\left[1 + (1.8975T^*)^{\frac{1}{2}}\right]^{-\frac{1}{6}}$	(9.56)
	[25]	$\sigma^*_{\text{BLSM}}(T^*) = 2^{\frac{1}{6}}\left[1 + (1.3229T^*)^{\frac{1}{2}}\right]^{-\frac{1}{6}}$	(9.57)
Barker–Henderson	[20]	$\sigma^*_{\text{BHVW}}(T^*) = \dfrac{0.3837 + 1.068/T^*}{0.4293 + 1/T^*}$	(9.58)
	[30]	$\sigma^*_{\text{BHMM}}(T^*) = 2^{\frac{1}{6}}\left\{1 - \exp\left[\sum_{i=1}^{5} a_i T^{*0.093(i-1)}\right]\right\}$ where $a_1 = -13.95687$, $a_2 = 24.13697$, $a_3 = -18.01784$, $a_4 = 6.355364$, $a_5 = -0.869877$	(9.59)
	[22]	$\sigma^*_{\text{BHAH}}(T^*) = 1.1154\left[1 + (0.5685T^*)^{\frac{1}{2}}\right]^{-\frac{1}{6}}$	(9.60)
	[31]	$\sigma^*_{\text{BHSA}}(T^*) = 2^{\frac{1}{6}}\left\{1 + \left[1 + \frac{T^* + a_2 T^{*2} + a_3 T^{*4}}{a_1}\right]^{\frac{1}{2}}\right\}^{-\frac{1}{6}}$ where $a_1 = 1.1287$, $a_2 = -0.05536$, $a_3 = 0.0007278$	(9.61)
	[32]	$\sigma^*_{\text{BHN}}(T^*) = \sigma_t \exp\left(b_1 \text{Ln}\, t + b_2 \text{Ln}^2 t + b_3 \text{Ln}^3 t\right)$ where $t = T^*/T_t$, $\sigma_t = 1.02851$, $T_t = 0.69$, $b_1 = -0.03208$, $b_2 = -0.00520$, $b_3 = 0.00015$	(9.62)

Herschbach [22] from equilibrium data and that obtained from self-diffusivity data [25] (curves 9 and 6 of Fig. 9.4, respectively), which explains the good results it furnishes. It should be emphasized that the merits of the B EHSD owe principally to its convincing theoretical basis, once its fundamentals rest on the transport phenomena domain.

Table 9.2. Explicit expressions for temperature- and density-dependent EHSD, reduced according to $\sigma^*_{\text{eff}}(T^*,\rho^*) = \sigma_{\text{eff}}(T^*,\rho^*)/\sigma_{\text{LJ}}$

Criterion	Ref.	Model	Eq.
Weeks–Chandler–Andersen	[30]	$\sigma^*_{\text{WCAMM}}(T^*,\rho^*) = \sigma^*_{\text{BHMM}}(T^*)\left[1 + S(\varphi)\, D(T^*)\right]$ where $\sigma^*_{\text{BHMM}}(T^*)$ is given by Eq. (9.59), and $S(\varphi) = \dfrac{1 - 5.224\varphi + 5.72\varphi^2 - 2.623\varphi^3 + 1.693\varphi^4}{(1-\varphi)^3}$ $\text{Ln}\, D(T^*) = a_6 + a_7 \text{Ln}\,(1/T^* + a_8)$, and $a_6 = -6.411461$, $a_7 = -0.551805$, $a_8 = 0.05$	(9.63)
	[22]	$\sigma^*_{\text{WCAAH}}(T^*,\rho^*) = 1.1137\left[1 + \left(\frac{T^*}{T^*_0}\right)^{\frac{1}{2}}\right]^{-\frac{1}{6}}$ $(T^*_0)^{-\frac{1}{2}} = 0.72157 + 0.04561\rho^* - 0.07468{\rho^*}^2 + 0.12344{\rho^*}^3$	(9.64)
	[31]	$\sigma^*_{\text{WCASA}}(T^*,\rho^*) = 2^{\frac{1}{6}}\left\{1 + \left[\frac{T^* + a_2 {T^*}^2 + a_3 {T^*}^4}{a_1\left(1 + a_4\rho^* + a_5{\rho^*}^2 + a_6{\rho^*}^3\right)}\right]^{\frac{1}{2}}\right\}^{-\frac{1}{6}}$ where $a_1 = 1.5001$, $a_2 = -0.03367$, $a_3 = 0.0003935$, $a_4 = -0.09835$, $a_5 = 0.04937$, $a_6 = -0.1415$	(9.65)
	[32]	$\sigma^*_{\text{WCAN}}(T^*,\rho^*) = \sigma^*_{\text{BHN}}(T^*)\left[d_1 + d_2 t^{d_7} + \frac{d_3 + d_4 t^{d_8}\rho^0}{1 + d_5\rho^0 + d_6\rho^{02}}\right]$ where $\sigma_{\text{BHN}}(T^*)$ is given by Eq. (9.62), and $d_1 = 0.975101$, $d_2 = 0.040766$, $d_3 = -0.014391$, $d_4 = -0.002316$, $d_5 = 0.075294$, $d_6 = -0.313557$, $d_7 = 0.018699$, $d_8 = 0.115357$, $\rho^0 = \rho^*\sigma_t^3$	(9.66)
Lado WCA	[22]	$\sigma^*_{\text{LWCAAH}}(T^*,\rho^*) = 1.1152\left[1 + \left(\frac{T^*}{T^*_0}\right)^{\frac{1}{2}}\right]^{-\frac{1}{6}}$ $(T^*_0)^{-\frac{1}{2}} = 0.73454 + 0.1025\rho^* - 0.1296{\rho^*}^2 + 0.15976{\rho^*}^3$	(9.67)

9.3.4 Free-Volume Theories for Transport Properties

An alternative procedure to treat transport properties of dense fluids is that of free volume, where the related coefficients depend on the relative expansion from an intrinsic molar volume V_i. The importance of the free-volume theories in this context owes to the following facts: (i) their equations are simple and

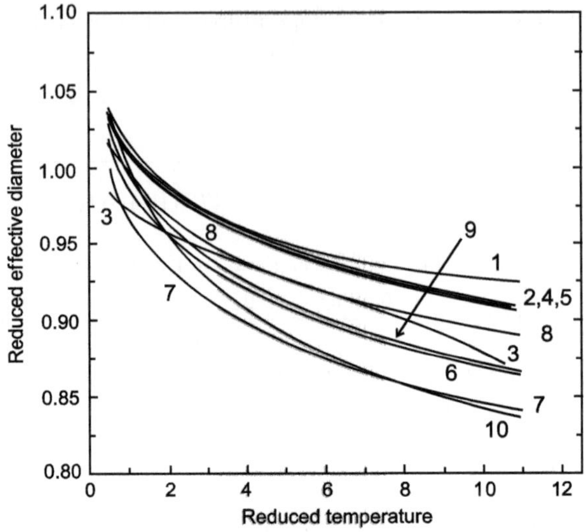

Fig. 9.4. Comparison between 10 temperature-dependent EHSD expressions: (1) BHVW, Eq. (9.58) [20]; (2) BHMM, Eq. (9.59) [30]; (3) BHSA, Eq. (9.61) [31]; (4) BHN, Eq. (9.62) [32]; (5) BHAH, Eq. (9.60) [22]; (6) BLSM, Eq. (9.57) [25]; (7) BS, Eq. (9.55) [24]; (8) SBT, Eq. (9.54) [27]; (9) BAH, Eq. (9.56) [22]; (10) HH, Eq. (9.53) [26]

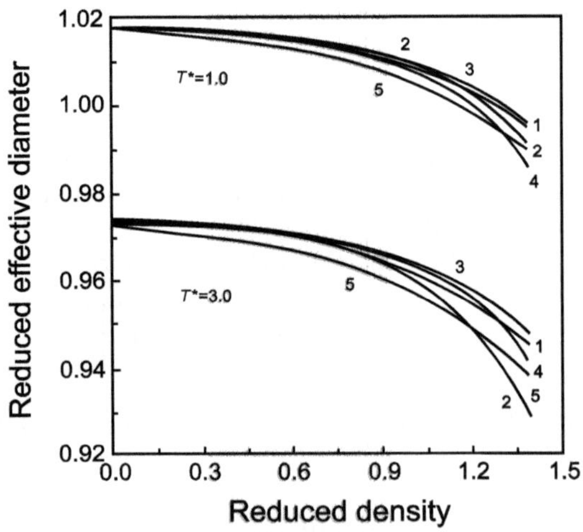

Fig. 9.5. Comparison between five temperature- and density-dependent EHSD expressions. Illustration of the density dependence: (1) WCAAH, Eq. (9.64) [22]; (2) WCAMM, Eq. (9.63) [30]; (3) WCASA, Eq. (9.65) [31]; (4) WCAN, Eq. (9.66) [32]; (5) LWCAAH, Eq. (9.67) [22]

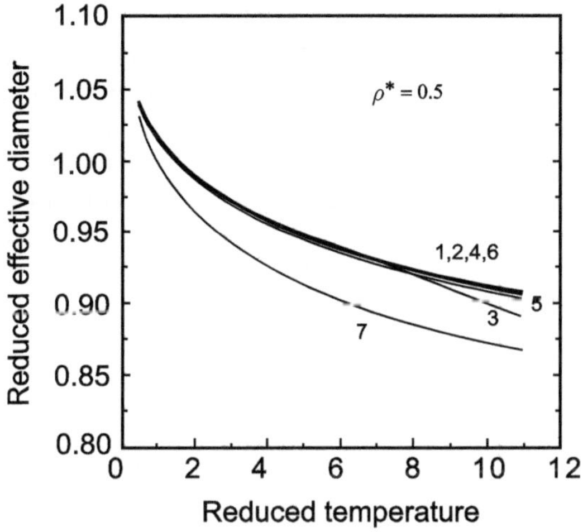

Fig. 9.6. Comparison between five temperature- and density-dependent EHSD expressions. Illustration of the temperature dependence (also shown are B and BH EHSDs): (1–5) same as in Fig. 9.5; (6) BHAH, Eq. (9.60) [22]; (7) BAH, Eq. (9.56) [22]

involve reduced number of parameters (usually between two and four); (ii) the parameters have, almost always, a precise physical meaning; (iii) they can be used over wide ranges of temperature and pressure; (iv) the main theories are based on the statistical mechanics, therefore giving them a solid theoretical background; and (v) they may be easily extended to multicomponent systems. There are many free-volume theories in literature concerning transport properties and choosing among them is sometimes puzzling, especially when they refer to the same conditions of applicability.

In this section, we present a brief description of the main theories, and discuss the concept of free volume with interest in this field—geometric free volume. It is usually identified with the thermodynamic free volume, although they are conceptually different quantities.

Relevant Free-Volume Theories

The free-volume models can be divided into two main groups. The first one correlates transport coefficients just with the molar free volume, $V_\mathrm{f} = V - V_i$:

$$D, \eta^{-1}, \lambda^{-1} = f(V_\mathrm{f}) \tag{9.68}$$

Those relations are only valid for fluids without attractive molecular interactions, e.g. HS and hard convex body fluids. The second group comprises the so-called hybrid models, which combine the concepts of free volume and activation energy, E_a, so embodying both attractive and repulsive interactions, and can be written as:

$$D, \eta^{-1}, \lambda^{-1} = f(V_f) \exp\left(-\frac{E_\mathrm{a}}{k_\mathrm{B}T}\right) \tag{9.69}$$

The major free-volume equations are the following: (i) Dymond–Hildebrand–Batschinski (DHB) [33, 34, 35], frequently adopted to describe non-polar substances with negligible attractive forces at moderate densities; (ii) Doolittle–Cohen–Turnbull (DCT) [36, 37], which accounts for the glass transition behaviour of fluids; (iii) Turnbull and Cohen (TC) [38], an improved form of their earlier expression; and (iv) Macedo and Litovitz (ML) [39], a milestone in the field of polymers. The dependence of the self-diffusivity, fluidity, and thermal resistivity ($Y = D, \eta^{-1}, \lambda^{-1}$) upon free volume predicted by those models may be compiled as follows:

$$\text{DHB}: \quad Y \propto V_f \tag{9.70}$$

$$\text{DCT}: \quad Y \propto \exp\left(-\frac{C_1}{V_f}\right) \tag{9.71}$$

$$\text{TC}: \quad Y \propto (C_2 + V_f)\ \exp\left(-\frac{C_3}{V_f}\right) \tag{9.72}$$

$$\text{ML}: \quad Y \propto \exp\left(-\frac{C_4}{V_f} - \frac{C_5}{k_\text{B}T}\right) \tag{9.73}$$

where C_1, C_2, C_3, C_4, and C_5 are constants of the models.

van Loef [40] presented a comprehensive investigation of the DHB models for compressed atomic and molecular liquids composed of linear, quasi-spherical and dipolar molecules, and some hydrocarbons. He studied the volume dependence of the transport coefficients of liquids along an isotherm over wide ranges of pressure and temperature, since in dense fluids they depend on density much more than on temperature. Such fact gives good reason to interpret transport properties in terms of a modified HS scheme based on free volume. The experimental data analysed appear to depend linearly on V in a substantial fraction of the liquid range, and it was evidenced that these dependencies, extrapolated linearly to $D = \eta^{-1} = 0$, intercept at the same intrinsic volume V_i:

$$\frac{1}{\eta} = B\left(\frac{V}{V_i} - 1\right) \quad \text{and} \quad D = D'\left(\frac{V}{V_i} - 1\right) \tag{9.74}$$

where B and D' are the specific proportionality constants of the models. This is consistent with the fact that these kinetic coefficients satisfy the classical Stokes–Einstein relation

$$D = \frac{k_\text{B}T}{3\pi\eta\sigma} \tag{9.75}$$

Such finding is illustrated in Fig. 9.7, where a three-dimensional representation of the fluidity and self-diffusivity of compressed liquid CO_2 against molar volume and temperature are given [40].

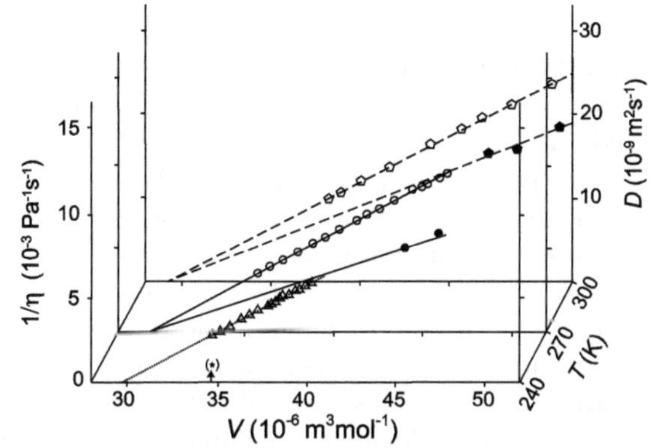

Fig. 9.7. Fluidity (unfilled symbols) and self-diffusion coefficient (filled symbols) of compressed liquid CO_2 as function of molar volume and temperature. (*) Molar volume of liquid CO_2 at 240 K along melting curve

For the thermal resistivity an analogous expression was analysed:

$$\frac{1}{\lambda} = C\left(\frac{V}{V_i'} - 1\right) \tag{9.76}$$

The fluidity and the thermal resistivity constants B and C for classical atomic and non-associated molecular liquids appeared to be interrelated through a square-root-mass law: $M^{-1/2}$ dependence for B, and $M^{1/2}$ for C. In contrast to B and C, a single diffusion constant D' that could describe liquid self-diffusivities over wide ranges of temperature and density apparently does not exist. Instead, a diffusion parameter D_T' is introduced to represent the volume dependence of D in liquids and dense gases along an isotherm. Comparing this parameter for several atomic and non-associated molecular liquids at a given temperature it appears that D_T' closely satisfies an $M^{-1/2}$ dependence. The transport properties of associated molecular liquids strongly depart from the aforementioned relationships, though from a comparison between the behaviour of associated and non-associated molecular liquids, qualitative information on association effects is obtained [40].

The works of Doolittle, Cohen, and Turnbull account for the diffusive transport in liquids and glasses and for the liquid–glass transition. Their free-volume model was developed for a dense HS fluid, in which the free volume is well defined, and then applied to interpret transport and glass transition behaviour of molecular systems. According to Cohen and Turnbull [37], the molecular transport in the HS liquid occurs by statistical redistribution of free volume, specifically by the formation of adjacent holes of sufficient size into which the diffusive sphere can jump. Starting from this assumption, they were able to demonstrate the equation found empirically by Doolittle in 1951, Eq. (9.71). An important deduction from the model is a crude correction for the effects of backscattering, neglected in the Enskog theory, since the diffusive motion is cancelled in those parts of the system which are locally most dense, i.e. where local free volume is smaller than a critical value.

Owing to the relative success of both free-volume and Eyring reaction-rate theories, Macedo and Litovitz [39] combined their desirable aspects into a hybrid equation for the viscosity of liquids. This expression, Eq. (9.73), was demonstrated to give useful description of the temperature and pressure dependence of viscosity of many molecular, hydrogen-bonded as well as"network" liquids. Furthermore, one of the most successful applications of Eq. (9.73) is the prediction of the solvent diffusion in polymers [41, 42, 43]. Chung [44] derived it from statistical-mechanical arguments, based upon the assumption that diffusion in liquids is controlled by two simultaneous events: (i) the existence of an adjacent local free volume of certain size into which the molecule can jump, and (ii) the acquisition of sufficient energy by the molecule in order to escape from the force field of its neighbours.

Liu et al. [45] have shown that those relevant models are particular cases of a generalized equation derived in their work. The authors clarified the validity and range of applicability of those equations, since they arise by introducing approximations duly established in the fundamental model devised by them. The theoretical basis for the DHB expression, Eq. (9.70), has also been given.

Geometrical Free Volume versus Thermodynamic Free volume

The free volume concept used in transport properties should be read as geometric free volume and it is conceptually different from thermodynamic free volume, this being strictly related to thermodynamic properties based on the generalized van der Waals theory. For van der Waals fluid, the two free volumes are identical.

Geometric free volume, $v_{\mathrm{gf}} = V_{\mathrm{gf}}/N$, is defined as the volume over which the centre of a sphere can move, when other spheres are frozen in positions [46, 47]. A related concept is cavity, $v_{\mathrm{c}} = V_{\mathrm{c}}/N$, sometimes called vacancy. Cavity is defined as a connected region, in a particular configuration, where there is space available for the addition of another sphere [47, 48]. The relationship between geometric free volume and cavity is [48]

$$V_{\mathrm{gf}} = C_0 V_{\mathrm{c}} \tag{9.77}$$

where C_0 is a constant, which is equal to 5 for liquid region. A general relationship between cavity and thermodynamic free volume, V_{f}, has been given in [47, 48]:

$$\mathrm{Ln}\left(\frac{V_{\mathrm{c}}}{V}\right) = \frac{S^{\mathrm{ex}}}{Nk_{\mathrm{B}}} + F = \mathrm{Ln}\left(\frac{V_{\mathrm{f}}}{V}\right) + F \tag{9.78}$$

where F is a function generally dependent on density for supercooled liquid and solid, while for stable liquid it can be taken as constant ($F = -2.6$) [48].

Taking into account these equations and following the work of Speedy and Reiss [47], we can assume a linear relation between free volume, geometric free volume, and cavity for the HS system:

$$\frac{V_\mathrm{f}}{V} = C_1 \frac{V_\mathrm{c}}{V} = C_2 \frac{V_\mathrm{gf}}{V} \tag{9.79}$$

where C_1 and C_2 are constants. Unfortunately, there is no sufficient simulation data to test Eq. (9.79). Bowles and Speedy [48] only reported some data points for the region beyond stable fluid. For liquid, we take Eq. (9.79) as an approximation. Obviously, it does not satisfy the dilute gas limit: when density approaches zero, all kind of molecular free volumes should reduce to V. Namely, geometric and thermodynamic free volume should be the same as density vanishes.

Although it is difficult to derive an expression for the geometric free volume interested in transport properties, the empirical equations given in Eqs. (9.78) and (9.79) suggest that the thermodynamic free volume (or entropy) does provide very useful information. Therefore, in the following some V_f models are presented and discussed.

Analysis of Free-Volume Expressions

It is crucial to have accurate expressions to estimate the free volume of the HS system. It is well known that the simplest van der Waals model ($V_\mathrm{f} = V - b$, where b is the co-volume) is only valid within limited density range, although its simplicity results in its frequency of utilization. In this section, some representative equations are briefly analysed and compared according to their density, molecular shape, and temperature dependencies. For a detailed discussion see [45].

In Table 9.3, six representative models for the repulsive compressibility factor and free-volume fraction are listed: (i) three earlier empirical models, Eqs. (9.80)–(9.82): vdW, Eyr-Hfd, and Gugg; (ii) three models obtained from Monte Carlo (MC) simulation results for the HS compressibility, Eqs. (9.83)–(9.85): CS, H-Wood, and M-Nia. We also note that Ishiwaka et al. [55] suggested empirically the same equation as Eq. (9.85), but using constants 2 and 2, instead of 1.88 and 2.48, respectively. In Chap. 3, some other models can be found.

As is shown in Chap. 3, for the EOSs included in Table 9.3, we know that the vdW equation (9.80) only gives acceptable results at low densities; Eq. (9.81) is completely unacceptable; Gugg and M-Nia models, Eqs. (9.82) and (9.85), offer good representations at reduced densities below approximately 0.7; finally, both Eqs. (9.83) and (9.84) (CS and H-Wood) perform very well, exhibiting excellent agreement with computer simulation data.

Figure 9.8 presents the density dependence of the free-volume fraction of the HS fluid equations, Eqs. (9.80)–(9.85). This figure shows that Gugg, CS, H-Wood, and M-Nia models (Eqs. (9.82), (9.83), (9.84) and (9.85), respectively) give very similar results for V_f/V in the whole density range. The vdW equation also gives acceptable results when $\rho^* < 0.2$. The Eyr-Hfd model only presents quantitative correct results at very high densities. This model was

Table 9.3. Expressions for the repulsive compressibility factor (Z^{r}) and free-volume fraction (V_{f}/V) of hard spherical molecules

van der Waals (vdW)

$$Z^{\mathrm{r}} = \frac{1}{1-4\varphi} \qquad \text{and} \qquad V_{\mathrm{f}}/V = 1-4\varphi \tag{9.80}$$

Eyring–Hirschfelder (Eyr-Hfd) [2, 49]

$$Z^{\mathrm{r}} = \frac{1}{1-\left(\frac{6}{\pi}\varphi\right)^{1/3}} \qquad \text{and} \qquad V_{\mathrm{f}}/V = 8\left[1-\left(\frac{6}{\pi}\varphi\right)^{\frac{1}{3}}\right]^3 \tag{9.81}$$

Guggenheim (Gugg) [50]

$$Z^{\mathrm{r}} = \frac{1}{(1-\varphi)^4} \qquad \text{and} \qquad V_{\mathrm{f}}/V = (1-\varphi)\exp\left[\frac{3\varphi(\varphi-2)}{2(1-\varphi)^2} - \frac{\varphi^3}{3(1-\varphi)^3}\right] \tag{9.82}$$

Carnahan and Starling (CS) [12]

$$Z^{\mathrm{r}} = \frac{1+\varphi+\varphi^2-\varphi^3}{(1-\varphi)^3} \qquad \text{and} \qquad V_{\mathrm{f}}/V = \exp\left[\frac{\varphi(3\varphi-4)}{(1-\varphi)^2}\right] \tag{9.83}$$

Heyes and Woodcock (H-Wood) [13]

$$Z^{\mathrm{r}} = 1+\frac{4\varphi}{(1-1.175\varphi)^2} \qquad \text{and} \qquad V_{\mathrm{f}}/V = \exp\left[-\frac{4\varphi}{1-1.175\varphi}\right] \tag{9.84}$$

Moshen-Nia et al. (M-Nia) [51]

$$Z^{\mathrm{r}} = \frac{1+2.48\varphi}{1-1.88\varphi} \qquad \text{and} \qquad V_{\mathrm{f}}/V = (1-1.88\varphi)^{1+\frac{2.48}{1.88}} \tag{9.85}$$

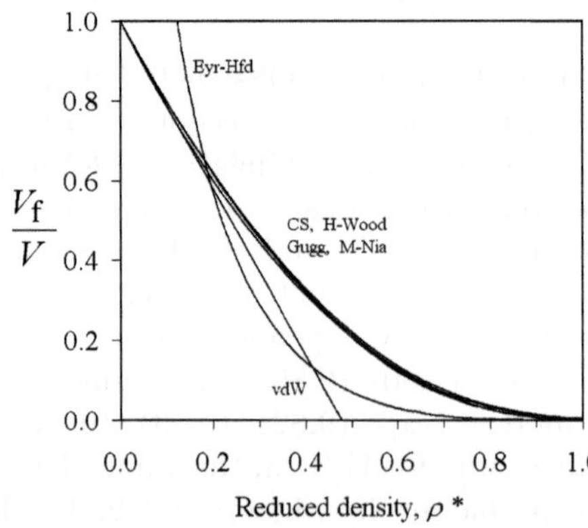

Fig. 9.8. Free-volume fraction versus reduced density, calculated by Eqs. (9.80)–(9.85) (see Table 9.3)

proposed for cubic regular liquids, and does not satisfy the necessary boundary condition: $\rho^* \to 0 \Rightarrow V_\mathrm{f}/V \to 1$. In fact, Eyr-Hfd achieves $V_\mathrm{f}/V \to 8$. In summary, Eqs. (9.82)–(9.85) can be employed as HS free-volume models.

To take account of the non-spherical character of real molecules, a shape parameter, α, is commonly introduced. Table 9.4 lists three models for illustration, Eqs. (9.86)–(9.88): Boub, Chien, and Cott [52, 53, 54]. Equation (9.86) has been derived from the MC simulation results for the Hard Convex Body fluid [52] and employed in the well-known BACK EOS, which is applicable for both non-polar and polar systems [56, 57]. Figure 9.9 shows the effect that α has upon V_f/V at two reduced densities according to Boub, Chien, and Cott models. It reveals the sensibility of Boub and Cott expressions to the shape factor, and that both exhibit similar trends; Chien model appears very insensitive, instead. Therefore, Eq. (9.88) (Cott) is recommended owing to its simplicity.

The models considered above, Eqs. (9.80)–(9.88), are only valid for HS or hard convex body fluids, where the molecular diameter, σ, or the close-packed molar volume, V_0, is temperature-independent. To account for the softness of real molecules, a temperature-dependent (and eventually also density-dependent) expression for σ is necessary, being well reported the large

Table 9.4. Expressions for the repulsive compressibility factor (Z^r) and free-volume fraction (V_f/V) of hard non-spherical molecules

Boublik (Boub) [52]

$$Z^\mathrm{r} = \frac{1 + (3\alpha - 2)\,\varphi + \left(3\alpha^2 - 3\alpha + 1\right)\varphi^2 - \alpha^2\varphi^3}{(1-\varphi)^3}$$

$$V_\mathrm{f}/V = (1-\varphi)^{1-\alpha^2} \exp\left[\frac{3\alpha\varphi^2 - \left(3\alpha + \alpha^2\right)\varphi}{(1-\varphi)^2}\right] \qquad (9.86)$$

Chien et al. (Chien) [53]

$$Z^\mathrm{r} = \frac{1 + \varphi + \varphi^2 - \varphi^3}{(1-\varphi)^3} + \frac{\alpha}{2}(C-1)\frac{3\varphi + 3C\varphi^2 - (C+1)\,\varphi^3}{(1-\varphi)^3}$$

$$V_\mathrm{f}/V = (1-\varphi)^{\frac{\alpha}{2}\left(1-C^2\right)} \exp\left[\frac{\varphi\,(3\varphi - 4)}{(1-\varphi)^2} + \frac{\alpha}{2}\frac{3\,(C-1)\,\varphi^2 - \left(C^2 + 3C - 4\right)\varphi}{(1-\varphi)^2}\right] \qquad (9.87)$$

where: C=1.078

Cotterman et al. (Cott) [54]

$$Z^\mathrm{r} = \frac{1 + (4\alpha - 3)\,\varphi + (3 - 2\alpha)\,\varphi^2 - \varphi^3}{(1-\varphi)^3}$$

$$V_\mathrm{f}/V = \exp\left[\frac{\alpha\varphi\,(3\varphi - 4)}{(1-\varphi)^2}\right] \qquad (9.88)$$

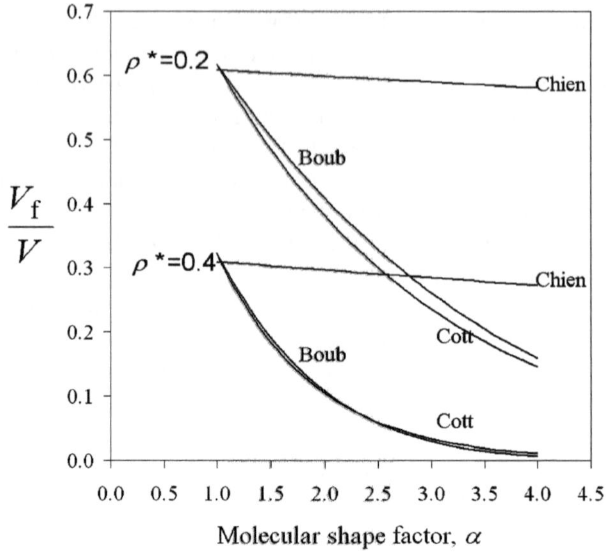

Fig. 9.9. Free-volume fraction versus molecular shape factor, calculated by Eqs. (9.86) to (9.88) (see Table 9.3) at two reduced densities

influence it has upon calculated results (e.g. [25, 28, 45, 58]). The EHSD method introduced in Sect. 9.3.3 is the most common tool adopted.

9.3.5 van der Waals Theory of Transport Properties

According to van der Waals (vdW) theory, the molecules exhibit a weak-long-range attractive energy and a hard-core repulsive energy. In real systems, intermolecular potential does have a steep repulsive part and the range of the attractive branch can be considered large relative to the average intermolecular separation at densities greater than critical density. Therefore, the attractive energy gives rise to a uniform energy surface where molecules move in straight lines between core collisions, which implies the vdW model becomes equivalent to the HS model. Such description should be approximately correct at temperatures greater than critical temperature, where the kinetic energy of molecules exceeds the attractive potential energy [59].

The vdW model is physically realistic and has been successfully applied to interpret and represent equilibrium properties of dense fluids. It is in fact a milestone in the field of thermodynamics. Such fact persuades and justifies the application of the HS model to transport coefficients, since the vdW theory of dynamic properties is equivalent to the former, provided the molecular core sizes are allowed to decrease as temperature increases to reflect the somewhat soft repulsive energy of real systems [59, 60].

To calculate the self-diffusivity, viscosity, or thermal conductivity of any substance at known conditions of temperature and density, it is only necessary to assign a value to the core size, σ, appearing in the appropriate HS expression. Nonetheless, owing to the uncertainties found in determining σ from equilibrium data, Dymond [33, 59, 60, 61] devised a methodology to test the applicability of the HS model by comparing calculated and experimental transport coefficients avoiding prior estimation or measurement of the core size. He treated data at fixed temperature, introducing the following reduced coefficients independent of molecular diameter:

$$D^* = \frac{\rho}{\rho_o}\frac{D}{D_o}\left(\frac{V}{V_o}\right)^{2/3} \tag{9.89}$$

$$\eta^* = \frac{\eta}{\eta_o}\left(\frac{V}{V_o}\right)^{2/3} \tag{9.90}$$

$$\lambda^* = \frac{\lambda}{\lambda_o}\left(\frac{V}{V_o}\right)^{2/3} \tag{9.91}$$

It is straightforward to evaluate these reduced quantities for experimental data, if the fluid is considered to be an assembly of hard spheres. Accordingly, taking into account Eqs. (9.21)–(9.23) for $\rho_o D_o$, η_o, λ_o, and Eq. (9.34) for V_o, the expressions required for the dimensionless experimental coefficients are

$$D^* = 5.030 \times 10^8 \frac{D}{V^{1/3}}\sqrt{\frac{M}{RT}} \tag{9.92}$$

$$\eta^* = 6.035 \times 10^8 \eta \frac{V^{2/3}}{\sqrt{MRT}} \tag{9.93}$$

$$\lambda^* = 1.936 \times 10^7 \lambda V^{2/3}\sqrt{\frac{M}{RT}} \tag{9.94}$$

where all the quantities are in SI units.

Finally, it is only necessary to evaluate theoretically Eqs. (9.89)–(9.91) to accomplish a comparison between isothermal experimental and predicted data. The analytic expressions needed to calculate $(\rho D)/(\rho_o D_o)$, η/η_o, and λ/λ_o depend solely on V/V_o (or ρ^*), and are presented in Sects. 9.4, 9.5, and 9.6; otherwise, computer experimental results from literature may be used directly. Therefore, D^*, η^*, and λ^* from HS theory may be plotted against $\log(V/V_o)$, while, at the same time, the experimental reduced coefficients, Eqs. (9.92)–(9.94), may be graphed as function of $\log V$. If both curves are superimposable laterally, then the HS theory does represent the density dependence of transport coefficients, and it is possible to establish their range of validity. Furthermore, V_o, or equivalently σ, can be determined from curve-fitting. It is important to emphasize that V_o or σ must be the same for D, η, and λ because it is a characteristic of the fluid. Therefore, it should be optimized simultaneously for the three properties. Nonetheless, if it is available from one property alone it becomes an input parameter for the remaining ones.

Dymond [61] analysed the LJ system to provide a theoretical justification for the correlation approach described. Results demonstrated that the equivalent core size for self-diffusion should be larger by about 5% than that for viscosity and thermal conductivity at a given temperature. It is not unreasonable that V_o may differ for systems of real molecules also, as different equivalent HS collision cross sections have been previously found for dilute particles interacting with LJ and inverse-power potentials.

According to the vdW theory, the deviations found at low and/or intermediate densities are a consequence of the non-uniformities in the attractive potential energy surface in real fluids at such densities.

Figure 9.10 illustrates such procedure for methane, krypton, argon and tetrafluoride, and shows that the HS theory represents very satisfactorily the density dependence of these data at temperatures above their critical values, from $V/V_o = 2$ to $V/V_o = 7.5$, which represents a density range from more than twice the critical density down to about two thirds the critical density. The molecular core sizes may be optimized using isothermal data. For instance, for methane, $\sigma = 0.360$, 0.355, 0.352, 0.350 nm at 200, 250, 300, 350 K, respectively, following expected trend.

9.3.6 Rough Hard Sphere Theory

The kinetic theory of dense gases due to Enskog [11] was developed for perfectly smooth hard spheres (SHS), i.e. particles are not allowed to exchange

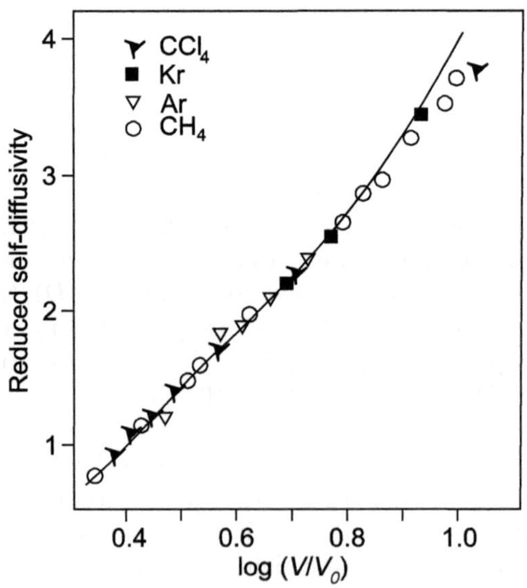

Fig. 9.10. Reduced self-diffusion coefficients against $\log(V/V_o)$ plotted in Dymond' fashion. (*Line*) MD results for HS (Eq. (9.114), Sect. 9.4.1); (*points*) experimental data [62, 63, 64, 65]

angular momentum upon collision. This assumption is expected to break down to some degree when applied to real fluid where particles of any shape may be found, especially for polyatomic molecules with non-isotropic intermolecular potentials. Therefore it is obligatory to modify Eqs. (9.24)–(9.26) (and those derived from them) to take into account the possibility of angular momentum transfer upon collision, that is the coupling of translational and rotational motions of molecules. Such evidence has driven Chandler [66] to the concept of the rough hard sphere (RHS) fluid, which is composed of spherical particles that collide impulsively and are capable of changing angular and linear momentum under their instantaneous collisions.

Chandler [66] finds that in the case of self-diffusion this coupling can be accounted for by including in the Enskog, and other derived equations, a frontal coefficient A_{D} which is rigorously independent of density and temperature:

$$D_{\mathrm{RHS}} = A_{\mathrm{D}} D_{\mathrm{SHS}} \quad (0 < A_{\mathrm{D}} \leq 1) \tag{9.95}$$

The roughness factor A_{D} is the unity only when the hard spheres are perfectly smooth, and it has been shown that the coupling between translational and rotational motions has the effect of reducing the value of the diffusion coefficient. The physical explanation of this phenomenon is that such coupling provides an additional mechanism for the relaxation of the velocity of the particle in the fluid [66]. As a result, the velocity autocorrelation function, Eq. (9.9), will decay to zero more rapidly if the coupling exists than if it does not, which implies lower diffusivity (see Eq. (9.8)).

The model is applicable for densities greater than about twice the critical density, given that a physically reasonable temperature- and perhaps density-dependent HS diameter is used. For dense liquids it will also not be useful if the molecular shape is highly non-spherical or if there exist attractive interactions between molecules (e.g. water, where strong hydrogen bond attractions play the crucial role in determining the arrangement of molecules). The most striking aspect of this analysis is the large impact produced even for molecules as spherical as carbon tetrachloride (factor of 2) [66].

The shear viscosity of a fluid also depends on whether the translational and rotational motions are coupled. By following the same analysis, Chandler [66] assumed that

$$\eta_{\mathrm{RHS}} = A_{\eta} \eta_{\mathrm{SHS}} \quad (A_{\eta} \geq 1) \tag{9.96}$$

where roughness constant A_{η} is the unity only when binary collisions do not allow for the transfer of translational and rotational energies. The author confirmed once more that the coupling should not be ignored even for liquid carbon tetrachloride, for which the proportionality constant found was $A_{\eta} = 1.74$.

Subsequent researches have shown that the roughness factors may be temperature- and/or density-dependent. For instance, strong density dependence was observed for mesitylene, 1,2-dichloroethane, and octamethylcyclotetrasiloxane [67]; for benzene, A_{D} exhibits a pronounced temperature- and density-dependence; in the case of acetonitrile, A_D is temperature-dependent, but continues independent of density [66]. When the proportionality constants display strong density dependence it might be concluded that the RHS model is invalid or that the core sizes are being incorrectly determined. In fact, small variations in the HS diameter have significant influence on the density dependence of the calculated diffusion coefficient at high densities [25, 60]. Arends et al. [68] measured self-diffusivities of ethylene, and their RHS results amount to a temperature-dependent coupling in such a way that A_D increases with decreasing temperature. They argue that the behaviour of $A_D(T)$ is due to the rising influence of the attractive part of the intermolecular potential when temperature is lowered, as is well known from modelling and molecular dynamics simulations of Lennard-Jones and Square-Well fluids [25, 69, 70, 71].

Bertucci and Flygare [72] investigated the application of the basic ideas of the HS model to the calculation of mutual diffusion coefficients, D_{12}. In order to explore the degree to which a rigorous HS treatment may be relaxed in the consideration of real fluid systems, they compared calculated and experimental D_{12} values for binary liquid mixtures, whose molecules vary from the nearly spherical carbon tetrachloride and cyclohexane to the fairly non-spherical derivatives of benzene. Their studies verified the idea of Chandler [66] that the translational and rotational motions of molecules must be accounted for. Mathematically,

$$D_{12}^{\mathrm{real}} \approx D_{12}^{\mathrm{RHS}} = A_D D_{12}^{\mathrm{HS}} \tag{9.97}$$

Bertucci and Flygare found, similar to Chandler, that such coupling has about a factor of two effect on the calculated diffusion coefficients. The parameter A_D is expected to vary from system to system since that coupling will certainly depend on the nature of the colliding particles.

The RHS theory of the liquid state has been frequently combined empirically and semi-empirically with several models for pure substances and binary mixtures, even with supercritical fluids [25, 58, 73, 74, 75, 76].

9.3.7 Excess Entropy-Based Scaling Laws

Another very interesting approach is to relate dynamic properties with fluid structural properties, which remains one of the most challenging tasks in the field of condensed matter. A first attempt in establishing a correspondence between reduced transport coefficients and reduced excess entropy (over the ideal-gas value) of dense fluids has been done by Rosenfeld [77, 78] and revealed some universal characteristics. Such excess entropy corresponding states consideration was originally motivated by the success of the variational HS thermodynamic perturbation theory for simple fluids in which the hard sphere diameter, or equivalently the excess entropy, was used to parametrize the structure of equilibrium fluids.

The excess entropy can be calculated knowing the compressibility factor, $Z = P/(\rho k_B T)$, which can be evaluated analytically for a given EOS analytic in φ, as follows:

$$\frac{S^{\mathrm{ex}}}{Nk_{\mathrm{B}}} = -\int_0^{\varphi} \frac{Z-1}{\varphi} \mathrm{d}\varphi \tag{9.98}$$

For instance, the Carnahan–Starling and Kolafa EOSs for HS systems (see Eq. (9.31)) give rise to [12, 79]

$$\frac{S^{\mathrm{ex}}}{Nk_{\mathrm{B}}} = -\varphi \frac{4-3\varphi}{(1-\varphi)^2} \tag{9.99}$$

and

$$\frac{S^{\mathrm{ex}}}{Nk_{\mathrm{B}}} = -\frac{5}{3}\mathrm{Ln}(1-\varphi) - \varphi \frac{4\varphi^2 - 33\varphi + 34}{6(1-\varphi)^2} \tag{9.100}$$

respectively, the latter derived by Heyes et al. [80].

A number of semi-empirical formulas of the kind

$$Y^* = a \ \exp\left(b \frac{S^{\mathrm{ex}}}{Nk_{\mathrm{B}}}\right) \tag{9.101}$$

where Y^* denotes reduced transport coefficients, have been proposed in literature for HS, OCP, SS, LJ, and real systems. See Sect. 9.1.1 for a brief description of these model fluids.

Rosenfeld [77, 81] defined the reduced (dimensionless) coefficients scaled by macroscopic parameters (density and temperature), namely a mean interparticle distance, $d = \rho^{-1/3}$, and the thermal velocity, $v = \sqrt{k_{\mathrm{B}}T/m}$. Thus the proposals are

$$D^* = D \frac{\rho^{\frac{1}{3}}}{\sqrt{k_{\mathrm{B}} \frac{T}{m}}}, \qquad \eta^* = \eta \frac{\rho^{\frac{-2}{3}}}{\sqrt{mk_{\mathrm{B}}T}}, \qquad \lambda^* = \lambda \frac{\rho^{\frac{-2}{3}}}{k_{\mathrm{B}}\sqrt{k_{\mathrm{B}} \frac{T}{m}}} \tag{9.102}$$

This form of reduced transport coefficients is also suggested by an elementary kinetic theory for a dense medium of particles with thermal velocities but with a mean free path between collisions which is of the order of the average interparticle distance. Because of the choice of macroscopic reduction parameters, excess entropy corresponding states relations can be applied directly to real materials.

In Figs. 9.11 and 9.12, D^* and η^* are plotted against excess entropy for purely repulsive potentials, namely HS, SS, and OCP, and to a good accuracy exhibit linear trends on logarithmic scale. Moreover, Rosenfeld [77] studied the effect of attractive forces by comparing results for SS and LJ systems, and found that the self-diffusion coefficient lie on nearly the same line (see Fig. 9.13). Therefore, the addition of an attractive component does not change the functional dependence of D^* upon S^{ex}, which indicates once more that for real materials it is enough to have an idea of the repulsive forces to interpret transport phenomena.

Later, Rosenfeld [81] extended the quasi-universal excess entropy scaling to dilute gases, where it is least expected to be observed. The analytic expressions he obtained for the Enskog domain are power-law dependence between

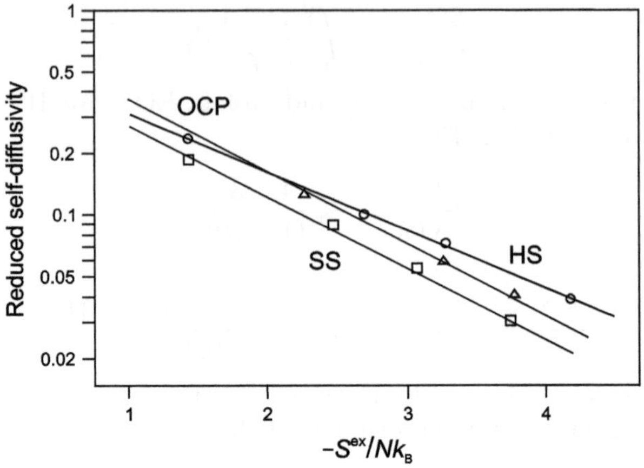

Fig. 9.11. Reduced self-diffusion coefficient versus excess entropy for hard spheres, soft spheres, and one-component plasma systems, according to Rosenfeld's approach

reduced transport coefficients and excess entropy. By considering in detail HS results, Rosenfeld [81] pointed out that the analytic form of the quasi-universal function is probably not exponential, yet an exponential form can provide good numerical representation of the true function. The simplified relations of the type of Eq. (9.101) are in fact semi-quantitative models established empirically by computer simulations of systems with well-defined potentials. However, in view of the absence of a unifying quantitative description of atomic transport in condensed matter, the excess entropy scaling is important for estimating unknown transport coefficients and for providing guidelines for theoretical analysis. Such a view point was also taken by Dzugutov [82], who proposed

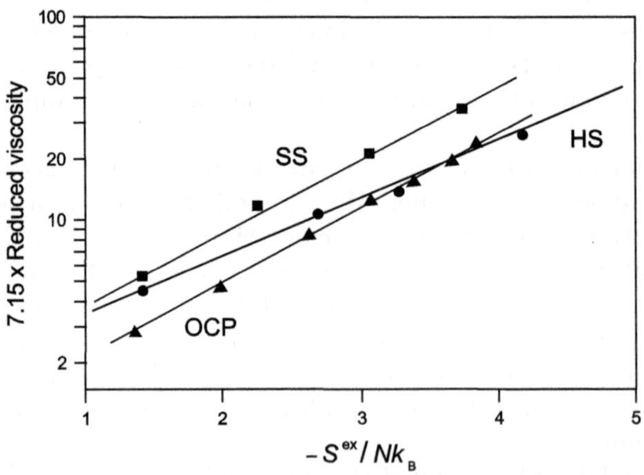

Fig. 9.12. Reduced viscosity coefficient versus excess entropy for hard spheres, soft spheres, and one-component plasma systems, according to Rosenfeld's approach

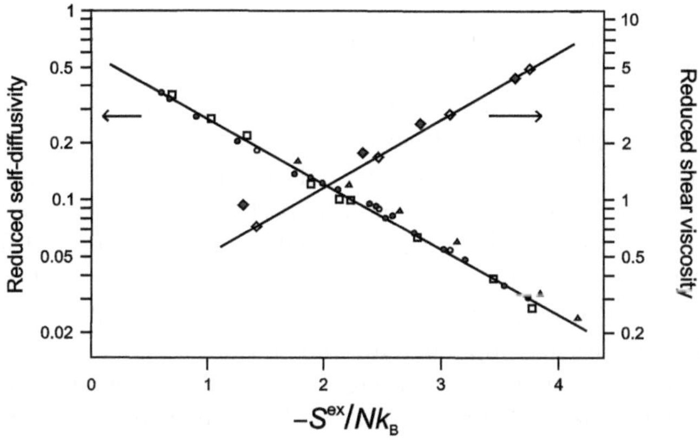

Fig. 9.13. Reduced self-diffusion and viscosity coefficients versus excess entropy for soft spheres and Lennard–Jones fluid, calculated by Rosenfeld's approach. (*unfilled symbols*) SS; (*filled symbols*) LJ

a variant of the entropy scaling, following a microscopic reduction approach tailored specifically for the diffusion coefficient of dense matter.

Dzugutov [82] assumed that energy and momentum transfer in liquids is mainly governed by the uncorrelated binary collisions described by the Enskog theory. Hence, the diffusion coefficient should be appropriately expressed in units of $\sigma^2\Gamma_\mathrm{E}$, where σ is the HS diameter that corresponds practically to the position of the first peak of the pair correlation function, $g(r)$, and $\Gamma_\mathrm{E} = 4\sigma^2 g(\sigma)\rho\sqrt{\pi k_\mathrm{B}T/m}$ is the collision frequency given by Enskog theory. Accordingly,

$$D^* = \frac{D}{\sigma^2\Gamma_\mathrm{E}} = \frac{D}{4\sigma^4 g(\sigma)\rho\sqrt{\pi k_\mathrm{B}\frac{T}{m}}} \tag{9.103}$$

Dzugutov assumed that atomic dynamics in condensed phases is dominated by the so-called cage effect, whereby diffusion is coupled to structural relaxations. In fact, since each atom finds itself in a cage formed by the immediate neighbours, its diffusive movement is determined by local density fluctuations, which modifies the surrounding atomic configuration. The frequency of the local structural relaxations, which defines the rate of cage diffusion, is obviously proportional to the number of accessible configurations. In an equilibrium system, the constraints imposed by the structural correlations reduce this configuration number by a factor of $\exp(S^\mathrm{ex})$. Therefore, D^* and $\exp(S^\mathrm{ex})$ must be connected by a universal linear relationship [82]:

$$D^* = a\ \exp\left(\frac{S^\mathrm{ex}}{Nk_\mathrm{B}}\right) \tag{9.104}$$

Furthermore, it was assumed that the excess entropy could be restricted to the two-particle approximation, independent of whether or not the potential energy is pairwise additive:

$$\frac{S^{\mathrm{ex}}}{Nk_{\mathrm{B}}} = -2\pi\rho \int_0^{\infty} \{g(r)\,\mathrm{Ln}\,[g(r)] - [g(r) - 1]\}\, r^2\, \mathrm{d}r \qquad (9.105)$$

In the case of multicomponent systems, Γ_{E}, σ, and S^{ex} for each species is derived from the relevant partial $g(\sigma)$.

Dzugutov's conjecture was tested by MD simulations with a diverse set of simple monatomic liquids: HS, LJ, one liquid characterized by predominantly icosahedral local order, and another one whose local order is topologically related to the primitive hexagonal lattice with only six nearest neighbours. Liquid Cu, liquid Pb, and a two-component LJ liquid model have also been simulated. The results accomplished are plotted in Fig. 9.14 and demonstrates convincingly an existent universal relationship. Also graphed for comparison are the classical Enskog theory results for HS, which deviates progressively more with decreasing excess entropy.

Bretonnet [83] showed, from an analysis of the HS fluid properties, that Eq. (9.104) is only legitimate over a restricted range of densities corresponding to around $S^{\mathrm{ex}} = -2.5$ in unit of Nk_{B}. Afterwards, he addressed a fundamental investigation to make more definite assessment about the reduction parameter to form the reduced diffusion coefficient, based on MD simulation of both excess entropy and diffusivity in liquid Rb and Cs along the liquid–vapour coexistence curve [84]. To look at in what extent it changes the exponential form of the scaling law connecting S^{ex} and D^*, he examined three other reduction parameters stemming from: (i) the principle of corresponding states

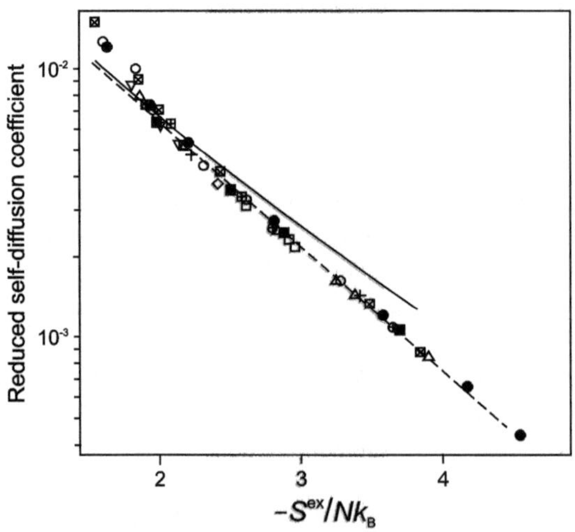

Fig. 9.14. Reduced self-diffusion coefficient of several fluid models versus excess entropy, calculated according to Dzugutov's approach. Symbols: results for HS, LJ, two-component LJ, liquid Pb, liquid Cu, icosahedral local order liquid, liquid with local order topologically related to the primitive hexagonal lattice with only six nearest neighbours, Ag diffusion in AgI. (*Full line*): Enskog theory results; (*dashed line*) linear fitting by Dzugutov [82] (see Sect. 9.4.1, Eq. (9.131))

(PCS), (ii) the quasi-harmonic Einstein model, which assumes that the nature of liquid and solid phases are similar, and (iii) the Green-Kubo formalism. The corresponding diffusion reduction parameters proposed are, respectively [84]:

$$D_{\mathrm{PCS}} = \sigma\sqrt{\varepsilon/m} \tag{9.106}$$

$$D_{\mathrm{Ein}} = \frac{d^3\omega_{\mathrm{E}}^2}{\sqrt{k_{\mathrm{B}}T/m}} \tag{9.107}$$

$$D_{\mathrm{GK}} = \frac{k_{\mathrm{B}}T}{mM_0\tau} \tag{9.108}$$

where σ and ε are the distance and energy characteristics to be specified for each interatomic potential; ω_{E} is the Einstein frequency at which an atom vibrates in the cage formed by its neighbours, and d is the interatomic separation in the crystalline phase; $\tau = \sqrt{2M_0/M_2}$, M_0 and M_2 being the first moments of the expansion of the Enskog contribution to the memory function, $M_{\mathrm{B}}(t)$, in even power series in time:

$$M_{\mathrm{B}}(t) = \sum_{n=0}^{\infty} \frac{(-1)^n}{(2n)!} M_{2n} t^{2n} \tag{9.109}$$

This memory function is commonly split into two components according to

$$M(t) = M_{\mathrm{B}}(t) + M_{\mathrm{MC}}(t) \tag{9.110}$$

Here, $M_{\mathrm{B}}(t)$ takes into account the uncorrelated binary collisions, which dominates at very short times, and $M_{\mathrm{MC}}(t)$ comprises the mode coupling effects

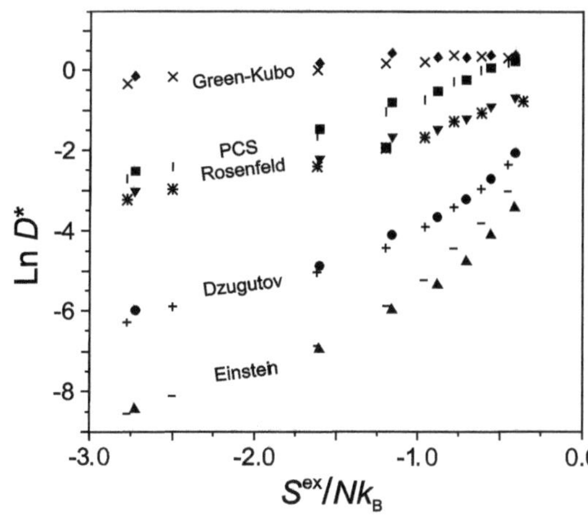

Fig. 9.15. Reduced self-diffusion coefficient of Rb (*filled symbols*) and Cs (*other symbols*) against excess entropy for several reduction principles

prevailing at long time. In Fig. 9.15 the results achieved with Eqs. (9.106)–(9.108) are compared with those due to Rosenfeld and Dzugutov, Eqs. (9.102) and (9.103), to demonstrate that the parameters involved in the $S^{\text{ex}} - D^*$ scaling law depend upon the particular reduction adopted. At least for the liquid alkali metals used in the calculations, Bretonnet [84] showed that the abovementioned quasi-universal relation holds effectively when Green-Kubo formalism is used to generate D^*.

In this section the main theories to interpret and represent the transport properties of HS and modified systems have been described. They are fundamental tools to analyse and comprehend Sects. 9.4–9.6, where diffusion, viscosity, and thermal conductivity coefficients are treated separately, as most theoretical approaches presented here are common to these coefficients.

9.4 Diffusion Coefficient

The diffusion coefficient in liquids and compressed fluids plays an important role in the design and simulation of many units involving mass transfer and its interaction with chemical reaction. The extent of mass transfer limitations within processes is frequently not well characterized, in part due to the lack of experimental data of the various constituents. Hence, it is fundamental to have reliable and accurate equations to calculate diffusivities at the operating conditions found in industrial plants. This fact justifies an individual section for the diffusion coefficient, where we describe and compare several models published in literature to assess its predictive or correlative capability. Accordingly, the self-diffusion and tracer diffusion coefficients of the HS fluid are treated firstly. Afterwards, its results are embodied in or extended to other systems such as SW, WCA, LJ, and real substances.

9.4.1 Models for the Diffusion Coefficient of the Hard Sphere System

As mentioned above, Sect. 9.4 initiates with the study of the self- and tracer diffusion coefficients of the HS system. Special attention is devoted to the indispensable corrections to the Enskog theory, which have been assessed by computer simulation since the pioneer investigation of Alder et al. [14]. The sets of MD data points most frequently adopted in the literature to develop empirical and semi-empirical models are compared and analysed, as well as the models expressions themselves. This subsection contains the raw material needed to examine modified HS systems. It is worth noting the behaviour of real substances is our final objective.

Self-Diffusion Coefficient of the Hard Sphere System

As discussed in Sect. 9.3.1, Enskog theory exhibits large deviations to MD self-diffusion coefficients of HS fluid in the intermediate and high-density regions.

Consequently, Eq. (9.24) must be corrected, which is frequently accomplished by parameterizing the hard sphere MD data according to

$$\rho D_{\mathrm{HS}} = F(\rho^*)\rho D_{\mathrm{E}} \tag{9.111}$$

$$\rho D_{\mathrm{HS}} = G(\rho^*)\rho_o D_o \tag{9.112}$$

$F(\rho^*)$ and $G(\rho^*)$ are correction functions, which according to their definition are related by

$$F(\rho^*) \equiv \left(\frac{\rho D_{\mathrm{HS}}}{\rho D_{\mathrm{E}}}\right) = \left(\frac{\rho D_{\mathrm{HS}}}{\rho_o D_o}\right)\left(\frac{\rho_o D_o}{\rho D_{\mathrm{E}}}\right) \equiv G(\rho^*)\left(\frac{\rho_o D_o}{\rho D_{\mathrm{E}}}\right) \tag{9.113}$$

Dymond [59] investigated the transport coefficients on the basis of the van der Waals model summarily described in Sect. 9.3.5, and fitted the exact data of Alder et al. [14] to a polynomial in $x \equiv (V/V_o)^{2/3}$. The expression provides good representation over the (V/V_o) range from 1.7 to 10, giving AAD = 1.5%:

$$\frac{\rho D_{\mathrm{HS}}}{\rho_o D_o} = -3.629 + 5.096x - 2.223x^2 + 0.4316x^3 - 0.03105x^4 \tag{9.114}$$

This function is plotted in Fig. 9.10 (line), reduced according to Eq. (9.89). The experimental self-diffusivities of CCl_4, Kr, Ar, and CH_4 plotted demonstrate that the HS theory is able to represent the density dependence of simple molecules.

Although the vdW model for transport properties has sound theoretical justification for gases at high-density and high-temperature conditions, one expects it may be extended to liquids due to the characteristic small intermolecular distances. Dymond [33] examined the region $1.5 < V/V_0 < 2$ or $0.370 < \varphi < 0.494$ using the computer results for hard spheres of Alder et al. [14], and fitted them to the equation

$$\frac{D_{\mathrm{HS}}}{D_o} = 1.271\left(1 - 1.384\frac{V_o}{V}\right) \quad \text{or} \quad \frac{D_{\mathrm{HS}}}{D_o} = 1.271\left(1 - \frac{\varphi}{\varphi_i}\right) \tag{9.115}$$

Here, $\varphi_i = 0.5350$ is the HS packing fraction at which diffusion ceases (the same for $V/V_o = 1.384$). Actually, Eq. (9.115) is a free-volume type equation where D_{HS} depends on the relative expansion from an intrinsic molar volume V_{i}, i.e. $D_{\mathrm{HS}} \propto (V - V_{\mathrm{i}}) = V_{\mathrm{f}}$ (see Sect. 9.3.4, Eq. (9.74)). It reproduces MD values to better than 2% from 1.5 to $2V_o$; at 2.5 and $3V_o$ it overestimates computer results by 3% and 6%, respectively.

In the region $1.6 < V/V_o < 6.6$, Dymond [85] corrected the Enskog theory and found the following alternative proportion to hold:

$$\frac{D_{\mathrm{HS}}}{D_o} \propto \frac{V_o}{V}\left[\left(\frac{V}{V_o}\right)^{2/3} - 1.256\right] \tag{9.116}$$

Later, Erkey et al. [74] used the RHS theory to interpret tracer diffusion coefficients of benzene and tetrachloride in sub and supercritical carbon dioxide. With that purpose, they first derived an expression for the self-diffusivity of HS system using the MD simulation data of Easteal et al. [86], which covers the range $1.5 \le V/V_o \le 4.0$. To relate D explicitly to molar volume and temperature, the following reduced coefficient is defined:

$$D^+ \equiv \left(\frac{D_{\mathrm{HS}}}{D_o}\right)\left(\frac{V}{V_o}\right) \tag{9.117}$$

From Enskog equation, Eq. (9.24), it is possible to conclude that D^+ only contains the volume dependence of the self-diffusion coefficient:

$$D^+ = \left(\frac{D_{\mathrm{HS}}}{D_{\mathrm{E}}}\right)\left(\frac{D_{\mathrm{E}}}{D_o}\right)\left(\frac{V}{V_o}\right) = \frac{D_{\mathrm{HS}}}{D_{\mathrm{E}}}\frac{1}{g(\sigma)}\frac{V}{V_o} \tag{9.118}$$

The right-hand side of Eq. (9.118) was fitted as $D^+ = 2.190[(V/V_o)^{2/3} - 1.271]$ over $1.7 \le V/V_o \le 4.0$ with AAD $= 1.32\%$. Taking into account the definition of D^+, the final model easily establishes (note the similarity with Eq. (9.116))

$$\frac{D_{\mathrm{HS}}}{D_o} = 2.190\frac{V_o}{V}\left[\left(\frac{V}{V_o}\right)^{2/3} - 1.271\right] \tag{9.119}$$

Speedy [87] proposed the following equation to represent 25 points from Alder et al. [14], Woodcock [88], and Easteal et al. [86, 89]:

$$D_{\mathrm{HS}} = D_0\left(1 - \rho^*/1.09\right)\left[1 + \rho^{*^2}\left(0.4 - 0.83\rho^{*^2}\right)\right] \tag{9.120}$$

where the constants 1.09, 0.4, and −0.83 are the regressed parameters. The first factor represents the cage constraining effects at high density, which leads eventually to a glass transition (i.e. diffusion ceases) when $\rho^* \to 1.09$ or $\varphi \to 0.571$. The second factor accounts for the hydrodynamic enhancement discussed in Sect. 9.3.1. Speedy's model has been the starting point of numerous researches on self- and mutual diffusion coefficients. It is shown in Figs. 9.16 and 9.17, along with data points compiled by the author.

Heyes [90, 91] presented a polynomial correlation with five parameters that provides similar results to Eq. (9.120), since it refers essentially to the same database:

$$\frac{\rho D_{\mathrm{HS}}}{\rho_o D_o} = 1.035 - 0.553\rho^* + 5.640\rho^{*^2} - 7.639\rho^{*^3} + 1.814\rho^{*^4} \tag{9.121}$$

It is important to emphasize that the MD data used by Speedy [87] are inconsistent with the final values reported by Alder et al. [14], as Figs. 9.16 and 9.17 confirm. Such fact implies significant discrepancies between models based on different data sources, which warrants a brief comment in the following.

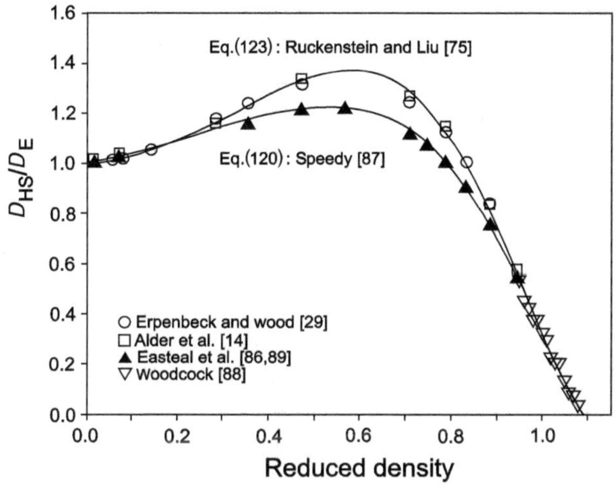

Fig. 9.16. Self-diffusion coefficient of the HS system normalized by the Enskog value against reduced density, $\rho^* = \rho\sigma^3$, from MD data and from models

Alder et al. [14], in the earlier work on the subject, calculated $D_{\text{HS}}(N)$ in systems of $N = 108$ and $N = 500$ spheres with a precision of about 1%. In addition, they have made correction to their points to extrapolate to the infinite system limit, $D_{\text{HS}}(\infty)$, and have included estimates for the effect of the truncation of the length of the calculations (the long-time tail). The corrections were large, in excess of 10% at some densities. Later, Easteal et al. [86, 89]

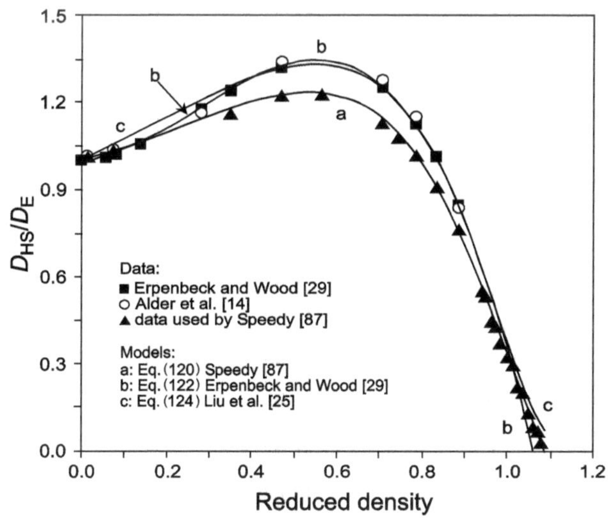

Fig. 9.17. Ratio of the self-diffusion coefficient of the HS system to the Enskog value against reduced density, $\rho^* = \rho\sigma^3$, from MD data and from models

carried out additional simulations in the dense fluid region and explicitly examined the system size dependence by computing the self-diffusion coefficient for N ranging from 128 to 4394. The implication of their results was that $D_{\mathrm{HS}}(N)$ is essentially number-independent provided that $N \geq 250$, which disproves the appropriateness of the Alder et al. corrections. Their conclusion was substantiated at one density, $\rho^* = 0.707$, by the study of systems of $N = 128, 250, 432, 1024, 2000$, and 4394 particles, in which $D_{\mathrm{HS}}(N)$ increased by only 1% on going from $N = 250$ to $N = 4394$, in etc. contrast to Alder–Gass–Wainwright's estimate that $D_{\mathrm{HS}}(\infty)$ is 11% greater than $D_{\mathrm{HS}}(500)$ at that density [14].

Based on these findings, Speedy [87] considered the values of $D_{\mathrm{HS}}(500)$ reported by Alder et al. [14] and of $D_{\mathrm{HS}}(512)$ computed by Woodcock [88], for states of the metastable fluid below the freezing volume, as reliable estimates of $D_{\mathrm{HS}}(\infty)$ without correction. The author combined these data with those of Easteal et al. [86, 89] to suggest Eq. (9.120) for the self-diffusion coefficient of the HS system.

Erpenbeck and Wood [92] presented calculations of the velocity autocorrelation functions for the HS fluid, extending the Alder–Gass–Wainwright studies to larger systems and more densities. The specific aim was to determine the nature of the long-time tail corrections and they have concluded that these could be quantitatively calculated by mode-coupling theory. Later, Erpenbeck and Wood [29] explicitly published ten self-diffusion coefficients and were compelled to directly contradict the researches of Easteal et al. [86, 89] and Speedy [87]. They found that there is indeed no apparent number dependence. It is the long-time tail correction that is number-dependent. Speedy simply ignored this correction without any justification, which led Erpenbeck and Wood [29] to advise against the use of his equation. Their final results approximate those originally given by Alder et al. [14], as Figs. 9.16 and 9.17 point out, though the precision is somewhat better.

Erpenbeck and Wood [29] represented their 10 MD self-diffusivities by a polynomial of the third degree in V_o/V, normalized by the Enskog coefficient:

$$\frac{D_{\mathrm{HS}}}{D_{\mathrm{E}}} = 1.0 + 0.054034\frac{V_o}{V} + 6.3656\left(\frac{V_o}{V}\right)^2 - 10.9425\left(\frac{V_o}{V}\right)^3 \qquad (9.122)$$

where coefficients 0.054034, 6.3656, and −10.9425 are optimized parameters, and gives $D_{\mathrm{HS}} = 0$ for $\varphi = 0.5557$ or $\rho^* = 1.061$. The model and data points fitted are plotted in Fig. 9.17 (curve b), along with Speedy's equation for comparison (curve a). It is noticeable the disagreement found in the range $0.2 \leq \rho^* \leq 1.0$, owing to the above-mentioned dissimilar data sources involved.

Harris [93] proposed another polynomial expression based on the data from Alder et al. [14], Woodcock [88], and Erpenbeck and Wood [29]. This correlation covers only the reduced density range 0.471–0.943, which forbids its use in the dilute region. Besides, it uses distinct and inconsistent sets of data, as has been mentioned above. The same recurred also with Amoros [94], who fitted the same MD data to a nine-degree polynomial.

Ruckenstein and Liu [75] systematically studied self-diffusion in gases and liquids, starting with the HS system. A new equation was obtained fitting MD data of Alder et al. [14], Woodcock [88], and Erpenbeck and Wood [29], and represents accurately D_{HS} in the whole density range:

$$\frac{D_{\mathrm{HS}}}{D_{\mathrm{E}}} = 1.0 + 0.94605\rho^{*^{1.5}} + 1.4022\rho^{*^{3}} - 5.6898\rho^{*^{5}} + 2.6626\rho^{*^{7}} \qquad (9.123)$$

This expression is shown in Fig. 9.16 together with the corresponding simulation data. Equation (9.123) clearly detaches from Speedy's model, Eq. (9.120), especially at intermediate densities, corroborating our previous discussion.

An alternative procedure to treat the diffusion behaviour of dense HS fluid is based on the free-volume theory exhibiting exponential dependence on V_{f} (Eq. (9.71), Sect. 9.3.4). Liu et al. [25] presented a reduced expression of this category, involving nothing but the concept of molecular volume and particularly adopted to be simple and accurate over the intermediate and high-density regions:

$$\frac{D_{\mathrm{HS}}}{D_0} = \exp\left(-\frac{0.75\rho^*}{1.2588 - \rho^*}\right) \qquad (9.124)$$

The right-hand side of Eq. (9.124) is equivalent to the DCT model, Eq. (9.71), established empirically by Doolittle [36] and later derived by Cohen and Turnbull [37]. Hence it is able to interpret the glass transition behaviour. The numerical factor 0.75 was previously fixed because it is intrinsic to the model. Using the 10 MD data points of Erpenbeck and Wood [29], the unique parameter was optimized giving $\rho_i^* = 1.2588$ with an average absolute deviation equal to 1.43% over $0.0566 \leq \rho^* \leq 0.8839$ (or $1.6 \leq V/V_o \leq 24.99$, or $0.02964 \leq \varphi \leq 0.4628$). Equation (9.124) may be seen in Fig. 9.17 (curve c) as function of reduced density. Its trend approximates Erpenbeck and Wood's model in $0.3 \leq \rho^* \leq 1.0$, while sets apart that of Speedy over almost all range.

Another formula for the HS self-diffusivities that also fits MD data quite well is due to Sigurgeirsson and Heyes [95]:

$$\frac{D_{\mathrm{HS}}}{D'_o} = 1.06\left(\frac{\varphi_i}{\varphi} - 1\right) \qquad (9.125)$$

where $D'_o = D_o\rho^*$ and $\varphi_i = 0.545$. This relation reproduces data within the simulation statistical uncertainty for $\varphi > 0.25$ or $\rho^* > 0.477$. It is worth noting that D'_o is independent of density, therefore Eqs. (9.115) and (9.125) are equivalent. Expressing both equations in the same coordinates, one verifies they have different frontal coefficients, although they exhibit quite similar zeros: $D_{\mathrm{HS}} = 0$ for $\varphi_i = 0.5350$ and $\varphi_i = 0.545$, respectively.

Heyes et al. [80] fitted their MD data to the following analytic form:

$$\frac{D_{\mathrm{HS}}}{D_{\mathrm{E}}} = 1 + 0.277645\varphi + 3.98964\varphi^2 + 26.49881\varphi^3 - 134.0015\varphi^4 + 110.7344\varphi^5 \qquad (9.126)$$

which predicts zero diffusion at $\varphi = 0.5602$ or $\rho^* = 1.070$. They presented an alternative equation analogous to that of Speedy, Eq. (9.120), to separate the crowding effect from the hydrodynamic enhancement one:

$$\frac{D_{\mathrm{HS}}}{D_{\mathrm{E}}} = \left(1 - \frac{\varphi}{0.5531}\right)^{0.04128} \left[1 - 0.160226\varphi + 13.8375\varphi^2 - 30.3387\varphi^3\right] \tag{9.127}$$

In Sect. 9.3.7 a distinct approach based on entropy scaled laws to describe transport properties in the specific field of condensed matter is presented, namely equations of the type of Eqs. (9.101) and (9.104). It is important to emphasize that those expressions are mainly for real substances, although authors generally include MD simulations for hard spheres in their databases for comparison.

Rosenfeld [77, 81] found the self-diffusion coefficient reduced according to Eq. (9.102) follows the quasi-universal relation:

$$D^*_{\mathrm{HS}} \cong D\frac{\rho^{1/3}}{\sqrt{k_{\mathrm{B}}T/m}} = 0.585 \exp\left(0.788\frac{S^{\mathrm{ex}}}{Nk_{\mathrm{B}}}\right) \tag{9.128}$$

which holds for quite disparate pair interactions. Recall that S^{ex} represents excess entropy over the ideal-gas value, Eqs. (9.98)–(9.100). Figure 9.13 shows Eq. (9.128) plotted together with MD data for SS and LJ systems, and evidences the good representation achieved by previous correlation. For the HS fluid, somewhat different exponential argument is given (see Fig. 9.11):

$$D^*_{\mathrm{HS}} = D\frac{\rho^{1/3}}{\sqrt{k_{\mathrm{B}}T/m}} = 0.585 \exp\left(0.65\frac{S^{\mathrm{ex}}}{Nk_{\mathrm{B}}}\right) \tag{9.129}$$

On the whole, it is important to emphasize these equations estimate diffusivities, which vary by about two orders of magnitude, within about 30%.

Later, Rosenfeld [81] investigated dilute gases also and discovered power-law dependence between D^* and excess entropy for inverse power potentials of the form $\phi = \varepsilon(\sigma/r)^\nu$ (see comment to Eq. (9.5) in Sect. 9.1.1). Between $\nu = 4$ and $\nu = \infty$ (hard spheres) the universal expression is

$$D^* = D\frac{\rho^{1/3}}{\sqrt{k_{\mathrm{B}}T/m}} = D'\left(-\frac{S^{\mathrm{ex}}}{Nk_{\mathrm{B}}}\right)^{-2/3} \tag{9.130}$$

the constant D' varying from 0.409 to 0.346 in that range of ν.

Throughout the liquid domain, Dzugutov [82] established another simple scaling law appropriately expressed in terms of microscopic reduction parameters (Eq. (9.103)):

$$D^* = \frac{D}{4\sigma^4 g(\sigma)\rho\sqrt{\pi k_{\mathrm{B}}T/m}} = 0.049 \exp\left(\frac{S^{\mathrm{ex}}}{Nk_{\mathrm{B}}}\right) \tag{9.131}$$

which obeyed equally well for HS, LJ, and other simple liquids, and also remains applicable to atomic transport in quasicrystal and to silver–ion diffusion in the solid-state ionic conductor α–AgI (see Fig. 9.14). Nonetheless, for liquid metals modelled by the embedded atom potentials, some scatter in the data is observed, especially for liquid silicon. In fact, for systems with isotropic, many-body interactions, it is necessary to include multiparticle contributions to the excess entropy, while for systems with anisotropic interatomic interactions (such as Si) the relation does not hold [96, 97, 98, 99].

Bretonnet [83] analysed the entropy dependence of the self-diffusivity in terms of positional correlations in an HS fluid, and concluded the exponential behaviour due to Dzugutov is correct over a restricted domain around $\rho^* = 0.7$, corresponding to $S^{\text{ex}}/Nk_{\text{B}} = -2.5$. From the analysis of the structural, thermodynamic, and transport properties of the HS fluid, a new semi-empirical expression was presented to predict the diffusion coefficient of simple dense fluids over large-density range. The formula has no adjustable parameters:

$$D = \frac{\sigma}{8}\left(\frac{\pi k_{\text{B}}T}{m}\right)^{1/2}\frac{(1-\varphi)^4}{\varphi\,(2+\varphi)}\exp\left(\frac{6\varphi}{1-\varphi}\right)\exp\left(\frac{S^{\text{ex}}}{Nk_{\text{B}}}\right) \tag{9.132}$$

σ is assumed to be the position of the first peak of the pair correlation function, $g(r)$.

Tracer Diffusion Coefficient of the Hard Sphere System

The tracer diffusion coefficient refers to the diffusion of a solute infinitely diluted in a solvent, being one of the most important transport properties. In fact, in some industrial processes, such as those encountered in supercritical fluid extraction, the systems can be considered simply as infinitely dilute. On the other hand, in most concentrated liquid mixtures, the mutual diffusion coefficients can be estimated as function of tracer diffusivities using Darken [100], Vignes [101], or other equations reviewed by Pertler et al. [102]. Following on, some equations devised for the tracer diffusion coefficient of the HS system are presented, due to the success of the methods based on HS theories to describe transport properties of single-component dense fluids.

Tracer diffusivities may be calculated from an extension of the Enskog method for self-diffusion, Eq. (9.24). Accordingly, at high density, coefficient D_{12}^{E} is related to the low-density one, D_{12}^{0}, by the unlike pair radial distribution function at contact, $g_{12}(\sigma_{12})$:

$$\frac{\rho_2 D_{12}^{\text{E}}}{\rho_2^0 D_{12}^0} = \frac{1}{g_{12}(\sigma_{12})} \tag{9.133}$$

$$D_{12}^0 = \frac{3}{8\rho_2^0\sigma_{12}^2}\left(\frac{k_{\text{B}}T}{2\pi m_{12}}\right)^{\frac{1}{2}} \tag{9.134}$$

$$m_{12} = \frac{m_1 m_2}{m_1+m_2} = \frac{0.001}{N_{\text{a}}}\frac{M_1 M_2}{M_1+M_2} \tag{9.135}$$

where subscripts "1" and "2" stand for solute and solvent, respectively, superscript "0" denotes ideal gas, $\sigma_{12} = (\sigma_1 + \sigma_2)/2$, m_{12} is the reduced mass of the system in kg, M_i is the molecular weight of component i in g/mol, and N_a is the Avogadro number. The pair distribution function at contact may be calculated by the approximated expression of Lebowitz [103] or by the equations due to Carnahan and Starling [104], Mansoori et al. [105], and Alder et al. [106], for instance. According to Carnahan and Starling [104],

$$g_{12}(\sigma_{12}) = \frac{\sigma_1 g_{22}(\sigma) + \sigma_2 g_{11}(\sigma)}{\sigma_1 + \sigma_2} \tag{9.136}$$

where $g_{ii}(\sigma)$ is given by

$$g_{ii}(\sigma) = \frac{1}{1-\varphi} + \frac{3y_i}{2(1-\varphi)^2} + \frac{y_i^2}{2(1-\varphi)^3} \tag{9.137}$$

and

$$\varphi = \sum_i \varphi_i = \sum_i \pi\rho_i^*/6 = \sum_i \pi\rho_i\sigma_i^3/6 \tag{9.138}$$

$$y_i = \frac{\sigma_i\varphi_j + \sigma_j\varphi_i}{\sigma_j} \tag{9.139}$$

An alternative expression is that of Mansoori et al. [105]:

$$g_{12}(\sigma_{12}) = \frac{1}{(1-\varphi_2)^3}\left(1 - \varphi_2 + \frac{2\varphi_2}{1+\sigma_2/\sigma_1}\right) \times \left(1 - \varphi_2 + \frac{\varphi_2}{1+\sigma_2/\sigma_1}\right) \tag{9.140}$$

which provide essentially the same results of that of Alder et al. [106].

Analogous to one-component systems, Eq. (9.133) also neglects both backscattering and vortex flow effects described in Sect. 9.3.1. Such deviations are investigated quantitatively by computer simulation studies for systems of a single test particle in a solvent, for selected size and mass ratios, to assess the necessary correction factor, henceforth identified by F_{12}. Therefore, the HS tracer diffusivity, D_{12}^{HS}, can be written as a modification of the Enskog equation:

$$D_{12}^{\mathrm{HS}} = \frac{D_{12}^{\mathrm{HS}}}{D_{12}^{\mathrm{E}}} D_{12}^{\mathrm{E}} = F_{12} D_{12}^{\mathrm{E}} \tag{9.141}$$

The correction factor generally depends on the reduced density of the solvent, $\rho_2^* = \rho_2\sigma_2^3$, and on the size and mass ratios:

$$F_{12} = F_{12}(\rho_2^*, \sigma_1/\sigma_2, m_1/m_2) \tag{9.142}$$

As shown by Alder et al. [106], the dependence of F_{12} on σ_1/σ_2 is complex. At low densities, it increases with increasing σ_1/σ_2 when $m_1/m_2 = 1$, while diminishes with increasing σ_1/σ_2 for $m_1/m_2 \leq 0.1$; at high densities, F_{12} always decreases as σ_1/σ_2 increases.

The correction factor should satisfy the following four restrictions:

(i) When diameters and masses coincide, F_{12} reduces to the self-diffusion correction factor, F_{22}, already presented and discussed in Sect. 9.4.1 (Eq. (9.111)):

$$\lim_{\substack{\sigma_1/\sigma_2 \to 1 \\ m_1/m_2 \to 1}} F_{12} = \frac{D_{22}^{\mathrm{HS}}}{D_{22}^{\mathrm{E}}} = F_{22}(\rho_2^*) = F_{22} \tag{9.143}$$

(ii) At low density the Enskog theory has to be recovered:

$$\lim_{\rho_2^* \to 0} F_{12}(\rho_2^*, \sigma_1/\sigma_2, m_1/m_2) = 1 \tag{9.144}$$

(iii) At low density the self-diffusion correction factor obtained from Eq. (9.143), F_{22}, has to obey Enskog theory:

$$\lim_{\rho_2^* \to 0} F_{22}(\rho_2^*) = 1 \tag{9.145}$$

(iv) The self-diffusion correction factor obtained from Eq. (9.143), F_{22}, should vanish at an intrinsic high-density ρ_s^*:

$$\lim_{\rho_2^* \to \rho_s^*} F_{22}(\rho_2^*) = 0 \tag{9.146}$$

Some expressions for F_{12} will be analysed in the following, namely those from Sung and Stell [107], Sun and Chen [108], Eaton and Akgerman [109], Liu and Ruckenstein [110], and Easteal and Woolf [111].

Sung and Stell [107] derived the following equation for the HS tracer diffusion coefficient:

$$F_{12} = \left(\frac{1 + \gamma_{\mathrm{B}}\ \sigma_2/\sigma_1}{1 + \gamma\ \sigma_2/\sigma_1} \right)^2 \frac{g_{12}(\sigma_{12})}{1 + 4\varphi_2 g_{12}(\sigma_{12})} + \frac{1 + \sigma_2/\sigma_1}{1 + \gamma\ \sigma_2/\sigma_1} \frac{D_{12}^{\mathrm{HS}}}{D_{12}^{\mathrm{E}}} \tag{9.147}$$

where γ and γ_{B} ($\gamma_{\mathrm{B}} \equiv \lim_{\rho_2 \to 0} \gamma$) are given by

$$\begin{aligned} &\frac{1}{1 - \gamma^3 \varphi_2} \left(1 - \gamma^3 \varphi_2 + \frac{2\gamma^3 \varphi_2}{1 + \gamma\ \sigma_2/\sigma_1} \right) \times \left(1 - \gamma^3 \varphi_2 + \frac{\gamma^3 \varphi_2}{1 + \gamma\ \sigma_2/\sigma_1} \right) \\ &= 1 + 4\varphi_2 \frac{1 - \varphi_2/2}{(1 - \varphi_2)^3} \end{aligned} \tag{9.148}$$

The first term on the right-hand side of Eq. (9.147) represents the Enskog limit, while the second is the hydrodynamic limit; the ratio $D_{12}^{\mathrm{HS}}/D_{12}^{\mathrm{E}}$ is given by [112]

$$\frac{D_{12}^{\mathrm{HS}}}{D_{12}^{\mathrm{E}}} = 2.881 \left(1 + \frac{\sigma_1}{\sigma_2} \right) \left(1 + \frac{m_2}{m_1} \right)^{-1/2} \frac{\varphi_2 g_{12}(\sigma_{12})}{\eta_2^{\mathrm{HS}}/\eta_o} \tag{9.149}$$

where $\eta_2^{\mathrm{HS}}/\eta_o$ is the ratio between the viscosities of an HS solvent and an ideal-gas. Taking into account Enskog formula, Eq. (9.25), it is possible to write

$$\frac{\eta_2^{\mathrm{HS}}}{\eta_0} = \frac{\eta_2^{\mathrm{HS}}}{\eta_2^{\mathrm{E}}} \times \frac{\eta_2^{\mathrm{E}}}{\eta_o} = \frac{\eta_2^{\mathrm{HS}}}{\eta_2^{\mathrm{E}}} \left[\frac{1}{g(\sigma_{22})} + 3.2\varphi_2 + 12.176\varphi_2^2 g(\sigma_{22}) \right] \tag{9.150}$$

(recall that $b\rho = 4\varphi$). Using the MD simulations of Alder et al. [14] for an HS fluid, Liu and Ruckenstein [110] obtained the following accurate expression, whose plot may be seen in Fig. 9.24 (Sect. 9.5.1):

$$\frac{\eta_2^{\mathrm{HS}}}{\eta_2^{\mathrm{E}}} = 1 + 0.007825 \rho_2^{*^{0.1}} \exp(6.0037 \rho_2^{*^3}) \tag{9.151}$$

For pure solvents, Eq. (9.147) combined with Eq. (9.149) leads to

$$F_{22} = \left(\frac{1 + \gamma_B^0}{1 + \gamma^0} \right)^2 \frac{g(\sigma_{22})}{1 + 4\varphi_2 g(\sigma_{22})} + \frac{8.15\varphi_2 g(\sigma_{22})}{(1 + \gamma^0)\, \eta_2^{HS}/\eta_o} \tag{9.152}$$

where superscript “0” refers to pure solvent and $\gamma_{\mathrm{B}}^0 = 1.1909$ (γ^0 should be calculated from Eq. (9.148) for pure solvent).

Equations (9.147) and (9.152) satisfy the zero density constraints (Eqs. (9.144) and (9.145), respectively). As curve 1 of Fig. 9.18 shows, Eq. (9.152) provides raw quantitative agreement with the MD simulations of Alder et al.

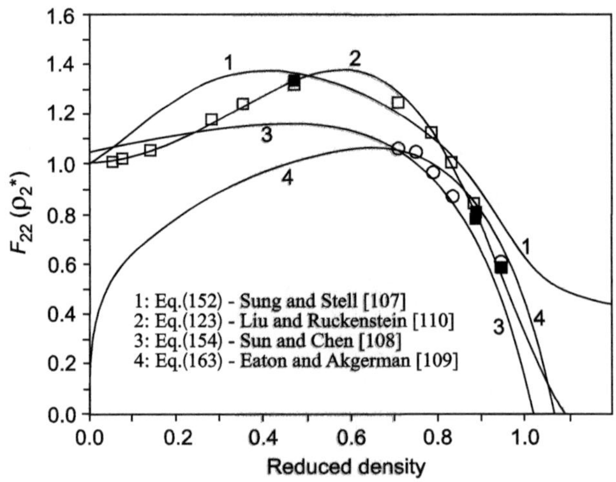

Fig. 9.18. Self-diffusion correction factor of the HS system as function of reduced density, $\rho_2^* = \rho_2 \sigma_2^3$. $F_{22}(\rho_2^*)$ is calculated by the limit of several tracer diffusion models, see Eq. (9.143). The Liu and Ruckenstein' model, Eq. (9.165), recovers the self-diffusion equation of Ruckenstein and Liu, Eq. (9.123). MD data: (■) Alder et al. [106]; (□) Alder et al. [14]; (○) Easteal and Woolf [111]

[14] in the dilute and middle-density ranges. Nonetheless, at high densities it leads to non-zero values, hence violating restriction given by Eq. (9.146).

Later, Sun and Chen [108] proposed another expression for F_{12} on the basis of the MD simulations of Herman and Alder [113] and Alder et al. [106], which cover the ranges $0.5 < \sigma_1/\sigma_2 < 1.6$, $0.5 < m_1/m_2 < 4.0$, and $1.5 < V_2/V_0 < 3.0$:

$$F_{12} = 1.0514 \frac{\sigma_{12}^2}{\sigma_1 \sigma_2} \left(\frac{2}{1 + m_2/m_1} \right)^{\frac{1}{2}} g_{12}(\sigma_{12}) \left(\frac{m_2}{m_1} \right)^{0.167} \times \left[1 - 0.9779 \rho_2^* \left(\frac{m_2}{m_1} \right)^{0.0165} \left(\frac{\sigma_1}{\sigma_2} \right)^{0.129} \right] \quad (9.153)$$

For pure solvent, Eq. (9.153) reduces to

$$F_{22}(\rho_2^*) = 1.0514\, g(\sigma_{22}) \times (1 - 0.9779 \rho_2^*) \quad (9.154)$$

which satisfies approximately the zero density limit (with a slightly different numerical factor, 1.0514) and the high-density restriction, but underestimates F_{22} in the middle-density range — see Fig. 9.18, curve 3. Furthermore, Eq. (9.154) predicts that $F_{22}/g(\sigma_{22})$ depends linearly on ρ_2^*, which is totally incorrect since $F_{22}/g(\sigma_{22}) = D_{\mathrm{HS}}/D_o$.

Easteal and Woolf [111] carried out MD simulations for tracer diffusion coefficients in HS systems and proposed the following empirical model for the molecular-weight ratio dependence of F_{12}/F_{22}:

$$\mathrm{Ln} \left(\frac{F_{12}}{F_{22}} \right) = a_0 + a_1\, \mathrm{Ln} \left(\frac{m_1}{m_2} \right) + a_2\, \mathrm{Ln}^2 \left(\frac{m_1}{m_2} \right) \quad (9.155)$$

where a_0, a_1, and a_2 depend upon both ρ^* and σ_1/σ_2.

Salim and Trebble [114] adopted the same expression for $\mathrm{Ln}(G_{12}/G_{22})$, where G_{12} and G_{22} are corrections to the ideal-gas values, i.e. $D_{12}^{HS} = G_{12} D_{12}^0$ and $\lim_{\substack{\sigma_1/\sigma_2 \to 1, \\ m_1/m_2 \to 1}} G_{12} = G_{22}(\rho_2^*) = G_{22}$. On the basis of the same MD data, Salim and Trebble published the following coefficients:

$$a_0 = -1.02564\, {\rho_2^*}^4\, \mathrm{Ln}(\sigma_1/\sigma_2) \quad (9.156)$$

$$a_1 = -0.24107 + 1.27589\, h_{12} - 1.35439\, h_{12}^2 + 0.62393\, h_{12}^3 \quad (9.157)$$

$$a_2 = -0.056 \quad (9.158)$$

where $h_{12} = {\rho_2^*}^2 (\sigma_1/\sigma_2)$, and selected the model of Speedy [87] for G_{22} (Eq. (9.120)). One can easily confirm that this $\mathrm{Ln}(G_{12}/G_{22})$ does not satisfy the low density restriction $\lim_{\rho_2^* \to 0} G_{12}(\rho_2^*, \sigma_1/\sigma_2, m_1, m_2)$, owing to Eqs. (9.157) and (9.158).

Eaton and Akgerman [109] proposed an alternative expression for the correction factor of the HS tracer diffusivity, based on the MD simulations of Easteal and Woolf [111]:

$$F_{12} = a\,\frac{V_o}{V_2}\left[\left(\frac{V_2}{V_o}\right)^{\alpha} - b_{12}\right] \tag{9.159}$$

where

$$\alpha = \frac{\sigma_2}{\sigma_1} - \frac{1}{3} \tag{9.160}$$

$$b_{12} = \left(\frac{m_2}{m_1}\right)^{0.03587}\left[0.6001 + 0.8491\left(\frac{\sigma_2}{\sigma_1}\right) - 0.244\left(\frac{\sigma_2}{\sigma_1}\right)^2\right] \tag{9.161}$$

and a is a function of the molecular size ratio (σ_1/σ_2). Using the same MD simulations, Liu and Ruckenstein [110] obtained

$$a = 1.689\,(\sigma_1/\sigma_2)^{2.0674} \tag{9.162}$$

For pure solvent, Eq. (9.159) gives rise to

$$F_{22}(\rho_2^*) = 1.1943\;g(\sigma_{22})\;(1.26\rho_2^{*^{1/3}} - 1.2052\rho_2^*) \tag{9.163}$$

An immediate shortcoming is that it vanishes at zero density instead of being unitary. Consequently, the model of Eaton and Akgerman [109], Eq. (9.159), is valid only at high densities (see Fig. 9.18, curve 4).

Liu and Ruckenstein [110] devised a new expression for F_{12}, on the basis of the MD simulations of Herman and Alder [113] and Alder et al. [106]. Their empirical model involves ρ_2^*, σ_1/σ_2, m_1/m_2, radial distribution functions at contact given by Eqs. (9.32) and (9.140), and some terms involved in the theoretical expressions of Sung and Stell [107] and Alder et al. [106], namely $\eta_2^{\mathrm{HS}}/\eta_0$ (Eq. (9.150)), $\eta_2^{\mathrm{HS}}/\eta_2^{\mathrm{E}}$ (Eq. (9.151)), and the quantity

$$\triangle = \left[\frac{\eta_2^{\mathrm{HS}}/\eta_o + 1.2/g\,(\sigma_{22})}{\eta_2^{\mathrm{HS}}/\eta_o + 1.2\,(0.5m_2/m_{12})^{1/2}\,(\sigma_{22}/\sigma_{12})^2\,g_{12}^{-1}\,(\sigma_{12})}\right]^{\frac{3}{2}} \tag{9.164}$$

Alder et al. [106] introduced $\triangle$ along with σ_1/σ_2 and m_1/m_2, to take into account a hydrodynamic vortex effect in F_{12}. The final equation is

$$\frac{F_{12}}{F_{22}} = 1.0 + \sum_{i=1}^{10} b_i H_i \tag{9.165}$$

The coefficients b_i and the functions H_i are listed in Table 9.5. The correction factor for self-diffusion, F_{22}, is calculated directly from Eq. (9.123), being represented by curve 2 in Fig. 9.18. The model obeys all restrictions given by Eqs. (9.143)–(9.146), and is applicable in the range $0.0 < \rho_2^* < 0.95$, $0.25 < \sigma_1/\sigma_2 < 1.0$, and $0.0002 < m_1/m_2 < 4.0$. It is not expected to provide reliable predictions at low (< 0.25) and high (> 1.0) molecular size ratios.

For the 58 MD data points of Herman and Alder [113] and Alder et al. [106] (the last ones extrapolated to an infinite number of spheres), the average absolute deviations found for the previous models are: $\mathrm{AAD}_{\mathrm{Sung-Stell}} = 72.62\%$,

Table 9.5. Coefficients and functions of the model of Liu and Ruckenstein [110] for the tracer diffusion of the HS system (Eq. 9.165) Note:

$$\chi = \rho_2^{*0.01} \quad \text{and} \quad \delta(\sigma_1/\sigma_2, m_1/m_2) = \begin{cases} 0, & \text{if } (\sigma_1/\sigma_2) \text{ and } (m_1/m_2) = 1 \\ 1, & \text{if } (\sigma_1/\sigma_2) \text{ or } (m_1/m_2) \neq 1 \end{cases}$$

i	b_i	H_i
1	−0.293977	$\chi\,\mathrm{Ln}(\sigma_1/\sigma_2)$
2	−0.106089	$\chi\,\mathrm{Ln}^2(\sigma_1/\sigma_2)\mathrm{Ln}(m_1/m_2)$
3	0.111318	$\chi\,\mathrm{Ln}(m_1/m_2)$
4	0.002865	$\mathrm{Ln}(\eta_2^{HS}/\eta_o)\mathrm{Ln}^2(m_1/m_2)$
5	−0.0069724	$\chi\,\mathrm{Ln}(m_1/m_2)(\sigma_2/\sigma_1)^{0.5}$
6	−0.0020652	$\chi\,\mathrm{Ln}(\sigma_1/\sigma_2)(m_2/m_1)^{0.5}$
7	3.21576	$\mathrm{Ln}^2[g(\sigma_{22})/g_{12}(\sigma_{12})]$
8	−0.334952	$\chi\,\mathrm{Ln}\Delta$
9	−0.26190	$\chi\,\mathrm{Ln}^2(\sigma_1/\sigma_2)(m_2/m_1)^{1/6}$
10	0.143989	$\delta(\sigma_1/\sigma_2, m_1/m_2)\mathrm{Ln}(\eta_2^{HS}/\eta_2^E)$

$\mathrm{AAD}_{\mathrm{Sun-Chen}} = 34.51\%$, and $\mathrm{AAD}_{\mathrm{Liu-Ruckenstein}} = 4.97\%$. In spite of the lengthy appearance of the Liu and Ruckenstein's model, its excellent performance is noticeable; on the contrary, Eqs. (9.147) and (9.153) provide unsatisfactory results.

Figures 9.19 and 9.20 illustrate the dependence of F_{12} on σ_1/σ_2 and m_1/m_2, respectively. Results achieved with previous models are graphed together with the MD simulations of Alder et al. [106]. Figure 9.19 shows that the theory of Sung and Stell [107], Eq. (9.147), and the empirical expression of Sun and Chen [108], Eq. (9.153), present wrong trends under some conditions. According to Fig. 9.20, Sung and Stell [107] overestimate F_{12} at high values

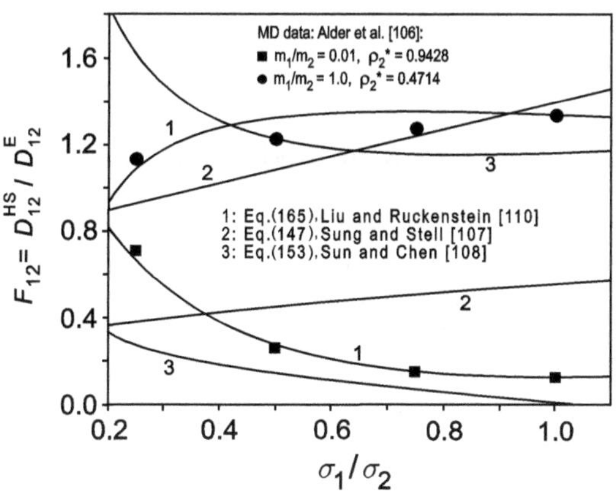

Fig. 9.19. Correction factor of the tracer diffusion coefficient of HS system versus molecular size ratio

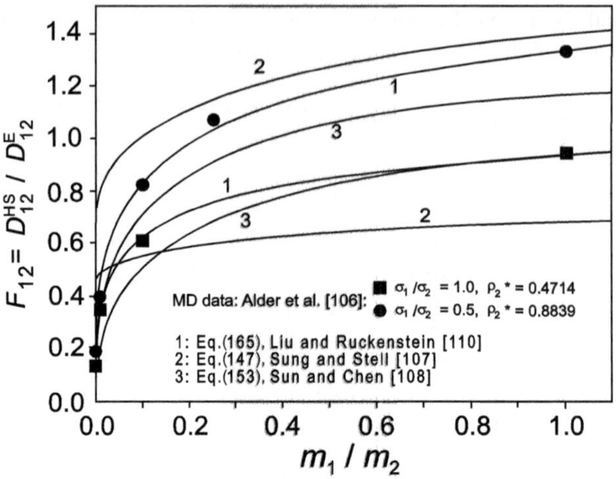

Fig. 9.20. Correction factor of the tracer diffusion coefficient of HS system versus molecular weight ratio

of σ_1/σ_2 and underestimate it at low values of σ_1/σ_2; on the other hand, Sun and Chen [108] always underestimate the correction factor.

9.4.2 Extension to Modified Hard Sphere Systems

In the preceding subsection, MD data and models for the diffusion coefficients of the HS fluid have been reviewed. At the moment, one intends to extend such results and principles to modified HS systems, where intermolecular repulsive interactions are not as hard as those of hard spheres, and attractive forces may be simultaneously present. Namely the diffusivities of the SW, WCA, LJ, and rough LJ fluids will be investigated, and above all their application to real substances. The self-diffusion coefficient is treated first, followed by tracer and intra diffusion coefficient.

Self-Diffusion Coefficient of Modified Hard Sphere Systems

This subsection initiates with the self-diffusion coefficient of modified HS systems. Both theoretical and real systems are covered. Most successful and interesting approaches to describe polyatomic and attractive fluids are presented.

Simple models for the SW fluid can be obtained coupling HS equations with a simple energy (temperature) term. Based on the memory function theory, Straub [71] has proved it is possible to correlate the self-diffusivities of both fluids with a simple equation

$$\frac{D_{\mathrm{SW}}}{D_{\mathrm{HS}}} = \frac{1}{1 + \alpha_{\mathrm{SW}}/T^*} \tag{9.166}$$

where $\alpha_{\rm SW}$ is a parameter, and $T^* = k_{\rm B}T/\varepsilon_{\rm SW}$ is the reduced temperature. Similar equations have been presented for the Sutherland potential and for the LJ fluid [2, 71].

Since the self-diffusion coefficient of the SW system has to consider both the concepts of molecular volume and activation energy, to agree with the hybrid principles of that fluid, the free-volume-based equation suggested by Macedo and Litovitz [39], Eq. (9.73), was chosen by Liu et al. [25] to relate $D_{\rm SW}$ and $D_{\rm HS}$:

$$D_{\rm SW} = D_{\rm HS} \exp\left(-\frac{\alpha_{\rm SW}\varepsilon_{\rm SW}}{k_{\rm B}T}\right) = D_{\rm HS} \exp\left(-\frac{\alpha_{\rm SW}}{T^*}\right) \tag{9.167}$$

Note this relation reduces to Eq. (9.166) if the exponential is replaced by its first-order Taylor approximation. The authors inserted their $D_{\rm HS}$ expression, Eq. (9.124), obtaining for the reduced SW self-diffusivity:

$$D^*_{\rm SW} = 0.21157 \exp\left(-\frac{0.75\rho^*}{1.2588 - \rho^*}\right) \exp\left(-\frac{0.81061}{T^*}\right) \tag{9.168}$$

the normalizing relation being:

$$D^*_{\rm SW} \equiv \rho\sigma^2 D_{\rm SW}/(k_{\rm B}T/m)^{1/2} \tag{9.169}$$

The unique parameter, $\alpha_{\rm SW} = 0.81061$, has been optimized using 123 MD data points in the range $0.02 < \rho^* < 0.90$ and $0.94 < T^* < \infty$, with AAD = 14.57%. Liu et al. [25] found important inconsistencies between data available for the SW fluid. If Eq. (9.166) is used instead of Eq. (9.167), then an AAD = 14.92% is found.

As has been referred to in Sect. 9.1.1, the LJ model can arise from a perturbation on the WCA fluid, by adding on its attractive part. The works of Kushick and Berne [69] and Straub [71] reveal that the differences between the repulsive and full LJ systems are not negligible, especially at low temperatures. An interesting approach to the LJ fluid embodies two steps: (i) First, devise models for the WCA self-diffusivities, appropriately validated with MD data; this step employs frequently the EHSD method:

$$D_{\rm WCA}(T^*, \rho^*) = D_{\rm HS}[\rho^*, \sigma_{\rm eff}(T^*, \rho^*)] \tag{9.170}$$

the reduced temperature being $T^* = k_{\rm B}T/\varepsilon_{\rm LJ}$. (ii) Then, relate $D_{\rm WCA}$ to the full LJ coefficient by means of simple temperature-dependent equations; generally, only one parameter is involved, being optimized with MD data:

$$D_{\rm LJ} = D_{\rm WCA} \times f(\alpha_{LJ}/T^*) \tag{9.171}$$

where $\alpha_{\rm LJ}$ is an adjustable parameter.

It is worth noting that many researchers usually apply the EHSD method directly to the LJ fluid and even to real fluids, due to the dominant role played by the repulsive interaction between molecules.

Straub [71] suggested an equation similar to Eq. (9.166) to relate LJ and WCA coefficients:

$$\frac{D_{\mathrm{LJ}}}{D_{\mathrm{WCA}}} = \frac{1}{1+\alpha_{\mathrm{LJ}}/T^*} \tag{9.172}$$

where $\alpha_{\mathrm{LJ}} = 0.23$ is a constant fitted to MD data. Based on the results of Kushick and Berne [69], Speedy et al. [24] proposed another equation:

$$D_{\mathrm{LJ}} = D_{\mathrm{HS}}\left(\rho^*, \sigma_{\mathrm{eff}}\right)\left(1 - \frac{\alpha_{\mathrm{LJ}}}{T^*}\right) \tag{9.173}$$

where $\alpha_{\mathrm{LJ}} = 0.14$. Furthermore, they used an alternative exponential form to express the attractive contribution:

$$D_{\mathrm{LJ}} = D_{\mathrm{HS}}\left(\rho^*, \sigma_{\mathrm{eff}}\right)\exp\left(-\frac{\alpha_{\mathrm{LJ}}}{T^*}\right) \tag{9.174}$$

where $\alpha_{\mathrm{LJ}} = 0.5$, together with the BS EHSD, Eq. (9.55) in Table 9.1, in the repulsive part. Nevertheless, only limited number of data points were tested.

Later, Enninghorst et al. [115] adopted Eq. (9.174) with $\alpha_{\mathrm{LJ}} = 0.25$ in conjunction with the D_{HS} of Speedy [87] and the EHSD of Speedy et al. [24] (Eqs. (9.120) and (9.55), respectively) to correlate their experimental data for isopentane and cyclopentane.

Following this philosophy, Liu et al. [25] and Silva et al. [58] derived new expressions for the self-diffusion coefficient of WCA, LJ, and real fluids. For repulsive LJ, they adopted their HS expression, Eq. (9.124), and the general two-parameter equation derived by Ben-Amotz and Herschbach [22] for the effective diameter:

$$\sigma^*_{\mathrm{eff}} = \frac{\sigma_{\mathrm{eff}}}{\sigma_{\mathrm{LJ}}} = \alpha^*_0\left[1 + \left(\frac{T^*}{T^*_0}\right)^{1/2}\right]^{\frac{-1}{6}} \tag{9.175}$$

Adopting the Boltzmann criterion and using MD self-diffusivities available for the WCA system (36 data points), their reduced EHSD equation, σ^*_{BSLM}, arises to Eq. (9.57) in Table 9.1.

Hence, the D_{WCA} model of Liu et al. [25] corresponds to Eqs. (9.124) and (9.57), being applicable in the range $0.74 \le T^* \le 5.10$ and $0.30 \le \rho^* \le 0.95$ with AAD = 3.33%. For the same data points, the deviation found with the usual equations of Speedy and co-workers, i.e. Eqs. (9.120) and (9.55), is 11.22% [24, 87].

To generate LJ self-diffusivities, Liu et al. [25] coupled their D_{WCA} model with an exponential energetic term, i.e. $D_{\mathrm{LJ}} = D_{\mathrm{WCA}}\exp(-\alpha_{\mathrm{LJ}}/T^*)$, and optimized parameter α_{LJ} fitting 435 data points available in literature for the LJ system. The obtained expression is

$$D^*_{\mathrm{LJ}} = \frac{0.2116\sqrt{T^*}}{\sigma^{*2}_{\mathrm{BLSM}}}\exp\left(-\frac{0.75\rho^*}{1.2588-\rho^*}\right)\exp\left(-\frac{0.27862}{T^*}\right) \tag{9.176}$$

where the reduced coefficient was defined as

$$D^*_{\mathrm{LJ}} \equiv \rho\sigma^2_{\mathrm{LJ}} D_{\mathrm{LJ}}/(\varepsilon_{\mathrm{LJ}}/m)^{1/2} \tag{9.177}$$

In the validity range $0.72 < T^* < \infty$ and $\rho^* \leq 1.22$, the average absolute deviation reported is only 5.62%, which is significantly smaller than those furnished by the exponential model proposed by Speedy [24], AAD=11.24%, and by an empirical equation with 13 parameters derived by Heyes [116], AAD=9.80%.

Liu et al. [25] seek to interpret the self-diffusion coefficient of real fluids in terms of their LJ model. Accordingly, Eqs. (9.176) and (9.57) were explicitly written as function of T and ρ, giving rise to the following two-parameter expression (LJ2):

$$D^{\mathrm{real}}_{\mathrm{LJ2}} = \frac{21.16}{\rho\sigma^2_{\mathrm{BLSM}}}\sqrt{\frac{1000RT}{M}}\exp\left\{-\frac{0.75\rho\sigma^3_{\mathrm{BLSM}}}{1.2588-\rho\sigma^3_{\mathrm{BLSM}}} - \frac{0.27862(\varepsilon_{\mathrm{LJ}}/k_{\mathrm{B}})}{T}\right\} \tag{9.178}$$

$$\sigma_{\mathrm{BLSM}} = \sigma_{\mathrm{LJ}} 2^{1/6}\left[1+\sqrt{\frac{1.3229T}{(\varepsilon_{\mathrm{LJ}}/k_{\mathrm{B}})}}\right]^{-1/6} \tag{9.179}$$

The diameter and the energy of the LJ potential function are the parameters, whose determination rests solely on the existence of experimental data; $R = 8.3144$ J mol^{-1} K^{-1} and M is molecular weight in g mol^{-1}. This model will be represented shortly by $D^{\mathrm{real}}_{\mathrm{LJ2}} = D(T, \rho; \sigma_{\mathrm{LJ}}, \varepsilon_{\mathrm{LJ}}/k_{\mathrm{B}})$. Results obtained for 40 substances, completing 2514 data points at the most, showed that the diffusive behaviour of almost all molecules can be interpreted by means of the two-parameter LJ model, and this assertion comprehends molecules as polar as acetonitrile and non-spherical as n-hexadecane. Alcohols and water, substances commonly identified with hydrogen bonding, exact a better description, which implies more parameters [58]. Table 9.6 compiles reported σ_{LJ} and $\varepsilon_{\mathrm{LJ}}/k_{\mathrm{B}}$ values, as well as individual deviations for $D^{\mathrm{real}}_{\mathrm{LJ2}}$. Global AADs are listed in Table 9.7 for both hydrogen- and non-hydrogen-bonding substances. For comparison, results achieved by the well-known models of Dymond-Hildbrand-Batschinski [33, 34, 35] and Lee and Thodos (LT) [117] are also included. A brief description of both models is presented below.

Silva et al. [58] used the same database and proposed generalized correlations, Eqs. (9.180)–(9.181), for the LJ force constants of molecules not covered by Liu et al. [25], and for which experimental self-diffusivities do not exist, according to the following procedure: (i) $D^{\mathrm{real}}_{\mathrm{LJ2}}$ is much more sensitive to σ_{LJ} than to $\varepsilon_{\mathrm{LJ}}/k_{\mathrm{B}}$, so this has been fixed to the simplest corresponding states correlation available [118], Eq. (9.180). (ii) Dimensional analysis and the empirical principle of corresponding states ascertain that critical constants reduced in terms of LJ parameters are universal [2]. For LJ molecules, this approximation establishes that σ_{LJ} is a function of $T_{\mathrm{c}}/P_{\mathrm{c}}$ (critical constants), which induced Silva et al. to build Fig. 9.21 using diameters optimized with fixed $\varepsilon_{\mathrm{LJ}}/k_{\mathrm{B}}$ of 34 real substances. The final correlations are

Table 9.6. Average absolute deviations and parameters of several models for the self-diffusion coefficient of real fluids

Substance	LJ2 Eqs. (9.178)+(9.179) Liu et al. [25]				LJ4 Eqs. (9.184)+(9.185) Silva et al. [58]					
	NDP	AAD (%)	σ_{LJ} (Å)	$\frac{\varepsilon_{LJ}}{k_B}$ (K)	NDP	AAD (%)	A_D	E_D (J/mol)	σ_{LJ} (Å)	T_D (K)
Argon	38	5.65	3.40744	123.55	38	4.028	0.9575	144.70	3.39771	1.1684E+02
Krypton	52	8.07	2.89870	511.92	52	6.222	0.3725	651.06	4.81369	2.6930E−03
Hydrogen	23	5.46	5.94111	3.45E−03	23	5.326	0.8963	−43.15	7.39111	2.1100E−04
Carbon disulphide	29	1.08	4.29901	376.51	29	1.075	1.0007	876.07	4.28933	3.0023E+02
Carbon dioxide	191	6.41	3.26192	500.71	191	6.150	1.1390	1496.10	7.37016	4.9080E−03
Methane	369	6.73	3.58484	167.15	369	5.943	0.9028	254.00	4.78869	1.2767E+00
Ethylene	118	5.80	4.04838	169.08	118	5.770	0.9870	340.25	4.04079	1.3697E+02
Ethane	65	3.01	4.17587	213.99	65	1.840	1.2428	769.65	3.94898	8.9234E+02
Propane	92	4.81	4.50412	457.99	92	2.566	1.2099	1147.69	4.35444	1.4156E+03
n-Butane										
n-Pentane										
n-Hexane	59	3.06	5.61841	434.76	59	2.702	0.7226	271.58	5.79056	1.6366E+02
n-Heptane	35	1.80	5.94356	404.05	35	1.693	1.1222	2702.70	12.15112	8.3640E−03
n-Octane	43	3.49	6.17328	478.32	43	2.938	0.4845	−627.36	6.91989	3.7666E+01
n-Nonane	14	1.07	6.43057	497.35	14	1.026	1.2737	3520.04	10.18732	2.2159E−01
n-Decane	20	1.12	6.71395	434.86	20	1.048	1.3971	3756.80	10.07872	4.1972E−01
n-Dodecane	10	0.44	7.00451	672.90	10	0.395	0.7094	−1172.54	6.65194	1.1514E+04
n-Tetradecane										
n-Hexadecane	25	5.36	7.36480	1669.19	25	2.099	0.3233	115.24	12.22689	3.7897E−01
n-$C_{30}H_{62}$										
n-$C_{50}H_{102}$										

n-$C_{78}H_{158}$										
n-$C_{154}H_{310}$										
Isopentane										
Cyclopentane										
Cyclohexane	39	2.59	5.73075	224.87	39	2.207	1.4797	1576.54	5.36688	8.7063E+02
Tetramethylsilane	42	6.82	5.53310	721.26	42	6.792	1.3890	2525.33	5.31579	2.1709E+03
Benzene	65	5.15	5.19165	308.43	65	4.629	1.1031	989.82	4.95708	7.5750E+02
Toluene	54	2.98	5.45450	350.74	54	1.570	1.0086	1032.20	5.60700	1.3709E+02
Chlorobenzene	15	1.00	5.56838	207.50	15	0.554	1.4322	2860.12	6.38257	1.0352E+01
Chloromethane	42	3.05	3.99520	402.67	42	2.514	0.5996	−267.01	5.20300	3.7767E+00
Dichloromethane	43	6.50	4.41862	411.02	43	3.299	1.1804	2191.59	4.65153	7.0850E+01
Chloroform	66	9.67	4.75692	505.30	66	8.693	1.7320	2118.01	4.43628	9.7730E+03
Carbon tetrachloride	37	3.48	5.29240	418.84	37	3.378	1.7590	2400.29	5.04569	1.2815E+03
Fluoromethane										
Trifluoromethane	38	7.89	4.05121	332.17	38	6.879	1.8983	1862.78	3.80259	1.9792E+03
Carbon tetrafluoride	53	6.50	4.33323	244.02	53	5.628	1.1608	894.60	7.21628	1.3263E−01
Chlorotrifluoromethane	115	6.47	4.37636	410.79	115	6.052	1.1278	1181.32	4.17242	1.5465E+03
Sulfur hexafluoride	38	8.48	4.76629	271.68	38	6.309	1.9830	2302.57	4.27221	1.4262E+04
Perfluorocyclobutane	59	3.82	6.20449	71.01	59	3.302	0.7068	−1082.69	6.88164	1.1597E+01
Ammonia	22	1.40	4.24397	4.46	22	0.910	1.5404	1498.65	10.17306	5.8000E−05
Acetone	13	0.95	4.67012	332.97	13	0.767	0.5775	−2249.42	4.40495	1.0484E+04
Acetic acid	8	0.16	4.47879	499.29	8	0.121	1.4831	3875.00	5.29186	1.1815E+01
Pyridine	55	8.97	5.03753	296.26	55	2.669	5.3936	5728.94	4.52573	9.9985E+03
Acetonitrile	60	3.49	4.02424	652.53	60	2.262	0.6656	527.85	11.88230	2.1300E−04
Methanol	106	35.52	3.79957	685.96	106	6.211	2.4728	5512.52	11.16008	1.2800E−04
Methanol-OD	51	17.85	2.74261	3309.05	51	5.498	6.5244	8567.08	4.80736	3.0560E+00
Ethanol	139	29.53	4.23738	1291.41	139	3.459	6.0393	9221.90	4.50700	8.7689E+01
Propanol	53	12.17	4.4919	2120.83	53	1.343	19.1215	12573.39	4.77935	2.4404E+02
Heavy water	41				41	3.439	7.7683	9848.88	2.36560	7.1174E+04
Water (supercritical)	34				34	5.461	0.0044	971.93	0.24834	1.8304E+01
Water (all states)	118	19.14	1.53091	3788.51	118	16.171				

Table 9.6. (Continued)

	LJC Eqs. (9.216)+(9.211)+(9.56) Yu and Gao [125]				RLJ Eqs. (9.205)+(9.123)+(9.56) Ruckenstein and Liu [75]					LJC Eqs. (9.192)+(9.193)+(9.186) Salim and Trebble [114]			
Substance	NDP	AAD (%)	N	σ_{LJ} ($\mathring{A}$)	NDP	AAD (%)	A_D	σ_{LJ} ($\mathring{A}$)	$\frac{\varepsilon_{LJ}}{k_B}$ (K)	AAD (%)	N	σ_{LJ} ($\mathring{A}$)	$\frac{\varepsilon_{LJ}}{k_B}$ (K)
Argon					72	4.81	1.0	3.4236	119.75				
Krypton													
Hydrogen													
Carbon disulphide	29	1.18	1.846	3.517	29	2.92	0.9314	4.2476	438.34				
Carbon dioxide	105	7.29	1.000	3.655	191	7.52	0.8669	3.4482	241.56				
Methane	59	3.02	1.102	3.547	369	5.96	1.0	3.6387	151.32	3.79	0.65	4.13	52.54
Ethylene	63	5.06	1.326	3.653	118	7.73	0.9393	3.9658	224.25				
Ethane					65	5.11	0.9354	4.1336	242.52	2.59	1.30	4.60	39.94
Propane										2.93	1.39	4.90	98.36
n-Butane										7.64	1.08	5.25	203.44
n-Pentane										5.08	1.19	5.69	138.71
n-Hexane	59	2.31	2.021	4.524	59	8.31	0.8292	5.5731	402.92				
n-Heptane													
n-Octane					43	5.22	0.7558	6.1092	451.68	3.02	1.32	6.74	101.30
n-Nonane													
n-Decane										2.20	1.38	7.16	135.41
n-Dodecane													
n-Tetradecane										6.29	1.60	7.69	267.22
n-Hexadecane	25	5.83	4.911	4.784	25	11.15	0.3944	7.5920	573.33	7.01	1.70	7.89	386.31
n-$C_{30}H_{62}$										6.37	2.10	9.37	684.20
n-$C_{50}H_{102}$										6.72	2.10	10.95	1211.39

n-$C_{78}H_{158}$										2.60	2.60	12.47	1474.90
n-$C_{154}H_{310}$										2.29	4.20	14.80	1831.00
Isopentane	22	5.93	1.480	4.817	22	5.73	0.8739	5.3987	365.60				
Cyclopentane	21	4.42	1.571	4.469	21	4.42	0.8909	5.1088	406.34				
Cyclohexane	39	2.06	1.351	5.026	39	2.64	0.8823	5.4805	440.05				
Tetramethylsilane	42	4.52	2.533	4.297	42	7.88	0.5637	5.6111	356.24				
Benzene	35	4.22	1.585	4.414	65	4.98	0.8823	5.0506	446.35				
Toluene	54	3.46	1.813	4.458	54	6.68	0.8524	5.3176	469.94				
Chlorobenzene													
Chloromethane	42	3.33	1.956	3.278	46	8.62	0.9123	4.0294	330.58				
Dichloromethane	43	8.33	1.888	3.627	43	9.46	0.8893	4.3936	404.99				
Chloroform	39	4.52	1.893	3.923	64	10.16	0.8789	4.7763	425.95				
Carbon tetrachloride	27	2.68	2.017	4.231	37	5.34	0.8925	5.2335	441.83				
Fluoromethane	57	8.12	1.369	3.293									
Trifluoromethane	75	5.36	1.636	3.430	38	9.61	0.8543	4.0968	237.67				
Carbon tetrafluoride					53	7.49	0.9673	4.4646	180.74				
Chlorotrifluoromethane	66	2.10	1.000	4.657	115	13.13	0.8898	4.4789	239.82				
Sulfur hexafluoride					38	13.50	0.8378	4.7229	253.08				
Perfluorocyclobutane	59	4.08	1.102	5.510	59	4.66	0.9673	5.6569	308.50				
Ammonia													
Acetone													
Acetic acid													
Pyridine	55	10.17	1.634	4.206	55	11.45	0.8646	4.8610	492.34				
Acetonitrile	65	5.89	2.303	3.232	60	6.12	0.8098	4.0970	433.18				
Methanol													
Methanol-OD													
Ethanol													
Propanol													
Heavy water													
Water (supercritical)													
Water (all states)													

Table 9.7. Global deviations for the self-diffusion coefficient of real fluids calculated by models described in this section. Non-hydrogen-bonding substances are from Table 9.6 for argon to acetonitrile. Hydrogen-bonding substances are from Table 9.6 for methanol to water

	LJ2 (2 parameters) Eqs. (9.178)+(9.179) Liu et al. [25]		**LJ4** (4 parameters) Eqs. (9.184)+(9.185) Silva et al. [58]		**LT** (4 parameters) Eq. (9.183) Lee and Thodos [117]		**DHB** (2 parameters) Eq. (9.182) Dymond [33]	
Substances	NDP	AAD(%)	NDP	AAD(%)	NDP	AAD(%)	NDP	AAD(%)
Non-hydrogen bonding	2047	5.45	2047	4.49	2047	5.16	2047	29.01
Hydrogen bonding	424	23.43	424	4.29	424	34.11	424	440.78
Total	2471	8.53	2471	4.45	2471	10.13	2471	99.66

	RLJ (1 parameter) Eqs. (9.205)+(9.123)+(9.56) Ruckenstein and Liu [75]		**LJ** (0 parameters) Eqs. (9.207)–(9.209) Zhu et al. [122]		**LJC** (2 parameters) Eqs. (9.216)+(9.211)+(9.56) Yu and Gao [125]		**LJC** (3 parameters) Eqs. (9.192)+(9.193)+(9.186) Salim and Trebble [114]	
Substances	NDP	AAD(%)	NDP	AAD(%)	NDP	AAD(%)	NDP	AAD(%)
Non-hydrogen bonding	1822	7.33	1226	18.51	1081	4.72	na	na
Total	1822	7.33	1226	18.51	1081	4.72	na	na

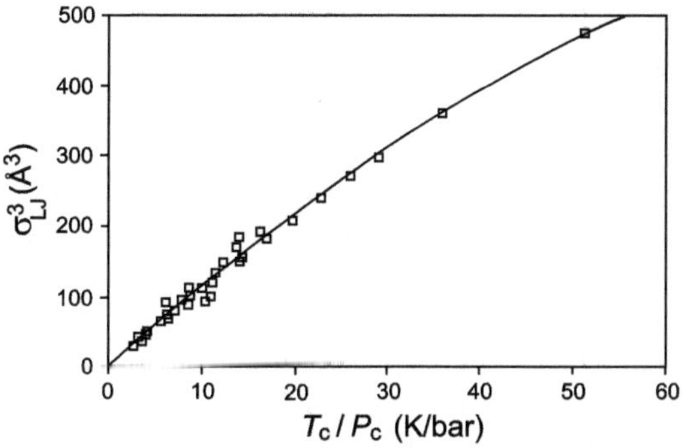

Fig. 9.21. Generalized correlation for the LJ diameter, Eq. (9.181). Data from Silva et al. [58]

$$\varepsilon_{\mathrm{LJ}}/k_{\mathrm{B}} = 0.774\ T_{\mathrm{c}} \tag{9.180}$$

$$\sigma_{\mathrm{LJ}}^3 = 0.17791 + 11.779\ (T_{\mathrm{c}}/P_{\mathrm{c}}) - 0.049029\ (T_{\mathrm{c}}/P_{\mathrm{c}})^2 \tag{9.181}$$

where σ_{LJ}, T_{c}, and P_{c} are in Å, K, and bar, respectively.

According to the original work of Batschinski [35] for real liquids, later reestablished by Hildebrand [34], the self-diffusivities will obey Eq. (9.74). In 1974, results obtained by Dymond for the HS system suggest that the preferred expression should be Eq. (9.115), which we recast as Eq. (9.182) after substituting D_0, Eq. (9.21), for hard spheres:

$$D_{\mathrm{DHB}}^{\mathrm{real}} = B\sqrt{T}(V - V_{\mathrm{D}}) \tag{9.182}$$

where B and V_{D} are adjustable constants for each real fluid.

Lee and Thodos [117] presented the following correlation for generalized predictions for all state conditions:

$$\frac{\tau}{D_{\mathrm{LT}}^{\mathrm{real}}\delta_t}10^5 = \exp(B_1 x^{B_3} + B_2 x^{B_4}) - 1 \tag{9.183}$$

Here, $\delta_{\mathrm{t}} = M^{1/2}T_{\mathrm{t}}^{-1/2}V_{\mathrm{lt}}^{-2/3}$ is a self-diffusivity parameter, T_{t} is the triple-point temperature, V_{lt} is liquid molar volume at the triple point, M is molecular weight, $\tau = T/T_{\mathrm{t}}$ is a normalized temperature, $x = \omega/\tau^{E_1\omega^{E_2}}$ is a density-temperature variable, $\omega = \rho/\rho_{\mathrm{lt}}$ is a normalized density, ρ_{lt} is liquid density at the triple point, $E_1 = 0.09$ and $E_2 = 3.1$ are established and found to conform to universal constants, and, finally, B_1, B_2, B_3, and B_4 are parameters obtained for each fluid. Although the LT model has been originally subjected to an ample validation (58 substances; 975 data points), it exhibits some important drawbacks: (i) it requires the evaluation of four parameters physically meaningless; (ii) the independent thermodynamic variables used (T and V) are normalized introducing the triple-point values as reference basis; (iii) its validation sharply underestimated some very important substances.

For example, they have used only 7 self-diffusivities for carbon dioxide and 88 for methane, while both Liu et al. [25] and Silva et al. [58] employed, respectively, 191 and 369 points.

Table 9.6 shows that the model of Liu et al. [25] (LJ2: Eqs. (9.178) and (9.179)) provides reliable representation for real non-hydrogen-bonding substances ($\mathrm{AAD}^{\mathrm{real}}_{\mathrm{LJ2}} = 5.45\%$). The average deviation is well inside the experimental accuracy (2–10%) and is comparable to that given by the LT four-parameter equation ($\mathrm{AAD}^{\mathrm{real}}_{\mathrm{LT}} = 5.16\%$). With respect to the DHB model, Eq. (9.182), its performance is completely unacceptable ($\mathrm{AAD}^{\mathrm{real}}_{\mathrm{DHB}} = 29.01\%$) and its interest should be confined to correlative purposes in narrow ranges of temperature and pressure. It is well known that it embodies the simplest free-volume expression (i.e. van der Waals) whose applicability is merely limited to low densities (see Sect. 9.3.4). The high-density region exacts for better equations, an example being that of Carnahan-Starling [12] obtained from Monte Carlo simulation results for the compressibility of the HS fluid. All models fail to represent hydrogen-bonding molecules ($\mathrm{AAD}^{\mathrm{real}}_{\mathrm{LJ2}} = 23.43\%$, $\mathrm{AAD}^{\mathrm{real}}_{\mathrm{LT}} = 34.11\%$, $\mathrm{AAD}^{\mathrm{real}}_{\mathrm{DHB}} = 440.78\%$), but it is curious to verify that LT equation is even more unsuitable though involving the double number of parameters.

Silva et al. [58] extended the LJ2 model to calculate self-diffusion coefficients of hydrogen-bonding substances, by adopting the rough hard sphere approach described in Sect. 9.3.6 and introducing a new parameter in the EHSD equation. First of all, as in Eq. (9.95), they introduced a roughness or shape factor (A_D) intended to encompass the effect of coupling between translational and rotational momentum exchange during collisions. Afterwards, they fixed $\alpha_0^* = 2^{1/6}$ in Eq. (9.175) and took T_0^* as new parameter to allow the molecules to personalize the temperature-dependence of their effective diameter. The equation thus achieved (LJ4 model) may be represented by $D^{\mathrm{real}}_{\mathrm{LJ4}} = A_\mathrm{D} D(T, \rho; \sigma_{\mathrm{LJ}}, \varepsilon_{\mathrm{LJ}}/k_\mathrm{B}, T_0^*)$ or, for convenience, $D^{\mathrm{real}}_{\mathrm{LJ4}} = A_\mathrm{D} D(T, \rho; \sigma, E_\mathrm{D}, T_\mathrm{D})$:

$$D^{\mathrm{real}}_{\mathrm{LJ4}} = 21.16 \frac{A_\mathrm{D}}{\rho \sigma^2_{\mathrm{LJ4}}} \sqrt{\frac{1000RT}{M}} \exp\left\{ -\frac{0.75\rho\sigma^3_{\mathrm{LJ4}}}{1.2588 - \rho\sigma^3_{\mathrm{LJ4}}} - \frac{E_\mathrm{D}}{RT} \right\} \tag{9.184}$$

$$\sigma_{\mathrm{LJ4}} = 2^{1/6}\sigma \left[1 + (T/T_\mathrm{D})^{\frac{1}{2}} \right]^{\frac{-1}{6}} \tag{9.185}$$

The individual deviations and parameters A_D, σ, E_D, and T_D found are given in Table 9.6 for the database compiled. Global averages for hydrogen- and non-hydrogen-bonding substances are shown in Table 9.7. Results demonstrate that the LJ4 model is capable to represent the self-diffusivities of all molecules studied, which includes polar, non-spherical and hydrogen-bonded substances, with average deviations well inside the experimental accuracy ($\mathrm{AAD}^{\mathrm{real}}_{\mathrm{LJ4}} = 4.45\%$ for 2471 data points). Particularly welcome is the improvement achieved in relation to the LT equation, Eq. (9.183), for hydrogen-bonding molecules, taking into account that both involve four parameters ($\mathrm{AAD}^{\mathrm{real,hyd}}_{\mathrm{LJ4}} = 4.29\%$ versus $\mathrm{AAD}^{\mathrm{real,hyd}}_{\mathrm{LT}} = 34.11\%$ for 424 data points).

Salim and Trebble [114] published a modified interacting-sphere model for HS chain and LJ chain fluids. Extension to real fluids is also given, being particularly beneficial for n-alkanes larger than n-triacontane. A generalized version of the model is also available and requires only the carbon number in order to predict n-alkane self-diffusivity.

For a chain molecule containing N hard sphere segments of diameter σ and mass m, it is assumed that the self-diffusion coefficient at ideal-gas condition, $D_{0,\mathrm{HSC}}$, is the same as that of a single hard sphere of diameter σ' and mass m', that is

$$D_{0,\mathrm{HSC}} = \frac{3}{8\rho_0\sigma'^2}\left(\frac{k_\mathrm{B}T}{\pi m'}\right)^{\frac{1}{2}} \tag{9.186}$$

where σ' and m' are calculated by

$$\sigma'^3 = N\sigma^3 \tag{9.187}$$

$$m' = Nm \tag{9.188}$$

to preserve total volume and mass of molecules. It is worth noting that Eq. (9.186) is equivalent to Eq. (9.21) for HS, since $\Omega^{(\mathrm{s},1)^*} = 1$ in this case. The reduced density interested for long-chained molecules is

$$\rho'^* = \rho\sigma'^3 = \rho N\sigma^3 \tag{9.189}$$

The self-diffusion coefficient determined according to the parametrization given by Eq. (9.112) allows us to view $G(\rho^*)$ as the probability of molecular transfer, since its value is the reciprocal of the probability of molecular collision. Once more, $G(\rho^*) = 1$ in the ideal-gas limit and $G(\rho^*) = 0$ in the metastable fluid range at the extreme high density. In Sect. 9.4.1 several expressions for the correction factor $G(\rho^*)$ are listed. Salim and Trebble [114] adopted Speedy's model, Eq. (9.120), for the HS system. For long-chained molecules composed of N hard sphere segments, statistically independent transfer events have been assumed, so the transfer probability is $[G(\rho'^*)]^N$. Therefore, the hard sphere chain (HSC) model is written as

$$\frac{D_\mathrm{HSC}}{D_{0,\mathrm{HSC}}} = [G(\rho'^*)]^N = \left\{\left(1-\frac{\rho'^*}{1.09}\right)\left[1+\rho'^{*2}\left(0.4-0.83\rho'^{*2}\right)\right]\right\}^N \tag{9.190}$$

For Lennard-Jones long-chained (LJC) molecules, Salim and Trebble [114] combined their HSC model with: (i) the EHSD of Speedy et al. [24], Eq. (9.55) in Table 9.1, and (ii) the exponential attractive term of the same authors (see Eq. (9.174)):

$$\frac{D_\mathrm{LJC}}{D_{0,\mathrm{HSC}}} = [G(\rho'^*)]^N \exp\left(-\frac{0.5}{T^*}\right) \tag{9.191}$$

The extension to real fluids is straightforward, if the LJC equation is explicitly written as function of T and ρ. In this case, $D_{\mathrm{LJC}}^{\mathrm{real}} = D_{\mathrm{LJC}}(T, \rho; N, \sigma'_{\mathrm{LJ}}, \varepsilon_{\mathrm{LJ}}/k_{\mathrm{B}})$, where N, σ'_{LJ}, and $\varepsilon_{\mathrm{LJ}}/k_{\mathrm{B}}$ are taken as parameters to fit to experimental data. Accordingly,

$$\frac{D_{\mathrm{LJC}}^{\mathrm{real}}}{D_{0,\mathrm{HSC}}} = \left\{\left(1 - \frac{\rho\sigma_{\mathrm{BS}}'^3}{1.09}\right)\left[1 + \left(\rho\sigma_{\mathrm{BS}}'^3\right)^2\left(0.4 - 0.83\left(\rho\sigma_{\mathrm{BS}}'^3\right)^2\right)\right]\right\}^N \times \exp\left(-\frac{0.5\varepsilon_{\mathrm{LJ}}/k_{\mathrm{B}}}{T}\right) \tag{9.192}$$

with

$$\sigma'_{\mathrm{BS}}(T^*) = \sigma'_{\mathrm{LJ}} 2^{\frac{1}{6}}\left[1 + (2T^*)^{\frac{1}{2}}\right]^{-1/6} \tag{9.193}$$

In Table 9.6, calculated results for the modified interacting-sphere model are listed, showing that it provides good representations even for long n-alkanes. Unfortunately, the authors do not indicate the number of data points used in calculations, so it is impossible to present global averages in Table 9.7.

For engineering problems, it is important to have generalized functions. Therefore, based on plots of self-diffusion with respect to carbon number, N_{C}, Salim and Trebble [114] generated the following correlations to predict the parameters involved:

$$N = 1 + 0.36337\,\mathrm{Ln}(N_{\mathrm{C}}), \quad \text{if } N_{\mathrm{C}} \leq 3 \tag{9.194}$$

$$N = -1.20548 + 2.88886\,\mathrm{Ln}(N_{\mathrm{C}}) - 1.10280\,\mathrm{Ln}^2(N_C) + 0.14733\,\mathrm{Ln}^3(N_{\mathrm{C}}), \quad \text{if } 4 \leq N_{\mathrm{C}} \leq 154 \tag{9.195}$$

$$\sigma'_{LJ}(\mathrm{nm}) = 0.40076 + 0.05158\,\mathrm{Ln}(N_C) + 0.03236\,\mathrm{Ln}^2(N_{\mathrm{C}}) \tag{9.196}$$

$$\varepsilon_{\mathrm{LJ}}/k_{\mathrm{B}}(\mathrm{K}) = 45.7211 - 77.5105\,\mathrm{Ln}(N_C) + 224.7692\,\mathrm{Ln}^2(N_{\mathrm{C}}) - 159.437\,\mathrm{Ln}^3(N_{\mathrm{C}}) + 53.98346\,\mathrm{Ln}^4(N_{\mathrm{C}}) - 5.52783\,\mathrm{Ln}^5(N_{\mathrm{C}}) \tag{9.197}$$

This proposed generalization offers significantly worse results for self-diffusivities. Therefore, results based on optimized parameters should always be preferred when data are available.

Ruckenstein and Liu [75] presented models for LJ and real compounds over the entire range of gaseous and liquid states. The parameters are the LJ force constants, σ_{LJ} and $\varepsilon_{\mathrm{LJ}}$, and a translational-rotational factor equivalent to that of the rough hard sphere theory, A_{D}. Generalized correlations are given for A_{D} and $\varepsilon_{\mathrm{LJ}}$. In the following a brief description of the models is given.

The self-diffusion coefficient may be expressed in terms of repulsive and soft attractive friction coefficients, ζ_{R} and ζ_{S}, as follows [119]:

$$D = \frac{k_{\mathrm{B}}T}{\zeta_{\mathrm{R}} + \zeta_{\mathrm{S}}} \tag{9.198}$$

For the HS fluid at low density, the following equation may be derived [119, 120]:

$$\zeta_{\mathrm{R}} = \frac{8}{3}\rho g(\sigma)\sigma^2\sqrt{\pi m k_{\mathrm{B}} T} \tag{9.199}$$

which combined with Eq. (9.198) gives rise to the famous Enskog equation, Eq. (9.24). Extension to higher densities demands the well-known corrections described in Sects. 9.3.1 and 9.4.1 (identified by $F(\rho^*)$ and $G(\rho^*)$ in Eqs. (9.111) and (9.112)) which Ruckenstein and Liu take into account with Eq. (9.123). Adding on the attractive contribution, ζ_{S}, and adopting an EHSD, σ_{eff}, the following equation for the LJ self-diffusion coefficient is obtained:

$$D_{\mathrm{LJ}} = \frac{k_{\mathrm{B}}T}{\frac{8}{3}\rho g(\sigma_{\mathrm{eff}})\sigma_{\mathrm{eff}}^2(\pi m k_{\mathrm{B}}T)^{\frac{1}{2}}F^{-1}(\rho^*) + \zeta_{\mathrm{S}}} \tag{9.200}$$

The soft friction coefficient of the LJ fluid was determined to be well approximated by

$$\zeta_{\mathrm{S}} = \frac{k_{\mathrm{B}}T}{D_0}\frac{0.4}{T^{*1.5}} = \frac{8}{3}\rho\sigma_{\mathrm{eff}}^2\sqrt{\pi m k_{\mathrm{B}}T}\frac{0.4}{T^{*1.5}} \tag{9.201}$$

The final model is obtained by substituting this result in Eq. (9.200), further combined with the EHSD of Ben-Amotz and Herschbach [22] according to Boltzmann criterion, Eq. (9.56) in Table 9.1. The radial distribution function at contact is that of Carnahan-Starling, Eq. (9.32). Thus, in the canonical reduced form,

$$D_{\mathrm{LJ}}^* = \frac{\sqrt{T^*}}{\frac{8\sqrt{\pi}}{3}\left[\sigma_{\mathrm{BAH}}^*(T^*)\right]^2\left[\frac{g(\sigma_{\mathrm{BAH}})}{F(\rho^*)} + \frac{0.4}{T^{*1.5}}\right]} \tag{9.202}$$

where $T^* = k_{\mathrm{B}}T/\varepsilon_{\mathrm{LJ}}$, and D_{LJ}^* is defined by Eq. (9.177). Ruckenstein and Liu [75] compiled a database with 411 LJ self-diffusion coefficients from literature, and achieved an average absolute deviation of 5.66%.

To extend Eq. (9.202) to real compounds, the RHS theory has been taken into account by multiplying a frontal parameter A_{D} independent of temperature and density (see Eq. (9.95)). Hence, $D_{\mathrm{RLJ}}^{\mathrm{real}} = A_{\mathrm{D}}D(T,\rho;\sigma_{\mathrm{LJ}},\varepsilon_{\mathrm{LJ}}/k_{\mathrm{B}})$, where RLJ denotes "rough LJ". As $D_{\mathrm{RLJ}}^{\mathrm{real}}$ is much more sensitive to σ_{LJ} than to $\varepsilon_{\mathrm{LJ}}/k_{\mathrm{B}}$, this last has been fixed by the relation between the energy parameter and the critical temperature, obtained from viscosity data [121]:

$$\frac{\varepsilon_{\mathrm{LJ}}}{k_{\mathrm{B}}} = \frac{T_{\mathrm{c}}}{1.2593} \tag{9.203}$$

It is worth noting that a change of 1% in the diameter may lead to a variation in the diffusion coefficient by 10–20%. Furthermore, since the coupling factor is expected to depend on the molecular size and shape, it has been correlated with the acentric factor, ω, using diffusion data:

$$A_{\mathrm{D}} = 0.9673 - 0.2527\omega - 0.70\omega^2 \tag{9.204}$$

Concluding, the model of Ruckenstein and Liu for real substances has only σ_{LJ} as adjustable parameter to fit to experimental data: $D_{\mathrm{RLJ}}^{\mathrm{real}} = A_{\mathrm{D}}(\omega) D(T, \rho; \sigma_{\mathrm{LJ}})$. Hence,

$$D_{\mathrm{RLJ}}^{\mathrm{real}} = A_{\mathrm{D}}(\omega) \frac{k_{\mathrm{B}} T}{\frac{8}{3} \rho \sigma_{\mathrm{BAH}}^2 (\pi m k_{\mathrm{B}} T)^{1/2} \left[\frac{g(\sigma_{\mathrm{BAH}})}{F(\rho^*)} + \frac{0.4}{T^{*1.5}} \right]} \tag{9.205}$$

Table 9.6 lists calculation results for 26 molecules with 1822 data points. For most compounds, the AAD is comparable to the experimental uncertainty, but higher deviations are found for long-chain or polar molecules. The global AAD for all points is 7.33% (see Table 9.7). For compounds not covered in their essay, the diameter may be estimated by the correlation [75]

$$\sigma_{\mathrm{LJ}} = 0.7889 V_{\mathrm{c}}^{1/3} \tag{9.206}$$

Nonetheless, the resulting predictive model is only acceptable for small non-polar and weakly polar molecules, otherwise deviations can attain more than 30%.

Later, Zhu et al. [122] developed another equation for the LJ system, with the ultimate purpose to represent pure and infinite dilute coefficients of real fluids. The model is based on the HS expression of Speedy [87], Eq. (9.120), where eight parameters have been introduced, and embodies the usual exponential attractive contribution (Eq. (9.174)):

$$\begin{aligned} D_{\mathrm{LJ}}^* = \frac{D_{\mathrm{LJ}}}{\sqrt{\sigma_{\mathrm{LJ}}^2 \varepsilon_{\mathrm{LJ}}/m}} &= \frac{3}{8\sqrt{\pi}} \frac{\sqrt{T^*}}{\rho^*} \left(1 - \frac{\rho^*}{1.029079 T^{*0.165377}} \right) \\ &\times \left[1 + \rho^{*0.126978} \left(\frac{0.596103\,(\rho^* - 1)}{0.539292\,(\rho^* - 1) + T^{*(0.400152 - 0.41054\rho^*)}} + 0.68856 \right) \right] \\ &\times \exp\left(-\frac{\rho^*}{2T^*} \right) \end{aligned} \tag{9.207}$$

This equation is consistent with the MD simulation results for the LJ system reported by Rowley and Painter [123] (182 data points), and can be used over $0 \leq \rho^* \leq 1.0$ and $0.8 \leq T^* \leq 4.0$ with AAD = 4.95%. The normalization of D_{LJ} is different from that of Eq. (9.177), because that includes a factor ρ^*. Please note, for comparison, that the LJ model of Liu et al. [25], Eq. (9.176), gives AAD=5.62% for 435 data points collected from different authors, and is applicable over a much larger temperature and density ranges: $0.72 < T^* < \infty$ and $\rho^* \leq 1.22$.

To apply Eq. (9.207) to real fluids, Zhu et al. [122] proposed relations to estimate the LJ force constants of each substance:

$$\varepsilon_{\mathrm{LJ}}/k_{\mathrm{B}}\ (\mathrm{K}) = \frac{T_{\mathrm{c}}}{T_{\mathrm{c}}^*} [1 + 0.47527332 \rho_{\mathrm{r}} + (0.06300484 + 0.12374707 \rho_{\mathrm{r}})\, T_{\mathrm{r}}] \tag{9.208}$$

$$\sigma_{\mathrm{LJ}}\ (\mathrm{cm}) = \left(\frac{\rho_c^*}{\rho_c}\right)^{\frac{1}{3}} [1 - 0.0368868\rho_r + (0.00006945 + 0.01089228\ \rho_r)\ T_r] \tag{9.209}$$

where $T_c^* = 1.313$ and $\rho_c^* = 0.31$ is the LJ critical point published by Johnson et al. [124], and T_c(K), ρ_c (cm^{-3}), and $\rho_r = \rho/\rho_c$, are the critical temperature, critical density, and reduced density, respectively. Once more, $T^* = k_B T/\varepsilon_{\mathrm{LJ}}$ and $\rho^* = \rho\sigma_{\mathrm{LJ}}^3$ are considered in Eq. (9.207). It has been validated with 1226 data points of 17 substances, giving rise to 18.51% (see Table 9.7). This model is totally predictive, being recommended when there are no parameters reported to utilize others equations or experimental self-diffusivities to optimize them previously.

To finish this subject on self-diffusion coefficient of modified HS systems, another model for polyatomic molecules due to Yu and Gao [125] will be described. It comprehends three contributions: a temperature-dependent HS part, a chain friction part, and a soft friction contribution.

Yu and Gao started by researching model fluids consisting of chains of freely tangent hard spheres. Accordingly, the MD simulations of hard sphere chains (HSC) composed of 2, 4, 8, and 16 particles carried out by Smith et al. [126] have been used to develop an equation for D_{HSC}. Similarly to Salim and Trebble [114], the coefficient of an HSC molecule in the ideal-gas condition is calculated by Eqs. (9.186)–(9.188). At higher densities, the correcting procedure analogous to that formulated for single spheres, Eq. (9.111), has been adopted:

$$\rho D_{\mathrm{HSC}} = F(N,\rho'^*)(\rho D_{\mathrm{E}}) = F(N,\rho'^*)\frac{\rho_0 D_{0,\mathrm{HSC}}}{g(\sigma')} \tag{9.210}$$

where $\rho'^* = \rho\sigma'^3 = \rho N\sigma^3$, Eq. (9.189), and the radial distribution function at contact is calculated by the Carnahan-Starling equation, Eq. (9.32). Taking into account the MD data mentioned above, the authors proposed the following correction factor:

$$F(N,\rho'^*) = F(\rho'^*) \times \exp\left[-0.06356(N-1) - 0.05212\tfrac{N-1}{N}\rho'^* - 1.9709\left(\tfrac{N-1}{N}\right)^2\rho'^*\right] \tag{9.211}$$

where constants −0.06356, −0.05212, and −1.9709 have been optimized, using the HS expression of Ruckenstein and Liu [75] for $F(\rho'^*)$, now rewritten in terms of ρ'^*:

$$F(\rho'^*) = 1 + 0.94605\rho'^{*1.5} + 1.4022\rho'^{*3} - 5.6898\rho'^{*5} + 2.6626\rho'^{*7} \tag{9.212}$$

The resulting HSC model was validated at reduced densities from 0.191 to 0.955 and for particles of length 2, 4, 8, and 16, giving rise to an average absolute deviation of 3.95%. Figure 9.22 exhibits the MD data along with the calculated values by Eqs. (9.210)–(9.212), pointing out the good fitting achieved.

The extension to the Lennard-Jones chain (LJC) fluid is accomplished according to the fundamentals of Rice and Gray [119], i.e. $D_{\mathrm{LJC}} = k_B T/\zeta$,

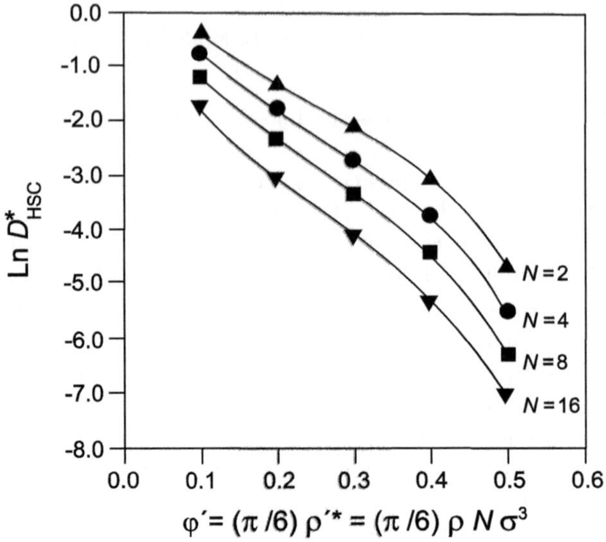

Fig. 9.22. Reduced self-diffusion coefficient of Yu and Gao [125] for the hard sphere chain fluid, Eqs. (9.210)–(9.212). MD data from Smith et al. [126]

where friction coefficient is modified to integrate a chain contribution (ζ_C) in addition to the common repulsive (ζ_R) and soft attractive (ζ_S) parts shown in Eq. (9.198):

$$\zeta_{LJC} = \zeta_R + \zeta_C + \zeta_S = \zeta_{HSC} + \zeta_S \tag{9.213}$$

where ζ_{HSC} can be obtained from Eq. (9.210):

$$\zeta_{HSC} = \frac{k_B T}{D_{HSC}} = \frac{k_B T}{D_{0,HSC}} \frac{g(\sigma')}{F(N, \rho'*)} \tag{9.214}$$

The attractive contribution is taken from that derived by Ruckenstein and Liu [75] for the LJ system, replacing D_0 in Eq. (9.201) by $D_{0,HSC}$:

$$\zeta_S = \frac{k_B T}{D_{0,HSC}} \frac{0.4}{T^{*1.5}} \tag{9.215}$$

The final model for D_{LJC} combines the friction coefficients and the EHSD of Ben-Amotz and Herschbach [22] according to Boltzmann criterion (Eq. (9.56) in Table 9.1)

$$D_{LJC} = \frac{D_{0,HSC}}{g(\sigma_{BAH})/F(N, \rho'^*) + 0.4/T^{*1.5}} \tag{9.216}$$

At first glance, its direct extension to real fluids comprises three parameters $(N, \sigma_{LJ}, and\ \varepsilon_{LJ}/k_B)$. Nonetheless, since D_{LJC} is much more sensitive to the segment diameter and chain length than to the energy, this has been fixed analogously to Eq. (9.203):

$$\frac{\varepsilon_{\mathrm{LJ}}}{k_{\mathrm{B}}}N = \frac{T_{\mathrm{c}}}{1.2593} \tag{9.217}$$

Hence, the LJC-based model for real fluids due to Yu and Gao [125] involves only two adjustable parameters, σ_{LJ} and N, and may be represented by $D_{\mathrm{LJC}}^{\mathrm{real}} = D(T, \rho; N, \sigma_{\mathrm{LJ}})$. It has been used to calculate the self-diffusion coefficients of 1081 data points of 22 polyatomic substances over wide temperature and pressure ranges, and the total AAD is 4.72% (see Table 9.7). Individual parameters and deviations obtained are compiled in Table 9.6. The equation has been compared with the rough LJ model of Ruckenstein and Liu [75], which performs slightly better (AAD = 4.48%) despite containing only one parameter. In Fig. 9.23, the self-diffusion coefficient of acetonitrile is plotted against pressure at constant temperature to point out the good correlation achieved by Yu and Gao's model.

Before proceeding to mutual diffusion coefficients, it is worth remembering that the entropy scaled laws introduced in Sect. 9.3.7, and later described by Eqs. (9.128)–(9.132), are mainly applicable to real substances in liquid domain. For instance, a series of simple melts, including Ag, Al, Au, Cd, Co, Cs, Cu, Fe, Ga, Hg, K, Li, Mg, Na, Ni, Pb, Pd, Pt, Rb, Rh, Sb, Sn, and Zn, have been studied by several authors (e.g. [82, 84, 128, 129]).

Tracer and IntraDiffusion Coefficients of Modified Hard Sphere Systems

In this section some models for the diffusion coefficient of LJ and real fluids are described, namely those by Liu et al. [76], Salim and Trebble [114], Liu

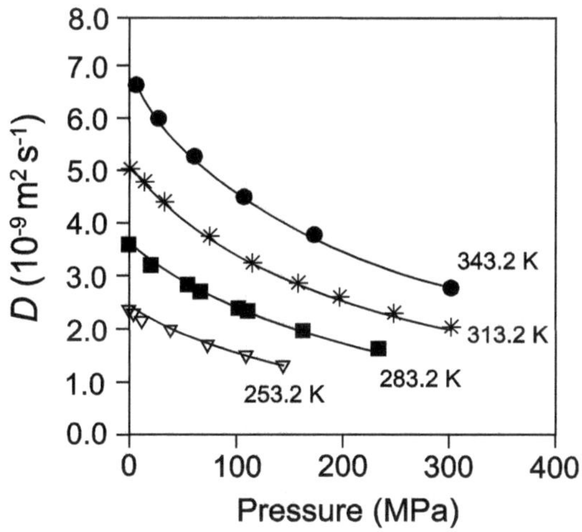

Fig. 9.23. Self-diffusion coefficient of acetonitrile plotted against pressure at different temperatures calculated by the model of Yu and Gao [125]. Experimental data from Hurle and Woolf [127]

and Ruckenstein [110], Zhu et al. [122], and Merzliak and Pfennig [130]. All of them have been grafted onto former HS equations discussed above, allowing us to enlighten how multicomponent expressions for real and LJ substances may be generated from simple model fluids.

The reliable results accomplished by the LJ2 model of Liu et al. [25] for the self-diffusion coefficient induced them to propose an empirical extension of Eq. (9.178) to calculate limiting mutual diffusion coefficients of real substances [76], $D_{12,\mathrm{TLJ2}}^{\mathrm{real}}$:

$$D_{12,\mathrm{TLJ2}}^{\mathrm{real}} = 21.16\sqrt{\frac{1000RT}{2M_{12}}}\frac{V_2}{N_{\mathrm{a}}\sigma_{12,\mathrm{BLSM}}^2}\exp\left(-\frac{0.75\rho_2^*}{1.2588-\rho_2^*}-\frac{0.27862}{T_{12}^*}\right) \tag{9.218}$$

where T_{12}^* and $\sigma_{12,\mathrm{BLSM}}$ are the reduced temperature and the BLSM-effective diameter, Eq. (9.57) in Table 9.1 or Eq. (9.179), calculated in terms of binary LJ parameters, $\sigma_{12,\mathrm{LJ}}$ and $\varepsilon_{12,\mathrm{LJ}}$:

$$T_{12}^* = \frac{T}{\varepsilon_{12,\mathrm{LJ}}/k_{\mathrm{B}}} \tag{9.219}$$

$$\sigma_{12,\mathrm{BLSM}} = \sigma_{12,\mathrm{LJ}}2^{\frac{1}{6}}\left[1+\sqrt{1.3229\ T_{12}^*}\right]^{\frac{-1}{6}} \tag{9.220}$$

M_{12} (g mol^{-1}) is the reduced molar mass of the system:

$$M_{12} = \frac{M_1M_2}{M_1+M_2}, \tag{9.221}$$

V_2 is the solvent molar volume, and ρ_2^* is the reduced number density of solvent calculated using effective diameter:

$$\rho_2^* = \rho_2\sigma_{2,\mathrm{eff}}^3 = \rho_2\sigma_{2,\mathrm{BLSM}}^3 = \frac{N_{\mathrm{a}}}{V_2}\sigma_{2,\mathrm{BLSM}}^3 \tag{9.222}$$

It is now possible to devise a predictive model and two correlations. The first one is established with the following energy and diameter combining rules:

$$\varepsilon_{12,\mathrm{LJ}}/k_{\mathrm{B}} = \frac{\sqrt{\sigma_{1,\mathrm{LJ}}^3\left(\varepsilon_{1,\mathrm{LJ}}/k_{\mathrm{B}}\right)\times\sigma_{2,\mathrm{LJ}}^3\left(\varepsilon_{2,\mathrm{LJ}}/k_{\mathrm{B}}\right)}}{\sigma_{12,\mathrm{LJ}}^3} \tag{9.223}$$

$$\sigma_{12,\mathrm{LJ}} = \frac{\sigma_{1,\mathrm{LJ}}+\sigma_{2,\mathrm{LJ}}}{2} \tag{9.224}$$

This model will henceforth be labelled by TLJ2, where "T" stands for Tracer. Two correlations appear when interaction parameters (k_{12}^{en} and k_{12}^{d}) are introduced in these combining rules. Accordingly, the correlation model with interaction energy parameter (TLJ2$_{\mathrm{en}}$) has been defined with the following combining rule for $\varepsilon_{12,\mathrm{LJ}}/k_{\mathrm{B}}$:

$$\varepsilon_{12,\mathrm{LJ}}/k_{\mathrm{B}} = (1 - k_{12}^{\mathrm{en}}) \frac{\sqrt{\sigma_{1,\mathrm{LJ}}^{3} (\varepsilon_{1,\mathrm{LJ}}/k_{\mathrm{B}}) \times \sigma_{2,\mathrm{LJ}}^{3} (\varepsilon_{2,\mathrm{LJ}}/k_{\mathrm{B}})}}{\sigma_{12,\mathrm{LJ}}^{3}} \tag{9.225}$$

where k_{12}^{en} is an adjustable parameter and $\sigma_{12,\mathrm{LJ}}$ is given by Eq. (9.224). The correlation model with interaction diameter parameter (TLJ2$_{\mathrm{d}}$) is defined with the following combining rule for $\sigma_{12,\mathrm{LJ}}$:

$$\sigma_{12,\mathrm{LJ}} = (1 - k_{12}^{\mathrm{d}}) \frac{\sigma_{1,\mathrm{LJ}} + \sigma_{2,\mathrm{LJ}}}{2} \tag{9.226}$$

where k_{12}^{d} is an adjustable parameter, and $\varepsilon_{12,\mathrm{LJ}}/k_{\mathrm{B}}$ is given by Eq. (9.223) with $\sigma_{12,\mathrm{LJ}}$ computed by Eq. (9.224), that is

$$\varepsilon_{12,\mathrm{LJ}}/k_{\mathrm{B}} = 8 \frac{\sqrt{\sigma_{1,\mathrm{LJ}}^{3} (\varepsilon_{1,\mathrm{LJ}}/k_{\mathrm{B}}) \times \sigma_{2,\mathrm{LJ}}^{3} (\varepsilon_{2,\mathrm{LJ}}/k_{\mathrm{B}})}}{(\sigma_{1,\mathrm{LJ}} + \sigma_{2,\mathrm{LJ}})^{3}} \tag{9.227}$$

With regard to TLJ2, the only input data required are temperature, solvent density, and, for both components, molecular weight and LJ force constants. The molecular weight and LJ force constants are available for a large number of substances in Table 9.6, or may be easily estimated from the knowing of the critical temperature and pressure, Eqs. (9.180) and (9.181). The calculation results in Table 9.8 show that the TLJ2 model is reliable for liquids and supercritical fluids (SCF) (with global AAD = 14.77%). Better performance for liquids (AAD = 9.58%) is expected in advance, since most self-diffusivity data used in the optimization of the LJ parameters refer to that state [25, 58]. Detailed results may be consulted in [76].

Concerning the one-parameter correlations TLJ2$_{\mathrm{en}}$ and TLJ2$_{\mathrm{d}}$, it has been found that both provide similar results, being irrelevant the type of correction performed (energy or diameter): from Table 9.8, AAD$_{\mathrm{en}}$ = 6.50% and AAD$_{\mathrm{d}}$ = 6.57%. One binary parameter is sufficient to accomplish good representations of tracer diffusivities of all systems studied by Liu et al. [76], which makes them confident correlations. The interaction parameters, k_{12}^{en} or k_{12}^{d}, needed to estimate diffusion coefficient may be taken from Table 9.3, page 250, of [76].

Concluding, the TLJ2 model is recommended to predict tracer coefficients with acceptable accuracy, if there is no experimental data available. On the other hand, when some few data points can be found, correlations TLJ2$_{\mathrm{en}}$ and TLJ2$_{\mathrm{d}}$ may be used with confidence, by first optimizing the binary interaction parameter embodied.

Salim and Trebble [114] derived a correlation for the tracer diffusion coefficient of real fluids, $D_{12,\mathrm{TLJC}}^{\mathrm{real}}$, based on their LJC expression, Eq. (9.192), for self-diffusivities of long-chained particles. Remember this model deals with molecules composed of N segments, whose diameters and masses are related by Eqs. (9.187) and (9.188). Label TLJC denotes Tracer LJC. Final model is generated introducing appropriate binary parameters and the correction factors ratio G_{12}/G_{22} in Eq. (9.192):

Table 9.8. Models for tracer diffusion coefficient of real substances (NDP = number of data points, SCF = supercritical fluids)

Label	TLJ2	$TLJ2_{en}$	$TLJ2_{d}$	TLJC	TRLJ	$TRLJ_{Ad}$	TLJ-Zhu
No. of parameters	0	1	1	1	0	1	0
Parameters	-	k_{12}^{en}	k_{12}^{d}	k_{12}^{en}	-	A_D	-
Main equation	(9.218)	(9.218)	(9.218)	(9.228)	(9.243)	(9.243)	(9.246)
Model equations	(9.218)–(9.224)	(9.218)–(9.222)+ (9.224)+(9.225)	(9.218)–(9.222)+ (9.226)+(9.227)	(9.228)– (9.238)	(9.243)–(9.245)+ (9.135)+(9.140)+ (9.165)+(9.203)+ (9.206)+(9.219)+ (9.240)+(9.241)	(9.243)+(9.135)+ (9.140)+(9.165)+ (9.203)+(9.206)+ (9.219)+(9.240)+ (9.241)	(9.246)– (9.249)+ (9.208)+ (9.209)
AAD (%)	SCF: 16.88 Liquid: 9.58 Total: 14.77	SCF: 7.16 Liquid: 4.88 Total: 6.50	SCF: 7.17 Liquid: 5.10 Total: 6.57	2.48	Gases: 6.42 SCF: 9.24 Liquid: 7.74 Total: 8.42	Gases: 4.18 SCF: 6.24 Liquid: 3.72 Total: 5.10	17.32
NDP in database	SCF: 734 Liquid: 299 Total: 1033	SCF: 734 Liquid: 299 Total: 1033	SCF: 734 Liquid: 299 Total: 1033	na	Gases: 157 SCF: 793 Liquid: 493 Total: 1443	Gases: 157 SCF: 793 Liquid: 493 Total: 1443	1141
No. of systems in database	77	77	77	17	120	120	74
Reference	[76]	[76]	[76]	[114]	[110]	[110]	[122]
Detailed results in original paper	AADs Table 9.3 P. 250	AADs + k_{12}^{en} Table 9.3 P. 250	AADs + k_{12}^{d} Table 9.3 P. 250	AADs + k_{12}^{en} Table 9.5 P. 249	AADs Table 9.5 P. 5494	AADs + A_D Table 9.5 P. 5494	AADs Table 9.7 P. 1151

$$D^{\text{real}}_{12,\text{TLJC}} = D^0_{12,\text{HSC}} \times \exp\left(-\frac{0.5\varepsilon_{12,\text{LJ}}/k_\text{B}}{T}\right)\frac{G_{12}}{G_{22}} \tag{9.228}$$
$$\times\left[\left(1-\rho_2\sigma'^3_{2,\text{BS}}/1.09\right)\left[1+(\rho_2\sigma'^3_{2,\text{BS}})^2\left(0.4-0.83\left(\rho_2\sigma'^3_{2,\text{BS}}\right)^2\right)\right]\right]^{N_2}$$

where N_2 is the parameter N of the solvent, $D^0_{12,\text{HSC}}$ is the ideal-gas binary diffusivity for the interested HSC system:

$$D^0_{12,\text{HSC}} = \frac{3}{8\rho_2\sigma'^2_{12,\text{BS}}}\left(\frac{k_\text{B}T}{2\pi m'_{12}}\right)^{\frac{1}{2}} \tag{9.229}$$

$$m'_{12} = \frac{m'_1 m'_2}{m'_1 + m'_2} = \frac{M'_1 M'_2}{M'_1 + M'_2} \tag{9.230}$$

$\sigma'_{2,\text{BS}}$ and $\sigma'_{12,\text{BS}}$ are the EHSD of the solvent and the effective binary diameter of the system, respectively, both calculated by the equation of Speedy et al. [24] (Eq. (9.55) in Table 9.1):

$$\sigma'_{2,\text{BS}}(T) = \sigma'_{2,\text{LJ}}2^{1/6}\left[1+\left(\frac{2T}{\varepsilon_{2,\text{LJ}}/k_\text{B}}\right)^{1/2}\right]^{\frac{-1}{6}} \tag{9.231}$$

$$\sigma'_{12,\text{BS}}(T) = \sigma'_{12,\text{LJ}}2^{1/6}\left[1+\left(\frac{2T}{\varepsilon_{12,\text{LJ}}/k_\text{B}}\right)^{1/2}\right]^{\frac{-1}{6}} \tag{9.232}$$

The binary parameters $\sigma'_{2,\text{LJ}}$ and $\varepsilon_{12,\text{LJ}}/k_\text{B}$ are determined in terms of pure component values by common combining rules:

$$\sigma'_{12,\text{LJ}} = \frac{\sigma'_{1,\text{LJ}} + \sigma'_{2,\text{LJ}}}{2} \tag{9.233}$$

$$\varepsilon_{12,\text{LJ}}/k_\text{B} = (1-k^{\text{en}}_{12})\sqrt{(\varepsilon_{1,\text{LJ}}/k_\text{B})\times(\varepsilon_{2,\text{LJ}}/k_\text{B})} \tag{9.234}$$

where k^{en}_{12} is a binary interaction parameter to fit to experimental data. For G_{12}/G_{22}, Salim and Trebble adopted the same dependence proposed by Easteal and Woolf [111] for F_{12}/F_{22}:

$$\text{Ln}(G_{12}/G_{22}) = a_0 + a_1\,\text{Ln}(m'_1/m'_2) + a_2\,\text{Ln}^2(m'_1/m'_2) \tag{9.235}$$

and used the coefficients they have optimized (see commentary to Eqs. (9.156)–(9.158)):

$$a_0 = -1.02564\;\rho'^{*4}_2\,\text{Ln}(\sigma'_{1,\text{BS}}/\sigma'_{2,\text{BS}}) \tag{9.236}$$

$$a_1 = -0.24107 + 1.27589\;h'_{12} - 1.35439\;h'^2_{12} + 0.62393\;h'^3_{12} \tag{9.237}$$

$$a_2 = -0.056 \tag{9.238}$$

where $h'_{12} = \rho_2'^{*^2}(\sigma'_{1,\mathrm{BS}}/\sigma'_{2,\mathrm{BS}})$, $\rho_2'^{*} = \rho_2\sigma'^3_{2,\mathrm{BS}}$, and $\sigma'_{1,\mathrm{BS}}$ is the BS EHSD of the solute calculated by Eq. (9.55) also. Note that G_{22} corresponds to $[G(\rho'^{*})]^N$ in Eqs. (9.190) and (9.191), being given by the last term in Eq. (9.228).

Salim and Trebble [114] tested their $D^{\mathrm{real}}_{12,\mathrm{TLJC}}$ correlation with infinite dilution mutual diffusivities of n-alkanes (17 binary systems) reported by Matthews and Akgerman [131] and Erkey and Akgerman [132]. The binary interaction parameter of the model, k^{en}_{12}, was optimized using pure component parameters N_i, $\sigma'_{i,\mathrm{LJ}}$, and $\varepsilon_{i,\mathrm{LJ}}/k_\mathrm{B}$ $(i = 1, 2)$ estimated by Eqs. (9.194)–(9.197). The correlation was able to match binary diffusivity data to the overall AAD of 2.48%. In Table 9.8, most important specifications and results of the model are listed. Detailed results may be consulted in [114].

Liu and Ruckenstein [110] derived an equation for the tracer diffusivities of real substances based on the rough Lennard-Jones approach. It will be denoted by initials TRLJ. They started by extending their LJ expression, Eq. (9.202), to binary mixtures at infinite dilution as

$$D_{12,\mathrm{LJ}} = \frac{k_\mathrm{B}T}{\frac{8}{3}\rho_2\sigma^2_{12,\mathrm{BAH}}(2\pi m_{12}k_\mathrm{B}T)^{1/2}\left[\frac{g_{12}(\sigma_{12,\mathrm{BAH}})}{F_{12}} + \frac{0.4}{T^{*1.5}_{12}}\right]} \tag{9.239}$$

where $\sigma_{12,\mathrm{BAH}}$ is the effective diameter of the system computed by the expression of Ben-Amotz and Herschbach [22] according to Boltzmann criterion, Eq. (9.56) in Table 9.1:

$$\sigma_{12,\mathrm{BAH}} = 1.1532\sigma_{12,\mathrm{LJ}}\left[1 + (1.8975T^*_{12})^{1/2}\right]^{-1/6} \tag{9.240}$$

m_{12} is given by Eq. (9.135), $g_{12}(\sigma_{12,\mathrm{BAH}})$ by Eq. (9.140), T^*_{12} by Eq. (9.219), and F_{12} by Eq. (9.165) (where diameters σ_1 and σ_2 are replaced by the BAH EHSDs, and $\rho^*_2 = \rho_2\sigma^3_{2,\mathrm{BAH}}$). Binary diameter and energy parameters are determined by Lorentz-Berthelot combining rules:

$$\sigma_{12,\mathrm{LJ}} = \frac{\sigma_{1,\mathrm{LJ}} + \sigma_{2,\mathrm{LJ}}}{2} \qquad \varepsilon_{12,\mathrm{LJ}}/k_\mathrm{B} = \sqrt{(\varepsilon_{1,\mathrm{LJ}}/k_\mathrm{B})(\varepsilon_{2,\mathrm{LJ}}/k_\mathrm{B})} \tag{9.241}$$

To take into account the polyatomic or non-spherical nature of real molecules, Eq. (9.239) is multiplied by a roughness factor independent of density and temperature, A_D, as described in Sect. 9.3.6. Furthermore, as has been pointed out in Sect. 9.4.1, Eq. (9.165) for F_{12} is valid in limited ranges of molecular size and mass ratios: $0.25 < \sigma_1/\sigma_2 < 1.0$, and $0.0002 < m_1/m_2 < 4.0$. Therefore, to enlarge the applicability of Eq. (9.239) to real systems outside these ranges, one factor empirically determined by fitting experimental data has been also introduced as follows:

$$\begin{aligned} D^{\mathrm{real}}_{12,\mathrm{TRLJ}} &= A_\mathrm{D}\left(\frac{\sigma_{1,\mathrm{LJ}}}{\sigma_{2,\mathrm{LJ}}}\right)^{0.42}\left(\frac{m_1}{m_2}\right)^{0.06} D_{12,\mathrm{LJ}} \\ &= A_\mathrm{D}\left(\frac{V_{\mathrm{C}1}}{V_{\mathrm{C}2}}\right)^{0.14}\left(\frac{m_1}{m_2}\right)^{0.06} D_{12,\mathrm{LJ}} \end{aligned} \tag{9.242}$$

The relation between diameter and critical volume is given by Eq. (9.206). Substituting Eq. (9.239) into (9.242), final model arises:

$$D_{12,\mathrm{TRLJ}}^{\mathrm{real}} = \frac{A_{\mathrm{D}}\left(V_{\mathrm{C1}}/V_{\mathrm{C2}}\right)^{0.14}\left(m_1/m_2\right)^{0.06} k_{\mathrm{B}}T}{\frac{8}{3}\rho_2\sigma_{12,\mathrm{BAH}}^2(2\pi m_{12}k_{\mathrm{B}}T)^{1/2}\left[\frac{g_{12}(\sigma_{12,\mathrm{BAH}})}{F_{12}} + \frac{0.4}{T_{12}^{*1.5}}\right]} \tag{9.243}$$

To generate a predictive equation, the correlations given in Eqs. (9.203) and (9.206) for molecular energy and size parameters, respectively, are used. For the coupling factor they use

$$A_{\mathrm{D}} = 1 - 0.0133\omega_{12} - 0.3662\omega_{12}^2 \tag{9.244}$$

where ω_{12} is the binary acentric factor. Equation (9.244) was obtained by Liu and Ruckenstein [110], after plotting the roughness factors optimized from experimental data against ω_{12}. The combining rule assumed is

$$\omega_{12} = (\omega_1 + \omega_2)/2 \tag{9.245}$$

The predictive model, TRLJ, has been validated with 1443 data points for 120 binary systems. Results accomplished are compiled in Table 9.7, and point out that its good performance compensates the model complexity. Actually, total average absolute deviation is only 8.42%. For most systems, predicted errors are comparable to experimental uncertainties. Nonetheless, Eq. (9.243) is not applicable to hydrogen-bonding systems.

When the translational-rotational factor, A_{D}, is taken as adjustable constant characteristic of each pair of components, a one-parameter correlation for tracer diffusivity arises. It is denoted by $\mathrm{TRLJ}_{\mathrm{Ad}}$ and may be represented mathematically as $D_{12,\mathrm{TRLJ}_{\mathrm{Ad}}}^{\mathrm{real}} = D(\rho, T; A_{\mathrm{D}})$. In this case, global AAD diminishes to 5.10%. Detailed results for both TRLJ and $\mathrm{TRLJ}_{\mathrm{Ad}}$ may be found in Liu and Ruckenstein [110] p. 5494.

The model of Zhu et al. [122] for the binary diffusion coefficient of real systems at infinite dilution (TLJ-Zhu) is simply obtained from their self-diffusion equation, Eq. (9.207), by just substituting $(\sigma_{\mathrm{LJ}}, \varepsilon_{\mathrm{LJ}}/k_{\mathrm{B}}, m)$ by $(\sigma_{12,\mathrm{LJ}}, \varepsilon_{12,\mathrm{LJ}}/k_{\mathrm{B}}, m_2)$:

$$\begin{aligned} D_{12,\mathrm{TLJ-Zhu}}^{\mathrm{real}} &= \frac{3}{8\sqrt{\pi}}\sqrt{\frac{\sigma_{12,\mathrm{LJ}}^2\varepsilon_{12,\mathrm{LJ}}}{m_2}}\frac{\sqrt{T_{12}^*}}{\rho_{12}^*}\left(1 - \frac{\rho_{12}^*}{1.029079 T_{12}^{*0.165377}}\right) \\ &\times \left[1 + \rho_{12}^{*0.126978}\left(\frac{0.596103\,(\rho_{12}^* - 1)}{0.539292\,(\rho_{12}^* - 1) + T_{12}^{*(0.400152 - 0.41054\rho^*)}} + 0.68856\right)\right] \\ &\times \exp\left(-\frac{\rho_{12}^*}{2T_{12}^*}\right) \end{aligned} \tag{9.246}$$

Here, T_{12}^* is calculated as before, whereas distinct reduced density is introduced, as $\sigma_{12,\mathrm{LJ}}$ is implied instead of $\sigma_{2,\mathrm{LJ}}$:

$$T^*_{12} = \frac{T}{\varepsilon_{12,\mathrm{LJ}}/k_\mathrm{B}} \qquad \text{and} \qquad \rho^*_{12} = \rho_2 \sigma^3_{12,\mathrm{LJ}} \tag{9.247}$$

The combining rules adopted to determine binary parameters are

$$\varepsilon_{12,\mathrm{LJ}}/k_\mathrm{B} = \sqrt{(\varepsilon_{1,\mathrm{LJ}}/k_\mathrm{B})(\varepsilon_{2,\mathrm{LJ}}/k_\mathrm{B})} \ \text{ and } \ \sigma_{12,\mathrm{LJ}} = (1 - k^\mathrm{d}_{12}) \frac{\sigma_{1,\mathrm{LJ}} + \sigma_{2,\mathrm{LJ}}}{2} \tag{9.248}$$

where $k^\mathrm{d}_{12} = 0.7926(\sigma_{1,\mathrm{LJ}} - \sigma_{2,\mathrm{LJ}})/(\sigma_{1,\mathrm{LJ}} + \sigma_{2,\mathrm{LJ}})$. The LJ parameters of the solvent are estimated by Eqs. (9.208) and (9.209), while those of the solute are evaluated as function of its critical temperature and pressure [133] by

$$\varepsilon_{1,\mathrm{LJ}}/k_\mathrm{B} = \frac{T_{1,\mathrm{C}}}{1.313} \qquad \text{and} \qquad \sigma_{1,\mathrm{LJ}} = \frac{0.13\varepsilon_{1,\mathrm{LJ}}}{P_{1,\mathrm{C}}} \tag{9.249}$$

These equations are based on the principle of corresponding states, and on the critical point computed by Johnson et al. [124] for the LJ fluid ($P^*_\mathrm{C} = 0.13$, $T^*_\mathrm{C} = 1.313$, $\rho^*_\mathrm{C} = 0.31$). Equation (9.246) has been validated with six solvents and 55 solutes, completing 74 systems and 1141 data points. Global AAD is 17.32%, a high value even for predictive equations (see Table 9.8).

To complete this section about the diffusion coefficient, a model by Merzliak and Pfennig [130] for intradiffusivities in LJ mixtures is described.

The model of Liu et al. [25] for self-diffusion, Eq. (9.176), has been chosen as starting point, being extended to LJ mixtures by introducing suitable combining and mixing rules which account for the different relation of particle size and mixture density, which depends on system conditions and the molecular parameters in the mixture. They not only describe the effective particle size and the particle–particle interaction energies, but also the effective densities, the effective reduced temperatures, and the inverse masses for each component of the mixture. The proposed equations for the intra-diffusion coefficients are the following:

$$D_{i,\mathrm{LJ}} = \frac{3}{8} \frac{1}{\rho \sigma^2_{\mathrm{mix},i}} \sqrt{\frac{k_\mathrm{B} T}{\pi m_{\mathrm{eff},i}}} \exp\left(-\frac{\theta_4 \rho^*_{\mathrm{eff},i}}{\theta_2 - \rho^*_{\mathrm{eff},i}}\right) \exp\left(-\frac{\theta_3}{T^*_{\mathrm{mix},i}}\right) \tag{9.250}$$

with

$$\rho^*_{\mathrm{eff},i} = \rho \sigma^a_{\mathrm{mix},i} \sigma^b_\mathrm{mix} \tag{9.251}$$

$$\sigma^a_{\mathrm{mix},i} = \sum_{j=1}^{n} x_j \sigma^{3-\theta_1}_{\mathrm{eff},j,i} \tag{9.252}$$

$$\sigma^b_\mathrm{mix} = \sum_{i=1}^{n} \sum_{j=1}^{n} x_i x_j \sigma^{\theta_1}_{\mathrm{eff},j,i} \tag{9.253}$$

$$\sigma^2_{\mathrm{mix},i} = \sum_{j=1}^{n} x_j \sigma^2_{\mathrm{eff},j,i} \tag{9.254}$$

$$\sigma_{\mathrm{eff},i,j} = 2^{1/6}\sigma_{\mathrm{LJ},i,j}\left[1+\left(\frac{T^*_{\mathrm{eff},i,j}}{\theta_5}\right)^{\frac{1}{2}}\right]^{\frac{-1}{6}} \tag{9.255}$$

$$T^*_{\mathrm{eff},i,j} = \frac{k_{\mathrm{B}}T}{\varepsilon_{\mathrm{LJ},i,j}} \tag{9.256}$$

$$T^*_{\mathrm{mix},i} = \sum_{j=1}^{n} x_j T^*_{\mathrm{eff},j,i} \tag{9.257}$$

$$m_{\mathrm{eff},i}^{-1/2} = \frac{1}{2}\sum_{j=1}^{n} x_j (m_i^{-1/2} + m_j^{-1/2}) \tag{9.258}$$

where indices $i, j = 1, \ldots, n$ number the components of the mixture, m_i is the particle mass, $\sigma_{\mathrm{LJ},i,i}$ and $\varepsilon_{\mathrm{LJ},i,i}$ are the LJ parameters of the pure component, $\sigma_{\mathrm{LJ},i,j}$ and $\varepsilon_{\mathrm{LJ},i,j}$ are the cross LJ parameters in the mixture, x_i is the mole fraction, ρ is the number density of the mixture, $\rho^*_{\mathrm{eff},i}$, $T^*_{\mathrm{eff},i,j}$, and $m_{\mathrm{eff},i}$ are the reduced density, the reduced temperature, and the effective mass corresponding to each component or interaction between the particles in the mixture, respectively. θ_1 to θ_5 are parameters of the model, which were fitted by least-squares to a database of intradiffusion coefficients with an average absolute deviation of 5.56%:

$$\begin{aligned} \theta_1 &= 2.159807 \\ \theta_2 &= 1.156846 \\ \theta_3 &= 0.414496 \\ \theta_4 &= 0.610344 \\ \theta_5 &= 0.564022 \end{aligned} \tag{9.259}$$

This database was obtained from MD simulations of liquid mixtures and pure substances over $100 < \rho\ (\mathrm{kg\,m^{-3}}) < 2580$ and $90 < T\ (\mathrm{K}) < 350$. The Lorentz-Berthelot combining rules were adopted in the calculations.

The model of Merzliak and Pfennig [130] was then applied to calculate 42 intra-diffusivities of LJ mixtures available in literature. The comparison between calculated and experimental data showed that Eq. (9.250) is able to predict coefficients with an accuracy of 13.08%.

The model of Merzliak and Pfennig for LJ intradiffusivities [130] concludes this section on diffusion coefficients. In the following part, the viscosity coefficient is focused and the importance of the HS fluid is once more highlighted. It will be shown that accurate HS models coupled with perturbation approaches can generate the viscosity of modified and real substances.

9.5 Viscosity

The similarity of modelling viscosity to that of diffusion coefficient can be seen from previous general discussions in Sect. 9.3. For liquids, Stokes–Einstein relationship, Eq. (9.75), correlates both transport properties. However, for real fluids, there is a major difference between viscosity and diffusivity over the entire gas and liquid range regarding the temperature effect. Diffusion coefficient always increases with raising temperature, in contrast with viscosity which increases in gases and decreases for liquids. This behaviour makes viscosity more difficult to model over the whole fluid range.

Viscosity of gases in low and intermediate density range can be described by kinetic theories, such as Enskog theory, Eq. (9.25). Nonetheless, as Fig. 9.1 illustrates, obligatory corrections must be contemplated even for the HS system, akin to those in Sect. 9.4.1 for the diffusion coefficient. Final extension of HS models to real substances implies additional modifications, which are the subject of this section.

9.5.1 Perturbation Approaches: Hard Sphere System

As for the diffusion coefficient, the HS fluid also plays a key role in modelling viscosity of real substances. Therefore, it is fundamental to have accurate models to interpret the behaviour of the HS viscosities, since they may be further extended to real systems by the EHSD method, which is established by the perturbation theory, and by coupling energetic terms.

Analogously to the diffusion coefficient, a general relation for the fluidity, $\psi \equiv 1/\eta$, of a real substance can be written as sum of a reference term and a perturbation contribution related to activation energy, E_ψ:

$$\mathrm{Ln}\ \psi = A_\psi + \mathrm{Ln}\ \psi_{\mathrm{ref}} - E_\psi/(k_\mathrm{B}T) \tag{9.260}$$

where A_ψ is a constant. A simple reference fluid is the HS system:

$$\mathrm{Ln}\ \psi = A_\psi + \mathrm{Ln}\ \psi_{\mathrm{HS}} - E_\psi/(k_\mathrm{B}T) \tag{9.261}$$

There are two extreme cases: (1) the reference is ignored and only the activation energy term is taken into account, which leads to the well-known Eyring's rate theory [134]; (2) the HS reference is the unique contribution, and in this case one obtains

$$\mathrm{Ln}\ \psi_{\mathrm{RHS}} = A_\psi + \mathrm{Ln}\ \psi_{\mathrm{HS}} \tag{9.262}$$

which corresponds in fact to the RHS model, Eq. (9.96), if $A_\psi = -\mathrm{Ln}\ A_\eta$.

Alternatively, one can adopt other fluid as reference system, such as the LJ fluid. In this case, the so-called rough LJ (RLJ) model arises:

$$\mathrm{Ln}\ \psi_{\mathrm{RLJ}} = A_\psi + \mathrm{Ln}\ \psi_{\mathrm{LJ}} \tag{9.263}$$

The RLJ theory has been already applied to interpret tracer and self-diffusion coefficients in Sect. 9.4, when the models of Ruckenstein and Liu [75,109], Eqs. (9.205) and (9.243), have been presented. From Sect. 9.3.3, the viscosity of a LJ fluid can be described by the EHSD method. Simultaneously, it may be coupled with a perturbation energetic term, as happens with hybrid models of the type of Eq. (9.69). Accordingly, fluidity may be expressed as

$$\psi_{\mathrm{LJ}} = \psi_{\mathrm{EHSD}}\ \exp[W/(k_{\mathrm{B}}T)] \equiv \psi_{\mathrm{HS}}\ [\sigma^*_{\mathrm{EHSD}}(T^*,\rho^*)]\ \exp[W/(k_{\mathrm{B}}T)] \quad (9.264)$$

where W is a parameter.

The vital importance of the HS fluid may be easily seen in both situations, Eqs. (9.262) and (9.264). Some models for the HS viscosity will be presented next.

Models for the Viscosity of the Hard Sphere System

Various efforts have been made to derive accurate models for the HS system and some theories have been published in literature. In the low-density range (gas), Enskog expression serves as a good example of a simple and strict theory. Nonetheless, at high densities it requires significant correction as Fig. 9.1 shows. On the other hand, there are some theories developed for very high density range, e.g. that proposed by Koo and Hess [135] based on the statistical mechanics theory of Kirkwood et al. [136, 137]. The whole development of this theory is lengthy, and here we only present the final result, which is surprisingly simple:

$$\eta_{\mathrm{HS}} = \frac{16}{135}(\pi m k_{\mathrm{B}}T)^{1/2}\rho^3\sigma^4\frac{\sqrt{g(\sigma)}}{1+\frac{1}{6}g'(\sigma)g^{-1}(\sigma)} \quad (9.265)$$

where $g'(\sigma) = [dg(r)/dr]_{r=\sigma}$. The major advantage of the theory is that it comprises a divergence at $1+\frac{1}{6}g'(\sigma)g^{-1}(\sigma)$. Using an EOS, the root found is $\rho^* \approx 1.003$, which is quite close to the simulation results [90, 138]. This model obviously fails at low density, since it predicts:

$$\eta_{\mathrm{HS}} \propto \rho^{*3} \quad (9.266)$$

which contradicts Enskog theory. Therefore, for practical applications, pure theoretical models are still far from being satisfactory [137], since transport properties consist analytically in N-body problems. Consequently, a simple way to surmount these obstacles is to recur to empirical and semi-empirical expressions intended to correct Enskog formulae, once reliable simulation data over wide density range are now available for the HS system [14, 90, 95, 126].

The Enskog viscosity, Eq. (9.25), may be corrected similarly to the diffusion coefficient, by parameterizing hard sphere MD data according to:

$$\eta_{\mathrm{HS}} = F(\rho^*)\eta_{\mathrm{E}} \quad (9.267)$$

$$\eta_{\mathrm{HS}} = G(\rho^*)\eta_0 \quad (9.268)$$

where $F(\rho^*)$ and $G(\rho^*)$ are correction functions related by

$$F(\rho^*) \equiv \left(\frac{\eta_{\mathrm{HS}}}{\eta_{\mathrm{E}}}\right) = \left(\frac{\eta_{\mathrm{HS}}}{\eta_0}\right)\left(\frac{\eta_0}{\eta_{\mathrm{E}}}\right) \equiv G(\rho^*)\left(\frac{\eta_0}{\eta_{\mathrm{E}}}\right) \quad (9.269)$$

and η_0 is the ideal-gas coefficient calculated by Eq. (9.22). Compare previous relations with Eqs. (9.111)–(9.113) for self-diffusion coefficient and Eq. (9.141) for tracer diffusivity. Dymond [33] investigated the viscosity over the liquid region. Within the uncertainty of the computer simulations of Alder et al. [14] for hard spheres he proposed the expression

$$\frac{\eta_0}{\eta_{\mathrm{HS}}} = 0.2195\left(\frac{V}{V_0} - 1.384\right) \tag{9.270}$$

to reproduce the MD values in the range $1.5 < V/V_0 < 3.0$. Note this equation is of the simplest free-volume type, as fluidity depends on the relative expansion from an intrinsic molar volume, i.e. $\eta_{\mathrm{HS}}^{-1} \propto (V - V_i) = V_f$; compare it with DHB model, Eq. (9.70), in Sect. 9.3.4.

The dense gas region has also been studied by Dymond [59, 60] on the basis of the van der Waals theory. He adopted the correlation scheme described in Sect. 9.3.5, involving reduced coefficients η^* given by Eqs. (9.90) and (9.93), to test the applicability of the HS model to interpret the viscosity of noble gases. He proposed the following polynomial in $x \equiv (V/V_0)^{2/3}$ to represent the HS model [59] over the V/V_0 range from 2 to 10:

$$\frac{\eta_0}{\eta_{\mathrm{HS}}} = 1.466 - 3.545x + 2.927x^2 - 0.9972x^3 + 0.1580x^4 - 0.00959x^5 \tag{9.271}$$

Later, Dymond and Assael [139] carried out a more extensive test of the van der Waals theory utilizing accurate experimental transport properties of monatomic fluids, namely neon, argon, krypton, and xenon. The procedure of Sect. 9.3.5 has been accomplished simultaneously for D, η, and λ because V_0 is in principle independent of each coefficient. Nonetheless, as has been referred to in that section, it is not unreasonable that they may differ [61]. The universal curve provided for the HS viscosity in the reduced volume range $1.5 < V/V_0 < 5.0$ is

$$\begin{aligned}\log\ \eta_{\mathrm{HS}}^* = \log\left[\frac{\eta_{\mathrm{HS}}}{\eta_0}\left(\frac{V}{V_0}\right)^{\frac{2}{3}}\right] &= 1.0945 - 9.26324\,(V/V_0)^{-1} \\ &+ 71.0385\,(V/V_0)^{-2} - 301.9012\,(V/V_0)^{-3} + 797.6900\,(V/V_0)^{-4} \\ &- 1221.9770\,(V/V_0)^{-5} + 987.5574\,(V/V_0)^{-6} - 319.4636\,(V/V_0)^{-7}\end{aligned} \tag{9.272}$$

Liu and Ruckenstein [110] derived the correlation given by Eq. (9.151) to represent the MD simulations of Alder et al. [14] in the range $\rho^* < 0.95$. Figure 9.24 illustrates the good fit achieved by their $F(\rho^*)$ equation.

Another correction function $F(\rho^*)$ was proposed later by Silva et al. [140]. The authors fitted the MD data of Alder et al. [14] and Smith et al. [126] with the following expression:

$$\begin{aligned}\frac{\eta_{\mathrm{HS}}}{\eta_{\mathrm{E}}} = \frac{\psi_{\mathrm{E}}}{\psi_{\mathrm{HS}}} = F(\rho^*) = 1 + 0.000324 \times \exp(2.927464 \\ + 12.49621\rho^* - 29.94134\rho^{*2} + 24.01674\rho^{*3})\end{aligned} \tag{9.273}$$

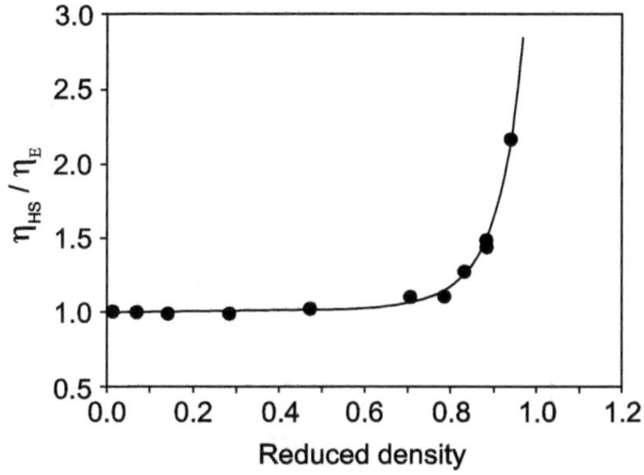

Fig. 9.24. Ratio between HS and Enskog viscosities against reduced density, $\rho^* = \rho\sigma^3$. Data from Alder et al. [14]; Model by Liu and Ruckenstein [110], Eq. (9.151)

which reproduces computer experiments with average absolute deviation equal to 2.8%.

A simple relation by Heyes and Sigurgeirsson [138] has been recently proposed:

$$\frac{\eta_0}{\eta_{\mathrm{HS}}} = \frac{\psi_{\mathrm{HS}}}{\psi_0} = G^{-1}(\rho^*) = (1 - \varphi/0.58)^{1.92} = (1 - \rho^*/1.1077)^{1.92} \qquad (9.274)$$

where

$$\psi_0^{-1} = 1.016 \frac{5}{16\sigma^2} \left(\frac{m k_{\mathrm{B}} T}{\pi} \right)^{\frac{1}{2}} \qquad (9.275)$$

The constants 0.58 and 1.92 have been fitted to the MD calculations of Sigurgeirsson and Heyes [95], and makes that Eq. (9.274) can reproduce simulation data within statistical uncertainty upto $\varphi \cong 0.51 - 0.52$.

9.5.2 Free-Volume Models

In the following, we focus our attention mainly on free-volume models due to two reasons: (1) although its strict theoretical base is still being investigated, the apparent physical background seems attractive; and (2) the theories developed for most cases are simple and present very reasonable correlations for both model (HS, SW, LJ, etc.) and real systems.

If we introduce a general free-volume equation, $f(V_{\mathrm{f}})$, for the HS fluidity in Eq. (9.261), then one gets an expression akin to Eq. (9.69):

$$\psi = A'_{\psi} f(V_{\mathrm{f}}) \, \exp\left(-\frac{E_{\psi}}{k_{\mathrm{B}} T} \right) \qquad (9.276)$$

where A'_{ψ} denotes $\exp(A_{\psi})$. It is worth noting that V_{f} should be seen as geometric free volume (read comments around Eqs. (9.77)–(9.79) in Sect. 9.3.4).

Consequently, assuming the simple relation given in Eq. (9.79), the thermodynamic free volume expressions discussed and compiled in Sect. 9.3.4 can be used (see Tables 9.3, 9.4, and Figs. 9.8, 9.9). In the simplest case, taking the ideal-gas as reference, $f(V_\mathrm{f}) = V$, which gives rise to the well-known reaction rate theory due to Eyring [134]:

$$\psi = A'_\psi V \exp\left(-\frac{E_\psi}{k_\mathrm{B}T}\right) \tag{9.277}$$

On the other hand, if the activation energy term is neglected, which corresponds to start directly from Eq. (9.262), a free-volume model is obtained. The Doolittle equation is a typical example [36]:

$$\mathrm{Ln}\ \psi = A + \frac{B}{V - V_0} \tag{9.278}$$

where A, B, and V_0 are adjustable parameters. This model has been re-established on the HS free-volume theory [37, 38]. A recent application can be found in Laghaei et al. [141]. A modification by Allal et al. [142] has also been proposed.

Another simple and successful model is that of Hildebrand–Batschinski [34], Eq. (9.74), which embodies the free-volume expression of vdW (Eq. (9.80) in Table 9.3). In fact, it relates the fluidity to the relative expansion from the molar volume of the intercept, V_0, at $\psi = 0$ obtained by linear extrapolation, which corresponds to the situation where molecules are so close that viscous flow ceases. Hildebrand-Batschinski model has been generalized by Przezdziecki and Sridhar [143] to give rise to an entirely predictive correlation for pure liquids. This may be taken as a very good example to emphasize the importance of the HS fluid to describe liquid structure and properties. The viscosity of a liquid substance, in cP, is given by [121, 143]

$$\eta = \frac{V_0}{E(V - V_0)} \tag{9.279}$$

where liquid volume is in $\mathrm{cm^3\,mol^{-1}}$. The parameters are defined below:

$$E = -1.12 + \frac{V_\mathrm{C}}{12.94 + 0.097M - 0.23P_\mathrm{C} + 0.0424T_\mathrm{f} - 11.58(T_\mathrm{f}/T_\mathrm{C})} \tag{9.280}$$

$$V_0 = 0.0085\omega T_\mathrm{C} - 2.02 + \frac{V_\mathrm{m}}{0.342(T_\mathrm{f}/T_\mathrm{C}) + 0.894} \tag{9.281}$$

where T_C is the critical temperature in K, P_C the critical pressure in bar, V_C the critical volume in $\mathrm{cm^3\,mol^{-1}}$, M the molecular weight in $\mathrm{g\,mol^{-1}}$, T_f is the freezing point temperature, ω the acentric factor, and V_m is the liquid volume at T_f. Therefore, if a predictive equation to calculate molar volume is available (e.g. from Reid et al. [121]) then Eqs. (9.279)–(9.281) may be used to compute liquid viscosities of pure substances, without any experimental coefficients a priori.

Reid et al. [121] tested the model of Przezdziecki and Sridhar [143] with more than 30 different organic compounds, including polar and non-polar molecules, and found that it accomplishes satisfactory predictions for most of them, in comparison with the best predictive models.

Another modification to the Hildebrand-Batschinski equation has been proposed by Ertl and Dullien [144], who introduced a third parameter, n, in Eq. (9.70) to improve results obtained. Accordingly,

$$\psi = B\left(\frac{V - V_0}{V_0}\right)^n \tag{9.282}$$

However, Eq. (9.282) has not been widely used due to its limited capacity. In fact, much valuable and successful results arise by coupling the activation energy term.

Despite the simplicity of the Hildebrand–Batschinski model, it only works reasonably well with non-polar or weakly polar substances [40], which prompts researchers to include an activation energy term like that in Eq. (9.276) for most applications. At the moment, we will present a simple expression by Liu et al. [145], which combines Eqs. (9.276) and (9.279) with minor modification.

If the free-volume model of Hildebrand-Batschinski, Eq. (9.74), is chosen as the reference part of Eq. (9.276), then $f(V_\mathrm{f}) \propto (V - V_0)/V_0$, where V_0 is a parameter. Hence, Eq. (9.276) becomes

$$\psi = A_\psi \frac{V - V_0}{V_0} \exp\left(-\frac{E_\psi}{k_\mathrm{B}T}\right) \tag{9.283}$$

In addition, according to Eyring [134], the activation energy can be related to the evaporation energy and further to the potential energy. Hence, taking account of the van der Waals approximation,

$$E_\psi = -a/V \tag{9.284}$$

where a is constant, the following simple model for liquid viscosity [145, 146] arises:

$$\psi = B\frac{V - V_0}{V_0} \exp\left(\frac{C}{VRT}\right) \tag{9.285}$$

where parameters V_0, B, and C are characteristic of each substance. The application of this equation to various liquids has been accomplished successfully [146, 147]. For 83 pure compounds, including non-polar and strongly polar substances, Eq. (9.285) gives an average absolute deviation of only 1.30% for 1044 data points. In comparison, the three-parameter Antoine equation,

$$\mathrm{Ln}\ \eta = A + \frac{B}{T - T_0} \tag{9.286}$$

gives rise to an AAD = 4.38% for the same database.

Another major advantage of the model is that it can be easily extended to mixtures [145]:

$$\text{Ln}\ \psi_{\text{mix}} = \sum_i \theta_i \ \text{Ln}\ \psi_i + \text{Ln}\frac{\widetilde{V}_{\text{mix}} - 1}{\prod_i (\widetilde{V}_i - 1)^{\theta_i}} + \frac{\sum_{(i<j)} x_i x_j G_{ij}}{V_{\text{mix}} RT} \tag{9.287}$$

where subscript "mix" stands for mixture, $\widetilde{V} = V/V_0$, $\theta_i = x_i V_i / V_{\text{mix}}$, and x_i are molar fractions. The mixing and combining rules needed to evaluate $V_{0,\text{mix}}$ are given by

$$V_{0,\text{mix}} = \sum_i \sum_j x_i x_j V_{0ij} \tag{9.288}$$

$$V_{0ij} = \frac{(1 - k_{ij}) \left(V_{0i}^{1/3} + V_{0j}^{1/3} \right)}{8} \tag{9.289}$$

Besides V_{0i} for each pure substance, the binary parameters k_{ij} and G_{ij} are employed. For all non-aqueous solutions, set $k_{ij} = 0$, hence only one binary parameter is involved. For symmetric systems (i.e. composed of similar components [145]), one can use a predictive model as $G_{ij} = 0$:

$$\text{Ln}\ \psi_{\text{mix}} = \sum_i \theta_i \ \text{Ln}\ \psi_i + \text{Ln}\frac{\widetilde{V}_{\text{mix}} - 1}{\prod_i (\widetilde{V}_i - 1)^{\theta_i}} \tag{9.290}$$

Extensive tests validating the model by Liu et al. [145] have shown that it performs better than other ones proposed in literature [145, 146, 147]. Database compiled includes aqueous solutions, where there exists a maximum in binary mixture viscosities. Table 9.9 summarizes the results obtained, from which two remarkable conclusions may be drawn: (1) for symmetric systems (i.e., components have similar size and polarity), the simplified free-volume model, Eq. (9.290), predicts excellently the mixture viscosity; (2) for strongly non-ideal aqueous solutions, which exhibit a maximum in viscosity, the modified free-volume model, Eq. (9.287), behaves much better than the Teja and

Table 9.9. Calculation results (AADs) for the viscosity of binary mixtures using the models of Liu et al. [145] and Teja and Rice [148]. The systems are identified in [145], NS is the number of systems, NDP is the number of data points, and NP is the number of parameters in Eq. (9.287)

System	NS	NDP	NP (Liu et al.)	AAD(%) (Liu et al.)	AAD(%) (Teja and Rice)
Symmetric	19	531	0	0.96	3.93
Asymmetric	40	1093	1	1.59	2.54
Aqueous	11	409	2	2.47	7.32
Overall	70	2033	–	1.60	3.90

Rice' method [148] based on the principle of corresponding states and recommended by Reid et al. [121]. A very recent essay by Aznarez et al. [149] reached the same conclusion for mixtures of 2-alkanols with tetraethyleneglycol dimethyl ether. For ternary systems, Eq. (9.287) predicts the solution viscosity remarkably well. The AAD for 23 ternary systems (560 data points) is only 2.76%.

Under the same assumption of Eq. (9.79), which established a relation between geometric and thermodynamic free volumes (V_{gf} and V_f, respectively), we may adopt the Carnahan-Starling expression, Eq. (9.83) in Table 9.3, to evaluate $f(V_f)$ in Eq. (9.276). At the end, an accurate model for the viscosity of compressed liquids can be obtained [146, 150].

In summary, the models described above pointed out that simple HS free-volume equations may be applied to various real systems when combined with an activation energy term. Free-volume (i.e. hard sphere) contribution dominates the liquid viscosity, as Eqs. (9.279) and (9.290) express.

9.5.3 Lennard-Jones-Based Models

As mentioned above, better models might be developed by taking the LJ fluid as reference to calculate fluidity, ψ_{LJ}, which can in turn be obtained by means of the EHSD method discussed in Sect. 9.3.3 and already applied to diffusion coefficients along Sect. 9.4.2. As an example, the model proposed recently by Silva et al. [140] will be briefly discussed in the following.

The EHSD method has been chosen by Silva et al. [140] to estimate the viscosity of the WCA fluid. With that purpose, the authors adopted their HS expression for η_{HS}, Eq. (9.273), and the effective diameter due to Souza and Ben-Amotz following Week-Chandler-Andersen criterion, Eq. (9.65) in Table 9.2. Accordingly,

$$\eta_{\mathrm{WCA}} = \eta_{\mathrm{HS}}\ [\sigma^*_{\mathrm{WCASA}}(T^*, \rho^*)] \tag{9.291}$$

The full LJ viscosity, η_{LJ}, is finally obtained from η_{WCA} by means of the temperature-dependent relationship given by:

$$\eta_{\mathrm{LJ}} = \eta_{\mathrm{WCA}} \left(\theta_1 + \frac{\theta_2}{T^*}\right) \tag{9.292}$$

where $\theta_1 = 0.9932$ and $\theta_2 = 0.2778$ for $\rho^* \geq 0.6$ and $T^* \geq 0.8$. It is worth noting that the last equations correspond to the theoretical procedure traduced by Eqs. (9.170) and (9.171) for the self-diffusion coefficient.

Silva et al. [140] have applied this model to simple non-polar substances, namely Ar, N_2, F_2, O_2, CO_2, and n-alkanes up to C_6. It involves only one adjustable parameter (σ_{LJ}), since the $\varepsilon_{\mathrm{LJ}}/k_{\mathrm{B}}$ parameter comprised in T^* has been fixed a priori by the equation:

$$\varepsilon_{\mathrm{LJ}}/k_{\mathrm{B}} = 0.77T_{\mathrm{C}} \tag{9.293}$$

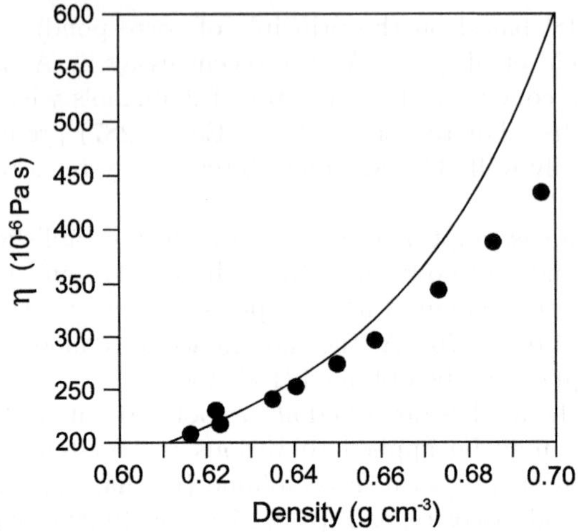

Fig. 9.25. Comparison between results obtained with Eq. (9.292) and experimental data of n-pentane at 303.15 K. Data from [151]

Note that the same assumption has been accounted for in Eqs. (9.180), (9.203), and (9.217) for diffusion coefficients [58, 75, 125]. For 11 substances, 994 data points totally, the average absolute deviation found is 3.66%.

In Figs. 9.25 and 9.26 the calculated viscosities of n-pentane at 303.15 and 383.15 K, respectively, are plotted against density, along with experimental

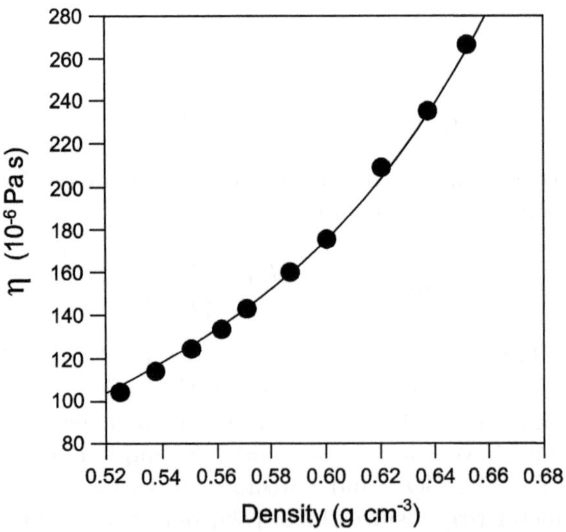

Fig. 9.26. Comparison between results obtained with Eq. (9.292) and experimental data of n-pentane at 383.15 K. Data from [151]

data. Equation (9.292) fails at 303.15 K, since data fall on the low-temperature–high-density region, $0.84 < \rho^* < 0.95$ and $T^* = 0.839$. This temperature is near the inferior limit of application of the model, despite being slightly higher. At 383.15 K (Fig. 9.26) the predictions accomplished are accurate, as $T^* = 1.059$ is clearly above the inferior limit of validity.

9.5.4 Alternative Approaches

There are several alternative approaches available to study the viscosity of liquids and dense gases. In the following we are going to enunciate some of them briefly.

The MET described in Sect. 9.3.2 has been successfully applied to the representation of viscosities. One may cite the good results achieved by two different and original methods reported by Mehdipour and Eslami [152], and Parsafar and Kalantar [153]. The first one has been tested with experimental data available for noble gases, N_2, O_2, F_2, CO, CO_2, CH_4, CF_2Cl_2, C_2H_6, and C_3H_8, and the average absolute deviation found was 3.03% for 3033 points examined. The method of Parsafar and Kalantar [153] is applicable at high densities, $\rho > \rho_C$, and was able to estimate viscosities of some hydrocarbons with $AAD < 1.6\%$ and maximum deviation equal to 2.0%. Moreover, it is appropriate for long-chain fluids.

The van der Waals theory of transport properties introduced in Sect. 9.3.5 provides good results for the viscosity of liquids and dense gases. It is able to represent the behaviour of molecules as different as noble gases, n-alkanes, aromatics, alcohols, refrigerants, and simple compounds [139, 154, 155, 156, 157, 158, 159, 160]. It is frequently combined with the RHS theory (see Sect. 9.3.6) to take into account the coupling between translational and rotational motions observed in real non-spherical particles. Accordingly, following Eq. (9.96) it is assumed that:

$$\eta_{\mathrm{real}} \approx \eta_{\mathrm{RHS}} = A_\eta \, \eta_{\mathrm{SHS}} \tag{9.294}$$

and the van der Waals' scheme should be implemented with corrected coefficient η/A_η substituted in the right-hand side of Eq. (9.93). In SI units, the reduced viscosity becomes:

$$\eta^* = 6.035 \times 10^8 \frac{\eta}{A_\eta} \frac{V^{\frac{2}{3}}}{\sqrt{MRT}} \tag{9.295}$$

where $V_0(T)$ and A_η are obtained from the horizontal and vertical shifts necessary to superimpose plots of $\log \eta^*$ versus $\log V$ from experiment and $\log \eta^*$ versus $\log(V/V_0)$ from the HS system. Whenever possible such fitting should comprehend all three properties simultaneously.

There are expressions for $V_0(T)$ available in literature for many compounds [139, 154, 155, 156, 157, 158, 159, 160] which may be substituted, for instance, in Eq. (9.272) to compute η_{SHS}, and then multiplied by A_η to determine the viscosity of the real fluid. The method has also been extended to mixtures of n-alkanes and refrigerants [156, 160].

Another successful approach to study the viscosity coefficient is that based on excess entropy (over the ideal-gas value) scaling laws introduced in Sect. 9.3.7 [77, 81, 161, 162, 163, 164].

Figure 9.12 illustrates the dependence of the reduced viscosity defined by Rosenfeld [77, 81], Eq. (9.102), upon $-S^{\mathrm{ex}}/Nk_{\mathrm{B}}$ for HS, SS, and OCP systems [77] (see Sect. 9.1.1 for a brief introduction to these pair potentials). The effect of the attractive forces upon η^* is pointed out in Fig. 9.13, where results for SS and LJ fluids are plotted together.

Rosenfeld [81] published the following quasi-universal relationship for η^*:

$$\eta^* = 0.2\ \exp\left(-0.8\frac{S^{\mathrm{ex}}}{Nk_{\mathrm{B}}}\right) \tag{9.296}$$

which has been obtained from plots of hundreds of simulations of strongly coupled simple fluids, $-S^{\mathrm{ex}}/Nk_{\mathrm{B}} > 1$, with quite disparate pair interactions (freezing corresponds to about $4 < -S^{\mathrm{ex}}/Nk_{\mathrm{B}} < 5$).

A new scaling law for viscosity has been reported by Li et al. [163, 164]. They defined the reduced coefficient as

$$\eta^* = \frac{\sigma}{m\Gamma_{\mathrm{E}}}\eta \tag{9.297}$$

where $\Gamma_{\mathrm{E}} = 4\sigma^2 g(\sigma)\rho\sqrt{\pi k_{\mathrm{B}}T/m}$ is the collision frequency, and assumed that S^{ex} can be restricted to the two-particle approximation, Eq. (9.105). The new relationship is

$$\eta^* = 0.035\ \exp\left(-0.55\frac{S^{\mathrm{ex}}}{Nk_{\mathrm{B}}}\right) \tag{9.298}$$

Korkmaz et al. [162] verified this equation agrees well with experimental viscosity of liquid Na, K, Rb, and Cs metals.

Recently, new scaling laws relating the viscosity of pure component and binary mixtures to their excess thermodynamic entropies have been proposed by Ali [161]. For pure component, the function is

$$\eta = 0.37\sqrt{mk_{\mathrm{B}}T}\ \rho^{2/3}\exp\left(-\frac{2}{3}\frac{S^{\mathrm{ex}}}{Nk_{\mathrm{B}}}\right) \tag{9.299}$$

It is worth noting that the unique difference between Eqs. (9.296) and (9.299) are the constants involved, i.e. 0.37 and 2/3 versus 0.2 and 0.8. For binary mixtures composed of identical mass and size particles at intermediate densities, where it is legitimate to accept $\rho_1 g_{11}(\sigma_{11}) \cong \rho_2 g_{22}(\sigma_2) \cong 1$, Ali proposed:

$$\eta = 0.37\sqrt{mk_{\mathrm{B}}T}\ \rho^{2/3}\exp\left(-\frac{2}{3}\frac{S_T^{\mathrm{ex}}}{Nk_{\mathrm{B}}}\right) \tag{9.300}$$

where S_T^{ex} is the total partial entropy of the system computed by molar fraction average:

$$S_T^{\text{ex}} = x_1 S_1^{\text{ex}} + x_2 S_2^{\text{ex}} \tag{9.301}$$

Both relations exhibited good agreement with MD simulation results.

Rosenfeld [81] investigated the excess entropy scaling and discovered it holds also in the dilute gas regime, where it would be least expected. Nonetheless, in this region power-law dependencies prevail instead of the exponential ones. For power potentials of the form $\phi = \varepsilon(\sigma/r)^\nu$, the universal expression between $\nu = 4$ and $\nu = \infty$ (hard spheres) is

$$\eta^* \equiv \eta \frac{\rho^{-2/3}}{\sqrt{mk_{\text{B}}T}} = \eta'(-S^{\text{ex}}/Nk_{\text{B}})^{-2/3} \tag{9.302}$$

where the constant η' varies from 0.264 to 0.288 in that range of ν. As has been referred to in Sect. 9.3.7, such finding emphasizes that the analytic form of the quasi-universal relationships is probably not exponential, yet an exponential form can provide good numerical representation for dense fluids.

9.5.5 Conclusions and Discussion

The HS fluid plays a major role in modelling the viscosity of various real substances. The lack of accurate theories for dense fluids is largely compensated by the very accurate simulation data over the entire density range published during the last years. This fact allows researchers to derive simple correlation models for hard spheres, further applied to real compounds by using common techniques such as the EHSD method.

Another way to modify an HS model is to combine it with an activation energy term. In any case, free-volume models seem very suitable for modelling. For instance, the simplest Hildebrand–Batsinski's relationship can be beautifully applied to many real systems, especially when it is coupled with the refered activation energy contribution.

The predominance of the HS contribution to the viscosity of real fluids is also corroborated by researches on excess entropy scaled laws. Calculations accomplished for diverse pair potential functions evidence the HS viscosity is major part of the complete (real) coefficient. This justifies once again that it may be taken as good reference in perturbation theories.

Future directions in this field can be: (a) seek for an accurate theory of the HS system. It is not only useful in practice, but also very important to understand the dynamic behaviour of the HS fluid. Due to the difficulties met by present theories, such as the kinetic one, new ideas are welcome. (b) There is a very interesting and significant area we have not explored in this chapter: supercooled fluid and glass. Once more, the HS theory will play a key role in this research. The topic needs to be addressed in a whole chapter, or even in an entire book, since both theory and computer simulations have to be taken into account. (c) For the conventional methods discussed in this chapter, there is still room for improvements. Simple and accurate models are always necessary in industrial applications. Additionally, more investigation is necessary around free-volume approach. For instance, the connection between geometric and thermodynamic free volumes, and the "real" dependence between them and the transport properties are challenging themes. In fact, the present successes of free-volume-based models are partly due to their parameters. A "serious" investigation should avoid these parameters in the future.

9.6 Thermal Conductivity

Likewise other transport properties, the thermal conductivity provides information on the intermolecular forces established between the molecules of a fluid, and is particularly essential for the design of efficient refrigeration equipment and separation processes involving not only material but also energy balances.

This section is dedicated to the thermal conductivity of liquids and dense gases. It opens with a brief review on the HS system, where the obligatory corrections to the Enskog theory are quantified. After that, the application of the most successful theories of transport to real substances is presented, and some expressions and methods are described as well.

9.6.1 Thermal Conductivity of the Hard Sphere System

Research on thermal conductivity is mostly developed to describe the behaviour of real fluids directly, in spite of investigating specific corrected models for hard spheres. Nonetheless, some expressions and fundamentals are compiled in the following.

The deviations of the Enskog theory to the MD transport properties of the HS fluid are discussed in Sect. 9.3.1, where it has been shown that they penalize predominantly self-diffusion and shear viscosity coefficients. As Fig. 9.1 illustrates, the errors introduced by molecular chaos approximation in the thermal conductivity are less than 10% over the whole density range, being barely perceptible within the few percent accuracy of the data [59].

The Enskog thermal conductivity, Eq. (9.26), may be corrected similarly to the transport properties dealt in previous sections, by parameterizing hard sphere MD data according to

$$\lambda_{\mathrm{HS}} = F(\rho^*)\lambda_{\mathrm{E}} \tag{9.303}$$

$$\lambda_{\mathrm{HS}} = G(\rho^*)\lambda_0 \tag{9.304}$$

where $F(\rho^*)$ and $F(\rho^*)$ are correction functions related by

$$F(\rho^*) \equiv \left(\frac{\lambda_{\mathrm{HS}}}{\lambda_{\mathrm{E}}}\right) = \left(\frac{\lambda_{\mathrm{HS}}}{\lambda_0}\right) \times \left(\frac{\lambda_0}{\lambda_{\mathrm{E}}}\right) \equiv G(\rho^*) \times \left(\frac{\lambda_0}{\lambda_{\mathrm{E}}}\right) \tag{9.305}$$

where λ_0 is the ideal-gas coefficient calculated by Eq. (9.23). Compare previous relations with Eqs. (9.111)–(9.113) for self-diffusion coefficient and Eq. (9.141) for tracer diffusivity.

Dymond [33] investigated the thermal conductivity over the liquid region. Within the uncertainty of the computer simulations of Alder et al. [14] he proposed the expression

$$\frac{\lambda_0}{\lambda_{\mathrm{HS}}} = 0.1611 \left(\frac{V}{V_0} - 1.217 \right) \tag{9.306}$$

to reproduce the MD values in the volume range from 1.5 to $2V_0$. At $V/V_0 = 2.5$ it underestimates computer results by 6%. It is worth noting that this equation is of the simplest free-volume type, as the thermal resistivity depends on the relative expansion from an intrinsic molar volume, i.e. $\lambda_{\mathrm{HS}}^{-1} \propto (V - V_{\mathrm{i}}) = V_{\mathrm{f}}$ – compare it with DHB model, Eq. (9.70), in Sect. 9.3.4.

The dense gas region has also been studied by Dymond [59, 60] on the basis of the van der Waals theory. He adopted the correlation scheme described in Sect. 9.3.5, involving reduced coefficients λ^* given by Eqs. (9.91) and (9.94), to test the applicability of the HS model to interpret the thermal conductivity of noble gases, and obtained excellent agreement for densities down to about two thirds the critical density. In Fig. 9.27 the variation of λ^* with the logarithm of reduced volume is shown. The solid line fits experimental data of neon,

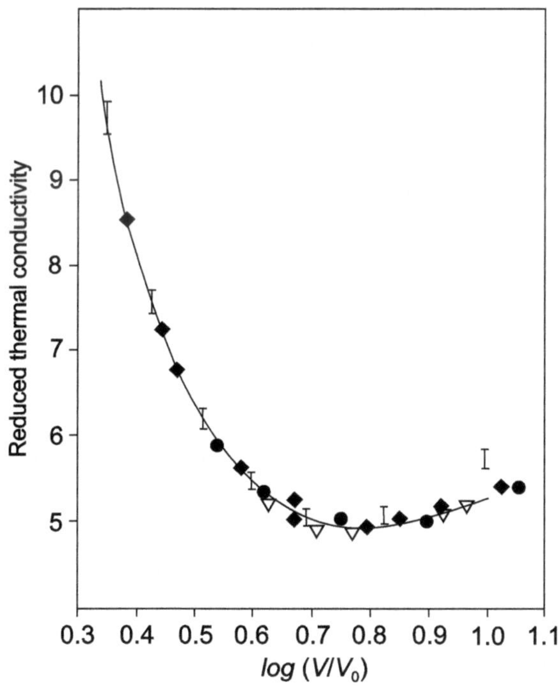

Fig. 9.27. Reduced thermal conductivity against $\log(V/V_0)$ plotted in Dymond' fashion (Eqs. (9.91) and (9.94)): HS results (*vertical lines* with error bar) and experimental data [165, 166, 167, 168, 169]; *curve*: from Eq. (9.307)

argon, and krypton, and vertical lines with error bar indicate MD values of the HS system. The curve is the following polynomial in $x \equiv (V/V_0)^{2/3}$, which is assumed to represent the HS model [59],

$$\frac{\lambda_0}{\lambda_{\mathrm{HS}}} = 0.0253 - 0.4746x + 0.5625x^2 - 0.1355x^3 + 0.01136x^4 \ , \qquad (9.307)$$

and provides good representation over the V/V_0 range from 2.5 to 7.

Later, Dymond and Assael [139] carried out a more extensive test of the van der Waals model using accurate experimental transport properties of monatomic fluids, namely neon, argon, krypton, and xenon. The procedure of the method has been accomplished simultaneously for D, η, and λ, since the molecular core sizes (V_0 or equivalently σ) should be characteristic of each fluid and independent of the coefficients. Nonetheless, as has been mentioned in Sect. 9.3.5, it is not unreasonable that they may differ [61]. The universal curve obtained for the reduced thermal conductivity is

$$\log \lambda^*_{\mathrm{HS}} = \log\left[\frac{\lambda_{\mathrm{HS}}}{\lambda_0}\left(\frac{V}{V_0}\right)^{\frac{2}{3}}\right] =$$

$$1.0655 - 3.538\left(\frac{V}{V_0}\right)^{-1} + 12.120\left(\frac{V}{V_0}\right)^{-2} - 12.469\left(\frac{V}{V_0}\right)^{-3} + 4.562\left(\frac{V}{V_0}\right)^{-4} \qquad (9.308)$$

This data fit covers the reduced volume range $1.5 < V/V_0 < 5$, and gives the closest agreement with the computed values for hard spheres. Therefore, it may be used to estimate λ_{HS}.

An alternative expression for the HS system was published by Heyes [91], who fitted the $\lambda_{\mathrm{HS}}/\lambda_{\mathrm{E}}$ values of Erpenbeck and Wood [92] to the following polynomial of the fourth degree in reduced density:

$$\frac{\lambda_{\mathrm{HS}}}{\lambda_{\mathrm{E}}} = 0.990 + 0.160\rho^* - 0.746\rho^{*2} + 1.215\rho^{*3} - 0.558\rho^{*4} \qquad (9.309)$$

Once the topic on thermal conductivity of the HS system comes to an end, it is now possible to analyse modifications to this model fluid with the final objective to interpret real substances. This is the subject of the next subsections.

9.6.2 Thermal Conductivity of Modified Hard Sphere Systems

In this subsection, the most frequently adopted models and approaches for the thermal conductivity of modified HS systems, together with final extensions to real fluids, are presented. Under the scope of this chapter, they consist essentially in the modified Enskog theory, free-volume models, and van der Waals theory. Some minor results are also mentioned for the excess entropy scaling laws approach.

Free-Volume Theory

As has been referred to in Sect. 9.3.4, the free-volume theories have been successfully applied in the field of transport properties, mostly viscosity and diffusivity. The simplest model is that of Dymond–Hildebrand–Batschinsky [33, 34, 35], Eq. (9.70), which expresses a linear relationship between transport coefficients and free volume.

Concerning thermal conductivity, Dymond [33] has arrived to such dependence between λ^{-1} and molar volume by applying the corrected Enskog theory in MD calculations for hard spheres over the liquid region—see Eq. (9.306) above. This result induced researchers to represent the thermal conductivity of liquids by Eq. (9.76), which in conjunction with Eq. (9.74) has been subjected to an ample validation by van Loef [40], who examined DHB model with classical monatomic, and linear, quasi-spherical and dipolar molecular liquids, as well as some hydrocarbons.

In the case of atomic liquids, van Loef [40] investigated λ values of compressed argon, krypton, and xenon along different isotherms, and found the thermal resistivity satisfies Eq. (9.76). The linear $V - \lambda^{-1}$ plot for liquid argon in the volume range up to $V/V_C = 0.6$ and temperatures between 85 and 145 K is shown by the circles in Fig. 9.28, from which proportionality constant $C = 10.0 \pm 0.5\,\text{mKW}^{-1}$ has been determined. It is worth noting the single slope found at different temperatures in a substantial fraction of the liquid range. The linear extrapolation to $1/\lambda = 0$ is represented by the dashed line, from which the intrinsic volume V_i' is calculated.

For the classical atomic liquids cited, constant C plotted against molecular weight on a log–log scale evidences a square-root dependence, $C \propto M^{1/2}$, and the volume intercept is a constant fraction of the critical volume, i.e. $V_i' = 0.21 V_C$.

Results for nitrogen and carbon dioxide are also shown in Fig. 9.28 to assess the behaviour of linear molecular liquids. For both N_2 in the range

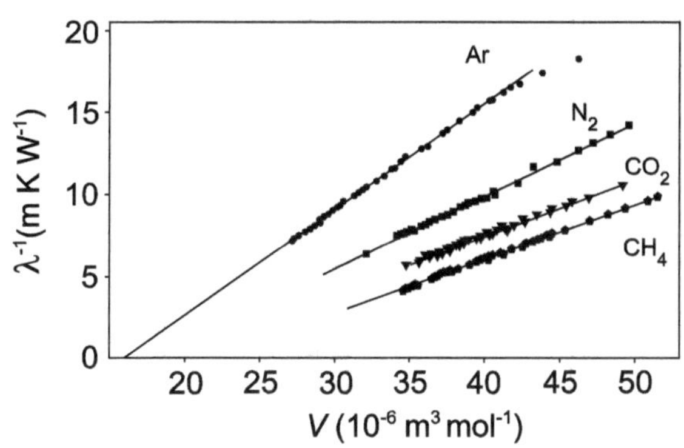

Fig. 9.28. Thermal resistivity of compressed liquid argon, nitrogen, carbon dioxide, and methane against molar volume. Experimental data [170, 171, 172, 173, 174, 175, 176, 177]

80–120 K and CO_2 in the range 230–290 K a linear relationship is observed with a single slope. Furthermore, van Loef [40] investigated O_2, F_2, N_2O, and CS_2 at constant temperature and concluded that the thermal resistivity vary linearly with molar volume with a slope that is within 10% constant in the liquid range up to $T/T_C = 0.9$. Generally, the values of C for these diatomic molecules are close to the $M^{1/2}$ relation established for monatomic liquids; CO_2 and N_2O exhibit C values about 30% smaller than the predicted ones.

The thermal resistivity of several compressed liquid hydrocarbons (e.g. CH_4 in Fig. 9.28) presented similar trends to the above-cited ones: λ^{-1} increases linearly with molar volume, and constant C obeys approximately the square root mass dependence observed in monatomic liquids. In fact, constant C of non-associated liquids with molecular mass in the range between 20 and 200 g mol^{-1} follows such dependence [40]. Apparently, the energy transfer is neither much dependent upon molecular shape nor strongly influenced by the internal motions of the molecules in a large part of the liquid range.

In Fig. 9.29 the thermal resistivity of compressed water is graphed as function of molar volume along six isotherms, to detach its abnormal behaviour. It is seen that λ increases with temperature at constant volume, and this behaviour is most pronounced at lower temperatures. Moreover, the linear trend is once more followed with approximately constant slope ($C = 1.1 \pm 0.1$ mKW^{-1}), and C is substantially smaller than that of an atomic liquid with equivalent mass. Similar result has been found for constant B in Eq. (9.74), which evidenced that compressed water is a strongly associated liquid even at elevated temperatures [40]. These results are consistent with the findings by Luck [178], who estimated the number of molecules per aggregate in the saturated liquid at $T = 423$ K to be nearly 20.

Modified Enskog Theory

The MET is one of the earliest attempts to correct the Enskog expressions, Eqs. (9.24)–(9.26), to represent transport properties of real systems accurately. The original approach by Hanley et al. [15] for the thermal conductivity of

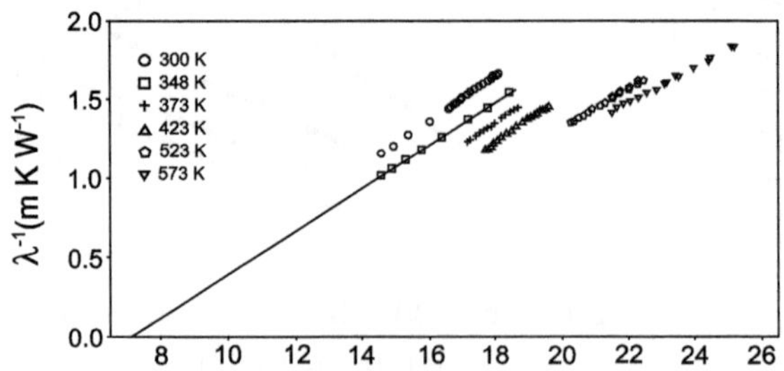

Fig. 9.29. Thermal resistivity of compressed water at six temperatures. Experimental data: [179, 180, 181, 182, 183]

dense fluids is described in Sect. 9.3.2. Recent contributions by Mehdipour and Eslami [152] and Parsafar and Kalantar [153] will be presented here due to their originality and good results achieved.

According to the MET, the co-volume and radial distribution function at contact appearing in Eqs. (9.24) to (9.26) are evaluated using real fluid data, namely b is redefined in terms of the second virial coefficient (see Eq. (9.40)), and $g(\sigma)$ is deduced from real compressibility isotherms. Last assumption implies the external pressure, P, in Eq. (9.28) should be replaced by the thermal pressure to give Eq. (9.38), to incorporate the attractive forces between molecules. The major limitation of the method is the lack of virial coefficients and thermal pressures for many fluids, which must be ultimately estimated from accurate EOSs.

Amorós et al. [184, 185] analysed the behaviour of the thermal pressure for several types of liquids and applied the MET to various substances. Moreover, Sheng et al. [186] calculated transport properties of dense gases using the Peng-Robinson EOS. Lee and Liu [187] derived a correlation to predict thermal pressure coefficients of liquids as function of molar volume and one reliable experimental value, which may be estimated by group contribution method when it is not available in the literature.

Mehdipour and Eslami [152] computed b and $g(\sigma)$ directly from an EOS proposed by Song et al. [188] and Ihm et al. [189], which is based on the statistical-mechanical perturbation theory:

$$\frac{P}{\rho k_{\mathrm{B}} T} = 1 + \frac{(B - B^{\mathrm{r}})\rho}{1 + 0.22\Gamma\beta\rho} + \frac{B^{\mathrm{r}}\rho}{1 - \Gamma\beta\rho} \tag{9.310}$$

where B is the second virial coefficient, B^{r} is the contribution of the repulsive forces to B, β is a temperature-dependent parameter analogous to the vdW co-volume, and Γ is an adjustable parameter. For most compounds studied, Γ values have been already reported [189]; for other molecules, it may be optimized using experimental PVT data; for noble gases, $\Gamma = 0.454$. The temperature-dependent parameters can be determined by specifying the intermolecular potential function. Mehdipour and Eslami [152] adopted the LJ and WCA pair potentials (see Sect. 9.1.1), hence:

$$B(T) = 2\pi \int_0^{\infty} (1 - \mathrm{e}^{-\phi_{\mathrm{LJ}}/(k_{\mathrm{B}}T)})\, r^2 \,\mathrm{d}r \tag{9.311}$$

$$B^{\mathrm{r}}(T) = 2\pi \int_0^{r_{\mathrm{m}}} (1 - \mathrm{e}^{-\phi_{\mathrm{WCA}}/(k_{\mathrm{B}}T)})\, r^2 \,\mathrm{d}r \tag{9.312}$$

$$\beta(T) = 2\pi \int_0^{r_{\mathrm{m}}} \left[1 - \left(1 + \frac{\phi_{\mathrm{WCA}}}{k_{\mathrm{B}}T}\mathrm{e}^{-\phi_{\mathrm{WCA}}/(k_{\mathrm{B}}T)}\right)\right] r^2 \,\mathrm{d}r \tag{9.313}$$

where $r_{\mathrm{m}} = 2^{1/6}\sigma_{LJ}$. Substituting Eq. (9.310) into Eq. (9.38) one obtains

$$b\,g(\sigma) = \frac{B + T(dB/dT) - B^{\mathrm{r}} - T(dB^{\mathrm{r}}/dT)}{1 + 0.22\Gamma\beta\rho} + \frac{B^{\mathrm{r}} + T(dB^{\mathrm{r}}/dT)}{1 - \Gamma\beta\rho} + \frac{\Gamma B^{\mathrm{r}}\rho T(d\beta/dT)}{(1 - \Gamma\beta\rho)^2} - \frac{0.22\Gamma\rho(B - B^{\mathrm{r}})T(d\beta/dT)}{(1 + 0.22\Gamma\beta\rho)^2} \tag{9.314}$$

Since in the dilute gas limit $g(\sigma) = 1$, Eq. (9.314) also satisfies classical Eq. (9.40), i.e., $b = B + T(dB/dT)$.

The MET based on Eqs. (9.314) and (9.40) was tested with experimental thermal conductivities of He, Ne, Ar, N_2, O_2, CO_2, CH_4, CF_2Cl_2, C_2H_6, and SF_6, and the average absolute deviation reported for 2587 data points was 2.66% [152]. It is worth noting the method is not applicable near the critical point, where large deviations have been found by the authors. Such behaviour is expected as has been already mentioned above.

Recently, Parsafar and Kalantar [153] developed a new method based on the MET to represent transport properties of dense fluids, applicable at $\rho > \rho_C$. Once more, the lack of experimental virial coefficients and thermal pressures for most substances make necessary to recur to accurate analytical EOSs. Their approach has been essentially tested and validated for viscosities. With respect to the thermal conductivity, the authors investigated specifically Xe, because of the abundance of λ and PVT measurements, and the existence of a precise EOSs [190, 191]. For long-chain organic molecules they suggest their modified linear isothermal regularity (MLIR) EOS, which has been tested with n-alkanes and their binary mixtures [192], and later with primary, secondary, and tertiary alcohols, ketones, and 1-carboxylic acids [193]. According to the MLIR EOS, the quantity $(Z/n - 1)/\rho^2$ is linear against ρ^2 for each isotherm of the chain fluid, i.e.

$$\frac{Z/n - 1}{\rho^2} = A_m + B_m n^2 \rho^2 \tag{9.315}$$

where n is the number of constituent groups of the chain molecule, and A_m and B_m are parameters predicted by a simple group contribution method described in their original essays [192, 193]. The MLIR EOS is an extension of the linear isothermal regularity (LIR) EOS that has been initially devised for spherically symmetrical molecules and their mixtures [194, 195]. From the MLIR EOS the density of several chain compounds has been calculated with AAD $< 1.2\%$, and the thermal pressure coefficient $(\partial P/\partial T)_\rho$ with AAD $< 1.1\%$ [192, 193].

The Enskog thermal conductivity, Eq. (9.26), may be recast as

$$\lambda_{\mathrm{E}} = \lambda_0 b\rho\,(1/Q + 1.2 + 0.755Q) \tag{9.316}$$

when the variable $Q \equiv b\rho g(\sigma)$ is introduced. Similarly, Eq. (9.28) may be rewritten as $Q = P/\rho k_{\mathrm{B}}T - 1$. Coefficient λ_0 may be estimated by Eq. (9.23) for dilute monatomic gases. On the contrary, polyatomic molecules demand an Eucken correction which leads to the following relation [2]:

$$\lambda_0 = \frac{\eta_0}{m}\left(C_V + (9/4)k_B\right) = \frac{5}{16\sigma^2}\left(\frac{k_B T}{\pi m}\right)^{\frac{1}{2}}\left(C_V + (9/4)k_B\right) \tag{9.317}$$

Note that when $C_V = (3/2)k_B$ is substituted here, Eq. (9.317) reduces to Eq. (9.23). According to Eqs. (9.27) and (9.317) for b and λ_0, respectively, Eq. (9.316) gives rise to:

$$\frac{\lambda Q}{\rho\sqrt{T}(C_V + \frac{9}{4}k_B)} = C_1 Q^2 + C_2 Q + C_3 \tag{9.318}$$

where parameters C_1, C_2, and C_3 are independent of temperature for hard particles; real fluids perform differently owing to the fact these parameters embody diameter σ. At high densities one may expect that $\lambda Q/[\rho\sqrt{T}(C_V + (9/4)k_B)]$ can be well represented by a quadratic function in Q for each isotherm of a dense fluid. Moreover, for monatomic fluids C_V is almost independent of temperature, therefore $\lambda Q/(\rho\sqrt{T})$ should also exhibit polynomial dependence of the second degree in Q. At high densities ($\rho > \rho_C$) the mentioned isotherms fall onto a single quadratic curve, which prefigures universal behaviour [153].

To take MET into account, Q is computed from an EOS containing thermal pressure (see Eq. (9.38)):

$$Q = \frac{1}{\rho k_B}\left(\frac{\partial P}{\partial T}\right)_\rho - 1 \tag{9.319}$$

and co-volume is redefined in terms of virial coefficients. Parsafar and Kalantar [153] involved second and third virial coefficients, B and C, in this step:

$$b = B + T\left(\frac{dB}{dT}\right) + C + T\left(\frac{dC}{dT}\right) \tag{9.320}$$

In Fig. 9.30 experimental data for Xe at nine temperatures, between 200 K and 600 K, are plotted in $Q - \lambda Q/(\rho\sqrt{T})$ coordinates, being possible to conclude they all lay in the same quadratic line (correlation coefficient $R > 0.999$). On the other hand, at low-density $\rho < \rho_C$ quadratic fits are no more representative, as Fig. 9.31 points out.

van der Waals Theory of Transport Properties

As mentioned in Sect. 9.3.5, the van der Waals theory for transport properties is equivalent to the HS model, which furnishes a realist description of the molecular interactions in the case of liquids and dense gases above critical density. It is frequently combined with the rough hard sphere (RHS) theory, briefly discussed in Sect. 9.3.6, to take into account the coupling between translational and rotational motions observed in real substances with non-spherical shape.

The van der Waals theory has been successfully applied to the thermal conductivity since its formulation in the early 1970s [59, 196]. Compounds as diverse as noble gases and simple molecules, alkanes, aromatics, n-alcohols,

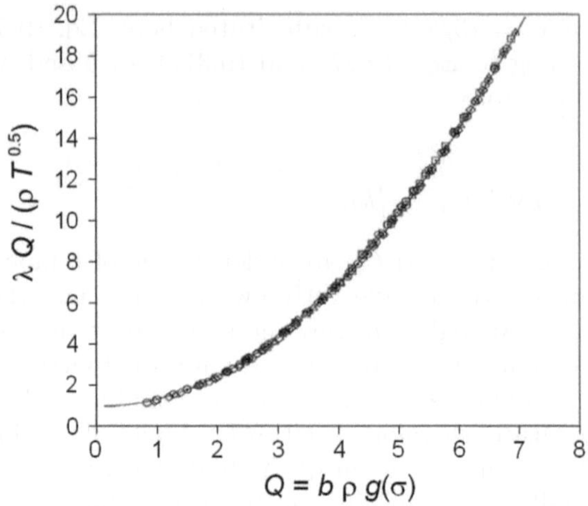

Fig. 9.30. Quadratic fit of $\lambda Q/(\rho\sqrt{T})$ against Q for xenon at $\rho > \rho_C$ and nine temperatures (200, 220, 240, 260, 280, 300, 400, 500, 600 K). Correlation coefficient, $R > 0.999$. Data from [191]

refrigerants, and even mixtures have been modelled since then [59, 60, 61, 139, 154, 155, 156, 157, 158, 159, 160, 196, 197, 198, 199, 200].

Following the optimization procedure described in Sect. 9.3.5, it is possible to determine the close-packed molar volume as a function of temperature, $V_0 =$

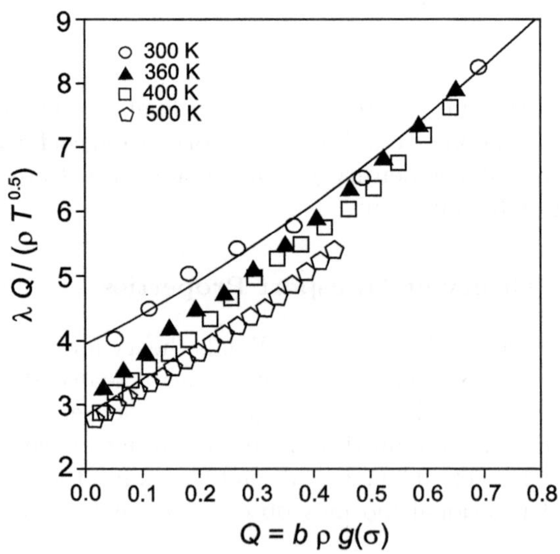

Fig. 9.31. Plot of $\lambda Q/(\rho\sqrt{T})$ against Q for xenon at $\rho < \rho_C$ and four temperatures. Data from [191]

$V_0(T)$, for molecules for which isothermal experimental data are available. The horizontal shift required to produce coincidence of the reduced HS transport property versus $\log(V/V_0)$ with corresponding plots of the reduced transport property versus $\log V$ for the real fluid allow us to get V_0 at each temperature.

For noble gases, Dymond and Assael [139] reported the following polynomial functions, in units of 10^{-6} m^3 mol^{-1} and K, for V_0:

$$\text{neon}: \qquad V_0 = 28.07115 - 0.123408T + 1.625 \times 10^{-4}T^2 \tag{9.321}$$

$$\begin{aligned}\text{argon}: \quad V_0 =& -8.25916 + 0.984981T - 0.0142467T^2 + 1.00518 \times 10^{-4}T^3 \\ & - 3.76457 \times 10^{-7}T^4 + 7.16878 \times 10^{-10}T^5 - 5.46113 \times 10^{-13}T^6\end{aligned} \tag{9.322}$$

$$\text{krypton}: \quad V_0 = 41.18 - 0.1699T + 4.00 \times 10^{-4}T^2 - 3.162 \times 10^{-7}T^3 \tag{9.323}$$

$$\text{xenon}: \quad V_0 = 57.6744 - 0.305577T + 1.00116 \times 10^{-3}T^2 - 1.15946 \times 10^{-6}T^3 \tag{9.324}$$

which may be substituted in an HS model to compute λ for each fluid. For consistency, Eq. (9.308) should be adopted because it has been used by the same authors.

Non-spherical molecules are not adequately traduced by the smooth hard sphere (SHS) expressions of Sect. 9.6.1. In these cases, the RHS theory of Chandler [66] is habitually extrapolated to the thermal conductivity, even without theoretical justification, by introducing a roughness parameter, A_λ, to account for the translational-rotational coupling of motions. Hence, similarly to Eqs. (9.95)–(9.97), the following relationship is assumed:

$$\lambda_{\text{real}} \approx \lambda_{\text{RHS}} = A_\lambda \, \lambda_{\text{SHS}} \tag{9.325}$$

Accordingly, for non-spherical fluids the van der Waals approach and the correlation scheme of Dymond and Assael should be implemented with corrected coefficients (D/A_{D}), (η/A_η), and (λ/A_λ) on the right-hand side of Eqs. (9.92)–(9.94), respectively. For instance, the reduced thermal conductivity expression to be used is, in SI units:

$$\lambda^* = 1.936 \times 10^7 V^{2/3} \frac{\lambda}{A_\lambda} \sqrt{\frac{M}{RT}} \tag{9.326}$$

Therefore, $V_0(T)$ and A_λ may be simply derived from the horizontal and vertical shifts necessary to superimpose plots of $\log \lambda^*$ versus $\log V$ from experiment, and $\log \lambda^*$ versus $\log(V/V_0)$ from the HS system. As in the case of spherical compounds, this fitting scheme should comprise, whenever it is possible, the three properties simultaneously.

For n-alkanes, Dymond, Assael and collaborators [139, 154] correlated the optimized parameters V_0 and A_λ with carbon number, N_C. The resulting equations are, in 10^{-6} m^3 mol^{-1} and K:

$$\begin{aligned}&\mathrm{CH_4 - C_4H_{10}}:\\ &V_0 = 45.822 - 6.1867T^{0.5} + 0.36879T - 0.007273T^{1.5}\\ &\quad +N_\mathrm{C}\left(2.17871T^{0.5} - 0.185198T + 0.00400369T^{1.5}\right)\\ &\quad +N_\mathrm{C}^2\left(6.95148 - 52.6436T^{-0.5}\right)\\ &\quad +N_\mathrm{C}^3\left(-7.801897 + 42.24493T^{-0.5} + 0.4476523T^{0.5} - 0.009573512T\right)\end{aligned} \tag{9.327}$$

$$\mathrm{C_5H_{12}}: \qquad V_0 = 81.1713 - 0.046169T \tag{9.328}$$

$$\begin{aligned}&\mathrm{C_6H_{14} - C_{16}H_{34}}:\\ &V_0 = 117.874 + 0.15(-1)^{N_\mathrm{C}} - 0.25275T + 5.48\times10^{-4}T^2 - 4.2464\times10^{-7}T^3\\ &\quad +(N_\mathrm{C}-6)(1.27 - 0.0009T)(13.27 + 0.025N_\mathrm{C})\end{aligned} \tag{9.329}$$

$$\begin{aligned}&\mathrm{CH_4 - C_{16}H_{34}}:\\ &A_\lambda = -18.8416N_\mathrm{C}^{-1.5} + 41.461N_\mathrm{C}^{-1} - 30.15N_\mathrm{C}^{-0.5}\\ &\quad +8.6907 + 1.3371\times10^{-3}N_\mathrm{C}^{2.5}\end{aligned} \tag{9.330}$$

These may be now substituted into Eq. (9.308) to compute λ_{SHS} of a particular fluid, and then multiplied by A_λ to determine its real-thermal conductivity, as Eq. (9.325) explains. For 960 points, included in the database compiled, only 1.5% of them deviate more than 5% from experimental values. Function $A_\lambda = A_\lambda(N_\mathrm{C})$ is graphed in Fig. 9.32 to point out the strong influence of carbon number, as it is theoretically expected in advance. In fact, angular momentum exchange should be particularly significant for fluids

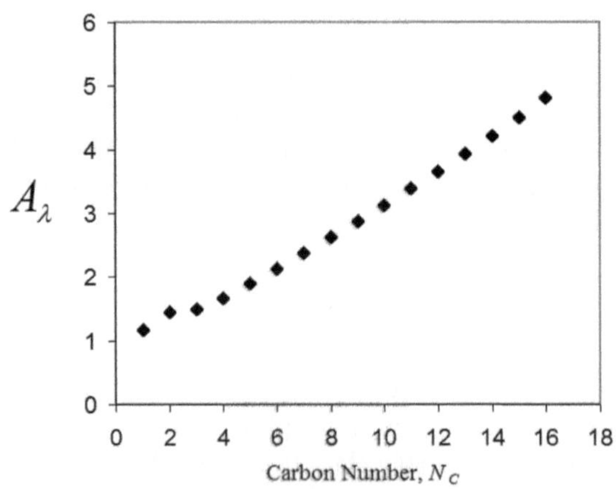

Fig. 9.32. Roughness factor for the thermal conductivity of n-alkanes as a function of carbon number, Eq. (9.330)

composed of large non-spherical molecules. It is worth noting that $A_\lambda \approx 1$ for methane and attains almost five for n-hexadecane, which implies a huge variation in the calculated conductivities.

The van der Waals theory has been extended to the thermal conductivity of n-alkanes [156] and refrigerants [160] mixtures. Introducing simple mixing rules, without adjustable parameters, for the characteristic volume, $V_{0,\mathrm{mix}}$, and roughness factor, $A_{\lambda,\mathrm{mix}}$, the method becomes purely predictive.

For instance, for n-alkanes Assael et al. [156] determined $V_{0,\mathrm{mix}}$ and $A_{\lambda,\mathrm{mix}}$ by mass fraction average of pure component values, and assumed zero mixing volumes in the density calculation. For 76 experimental data within wide ranges of temperature and pressure no predictions deviate more than 5%. Later, Teja et al. [160] investigated three binary refrigerant mixtures and adopted the same mixing rules though based on molar fraction averages. The solutions are R22+R152a, R12+R22, and R32+R134a, and comprehend 268 data points totally. They achieved AAD = 2.56% and maximum deviation equal to 6.60%.

Additional researches with refrigerants and alcohols evidenced that the roughness factors may be weak functions of temperature [158, 159, 160, 198]. For instance, Assael et al. [159] analysed the viscosity and thermal conductivity of 13 liquid halogenated ethane refrigerants, and observed such dependence for the coupling factor A_λ of R114, R115, R124, R125, R134a, and R152a. The relationship they proposed to account for the temperature effect is simply linear:

$$A_\lambda = C_1 + C_2\,(T/T_C - 1) \tag{9.331}$$

where C_1 and C_2 are fitted constants. For illustration, the coupling factor of four refrigerant compounds are shown in Fig. 9.33 as function of reduced temperature, T/T_C, where it is possible to observe variations in the range

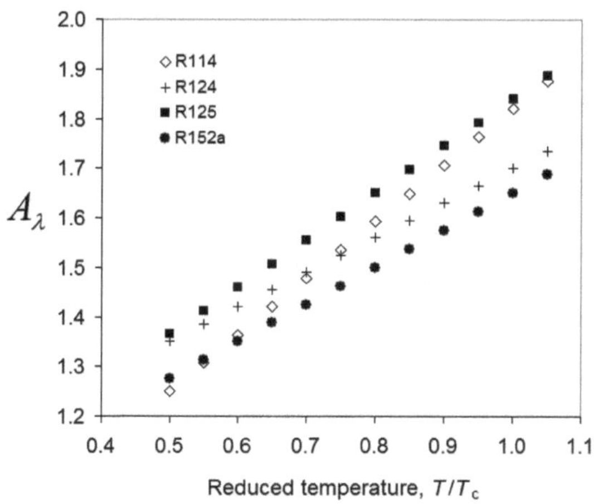

Fig. 9.33. Roughness factor for the thermal conductivity of some refrigerant molecules plotted against reduced temperature. Data from [159]

1.2–1.9. In that paper, Assael et al. [159] fixed $A_\eta = 1$ and determined $V_0 = V_0(T)$ from viscosity data available in literature. Then, $A_\lambda(T)$ was optimized using previous V_0 values and thermal conductivity data.

Teja et al. [160] proposed two simplified and systematic methods to obtain V_0, A_η, and A_λ of any substance, which avoids the original series of calculations necessary to extract them from experimental data by Dymond–Assael's procedure. A_D was not included, since they investigated refrigerants for which no self-diffusivities were available. However it is trivial to incorporate it in the calculations. The first iterative method may be summarized as follows:

1. Fix $A_\eta = 1$ or other realistic value.
2. Calculate $V_0 = V_0(T)$ from experimental viscosities using Dymond's method.
3. Compute deviation $\Delta_\eta = (\eta_{\text{calc}} - \eta_{\text{exp}})/\eta_{\text{exp}}$.
4. Determine A_λ using previous V_0 and experimental thermal conductivities.
5. Compute deviation $\Delta_\lambda = (\lambda_{\text{calc}} - \lambda_{\text{exp}})/\lambda_{\text{exp}}$.
6. Repeat steps 1–5 to minimize function $\sum \Delta_\eta^2 + \sum \Delta_\lambda^2$.

The second method embodies fundamentally an EHSD approach. Actually, $V_0(T)$ is directly evaluated from Eq. (9.34) using the EHSD of Heyes [201]:

$$V_0 = V_0(T) = N\sigma_{\text{EHSD}}^3(T)/\sqrt{2} \tag{9.332}$$

$$\frac{\sigma_{\text{EHSD}}}{\sigma_{\text{LJ}}} = 1.0217\,\frac{1 - 0.0178(T^*)^{-1.256}}{T^{*1/12}} \tag{9.333}$$

where $T^* = T/(\varepsilon_{\text{LJ}}/k_{\text{B}})$ as usually. The LJ force constants may be obtained from an LJ EOS and then substituted in Eq. (9.333) to generate σ_{EHSD}. After that, V_0 is calculated by Eq. (9.332). The determination of the roughness factors is straightforward: A_η is optimized using previous V_0 and experimental viscosities; A_λ is optimized using the same V_0 and experimental thermal conductivities. Teja et al. [160] showed that this technique can be used to correlate as well as predict the transport properties of refrigerants.

Gurova et al. [200] measured and correlated the thermal conductivity of 1,1-difluoroethane (HFC-152a) at five temperatures between 211.65 and 293.65 K. They used a modification of SHS (vdW) theory that corrects the value of the close-packed volume, instead of introducing roughness factors. This modification was previously reported [202, 203] and is linear in V/V_0, i.e.

$$\lambda^* = 1.936 \times 10^7 \lambda V^{2/3} \sqrt{\frac{M}{RT}}\,(C_1 + C_2(V/V_0)) \tag{9.334}$$

where C_1 and C_2 are empirically fitted constants (compare Eqs. (9.334) and (9.326)). For HFC-152a, Gurova et al. [200] obtained $C_1 = -0.321$, $C_2 = 0.8295$, and V_0 ($\text{m}^3\ \text{mol}^{-1}$) $= 4.4895 - 0.001877\ T(K)$, being able to reproduce 38 experimental values with a total uncertainty of 1.1% at a 95% confidence level. The application of the model to several refrigerants was described elsewhere [199].

Excess Entropy Scaling Laws

The interpretation of transport properties in terms of excess entropy scaling laws has been applied essentially to self-diffusion and viscosity coefficients [77, 78, 81, 82, 83, 84, 128, 129, 161, 162, 204] since its appearance in 1977. A general introduction to this research is provided in Sect. 9.3.7.

Rosenfeld [81] proposed the following quasi-universal relationship for λ reduced according to Eq. (9.102):

$$\lambda^* \cong 1.5\ \exp\left(\frac{-0.5\ S^{\mathrm{ex}}}{Nk_\mathrm{B}}\right) \tag{9.335}$$

which has been obtained from plots of hundreds of simulations of strongly coupled simple fluids, $-S^{\mathrm{ex}}/Nk_\mathrm{B} > 1$, with quite disparate pair interactions (freezing corresponds to about $4 < -S^{\mathrm{ex}}/Nk_\mathrm{B} < 5$). Remember that S^{ex} represents excess entropy over the ideal-gas value (Eqs. (9.98)–(9.100)).

Additionally, Rosenfeld [81] discovered that the excess entropy scaling holds in the dilute gas regime also, where it would be least expected, although power-law dependences prevail instead of the exponential ones. For power potentials of the form $\phi = \varepsilon(\sigma/r)^\nu$ (see comment to Eq. (9.5) in Sect. 9.1.1), the universal expression between $\nu = 4$ and $\nu = \infty$ (hard spheres) is:

$$\lambda^* = \lambda \frac{\rho^{-2/3}}{k_\mathrm{B}\sqrt{k_\mathrm{B}T/m}} = \frac{4}{15}\lambda'(-S^{\mathrm{ex}}/Nk_\mathrm{B})^{-2/3} \tag{9.336}$$

where constant λ' varies from 0.264 to 0.288 in that range of ν. As has been already referred to in Sect. 9.3.7, such finding is particularly interesting since it emphasizes that the analytic form of the quasi-universal relationships is probably not exponential, yet an exponential form can provide good numerical representation for dense fluids.

With this subject dedicated to thermal conductivity, the individual study of transport properties carried out in this chapter comes to an end. To conclude, some final remarks are given in the next section.

9.7 Conclusions and Final Remarks

Although the hard sphere system is the simplest model fluid, it provides the most satisfactory basis for the interpretation, correlation, and prediction of transport properties of real-dense fluids, namely liquids, compressed gases, and supercritical fluids. Up till now, no rigorous theory is established due to the complex many-body interactions involved, and because of the lack of detailed information on the pair interaction potential energy function for all but the simplest molecules.

Such limitations induce researchers to fit empirical and semi-empirical equations to the transport coefficients of model systems computed by equilibrium and non-equilibrium molecular dynamics. Afterwards, following the methods described in previous sections those expressions are extended to real substances, allowing us to assess the extent to which a hypothetical fluid approximates a real one.

As has been demonstrated throughout the chapter, the HS fluid may be taken as reliable reference system in perturbation theories to provide more realistic models for the diffusion coefficient, viscosity, and thermal conductivity of real substances. Three important examples are: (i) the effective hard sphere diameter method, which takes into account the softness of repulsive potential; (ii) the coupling of activation energy terms, which accounts for the effect of the attractive forces found in SW, LJ, and real compounds; and (iii) the rough hard sphere theory, which considers the non-isotropic potentials governing the interactions of polyatomic molecules.

In this chapter several generalized approaches have been presented for the three coefficients studied. The free-volume theories accomplish good representations for D, η, and λ, while more investigation around the connection between geometric free volume and thermodynamic free volume is necessary.

In the past years, variations of the classical modified Enskog theory have been also applied successfully to the three transport coefficients, although larger databases should be tested, particularly for self-diffusivities.

Special attention has been devoted to the van der Waals theory of transport. The model is realistic at densities greater than the critical density, as the molecular trajectory in real systems is determined predominantly by repulsive interactions. The advantage of such HS-based model is that it may be easily applied to D, η, and λ simultaneously, from which common molecular core sizes are optimized from experimental data at constant temperature. Nonetheless, it has been shown that for diffusion larger close-packed molar volumes may be obtained in comparison to those for viscosity and thermal conductivity.

Another interesting investigation in the field of condensed matter dynamics is the relationship between transport coefficients and structural properties. Such exploration acknowledges the fundamental works by Rosenfeld and Dzugutov, which connect the diffusivity and viscosity to the excess entropy defined through the radial distribution function.

Besides the generalized approaches presented in this chapter, it is worth mentioning unifying methods based on theoretical relationships between pairs of transport properties. Accordingly, if accurate expressions are known for a particular coefficient, new models may be suggested for the remaining ones. Typical examples are the well-known Stokes–Einstein (Eq. (9.75)) and Eucken [3, 205] relationships. The first one was derived for a macroscopic spherical particle suspended in a continuum fluid. The second connects the thermal conductivity and viscosity of monatomic gases. Generalized expressions for both relations may be found in literature (e.g. [206, 207]).

To conclude, this chapter emphasizes the importance of the hard sphere system to model the transport properties of dense fluids, as it has been successfully employed to correlate and predict the coefficients of real substances and their mixtures.

Acknowledgement

We acknowledge Simão P. Cardoso for his contribution in the preparation of some tables and figures. We also express our gratitude to Ángel Mulero for all support and collaboration conceded.

References

1. R.B. Bird, W.E. Stewart, E.N. Lightfoot: *Transport Phenomena.* (Ed. John Wiley & Sons, Singapore 1960)
2. J.O. Hirschfelder, C.F. Curtiss, R.B. Bird: *Molecular Theory of Gases and Liquids.* (Ed. John Wiley & Sons, New York 1967)
3. S. Chapman, T.G. Cowling: *The Mathematical Theory of Non Uniform Gases.* (Ed. Cambridge University Press, Cambridge 1970)
4. J.D. Weeks, D. Chandler, H.C. Andersen: J. Chem. Phys. **54**, 5237 (1971)
5. F. Vesely: *Computerexperimente na Flüssigkeitsmodellen.* (Physik Verlag, Weinheim 1978)
6. Hoover, W.G.: *Molecular Dynamics.* Lect. Notes Phys. **258** (1986)
7. W.G. Hoover: *Computational Statistical Mechanics.* (Ed. Elsevier, Amsterdam 1991)
8. D.J. Evans, G. Morriss: *Statistical Mechanics of Nonequilibrium Liquids.* (Ed. Academic, New York 1990)
9. J.M. Haile: *Molecular Dynamics Simulation - Elementary Methods.* (Ed. John Wiley & Sons, USA 1997)
10. M.P. Allen, D.J. Tildesley: *Computer Simulation of Liquids.* (Ed. Oxford University Press, USA 2000)
11. D. Enskog: Kgl. Svenska Ventensk. Handl. **63** (4) (1922)
12. N.F. Carnahan, K.E. Starling: J. Chem. Phys. **51**, 635 (1969)
13. D.M. Heyes, L.V. Woodcock: Mol. Phys. **59**, 1369 (1986)
14. B.J. Alder, D.M. Gass, T.E. Wainwright: J. Chem. Phys. **53**, 3813 (1970)
15. H.J.M. Hanley, R.D. McCarty, E.G.D. Cohen: Physica **60**, 3222 (1972)
16. E.A. Mason, L. Monchick: J. Chem. Phys. **36**, 1622 (1962)
17. J.D. Kirkwood, E.M. Boggs: J. Chem. Phys. **10**, 394 (1942)
18. J.A. Barker, D. Henderson: J. Chem. Phys. **47**, 4714 (1967)
19. H.C. Andersen, J.D. Weeks, D. Chandler: Phys. Rev. A **4**, 1597 (1971)
20. L. Verlet, J.J. Weis: Phys. Rev. A **5**, 939 (1972)
21. F. Lado: Mol. Phys. **52**, 871 (1984)

22. D. Ben-Amotz, D.R. Herschbach: J. Phys. Chem. **94**, 1038 (1990)
23. F.C. Andrews: J. Chem. Phys. **64**, 1941 (1976)
24. R.J. Speedy, F.X. Prielmeier, T. Vardag, E.W. Lang, H.D. Ludemann: Mol. Phys. **66**, 577 (1989)
25. H. Liu, C.M. Silva, E.A. Macedo: Chem. Eng. Sci. **53**, 2403 (1998)
26. K.D. Hammonds, D.M. Heyes: J. Chem. Soc. Faraday Trans. II **84**, 705 (1988)
27. T.F. Sun, J. Bleazard, A.S. Teja: J. Phys. Chem. **98**, 1306 (1994)
28. C.M. Silva, H. Liu, E.A. Macedo: Ind. Eng. Chem. Res. **37**, 221 (1998)
29. J.J. Erpenbeck, W.W. Wood: Phys. Rev. A **43**, 4254 (1991)
30. Y. Miyano, H. Masuoka: Fluid Phase Equilib. **18**, 131 (1984)
31. L.E.S. de Souza, D. Ben-Amotz: Mol. Phys. **78**, 137 (1993)
32. I. Nezbeda: Fluid Phase Equilib. **87**, 237 (1993)
33. J.H. Dymond: J. Chem. Phys. **60**, 969 (1974)
34. J.H. Hildebrand: Science **174**, 490 (1971)
35. A.J. Batschinski: Z. Phys. Chem. **84**, 643 (1913)
36. A.K. Doolittle: J. Appl. Phys. **22**, 1471 (1951)
37. M.H. Cohen, D. Turnbull: J. Chem. Phys. **31**, 1164 (1959)
38. D. Turnbull, M.H. Cohen: J. Chem. Phys. **52**, 3038 (1970)
39. P.B. Macedo, T.A. Litovitz: J. Chem. Phys. **42**, 245 (1965)
40. J.J. van Loef: Physica **103B**, 133 (1981)
41. H. Fujita: Adv. Polym. Sci. **3**, 1 (1961)
42. H. Fujita: Chem. Eng. Sci. **48**, 3037 (1993)
43. J.M. Zielinski, J.L. Duda: AIChE J. **38**, 405 (1992)
44. H.S. Chung: J. Chem. Phys. **44**, 1362 (1966)
45. H. Liu, C.M. Silva, E.A. Macedo: Fluid Phase Equilib. **202**, 89 (2002)
46. W.G. Hoover, N.E. Hoover, K. Hansen: J. Chem. Phys. **70**, 1837 (1979)
47. R.J. Speedy, H. Reiss, Mol. Phys. **72**, 999 (1991)
48. R. Bowles, R. J. Speedy, Mol. Phys. **83**, 113 (1994)
49. J.H. Vera, J.M. Prausnitz: Chem. Eng. J. **3**, 1 (1972)
50. E.A. Guggenheim: Mol. Phys. **9**, 199 (1965)
51. M. Moshen-Nia, H. Modaress, G.A. Mansoori: *SPE Paper N° 26667.* (Proc. 1993 Annual Meeting Soc. Pet. Eng., SPE. Richardson, TX 1993)
52. T. Boublik: J. Chem. Phys. **63**, 4084 (1975)
53. C.H. Chien, R.A. Greenkorn, K.C. Chao: AIChE J. **29**, 560 (1983)
54. R.L. Cotterman, B.J. Schwarz, J.M. Prausnitz: AIChE J. **32**, 1787 (1986)
55. T. Ishikawa, W.K. Chung, B.C.Y. Lu: AIChE J. **26**, 372 (1980)
56. S.S. Chen, A. Kreglewski: Ber. Bunsenges. Phys. Chem. **81**, 1049 (1977)
57. M.J. Lee, K.C. Chao: AIChE J. **34**, 825 (1988)
58. C.M. Silva, H. Liu, E.A. Macedo: Chem. Eng. Sci. **53**, 2423 (1998)
59. J.H. Dymond: Physica **75**, 100 (1974)
60. J.H. Dymond: Chem. Soc. Rev. **14**, 317 (1985)
61. J.H. Dymond: Int. J. Thermophys. **18**, 303 (1997)
62. P. Oosting, N.J. Trappeniers: Physica **51**, 418 (1971)
63. A. Hasman: Neutron quasi-elastic scattering studies on fluid Argon. PhD Thesis, Delft University of Technology, Delft (1971)
64. F. Khoury, R.J. Kobayashi: J. Chem. Phys. **55**, 2439 (1971)
65. L. Durbin, R.J. Kobayashi: J. Chem. Phys. **37**, 1643 (1962)
66. D. Chandler: J. Chem. Phys. **62**, 1358 (1975)
67. A.J. Easteal, L.A. Woolf: Physica **124B**, 182 (1984)

68. B. Arends, K.O. Prins, N.J. Trappeniers: Physica **107A**, 307 (1981)
69. J. Kushick, B.J. Berne: J. Chem. Phys. **59**, 3732 (1973)
70. W.E. Alley, B.J. Alder: J. Chem. Phys. **63**, 3764 (1975)
71. J.E. Straub: Mol. Phys. **76**, 373 (1992)
72. S.J. Bertucci, W.H. Flygare: J. Chem. Phys. **63**, 1 (1975)
73. S.H. Chen: Chem. Eng. Sci. **38**, 655 (1983)
74. C. Erkey, H. Gadalla, A. Akgerman: J. Supercr. Fluids **3**, 180 (1990)
75. E. Ruckenstein, H. Liu: Ind. Eng. Chem. Res. **36**, 3927 (1997)
76. H. Liu, C.M. Silva, E.A. Macedo: Ind. Eng. Chem. Res. **36**, 246 (1997)
77. Y. Rosenfeld: Phys. Rev. A **15**, 2545 (1977)
78. Y. Rosenfeld: Chem. Phys. Lett. **48**, 467 (1977)
79. T. Boublik, I. Nezbeda: Coll. Czech. Chem. Commun. **51**, 2301 (1986)
80. D.M. Heyes, M.J. Cass, J.G. Powles, W.A.B. Evans: J. Phys. Chem. B **111**, 1455 (2007)
81. Y. Rosenfeld: J. Phys. Condens. Matter **11**, 5415 (1999)
82. M. Dzugutov: Nature (London) **381**, 137 (1996)
83. J.L. Bretonnet: J. Chem. Phys. **117**, 9370 (2002)
84. J.L. Bretonnet: J. Chem. Phys. **120**, 11100 (2004)
85. J.H. Dymond: J. Chem. Soc. Far. Trans. II **68**, 1789 (1972)
86. A.J. Easteal, L.A. Woolf, D.L. Jolly: Physica **121A**, 286 (1983)
87. R.J. Speedy: Mol. Phys. **62**, 509 (1987)
88. L.V. Woodcock: Ann. N.Y. Acad. Sci. **371**, 274 (1981)
89. A.J. Easteal, L.A. Woolf, D.L. Jolly: Physica **127A**, 344 (1984)
90. D.M. Heyes: J. Chem. Soc. Faraday Trans. II **83**, 1985 (1987)
91. D.M. Heyes: Phys. Rev. B **37**, 5677 (1988)
92. J.J. Erpenbeck, W.W. Wood: Phys. Rev. A **32**, 412 (1985)
93. K.R. Harris: Mol. Phys. **77**, 1153 (1992)
94. J. Amoros: Mol. Phys. **83**, 771 (1994)
95. H. Sigurgeirsson, D.M. Heyes: Mol. Phys. **101**, 469 (2003)
96. I. Yokoyama: Physica B **254**, 172 (1998)
97. I. Yokoyama: Physica B **269**, 244 (1999)
98. J.J. Hoyt, M. Asta, B. Sadigh: Phys. Rev. Lett. **85**, 594 (2000)
99. M. Dzugutov: Phys. Rev. E **65**, 032501 (2002)
100. L.S. Darken: Am. Inst. Min. Metall. Pet. Eng. **175**, 184 (1948)
101. A. Vignes: Ind. Eng. Chem. Fundam. **5**, 189 (1966)
102. M. Pertler, E. Blass, G.W. Stevens: AIChE J. **42**, 910 (1996)
103. J.L. Lebowitz: Phys. Rev. A **895**, 133 (1964)
104. N.F. Carnahan, K.E. Starling: *Abstracts of Invited Lectures for the van der Waals Centennial Conference on Statistical Mechanics.* (Ed. North-Holland Publications Co, Amsterdam 1973)
105. G.A. Mansoori, N.F. Carnahan, K.E. Starling, T.W. Leland Jr.: J. Chem. Phys. **54**, 1523 (1971)
106. B.J. Alder, W.E. Alley, J.H. Dymond: J. Chem. Phys. **61**, 1415 (1974)
107. W. Sung, G. Stell: J. Chem. Phys. **80**, 3350 (1984)
108. C.K.J. Sun, S.H. Chen: AIChE J. **31**, 1510 (1985)
109. A.P. Eaton, A. Akgerman: Ind. Eng. Chem. Res. **36**, 923 (1997)
110. H. Liu, E.A. Ruckenstein: Ind. Eng. Chem. Res. **36**, 5488 (1997)
111. A.J. Easteal, L.A. Woolf: Chem. Phys. Lett. **167**, 329 (1990)
112. C.K.J. Sun, S.H. Chen: AIChE J. **31**, 1904 (1985)

113. P.T. Herman, B.J. Alder: J. Chem. Phys. **56**, 987 (1972)
114. P.H. Salim, M.A. Trebble: J. Chem. Soc., Faraday Trans. **91**, 245 (1995)
115. A. Enninghorst, F.D. Wayne, M.D. Zeidler: Mol. Phys. **88**, 437 (1996)
116. D.M. Heyes: J. Chem. Soc., Faraday Trans. II, **79**, 1741 (1983)
117. H. Lee, G. Thodos: Ind. Eng. Chem. Res. **27**, 992 (1988)
118. L.S. Tee, S. Gotoh, W.E. Stewart: Ind. Eng. Chem. Fundam. **5**, 356 (1966)
119. S.A. Rice, P. Gray: *The Statistical Mechanics of Simple Liquids.* (Ed. Interscience Publishers. New York 1965)
120. E. Helfand: Phys. Fluids **4**, 681 (1961)
121. R.C. Reid, J.M. Prausnitz, B.E. Poling: *The Properties of Gases and Liquids.* (Ed. McGraw-Hill Book Company. 4th ed., New York 1987)
122. Y. Zhu, X. Lu, J. Zhou, Y. Wang, J. Shi: Fluid Phase Equilib. **194-197**, 1141 (2002)
123. R.L. Rowley, M.M. Painter: Int. J. Therm. **18**, 1109 (1997)
124. J. Johnson, J.A. Zollweg, K.E. Gubbins: Mol. Phys. **78**, 591 (1993)
125. Y.-X. Yu, G.-H. Gao: Fluid Phase Equilib. **166**, 111 (1999)
126. S.W. Smith, C.K. Hall, B.D. Freeman: J. Chem. Phys. **102**, 1057 (1995)
127. R.L. Hurle, L.A. Woolf: J. Chem. Soc. Faraday Trans. I **178**, 2233 (1982)
128. I. Yokoyama: Physica B **291**, 145 (2000)
129. G.X. Li, C.S. Liu, Z.G. Zhu: Phys. Rev. B **71**, 094209 (2005)
130. T. Merzliak, A. Pfenning: Mol. Simulation **30**, 459 (2004)
131. M.A. Matthews, A. Akgerman: AIChE J. **33**, 881 (1987)
132. C. Erkey, A. Akgerman: AIChE J. **35**, 443 (1989)
133. J. Zhou, X. Lu, Y. Wang, J. Shi: Fluid Phase Equilib. **172**, 279 (2000)
134. S. Glasstone, K.J. Laidler, H. Eyring: *The Theory of Rate Processes*, (McGraw-Hill, New York 1941)
135. H.-M. Koo, S. Hess: Z. Naturforsch. **42a**, 231 (1987)
136. J.G. Kirkwood, F.P. Buff, M.S. Green: J. Chem. Phys. **17**, 988 (1949)
137. K. Stephan, K. Lucas: *Viscosity of Dense Fluids* (Plenum, New York 1979)
138. D.M. Heyes, H. Sigurgeirsson: J. Rheol. **48**, 223 (2004)
139. J.H. Dymond, M.J. Assael: Modified hard-spheres scheme, in *Transport Properties of Fluids: Their Correlation, Prediction and Estimation*, J. Millat, J.H. Dymond, C.A. Nieto de Castro, eds. (Cambridge University Press, New York 1996)
140. F.D.C. Silva, L.A.F. Coelho, F.W. Tavares, M.J.E.M. Cardoso: Inter. J. Quant. Chem. **95**, 79 (2003)
141. R. Laghaei, A. E. Nasrabad, B. C. Eu: J. Phys Chem B **109**, 5873 (2005)
142. A. Allal, C. Boned, A. Baylaucq: Phys. Rev. E **64**, 11203 (2001)
143. J. W. Przezdziecki and T. Shridhar: AIChE J. **31**, 333 (1983)
144. H. Ertl, F.A.L. Dullien: J. Phys. Chem. **77**, 3007 (1973)
145. H. Liu, W. Wang, C.-H. Chang: IEC Res. **30**, 1617 (1991)
146. H. Liu: Viscosity of liquids and its measurement under high pressure. PhD Thesis, Tinajin University, 1988.
147. C.-H. Chang, H. Liu: J. Chem. Ind. and Eng. (China) **5**, 276 (1990)
148. A.S. Teja, P. Rice: IEC Fundam. **20**, 77 (1981)
149. S. Aznarez, M.M.E.F. de Ruiz Holgado, E.L. Arancibia: J. Mol. Liquids **124**, 78 (2006)
150. H. Liu: Petrol. Chem. Ind. (China) **22**, 32 (1993)
151. F. Audonnet, A.A.H. Pádua: Fluid Phase Equilib. **181**, 147 (2001)

152. N. Mehdipour, H. Eslami: Int. J. Thermophys. **41**, 949 (2002)
153. G.A. Parsafar, Z. Kalantar: Fluid Phase Equilib. **253**, 108 (2007)
154. M.J. Assael, J.H. Dymond, M. Papadaki, P.M. Patterson: Int. J. Thermophys. **13**, 269 (1992)
155. M.J. Assael, J.H. Dymond, M. Papadaki, P.M Patterson: Fluid Phase Equilib. **75**, 245 (1992)
156. M.J. Assael, J.H. Dymond, M. Papadaki, P.M. Patterson: Int. J. Thermophys. **13**, 659 (1992)
157. M.J. Assael, J.H. Dymond, P.M. Patterson: Int. J. Thermophys. **13**, 895 (1992)
158. M.J. Assael, J.H. Dymond, S.K. Polimatidou: Int. J. Thermophys. **15**, 189 (1994)
159. M.J. Assael, N.K. Dalaouti, J.H. Dymond, E.P. Feleki: Int. J. Thermophys. **21**, 367 (2000)
160. A.S. Teja, R.L. Smith, Jr., R.K. King, T.F. Sun: Int. J. Thermophys. **20**, 149 (1999)
161. Sk. M. Ali: J. Chem. Phys. **124**, 144504-1 (2006)
162. S. Korkmaz, U.U.N. Yazar, S.D. Korkmaz: Fluid Phase Equilib. **249**, 159 (2006)
163. G.X. Li, C.S. Liu, Z.G. Zhu: Chin. Phys. Lett. **21**, 2489 (2004)
164. G.X. Li, C.S. Liu, Z.G. Zhu: J.Non-Cryst. Solids **351**, 946 (2005)
165. R. Tufeu, B. Leneindre, P. Bury: Compt. Rend. Acad. Sci. Paris **271B**, 589 (1970)
166. J.V. Sengers, W.T. Bolk, C.J. Stigter: Physica **30**, 1018 (1964)
167. A. Michels, J.V. Sengers, L.J.M. Van de Klundert: Physica **29**, 149 (1963)
168. B. Leneindre, P. Bury, R. Tufeu, P. Johannin, B. Vodar: Proc. 7th Conf. Therm. Cond. N.B.S., Washington (1968) 579
169. R. Tufeu, B. Leneindre, P. Bury: Compt. Rend. Acad. Sci. Paris **273B**, 61 (1971)
170. H.J.M. Hanley, R.D. McCarty, W.M. Haynes: J. Phys. Chem. Ref. Data **3**, 979 (1974)
171. S.A. Ulybin, S.S. Bakulin: Thermal Engin. **24**, 68 (1977)
172. A.L. Gosman: Thermodynamic properties of argon from the triple point to 300 K at pressures to 1000 atmospheres (NSRD-NBS 27, Washington 1969)
173. N.B. Vargaftik: *Tables on the Thermophysical Properties of Liquids and Gases*, 2nd Ed. (John Wiley & Sons, New York 1975)
174. N. Mani, J.E.S. Venart: Adv. Cryog. Eng. **18**, 280 (1973)
175. H.J.M. Hanley, W.M. Haynes, R.D. McCarty: J. Phys. Chem. Ref. Data **6**, 597 (1977)
176. R.D. Goodwin: The thermophysical properties of methane, from 90 to 500 K at pressures to 700 bar (NBS 653, Washington 1974)
177. E.T.S. Huang, G.W. Swift, F. Karata: AIChE J. **13**, 846 (1967)
178. W.A.P. Luck, in The Hydrogen Bond, vol. III, P. Schuster, G. Zundel, C. Sandorfy, eds. (North Holland, Amsterdam, 1976) p. 1369.
179. A. Nagashima: J. Phys. Chem. Ref. Data **6**, 1133 (1977)
180. B. Le Neindre, P. Bury, R. Tufieu, B. Vodar: J. Chem. Eng. Data **21**, 265 (1975)
181. D.L. Bewilogua, C. Gladun, K. Busch: J. Low. Temp. Phys. **4**, 299 (1971)
182. T. Grindley, J.E. Lind Jr.: J. Chem. Phys. **54**, 3983 (1971)
183. P.W. Bridgman: *The Physics of High Pressure* (Bells and Sons, London 1949)

184. J. Amorós, M.J. Maeso, E. Villar: Int. J. Thermophys. **13**, 907 (1992)
185. J. Amorós, J.R. Solana, E. Villar: Mater. Chem. Phys. **10**, 557 (1984)
186. W. Sheng, G.-J. Chen, H.-C. Lu: Int. J. Thermophys. **10**, 133 (1989)
187. H-Y Lee, G. Liu: Fluid Phase Equilib. **105**, 141 (1995)
188. Y. Song, E.A. Mason: J. Chem. Phys. **91**, 7840 (1989)
189. G. Ihm, Y. Song, E.A. Mason: J. Chem. Phys. **94**, 3839 (1991)
190. O. Sifner, J. Klomfar: J. Phys. Chem. Ref. Data **23**, 63 (1994)
191. http://webbook.nist.gov
192. G.A. Parsafar, Z. Kalantar: Iran. J. Chem. Chem. Eng. **22**, 1 (2003)
193. G.A. Parsafar, Z. Kalantar: Fluid Phase Equilibria **234**, 19 (2005)
194. G.A. Parsafar, E.A. Mason: J. Phys. Chem. **97**, 9048 (1993)
195. G.A. Parsafar, E.A. Mason: J. Phys. Chem. **98**, 1962 (1994)
196. J.H. Dymond: Transport properties in dense fluids. Proc. 6*th Symp. Thermophys. Prop. A.S.M.E.*, New York (1973) 143.
197. M.J. Assael, E. Karagiannidis, S.K. Polimatidou: High Temp. - High Press. **25**, 259 (1993)
198. M.J. Assael, J.H. Dymond, S.K. Polimatidou: Int. J. Thermophys. **16**, 761 (1995)
199. A.N. Gurova: Thermal Conductivity of Environmentally Acceptable Refrigerants. PhD Thesis, University of Lisbon, Lisbon (1996)
200. A.N. Gurova, U.V. Mardolcar, C.A. Nieto de Castro: Int. J. Thermophys. **20**, 63 (1999)
201. D.M. Heyes: J. Chem. Soc. Faraday Trans. **84**, 705 (1988)
202. J.C.G. Calado, U.V.Mardolcar, C.A. Nieto de Castro, H.M. Roder, W.A. Wakeham: Physica A **143**, 314 (1987)
203. U.V. Mardolcar, C.A. Nieto de Castro: Ber. Bunsenges. Phys. Chem. **91**, 152 (1987)
204. C. Kaur, U. Harbola, S.P. Das: J. Chem. Phys. **123**, 034501-1 (2005)
205. A. Eucken: Phys. Z. **14**, 323 (1913)
206. B.C. Eu: J. Phys. Chem. A **110**, 831 (2006)
207. T.H. Chung, L.L. Lee, K.E. Starling: Int. J. Thermophys. **1**, 397 (1980)

10

Kinetic Theory for Binary Granular Mixtures at Low Density

V. Garzó

Departamento de Física, Universidad de Extremadura, E-06071 Badajoz, Spain
vicenteg@unex.es

Many features of granular media can be modeled as a fluid of hard spheres with inelastic collisions. Under rapid flow conditions, the macroscopic behavior of grains can be described through hydrodynamic equations. At low density, a fundamental basis for the derivation of the hydrodynamic equations and explicit expressions for the transport coefficients appearing in them is provided by the Boltzmann kinetic theory conveniently modified to account for inelastic binary collisions. The goal of this chapter is to give an overview of the recent advances made for binary granular gases by using kinetic theory tools. Some of the results presented here cover aspects such as transport properties, energy nonequipartition, instabilities, segregation or mixing, non-Newtonian behavior, etc. In addition, comparison of the analytical results with those obtained from Monte Carlo and molecular dynamics simulations is also carried out, showing the reliability of kinetic theory to describe granular flows even for strong dissipation.

10.1 Introduction

Granular systems have attracted the attention of the physics community in the past few years, in part because the behavior of these systems under many conditions exhibit a great similarity to ordinary fluids [1]. These conditions include rapid, dilute flows where the dominant transfer of momentum and energy is through binary collisions of the grains. The main difference from ordinary fluids is the absence of energy conservation, leading to both obvious and subtle modifications of the usual macroscopic balance equations as well as the constitutive equations for the irreversible fluxes. However, in spite of the utility of hydrodynamics to describe rapid granular flows, there are still some open questions about its domain of validity and the associated constitutive equations appearing in the hydrodynamic equations [2].

To isolate the effects of such collisional dissipation from other important properties of granular media, an idealized microscopic model system is

Garzó, V.: *Kinetic Theory for Binary Granular Mixtures at Low Density*. Lect. Notes Phys. **753**, 493–540 (2008)
DOI 10.1007/978-3-540-78767-9_10

usually considered: a system composed by smooth hard spheres with inelastic collisions. As in the elastic case, the collisions are specified in terms of the change in relative velocity at contact but with a decrease in the magnitude of the normal component measured by a positive coefficient of restitution $\alpha \leq 1$. This parameter distinguishes the granular fluid ($\alpha < 1$) from the ordinary fluid ($\alpha = 1$). Given that the hard sphere system with elastic collisions has been widely studied for both equilibrium and nonequilibrium statistical mechanics [3], it is tempting to apply the same methods for the case of inelastic collisions. However, some care is warranted in translating properties of ordinary fluids to granular fluids. In this presentation, a kinetic theory description based on the Boltzmann kinetic equation (which applies at sufficiently low density) will be considered as the appropriate tool to study granular flows from a microscopic point of view.

Although many efforts have been devoted in the past few years to the understanding of granular fluids, the derivation of the form of the constitutive equations with explicit expressions for the transport coefficients is still a subject of interest and controversy. The conditions to obtain a hydrodynamic description are expected to be similar to those for normal fluids. For a given initial state there are two stages of evolution. First, during the kinetic stage there is rapid velocity relaxation to a "universal" velocity distribution that depends on the average local density, temperature, and flow velocity. Subsequently, the hydrodynamic stage is described through a slower evolution of these local hydrodynamic fields as they approach uniformity. The solution to the Boltzmann equation in this second stage is said to be *normal*, where all space and time dependence of the distribution function occurs through the macroscopic hydrodynamic fields. The Chapman–Enskog method [4] provides a constructive means to obtain an approximation to such a solution for states whose spatial gradients are not too large. In this case, the explicit form of this normal solution is given as a perturbation expansion in the spatial gradients of the fields. This solution is then used to evaluate the fluxes in the macroscopic balance equations in terms of these gradients. To lowest order the balance equations become the granular Euler equations while to second order they are the granular Navier–Stokes equations. In carrying out this analysis, explicit forms for the transport coefficients are obtained as functions of the coefficient of restitution and other parameters of the system. In this general context, the study of hydrodynamics for granular gases is the same as that for ordinary fluids.

The derivation of hydrodynamics from the inelastic Boltzmann equation has been widely covered in the case of a monodisperse gas where the particles are of the same mass and size. As for elastic collisions, the transport coefficients are given in terms of the solutions of linear integral equations [5, 6], which are approximately solved by using Sonine polynomial expansions. The estimates for the transport coefficients provided by the Sonine solution compare in general quite well with both direct Monte Carlo simulation (DSMC) of the Boltzmann equation and molecular dynamics (MD) simulation of the gas, even for relatively strong degrees of dissipation [7, 8, 9, 10]. This good agreement supports the formal theoretical analysis and the claim that hydrodynamics is not limited to nearly elastic particles [11, 12, 13].

Nevertheless, a real granular system is generally characterized by some degrees of polydispersity in density and size, which leads to phenomena very often observed in nature and experiments, such as separation or segregation. Needless to say, the study of granular mixtures is much more complicated than for a monodisperse gas since not only the number of transport coefficients in a multicomponent system is higher than that of a single gas, but also they depend on parameters such as masses, sizes, composition as well as several independent coefficients of restitution α_{ij}. Due to these difficulties, studies for multicomponent gases are more scarce in the literature. Many of the previous attempts [14, 15, 16, 17, 18, 19, 20] to derive hydrodynamics from kinetic theory were carried out in the quasi-elastic limit where the equipartition of energy can be considered as an acceptable assumption. In addition, according to this level of approximation, the inelasticity is only accounted for by the presence of a sink term in the energy balance equation, so that the expressions for the transport coefficients are the same as those obtained for ordinary fluids. However, the theoretical prediction of the failure of energy equipartition in multicomponent granular gases [21] has been confirmed by computer simulations [22, 23, 24, 25, 26, 27, 28, 29, 30, 31], and even observed in real experiments [32, 33].

Although the possibility of nonequipartition was already pointed out many years ago [34], it has not been until recently that a systematic study of the effect of nonequipartition on the Navier–Stokes hydrodynamic equations has been carefully carried out [35, 36]. These new equations and associated transport coefficients provide a somewhat more stringent test of the analysis since the parameter space is much larger. As in the monodisperse case, explicit expressions for the transport coefficients requires also to consider Sonine polynomial expansions. The numerical accuracy of this Sonine expansion has been confirmed by comparison with Monte Carlo simulations of the Boltzmann equation in the cases of the shear viscosity [37] and the tracer diffusion [38] coefficients. Exceptions to this agreement are extreme mass or size ratios and strong dissipation, although these discrepancies between theory and simulation diminish as one considers more terms in the Sonine polynomial approximation [38].

The explicit knowledge of the Navier–Stokes transport coefficients allows quantitative application of the nonlinear hydrodynamic equations to a number of interesting problems for granular mixtures, such as to quantify the violation of the Einstein relation [39, 40] or the Onsager reciprocal relations [36], the stability analysis of the homogeneous cooling state [36], and segregation induced by a thermal gradient [41]. In all the cases, the analysis clearly shows the important role played by the inelasticity in the different physical situations.

The analogy between rapid granular flow and ordinary fluids can also be extended to many other transport situations. A particularly simple case, allowing detailed analysis even in far from equilibrium conditions is the simple or uniform shear flow (USF) problem. Macroscopically, it is characterized by uniform density and temperature and a constant mean velocity profile. This is a well-known nonequilibrium problem widely studied, for both granular monodisperse [13, 42, 43, 44, 45, 46, 47, 48, 49, 50, 51, 52, 53, 54, 55, 56] and ordinary gases [57]. Nevertheless, the nature of this state is quite different in each system. While for elastic fluids the temperature increases monotonically in time due to viscous heating, a steady state is possible for granular media when the effect of the viscous heating is exactly compensated by the dissipation in collisions. Thus, in the steady state, there is an intrinsic connection between the shear field and dissipation, so that the collisional cooling sets the strength of the velocity gradient. As a consequence, the USF state is inherently non-Newtonian and the rheological properties of the system cannot be obtained from the Navier-Stokes description, at least for finite dissipation [58].

The aim of this chapter is to offer a short review of recent results obtained for binary granular mixtures from the Boltzmann kinetic theory. It is structured as follows. The Boltzmann kinetic equation for a granular binary mixture and its associated macroscopic balance equations are introduced in Sect. 10.2. Section 10.3 deals with the solutions to the Boltzmann equation for homogeneous states in the free cooling case as well as when the mixture is heated by an external thermostat. The Chapman–Enskog method around the local version of the homogenous distributions obtained in Sect. 10.2 is applied in Sect. 10.3 to get the form of the Navier–Stokes hydrodynamic equations. Theoretical results for the diffusion and shear viscosity transport coefficients are compared with simulation data in Sect. 10.5, while the Einstein and the Onsager relations for granular mixtures are analyzed in Sects. 10.6 and 10.7, respectively. The dispersion relations for the hydrodynamic equations linearized about the homogeneous cooling state are obtained in Sect. 10.8, showing that the homogeneous reference state is unstable to long-wavelength perturbations. The conditions for stability are identified as functions of the wave vector, the dissipation, and the parameters of the mixture. Segregation due to thermal diffusion is studied in Sect. 10.9 by using the Navier–Stokes description. A new criterion for segregation is found that is consistent with recent experimental results. Section 10.10 deals with the USF problem for a granular mixture. Finally, the paper concludes in Sect. 10.11 with a discussion of the results presented here.

Before ending this section, I remark that the present account is a personal perspective based on the author's work and that of his collaborators so that no attempt is made to include the extensive related work of many others in this field. The references given are selective and apologies are offered at the outset to the many other important contributions not recognized explicitly.

10.2 Boltzmann Kinetic Equation for Binary Mixtures of Inelastic Hard Spheres

Consider a binary granular mixture composed by smooth inelastic disks ($d=2$) or spheres ($d=3$) of masses m_1 and m_2, and diameters σ_1 and σ_2. The inelasticity of collisions among all pairs is characterized by three independent constant coefficients of restitution α_{11}, α_{22}, and $\alpha_{12}=\alpha_{21}$, where $\alpha_{ij}\leq 1$ is the coefficient of restitution for collisions between particles of species i and j. Since the spheres are assumed to be perfectly smooth, only the translational degrees of freedom of grains are affected by dissipation. In the low-density regime, a simultaneous interaction of more than two particles is highly unlike and so can be neglected. Consequently, in a dilute gas the interactions among the particles reduce to a succession of *binary* collisions. At this level of description, all the relevant information on the state of the system is contained in the one-body velocity distribution functions $f_i(\mathbf{r},\mathbf{v};t)$ $(i=1,2)$ defined so that $f_i(\mathbf{r},\mathbf{v};t)\mathrm{d}\mathbf{r}\mathrm{d}\mathbf{v}$ is the most probable (or average) number of particles of species i which at time t lie in the volume element $\mathrm{d}\mathbf{r}$ centered at the point $\mathbf{r}$ and moving with velocities in the range $\mathrm{d}\mathbf{v}$ about $\mathbf{v}$. For an inelastic gas, the distributions $f_i(\mathbf{r},\mathbf{v};t)$ $(i=1,2)$ for the two species satisfy the coupled nonlinear Boltzmann equations [59, 60]

$$\left(\partial_t+\mathbf{v}\cdot\nabla+\frac{\mathbf{F}_i}{m_i}\cdot\frac{\partial}{\partial\mathbf{v}}+\mathcal{F}_i\right)f_i(\mathbf{r},\mathbf{v},t)=\sum_{j=1}^{2}J_{ij}\left[\mathbf{v}|f_i(t),f_j(t)\right]\,, \tag{10.1}$$

where the Boltzmann collision operator $J_{ij}\left[\mathbf{v}|f_i,f_j\right]$ is

$$J_{ij}\left[\mathbf{v}_1|f_i,f_j\right]=\sigma_{ij}^{d-1}\int \mathrm{d}\mathbf{v}_2\int \mathrm{d}\widehat{\boldsymbol{\sigma}}\,\Theta(\widehat{\boldsymbol{\sigma}}\cdot\mathbf{v}_{12})(\widehat{\boldsymbol{\sigma}}\cdot\mathbf{v}_{12})$$
$$\times\left[\alpha_{ij}^{-2}f_i(\mathbf{r},\mathbf{v}_1',t)f_j(\mathbf{r},\mathbf{v}_2',t)-f_i(\mathbf{r},\mathbf{v}_1,t)f_j(\mathbf{r},\mathbf{v}_2,t)\right]\,. \tag{10.2}$$

In Eq. (10.2), d is the dimensionality of the system, $\sigma_{ij}=\left(\sigma_i+\sigma_j\right)/2$, $\widehat{\boldsymbol{\sigma}}$ is an unit vector along the line of centers, Θ is the Heaviside step function, and $\mathbf{v}_{12}=\mathbf{v}_1-\mathbf{v}_2$ is the relative velocity. The primes on the velocities denote the initial values $\{\mathbf{v}_1',\mathbf{v}_2'\}$ that lead to $\{\mathbf{v}_1,\mathbf{v}_2\}$ following a binary (restituting) collision:

$$\begin{aligned}\mathbf{v}_1'&=\mathbf{v}_1-\mu_{ji}\left(1+\alpha_{ij}^{-1}\right)(\widehat{\boldsymbol{\sigma}}\cdot\mathbf{v}_{12})\widehat{\boldsymbol{\sigma}}\,,\\ \mathbf{v}_2'&=\mathbf{v}_2+\mu_{ij}\left(1+\alpha_{ij}^{-1}\right)(\widehat{\boldsymbol{\sigma}}\cdot\mathbf{v}_{12})\widehat{\boldsymbol{\sigma}}\,,\end{aligned} \tag{10.3}$$

where $\mu_{ij}\equiv m_i/\left(m_i+m_j\right)$. In addition, $\mathbf{F}_i$ denotes an external *conservative* force acting on species i (such as a gravity field) and $\mathcal{F}_i$ is an operator representing a possible effect of an external *nonconservative* forcing which injects energy into the system to compensate for the energy dissipated by collisional cooling. This type of force acts as a *thermostat* that tries to mimics a thermal bath. Some explicit forms for the operator $\mathcal{F}_i$ will be chosen later.

The relevant hydrodynamic fields for the mixture are the number densities n_i, the flow velocity $\mathbf{u}$, and the temperature T. They are defined in terms of moments of the velocity distribution functions f_i as

$$n_i = \int d\mathbf{v}\, f_i(\mathbf{v}) \,, \tag{10.4}$$

$$\rho\mathbf{u} = \sum_{i=1}^{2} m_i \int d\mathbf{v}\, \mathbf{v} f_i(\mathbf{v}) \,, \tag{10.5}$$

$$nT = p = \sum_{i=1}^{2} n_i T_i = \sum_{i=1}^{2} \frac{m_i}{d} \int d\mathbf{v}\, V^2 f_i(\mathbf{v}) \,, \tag{10.6}$$

where $\mathbf{V} = \mathbf{v} - \mathbf{u}$ is the peculiar velocity, $n = n_1 + n_2$ is the total number density, $\rho = m_1 n_1 + m_2 n_2$ is the total mass density, and p is the pressure. Furthermore, the third equality of Eq. (10.6) defines the kinetic temperatures T_i for each species, which measure their mean kinetic energies.

The collision operators conserve the particle number of each species and the total momentum, but the total energy is not conserved:

$$\int d\mathbf{v}\, J_{ij}[\mathbf{v}|f_i, f_j] = 0 \,, \tag{10.7}$$

$$m_i \int d\mathbf{v}\, \mathbf{v} J_{ij}[\mathbf{v}|f_i, f_j] + m_j \int d\mathbf{v}\, \mathbf{v} J_{ji}[\mathbf{v}|f_j, f_i] = \mathbf{0} \,, \tag{10.8}$$

$$\sum_{i=1}^{2}\sum_{j=1}^{2} m_i \int d\mathbf{v} V^2 J_{ij}[\mathbf{v}|f_i, f_j] = -dnT\zeta \,. \tag{10.9}$$

In Eq. (10.9), ζ is identified as the total "cooling rate" due to collisions among all species. It measures the rate of energy loss due to dissipation. At a kinetic level, it is also convenient to introduce the "cooling rates" ζ_i for the partial temperatures T_i. They are defined as

$$\zeta_i = \sum_{j=1}^{2} \zeta_{ij} = -\frac{m_i}{dn_i T_i} \sum_{j=1}^{2} \int d\mathbf{v}\, V^2 J_{ij}[\mathbf{v}|f_i, f_j] \,, \tag{10.10}$$

where the second equality defines the quantities ζ_{ij}. The total cooling rate ζ can be written in terms of the partial cooling rates ζ_i as

$$\zeta = T^{-1} \sum_{i=1}^{2} x_i T_i \zeta_i \,, \tag{10.11}$$

where $x_i = n_i/n$ is the mole fraction of species i.

From Eqs. (10.7)–(10.9), the macroscopic balance equations for the number densities n_i, the total momentum density $\rho\mathbf{u}$, and the energy density $(d/2)nT$ can be obtained. They are given, respectively, by [35]

$$D_t n_i + n_i \nabla \cdot \mathbf{u} + \frac{\nabla \cdot \mathbf{j}_i}{m_i} = 0 , \tag{10.12}$$

$$\rho D_t \mathbf{u} + \nabla \cdot \mathsf{P} = \sum_{i=1}^{2} n_i \mathbf{F}_i , \tag{10.13}$$

$$D_t T - \frac{T}{n} \sum_{i=1}^{2} \frac{\nabla \cdot \mathbf{j}_i}{m_i} + \frac{2}{dn} \left(\nabla \cdot \mathbf{q} + \mathsf{P} : \nabla \mathbf{u} - \sum_{i=1}^{2} \frac{\mathbf{F}_i \cdot \mathbf{j}_i}{m_i} \right) = -(\zeta - \xi) T . \tag{10.14}$$

In the above equations, $D_t = \partial_t + \mathbf{u} \cdot \nabla$ is the material derivative,

$$\mathbf{j}_i = m_i \int d\mathbf{v}\, \mathbf{V}\, f_i(\mathbf{v}) \tag{10.15}$$

is the mass flux for species i relative to the local flow,

$$\mathsf{P} = \sum_{i=1}^{2} m_i \int d\mathbf{v}\, \mathbf{V}\mathbf{V}\, f_i(\mathbf{v}) \tag{10.16}$$

is the total pressure tensor, and

$$\mathbf{q} = \sum_{i=1}^{2} \frac{m_i}{2} \int d\mathbf{v}\, V^2 \mathbf{V}\, f_i(\mathbf{v}) \tag{10.17}$$

is the total heat flux. On the right-hand side of the temperature equation (10.14), the source term ξ (measuring the rate of heating due to the external thermostat) is given by

$$\xi = -\frac{1}{dnT} \sum_{i=1}^{2} m_i \int d\mathbf{v} V^2 \mathcal{F}_i f_i(\mathbf{v}). \tag{10.18}$$

In the balance equations (10.12)–(10.14) it is assumed that the external driving thermostat does not change the number of particles of each species or the total momentum, i.e.,

$$\int d\mathbf{v}\, \mathcal{F}_i f_i(\mathbf{v}) = 0 , \tag{10.19}$$

$$\sum_{i=1}^{2} m_i \int d\mathbf{v}\, \mathbf{v}\, \mathcal{F}_i f_i(\mathbf{v}) = \mathbf{0}. \tag{10.20}$$

The utility of the balance equations (10.12)–(10.14) is limited without further specification of the fluxes and the cooling rate, which in general have a complex dependence on space and time. However, for sufficiently large space and time scales, one expects that the system reaches a hydrodynamic regime in which all the space and time dependence is given entirely through a functional dependence on the six hydrodynamic fields n_i, $\mathbf{u}$, and T. The corresponding functional dependence of $\mathbf{j}_i$, P, and $\mathbf{q}$ on these fields are called constitutive equations and define the transport coefficients of the mixture. The primary feature of a hydrodynamic description is the reduction of the description from many microscopic degrees of freedom to a set of equations involving only six local fields. At a kinetic level, the constitutive equations are obtained when one admits the existence of a *normal* solution to the Boltzmann equation where the velocity distribution functions depend on $\mathbf{r}$ and t only through their functional dependence on the fields, namely,

$$f_i(\mathbf{r}, \mathbf{v}_1, t) = f_i[\mathbf{v}_1 | n_i(\mathbf{r}, t), T(\mathbf{r}, t), \mathbf{u}(\mathbf{r}, t)]. \tag{10.21}$$

This normal solution is generated by the Chapman–Enskog method [4] conveniently adapted to dissipative dynamics. Since the method is based on an expansion around the local version of the homogeneous state, let us characterize it before considering inhomogeneous solutions.

10.3 Homogeneous States

In this section we are interested in spatially homogeneous isotropic states. In this case, we assume that the magnitude of the conservative external forces is at least of first order in the spatial gradients (i.e., $\mathbf{F}_i = \mathbf{0}$), so that Eq. (10.1) becomes

$$(\partial_t + \mathcal{F}_i) f_i = \sum_j J_{ij}[f_i, f_j]. \tag{10.22}$$

For elastic collisions ($\alpha_{ij} = 1$) and in the absence of external forcing ($\mathcal{F}_i = 0$), it is well known that the long-time solution to (10.22) is a Maxwellian distribution for each species at the same temperature T. However, if the particles collide inelastically ($\alpha_{ij} < 1$) and $\mathcal{F}_i = 0$, a steady state is not possible in uniform situations since the temperature decreases monotonically in time. This state is usually referred to as the homogeneous cooling state (HCS). In this situation, since n_i is uniform and $\mathbf{u} = \mathbf{0}$, the normal (hydrodynamic) solution to f_i requires that all its time dependence occurs only through the temperature $T(t)$. Consequently, $f_i(v, t)$ must be of the form [21]

$$f_i(\mathbf{v}, t) = n_i v_0^{-d}(t) \Phi_i \left(v / v_0(t) \right) , \tag{10.23}$$

where $v_0(t) = \sqrt{2T(t)(m_1 + m_2)/\left(m_1 m_2\right)}$ is a thermal speed defined in terms of the temperature $T(t)$ of the mixture. The balance equations (10.12)–(10.14) to this order become $\partial_t x_i = \partial_t \mathbf{u} = 0$, and $T^{-1}\partial_t T = -\zeta$, where the cooling rate ζ is determined by Eq. (10.9). In addition, from Eqs. (10.10) and (10.22) (when $\mathcal{F}_i = 0$) one can derive the time evolution for the temperature ratio $\gamma = T_1(t)/T_2(t)$:

$$\partial_t \ln \gamma = \zeta_2 - \zeta_1. \tag{10.24}$$

The fact that the distributions f_i depend on time *only* through $T(t)$ necessarily implies that the temperature ratio γ must be independent of time and so, Eq. (10.24) gives the condition

$$\zeta_1 = \zeta_2 = \zeta. \tag{10.25}$$

In the elastic case, where f_i is a Maxwellian distribution, the above condition yields $T_1(t) = T_2(t) = T(t)$ and the energy equipartition applies. However, in the inelastic case, the equality of the cooling rates leads to different values for the partial temperatures, even if one considers the Maxwellian approximation to f_i. Nevertheless, the constancy of γ assures that the time dependence of the distributions is entirely through $T(t)$, and in fact the partial temperatures can be expressed in terms of the global temperature as

$$T_1(t) = \frac{\gamma}{1 + x_1(\gamma - 1)} T(t), \quad T_2(t) = \frac{1}{1 + x_1(\gamma - 1)} T(t). \tag{10.26}$$

Just as for the single gas case [59, 61], the exact form of Φ_i has not yet been found, although a good approximation for thermal velocities can be obtained from an expansion in Sonine polynomials [21]. In the leading order, Φ_i is given by

$$\Phi_i(v^*) \to \left(\frac{\theta_i}{\pi}\right)^{d/2} \mathrm{e}^{-\theta_i v^{*2}} \left[1 + \frac{c_i}{4}\left(\theta_i v^{*4} - (d+2)\theta_i v^{*2} + \frac{d(d+2)}{4}\right)\right]. \tag{10.27}$$

Here, $v^* \equiv v/v_0$,

$$\theta_i = \frac{m_i}{\gamma_i} \sum_{j=1}^{2} m_j^{-1}, \tag{10.28}$$

and $\gamma_i = T_i/T$. The coefficients c_i (which measure the deviation of Φ_i from the reference Maxwellian) are determined consistently from the Boltzmann equation. The explicit form of the approximation (10.27) provides detailed predictions for the temperature ratio T_1/T_2 (through calculation of the cooling rates) and for the cumulants c_i as functions of the mass ratio, size ratio, composition, and coefficients of restitution [21]. The numerical accuracy of this truncated Sonine expansion has been confirmed by comparison with Monte Carlo [62] and MD [63] simulations.

As mentioned in the Introduction, the existence of different temperatures for each species has been observed in real experiments of driven steady states. These states are achieved from external forces that do work at the same rate as collisional cooling. In experiments this is accomplished by vibrating the system so that it is locally driven at walls. Far from these walls a steady state is reached whose properties are expected to be insensitive to the details of the driving forces. Due to the technical difficulties involved in incorporating oscillating boundary conditions, it is usual to introduce external forces (or thermostats) acting locally on each particle. These forces are represented by the operator $\mathcal{F}_i$ in Eq. (10.22) and depend on the state of the system. Two types of thermostats have been usually considered in the literature. One is a deterministic thermostat widely used in nonequilibrium MD simulations for ordinary fluids [64, 65]. The force is similar to a Stokes law drag force, linear in the velocity, but with the opposite sign so that it heats rather than cools the system. In this case, $\mathcal{F}_i$ is given by [37, 66]

$$\mathcal{F}_i f_i(\mathbf{v}) = \frac{1}{2}\zeta_i \frac{\partial}{\partial \mathbf{v}} [\mathbf{v} f_i(\mathbf{v})] , \tag{10.29}$$

where the friction constant in the force has been adjusted to get a constant temperature in the long-time limit. It must be remarked that the corresponding Boltzmann equation (10.22) for this Gaussian thermostat force is formally identical with the Boltzmann equation in the HCS (i.e., with $\mathcal{F}_i = 0$) when both equations are written in terms of the reduced distributions $\Phi_i(v^*)$ [66]. In particular, the dependence of γ on the parameters of the system is the same with and without the Gaussian thermostat.

A second method of driving the system is by means of a stochastic Langevin force representing Gaussian white noise [67]. The corresponding operator $\mathcal{F}_i$ has a Fokker–Planck form [61]

$$\mathcal{F}_i f_i(\mathbf{v}) = -\frac{1}{2}\frac{T_i}{m_i}\zeta_i \left(\frac{\partial}{\partial \mathbf{v}}\right)^2 f_i(\mathbf{v}) , \tag{10.30}$$

where for simplicity the covariance of the stochastic acceleration has been taken to be the same for each species [68, 69]. This requirement gives the steady-state condition

$$\frac{T_1}{m_1}\zeta_1 = \frac{T_2}{m_2}\zeta_2 . \tag{10.31}$$

The cooling rates ζ_i are no longer equal, in contrast to the HCS, and the dependence of γ on the control parameters is different as well [63]. The procedure for determining the temperature ratio and the cumulants c_i is the same as in the HCS state since the steady-state distribution Φ_i can also be represented as an expansion of the form (10.27) and the coefficients are now determined from the solution to the Boltzmann equation (10.22). The condition (10.31) gives the corresponding equation for the temperature ratio.

Figure 10.1 illustrates the differences between the HCS and the stochastic steady state at the level of the temperature ratio T_1/T_2. We have considered mixtures constituted by spheres ($d = 3$) of the same material [and so, $\alpha_{ij} = \alpha$, and $m_1/m_2 = (\sigma_1/\sigma_2)^3$] and equal volumes of large and small particles [i.e., $x_2 = (\sigma_1/\sigma_2)^3 x_1$]. Here, for simplicity, the cooling rates have been analytically estimated by using Maxwellians (namely, by taking $c_i = 0$) for the distributions $f_i(\mathbf{v})$:

$$\zeta_i = \sum_{j=1}^{2} \zeta_{ij} \to \frac{4\pi^{(d-1)/2}}{d\Gamma\left(\frac{d}{2}\right)} v_0 \sum_{j=1}^{2} n_j \mu_{ji} \sigma_{ij}^{d-1} \left(\frac{\theta_i + \theta_j}{\theta_i \theta_j}\right)^{1/2} \\ \times (1+\alpha_{ij}) \left[1 - \frac{\mu_{ji}}{2}(1+\alpha_{ij})\frac{\theta_i + \theta_j}{\theta_j}\right] . \tag{10.32}$$

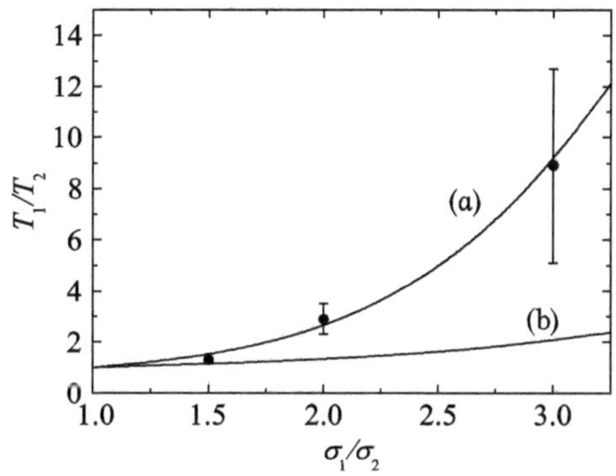

Fig. 10.1. Temperature ratio T_1/T_2 versus the size ratio σ_1/σ_2 for $\alpha_{ij} \equiv \alpha = 0.78$ in the case of mixtures constituted by particles of the same material and equal total volumes of large and small particles. The lines are the kinetic theory results in (**a**) the stochastic driving case and (**b**) the free cooling case, while the points refer to MD simulations [70]

Simulation data recently obtained from MD simulations in agitated mixtures have also been included [70]. The experimental value of the (common) coefficient of restitution is $\alpha = 0.78$. While a good agreement between kinetic theory and MD simulations is found when the gas is assumed to be driven by the stochastic thermostat, significant discrepancies appear in the undriven (HCS) case, especially as the size ratio σ_1/σ_2 increases. These results contrast with the comparison made by Brey et al. [71, 72] for agitated mixtures in the tracer limit ($x_1 \to 0$), where the predictions of the temperature ratio from kinetic theory based on the condition $\zeta_1 = \zeta_2$ compare quite well with MD simulations. However, for the cases studied in [71, 72], conditions (10.25) and (10.31) yield quite similar results for the dependence of T_1/T_2 on the parameters of the system. The good agreement found in Fig. 10.1 between MD simulations for agitated mixtures and kinetic theory suggests that the stochastic driving condition can be considered as a plausible first approximation for qualitative comparisons with experimental results [32, 33].

10.4 Navier–Stokes Hydrodynamic Equations

We now consider a spatially inhomogeneous state created either by initial preparation or by boundary conditions. We are interested in a hydrodynamic description where the state of the system is completely specified through their hydrodynamic fields. This implies that the latter dominate over other excitations for times large compared to the mean free time and for wavelengths large compared to the mean free path. The hydrodynamic regime is characterized by the existence of a *normal* solution to the Boltzmann equation which can be explicitly obtained by means of the Chapman–Enskog method [4]. For small spatial variations, the functional dependence (10.21) of the normal solution can be made local in space and time through an expansion in gradients of the fields:

$$f_i = f_i^{(0)} + \epsilon\, f_i^{(1)} + \cdots \,, \tag{10.33}$$

where each factor of ϵ means an implicit gradient of a hydrodynamic field. The reference distribution function $f_i^{(0)}(\mathbf{r},\mathbf{v},t)$ is the *local* version of the homogeneous distribution (10.23), namely, it is obtained from the homogeneous distribution by replacing the temperature, densities, and flow velocity by their nonequilibrium values:

$$f_i^{(0)}(\mathbf{r},\mathbf{v},t) = n_i(\mathbf{r},t) v_0^{-d}(T(\mathbf{r},t)) \Phi_i\left(V/v_0(T(\mathbf{r},t))\right) \,, \tag{10.34}$$

where $\mathbf{V} = \mathbf{v} - \mathbf{u}(\mathbf{r},t)$. The time derivatives of the fields are also expanded as $\partial_t = \partial_t^{(0)} + \epsilon\partial_t^{(1)} + \cdots$. The coefficients of the time derivative expansion are identified from the balance equations (10.12)–(10.14) with a representation of the fluxes and the cooling rate in the macroscopic balance equations as a similar series through their definitions as functionals of f_i.

This is the usual Chapman–Enkog method for solving kinetic equations [4, 57, 73]. Nevertheless, the complexity introduced by the energy dissipation in collisions has led to the introduction by some authors [14, 15, 16, 17, 18, 19, 20] of some additional approximations, restricting the validity of most of the results to the small inelasticity limit. Only very recently, explicit expressions for the fluxes to first order in the gradients as explicit functions of the coefficients of restitution have been obtained [35, 37, 74]. To Navier–Stokes order, the constitutive equations for mass, momentum, and heat fluxes are given, respectively, by

$$\mathbf{j}_1^{(1)} = -\frac{m_1 m_2 n}{\rho} D\nabla x_1 - \frac{\rho}{p} D_p \nabla p - \frac{\rho}{T} D' \nabla T + \sum_{i=1}^{2} \chi_{1i} \mathbf{F}_i, \quad \mathbf{j}_2^{(1)} = -\mathbf{j}_1^{(1)} \,, \tag{10.35}$$

$$P_{k\ell}^{(1)} = p\,\delta_{k\ell} - \eta\left(\nabla_\ell u_k + \nabla_k u_\ell - \frac{2}{d}\delta_{k\ell}\nabla\cdot\mathbf{u}\right), \tag{10.36}$$

$$\mathbf{q}^{(1)} = -T^2 D'' \nabla x_1 - L\nabla p - \lambda\nabla T + \sum_{i=1}^{2} \kappa_i \mathbf{F}_i. \tag{10.37}$$

The transport coefficients in these equations are the diffusion coefficient D, the pressure diffusion coefficient D_p, the thermal diffusion coefficient D', the mobility coefficient χ_{ij}, the shear viscosity η, the Dufour coefficient D'', the pressure energy coefficient L, the thermal conductivity λ, and the coefficient κ_i. These coefficients are defined as

$$D = -\frac{\rho}{d m_2 n} \int \mathrm{d}\mathbf{v}\, \mathbf{V} \cdot \boldsymbol{\mathcal{A}}_1 \,, \tag{10.38}$$

$$D_p = -\frac{m_1 p}{d\rho} \int d\mathbf{v}\, \mathbf{V} \cdot \boldsymbol{\mathcal{B}}_1 \,, \tag{10.39}$$

$$D' = -\frac{m_1 T}{d\rho} \int d\mathbf{v}\, \mathbf{V} \cdot \boldsymbol{\mathcal{C}}_1 \,, \tag{10.40}$$

$$\chi_{ij} = \frac{1}{d} \int d\mathbf{v}\, \mathbf{V} \cdot \boldsymbol{\mathcal{E}}_{ij} \,, \tag{10.41}$$

$$\eta = -\frac{1}{(d-1)(d+2)} \sum_{i=1}^{2} m_i \int d\mathbf{v}\, \mathbf{V}\mathbf{V} : \boldsymbol{\mathcal{D}}_i \,, \tag{10.42}$$

$$D'' = -\frac{1}{dT^2} \sum_{i=1}^{2} \frac{m_i}{2} \int d\mathbf{v}\, V^2 \mathbf{V} \cdot \boldsymbol{\mathcal{A}}_i \,, \tag{10.43}$$

$$L = -\frac{1}{d} \sum_{i=1}^{2} \frac{m_i}{2} \int d\mathbf{v}\, V^2 \mathbf{V} \cdot \boldsymbol{\mathcal{B}}_i \,, \tag{10.44}$$

$$\lambda = -\frac{1}{d} \sum_{i=1}^{2} \frac{m_i}{2} \int d\mathbf{v}\, V^2 \mathbf{V} \cdot \boldsymbol{\mathcal{C}}_i \,, \tag{10.45}$$

$$\kappa_i = \frac{1}{d} \sum_{j=1}^{2} \frac{m_i}{2} \int d\mathbf{v}\, V^2 \mathbf{V} \cdot \boldsymbol{\mathcal{E}}_{ij} \,. \tag{10.46}$$

Here, $\boldsymbol{\mathcal{A}}_i(\mathbf{V})$, $\boldsymbol{\mathcal{B}}_i(\mathbf{V})$, $\boldsymbol{\mathcal{C}}_i(\mathbf{V})$, $\boldsymbol{\mathcal{D}}_i(\mathbf{V})$, and $\boldsymbol{\mathcal{E}}_{ij}(\mathbf{V})$ are functions of the peculiar velocity and the hydrodynamic fields. They obey the following set of coupled linear integral equations:

$$\left[-c_\zeta^{(0)}\left(T\partial_T + p\partial_p\right) + \mathcal{F}_1^{(0)} + \mathcal{L}_1\right] \boldsymbol{\mathcal{A}}_1 + \mathcal{M}_1 \boldsymbol{\mathcal{A}}_2 = \mathbf{A}_1 + \left(\frac{\partial c_\zeta^{(0)}}{\partial x_1}\right)_{p,T} \times \left(p\boldsymbol{\mathcal{B}}_1 + T\boldsymbol{\mathcal{C}}_1\right) \,, \tag{10.47a}$$

$$\left[-c_\zeta^{(0)}\left(T\partial_T + p\partial_p\right) + \mathcal{F}_2^{(0)} + \mathcal{L}_2\right] \boldsymbol{\mathcal{A}}_2 + \mathcal{M}_2 \boldsymbol{\mathcal{A}}_1 = \mathbf{A}_2 + \left(\frac{\partial c_\zeta^{(0)}}{\partial x_1}\right)_{p,T} \times \left(p\boldsymbol{\mathcal{B}}_2 + T\boldsymbol{\mathcal{C}}_2\right) \,, \tag{10.47b}$$

$$\left[-c_\zeta^{(0)}\left(T\partial_T + p\partial_p\right) + \mathcal{F}_1^{(0)} + \mathcal{L}_1 - 2c_\zeta^{(0)}\right] \boldsymbol{\mathcal{B}}_1 + \mathcal{M}_1 \boldsymbol{\mathcal{B}}_2 = \mathbf{B}_1 + \frac{T c_\zeta^{(0)}}{p} \boldsymbol{\mathcal{C}}_1 \,, \tag{10.48a}$$

$$\left[-c_\zeta^{(0)}\left(T\partial_T + p\partial_p\right) + \mathcal{F}_2^{(0)} + \mathcal{L}_2 - 2c_\zeta^{(0)}\right] \boldsymbol{\mathcal{B}}_2 + \mathcal{M}_2 \boldsymbol{\mathcal{B}}_1 = \mathbf{B}_2 + \frac{T c_\zeta^{(0)}}{p} \boldsymbol{\mathcal{C}}_2 \,, \tag{10.48b}$$

$$\left[-c_\zeta^{(0)}\left(T\partial_T+p\partial_p\right)+\mathcal{F}_1^{(0)}+\mathcal{L}_1-\frac{1}{2}c_\zeta^{(0)}\right]\boldsymbol{\mathcal{C}}_1+\mathcal{M}_1\boldsymbol{\mathcal{C}}_2=\mathbf{C}_1-\frac{pc_\zeta^{(0)}}{2T}\boldsymbol{\mathcal{B}}_1\,, \tag{10.49a}$$

$$\left[-c_\zeta^{(0)}\left(T\partial_T+p\partial_p\right)+\mathcal{F}_2^{(0)}+\mathcal{L}_2-\frac{1}{2}c_\zeta^{(0)}\right]\boldsymbol{\mathcal{C}}_2+\mathcal{M}_2\boldsymbol{\mathcal{C}}_1=\mathbf{C}_2-\frac{pc_\zeta^{(0)}}{2T}\boldsymbol{\mathcal{B}}_2\,, \tag{10.49b}$$

$$\left[-c_\zeta^{(0)}\left(T\partial_T+p\partial_p\right)+\mathcal{F}_1^{(0)}+\mathcal{L}_1\right]\boldsymbol{\mathcal{D}}_1+\mathcal{M}_1\boldsymbol{\mathcal{D}}_2=\mathsf{D}_1\,, \tag{10.50a}$$

$$\left[-c_\zeta^{(0)}\left(T\partial_T+p\partial_p\right)+\mathcal{F}_2^{(0)}+\mathcal{L}_2\right]\boldsymbol{\mathcal{D}}_2+\mathcal{M}_2\boldsymbol{\mathcal{D}}_1=\mathsf{D}_2\,, \tag{10.50b}$$

$$\left[-c_\zeta^{(0)}\left(T\partial_T+p\partial_p\right)+\mathcal{F}_1^{(0)}+\mathcal{L}_1\right]\boldsymbol{\mathcal{E}}_{11}+\mathcal{M}_1\boldsymbol{\mathcal{E}}_{21}=\mathbf{E}_{11}\,, \tag{10.51a}$$

$$\left[-c_\zeta^{(0)}\left(T\partial_T+p\partial_p\right)+\mathcal{F}_1^{(0)}+\mathcal{L}_1\right]\boldsymbol{\mathcal{E}}_{12}+\mathcal{M}_1\boldsymbol{\mathcal{E}}_{22}=\mathbf{E}_{12}. \tag{10.51b}$$

Here, we have introduced the linearized Boltzmann collision operators

$$\mathcal{L}_1X=-\left(J_{11}[f_1^{(0)},X]+J_{11}[X,f_1^{(0)}]+J_{12}[X,f_2^{(0)}]\right)\,, \tag{10.52}$$

$$\mathcal{M}_1X=-J_{12}[f_1^{(0)},X], \tag{10.53}$$

with a similar definition for $\mathcal{L}_2$ and $\mathcal{M}_2$. In addition, $c_\zeta^{(0)}=\zeta^{(0)}$ in the undriven case while $c_\zeta^{(0)}=0$ in the driven case, where $\zeta^{(0)}$ is given by Eq. (10.9) to zeroth order, i.e.,

$$\zeta^{(0)}=-\frac{1}{dnT}\sum_{i=1}^{2}\sum_{j=1}^{2}m_i\int d\mathbf{v}V^2J_{ij}[\mathbf{v}|f_i^{(0)},f_j^{(0)}]\,. \tag{10.54}$$

In Eqs. (10.47a)–(10.51b) we have also introduced the operators

$$\mathcal{F}_i^{(0)}X=\frac{1}{2}\zeta_i^{(0)}\frac{\partial}{\partial\mathbf{V}}\cdot(\mathbf{V}X),\quad \text{(Gaussian thermostat)}\,, \tag{10.55a}$$

$$\mathcal{F}_i^{(0)}X=-\frac{1}{2}\frac{T_i}{m_i}\zeta_i^{(0)}\left(\frac{\partial}{\partial\mathbf{V}}\right)^2X,\quad \text{(stochastic thermostat)}\,, \tag{10.55b}$$

and the quantities

$$\mathbf{A}_i(\mathbf{V})=-\left(\frac{\partial}{\partial x_1}f_i^{(0)}\right)_{p,T}\mathbf{V}\,, \tag{10.56}$$

$$\mathbf{B}_i(\mathbf{V})=-\frac{1}{p}\left[f_i^{(0)}\mathbf{V}+\frac{nT}{\rho}\left(\frac{\partial}{\partial\mathbf{V}}f_i^{(0)}\right)\right]\,, \tag{10.57}$$

$$\mathbf{C}_i(\mathbf{V})=\frac{1}{T}\left[f_i^{(0)}+\frac{1}{2}\frac{\partial}{\partial\mathbf{V}}\cdot\left(\mathbf{V}f_i^{(0)}\right)\right]\mathbf{V}\,, \tag{10.58}$$

$$\mathsf{D}_i(\mathbf{V}) = \mathbf{V}\frac{\partial}{\partial \mathbf{V}} f_i^{(0)} - \frac{1}{d}\left(\mathbf{V}\cdot\frac{\partial}{\partial \mathbf{V}} f_i^{(0)}\right)\mathsf{I}\,, \tag{10.59}$$

$$\mathbf{E}_{ij}(\mathbf{V}) = -\left(\frac{\partial}{\partial \mathbf{V}} f_i^{(0)}\right)\frac{1}{m_j}\left(\delta_{ij} - \frac{\rho_j}{\rho}\right)\,. \tag{10.60}$$

Here, I is the unit tensor in d dimensions and $\rho_i = m_i n_i$ is the mass density of species i. Upon writing Eqs. (10.47a)–(10.51b) use has been made of the fact that there is no contribution to the cooling rate at this order, i.e., $\zeta^{(1)} = 0$. As a consequence, $\mathcal{F}_i^{(1)} = 0$. The property $\zeta^{(1)} = 0$ is special of the low-density Boltzmann kinetic theory (since $f_i^{(1)}$ does not contain any contribution proportional to $\nabla\cdot\mathbf{u}$), but such terms occur at higher densities [75, 76]. Note that in the particular case of the gravitational force $\mathbf{F}_i = m_i\mathbf{g}$ (where $\mathbf{g}$ is the gravity acceleration), the combination $m_1\mathbf{E}_{11} + m_2\mathbf{E}_{12} = \mathbf{0}$. This leads to $\boldsymbol{\mathcal{E}}_{ij} = \mathbf{0}$, and so there are no contributions to the mass and heat fluxes coming from the external conservative forces.

In summary, the solutions to the Boltzmann equations to first order in the spatial gradients are given by [35]

$$f_i = f_i^{(0)} + \boldsymbol{\mathcal{A}}_i\cdot\nabla x_1 + \boldsymbol{\mathcal{B}}_i\cdot\nabla p + \boldsymbol{\mathcal{C}}_i\cdot\nabla T + \boldsymbol{\mathcal{D}}_i : \nabla\mathbf{u} + \sum_j \boldsymbol{\mathcal{E}}_{ij}\cdot\mathbf{F}_j. \tag{10.61}$$

The solution to zeroth order is obtained from Eq. (10.22) while the functions $\{\boldsymbol{\mathcal{A}}_i, \boldsymbol{\mathcal{B}}_i, \boldsymbol{\mathcal{C}}_i, \boldsymbol{\mathcal{D}}_i, \boldsymbol{\mathcal{E}}_{ij}\}$ are determined from the integral equations (10.47a)–(10.51b). Once these equations are solved, the Navier–Stokes transport coefficients are obtained from Eqs. (10.38)–(10.46) and the mass, momentum, and heat fluxes are explicitly known. These fluxes, together with the macroscopic balance equations (10.12)–(10.14), provide the closed set of Navier–Stokes order hydrodynamic equations for a granular binary mixture. All these results are still formally exact and valid for arbitrary values of the coefficients of restitution.

However, explicit expressions for the Navier–Stokes transport coefficients require to solve the integral equations (10.47a)–(10.51b). Accurate approximations for $\{\boldsymbol{\mathcal{A}}_i, \boldsymbol{\mathcal{B}}_i, \boldsymbol{\mathcal{C}}_i, \boldsymbol{\mathcal{D}}_i, \boldsymbol{\mathcal{E}}_{ij}\}$ may be obtained using low-order truncation of expansions in a series of Sonine polynomials. The polynomials are defined with respect to a Gaussian weight factor whose parameters are chosen such that the leading term in the expansion yields the exact moments of the entire distribution with respect to 1, $\mathbf{v}$, and v^2. The procedure is similar to the one followed for elastic collisions [4] and yields explicit expressions for the transport coefficients in terms of the parameters of the mixture [35, 36, 74].

10.5 Comparison with Monte Carlo Simulations

As said before, the expressions derived for the Navier–Stokes transport coefficients are obtained by considering two different approximations. First, since the deviation of $f_i^{(0)}$ from its Maxwellian form is quite small in the region of thermal velocities, one uses the distribution (10.27) as a trial function for $f_i^{(0)}$. Second, one only considers the leading terms of an expansion of the distribution $f_i^{(1)}$ in Sonine polynomials. Both approximations allow one to offer a simplified kinetic theory for a granular binary mixture. To assess the degree of accuracy of these predictions, one resorts to numerical solutions of the Boltzmann equation, such as those obtained from the direct simulation Monte Carlo (DSMC) method [77]. Although the method was originally devised for normal fluids, its extension to granular gases is straightforward [66]. In this section we provide some comparisons between theory and numerical solutions of the Boltzmann equation by means of the DSMC method in the cases of the diffusion coefficient D (in the tracer limit) and the shear viscosity coefficient η of a heated gas. Let us study each coefficient separately.

10.5.1 Tracer Diffusion Coefficient

We consider a free granular mixture ($\mathcal{F}_i = 0$) in which one of the components of the mixture (say, for instance, species 1) is present in tracer concentration ($x_1 \to 0$). In this situation the diffusion coefficient of impurities in a granular gas undergoing homogeneous cooling state can be measured in simulations from the mean square displacement of the tracer particle after a time interval t [78]:

$$\frac{\partial}{\partial t}\langle |\mathbf{r}(t) - \mathbf{r}(0)|^2 \rangle = \frac{2dD}{n_2} , \qquad (10.62)$$

where $|\mathbf{r}(t) - \mathbf{r}(0)|$ is the distance traveled by the impurity from $t = 0$ until time t. The relation (10.62) written in appropriate dimensionless variables to eliminate the time dependence of $D(t)$ can be used to measure by computer simulations the diffusion coefficient [8, 38].

If the hydrodynamic description (or normal solution in the context of the Chapman–Enskog method) applies, then the diffusion coefficient $D(t)$ depends on time only through its dependence on the temperature $T(t)$. Dimensional analysis shows that $D(t) \propto \sqrt{T(t)}$. In this case, after a transient regime, the reduced diffusion coefficient $D^* = (m_1 m_2/\rho) D(t) \nu_0(t)/T(t)$ achieves a time-independent value. Here, $\nu_0(t) = n\sigma_{12}^{d-1} v_0(t) \propto \sqrt{T(t)}$ is an effective collision frequency for hard spheres. The fact that D^* reaches a constant value for times large compared with the mean free path is closely related with the validity of a hydrodynamic description for the system. In addition, as has been recently shown [79], the dependence of D^* on the mass ratio m_1/m_2 and the coefficient of restitution α_{12} is only through the effective mass $m_1^* = m_1 + (m_1 + m_2)(1 - \alpha_{12})/(1 + \alpha_{12})$.

The dependence of D^* on the common coefficient of restitution $\alpha_{ij} \equiv \alpha$ is shown in Fig. 10.2 in the case of hard disks ($d = 2$) for three different systems. The symbols refer to DSMC simulations while the lines correspond to the kinetic theory results obtained in the first Sonine approximation [35, 74]. MD results, reported in [8], when impurities and particles of the gas are

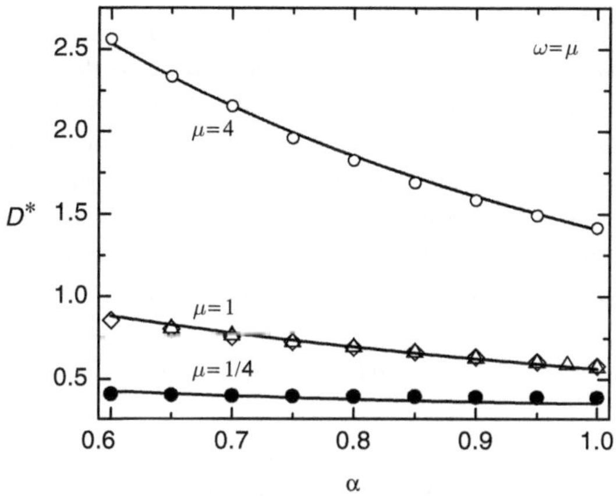

Fig. 10.2. Plot of the reduced diffusion coefficient D^* as a function of the (common) coefficient of restitution α for binary mixtures with $\omega = \mu$ in the case of a two-dimensional system ($d = 2$). Here, $\omega \equiv \sigma_1/\sigma_2$ and $\mu \equiv m_1/m_2$. The symbols are computer simulation results obtained from the mean square displacement and the lines are the theoretical results obtained in the first Sonine approximation. The DSMC results correspond to $\mu = 1/4$ (●), $\mu = 4$ (○), and $\mu = 1$ (◇). Molecular dynamics results reported in [8] for $\mu = 1$ (△) have also been included

mechanically equivalent have also been included. We observe that in the latter case MD and DSMC results are consistent among themselves in the range of values of α explored. This good agreement gives support to the applicability of the inelastic Boltzmann equation beyond the quasielastic limit. It is apparent that the agreement between the first Sonine approximation and simulation results is excellent when impurities and particles of the gas are mechanically equivalent and when impurities are much heavier and/or much larger than the particles of the gas (Brownian limit). However, some discrepancies between simulation and theory are found with decreasing values of the mass ratio $\mu \equiv m_1/m_2$ and the size ratio $\omega \equiv \sigma_1/\sigma_2$. These discrepancies are not easily observed in Fig. 10.2 because of the small magnitude of D^* for $\mu = 1/4$. For these systems, the second Sonine approximation [38] improves the qualitative predictions over the first Sonine approximation for the cases in which the gas particles are heavier and/or larger than impurities. This means that the Sonine polynomial expansion exhibits a slow convergence for sufficiently small values of the mass ratio μ and/or the size ratio ω. This tendency is also present in the case of elastic systems [80].

10.5.2 Shear Viscosity Coefficient of a Heated Gas

The shear viscosity η is perhaps the most widely studied transport coefficient in granular fluids. This coefficient can be measured in computer simulations in the special hydrodynamic state of uniform shear flow (USF). At a macroscopic level, this state is characterized by constant partial densities n_i, uniform temperature T, and a linear flow velocity profile $u_{1,k} = u_{2,k} = a_{k\ell} r_\ell$, $a_{k\ell} = a\delta_{kx}\delta_{\ell y}$, a being the constant shear rate. In this state, the temperature changes in time due to the competition between two mechanisms: on the one hand, viscous heating and, on the other, energy dissipation in collisions. In addition, the mass and heat fluxes vanish by symmetry reasons and the (uniform) pressure tensor is the only nonzero flux of the problem. The relevant balance equation is that for temperature, Eq. (10.14), which reduces to

$$\partial_t T + \frac{2}{dn} a P_{xy} = -(\zeta - \xi) T \,, \tag{10.63}$$

where

$$P_{xy} = \sum_{i=1}^{2} m_i \int d\mathbf{V}\, V_x V_y f_i(\mathbf{V}) \tag{10.64}$$

is the xy-element of the pressure tensor.

For a granular fluid under USF and in the absence of a thermostatting force ($\xi = 0$), the energy balance equation (10.63) leads to a steady state when the viscous heating effect is exactly balanced by the collisional cooling. This situation will be analyzed in Sect. 10.10. However, if for instance the mixture is heated by the Gaussian thermostat (10.29) (with $\mathbf{v} \to \mathbf{V}$), then the viscous heating still prevails so that the temperature increases in time. In this case, the collision frequency $\nu_0(t) \propto \sqrt{T(t)}$ also grows with t and hence the reduced shear rate $a^*(t) = a/\nu_0(t)$ (which is the relevant nonequilibrium parameter of the problem) monotonically decreases in time. Under these conditions, the system asymptotically achieves a regime described by linear hydrodynamics and the (reduced) shear viscosity $\eta^* = [\nu_0(t)/nT(t)]\eta(t)$ can be measured as

$$\eta^* = - \lim_{t \to \infty} \frac{P_{xy}^*}{a^*} \,, \tag{10.65}$$

where $P_{xy}^* = P_{xy}/nT$. This procedure allows one to identify the shear viscosity of a granular mixture *excited* by the Gaussian external force (10.29) and compare it with the predictions given by the Chapman–Enskog method.

In Fig. 10.3, we plot the ratio $\eta^*(\alpha)/\eta^*(1)$ versus the mass ratio m_1/m_2 in the case of hard spheres ($d = 3$) for $\sigma_1/\sigma_2 = 1$, $x_1 = \frac{1}{2}$, and three different values of the (common) coefficient of restitution $\alpha_{ij} \equiv \alpha$. Here, $\eta^*(1)$ refers to the elastic value of the shear viscosity coefficient. Again, the symbols represent the simulation data obtained by numerically solving the Boltzmann equation [37], while the lines refer to the theoretical results obtained from the Boltzmann equation in the first Sonine approximation. We see that in general the deviation of $\eta^*(\alpha)$ from its functional form for elastic collisions is quite important. This tendency becomes more significant as the mass disparity increases. The agreement between the first Sonine approximation and simulation is seen to be in general excellent. This agreement is similar to the one previously found in the monocomponent case [7, 10, 81]. At a quantitative level,

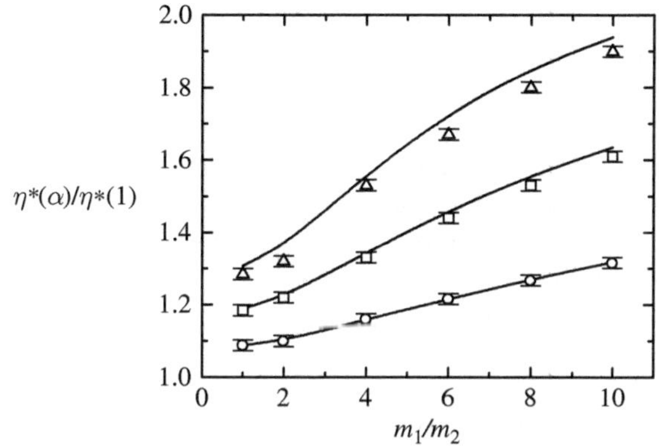

Fig. 10.3. Plot of the ratio $\eta^*(\alpha)/\eta^*(1)$ as a function of the mass ratio m_1/m_2 for $\sigma_1/\sigma_2 = n_1/n_2 = 1$ and three different values of the (common) coefficient of restitution α: $\alpha = 0.9$ (*circles*), $\alpha = 0.8$ (*squares*), and $\alpha = 0.7$ (*triangles*). The lines are the theoretical predictions and the symbols refer to the results obtained from the DSMC method

the discrepancies between theory and simulation tend to increase as the coefficient of restitution decreases, although these differences are quite small (say, for instance, around 2% at $\alpha = 0.7$ in the disparate mass case $m_1/m_2 = 10$). The influence of the size ratio on the shear viscosity is shown in Fig. 10.4 for $m_1/m_2 = 4$ and $x_1 = \frac{1}{2}$ [37]. We observe again a strong dependence of the shear viscosity on dissipation. However, for a given value of α, the influence

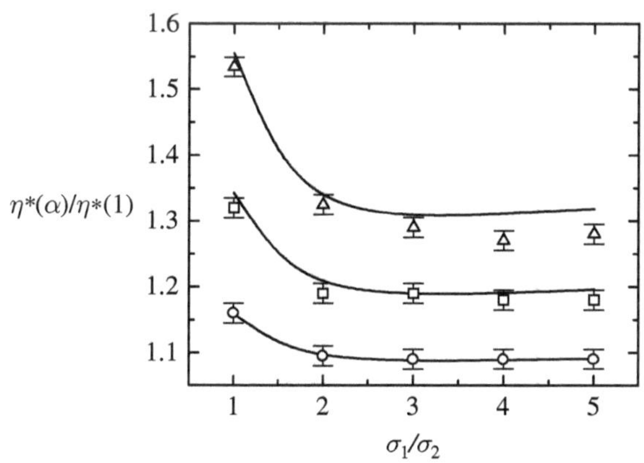

Fig. 10.4. Plot of the ratio $\eta^*(\alpha)/\eta^*(1)$ as a function of the size ratio σ_1/σ_2 for $m_1/m_2 = 4$, $n_1/n_2 = 1$ and three different values of the (common) coefficient of restitution α: $\alpha = 0.9$ (*circles*), $\alpha = 0.8$ (*squares*), and $\alpha = 0.7$ (*triangles*). The lines are the theoretical predictions and the symbols refer to the results obtained from the DSMC method

of σ_1/σ_2 on η^* is weaker than the one found before in Fig. 10.3 for the mass ratio. The agreement for both $\alpha = 0.9$ and $\alpha = 0.8$ is quite good, except for the largest size ratio at $\alpha = 0.8$. These discrepancies become more significant as the dissipation increases, especially for mixtures of particles of very different sizes. In summary, according to the comparison carried out in Figs. 10.3 and 10.4, one can conclude that the agreement between theory and simulation extends over a wide range values of the coefficient of restitution, indicating the reliability of the first Sonine approximation for describing granular flows beyond the quasielastic limit.

10.6 Einstein Relation in Granular Gases

The results presented in Sect. 10.5 give some support to the validity of the hydrodynamic description to granular fluids. However, in spite of this support some care is warranted in extending properties of normal fluids to those with inelastic collisions. Thus, for elastic collisions, in the case of an *impurity* (tracer) particle immersed in a gas the response to an external force on the impurity particle leads to a mobility coefficient proportional to the diffusion coefficient. This is the usual Einstein relation [78], which is a consequence of the fluctuation-dissipation theorem. A natural question is whether the Einstein relation also applies for granular fluids.

To analyze it, let us consider the tracer limit ($x_1 \to 0$) and assume that the current of impurities $\mathbf{j}_1^{(1)}$ is only generated by the presence of a weak concentration gradient ∇x_1 and/or a weak external field $\mathbf{F}_1$ acting only on the impurity particles. Under these conditions, Eq. (10.35) becomes

$$\mathbf{j}_1^{(1)} = -m_1 D \nabla x_1 + \chi_{11} \mathbf{F}_1. \tag{10.66}$$

The Einstein ratio ϵ' between the diffusion coefficient D and the mobility coefficient χ_{11} is defined as

$$\epsilon' = m_1 x_1 \frac{D}{T \chi_{11}}\ , \tag{10.67}$$

where $T \simeq T_2$ in the tracer limit. For elastic collisions, the Chapman–Enskog results yield $\epsilon' = 1$. However, at finite inelasticity the relationship between D and χ_{11} is no longer simple and, as expected, the Chapman–Enskog expressions for D and χ_{11} in the case of an *unforced* granular gas [39] clearly show that $\epsilon' \neq 1$. This means that the Einstein relation does not apply in granular gases. The deviations of the (standard) Einstein ratio ϵ' from unity has three distinct origins: the absence of the Gibbs state (non-Gaussianity of the distribution function of the HCS), time evolution of the granular temperature, and the occurrence of different kinetic temperatures between the impurity and gas particles. The second source of discrepancy can be avoided if the system is driven by an external energy input to achieve a stationary state. With respect to the third reason of violation, this could also be partially eliminated if the temperature of the gas T is replaced by the temperature of the impurity T_1 in the usual Einstein relation (10.67). This change yields the *modified* Einstein ratio

$$\epsilon = m_1 x_1 \frac{D}{T_1 \chi_{11}}. \tag{10.68}$$

As a consequence, the only reason for which $\epsilon \neq 1$ is due to the non-Maxwellian behavior of the HCS distribution. Given that the deviations of the gas distribution $f_2^{(0)}$ from its Maxwellian form are small [21], the discrepancies of ϵ from unity could be difficult to detect in computer simulations. This conclusion agrees with recent MD simulations [82] of granular mixtures subjected to the stochastic driving of the form (10.30), where no deviations from the (modified) Einstein relation $\epsilon = 1$ have been observed for a wide range of values of the coefficients of restitution and parameters of the system.

To illustrate the influence of dissipation on the Einstein ratio more generally, in Figs. 10.5 and 10.6 the Einstein ratio as given by (10.68) is plotted versus the coefficient of restitution α_{12} for $\sigma_1/\sigma_2 = 1$ and different values of the mass ratio m_1/m_2 and the coefficient of restitution α_{22}. The results obtained by using the Gaussian thermostat (10.29) are shown in Fig. 10.5, while Fig. 10.6 corresponds to the results derived when the system is heated by the stochastic thermostat (10.30) [40]. We observe that in general $\epsilon \neq 1$, although its value is very close to unity, especially in the case of the stochastic thermostat, where the deviations from the Einstein relation are smaller than 1%. However, in the case of the Gaussian thermostat the deviations from unity are about 8%, which could be detected in computer simulations. Figures 10.5 and 10.6 also show the fact that the transport properties are affected by the

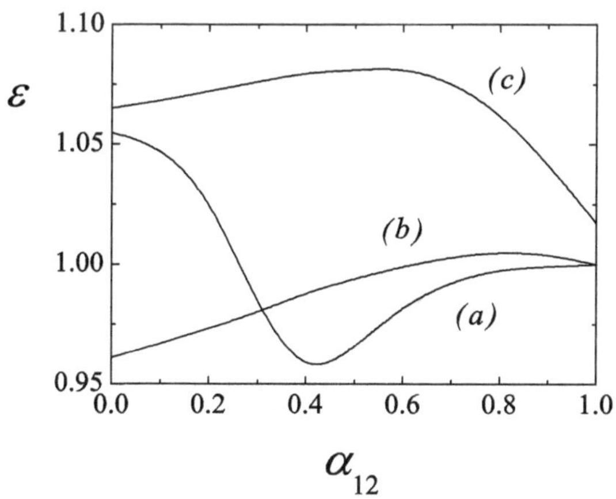

Fig. 10.5. Plot of the modified Einstein ratio ϵ versus the coefficient of restitution α_{12} for the Gaussian thermostat in the cases: (**a**) $\alpha_{22} = \alpha_{12}$, $m_1/m_2 = 5$, and $\sigma_1/\sigma_2 = 1$; (**b**) $\alpha_{22} = \alpha_{12}$, $m_1/m_2 = 0.5$, and $\sigma_1/\sigma_2 = 1$; and (**c**) $\alpha_{22} = 0.5$, $m_1/m_2 = 10$, and $\sigma_1/\sigma_2 = 1$

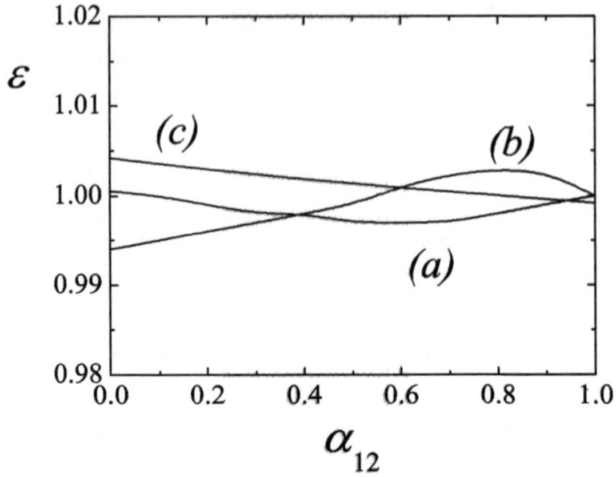

Fig. 10.6. Plot of the Einstein ratio ϵ versus the coefficient of restitution α_{12} for the stochastic thermostat in the cases: (**a**) $\alpha_{22} = \alpha_{12}$, $m_1/m_2 = 5$, and $\sigma_1/\sigma_2 = 1$; (**b**) $\alpha_{22} = \alpha_{12}$, $m_1/m_2 = 0.5$, and $\sigma_1/\sigma_2 = 1$; and (**c**) $\alpha_{22} = 0.5$, $m_1/m_2 = 10$, and $\sigma_1/\sigma_2 = 1$

thermostat introduced so that the latter does not play a neutral role in the problem [81].

10.7 Onsager's Reciprocal Relations in Granular Gases

In the usual language of the linear irreversible thermodynamics for *ordinary* fluids [83], the constitutive equations (10.35) and (10.37) for the mass flux and heat flux in the absence of external forces can be written as

$$\mathbf{j}_i^{(1)} = -\sum_j L_{ij} \left(\frac{\nabla \mu_j}{T} \right)_T - L_{iq} \frac{\nabla T}{T^2} - C_p \nabla p \,, \tag{10.69}$$

$$\mathbf{J}_q^{(1)} = -L_{qq} \nabla T - \sum_i L_{qi} \left(\frac{\nabla \mu_i}{T} \right)_T - C_p' \nabla p \,, \tag{10.70}$$

where

$$\mathbf{J}_q^{(1)} \equiv \mathbf{q}^{(1)} - \frac{d+2}{2} T \sum_i \frac{\mathbf{j}_i^{(1)}}{m_i} \tag{10.71}$$

and

$$\left(\frac{\nabla \mu_i}{T} \right)_T = \frac{1}{m_i} \nabla \ln(x_i p), \tag{10.72}$$

μ_i being the chemical potential per unit mass. In Eqs. (10.69) and (10.70), the coefficients L_{ij} are the so-called Onsager phenomenological coefficients and the coefficients C_p and C_p' can be expressed in terms of the transport coefficients associated with the heat and mass fluxes. For *elastic* fluids, Onsager showed [83] that time reversal invariance of the underlying microscopic equations of motion implies important constraints on the above set of transport coefficients, namely

$$L_{ij} = L_{ji}, \quad L_{iq} = L_{qi}, \quad C_p = C_p' = 0 . \tag{10.73}$$

The first two symmetries are called reciprocal relations as they relate transport coefficients for different processes. The last two are statements that the pressure gradient does not appear in any of the fluxes, even though it is admitted by symmetry. Even for a one-component fluid, Onsager's theorem is significant as it leads to a *new* contribution to the heat flux proportional to the density gradient [5]. Since there is no time reversal symmetry for granular fluids, Eq. (10.73) cannot be expected to apply. However, since explicit expressions for all transport coefficients are at hand, the quantitative extent of the violation can be explored.

To make connection with the expressions (10.35) and (10.37) for the mass and heat fluxes, respectively, it is first necessary to transform Eqs. (10.69)–(10.71) to the variables x_1, p, and T. Since $\nabla x_1 = -\nabla x_2$, Eq. (10.72) implies

$$\frac{(\nabla\mu_1)_T - (\nabla\mu_2)_T}{T} = \frac{n\rho}{\rho_1\rho_2}\left[\nabla x_1 + \frac{n_1 n_2}{n\rho}(m_2 - m_1)\nabla \ln p\right] . \tag{10.74}$$

The coefficients $\{L_{ij}, L_{iq}, L_{qi}, L_{qq}, C_p, C_p'\}$ then can be easily obtained in terms of the Navier–Stokes transport coefficients introduced in Sect. 10.4. The result is [36]

$$L_{11} = -L_{12} = -L_{21} = \frac{m_1 m_2 \rho_1 \rho_2}{\rho^2} D, \quad L_{1q} = \rho T D', \tag{10.75}$$

$$L_{q1} = -L_{q2} = \frac{T^2 \rho_1 \rho_2}{n\rho} D'' - \frac{d+2}{2}\frac{T \rho_1 \rho_2}{\rho^2}(m_2 - m_1) D , \tag{10.76}$$

$$L_{qq} = \lambda - \frac{d+2}{2}\rho \frac{m_2 - m_1}{m_1 m_2} D' , \tag{10.77}$$

$$C_p \equiv \frac{\rho}{p} D_p - \frac{\rho_1 \rho_2}{p\rho^2}(m_2 - m_1) D, \tag{10.78}$$

$$C_p' \equiv L - \frac{d+2}{2}\frac{T}{p}\frac{m_2 - m_1}{m_1 m_2} C_p - \frac{n_1 n_2}{n p \rho} T^2 (m_2 - m_1) D'' . \tag{10.79}$$

Onsager's relation $L_{12} = L_{21}$ holds since the diffusion coefficient D is symmetric under the change $1 \leftrightarrow 2$ [35]. However, in general $L_{1q} \neq L_{q1}$, $C_p \neq 0$, and $C_p' \neq 0$ [36]. The Chapman–Enskog results [35] show that there are only two limit cases for which $L_{1q} - L_{q1} = C_p = C_p' = 0$: (i) the elastic limit ($\alpha_{ij} = 1$) with arbitrary values of masses, sizes, and composition and (ii) the case of mechanically equivalent particles with arbitrary values of the (common) coefficient of restitution $\alpha \equiv \alpha_{ij}$. Beyond these limit cases, Onsager's relations do not apply. At macroscopic level the origin of this failure is due to the cooling of the reference state as well as the occurrence of different kinetic temperatures for both species.

10.8 Linearized Hydrodynamic Equations and Stability of the Homogeneous Cooling State

As shown in Sect. 10.4, the Navier-Stokes constitutive equations (10.35)–(10.37) have been expressed in terms of a set of experimentally accessible fields such as the composition of species 1, x_1, the pressure p, the mean flow field $\mathbf{u}$, and the granular temperature T. In terms of these variables and in the absence of external forces, the macroscopic balance equations (10.12)–(10.14) become

$$D_t x_1 + \frac{\rho}{n^2 m_1 m_2} \nabla \cdot \mathbf{j}_1 = 0 \,, \tag{10.80}$$

$$D_t p + p\nabla \cdot \mathbf{u} + \frac{2}{d}\left(\nabla \cdot \mathbf{q} + \mathsf{P} : \nabla \mathbf{u}\right) = -\zeta p \,, \tag{10.81}$$

$$D_t \mathbf{u} + \rho^{-1} \nabla \cdot \mathsf{P} = 0 \,, \tag{10.82}$$

$$D_t T - \frac{T}{n} \sum_i \frac{\nabla \cdot \mathbf{j}_i}{m_i} + \frac{2}{dn}\left(\nabla \cdot \mathbf{q} + \mathsf{P} : \nabla \mathbf{u}\right) = -\zeta T \,. \tag{10.83}$$

When expressions (10.35)–(10.37) for the fluxes and the cooling rate $\zeta \to \zeta^{(0)}$ are substituted into the above exact balance equations (10.80)–(10.83) one gets a closed set of hydrodynamic equations for $x_1, \mathbf{u}, T$, and p. These are the Navier–Stokes hydrodynamic equations for a binary granular mixture:

$$D_t x_1 = \frac{\rho}{n^2 m_1 m_2} \nabla \cdot \left(\frac{m_1 m_2 n}{\rho} D \nabla x_1 + \frac{\rho}{p} D_p \nabla p + \frac{\rho}{T} D' \nabla T \right) \,, \tag{10.84}$$

$$\begin{aligned}(D_t + \zeta)\, p + \frac{d+2}{d} p \nabla \cdot \mathbf{u} &= \frac{2}{d} \nabla \cdot \left(T^2 D'' \nabla x_1 + L \nabla p + \lambda \nabla T \right) \\ &\quad + \frac{2}{d} \eta \left(\nabla_\ell u_k + \nabla_k u_\ell - \frac{2}{d} \delta_{k\ell} \nabla \cdot \mathbf{u} \right) \nabla_\ell u_k \,,\end{aligned} \tag{10.85}$$

$$\begin{aligned}(D_t + \zeta)\, T + \frac{2}{dn} p \nabla \cdot \mathbf{u} &= -\frac{T}{n} \frac{m_2 - m_1}{m_1 m_2} \nabla \cdot \left(\frac{m_1 m_2 n}{\rho} D \nabla x_1 + \frac{\rho}{p} D_p \nabla p \right. \\ &\quad \left. + \frac{\rho}{T} D' \nabla T \right) + \frac{2}{dn} \nabla \cdot \left(T^2 D'' \nabla x_1 + L \nabla p + \lambda \nabla T \right) \\ &\quad + \frac{2}{dn} \eta \left(\nabla_\ell u_k + \nabla_k u_\ell - \frac{2}{d} \delta_{k\ell} \nabla \cdot \mathbf{u} \right) \nabla_\ell u_k \,,\end{aligned} \tag{10.86}$$

$$D_t u_\ell + \rho^{-1} \nabla_\ell p = \rho^{-1} \nabla_k \eta \left(\nabla_\ell u_k + \nabla_k u_\ell - \frac{2}{d} \delta_{k\ell} \nabla \cdot \mathbf{u} \right) \,. \tag{10.87}$$

For the chosen set of fields, $n = p/T$ and $\rho = p\left[(m_1 - m_2)\, x_1 + m_2\right]/T$. These equations are exact to second order in the spatial gradients for a low-density Boltzmann gas. Note that in Eqs. (10.84)–(10.87) the second-order contributions to the cooling rate have been neglected. These second-order terms have been calculated for a monocomponent fluid [5] and are found to be very small relative to corresponding terms from the fluxes. Consequently, they have not been considered in the hydrodynamic equations (10.84)–(10.87).

One of the main peculiarities of the granular gases (in contrast to ordinary fluids) is the existence of nontrivial solutions to the Navier–Stokes equations (10.84)–(10.87), even for spatially homogeneous states,

$$\partial_t x_{1H} = 0, \quad \partial_t \mathbf{u}_H = \mathbf{0} \,, \tag{10.88}$$

$$\left[\partial_t + \zeta\left(x_{1H}, T_H, p_H\right)\right] T_H = 0, \qquad \left[\partial_t + \zeta\left(x_{1H}, T_H, p_H\right)\right] p_H = 0, \tag{10.89}$$

where the subscript H denotes the homogeneous state. Since the dependence of the cooling rate $\zeta\left(x_{1H}, T_H, p_H\right)$ on x_{1H}, T_H, p_H is known [21, 74], these first-order nonlinear equations can be solved for the time dependence of the homogeneous state. The result is the familiar Haff cooling law for $T(t)$ at constant density [1, 84]:

$$T_H(t) = \frac{T_H(0)}{\left[1 + \zeta(0)\frac{t}{2}\right]^2}. \tag{10.90}$$

As said before, each partial temperature $T_i(t)$ has the same time dependence but with a different value [21],

$$T_{1H}(t) = \frac{\gamma}{1 + x_1(\gamma - 1)} T_H(t), \quad T_{2H}(t) = \frac{1}{1 + x_1(\gamma - 1)} T_H(t) \,, \tag{10.91}$$

where $\gamma = T_{1H}(t)/T_{2H}(t)$ is the time-independent temperature ratio.

Nevertheless, the homogeneous cooling state (HCS) is unstable to sufficiently long-wavelength perturbations. For systems large enough to support such spontaneous fluctuations, the HCS becomes *inhomogeneous* at long times. This feature was first observed in MD simulations of free monocomponent gases [85, 86]. In MD simulations the inhomogeneities may grow by the formation of clusters, ultimately aggregating to a single large cluster [87]; if cluster growth is suppressed, a vortex field may grow to the system size where periodic boundary conditions can induce a transition to a state with macroscopic shear. The mechanism responsible for the growth of inhomogeneities can be understood at the level of the Navier–Stokes hydrodynamics, where *linear* stability analysis shows two shear modes and a heat mode to be unstable [5, 7, 84, 88].

The objective here is to extend this analysis to the case of a binary mixture. To do that, we perform a linear stability analysis of the nonlinear hydrodynamic equations (10.84)–(10.87) with respect to this HCS for small initial spatial perturbations. For ordinary fluids such perturbations decay in time according to the hydrodynamic modes of diffusion (shear, thermal, mass) and damped sound propagation [4, 89, 90]. For inelastic collisions, the analysis is for fixed coefficients of restitution in the long-wavelength limit. As will be seen below, the corresponding modes for a granular mixture are then quite different from those for ordinary mixtures. In fact, an alternative study with fixed long wavelength and coefficients of restitution approaching unity yields the usual ordinary fluid modes. Consequently, the nature of the hydrodynamic modes is nonuniform with respect to the inelasticity and the wavelength of the perturbation.

Let us assume that the deviations $\delta y_\alpha(\mathbf{r},t)=y_\alpha(\mathbf{r},t)-y_{H\alpha}(t)$ are small. Here, $\delta y_\alpha(\mathbf{r},t)$ denotes the deviation of $\{x_1,\mathbf{u},T,p\}$ from their values in the HCS. If the initial spatial perturbation is sufficiently small, then for some initial time interval these deviations will remain small and the hydrodynamic equations (10.84)–(10.87) can be linearized with respect to $\delta y_\alpha(\mathbf{r},t)$. This leads to a set of partial differential equations with coefficients that are independent of space but which depend on time. As in the monocomponent case [5, 88], this time dependence can be eliminated through a change in the time and space variables, and a scaling of the hydrodynamic fields. We introduce the following dimensionless space and time variables:

$$\tau=\int_0^t \mathrm{d}t'\,\nu_{0H}(t'),\quad \mathbf{s}=\frac{\nu_{0H}(t)}{v_{0H}(t)}\mathbf{r}\,, \tag{10.92}$$

where $\nu_{0H}(t)$ is an effective collision frequency for hard spheres and $v_{0H}=\sqrt{2T_H(m_1+m_2)/m_1m_2}$. Since $\{x_{1H},\mathbf{u}_H,T_H,p_H\}$ are evaluated in the HCS, then Eqs. (10.88) and (10.89) hold. A set of Fourier-transformed dimensionless variables are then introduced as

$$\rho_{\mathbf{k}}(\tau)=\frac{\delta x_{1\mathbf{k}}(\tau)}{x_{1H}},\quad \mathbf{w}_{\mathbf{k}}(\tau)=\frac{\delta\mathbf{u}_{\mathbf{k}}(\tau)}{v_{0H}(\tau)},\quad \theta_{\mathbf{k}}(\tau)=\frac{\delta T_{\mathbf{k}}(\tau)}{T_H(\tau)},\quad \Pi_{\mathbf{k}}(\tau)=\frac{\delta p_{\mathbf{k}}(\tau)}{p_H(\tau)}\,, \tag{10.93}$$

where $\delta y_{\alpha\mathbf{k}}\equiv\{\delta x_{1\mathbf{k}},\delta\mathbf{u}_{\mathbf{k}},\delta T_{\mathbf{k}},\delta p_{\mathbf{k}}\}$ is defined as

$$\delta y_{\alpha\mathbf{k}}(\tau)=\int \mathrm{d}\mathbf{s}\;\mathrm{e}^{-\mathrm{i}\mathbf{k}\cdot\mathbf{s}}\delta y_\alpha(\mathbf{s},\tau). \tag{10.94}$$

Note that here the wave vector $\mathbf{k}$ is dimensionless.

In terms of the above variables, the transverse velocity components $\mathbf{w}_{\mathbf{k}\perp}=\mathbf{w}_{\mathbf{k}}-(\mathbf{w}_{\mathbf{k}}\cdot\widehat{\mathbf{k}})\widehat{\mathbf{k}}$ (orthogonal to the wave vector $\mathbf{k}$) decouple from the other four modes and hence can be obtained more easily. They obey the equation

$$\left(\frac{\partial}{\partial\tau}-\frac{\zeta^*}{2}+\eta^*k^2\right)\mathbf{w}_{\mathbf{k}\perp}=\mathbf{0}, \tag{10.95}$$

where $\zeta^*=\zeta_H/\nu_{0H}$ and

$$\eta^*=\frac{\nu_{0H}}{\rho_H v_{0H}^2}\eta\,, \tag{10.96}$$

where $\rho_H=m_1n_{1H}+m_2n_{2H}$. The solution for $\mathbf{w}_{\mathbf{k}\perp}(\tau)$ reads

$$\mathbf{w}_{\mathbf{k}\perp}(\tau)=\mathbf{w}_{\mathbf{k}\perp}(0)\exp[s_\perp(k)\tau]\,, \tag{10.97}$$

where

$$s_{\perp}(k) = \frac{1}{2}\zeta^* - \eta^* k^2. \tag{10.98}$$

This identifies $d-1$ shear (transversal) modes. We see from Eq. (10.98) that there exists a critical wave number $k_{\perp}^{\mathrm{c}}$ given by

$$k_{\perp}^{\mathrm{c}} = \left(\frac{\zeta^*}{2\eta^*}\right)^{1/2}. \tag{10.99}$$

This critical value separates two regimes: shear modes with $k \geq k_{\perp}^{\mathrm{c}}$ always decay, while those with $k < k_{\perp}^{\mathrm{c}}$ grow exponentially.

The remaining modes are called longitudinal modes. They correspond to the set $\{\rho_{\mathbf{k}}, \theta_{\mathbf{k}}, \Pi_{\mathbf{k}}, w_{\mathbf{k}||}\}$ where the longitudinal velocity component (parallel to $\mathbf{k}$) is $w_{\mathbf{k}||} = \mathbf{w}_{\mathbf{k}} \cdot \widehat{\mathbf{k}}$. These modes are the solutions of the linear equation [36]

$$\frac{\partial \delta z_{\alpha \mathbf{k}}(\tau)}{\partial \tau} = \left(M_{\alpha\beta}^{(0)} + ikM_{\alpha\beta}^{(1)} + k^2 M_{\alpha\beta}^{(2)}\right) \delta z_{\beta \mathbf{k}}(\tau), \tag{10.100}$$

where $\delta z_{\alpha \mathbf{k}}(\tau)$ now denotes the four variables $\left\{\rho_{\mathbf{k}}, \theta_{\mathbf{k}}, \Pi_{\mathbf{k}}, w_{\mathbf{k}||}\right\}$. The matrices in Eq. (10.100) are given by

$$\mathsf{M}^{(0)} = \begin{pmatrix} 0 & 0 & 0 & 0 \\ -x_1 \left(\frac{\partial \zeta^*}{\partial x_1}\right)_{T,p} & \frac{1}{2}\zeta^* & -\zeta^* & 0 \\ -x_1 \left(\frac{\partial \zeta^*}{\partial x_1}\right)_{T,p} & \frac{1}{2}\zeta^* & -\zeta^* & 0 \\ 0 & 0 & 0 & \frac{1}{2}\zeta^* \end{pmatrix}, \tag{10.101}$$

$$\mathsf{M}^{(1)} = \begin{pmatrix} 0 & 0 & 0 & 0 \\ 0 & 0 & 0 & -\frac{2}{d} \\ 0 & 0 & 0 & -\frac{d+2}{d} \\ 0 & 0 & -\frac{1}{2}\frac{\mu_{12}}{x_1\mu + x_2} & 0 \end{pmatrix}, \tag{10.102}$$

$$\mathsf{M}^{(2)} = \begin{pmatrix} -D^* & -x_1^{-1}D'^* & -x_1^{-1}D_p^* & 0 \\ M_{21}^{(2)} & M_{22}^{(2)} & M_{23}^{(2)} & 0 \\ -\frac{2}{d}x_1 D''^* & -\frac{2}{d}\lambda^* & -\frac{2}{d}L^* & 0 \\ 0 & 0 & 0 & -\frac{2}{d}(d-1)\eta^* \end{pmatrix}, \tag{10.103}$$

where

$$M_{21}^{(2)} = -x_1 \left(\frac{2}{d}D''^* - \frac{1-\mu}{x_1\mu + x_2}D^*\right), \tag{10.104}$$

$$M_{22}^{(2)} = \frac{1-\mu}{x_1\mu + x_2}D'^* - \frac{2}{d}\lambda^*, \tag{10.105}$$

$$M_{23}^{(2)} = -\frac{2}{d}L^* + \frac{1-\mu}{x_1\mu + x_2}D_p^*. \tag{10.106}$$

In these equations, $\mu = m_1/m_2$, $x_i = n_{iH}/n_H$, and we have introduced the reduced Navier–Stokes transport coefficients[1]

$$D^* = \frac{\nu_{0H}}{n_H v_{0H}^2} D, \quad D_p^* = \frac{\rho_H^2 \nu_{0H}}{m_1 m_2 n_H^2 v_{0H}^2} D_p, \quad D'^* = \frac{\rho_H^2 \nu_{0H}}{m_1 m_2 n_H^2 v_{0H}^2} D', \tag{10.107}$$

$$D''^* = \frac{\nu_{0H} T_H}{n_H v_{0H}^2} D'', \quad L^* = \frac{\nu_{0H}}{v_{0H}^2} L, \quad \lambda^* = \frac{\nu_{0H}}{n_H v_{0H}^2} \lambda. \tag{10.108}$$

The longitudinal modes have the form $\exp[s_n(k)\tau]$ with $n = 1, 2, 3, 4$, where $s_n(k)$ are the eigenvalues of the matrix $\mathsf{M}(k) = \mathsf{M}^{(0)} + ik\mathsf{M}^{(1)} + k^2\mathsf{M}^{(2)}$, namely, they are the solutions of the quartic equation

$$\det|\mathsf{M} - s\mathsf{I}| = 0. \tag{10.109}$$

The solution to (10.109) for arbitrary values of k is quite intricate. It is instructive to consider first the solutions to these equations in the extreme long-wavelength limit, $k = 0$. In this case, they are found to be the eigenvalues of the matrix of $\mathsf{M}^{(0)}$:

$$s_n^{(0)} = \left(0, 0, -\frac{1}{2}\zeta^*, \frac{1}{2}\zeta^*\right). \tag{10.110}$$

Hence, at asymptotically long wavelengths ($\mathbf{k} = 0$) the spectrum of the linearized hydrodynamic equations (both transverse and longitudinal) is comprised of a decaying mode at $-\zeta^*/2$, a two-fold degenerate mode at 0, and a d-fold degenerate unstable mode at $\zeta^*/2$. Consequently, some of the solutions are unstable. The two zero eigenvalues represent marginal stability solutions, while the negative eigenvalue gives stable solutions. For general initial perturbations all modes are excited. These modes correspond to evolution of the fluid due to uniform perturbations of the HCS, i.e., a global change in the HCS parameters. The unstable modes are seen to arise from the initial perturbations $w_{\mathbf{k}\perp}(0)$ or $w_{\mathbf{k}||}(0)$. The marginal modes correspond to changes in the composition at fixed pressure, density, and velocity, and to changes in $\Pi_{\mathbf{k}} - \theta_{\mathbf{k}}$ at constant composition and velocity. The decaying mode corresponds to changes in the temperature or pressure for $\Pi_{\mathbf{k}} = \theta_{\mathbf{k}}$. The unstable modes may appear trivial since they are due entirely to the normalization of the fluid velocity by the time-dependent thermal velocity. However, this normalization is required by the scaling of the entire set of equations to obtain time-independent coefficients.

The real parts of the modes $s_\perp(k)$ and $s_n(k)$ are illustrated in Fig. 10.7 in the case of hard spheres ($d = 3$) for $\alpha \equiv \alpha_{ij} = 0.9$, $\sigma_1/\sigma_2 = 1$, $x_1 = 0.2$, and $m_1/m_2 = 4$. The $k = 0$ values correspond to five hydrodynamic modes with two different degeneracies. The shear mode degeneracy remains at finite k but the other is removed at any finite k. At sufficiently large k a pair of real modes

[1] Note that the definition for the reduced diffusion coefficient D^* given here differs from the one introduced in Sect. 10.5.1.

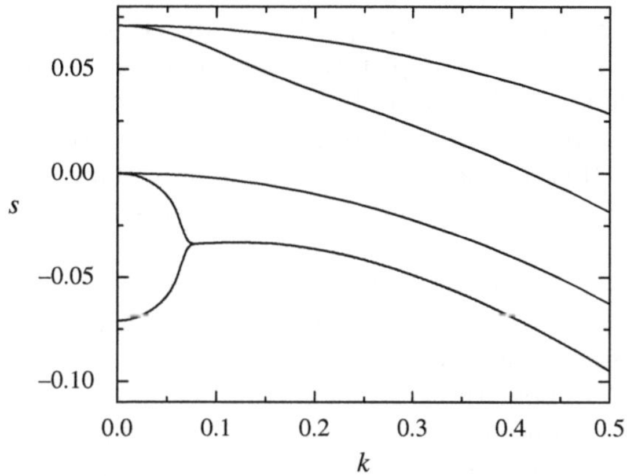

Fig. 10.7. Dispersion relations for $\alpha = 0.9$, $x_1 = 0.2$, $\omega = 1$, and $\mu = 4$

become equal and become a complex conjugate pair at all larger wave vectors, like a sound mode. The smallest of the unstable modes is that associated with the longitudinal velocity, which couples to the scalar hydrodynamic fields. It becomes negative at a wave vector smaller than that of Eq. (10.99) and gives the threshold for development of spatial instabilities.

The results obtained here for mixtures show no new surprises relative to the case for a monocomponent gas [5, 84, 88], with only the addition of the stable mass diffusion mode. Of course, the quantitative features can be quite different since there are additional degrees of freedom with the parameter set $\{x_{1H}, m_1/m_2, \sigma_1/\sigma_2, \alpha_{ij}\}$. Also, the manner in which these linear instabilities are enhanced by the nonlinearities may be different from that for the one component case [91].

10.9 Segregation in Granular Binary Mixtures: Thermal Diffusion

The analysis of the linearized hydrodynamic equations for a granular binary mixture has shown that the resulting equations exhibit a long-wavelength instability for d of the modes. These instabilities lead to the spontaneous formation of velocity vortices and density clusters when the system evolves freely. A phenomenon related with the density clustering is the separation or species segregation. Segregation and mixing of dissimilar grains is perhaps one of the most interesting problems in agitated granular mixtures. In some processes it is a desired and useful effect to separate particles of different types, while in other situations it is undesired and can be difficult to control. A variety of mechanisms have been proposed to describe the separation of particles of two sizes in a mixture of vertically shaken particles. Different mechanisms include void filling, static compressive force, convection, condensation, thermal diffusion, interstitial gas forcing, friction, and buoyancy [92]. However, in spite of the extensive literature published in the past few years on this subject, the problem is not completely understood yet. Among the different competing mechanisms, thermal diffusion becomes one of the most relevant at large shaking amplitude where the sample of macroscopic grains resembles a granular gas. In this regime, binary collisions prevail and kinetic theory can be quite useful to analyze the physical mechanisms involved in segregation processes.

Thermal diffusion is caused by the relative motion of the components of a mixture because of the presence of a temperature gradient. Due to this motion, concentration gradients subsequently appear in the mixture producing diffusion that tends to oppose those gradients. A steady state is finally achieved in which the separation effect arising from thermal diffusion is compensated by the diffusion effect. In these conditions, the so-called thermal diffusion factor Λ_{ij} characterizes the amount of segregation parallel to the temperature gradient. In this section, the thermal diffusion factor is determined from the Chapman–Enskog solution described before.

To make some contact with experiments, let us assume that the binary granular mixture is in the presence of the gravitational field $\mathbf{g} = -g\hat{\mathbf{e}}_z$, where g is a positive constant and $\hat{\mathbf{e}}_z$ is the unit vector in the positive direction of the z axis. In experiments [70], the energy is usually supplied by vibrating horizontal walls so that the system reaches a steady state. Here, instead of considering oscillating boundary conditions, particles are assumed to be heated by the action of the stochastic driving force (10.30), which mimics a thermal bath. As said above, although the relation between this driven idealized method with the use of locally driven wall forces is not completely understood, it must be remarked that in the case of boundary conditions corresponding to a sawtooth vibration of one wall, the condition to determine the temperature ratio coincides with the one derived from the stochastic force [63]. The good agreement between theory and simulation found in Fig. 10.1 for the temperature ratio confirms this expectation.

The thermal diffusion factor Λ_{ij} $(i \neq j)$ is defined at the steady state in which the mass fluxes $\mathbf{j}_i$ vanish. Under these conditions, the factor Λ_{ij} is given through the relation [93]

$$-\Lambda_{ij} \nabla \ln T = \frac{1}{x_i x_j} \nabla x_i, \quad \Lambda_{ij} + \Lambda_{ji} = 0. \tag{10.111}$$

The physical meaning of Λ_{ij} can be described by considering a granular binary mixture held between plates at different temperatures T (top plate) and T' (bottom plate) under gravity. For concreteness, we will assume that gravity and thermal gradient point in parallel directions, i.e., the bottom is hotter than the top ($T' > T$). In addition, without loss of generality, we also assume that $\sigma_1 > \sigma_2$. In the steady state, Eq. (10.111) describes how the thermal field is related to the composition of the mixture. Assuming that Λ_{12} is constant over the relevant ranges of temperature and composition, integration of Eq. (10.111) yields

$$\ln \frac{x_1 x_2'}{x_2 x_1'} = \Lambda_{12} \ln \frac{T'}{T}, \tag{10.112}$$

where x_i refers to the mole fraction of species i at the top plate and x_i' refers to the mole fraction of species i at the bottom plate. Consequently, according to Eq. (10.112), if $\Lambda_{12} > 0$, then $x_1' < x_1$, while if $\Lambda_{12} < 0$, then $x_1' > x_1$. In summary, when $\Lambda_{12} > 0$, the larger particles accumulate at the top of the sample (cold plate), while if $\Lambda_{12} < 0$, the larger particles accumulate at the bottom of the sample (hot plate). The former situation is referred to as the Brazil-nut effect (BNE) while the latter is called the reverse Brazil-nut effect (RBNE).

The RBNE was first observed by Hong et al. [94] in MD simulations of vertically vibrated systems. They proposed a very simple segregation criterion that was later confirmed by Jenkins and Yoon [95, 96] by using kinetic theory. More recently, Breu et al. [97] have experimentally investigated conditions under which the large particles sink to the bottom and claim that their experiments confirm the theory of Hong et al. [94] provided a number of conditions are chosen carefully. In addition to the vertically vibrated systems, some works have also focused in the last few years on horizontally driven systems showing some similarities to the BNE and its reverse form [98]. However, it is important to note that the criterion given in [94] is based on some drastic assumptions: elastic particles, homogeneous temperature, and energy equipartition. These conditions preclude a comparison of the kinetic theory derived here with the above simulations.

Some theoretical attempts to assess the influence of non-equipartition on segregation have been recently published. Thus, Trujillo et al. [99] have derived an evolution equation for the relative velocity of the intruders starting from the kinetic theory proposed by Jenkins and Yoon [95, 96], which applies for weak dissipation. They use constitutive relations for partial pressures that take into account the breakdown of energy equipartition between the two species. However, the influence of temperature gradients, which exist in the vibro-fluidized regime, is neglected in [99] because it is assumed that the pressure and temperature are constant in the absence of the intruder. A more refined theory has recently been provided by Brey et al. [71, 72] in the case of a single intruder in a vibrated granular mixture under gravity. The theory displayed in this section covers some of the aspects not accounted for in the previous theories [71, 72, 95, 96, 99] since it is based on a kinetic theory [35] that goes beyond the quasielastic limit [95, 96, 99] and applies for arbitrary composition x_1 (and so, it reduces to the results obtained in [71, 72] when $x_1 \to 0$). This allows one to assess the influence of composition and dissipation on thermal diffusion in bi-disperse granular gases without any restriction on the parameter space of the system.

To determine the dependence of the coefficient Λ_{12} on the parameters of the mixture, we consider a non-convecting ($\mathbf{u} = \mathbf{0}$) steady state with only gradients along the vertical direction (z-axis). In this case, the mass balance equation (10.12) yields $\mathbf{j}_1 = \mathbf{j}_2 = \mathbf{0}$, while the momentum equation (10.13) gives

$$\frac{\partial p}{\partial z} = -\rho g. \tag{10.113}$$

To first order in the spatial gradients, the constitutive equation for the mass flux $j_{1,z}$ is given by Eq. (10.35), i.e.,

$$j_{1,z} = -\frac{m_1 m_2 n}{\rho} D \frac{\partial x_1}{\partial z} - \frac{\rho}{p} D_p \frac{\partial p}{\partial z} - \frac{\rho}{T} D' \frac{\partial T}{\partial z} , \tag{10.114}$$

where the susceptibility coefficient $\chi_{ij} = 0$ in the particular case of the gravitational force. The condition $j_{1,z} = 0$ yields

$$\frac{\partial x_1}{\partial z} = \frac{\rho^3}{m_1 m_2 n p} \frac{D_p}{D} g - \frac{\rho^2}{m_1 m_2 p} \frac{D'}{D} \frac{\partial T}{\partial z} , \tag{10.115}$$

where use has been made of Eq. (10.113). Substitution of Eq. (10.115) into Eq. (10.111) leads to

$$\Lambda_{12} = \frac{n\rho^2}{\rho_1 \rho_2} \frac{D' - D_p g^*}{D} , \tag{10.116}$$

where

$$g^* \equiv \frac{\rho g}{n\left(\partial T/\partial z\right)} < 0 \tag{10.117}$$

is the reduced gravity acceleration. Since the mutual diffusion coefficient D is positive [35, 74], the sign of Λ_{12} is determined by the sign of the quantity $D' - D_p g^*$. This result is general since it goes beyond the regime of density considered.

To gain some insight into the explicit dependence of D' and D_p on the parameter space of the system, one has to resort to a kinetic theory description. For a low-density gas, the expressions of the coefficients D' and D_p in the first Sonine approximation are given by

$$D' = 0, \quad D_p = \frac{\rho_1 p}{\rho^2 \nu} \frac{x_2}{x_2 + x_1 \gamma} \left(\frac{\gamma}{\mu} - 1\right) , \tag{10.118}$$

where ν is the (positive) collision frequency [74]

$$\nu = \frac{2\pi^{(d-1)/2}}{d\Gamma\left(\frac{d}{2}\right)} n \sigma_{12}^{d-1} v_0 (1+\alpha_{12}) \left(\frac{\theta_1 + \theta_2}{\theta_1 \theta_2}\right)^{1/2} (x_2 \mu_{21} + x_1 \mu_{12}) . \tag{10.119}$$

Given that the driving stochastic term does not play a neutral role in the transport, it must be remarked that the expressions for the transport coefficients obtained in the driven case slightly differ from the ones derived in the free cooling case [35, 74].

Consequently, according to Eqs. (10.116) and (10.118), the sign of Λ_{12} is the same as that of the pressure diffusion coefficient D_p. The condition $\Lambda_{12} = 0$

(or equivalently, $D_p = 0$) provides the criterion for the transition from BNE to RBNE. Equation (10.118) shows that the sign of D_p is determined by the value of the control parameter

$$\theta \equiv \frac{\gamma}{\mu} = \frac{m_2 T_1}{m_1 T_2}. \tag{10.120}$$

This parameter gives the mean square velocity of the large particles relative to that of the small particles. Thus, when $\theta > 1$ ($\theta < 1$), the thermal diffusion factor is positive (negative), which leads to BNE (RBNE). The criterion for the transition condition from BNE to RBNE is $\theta = 1$, i.e.,

$$\frac{m_1}{m_2} = \frac{T_1}{T_2}. \tag{10.121}$$

In the case of equal granular temperatures (energy equipartition), $\theta \to \mu^{-1}$ and so segregation is predicted for particles that differ in mass, no matter what their diameters may be [95, 96]. It must be remarked that, due to the lack of energy equipartition, the condition $\theta = 1$ is rather complicated since it involves all the parameter space of the system. In particular, even when the species differ only by their respective coefficients of restitution they also segregate when subject to a temperature gradient. This is a novel pure effect of inelasticity on segregation [41, 100]. On the other hand, the criterion (10.121) for the transition BNE$\Longleftrightarrow$RBNE is the same as the one found previously in [99] when α_{ij} is close to 1 and in [71, 72] in the intruder limit case ($x_1 \to 0$). However, as said before, the results obtained here are more general since they cover all the range of the parameter space of the system.

To illustrate size segregation driven by thermal diffusion, we consider mixtures constituted by spheres ($d = 3$) of the same material and equal total volumes of large and small particles. In this case, $m_1/m_2 = (\sigma_1/\sigma_2)^3$ and $x_2/x_1 = (\sigma_1/\sigma_2)^3$. Figure 10.8 shows the phase diagram BNE/RBNE for this kind of systems. The data points represent the simulation results obtained by Schröter et al. [70] for $\alpha = 0.78$ in agitated mixtures constituted by particles of the same density. To the best of my knowledge, this is one of the few experiments in which thermal diffusion has been isolated from the remaining segregation mechanisms [92]. Our results show that, for a given value of the coefficient of restitution, the RBNE is dominant at small diameter ratios. However, since nonequipartition grows with increasing diameter ratio, the system shows a crossover to BNE at sufficiently large diameter ratios. This behavior agrees qualitatively well with the results reported in [70] at large shaking amplitudes, where thermal diffusion becomes the relevant segregation mechanism. At a quantitative level, we observe that the results are also consistent with the simulation results reported in [70] when periodic boundary conditions are used to suppress convection since they do not observe a change back to BNE for diameter ratios up to 3 (see red squares in Fig. 10.11 of [70]). Although the parameter range explored in MD simulations is smaller

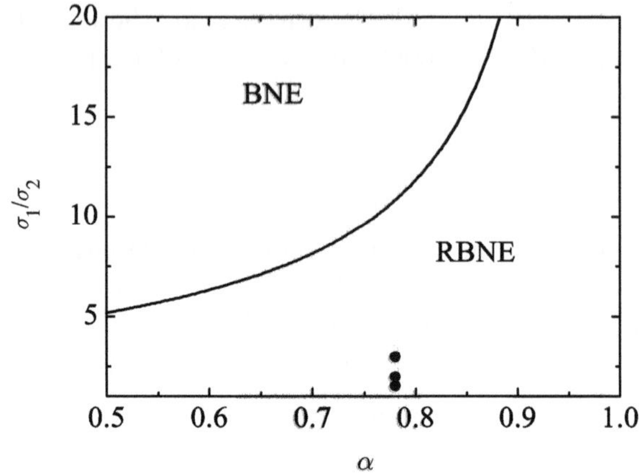

Fig. 10.8. Phase diagram for BNE/RBNE for mixtures constituted by spheres ($d = 3$) of the same mass density and equal total volumes of large and small particles. The data points represent the MD simulation results [70] for $\alpha = 0.78$ when convection is suppressed. Points below (above) the curve correspond to RBNE (BNE)

than the one analyzed here, one is tempted to extrapolate the simulation data presented in [70] to roughly predict the transition value of the diameter ratio at $\alpha = 0.78$ (which is the value of the coefficient of restitution considered in the simulations). Thus, if one extrapolates from the simulation data at the diameter ratios of $\sigma_1/\sigma_2 = 2$ and $\sigma_1/\sigma_2 = 3$, one sees that the transition from RBNE to BNE might be around $\sigma_1/\sigma_2 = 10$, which would quantitatively agree with the results reported in Fig. 10.8. Figure 10.8 also shows that the BNE is completely destroyed in the quasielastic limit ($\alpha \simeq 1$).

Let us now investigate the influence of composition on segregation. Figure 10.9 shows a typical phase diagram in the three-dimensional case for $\alpha_{ij} \equiv \alpha = 0.7$ and three different values of the mole fraction x_1. The lines separate the regimes between BNE and RBNE. We observe that the composition of the mixture has significant effects in reducing the BNE as the concentration of larger particles increases. In addition, for a given value of composition, the transition from BNE to RBNE may occur following two paths: (i) along the constant mass ratio m_2/m_1 with increasing size ratio σ_1/σ_2, and (ii) along the constant size ratio with increasing mass ratio m_2/m_1. The influence of dissipation on the phase diagrams BNE/RBNE is illustrated in Fig. 10.10 for $d = 3$ in the case of an equimolar mixture ($x_1 = \frac{1}{2}$) and three values of the (common) coefficient of restitution α. We observe that the role played by inelasticity is quite important since the regime of RBNE increases significantly with dissipation. Similar results are found for other values of composition.

In summary, thermal diffusion (which is the relevant segregation mechanism in agitated granular mixtures at large shaking amplitudes) can be analyzed by the Boltzmann kinetic theory. This theory is able to explain some of the experimental and/or MD segregation results [70] observed within the range

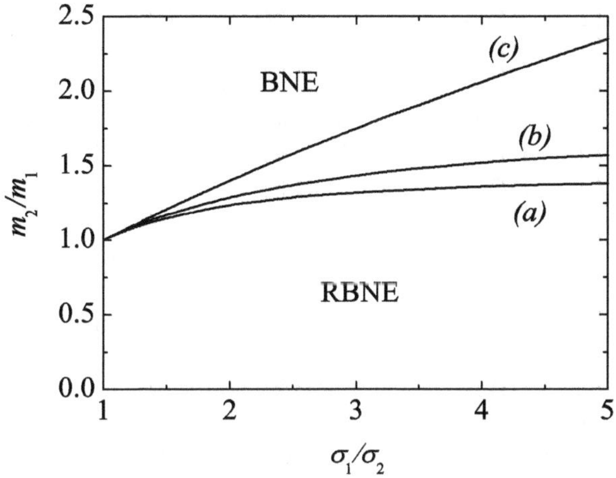

Fig. 10.9. Phase diagram for BNE/RBNE in three dimensions for $\alpha_{ij} = 0.7$ and three values of composition: (**a**) $x_1 = 0$, (**b**) $x_1 = 0.3$, and (**c**) $x_1 = 0.7$. Points below (above) each curve correspond to RBNE (BNE)

of parameter space explored. A more quantitative comparison in the dilute regime with MD simulations is needed to show the relevance of the Boltzmann equation to analyze segregation driven by a thermal gradient. As said before, comparison with MD simulations in the tracer limit case ($x_1 \to 0$) [71, 72] for a dilute gas has shown the reliability of the inelastic Boltzmann equation to describe segregation. In this context, one expects that the same agreement observed before in the intruder case [71, 72] is maintained when x_1 is different from zero.

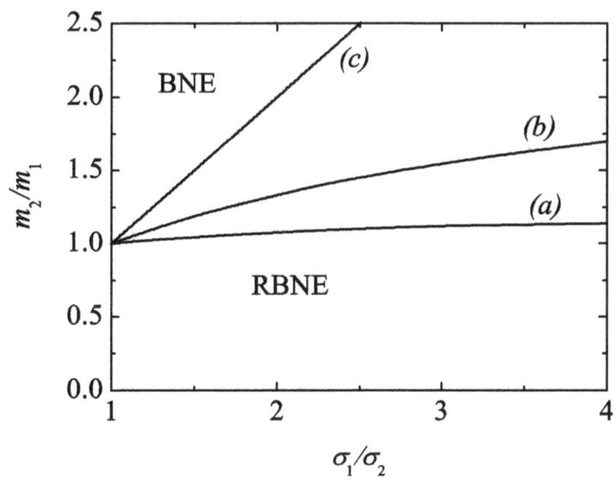

Fig. 10.10. Phase diagram for BNE/RBNE in three dimensions for $x_1 = 1/2$ and three values of the (common) coefficient of restitution: (**a**) $\alpha = 0.9$, (**b**) $\alpha = 0.8$, and (**c**) $\alpha = 0.5$. Points below (above) each curve correspond to RBNE (BNE)

10.10 Steady States: Uniform Shear Flow

In the preceding sections, the Navier–Stokes equations (constitutive equations that are linear in the hydrodynamic gradients) have been shown to be quite useful to describe appropriately several problems in granular mixtures. However, under some circumstances large gradients occur and more complex constitutive equations are required. The need for more complex constitutive equations does not signal a breakdown of hydrodynamics [101], only a failure of the Navier–Stokes approximation [102]. Although in this case the Chapman-Enskog method can be carried out to second order in gradients (Burnett order), it is likely that failure of the Navier-Stokes description signals the need for other methods to construct the normal solution that are not based on a small gradient expansion.

One of the most interesting problems in granular fluids is the simple or uniform shear flow (USF) [13, 42]. As described in Sect. 10.5, this state is characterized by uniform density and temperature and a simple shear with the local velocity field given by $u_{1,x} = u_{2,x} = ay, u_y = u_z = 0$, where a is the constant shear rate. The USF is a well-known nonequilibrium problem widely studied, for both granular [42, 43, 44, 45, 46, 47, 48, 49, 50, 51, 52, 53, 54, 55, 56, 103, 104, 105, 106, 107, 108, 109, 110, 111, 112, 113] and conventional [57, 114] gases. However, the nature of this state is quite different in both systems since a steady state is achieved for granular fluids when viscous (shear) heating is compensated for by energy dissipation in collisions:

$$aP_{xy} = -\frac{d}{2}nT\zeta. \tag{10.122}$$

This steady state is what we want to analyze in this section. The balance equation (10.122) shows the intrinsic connection between the shear field and dissipation in the system. This contrasts with the description of USF for elastic fluids where a steady state is not possible unless an external thermostat is introduced [57]. Note that the hydrodynamic steady shear flow state associated with the condition (10.122) is inherently beyond the scope of the Navier–Stokes or Newtonian hydrodynamic equations [58]. The reason for this is the existence of an internal mechanism, collisional cooling, that sets the strength of the velocity gradient in the steady state. For normal fluids, this scale is set by external sources (boundary conditions, driving forces) that can be controlled to admit the conditions required for Navier–Stokes hydrodynamics. In contrast, collisional cooling is fixed by the mechanical properties of the particles making up the fluid. This observation is significant because it prevents the possibility of measuring the Newtonian shear viscosity for granular fluids in the steady USF [105, 106, 107, 108, 109, 110, 111, 112, 113]. More generally, it provides a caution regarding the simulation of other steady states to study Navier–Stokes hydrodynamics when the gradients are strongly correlated to the collisional cooling [58].

From a microscopic point of view, the simple shear flow problem becomes spatially uniform in the local Lagrangian frame moving with the flow velocity $\mathbf{u}$. In this frame [57, 115, 116], the velocity distribution functions adopt the form: $f_i(\mathbf{r},\mathbf{v}) \to f_i(\mathbf{V})$, where $V_k = v_k - a_{k\ell}r_\ell$ is the peculiar velocity. Here, $a_{k\ell} = a\delta_{kx}\delta_{\ell y}$. Under these conditions, the set of Boltzmann kinetic equations (with $\mathcal{F}_i = 0$) for an isolated system reads

$$-aV_y\frac{\partial}{\partial V_x}f_i(\mathbf{V}) = \sum_{j=1}^{2} J_{ij}[\mathbf{V}|f_i,f_j]\,, \quad (i=1,2). \tag{10.123}$$

The most relevant transport properties in a shear flow problem are obtained from the pressure tensor $\mathsf{P} = \mathsf{P}_1 + \mathsf{P}_2$, where P_i is the partial pressure tensor of the species i given by

$$P_{i,k\ell} = m_i\int d\mathbf{V}\,V_kV_\ell f_i(\mathbf{V}). \tag{10.124}$$

The trace of P_i defines the partial temperatures T_i as $T_i = \mathrm{Tr}\mathsf{P}_i/dn_i$. As said before, these temperatures measure the mean kinetic energy of each species. The elements of the pressure tensor P_i can be obtained by multiplying the Boltzmann equation (10.123) by $m_i\mathbf{V}\mathbf{V}$ and integrating over $\mathbf{V}$. The result is

$$a_{km}P_{i,m\ell} + a_{\ell m}P_{i,mk} = \sum_{j=1}^{2} A_{ij,k\ell}\,, \tag{10.125}$$

where we have introduced the collisional moments A_{ij} as

$$A_{ij,k\ell} = m_i\int d\mathbf{V}V_kV_\ell J_{ij}[\mathbf{V}|f_i,f_j]. \tag{10.126}$$

From Eq. (10.125), in particular, one gets the balance equation for the partial temperature T_i

$$aP_{i,xy} = -\frac{d}{2}p_i\zeta_i\,, \tag{10.127}$$

where $p_i = n_iT_i$ is the partial pressure of species i and ζ_i is defined by Eq. (10.10). According to Eq. (10.127), the (steady) partial temperature in the simple shear flow problem can be obtained by equating the viscous heating term $a|P_{i,xy}|$ to the collisional cooling term $(d/2)p_i\zeta_i$.

The determination of A_{ij} requires the knowledge of the velocity distribution functions f_i. This is quite a formidable task, even in the monocomponent case [43, 44, 45, 46, 47, 48, 49, 50, 51, 52, 53, 54, 55, 56]. However, as in the elastic case, one expects to get a good estimate of A_{ij} by using Grad's approximation [73]:

$$f_i(\mathbf{V}) \to f_{i,M}(\mathbf{V})\left(1 + \frac{m_i}{2T_i}C_{i,k\ell}V_kV_\ell\right)\,, \tag{10.128}$$

where $f_{i,M}$ is a Maxwellian distribution at the temperature of the species i, i.e.,

$$f_{i,M}(\mathbf{V}) = n_i \left(\frac{m_i}{2\pi T_i}\right)^{d/2} \exp\left(-\frac{m_i V^2}{2T_i}\right). \tag{10.129}$$

As happens in the case of homogeneous states, in general the three temperatures T, T_1, and T_2 are different in the inelastic case. For this reason we choose the parameters in the Maxwellians so that it is normalized to n_i and provides the exact second moment of f_i. The Maxwellians $f_{i,M}$ for the two species can be quite different due to the temperature differences. This aspect is essential in a two-temperature theory and has not been taken into account in most of the previous studies [14, 15, 16, 17, 18, 19, 20, 103, 104, 105, 106, 107]. The coefficient C_i can be identified by requiring the moments with respect to $\mathbf{VV}$ of the trial function (10.128) to be the same as those for the exact distribution f_i. This leads to

$$\mathsf{C}_i = \frac{\mathsf{P}_i}{p_i} - \mathsf{I} \tag{10.130}$$

With this approximation, the Boltzmann collisional moments A_{ij} can be explicitly evaluated. The result is [117, 118]

$$\begin{aligned}\mathsf{A}_{ij} = &-\frac{2\pi^{(d-1)/2}}{d\Gamma(d/2)} m_i n_i n_j \mu_{ji} \sigma_{ij}^{d-1} \left(\frac{2T_i}{m_i} + \frac{2T_j}{m_j}\right)^{3/2} (1+\alpha_{ij}) \\ &\times \left\{\mu_{ji}\left[\frac{T_j - T_i}{(m_j/m_i)T_i + T_j} + \frac{1-\alpha_{ij}}{2}\right]\mathsf{I} + \frac{1}{1+(m_i T_j/m_j T_i)}\right. \\ &\times \left.\left[\frac{\mathsf{C}_i - \mathsf{C}_j}{1+(m_j T_i/m_i T_j)} + \frac{d+3}{2(d+2)}\lambda_{ij}\left(\mathsf{C}_i + \frac{m_i T_j}{m_j T_i}\mathsf{C}_j\right)\right]\right\}, \end{aligned} \tag{10.131}$$

where

$$\lambda_{ij} = 2\mu_{ji}\frac{T_j - T_i}{(m_j/m_i)T_i + T_j} + \frac{\mu_{ji}}{d+3}(2d+3-3\alpha_{ij}). \tag{10.132}$$

The partial cooling rates ζ_i can be easily obtained from Eqs. (10.10) and (10.131).

Substitution of Eq. (10.131) into the set of equations (10.125) allows one to get the partial pressure tensor P_i in terms of the temperature ratio $\gamma = T_1/T_2$ and the parameters of the mixture. The temperature ratio can be obtained from Eq. (10.127) as

$$\gamma = \frac{x_2 \zeta_2 P_{1,xy}}{x_1 \zeta_1 P_{2,xy}}. \tag{10.133}$$

When the expressions of P_i and ζ_i are used in Eq. (10.133), one gets a *closed* equation for the temperature ratio γ, that can be solved numerically. In Fig. 10.11 we plot γ versus the diameter ratio σ_1/σ_2 for a two-dimensional ($d=2$) granular gas with $x_1 = 1/2$ and two different values of α. The symbols refer to the simulation data obtained from the DSMC method [119]. Here, we have assumed that the disks are made of the same material, and hence $\alpha_{ij} = \alpha$ and $m_1/m_2 = (\sigma_1/\sigma_2)^2$. The dependence of γ on σ_1/σ_2 obtained in the homogeneous steady state driven by the stochastic thermostat (10.30) is

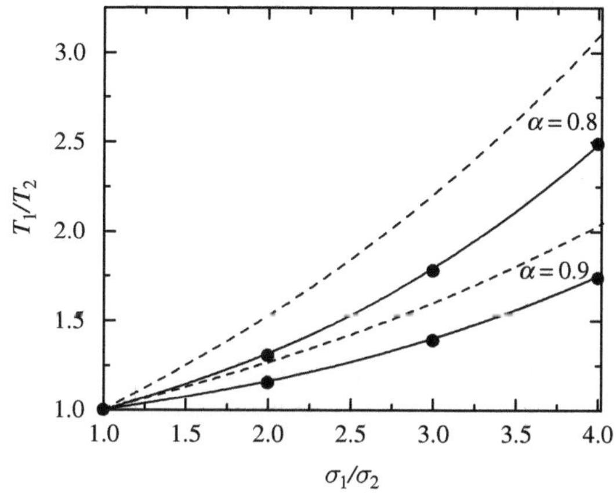

Fig. 10.11. Plot of the temperature ratio T_1/T_2 as a function of the size ratio $\sigma_1/\sigma_2 = (m_1/m_2)^{1/2}$ for a two-dimensional system in the case $x_1 = 1/2$ and two different values of the (common) coefficient of restitution: $\alpha = 0.9$ and $\alpha = 0.8$. The *solid lines* are the theoretical predictions based on Grad's solution, while the symbols refer to the DSMC results. The *dashed lines* correspond to the results obtained from the stochastic thermostat condition (10.31)

also included for comparison. It is clearly seen that the kinetic theory results based on Grad's solution agree very well with simulation data, even for quite large values of the size ratio. In addition, the thermostat results overestimate the simulation ones (especially for large mass ratio), showing that the properties of the system are not insensitive to the way at which the granular gas is driven.

Let us now consider the transport coefficients. To analyze the rheological properties in the steady state, it is convenient to introduce dimensionless quantities. As usual [105, 106, 107], for a low-density gas we introduce the reduced pressure p^* and the reduced shear viscosity η^* as

$$p^* = \frac{p\nu^2}{\rho_1 v_0^2 a^2}, \tag{10.134}$$

$$\eta^* = \frac{\eta\nu^2}{\rho_1 v_0^2 a}, \tag{10.135}$$

where $\eta = -P_{xy}/a$ is the non-Newtonian shear viscosity, $P_{xy} = P_{1,xy} + P_{2,xy}$; and $\nu = [\pi^{(d-1)/2}/\Gamma(d/2)]n\sigma_{12}^{d-1}v_0$.

In Figs. 10.12 and 10.13, we plot p^* and η^*, respectively, as functions of the mass ratio $\mu = m_1/m_2$ for an *equal-size* ($\sigma_1 = \sigma_2$) binary mixture of disks ($d = 2$) with $x_1 = 1/2$ and $\alpha = 0.9$. We have also included the predictions for p^* and η^* given by the kinetic theory but taking the expression of γ derived when the system is driven by the stochastic thermostat [69]. We observe again in both figures an excellent agreement between the Boltzmann theory based

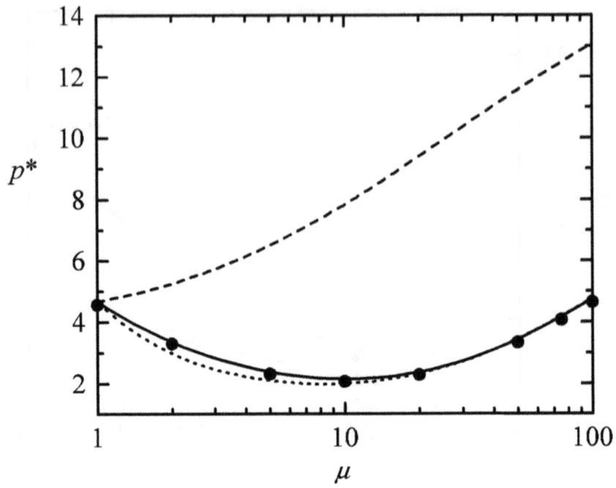

Fig. 10.12. Plot of the reduced pressure p^* versus the mass ratio $\mu = m_1/m_2$ for a two-dimensional system with $\sigma_1 = \sigma_2$, $x_1 = 1/2$, and $\alpha = 0.9$. The *solid line* corresponds to the theoretical predictions derived from Grad's solution, the *dotted line* refers to the latter theory but using the expression of T_1/T_2 obtained from the stochastic thermostat condition (10.31), and the *dashed line* is the result obtained from Grad's solution by assuming the equality of the partial temperatures ($\gamma = 1$). The symbols are the DSMC results

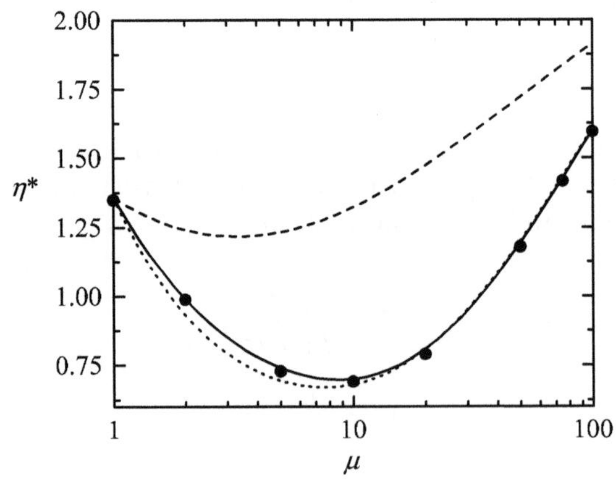

Fig. 10.13. Plot of the reduced shear viscosity η^* versus the mass ratio $\mu = m_1/m_2$ for a two-dimensional system with $\sigma_1 = \sigma_2$, $x_1 = 1/2$, and $\alpha = 0.9$. The *solid line* corresponds to the theoretical predictions derived from Grad's solution, the *dotted line* refers to the latter theory but using the expression of T_1/T_2 obtained from the stochastic thermostat condition (10.31), and the *dashed line* is the result obtained from Grad's solution by assuming the equality of the partial temperatures ($\gamma = 1$). The symbols are the DSMC results

on Grad's solution and the DSMC results, even for very disparate values of the mass ratio. With respect to the influence of energy nonequipartition, Fig. 10.12 shows that p^* presents a *non-monotonic* behavior with the mass ratio, whereas the theoretical predictions with the equipartition assumption *monotonically* increase with μ. In the case of the shear viscosity, as seen in Fig. 10.13, both theories (with and without energy nonequipartition) predict a *non-monotonic* dependence of η^* on μ. However, at a quantitative level, the influence of energy nonequipartition is quite significant over the whole range of mass ratios considered. The *non-monotonic* dependence of p^* and η^* on μ obtained here from the Boltzmann kinetic theory also agrees qualitatively well with MD simulations carried out for bidisperse dense systems [105, 106, 107]. Thus, for instance, the minimum values of p^* and η^* are located close to $\mu = 10$ in both dilute and dense cases. Moreover, the predictions for the transport properties given from the present theory by taking the stochastic thermostat expression of γ are quite close to those obtained from the actual value of γ, especially for large mass ratios.

10.11 Summary and Concluding Remarks

The primary objective of this review has been to derive the Navier–Stokes hydrodynamic equations of a binary mixture of granular gases from the (inelastic) Boltzmann kinetic theory. The Chapman–Enskog method [4, 73] is used to solve the Boltzmann equation up to the first order in the spatial gradients, and the associated transport coefficients are given in terms of the solutions of a set of linear integral equations. These equations have been approximately solved by taking the leading terms in a Sonine polynomial expansion. Comparison with controlled numerical simulations in some idealized conditions shows quite a good agreement between theory and simulation even for strong dissipation. This supports the idea that the hydrodynamic description (derived from kinetic theory) appears to be a powerful tool for analysis and predictions of rapid flow gas dynamics of polydisperse systems [12].

The reference state in the Chapman–Enskog expansion has been taken to be an exact solution of the uniform Boltzmann equation. An interesting and important result of this solution [21] is that the partial temperatures (which measure the mean kinetic energy of each species) are different. This does not mean that there are additional degrees of freedom since the partial temperatures can be expressed in terms of the global temperature. This is confirmed by noting that Haff's cooling law [1] (in the free cooling case) is the hydrodynamic mode at long wavelengths and MD simulations confirm that the global temperature dominates after a transient period of a few collision times [63]. In this case, only the global temperature should appear among the hydrodynamic fields. Nevertheless, the species temperatures play a new and interesting secondary role [36]. For an ordinary (molecular) gas, there is a rapid velocity relaxation in each fluid cell to a local equilibrium state on the time scale of a few collisions (e.g., as illustrated by the approach to Haff's law). Subsequently, the equilibration among cells occurs via the hydrodynamic equations. In each cell the species velocity distributions are characterized by the species temperatures. These are approximately the same due to equipartition, and the hydrodynamic relaxation occurs for the single common temperature [73]. A similar rapid velocity relaxation occurs for granular gases in each small cell, but to a universal state different from local equilibrium and one for which equipartition no longer occurs. Hence, the species temperatures T_i are different from each other and from the overall temperature T of the cell. Nevertheless, the time dependence of all temperatures (in the free cooling case) is the same in this and subsequent states, i.e., they are proportional to the global temperature. This implies that the species temperatures do not provide any new dynamical degree of freedom at the hydrodynamic stage. However, they still characterize the shape of the partial velocity distributions and affect the quantitative averages calculated with these distributions. The transport coefficients for granular mixtures therefore have new quantitative effects arising from the time-independent temperature ratios for each species [35]. This view contrasts with some recent works [120, 121, 122], where additional equations for each species temperature have been included among the hydrodynamic set. However, as mentioned before, this is an unnecessary complication, describing additional kinetics beyond hydrodynamics that is relevant only on the time scale of a few collisions.

Another important issue discussed here has been the applicability of the Navier–Stokes transport coefficients since their expressions are not restricted to weak inelasticity [12]. However, the Navier–Stokes hydrodynamic equations themselves may or may not be limited with respect to inelasticity, depending on the particular states studied. The Chapman–Enskog method assumes that the relative changes of the hydrodynamic fields over distances of the order of the mean free path are small. In the case of ordinary fluids this can be controlled by the initial or boundary conditions. For granular gases the situation is more complicated since in some cases (e.g., steady states such as the simple shear flow problem [58]) the boundary conditions imply a relationship between dissipation and gradients so that both cannot be chosen independently. In these cases, the Navier–Stokes approximation only holds for nearly elastic particles. However, the transport coefficients characterizing the Navier–Stokes hydrodynamic equations are nonlinear functions of the coefficients of restitution, regardless the applicability of those equations.

In spite of the above cautions, the Navier–Stokes approximation is appropriate and accurate for a wide class of flows. One group refers to spatial perturbations of the homogeneous cooling state (HCS) for an isolated system. Both MD and DSMC simulations [7] have confirmed the dependence of the Navier–Stokes transport coefficients on the coefficient of restitution, and application of the Navier–Stokes hydrodynamics with these coefficients to describe cluster formation has also been confirmed quantitatively [91]. The same kinetic theory results apply to driven systems as well. This is so since the reference state is a *local* HCS whose parameters vary throughout the system to match the physical values in each cell. Examples include application of Navier–Stokes hydrodynamics from kinetic theory to symmetry breaking and density/temperature profiles in vertical vibrated gases, for comparison with simulation [123, 124]. Similar comparison with Navier–Stokes hydrodynamics of the latter and of supersonic flow past a wedge in real experiments has been given [125, 126, 127, 128], showing both qualitative and quantitative agreement. In summary, the Navier–Stokes equations with the constitutive equations presented here remain an important and useful description for a wide class of granular flow, although more limited than for normal gases.

The explicit knowledge of the transport coefficients and the cooling rate allows one to make some applications of the Navier–Stokes hydrodynamic equations. One of them has been to obtain the linear hydrodynamic equations for small perturbations of the homogenous cooling state. The resulting equations exhibit a long wavelength instability for three of the modes. This is quite similar to the case of a monocomponent granular gas [5, 84, 88], and in fact the same modes are unstable here. The additional diffusion mode for two species behaves as for a normal fluid.

On the other hand, the constitutive equations for the mass and heat fluxes of a granular binary mixture differ from those obtained for ordinary fluids [4]. This is because the usual restrictions of irreversible thermodynamics no longer apply. These restrictions include Onsager's reciprocal relations among various transport coefficients and the extent to which these are violated has also been shown here. Another application of the Navier–Stokes equations has been to assess the violation of the Einstein relation between the diffusion and mobility coefficients. In the undriven case, the analysis shows that this violation is due to three independent reasons [39]: the absence of the Gibbs state, the cooling of the reference state, and the occurrence of different temperatures for the particle and surrounding fluid. However, when the mixture is subjected to stochastic driving, a modified Einstein relation suggested by recent MD simulations [82] has also been analyzed. In this case, the results show that the deviations of the (modified) Einstein ratio from unity are in general very small (less than 1%), in agreement with MD simulations [82].

Thermal diffusion becomes the relevant segregation mechanism in agitated granular mixtures at large shaking amplitudes. In these conditions, the use of the Boltzmann kinetic theory for low-density gases appears justified to understand the influence of thermal gradient on segregation phenomena. The thermal diffusion factor in a heated granular mixture has been explicitly evaluated from the Chapman–Enskog solution to the Boltzmann equation. The results show that the criterion for the transition Brazil-nut effect $\Longleftrightarrow$reverse Brazil-nut effect is provided by the control parameter $\theta = m_2T_1/m_1T_2$ [41, 71, 72, 99]. Given that the energy equipartition is broken, the condition $\theta = 1$ is quite complex since it involves all the parameters of the system: composition, masses, sizes, and coefficients of restitution. The Boltzmann kinetic theory results agree qualitatively well with recent MD simulations [70] within the range of parameter space analyzed.

The hydrodynamic description also seems to be justified in the case of steady states that are inherently beyond the scope of the Navier–Stokes hydrodynamic equations. The reason for this non-Newtonian behavior is the existence of an internal mechanism, collisional cooling, that sets the scale of the spatial gradients in the steady state. For ordinary fluids, this scale can be externally controlled by external sources so that the conditions for Navier–Stokes hydrodynamics apply. On the other hand, for granular gases, collisional cooling is fixed by the mechanical properties of the particles of the system and so the gas can depart from the Navier–Stokes description. One well-known example of steady states is the simple or uniform shear flow (USF). However, in spite of the extensive prior work on USF for granular fluids [43, 44, 45, 46, 47, 48, 49, 50, 51, 52, 53, 54, 55, 56, 105, 106, 107, 108, 109, 110, 111, 112, 113], the inherent non-Newtonian character of this state has not been conveniently taken into account. In fact, MD simulations of steady USF have been used for granular fluids to measure the Newtonian or Navier–Stokes shear viscosity. The results derived here from Grad's solution and DSMC simulations show that USF is an ideal testing ground for the study of rheology since any choice of the shear rate and the coefficients of restitution α_{ij} will provide non-Newtonian effects. It is one of the fascinating features of granular fluids that phenomena associated with complex fluids are more easily accessible than for simple atomic fluids [12, 129].

Hydrodynamics derived from hard-sphere models have found widespread use in the description of numerous industrial processes involving solid particles. Of particular relevance are high-speed, gas-solid flows as found in pneumatic conveyors (of ores, chemicals, grains, etc.) and fluidized beds (for fluid catalytic cracking, power generation, granulation of pharmaceutical powders, synthesis of fine chemicals like titania, etc.). Such descriptions are now standard features of commercial and research codes. Those codes rely upon accurate transport properties, and a first-order objective is to assure this accuracy from a careful theoretical treatment. As shown in this review, the price of this approach, in contrast to more phenomenological approaches, is an increasing complexity of the expressions as the systems become more complex.

The analysis carried out in this presentation has been focused on mixtures in the dilute regime, where the collisional transfer contributions to the transport coefficients are neglected and only their kinetic contributions are considered. A further step is to develop a theory for moderately dense granular mixtures. This will provide a fundamental basis for the application of hydrodynamics under realistic conditions. Possible extension of the present Boltzmann kinetic theory to higher densities can be done in the context of the Enskog kinetic equation [73]. Preliminary results [130] have been restricted to the uniform shear flow state to get directly the shear viscosity coefficient of a heated granular mixture. The extension of this study [130] to states with gradients of concentration, pressure, and temperature is somewhat intricate due to subtleties associated with the spatial dependence of the pair correlations functions considered in the revised Enskog theory. A future work is to extend the results derived for moderately dense mixtures of smooth *elastic* hard spheres [131] to inelastic collisions. This would allow us to assess the influence of density on the different problems addressed in this review. Of course, the precise expressions for transport coefficients in this case will be even more complex than for a dilute gas due to the expanded parameter space. However, this complexity is not a problem for implementation in a code.

As shown along this overview, granular mixtures exhibit a wide range of interesting phenomena for which the Navier–Stokes hydrodynamic equations can be considered as an accurate and practical tool. However, due to their complexity, many of their features are not fully understood. Kinetic theory and hydrodynamics (in the broader sense) can be expected to provide some insight into the understanding of such complex materials.

Acknowledgements

I acknowledge J. W. Dufty, J. M. Montanero, and A. Santos in their roles as collaborators and critics for much of the material discussed here. Partial support of the Ministerio de Ciencia y Tecnología (Spain) through Grant No. FIS2004-01399 (partially financed by FEDER funds) and from the European Community's Human Potential Programme HPRN-CT-2002-00307 (DYGLAGEMEM) is also acknowledged.

References

1. P.K. Haff: J. Fluid Mech. **134**, 187 (1983)
2. L.P. Kadanoff: Rev. Mod. Phys. **71**, 435 (1999)
3. See for instance, J.-P. Hansen, I.R. McDonald: *Theory of Simple Liquids* (Academic Press, London, 1986)
4. S. Chapman, T. G. Cowling: *The Mathematical Theory of Nonuniform Gases* (Cambridge University Press, Cambridge, 1970)
5. J.J. Brey, J.W. Dufty, C.S. Kim, A. Santos: Phys. Rev. E **58**, 4638 (1998)
6. J.J. Brey, D. Cubero: Hydrodynamics transport coefficients of granular gases. In: Pöschel, T., Luding, S. (eds.): *Granular Gases*. Lect. Notes Phys. **564**, 59–78. Springer, Berlin (2001)
7. J.J. Brey, M.J. Ruiz-Montero, D. Cubero: Europhys. Lett. **48**, 359 (1999)
8. J.J. Brey, M.J. Ruiz-Montero, D. Cubero, R. García-Rojo: Phys. Fluids **12**, 876 (2000)
9. J.J. Brey, M.J. Ruiz-Montero: Phys. Rev. E **70**, 051301 (2004)
10. J.M. Montanero, A. Santos, V. Garzó: DSMC evaluation of the Navier-Stokes shear viscosity of a granular fluid. In: *Rarefied Gas Dynamics 24*, ed by M. Capitelli (American Institute of Physics 762, 2005), pp 797–802 and preprint cond-mat/0411219
11. J.W. Dufty: J. Phys.: Condens. Matt. **12**, A47 (2000)
12. J.W. Dufty: Kinetic theory and hydrodynamics for rapid granular flow - A perspective. In: *Recent Research Develop. in Stat. Phys.* **2**, 21 (2002) and preprint cond-mat/0108444

13. I. Goldhirsch: Annu. Rev. Fluid Mech. **35**, 267 (2003)
14. J.T. Jenkins, F. Mancini: Phys. Fluids A **1**, 2050 (1989)
15. P. Zamankhan: Phys. Rev. E **52**, 4877 (1995)
16. B. Arnarson, J.T. Willits: Phys. Fluids **10**, 1324 (1998)
17. J.T. Willits, B. Arnarson: Phys. Fluids **11**, 3116 (1999)
18. M. Alam, J.T. Willits, B. Arnarson, S. Luding: Phys. Fluids **14**, 4085 (2002)
19. B. Arnarson, J.T. Jenkins: Phys. Fluids **16**, 4543 (2004)
20. D. Serero, I. Goldhirsch, S.H. Noskowicz, M.-L. Tan: J. Fluid Mech. **554**, 237 (2006)
21. V. Garzó, J.W. Dufty: Phys. Rev. E **60**, 5706 (1999)
22. J.M. Montanero, V. Garzó: Gran. Matt. **4**, 17 (2002)
23. A. Barrat, E. Trizac: Gran. Matt. **4**, 57 (2002)
24. S.R. Dahl, C.M. Hrenya, V. Garzó, J.W. Dufty: Phys. Rev. E **66**, 041301 (2002)
25. R. Pagnani, U.M.B. Marconi, A. Puglisi: Gran. Matt. **66**, 051304 (2002)
26. D. Paolotti, C. Cattuto, U.M.B. Marconi, A. Puglisi: Gran. Matt. **5**, 75 (2003)
27. P. Krouskop, J. Talbot: Phys. Rev. E **68**, 021304 (2003)
28. H. Wang, G. Jin, Y. Ma: Phys. Rev. E **68**, 031301 (2003)
29. J.J. Brey, M.J. Ruiz-Montero, F. Moreno: Phys. Rev. Lett. **95**, 098001 (2005)
30. J.J. Brey, M.J. Ruiz-Montero, F. Moreno: Phys. Rev. E **73**, 031301 (2006)
31. M. Schröter, S. Ulrich, J. Kreft, J.B. Swift, H.L. Swinney: Phys. Rev. E **74**, 011307 (2006)
32. R.D. Wildman, D.J. Parker: Phys. Rev. Lett. **88**, 064301 (2002)
33. K. Feitosa, N. Menon: Phys. Rev. Lett. **88**, 198301 (2002)
34. J.T. Jenkins, F. Mancini: J. Appl. Mech. **54**, 27 (1987)
35. V. Garzó, J.W. Dufty: Phys. Fluids **14**, 1476 (2002)
36. V. Garzó, J.M. Montanero, J.W. Dufty: Phys. Fluids **18**, 083305 (2006)
37. J.M. Montanero, V. Garzó: Phys. Rev. E **67**, 021308 (2003)
38. V. Garzó, J.M. Montanero: Phys. Rev. E **69**, 021301 (2004)
39. J.W. Dufty, V. Garzó: J. Stat. Phys. **105**, 723 (2001)
40. V. Garzó: Physica A **343**, 105 (2004)
41. V. Garzó: Europhys. Lett. **75**, 521 (2006)
42. C.S. Campbell: Annu. Rev. Fluid Mech. **22**, 57 (1990)
43. C.K.K. Lun, S.B. Savage, D.J. Jeffrey, N. Chepurniy: J. Fluid Mech. **140**, 223 (1984)
44. J.T. Jenkins, M.W. Richman: J. Fluid Mech. **192**, 313 (1988)
45. C.S. Campbell, J. Fluid Mech. **203**, 449 (1989)
46. M.A. Hopkins, H.H. Shen: J. Fluid Mech. **244**, 477 (1992)
47. P.J. Schmid, H.K. Kytömaa: J. Fluid Mech. **264**, 255 (1994)
48. C.K.K. Lun, A.A. Bent: J. Fluid Mech. **258**, 335 (1994)
49. I. Goldhirsch, M.L. Tan: Phys. Fluids **8**, 1752 (1996)
50. N. Sela, I. Goldhirsch, S.H. Noskowicz: J. Fluid Mech. **8**, 2337 (1997)
51. J.J. Brey, M.J. Ruiz-Montero, F. Moreno: Phys. Rev. E **55**, 2846 (1997)
52. C.-S. Chou M.W. Richman: Physica A **259**, 430 (1998)
53. C.-S. Chou: Physica A **287**, 127 (2000)
54. C.-S. Chou: Physica A **290**, 341 (2001)
55. J.M. Montanero, V. Garzó, A. Santos, J.J. Brey: J. Fluid Mech. **389**, 391 (1999)
56. A. Astillero, A. Santos: Phys. Rev. E **72**, 031309 (2005)
57. V. Garzó, A. Santos: *Kinetic Theory of Gases in Shear Flows. Nonlinear Transport* (Kluwer Academic, Dordrecht, 2003)

58. A. Santos, V. Garzó, J. W. Dufty: Phys. Rev. E **69**, 061303 (2004)
59. A. Goldshtein, M. Shapiro: J. Fluid Mech. **282**, 75 (1995)
60. J.J. Brey, J.W. Dufty, A. Santos: J. Stat. Phys. **87**, 1051 (1997)
61. T.P.C. van Noije, M.H. Ernst: Gran. Matt. **1**, 57 (1998)
62. J.M. Montanero, V. Garzó: Gran. Matt. **4**, 17 (2002)
63. S.R. Dahl, C.M. Hrenya, V. Garzó, J.W. Dufty: Phys. Rev. E **66**, 041301 (2002)
64. D.J. Evans, G.P. Morriss: *Statistical Mechanics of Nonequilibrium Liquids* (Academic Press, London, 1990)
65. W.G. Hoover: *Computational Statistical Mechanics* (Elsevier, Amsterdam, 1991)
66. J.M. Montanero, A. Santos: Gran. Matt. **2**, 53 (2000)
67. D.R.M. Williams, F.C. McKintosh: Phys. Rev. E **54**, R9 (1996).
68. C. Henrique, G. Batrouni, D. Bideau: Phys. Rev. E **63**, 011304 (2000)
69. A. Barrat, E. Trizac: Gran. Matt. **4**, 52 (2002)
70. M. Schröter, S. Ulrich, J. Kreft, J.B. Swift, H.L. Swinney: Phys. Rev. E **74**, 011307 (2006)
71. J.J. Brey, M.J. Ruiz-Montero, F. Moreno: Phys. Rev. Lett. **95**, 098001 (2005)
72. J.J. Brey, M.J. Ruiz-Montero, F. Moreno: Phys. Rev. E **73**, 031301 (2006)
73. J. Ferziger, H. Kaper: *Mathematical Theory of Transport Processes in Gases* (North-Holland, Amsterdam, 1972)
74. V. Garzó, J.M. Montanero: J. Stat. Phys. **129**, 27 (2007)
75. V. Garzó, J.W. Dufty: Phys. Rev. E **59**, 5895 (1999)
76. J.F. Lutsko: Phys. Rev. E **72**, 021306 (2005)
77. G.A. Bird: *Molecular Gas Dynamics and the Direct Simulation Monte Carlo of Gas Flows* (Clarendon, Oxford, 1994)
78. J.A. McLennan: *Introduction to Nonequilibrium Statistical Mechanics* (Prentice-Hall, New Jersey, 1989)
79. A. Santos, J.W. Dufty: Phys. Rev. Lett. **97**, 058001 (2006)
80. M. López de Haro, E.G.D. Cohen: J. Chem. Phys. **80**, 408 (1984)
81. V. Garzó, J.M. Montanero: Physica A **313**, 336 (2002)
82. A. Barrat, V. Loreto, A. Puglisi: Physica A **334**, 513 (2004)
83. S.R. de Groot, P. Mazur: *Nonequilibrium Thermodynamics* (Dover, New York, 1984)
84. N. Brilliantov, T. Pöschel: *Kinetic Theory of Granular Gases* (Oxford University Press, Oxford, 2004)
85. I. Goldhirsch, G. Zanetti: Phys. Rev. Lett. **70**, 1619 (1993)
86. I. Goldhirsch, M.L. Tan, G. Zanetti: J. Sci. Comput. **8**, 1 (1993)
87. S. Luding, H.J. Herrmann: Chaos **9**, 673 (1999)
88. V. Garzó: Phys. Rev. E **72**, 021106 (2005)
89. P. Résibois, M. de Leener: *Classical Kinetic Theory of Fluids* (John Wiley, NY, 1977)
90. J.P. Boon, S. Yip: *Molecular Hydrodynamics* (Dover, NY, 1980)
91. J.J. Brey, M.J. Ruiz-Montero, D. Cubero: Phys. Rev. E **60**, 3150 (1999)
92. See, for instance, A. Kudrolli: Rep. Prog. Phys. **67**, 209 (2004)
93. J. M. Kincaid, E.G.D. Cohen, M. López de Haro: J. Chem. Phys. **86**, 963 (1983)
94. D.C. Hong, P.V. Quinn, S. Luding: Phys. Rev. Lett. **86**, 3423 (2001)
95. J.T. Jenkins, D.K. Yoon: Phys. Rev. Lett. **88**, 194301 (2002)
96. D.K. Yoon, J.T. Jenkins: Phys. Fluids **18**, 073303 (2006)

97. A.P.J. Breu, H.M. Ensner, C.A. Kruelle, I. Rehberg: Phys. Rev. Lett. **90**, 014302 (2003)
98. T. Schautz, R. Brito, C.A. Kruelle, I. Rehberg: Phys. Rev. Lett. **95**, 028001 (2005) and references therein
99. L. Trujillo, M. Alam, H.J. Herrmann: Europhys. Lett. **64**, 190 (2003)
100. D. Serero, I. Goldhirsch, S.H. Noskowicz, M.-L. Tan: J. Fluid Mech. **554**, 237 (2006)
101. M.L. Tan, I. Goldhirsch: Phys. Rev. Lett. **81**, 3022 (1998)
102. J.W. Dufty, J.J. Brey: Phys. Rev. Lett. **82**, 4566 (1999)
103. J.T. Willits, B. Arnarson: Phys. Fluids **11**, 3116 (1999)
104. M. Alam, J.T. Willits, B. Arnarson, S. Luding: Phys. Fluids **14**, 4085 (2002)
105. M. Alam, S. Luding: Gran. Matt. **4**, 139 (2002)
106. M. Alam, S. Luding: J. Fluid Mech. **476**, 69 (2003)
107. M. Alam, S. Luding: Phys. Fluids **17**, 063303 (2005)
108. M. Alam, C.M. Hrenya: Phys. Rev. E **63**, 0613018 (2001)
109. R. Clelland, C.M. Hrenya: Phys. Rev. E **65**, 031301 (2002)
110. S.R. Dahl, R. Clelland, C.M. Hrenya: Phys. Fluids **14**, 1972 (2002)
111. S.R. Dahl, R. Clelland, C.M. Hrenya: Powder Tech., **138**, 7 (2003)
112. S.R. Dahl, C.M. Hrenya: Phys. Fluids **16**, 1 (2004)
113. H. Iddir, H. Arastoopour, C.M. Hrenya: Powder Tech. **151**, 117 (2005)
114. H.J.M. Hanley (ed.) *Nonlinear Fluid Behavior*, (North-Holland, Amsterdam, 1983)
115. A.W. Lees, S.F. Edwards: J. Phys. C **5**, 1921 (1972)
116. J.W. Dufty, A. Santos, J.J. Brey, R.F. Rodríguez: Phys. Rev. A **33**, 459 (1986)
117. J.M. Montanero, V. Garzó: Physica A **310**, 17 (2002)
118. V. Garzó: Phys. Rev. E **66**, 021308 (2002)
119. V. Garzó, J.M. Montanero: Gran. Matt. **5**, 165 (2003)
120. L. Huilin, D. Gidaspow, E. Manger: Phys. Rev. E **64**, 061301 (2001)
121. M.F. Ramahan, J. Naser, P.J. Witt: Powder Tech. **138**, 82 (2003)
122. J.E. Galvin, S.R. Dahl, C.M. Hrenya: J. Fluid Mech. **528**, 207 (2005)
123. J.J. Brey, M. J. Ruiz-Montero, F. Moreno, R. García-Rojo: Phys. Rev. E **63**, 061305 (2001)
124. J.J. Brey, M.J. Ruiz-Montero, F. Moreno, R. García-Rojo: Phys. Rev. E **65**, 061302 (2002)
125. X. Yang, C. Huan, D. Candela, R.W. Mair, R.L. Walsworth: Phys. Rev. Lett. **88**, 044301 (2002)
126. C. Huan, X. Yang, D. Candela, R.W. Mair, R.L. Walsworth: Phys. Rev. E **69**, 041302 (2004)
127. C. Bizon, M.D. Shattuck, J.B. Swift, H.L. Swinney: Phys. Rev. E **60**, 4340 (1999)
128. E.C. Rericha, C. Bizon, M.D. Shattuck, H.L. Swinney: Phys. Rev. Lett. **88**, 014302 (2002)
129. I. Goldhirsch: Granular gases: Probing the boundaries of hydrodynamics. In: Pöschel, T., Luding, S. (eds.): *Granular Gases*, Lect. Notes Phys. **564**, 79–99. Springer, Berlin (2001)
130. V. Garzó, J.M. Montanero: Phys. Rev. E **68**, 041302 (2003)
131. M. López de Haro, E.G.D. Cohen, J.M. Kincaid: J. Chem. Phys. **78**, 2746 (1983)

Index

MIX
Papier aus verantwortungsvollen Quellen
Paper from responsible sources
FSC® C105338

If you have any concerns about our products,
you can contact us on
ProductSafety@springernature.com

In case Publisher is established outside the EU,
the EU authorized representative is:
Springer Nature Customer Service Center GmbH
Europaplatz 3, 69115 Heidelberg, Germany

Printed by Libri Plureos GmbH
in Hamburg, Germany